Sebastian Basterrech, Technical University of Ostrava, Czech Republic
Witold Pedrycz, University of Alberta, Canada

National Advisory Committee

Ankush Mittal, Dehradun
Aruna Tiwari, IIT Indore
Ashish Verma, IIT Guwahati
Ashok Deshpande, Pune
Ashok Kumar Singh, DST, New Delhi
B.K. Das, Delhi University, India
C. Thangaraj, IIT Roorkee, India
D. Gnanaraj Thomas, Madras Christian College, Chennai
D. Nagesh Kumar, IISc., Bangalore
Debasish Ghose, IISc., Bangalore
Deepti, AHEC, IIT Roorkee
Dharmdutt, Saharanpur Campus, IIT Roorkee
Ghanshyam Singh Thakur, MANET Bhopal
Himani Gupta, IIFT, Delhi
Kanti S. Swarup, IIT Chennai
M.P. Biswal, IIT Kharagpur
Manoj Kumar Tiwari, IIT Kharagpur
N.R. Pal, ISI, Kolkata
Nirupam Chakraborti, IIT Kharagpur
P.C. Jha, Delhi University
Pankaj Gupta, Delhi University
Punam Bedi, University of Delhi
Raju George, (IIST), Trivandrum
Rama Mehta, NIH IIT Roorkee
Rama Sushil, Dehradun
Ravindra Gudi, IIT Bombay
Steven Fernendes, Sahyadri College of Engineering & Management, Mangaluru,
Karnataka

Contents

About the Editors

Dr. Millie Pant is an associate professor in the Department of Paper Technology, Indian Institute of Technology, Roorkee (IIT Roorkee) in India. A well-known figure in the field of swarm intelligence and evolutionary algorithms, she has published several research papers in respected national and international journals.

Dr. Kanad Ray is a professor of Physics at the Department of Physics at the Amity School of Applied Sciences, Amity University Rajasthan (AUR), Jaipur. In an academic career spanning over 19 years, he has published and presented research papers in several national and international journals and conferences in India and abroad. He has authored a book on the Electromagnetic Field Theory. Dr. Ray's current research areas of interest include cognition, communication, electromagnetic field theory, antenna and wave propagation, microwave, computational biology, and applied physics.

Dr. Tarun K. Sharma has a Ph.D. in artificial intelligence as well as MCA and MBA degrees and is currently associated with the Amity University Rajasthan (AUR) in Jaipur. His research interests encompass swarm intelligence, nature-inspired algorithms, and their applications in software engineering, inventory systems, and image processing. He has published more than 60 research papers in international journals and conferences. He has over 13 years of teaching experience and has also been involved in organizing international conferences. He is a certified internal auditor and a member of the Machine Intelligence Research (MIR) Labs, WA, USA and Soft Computing Research Society, India.

Dr. Sanyog Rawat is presently associated with the Department of Electronics and Communication Engineering, SEEC, Manipal University Jaipur, Jaipur, India. He holds a B.E. in Electronics and Communication, an M.Tech. in Microwave Engineering and Ph.D. in Planar Antennas. Dr. Rawat has been involved in organizing various workshops on "LabVIEW" and antenna designs and simulations using FEKO. He has taught various subjects, including electrical science, circuits and system, communication system, microprocessor systems, microwave devices, antenna theory and design, advanced microwave engineering and digital circuits.

Dr. Anirban Bandyopadhyay is a Senior Scientist in the National Institute for Materials Science (NIMS), Tsukuba, Japan. Ph.D. from Indian Association for the Cultivation of Science (IACS), Kolkata 2005, on supramolecular electronics. During 2005–2008, he was selected as Independent Researcher, ICYS Research Fellow in the International Center for Young Scientists (ICYS), NIMS, Japan, he worked on brain-like bio-processor building. In 2007, he started as permanent Scientist in NIMS, working on the cavity resonator model of human brain and brain-like organic jelly. During 2013–2014, he was a visiting professor in Massachusetts Institute of Technology (MIT), USA. He has received many honors such as Hitachi Science and Technology award 2010, Inamori Foundation award 2011–2012, Kurata Foundation Award, Inamori Foundation Fellow (2011–), Sewa Society International member, Japan, etc.

Geometric phase space model of a human brain argues to replace Turing tape with a fractome tape and built a new geometric-musical language which is made to operate that tape. Built prime metric has to replace space–time metric. Designed and built multiple machines and technologies include: (1) angstrom probe for neuron signals, (2) dielectric imaging of neuron firing, (3) single protein and its complex structure's resonant imaging, and (4) fourth circuit element Hinductor. A new frequency fractal model is built to represent biological machines. His group has designed and synthesized several forms of organic brain jelly (programmable matter) that learns, programs, and solves problems by itself for futuristic robots during 2000–2014, also several software simulators that write complex codes by itself.

Modified Critical Path and Top-Level Attributes (MCPTL)-Based Task Scheduling Algorithm in Parallel Computing

Ranjit Rajak, Diwakar Shukla and Abdul Alim

Abstract The parallel processing provides speedup to the computational tasks usually not solved efficiently by a sequential machine. It is basically, a transformation of an application program into task graph where tasks of the graph are allocated to the available processors. In this paper, we have proposed a new task scheduling algorithm which is a modified version of Critical Path and Top-Level Attributes-based Scheduling (CPTL) algorithm. This new algorithm is known as MCPTL algorithm. The proposed algorithm excludes the communication time between the tasks during allocations onto the processors. The MCPTL algorithm gives minimum scheduling length as compared the CPTL algorithm and also performance study is done with heuristic algorithms which is based on performance metrics, Scheduling Length Ratio (SLR), Speedup (SP), Efficiency (EFF), Load Balancing (LB), and Cost (C).

Keyword Task scheduling · Parallel computing · Scheduling length DAG · Speedup

1 Introduction

Parallel computing is defined as a collection of processing elements (PE), where each one is connected with each other through a high-speed interconnection network.

R. Rajak (✉) · D. Shukla · A. Alim
Department of Computer Science & Applications,
Dr. Harisingh Gour Central University, Sagar, Madhya Pradesh, India
e-mail: ranjit.jnu@gmail.com

D. Shukla
e-mail: diwakarshukla@rediffmail.com

A. Alim
e-mail: abdulaleem1990@gmail.com

© Springer Nature Singapore Pte Ltd. 2018
M. Pant et al. (eds.), *Soft Computing: Theories and Applications*,
Advances in Intelligent Systems and Computing 583,
https://doi.org/10.1007/978-981-10-5687-1_1

1

It is a platform where one can solve many scientific computation problems relating to bioinformatics, data mining, astronomy, and industries. These problems have a large number of tasks where each task interacts and communicate with each other. Task scheduling problem in parallel computing environment considered as NP-hard [1]. It is also known as a task scheduling optimization problem, which can fit in many fields [2] like economics, project management, production management, etc.

Task scheduling is a process to map the tasks onto the available processors so that it consumes minimum parallel execution time or scheduling length. There are a number of task scheduling algorithms have been developed to solve this problem. Each algorithm has own advantages and disadvantages. But the primary objective of each is same, i.e., to minimize scheduling length.

There are a number of characteristics [3] for classifying the task scheduling like finite number of tasks, precedence constraint between the tasks, the number of homogenous processors, their connectivity etc.

Generally, there are two types of task scheduling: static and dynamic. In static task scheduling, the number of tasks, their precedence constraint, the number of finite processors, and their connectivity are known in advanced or known at compilation time. There are two primary objectives [4] of the static task scheduling: to minimize parallel execution time of the task graph and communication time between the tasks. It is also known as deterministic scheduling.

In case of dynamic task scheduling, the information is not known in advance and is known at the time of execution. The goal [4] is to maximize processor utilization. It is also known as nondeterministic scheduling. We have considered that the tasks to be non-preemptive.

In parallel computing, the processors are either homogeneous or heterogeneous in nature. The homogeneous processors are those operating at equal speed; while heterogeneous processors operate at varying speed. In this paper, we have considered homogenous processors.

The task scheduling of an application program is generally represented by Directed Acyclic Graph (DAG). Figure 1 showing the mapping of an application program into task scheduling in a multiprocessor system. An application program is decomposed into a number of subtasks, and these subtasks are represented by a DAG, later, they are allocated to the number processors with hold precedence constraint between tasks.

This paper presents a new task scheduling algorithm which is based on two priority attributes: critical path and top-level (t-level). The suggested MCPTL algorithm does not use communication time, while allocating tasks to the available processors.

The proposed algorithm gives minimum parallel execution time, i.e., scheduling length. The content herein incorporates two DAG models for comparison of proposed algorithm and heuristic algorithms.

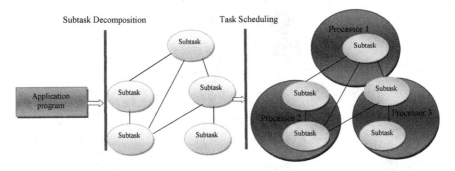

Fig. 1 Mapping application program to task scheduling [5]

2 Model of Task scheduling

A task scheduling environment consists of Application Model, System Model, DAG Attributes, and Objective Function.

2.1 Application Model

The task scheduling is represented using Directed Acyclic Graph (DAG), which is a weighted graph defined as $G_1 = (V, E, C_e, E_v)$ where V is a collection of the n tasks, and $V = \{T_1, T_2, T_3, ..., T_n\}$. E is a collection of edges, C_e is the communication time between two dependent tasks T_i and T_j. That is., $C_e(T_i, T_j)$. E_v is the execution time of each task T_i denoted as $E_v(T_i)$.

The DAG also consists of an entry and an exit tasks. The entry task T_{entry} is the first task denoted as having no predecessor task. Similarly, an exit task T_{exit} is the final task of the DAG and have no any successor tasks.

The DAG model of an application program is in Fig. 2.

2.2 System Model

Consider a parallel computing machine consisting of N_p number of homogenous processors fully connected. It is defined by $G_2 = (P, L)$ where P is a set of identical processors. That is, $P = \{P_1, P_2, P_3, ..., N_p\}$ and L is a set of communication links between the processors in full duplex mode (Fig. 3).

Here, we have not considered the communication time during the allocations of the tasks on the processors.

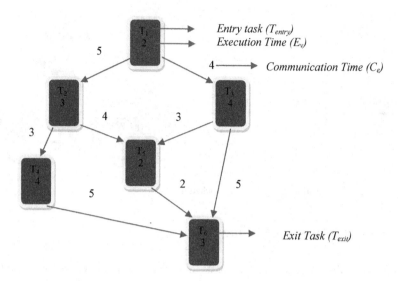

Fig. 2 DAG Model with 6 tasks [6]

Fig. 3 Fully connected
homogeneous processors [10]

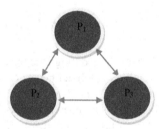

2.3 DAG Scheduling Attributes

The section contains some existing scheduling attributes, which will be used in proposed scheduling algorithm. It is necessary to discuss here for DAG scheduling. These are *Critical Path, T-level, Earliest Start Time, and Earliest Finished Time*.

Critical Path (CP) [5, 7] is the longest path from an entry task (T_{entry}) to an exit task (T_{exit}). It includes execution time of the tasks and communication time between the tasks. It is defined as follows:

$$CP = \max_{\text{path} \in G1} \{\text{Length}(\text{Path})\}$$

$$\text{Length}(\text{Path}) = \sum_{T_i \in V} E_v(T_i) + \sum_{e \in E} C_e(e_{ij})$$

T-level [8] is the longest path from an entry task (T_{entry}) to task T_i and it excludes the execution time of T_i. It is defined recursively as

$$T\text{-level}(T_i) = \begin{cases} 0 & \text{if} \quad T_i = T_{\text{entry}} \\ \max_{T_j \in \text{pred}(T_i)} \{T\text{-level}(T_j) + E_v(T_j) + C_e(T_j, T_i)\}, & \text{otherwise} \end{cases}$$

Earliest Start Time (EST) [9] for task T_i on processors P_j is defined as

$$\text{EST}(T_i, P_j) = \begin{cases} 0 & \text{if} \quad T_i = T_{\text{entry}} \\ \max_{T_j \in \text{pred}(T_i)} \{\text{EFT}(T_j, P_j) + C_e(T_j, T_i)\}, & \text{otherwise} \end{cases}$$

Note that pred(T_i) is the immediate predecessor of task T_i.

The proposed algorithm has excluded the communication time between the tasks during allocation onto processors. For this, $C_e(T_j, T_i)$ is excluded from EST and it is rewritten as Modified EST (MEST):

$$\text{MEST}(T_i, P_j) = \begin{cases} 0 & \text{if} \quad T_i = T_{\text{entry}} \\ \max_{T_j \in \text{pred}(T_i)} \{\text{EFT}(T_j, P_j)\}, & \text{otherwise} \end{cases}$$

Earliest Finished Time (EFT) [9] of the task T_i on processor P_j is defined as

$$\text{EFT}(T_i, P_j) = \text{MEST}(T_i, P_j) + W(T_i, P_j)$$

where $W(T_i, P_j)$ is the execution time of T_i on processor P_j.

2.4 Objective Function

The task scheduling algorithm is used to minimize *the overall execution time or scheduling length (SL)* on the multiprocessor system. An objective function can be defined as follows:

$$\text{SL} = \text{Max}\{\text{EFT}(T_{\text{exit}}, P_j)\}$$

where T_{exit} is the last task of DAG and P_j $\{j = 1, 2, 3, ..., N_p\}$ is a particular processor on which the last task will execute.

3 Algorithm Design

The section discusses the content of proposed algorithm which is a modified version of CPTL Algorithm [10].

3.1 Task Priority

Each task of given DAG has a priority based on two well-known task scheduling attributes critical path (CP) and top-level (T-level) as discussed in the previous section. A new attribute of each task using above attributes has been introduced as CPT attribute [10] which is the difference between critical path and top-level. Its computation is:

$\mathrm{CPT}(T_i) = \mathrm{CP} - T\text{-level}(T_i)$, where T_i is the tasks of given DAG and $T_i = \{T_1, T_2, T_3, \ldots, T_n\}$.

3.2 Proposed MCPTL Algorithm

The CPTL algorithm [10] uses CPT task priority attribute for calculating the priority of each task for a given DAG. These tasks are sorted in nonincreasing order as per CPT attribute value and inserted into a list. The highest priority task is removed from the list and allocated to the available processor as per two conditions, find minimum EST and satisfied precedence constraint (PC) [7]. Here, during allocation of the tasks to the processor, the communication time is considered between the tasks.

The MCPTL algorithm uses same priority attributes for calculating priority of the tasks but not consider communication time between the tasks during allocation of the tasks to processors and for this EST is modified as MEST which is used in the proposed algorithm. This will reduce the scheduling length of the proposed algorithm.

MCPTL algorithm

Step 1	Input: A DAG of finite number of task (T_n)
Step 2	Find CPT attribute of each task using following formula $\mathrm{CPT}(T_i) = \mathrm{CP} - T\text{-level}(T_i)$
Step 3	Sort the tasks in a list (LT) by nonincreasing order of CPT values
Step 4	**While** the unscheduled task in LT **do** Select the highest priority task T_i from LT
Step 5	**For** each processor P_j, $j = 1$ to N_p **do**
Step 6	**If** T_i is T_{entry} **then**

(continued)

(continued)

	T_i is allocated to available processor
Step 7	**Else** if T_i is satisfied PC **then** Compute MEST (T_i, P_j) and Compute EFT(T_i, P_j) Assign task T_i on processor P_j with minimum EFT

Step 8	**Else if** T_i is not satisfied PC **then**
	T_i is inserted to end of LT
	End For
	End While

Step 9	SL = Max$\{$EFT$(T_{\text{exit}}, P_j)\}$
Step 10	**End**

4 Performance Evaluation

The performance analysis of MCPTL algorithm is done based on comparison metrics and using two DAG models with 9 and 11 tasks. We have considered that all the tasks are of non-preemptive nature.

4.1 Comparison Metrics

The comparison metrics are used to analysis the proposed algorithm and heuristics algorithms. The metrics are Scheduling Length (SL), Scheduling Length Ration (SLR), Speedup (SP), Efficiency (EFF), Load Balancing (LB), and Cost (C) (Table 1).

4.2 Numerical Examples

To measure the performance of the proposed algorithm MCPTL, two models DAG1 [7] and DAG2 [6] with 9 and 11 tasks are considered on four homogenous processors (Figs. 4, 5) and (Tables 2, 3).

This is to observe that the scheduling length of MCPTL algorithm is minimum as compared to CPTL and heuristic algorithms. In case of DAG1, the MCPTL algorithm generates scheduling length 11 which is provides better as compared to other algorithms shown in Table 4. The proposed algorithm also provides better for other metrics like speedup, efficiency, load balancing, scheduling length ratio, and

Table 1 Comparison metrics

S. No	Metrics	Brief descriptions
1.	Scheduling length (SL)	It is the finished time of last task of given DAG on available processor. That is, $SL = \text{Max}\{EFT(T_i, P_j)\}$, where T_i is T_{exit} and P_j is available processor
2.	Scheduling length ratio (SLR) [11]	It is defined as a ratio of scheduling length and critical path. That is, $$SLR = \frac{SL}{CP}$$
3.	Speedup (SP) [12]	It is ratio of sum of sequential time for the tasks on single processor and parallel time or scheduling length (SL) on multiprocessors system. That is, $$SP = \frac{\sum_{i=1}^{n} ET_{(T_i)}}{SL}$$
4.	Efficiency (EFF) [13]	It is the ratio of Speedup and number of processors is used. $EFF = \frac{SP}{N_p}$, where N_p is number of processors is used
5.	Load balancing (LB) [14]	It is defined as the ratio of scheduling and average, where average is the ratio of sum of processing time of each processors and number of processors are used in scheduling. That is $LB = \frac{SL}{AVG}$, $AVG = \frac{\sum_{j=1}^{N_p} PT[P_j]}{N_p}$ where $PT[P_j]$ is execution time of processor P_j
6.	Cost (C) [13]	It is defined as the product of number of processor used and parallel time or scheduling length. That is, $C = N_p \times SL$

Fig. 4 DAG1 with 9 tasks

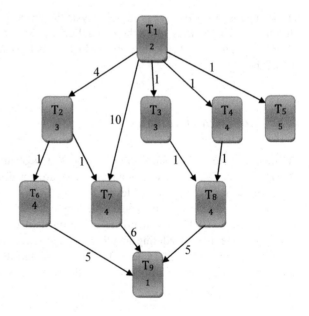

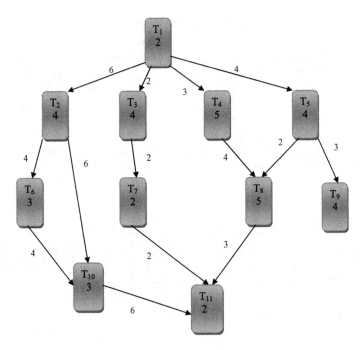

Fig. 5 DAG2 with 11 tasks

Table 2 CPT (T_i) attribute for DAG1 [10]

Tasks(T_i)	T-Level(T_i)	CPT(T_i) = CP − T-level(T_i)
T_1	0	23
T_2	6	17
T_3	3	20
T_4	3	20
T_5	3	20
T_6	10	13
T_7	12	11
T_8	8	15
T_9	22	1

Table 3 Computing CPT (T_i) for DAG2 [10]

Tasks(T_i)	T-Level(T_i)	CPT(T_i) = CP − T-level(T_i)
T_1	0	34
T_2	8	26
T_3	4	30
T_4	5	29
T_5	6	28
T_6	16	18
T_7	10	24
T_8	14	20
T_9	13	21
T_{10}	23	11
T_{11}	32	2

Table 4 Results of task scheduling algorithms for DAG1 and DAG2

Scheduling algorithms	DAG1						DAG2					
	Performance metrics						Performance metrics					
	SL	SP	EFF	LB	SLR	C	SL	SP	EFF	LB	SLR	C
MCPTL	11	2.77	0.68	1.22	0.478	44	14	2.710	0.67	1.272	0.411	56
CPTL [10]	17	1.77	0.44	1.89	0.730	68	22	1.728	0.432	1.375	0.648	88
HLFET [6]	19	1.57	0.39	1.58	0.826	76	22	1.728	0.432	1.571	0.648	88
MCP [6]	20	1.50	0.37	1.50	0.869	80	24	1.583	0.395	1.655	0.706	96
ETF [6]	19	1.57	0.39	1.42	0.732	76	25	1.520	0.380	1.666	0.732	100
DLS [6]	19	1.57	0.39	1.42	0.732	76	24	1.583	0.395	1.846	0.706	96

	0	1	2	3	4	5	6	7	8	9	10	11
P1	T_1	T_1	T_3	T_3	T_3	T_7	T_7	T_7	T_7			
P2			T_4	T_4	T_4	T_4	T_8	T_8	T_8	T_8	T_9	
P3			T_5	T_5	T_5	T_5	T_5					
P4			T_2	T_2	T_2	T_6	T_6	T_6	T_6			

Fig. 6 Scheduling length of DAG1 is 11 units

0	1	2	3	4	5	6	7	8	9	10	11	12	13	14
P1	T_1	T_1	T_3	T_3	T_3	T_3	T_7	T_7						
P2			T_4	T_4	T_4	T_4	T_4	T_8	T_8	T_8	T_8	T_8	T_{11}	T_{11}
P3			T_5	T_5	T_5	T_5	T_9	T_9	T_9	T_9				
P4			T_2	T_2	T_2	T_2	T_6	T_6	T_6	T_{10}	T_{10}	T_{10}		

Fig. 7 Scheduling length of DAG2 is 14 units

Fig. 8 Scheduling length

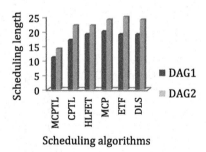

Scheduling algorithms

Fig. 9 Speedup

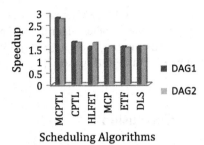

Fig. 10 Efficiency

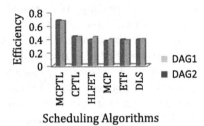

Fig. 11 Load balancing

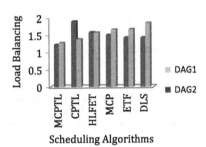

Fig. 12 Scheduling length ratio

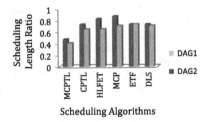

Fig. 13 Cost

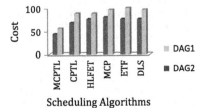

cost. Similarly for DAG2, provides minimum scheduling length 14 which is better as compared to other algorithms. The cost metrics and load balancing are also minimum for proposed algorithm. We have drawn various graphs-based result shown in Table 4 (Figs. 6, 7, 8, 9, 10, 11, 12, and 13).

5 Conclusion

The content has suggested a new algorithm for task scheduling in the environment of parallel computing. It is modified version of Critical Path and Top-Level attributes-based Task Scheduling Algorithm for DAG (CPTL) algorithm. The suggested hereby named as MCPTL algorithm which is free from communication time between the tasks. After testing on two different models of DAGs and four homogenous processors, it is found that it reduces scheduling length. The MCPTL algorithm is not only better scheduling length but it also gives minimum load balancing, less cost, less SLR as compared to CPTL and heuristic algorithms like HLFET, MCP, EFT, and DLS algorithms. The proposed algorithm is better in terms of speedup and efficiency as compared to others.

References

1. Pinedo, M.L.: Scheduling: theory, algorithms, and systems, 3rd Edn. Springer, Berlin (2008)
2. Noronha, S., Sharma, V.: Knowledge-based approaches for scheduling problems: a survey. IEEE Trans. Knowl. Data Engin. 3(2), 160–171 (1991)
3. Mostafa, R.M., Medhat, H.A.: Hybrid algorithms for multiprocessor task scheduling. Int. J. Comput. Sci. Issues 8(3), 79–89 (2011)
4. Rajaraman, V., Murthy, C.S.R.: Parallel computers: Architecture and programming, PHI Publication, (2012)
5. Sinnen, O.: Task scheduling for parallel systems, Wiley-Interscience Publication (2007)
6. Rajak, R.: Comparison of BNP class of scheduling algorithms based on metrics, GESJ Comput. Sci. Telecomun. 34(2) 35–44 (2012)
7. Kwok, Y.K., Ahmad, I.: Static scheduling algorithms for allocating directed task graphs to multiprocessors. ACM Comput. Surv. 31(4), 406–471 (1999)
8. Rahmani, A.M., Vahedi, M.A.: A novel task scheduling in multiprocessor systems with genetic algorithm by using elitism stepping method, science and research branch, Tehran, Iran 26 May 2008
9. Zhou, G., Xu Y., Tian, S., Zhao, H.: A genetic-based task scheduling algorithms on heterogeneous computing systems to minimize makespan. J. Converg. Inf. Technol. (JCIT) 8 (5) (2013)
10. Rajak, N., Rajak, R., Dixit, A.: A critical-path and top-level attributes based task scheduling algorithm for DAG (CPTL). Int. J. New Comput. Archit. Their Appl. Hong Kong, 4(4), 130–136 (2014)

11. Topcuoglu, H., Wu, M.Y.: Performance effective and low complexity task scheduling for heterogeneous computing. IEEE Trans. Parallel Distrib. Comput. **13**(3), 260–274 (2002)

12. Quinn, M.J.: Parallel programming in C with MPI and OpenMP, Tata McGraw-Hill, Edition (2003)

13. Gramma, A., Kumar, V., Gupta, A.: Introduction to parallel computing, Pearson Edition (2009)

14. Omara, F.A., Arafa, M.M.: Genetic algorithm for task scheduling problem. J. Parallel Distrib. Comput. **70**, 13–22 (2010)

Critical Path Problem for Scheduling Using Genetic Algorithm

Harsh Bhasin and Nandeesh Gupta

Abstract The critical path problem, in Software Project Management, finds the longest path in a Directed Acyclic Graph. The problem is immensely important for scheduling the critical activities. The problem reduces to the longest path problem, which is NP as against the shortest path problem. The longest path is an important NP-hard problem, which finds its applications in many other areas like graph drawing, sequence alignment algorithms, etc. The problem has been dealt with using Computational Intelligence. The paper presents the state of the art. The applicability of Genetic Algorithms in longest path problem has also been discussed. This paper proposes a novel Genetic Algorithm-based solution to the problem. This algorithm has been implemented and verified using benchmarks. The results are encouraging.

Keywords Longest path problem · Genetic algorithms · NP-Hard problems
Heuristics

1 Introduction

Handling non-deterministic polynomial (NP) time problems is one of the most precarious tasks. The problems which can be solved in polynomial time by deterministic algorithms are referred to as polynomial time problems or P problems [1].

H. Bhasin (✉)
School of Computer and Systems Sciences,
Jawaharlal Nehru University, New Delhi, India
e-mail: i_harsh_bhasin@yahoo.com

N. Gupta
Department of Computer Science,
YMCA University of Science and Technology,
Faridabad, India
e-mail: guptanandeesh@gmail.com

© Springer Nature Singapore Pte Ltd. 2018
M. Pant et al. (eds.), *Soft Computing: Theories and Applications*,
Advances in Intelligent Systems and Computing 583,
https://doi.org/10.1007/978-981-10-5687-1_2

Finding shortest path in a given graph is an example of P-type problem. NP problems can be solved by non-deterministic algorithms in polynomial time [2]. These problems are further divided into two types: NP-complete and NP-hard.

In NP-complete problems, solution can be verified in polynomial time. Hamiltonian cycle problem is an example of NP-complete problem. Problems which can neither be solved in polynomial time by deterministic algorithms, nor their solutions can be verified in polynomial time are called NP-hard problems. The longest path problem is an example of NP-hard problem.

Generally, NP-hard problems are optimization problems. However, global optimization is an illusion. Therefore, soft computing approaches like Genetic Algorithms (GAs) are used to handle such problems. GAs are heuristic search processes that are based on survival of the fittest [3].

The paper presents a GA-based solution for the critical path problem in software planning and management. The approach has been implemented and the results are encouraging.

The organization of the paper is as follows: Section 2 presents the review. Section 3 explains the basics of GA. Section 4 presents the background. Section 5 presents the proposed work. The next section presents the results and the last section concludes.

2 Review

As stated earlier, longest path problem is an NP-hard problem, which finds its application in various fields. In order to access the state of the art and place the proposed technique in the right perspective, a comprehensive literature review has been carried out. The review has been carried out in accordance with the guidelines proposed by Kitchenham [4]. This section has been divided into various subsections, which are as follows:

1. Research questions
2. Review methodology
3. Review

Research questions: In order to carry out a meaningful review, appropriate research question should be crafted. The review, in turn, should be able to answer the research questions. The aim of this review is to answer the following research questions.

RQ. 1. What is the trend in research in NP problems notably the longest path problem?

RQ. 2. What are the existing techniques used to tackle the problem?

Review methodology: the present study intends to summarize the work concerning NP problems in general and longest path problem in particular. The study also explores the applicability of Diploid Genetic Algorithms (DGAs) in longest path problem.

The databases that have searched while carrying out the review are as follows:

1. ACM Digital Library
2. Science Direct
3. IEEE
4. Wiley
5. Springer

From amongst these, the journals having high impact factor were selected. The papers which have been considered important in the topic have also been selected even if they were not from above journals.

Initially, all the papers related to longest path problem were included in the search, from those papers, the papers which were related to computer science discipline were selected, a further search was carried out by reading the abstract.

The selected papers were filtered according to the proposed technique. The results of the review have been presented in the next section.

The data was collected in a scientific way and the prime criterion was the quality of a paper. The summary of the review has been presented in Table 1.

Answer to RQ1. The search was carried out as follows. The keyword that was used to search the relevant articles was <longest path problem>. This was followed by applying various criteria to filter out the irrelevant papers. English was selected as the language for the papers. The topics selected were <Computer Science>, <AI> and <Theoretical Computer Science>. The year was selected as follows: the beginning and the end years were set same, For example, to select the papers of 2015, the beginning and the end year was 2015. Thus resulting in a total of 1209 articles and 12,496 chapters, in Springer link. Similar searches were carried out in

Table 1 Review of techniques to solve longest path problem

Papers	Proposed work
[5, 6]	These papers relate travelling salesman problem with the longest path problem and considers it to be a special case of longest path problem. It also discusses some of the algorithms from longest path problem
[7]	The paper applies some algorithms on super classes of interval and permutation graph and discusses the issue concerning the comparability of the graphs
[8]	The paper presents an approximation algorithm from the problem. The complexity of the proposed algorithm is $2^\wedge O(\log^{1-\varepsilon} n)$
[9]	The paper uses the concept of Hamiltonian Path problem to handle the longest path. It explores weighted trees, block graphs, ptolemaic graphs, and cacti and then solves the problem
[10]	The paper proposes $O(n)$ algorithm for finding the longest path in a bipartite graph
[11]	The paper proposes an $O(n^6)$ algorithm for finding the longest path in a biconvex graph and an $O(n^4)$ approximation algorithm for the same problem
[12]	The paper shows that interval graphs can be used to solve the longest path problem in polynomial time

other databases. However, it was found that most of them were not related to the core problem itself.

The total number of papers finally selected was 8.

Answer to RQ2. The critical path problem reduces to the longest path problem, using problem reduction approach. The problem reduces to that of finding a simple path of longest length which does not have any repeated vertex. Though the shortest path problem can be solved in polynomial time, this version of the longest path problem is an NP-hard problem.

The researches have been able to find a solution for Directed Acyclic Graph (DAG) which is used in many applications such as compiler, but the solution for undirected graphs still eludes the fraternity.

It was earlier suggested that if a graph having all the weight edges is given, then the procedure used to find the shortest path can also be used to find the longest path by simply converting all the weights to negative. However, the selection would not work as the susceptibility of creation of negative length cycles is high.

The longest path terminating at a given vertex can also be found by its incoming neighbors and incrementing the maximum length recorded by those neighbors. In the case, where the given vertex does not have any neighbor, the length of the longest path is Zero.

The problem has also been handled using approximation algorithms. The approximation ratio of the proposed solution is approximately O ($N/ \log N$) using color coding technique.

Except for the above, the technique has also been applied on other NP-complete problems. The review considered all those problems in which GA were applied. [13–15].

3   Genetic Algorithm

Genetic Algorithm (GA) is a heuristic search process which is based on the theory of survival of the fittest [3, 16]. It is generally used in optimization problems. The process starts with the creation of a population of biological units called chromosomes. Chromosomes further consist of a number of smaller units called cells. These cells may be binary, decimal, or even hexadecimal. Fig. 1 shows a population having cells which are binary.

Fig. 1 A binary Population

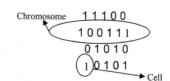

This work uses binary chromosomes which then together form a binary population. There are many genetic operators, some of them are: crossover, mutation, and roulette wheel selection. These genetic operators help us modify the parent chromosomes, thus helping us to reach the requisite solution.

1. **Crossover.** It helps to produce new chromosomes (child chromosomes) with traits of both the parent chromosomes. The crossover population merges the two parents for producing the child. There are many types of crossover: One-point, Two-point, Multipoint, and Uniform crossover.

   The One-point crossover is the simplest crossover in which a random point (R) is selected in the parent chromosomes. Now, the child is produced by merging the part to the left of R in first chromosome and that to the right in second chromosome. Similarly, the other child will consist of the cells of first parent after R and cells of second parent before R.

   The Two-point crossover is done by amalgamating the parent chromosomes wherein two random points are selected on basis of which the formation of new chromosome takes place.

2. **Mutation.** It is an operation which is carried out so as to break the local maxima. One of the ways of implementing mutation is by simply complementing a randomly chosen bit of a random chromosome.

3. **Selection.** It is a process in which some chromosomes having higher fitness values are selected from the population and then these chromosomes are replicated so as to form a new population with a majority of chromosomes having high fitness values, thus increasing the final solution.

Genetic algorithms have been successfully used for optimization problems, particularly in NP problems. This work explores the use of genetic algorithms in one such NP-hard problem called longest path problem.

4   Background

In computation theory problems can be broadly divided into two parts namely P and NP. The first class of problems is those for which there exists a deterministic Turing machine (Enigma machine) which can accomplish the task in polynomial time. Examples of such problem are linear search and most of the sorting algorithms. For the other class of problems, there exists a non-deterministic Turing machine which can accomplish the given task in polynomial time [1].

These problems are further of two types namely NP-complete and NP-hard. In NP-complete problems; though they are non-deterministic algorithms for solving problem, however, if the solution is given it is not difficult to verify whether it is correct in polynomial time. These are generally decision-type problems. The other kind of problems neither has an algorithm which takes place in polynomial time.

These problems are generally optimization problems. The critical path problem is one such problem.

The problem reduces to the longest path problem, which calls for finding out the longest path in the graph. There are two versions of the problem one is the NP-complete and the other is NP-hard.

In the first version, the problem is to find out there exists a path having K vertices? The problem is NP complete as if a path is given it can be easily said whether the solution is correct or not. The other version of the problem calls for finding out the longest path in the given graph, this version is NP-hard. During the literature review, it was found that the NP completeness of the first problem can be easily proved by converting it into Hamiltonian cycle problem [17].

As per the literature review, it was found that the approximation algorithms using color coding, though has minimum complexity it has, as yet not been verified on all the possible benchmarks.

Applications:

1. **Critical Path Method for scheduling on activity.** Critical Path Method is a technique that helps us to plan various projects consisting of numerous activities. Usually, some of these activities are dependent on others, i.e., in order to begin with some of our activities, we need to finish some others, thus making it a complex project. It also helps us to find the total time or the maximum time taken to complete our project and also points out more prior activities. Thus, helping us to determine various factors such as cost and speed [18].

2. **Layered Graph Drawing.** Longest path problem helps in layered graph drawing. It is a technique of drawing graphs where in, the edges are generally directed downwards and the vertices are generally in some particular horizontal row, such graphs reduce the possibility of getting the edges which cross and also reduce the number of inconsistently oriented edges. Both of which are NP-hard problems [19].

3. **Sequence Alignment Algorithms.** The longest path problem has also been used by some researchers to handle one of the most important problems in Bioinformatics that is Sequence alignment [1, 20].

5   Proposed Work

The longest path problem, as stated earlier is an NP-Hard problem. The work proposes a novel solution to solve the problem. The procedure has been explained in this section. The technique has been implemented and the results are encouraging. The results have been presented in the next section.

The application of GA to a problem requires an effective problem reduction approach. The first step of GA is population generation. The population consists of chromosomes. The number of cells in a chromosome is determined as follows.

For the graph with N vertices, the nearest power of two is found. Let this power be K then,

$$N \leq 2^K \tag{1}$$

Now each chromosome consists of N multiplied by K ($N \times K$) number of cells. This is done so that a chromosome can be divided into N parts each containing K cells. These K cells (a binary number) is converted into a decimal number and its modulus with n is then taken. The number obtained, if same as that obtained earlier (in any set of the same chromosome), then the bits are re-randomized. This procedure essentially generates a permutation of the n numbers. However, the number of chromosomes would determine the number of permutations. These limited permutations would lead us to the best (or almost best) solution. This logic also justifies the need of using GA.

This is followed by assigning fitness to each chromosome. The fitness is assigned as follows. The sequence generated (essentially a path, in the given graph), has some cost. This cost can be obtained by the given graph. In the implementation, the cost matrix of the graph is given as the input. The fitness of a chromosome is directly related to the cost. A chromosome having more cost should have more fitness. The formula employed to find the fitness is as follows.

Fitness $= K \times$ (The distance of the path determined by chromosome, where K is a parameter)

$$\tag{2}$$

This is followed by Roulette Wheel Selection. The selection procedure has been explained in the preceding sections. The work uses one-point crossover and two-point crossover but the results of the two-point crossover have been presented in the following section. This is because on an average one-point crossover does not show any better result as compared Fig. 2.

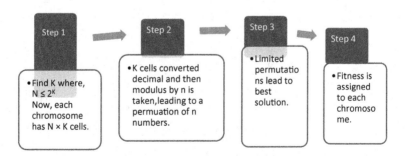

Fig. 2 Proposed algorithm for the longest path problem

The crossovered population is then made to undergo mutation. The stopping criterion is the number of generations (as stated in the next section) or the stagnation in the average fitness value.

6   Result and Conclusion

The problem has been implemented in MATLAB. The code has been tested on a few graphs of moderate size. The size of the graphs for which the algorithm was tested was 5, 7, 8, 9, and 10. The number of generations was restrained owing to the limited time. However, the results were encouraging. Table 2 gives the summary of the results and the actual answer.

Initial Population: 20
Number of generations (Maximum): 50
Crossover: Mentioned in the previous section.
Mutation: Mentioned in the previous section.
Selection: Roulette wheel.

The graphs were selected keeping in consideration the completeness of the path and feasibility of maximum number of paths Figs. 3 and 4.

Table 2 Results

n	Path found by the algorithm	Actual
5	34	34
5	32	34
5	34	34
5	30	34
5	29	34
7	43	47
7	42	47
7	47	47
7	45	47
7	47	47
9	62	65
9	62	65
9	62	65
9	62	65
9	62	65
10	90	90
10	82	90

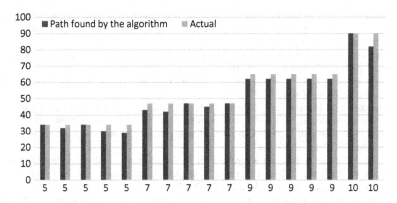

Fig. 3 Graphical representation of the results

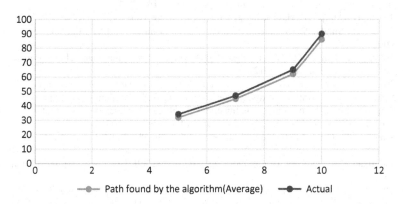

Fig. 4 Comparison of average cost in various trials

The work implements the proposed algorithm for the critical path problem. The problem is now being tested for larger population. Also, DGA is being applied to solve the problem. An extensive literature review of DGA has already been carried out [21]. The technique has also been applied to dynamic travelling salesman problem [22].

References

1. Bhasin, H.: Algorithm Analysis and Design. Oxford University Press (2015)
2. Cormen, L., Rivest, S.: Introduction to algorithms. Second Edition, Prentice Hall of India (1990)
3. Goldberg, D.E.: Genetic algorithms in search, Optimization and Machine Learning. Addison-Wesley, Reading, MA (1989)

4. Kitchenham, B.A., Charters, S.: Guidelines for performing systematic literature reviews in software engineering. Technical Report EBSE-2007–01, School of Computer Science and Mathematics, Keele University (2007)
5. Hardgrave, W.W., Nemhauser, G.L.: On the relation between the traveling-salesman and the longest-path problems. Oper. Res. **10**, 647 (1962)
6. Bhasin, H., Singla, N.: Harnessing cellular automata and genetic algorithms to solve travelling salesman problem. ICICT, 72–77 (2012)
7. Ioannidou, K., Mertzios, G.B., Nikolopoulos, G.B.: The longest path problem is polynomial on interval graphs. In: MFCS'09, Springer-Verlag, Berlin, Heidelberg (2009)
8. Karger, D., Motwani, R., Ramkumar, G.: On approximating the longest path in a graph. Algorithmica **18**(1), 82–98 (1997)
9. Uehara, R., Uno, Y.: Efficient algorithms for the longest path problem. In: ISAAC (2004)
10. Uehara, R., Valiente, G.: Linear structure of bipartite permutation graphs and the longest path problem. Inf. Process. Lett. **103**(2), 71–77 (2007)
11. Ghosh, E.: Hamiltonicity and longest path problem on special classes of graphs. Research Thesis, Indian Institute of Technology, Madras (2011)
12. Ioannidou, K., Nikolopoulos, S.D.: The longest path problem is polynomial on cocomparability graphs. Lect. Notes Comput. Sci. Springer **6410**, 27–38
13. Bhasin, H., Ahuja, G.: Harnessing genetic algorithm for vertex cover problem. IJCSE (2012)
14. Bhasin, H., Singla, N.: Genetic based algorithm for N-puzzle problem. Int. J. Comput. Appl. **51**(22) (2012)
15. Bhasin, H., Mahajan, R.: Genetic algorithms based solution to maximum clique problem, Int. J. Comput. Sci. Eng. **4**(8) (2012)
16. Bhasin, H., Behal, G., Aggarwal, N., Saini, R.K., Choudhary, S.: On the applicability of diploid genetic algorithms in dynamic environments. Soft Computing, 1–8. Springer (2015)
17. Karp, R.: Reducibility among combinatorial problems. Complexity of computer computations. In: Miller, R., Thatcher, J. (eds.) Plenum Press pp. 85–103 (1970)
18. Senior, B.A.: Critical Path Method Implementation Drawbacks: A Discussion Using Action Theory
19. Suderman, M.: Layered Graph Drawing. ACM, McGill University, Montreal, Quebue, Canada (2005)
20. Attwood, T.K., et al.: Introduction to Bioinformatics. Pearson Publications (2009)
21. Bhasin et al., H.: On the applicability of Diploid Genetic Algorithm. AI and society. Springer **31**(2), 265–274 (2016)
22. Bhasin, H., et al.: On the applicability of Diploid Genetic Algorithm in Dynamic Environments. Soft Computing, Springer, First Online, 1–10 (2015)

A Robust Digital Image Watermarking Algorithm Using DWT and SVD

Bandana Yadav, Ashish Kumar and Yogendera Kumar

Abstract A robust digital image watermarking method based on Discrete Wavelet Transform (DWT) and Singular Value Decomposition (SVD) is proposed in the present work. In this method, first, the original image of size 256×256 is DWT decomposed into the third level using Haar wavelet providing the four sub-bands LL3, LH3, HL3, and HH3. After that, SVD is applied on these sub-bands to get the diagonal matrices of singular values. The watermark image is then embedded in these singular values of the four sub-bands. Proposed algorithm is simulated using MATLAB v. 2013 and the results show that the PSNR value obtained is 84.25 which is in the range of 0.1–0.11 (of the scale factor). The PSNR value obtained for the current work is better compared to the previous approaches. Furthermore, the obtained results also show that using the present method the watermark image can be extracted properly even when the watermarked image is under various attacks like rotation, motion blur, Gaussian noise, gamma correction, rescaling, cropping, Gaussian blur, contrast adjustment, histogram equalization etc.

Keywords Digital image watermarking · Discrete wavelet transform (DWT) Haar wavelet · Singular value decomposition (SVD) · Median filter

B. Yadav (✉) · Y. Kumar
School of Electrical, Electronics and Communication Engineering, Galgotias University,
Greater Noida 201301, UP, India
e-mail: vandana91915@gmail.com

Y. Kumar
e-mail: yogiuor@gmail.com

A. Kumar
Department of Electronics and Communication Engineering, Bennett University,
Greater Noida 201310, UP, India
e-mail: akumar.1june@gmail.com

© Springer Nature Singapore Pte Ltd. 2018
M. Pant et al. (eds.), *Soft Computing: Theories and Applications*,
Advances in Intelligent Systems and Computing 583,
https://doi.org/10.1007/978-981-10-5687-1_3

25

1   Introduction

Advancements in digital multimedia technology and the popularity of internet have given rise to the abundant use and distribution of digital images. Since, these images are prone to the attacks from other users especially hackers as they can use and distribute these without permission from the owner, the safeguarding of the ownership rights has become important. Digital watermarking has been able to provide a solution to this type of problems [1, 2]. The ownership right (copyright protection) is provided by embedding some certifiable information in the original digital image, such as owner's name, company's logo, some types of copyright information etc. This information embedded in the digital multimedia can be extracted from or detected for security purposes. The embedded information (usually known as a watermark structure), a method to embed (embedding algorithm) and a method to extract (extraction algorithm) are the major components of a digital watermarking algorithm [3]. Although, hiding data in the form of water mark was reported as early as 1282 in Italy in plain paper and later used in currency notes, but the term digital watermarking appeared in 1980s and became popular in early 1990s. This term was probably first used by Komatsu and Tominaga [4] and later on Tirkel et al. [5] made it popular by presenting two techniques to hide data in images. Various other techniques have been proposed after that [2].

Digital watermarking techniques can be spatial domain based, frequency domain based, or both [6]. Watermarking techniques based on spatial domain approaches, e.g., LSB, correlation-based technique, predictive coding, etc., were easy in implementation, but were not able to withstand the attacks like low-pass filtering and lossy compression [7, 8]. Frequency domain-based approaches like Discrete Cosine Transform (DCT), Discrete Fourier Transform (DFT), and Discrete Wavelet Transform (DWT) provide better results than spatial domain approaches against attacks like low-pass filtering, contrast adjustment, brightness, etc., but these approaches show lack of resistance against geometric attacks. Although, the problem of lack of resistance against geometric attacks is removed by using DFT watermarking techniques, but the problem with DFT watermarking techniques is that if any change is made in transform coefficients, it affects the entire image [9, 10]. As DWT shows a multi-resolution description of the image, the watermarking techniques based on DWT appeared as a solution to the above-mentioned problems, but these techniques are not robust against the attacks of geometric distortions like scaling, rotation, translation, etc. Singular value decomposition (SVD) is another type of transform explored for digital watermarking. SVD is a powerful tool of linear algebra already used in image compression and other signal processing applications. An important mathematical property of SVD is that the slight variations of singular values do not affect the visual perception of the image. To achieve better transparency and robustness in the watermark embedding process, this property of SVD plays an important role. Several hybrid DWT and SVD watermarking techniques are reported [11–16] as these are preferred to the other approaches. The problem with above-mentioned techniques is that they are not able to provide good robustness with the better imperceptibility and, also, the information of the original image is not retained. Hence, there is need of a watermarking technique which is able to overcome the above-mentioned problems. A more robust watermarking algorithm compared to the previously proposed approaches based on third-level decomposition of DWT and SVD is proposed in the present work.

2 Review of Watermarking Methods Based on DWT and SVD

2.1 Watermarking Methods Based on DWT

DWT based approaches are used in many signal processing applications. Image watermarking is one such area. DWT shows the multi-resolution analysis property [17]. It decomposes the image in the constant bandwidth of frequency on a logarithmic scale. An image is decomposed by applying DWT (e.g., using Haar wavelet) into four frequency sub-bands namely LL (Low–Low or approximation), LH (Low–High or horizontal), HL (High–Low or vertical), and HH (High–High or diagonal) at the first level. Further decomposition of LL sub-band gives LL2, LH2, HL2, and HH2 at the second level. Figure 1 shows the LL3, LH3, HL3, and HH3 after third-level DWT decomposition. Usually, the low-frequency component (LL) represents the smooth variations in color whereas the high-frequency components (LH, HL, and HH) represent the sharp variations. In other words, the base of an image is constituted by the LL, and the edges which give the details are constituted by LH, HL, and HH (the high-frequency components). The detailed image is refined by adding LH, HL, and HH upon LL [18].

Various watermarking algorithms based on DWT and reported in the literature have put stress on embedding a visual watermark in low as well as high frequencies. As a result, a robust watermarking method is obtained which is able to withstand various types of attacks. If the watermark image is embedded in the low frequencies it is able to withstand the attacks having low-pass characteristics like filtering, lossy compression, geometric distortions, etc., but at the same time it makes the watermarked image more sensitive to modifications of the image histogram, such as gamma correction, contrast/brightness adjustment, histogram equalization, etc.

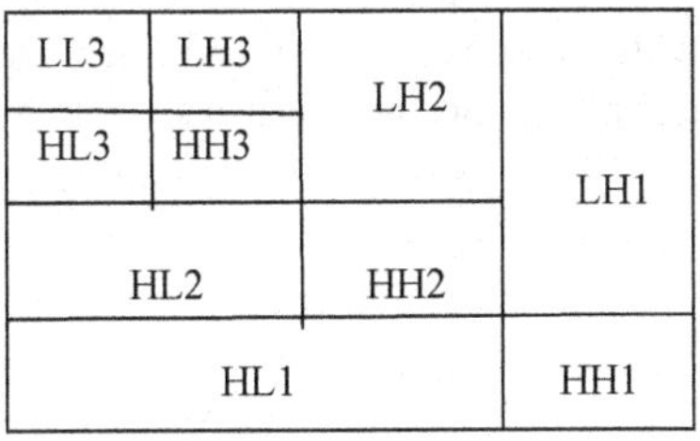

Fig. 1 Discrete wavelet transform-based third-level decomposition

LL3	LH3	LH2	LH1
HL3	HH3		
HL2		HH2	
HL1			HH1

If the watermark image is embedded in the middle and high frequencies then it is less robust to attacks such as low-pass filtering, lossy compression, and small geometric deformations of the image, but it makes the watermarked image more robust with respect to attacks like noise adding, and nonlinear deformations of the gray scale [12]. Thus the watermark image has to be embedded in all sub-bands (LL, LH, HL, and HH) to make the watermarked image able to withstand various types of attacks.

2.2 Watermarking Methods Based on SVD

A matrix is transformed into three matrices in the SVD transform. These three matrices are similar in size to the original matrix. According to linear algebra, an array of nonnegative scalar entries which can be considered as a matrix is represented as an image. If a square image is denoted by A, where $A \in R_{n \times n}$, R is the real number domain, then SVD of A is given by [19]:

$$A = USV^T$$

$U \in R_{n \times n}$ and $V \in R_{n \times n}$ are orthogonal matrices and $S \in R_{n \times n}$ is a diagonal matrix given by,

$$S = \begin{bmatrix} \sigma_1 & & & \\ & \sigma_2 & & \\ & & \ddots & \\ & & & \sigma_n \end{bmatrix}$$

Here σ's (diagonal elements) are known as singular values which satisfy

$$\sigma_1 \geq \sigma_2 \geq \sigma_r \geq \sigma_{r+1} = \sigma_{r+2} \ldots = \sigma_n = 0 \tag{1}$$

Various properties of SVD and hence its usefulness in digital image watermarking has been discussed in [20–22].

The approaches based on SVD are usually working on the principle of finding singular value decomposition of the image and then modifying the singular values to place the watermark in it. Recently, techniques based on SVD have been mixed with other types of techniques including DCT and DWT based. These hybrid techniques are getting very popular in digital image watermarking applications.

2.3 Median Filter

The median of the extracted watermark images from all four sub-bands which are much similar to the original watermark image is taken by using the median filter.

3 Proposed Watermarking Method Based on DWT and SVD

3.1 Process of Watermark Embedding

1. A gray scale image as a host image 'k' having size 256×256 and a to be embedded watermark image 'h' of size 144×144 are taken.
2. Haar wavelet-based third-level DWT decomposition on the host image is performed. It gives four sub-bands LL3, LH3, HL3, and HH3 (approximation, horizontal, vertical, and diagonal) respectively.
3. Singular value decomposition (SVD) has been applied to all four sub-bands (LL3, LH3, HL3, and HH3) given as

$$k^n = U^n S^n V^{nT}, \quad n = 1, 2, 3, 4 \tag{2}$$

n represents four sub-bands.

4. Singular values in all four sub-bands with the watermark image have been modified and then SVD is applied to them, described as

$$S^n + \alpha W = U_w^n S_w^n V_w^{nT} \tag{3}$$

α is the scale factor which is used to control the strength of the watermark to be embedded.

5. Four sets of modified sub-bands are obtained, i.e.,

$$k^{*n} = U^n S_w^n V^{nT}, \quad n = 1, 2, 3, 4 \tag{4}$$

where n represents four sub-bands.

6. Inverse discrete wavelet transform (IDWT) is performed on all the modified sub-bands to get the watermarked image k_w.

3.2   *Watermark Extraction Process*

7. Haar wavelet-based third-level DWT decomposition is applied on the water-marked image (similar to host image) k_w^* to obtain four sub-bands LL3, LH3, HL3, and HH3 respectively.

8. Singular value decomposition (SVD) is applied to all four sub-bands i.e.,

$$k_w^{*n} = U^{*n} S_w^{*n} V^{*n}, \quad n = 1, 2, 3, 4 \tag{5}$$

n represents four sub-bands.

9. Compute the value

$$D^{*n} = U_w^n S_w^{*n} V_w^{nT}, \quad n = 1, 2, 3, 4 \tag{6}$$

10. Watermark image from each sub-band is extracted, i.e.,

$$W^{*n} = \frac{D^{*n} - S^n}{\alpha}, \quad n = 1, 2, 3, 4 \tag{7}$$

11. To minimize the effect of negative values appearing in watermark image due to SVD effect, apply average function given as

$$W(i,j) = \frac{W(i,j-1) + W(i,j+1)}{2} \tag{8}$$

where i, j are the index values of the pixel intensity of watermark image.

12. Apply median filter to watermark images which are present in all four sub-bands to get the watermark (which are much similar to original watermark)

$$W_m = \text{median}(W_n) \tag{9}$$

n represents four sub-bands.

4   Results

The proposed technique is applied on the 256×256 size host images and the watermark image taken is of size 144×144. The proposed DWT-SVD-based watermarking algorithm is tested using several samples. The DWT decomposition is performed using the Haar wavelet and then SVD is applied. The peak signal-to-noise ratio (PSNR) of the watermarked image and the correlation coefficients of the extracted watermark are also calculated. Here, PSNR describes the imperceptibility and the correlation coefficients describe the robustness. Several results have been obtained for the watermarked image without attacks and also with attacks like Gaussian noise, Gamma correction, Average filtering, and Contrast adjustment, etc., on watermarked images. Figure 2 shows the 256×256 grayscale image Lena, the 144×144 grayscale visual watermark, the watermarked image, and the extracted watermarks. The quality of the watermarked image is calculated using PSNR value expressed as

$$PSNR(db) = 10\log \frac{255^2}{MSE} \tag{10}$$

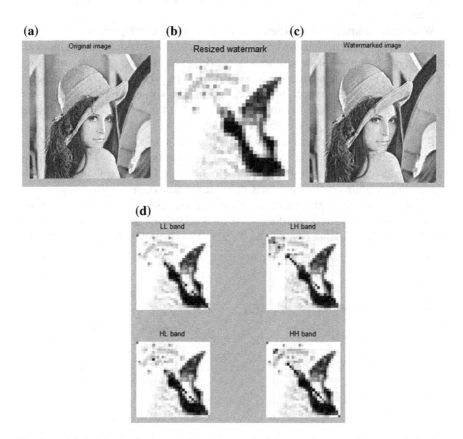

Fig. 2 **a** Original image, **b** watermark, **c** watermarked image, **d** extracted watermark in four sub-bands

where Mean Square Error (MSE) is calculated as

$$\text{MSE} = \frac{1}{N \times N} \sum_{x=0}^{N-1} \sum_{y=0}^{N-1} \left[\hat{f}(x,y) - f(x,y) \right]^2 \tag{11}$$

where $N \times N$ is the image size, $\hat{f}$ is the reconstructed image, and f is the original image. The robustness of the extracted watermark against different attacks is calculated using Pearson's Correlation Coefficient formula which is expressed as

$$\text{PCC} = \frac{n \sum(xy) - (\sum x)(\sum y)}{\sqrt{[n \sum x^2 - (\sum x^2)][n \sum y^2 - (\sum y^2)]}} \tag{12}$$

where x and y are original and extracted watermark.

Table 1 shows the MSE and PSNR values of the watermarked images for the different scale factors.

Watermarked images with different attacks are shown in Fig. 3a–k.

Table 2 shows the correlation coefficient values of the extracted watermark in all four sub-bands and also shows the watermark which is extracted after applying median filter to four sub-bands and denoted by Wm. The quality of constructed watermarks can be evaluated subjectively through a visual comparison with the reference watermark. It is obvious from Table 2 that the watermarks constructed from the four sub-bands look different for each attack.

Figure 4a–k shows the extracted watermark after attacks.

These results are compared with the technique mentioned in [13] as shown in Tables 3 and 4 which contain PSNR values in the Scale Factor (SF) range of 0.01–0.09 and correlation coefficient values of extracted watermark respectively.

Some other results with attacks like Poisson noise, Gaussian blur, Motion blur, Uniform noise, and Rescaling which shows the good correlation coefficient values of extracted watermark are found to be 0.9695, 0.8967, 0.9510, 0.9730, and 0.9812 respectively.

Table 1 MSE and PSNR values

Scale factor (α)	MSE	PSNR
0.01	0.00024	84.25
0.03	0.095	58.31
0.05	0.565	50.6
0.07	2.054	45
0.09	5.526	40.71
0.11	10.77	37.81

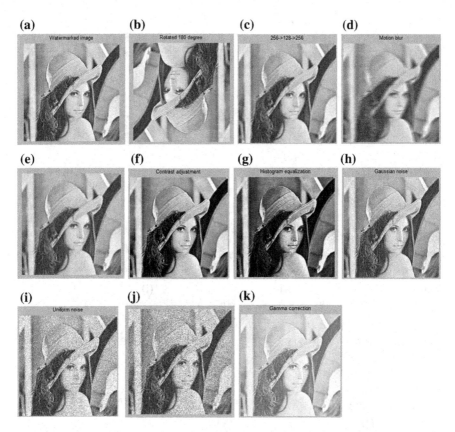

Fig. 3 Watermarked image **a** cropped image, **b** rotated by 180°, **c** rescaling 256 → 128 → 256, **d** motion blur, **e** Gaussian blur, **f** contrast adjustment, **g** histogram equalization, **h** Gaussian noise, **i** uniform noise, **j** Poisson noise, **k** gamma correction

Table 2 Correlation coefficient values of the extracted watermark in all sub-bands with median filter

Attacks	LL	LH	HL	HH	Wm
Crop	0.5752	0.5049	0.8491	0.8748	0.9323
Rotate	0.9649	0.9808	0.9697	0.9769	0.9834
Rescaling	0.9627	0.9670	0.9686	0.9645	0.9812
Motion blur	0.8515	0.9025	0.6075	0.8635	0.8967
Gaussian blur	0.9674	0.9175	0.9268	0.9193	0.9695
Contrast adjustment	0.9037	0.9685	0.9652	0.9650	0.9541
Histogram equalization	0.7204	0.8665	0.8810	0.9463	0.9323
Gaussian noise	0.9610	0.9735	0.9582	0.9797	0.9821
Uniform noise	0.9241	0.7123	0.9356	0.7343	0.9730
Poisson noise	0.8315	0.7363	0.8236	0.8254	0.9510
Gamma correction	0.7602	0.7991	0.9402	0.8891	0.9198

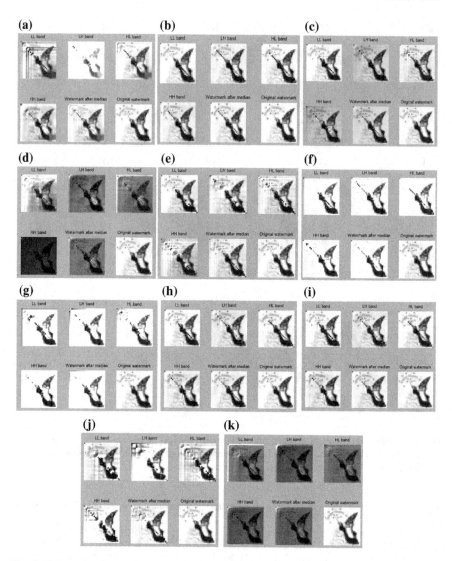

Fig. 4 Extracted watermark from attacks **a** crop, **b** rotate, **c** rescaling, **d** motion blur, **e** Gaussian blur, **f** contrast adjustment, **g** histogram equalization, **h** Gaussian noise, **i** uniform noise, **j** Poisson noise, **k** gamma correction

Table 3 Comparison of PSNR for G&E [12], L&T [11], L&T [13], and proposed algorithm

SF value	0.01	0.03	0.05	0.07	0.09
G&E [10]	37.80	36.79	35.29	33.27	32.26
L&T [9]	51.50	51.26	50.05	47.84	45.56
L&T [11]	51.14	51.14	50.89	49.52	47.49
Proposed	84.25	58.31	50.6	45	40.71

Table 4 Comparison of correlation coefficients for G&E [12], L&T [11], L&T [13] and proposed algorithm

Method	Crop (CR)	Rotate (RO)	CA	HE	GC	GN
G&E [12]	0.7063	0.9091	0.9759	0.9700	0.9989	0.9377
L&T [11]	0.9578	0.9444	0.9848	0.9780	0.9871	0.8953
L&T [13]	0.9843	0.9897	0.9958	0.9890	0.9994	0.9756
Proposed	0.9323	0.9834	0.9541	0.9323	0.9198	0.9821

5 Conclusion

In the present work, Haar wavelet-based digital image watermarking algorithm is developed. MATLAB v. 2013 has been used for the simulation purpose in the proposed work. In this work, Original image of size 256×256 is decomposed into third level of DWT which gives four sub-bands. SVD is applied on these four sub-bands which give diagonal matrices of singular values. Then, the watermark image is embedded in these singular values of four sub-bands. The obtained results show 84.25 as PSNR value in range of 0.1–0.11 (of the scale factor). The proposed work gives better PSNR value compared to the previous approaches.

References

1. Podilchuk, C.I., Delp, E.J.: Digital watermarking: algorithms and applications. IEEE Signal Proc. Mag. 33–46 (2007)
2. Cox, I., Miller, M., Bloom, J., Fridrich, J., Kalke, T.: Digital Watermarking and Steganography, 2nd edn. Morgan Kaufmann, Burlington, MA (2007)
3. Hartung, F., Kutter, M.: Multimedia watermarking techniques. Proc. IEEE **87**(7), 1079–1107 (1999)
4. Komatsu, N., Tominaga, H.: Authentication System Using Concealed Image in Telematics. Memoirs of the School of Science and Engineering, Waseda University, **52**, 45–60 (1988)
5. Tirkel, A.Z., Rankin, G.A., Schyndel, R.M.V., Ho, W.J., Mee, N.R.A., Osborne, C.F.: Electronic watermark. In: Digital Image Computing, Technology and Applications (DICTA'93), 666–673 (1993)
6. Swanson, M.D., Kobayashi, M., Tewfik, A.H.: Multimedia data embedding and watermarking technologies. Proc. IEEE **86**(6), 1064–1087 (1998)
7. Mukherjee, D., Maitra, S., Acton, S.: Spatial domain digital image watermarking of multimedia objects for buyer authentication. IEEE Trans. Multimedia **6**(1), 1–15 (2004)
8. Bender, W., Gruhl, D., Morimoto, N.: Techniques for data hiding. IBM Syst. J. **35**(3–4), 313–336 (1996)
9. Muhermagic, E., Furt, B.: Survey of Watermarking Techniques and Applications (2004)
10. Mohammed, G.N., Yasin, A., and Zeki, A.M.: Digital image watermarking, analysis of current methods. In: International Conference on Advanced Computer Science Applications and Technologies, 324–329 (2012)
11. Liu, R., Tan, T.: A SVD-based watermarking scheme for protecting rightful ownership. IEEE Trans. Multimedia. **4**(1), 121–128 (2002)

12. Ganic, E., Eskicioglu, A.M.: Robust DWT-SVD domain image watermarking: embedding data in all frequencies. In: MM&SEC'04, 166–174 (2004)
13. Lai, C.C., Tsai, C.C.: Digital image watermarking using discrete wavelet transform and singular value decomposition. IEEE Trans. Instrum. Meas. **49**, 3060–3063 (2010)
14. Ansari, I.A., Pant, M., Ahn, C.W.: Robust and false positive free watermarking in IWT domain using SVD and ABC. Eng. Appl. Artif. Intell. **49**, 114–125 (2016)
15. Ali, M., Ahn, C.W.: An optimized watermarking technique based on self-adaptive DE in DWT–SVD transform domain. Sig. Process. **94**, 545–556 (2014)
16. Agarwal, C., Mishra, A., Sharma, A., Bedi, P.: Optimized gray-scale image watermarking using DWT–SVD and firefly algorithm. Expert Syst. Appl. **41**(17), 7858–7867 (2014)
17. Mallat, S.: The theory for multiresolution signal decomposition: the wavelet representation. IEEE Trans. Pattern Anal. Mach. Intell. **11**(7), 654–693 (1989)
18. Muhit, A.A., Islam, M.S., Othman, M.: VLSI implementation of discrete wavelet transform (DWT) for image compression. In: 2nd International Conference on Autonomous Robots and Agents, pp. 391–395 (2005)
19. Bhatnagar, G., Raman, B.: A new robust reference watermarking scheme based on DWT-SVD. Comput. Stand. Interfaces **31**(5), 1002–1013 (2009)
20. Chang, C.C., Tsai, P., Lin, C.C.: SVD-based digital image watermarking scheme. Pattern Recogn. Lett. **26**(10), 1577–1586 (2005)
21. Zhou, Z., Tang, B., and Liu, X.: A block-SVD based image watermarking method. In: The 6th World Congress on Intelligent Control and Automation, **2**, 10347–10351 (2006)
22. Zhou, B., Chen, J.: A geometric distortion resilient image watermarking algorithm based on SVD. Chin. J. Image Graph. **9**, 506–512 (2004)

A Novel and Secure Multiparty Key Exchange Scheme Using Trilinear Pairing Map Based on Elliptic Curve Cryptography

Manoj Kumar and Pratik Gupta

Abstract Elliptic curves have been broadly studied for more than hundred years. Recently they have become a tool in various important applied fields such as coding theory, pseudorandom bit generation, number theory algorithms, etc. Actually, elliptic curve cryptography (ECC) is an alternative technique for conventional asymmetric cryptography like RSA, DSA, and Diffie–Hellman key exchange scheme. Instead of larger key size, ECC uses smaller key size to provide the highest strength-per-bit of any cryptographic system known today. This results in faster computations, lower power consumption, and less memory allocations. Another benefit of using ECC is that authentication schemes based on ECC are much secure even if a small key size is used. ECC also provides a methodology to obtain high speed, efficient, and scalable implementations of protocols for authentication and key agreement. In the present paper, we have discussed a trilinear pairing map on finitely generated free R-modules with rank three, where R is a commutative ring with unity. A trilinear pairing on an elliptic curve is constructed and we used this pairing map to a multiparty key exchange scheme. Since the secret shared key generated among the members of the group is constructed by the contribution of each member of the group, it increases the security of the proposed scheme.

Keywords Elliptic curve · Free modules · Bilinear pairing map
Authentication · Torsion points · Finite field · Jacobian projective transformation

Mathematics Subject Classification 94A60 · 14G50

M. Kumar (✉) · P. Gupta
Department of Mathematics and Statistics, Gurukula Kangri Vishwavidyalaya,
Haridwar 249404, Uttrakhand, India
e-mail: sdmkg1@gmail.com

P. Gupta
e-mail: pratikgupta1810@gmail.com

© Springer Nature Singapore Pte Ltd. 2018
M. Pant et al. (eds.), *Soft Computing: Theories and Applications*,
Advances in Intelligent Systems and Computing 583,
https://doi.org/10.1007/978-981-10-5687-1_4

1   Introduction

Very recently the authors, Kumar and Gupta [1] obtained cryptographic schemes based on elliptic curves over the ring $Z_p[i]$. In the present paper, we introduced a trilinear pairing map on finitely generated free R-modules with rank three, where R is a commutative ring with unity. We used this pairing map to generate a secret shared key for group communication. In the recent years, pairing-based crypto-graphic schemes on elliptic curve have been a very needful domain of research in cryptography. The concept of pairing in cryptography was first introduced by Weil [2]. Generally, pairings map use of pair of points on an elliptic curve into the multiplicative group of a finite field. The use of pairings by the publication of the paper of Joux [3] in cryptography has developed at an extraordinary pace. The identity-based encryption scheme of Boneh and Franklin [4] and, the short signa-ture scheme of Boneh et al. [5] are important applications of pairings in cryptog-raphy. In past four decades, pairing maps are continuously studied by several researchers [6–11].

Let E with $y^2 = x^3 + ax + b$ be an elliptic curve defined over a finite field F. Then, we know that [12–14] each elliptic curve point can be described by two coordinates $x, y \in F$. In this case, we say that elliptic curve points belong to two-dimensional affine planes $A_F^2 = \{(x, y) \in F \times F\}$. Suppose the coordinates (x, y) of the affine plane $A_F^2 = \{(x, y) \in F \times F\}$ are mapped to the coordinates (X, Y, Z) of projective plane $P_F^3 = \{(X, Y, Z) \in F \times F \times F\}$ as

$$(X, Y, Z) = \left(x.Z^c, y.Z^d, 1\right) \text{ or } x = X/Z^c \text{ and } y = Y/Z^d \tag{1}$$

where c, d are integers.

After applying the Jacobian projective transformation with $c = 2$ and $d = 3$, elliptic curve E can be rewritten as $E : Y^2 = X^3 + aXZ^4 + bZ^6$.

If $P_1 = (X_1, Y_1, Z_1)$ and $P_2 = (X_2, Y_2, Z_2)$ are two distinct points on the pro-jective plane then their point addition $(P_3 = (X_3, Y_3, Z_3) = P_1 + P_2)$ and point doubling $(P_3 = 2P_1)$ can be described as follows.

1.1   Addition of Points on Projective Plane

Case-I: If $x_1 \neq x_2$ then we have

$$\lambda = \frac{y_2 - y_1}{x_2 - x_1} = \frac{\frac{Y_2}{Z_2^d} - \frac{Y_1}{Z_1^d}}{\frac{X_2}{Z_2^c} - \frac{X_1}{Z_1^c}} = \frac{\left(Y_2 Z_1^d - Y_1 Z_2^d\right) Z_2^c Z_1^c}{\left(X_2 Z_1^c - X_1 Z_2^c\right) Z_2^d Z_1^d}$$

It is obvious from above expression that λ exists because $x_1 \neq x_2$. Now the point P_3 can be calculated as

$$x_3 = \lambda^2 - x_1 - x_2 = \left(\frac{(Y_2 Z_1^d - Y_1 Z_2^d) Z_2^c Z_1^c}{(X_2 Z_1^c - X_1 Z_2^c) Z_2^d Z_1^d}\right)^2 - \frac{X_1 Z_2^c + X_2 Z_1^c}{Z_2^c Z_1^c}$$

$$= \frac{(Y_2 Z_1^d - Y_1 Z_2^d) Z_2^{3c} Z_1^{3c} - (X_1 Z_2^c + X_2 Z_1^c)(X_2 Z_1^c - X_1 Z_2^c)^2 Z_2^{2d} Z_1^{2d}}{(X_2 Z_1^c - X_1 Z_2^c) Z_2^{2d+c} Z_1^{2d+c}}$$

$y_3 = \lambda(x_1 - x_3) - y_1$

$$= \frac{(Y_2 Z_1^d - Y_1 Z_2^d) Z_2^c Z_1^c}{(X_2 Z_1^c - X_1 Z_2^c) Z_2^d Z_1^d} \left(\frac{X_1}{Z_1^c} - \frac{(Y_2 Z_1^d - Y_1 Z_2^d)^2 Z_2^{3c} Z_1^{3c} - (X_1 Z_2^c + X_2 Z_1^c)(X_2 Z_1^c - X_1 Z_2^c)^2 Z_2^{2d} Z_1^{2d}}{(X_2 Z_1^c - X_1 Z_2^c) Z_2^{2d+c} Z_1^{2d+c}}\right) - \frac{Y_1}{Z_1^d}$$

$$= \frac{(Y_2 Z_1^d - Y_1 Z_2^d) Z_2^c Z_1^c}{(X_2 Z_1^c - X_1 Z_2^c) Z_2^d Z_1^d} \left(\frac{(2X_1 Z_2^c + X_2 Z_1^c)(X_2 Z_1^c - X_1 Z_2^c)^2 Z_2^{2d} Z_1^{2d} - (Y_2 Z_1^d - Y_1 Z_2^d)^2 Z_2^{3c} Z_1^{3c}}{(X_2 Z_1^c - X_1 Z_2^c) Z_2^{2d+c} Z_1^{2d+c}}\right) - \frac{Y_1}{Z_1^d}$$

$$= \frac{(Y_2 Z_1^d - Y_1 Z_2^d)(2X_1 Z_2^c + X_2 Z_1^c)(X_2 Z_1^c - X_1 Z_2^c)^2 Z_2^{2d+c} Z_1^{2d+c} - (Y_2 Z_1^d - Y_1 Z_2^d)^3 Z_2^{4c} Z_1^{4c}}{(X_2 Z_1^c - X_1 Z_2^c) Z_2^{3d+c} Z_1^{3d+c}} - \frac{Y_1}{Z_1^d}$$

$$= \frac{((Y_2 Z_1^d - Y_1 Z_2^d)(2X_1 Z_2^c + X_2 Z_1^c) - Y_1 Z_2^d(X_2 Z_1^c - X_1 Z_2^c))(X_2 Z_1^c - X_1 Z_2^c)^2 Z_2^{2d} Z_1^{2d} - (Y_2 Z_1^d - Y_1 Z_2^d)^3 Z_2^{3c} Z_1^{3c}}{(X_2 Z_1^c - X_1 Z_2^c) Z_2^{3d} Z_1^{3d}}$$

Using (1) and Jacobian projective transformation with $c = 2$ and $d = 3$, P_3 is given by

$$X_3 = (Y_2 Z_1^3 - Y_1 Z_2^3)^2 - (X_1 Z_2^2 + X_2 Z_1^2)(X_2 Z_1^2 - X_1 Z_2^2)^2,$$

$$Y_3 = ((Y_2 Z_1^3 - Y_1 Z_2^3)(2X_1 Z_2^2 + X_2 Z_1^2) - Y_1 Z_2^3(X_2 Z_1^2 - X_1 Z_2^2)) - (Y_2 Z_1^3 - Y_1 Z_2^3)^3,$$

and

$$Z_3 = (X_2 Z_1^2 - X_1 Z_2^2) Z_2 Z_1.$$

Case-II: If $x_1 = x_2$ then we have $P_3 = P_1 + P_2 = O$, where O is the point at infinity of the elliptic curve E in projective coordinates. It can be easily seen that for Jacobian projective coordinates, the point at infinity has the form $(1, 1, 0)$.

1.2 Point Doubling on Projective Plane

For point doubling, we can take $P_1 = P_2$ then $P_3 = P_1 + P_2 = 2P_1 = (X_3, Y_3, Z_3)$. We have

$$\lambda = \frac{3x_1^2 + a}{2y_1} = \frac{3X_1^2 Z_1^d + a Z_1^{2c+d}}{2Z_1^{2c} Y_1}$$

Evidently, λ exists if $y_1 \neq 0$

So we get

$$x_3 = \lambda^2 - 2x_1 = \frac{(3X_1^2 + aZ_1^{2c})^2 Z_1^{2d}}{4Z_1^{4c} Y_1^2} - 2\frac{X_1}{Z_1^c} = \frac{(3X_1^2 + aZ_1^{2c})^2 Z_1^{2d} - 8Z_1^{3c} X_1 Y_1^2}{4Z_1^{4c} Y_1^2}$$

$$\begin{aligned}
y_3 &= \lambda(x_1 - x_3) - y_1 = \lambda\left(3x_1 - \lambda^2\right) - y_1 \\
&= \frac{\left(3X_1^2 + aZ_1^{2c}\right)^2 Z_1^d}{2Z_1^{2c} Y_1}\left(3\frac{X_1}{Z_1^c} - \frac{\left(3X_1^2 + aZ_1^{2c}\right)^2 Z_1^{2d}}{4Z_1^{4c} Y_1^2}\right) - \frac{Y_1}{Z_1^d} \\
&= \frac{12X_1 Y_1^2\left(3X_1^2 + aZ_1^{2c}\right)^2 Z_1^{3c+2d} - \left(3X_1^2 + aZ_1^{2c}\right)^3 Z_1^{4d} - 8Z_1^{6c} Y_1^4}{8Z_1^{6c+d} Y_1^3}
\end{aligned}$$

Using (1) and Jacobian projective transformation with $c = 2$ and $d = 3$, the doubling of point P_1 is given by $P_3 = (X_3, Y_3, Z_3)$ where

$$X_3 = (3X_1^2 + aZ_1^4)^2 - 8X_1 Y_1^2,$$
$$Y_3 = 12X_1 Y_1^2(3X_1^2 + aZ_1^4) - (3X_1^2 + aZ_1^4)^3 - 8Y_1^4,$$

and

$$Z_3 = 2Z_1 Y_1.$$

Point subtraction can be performed as $P_3 = P_1 - P_2 = P_1 + (-P_2)$ where $-P_2$ is the additive inverse of P_2 and $-P_2 = (X_2, -Y_2, Z_2)$.

Here, it is remarkable that we are no need of division and multiplication operations for calculating elliptic curve point P_3 on the projective plane.

2 Construction of a Trilinear Pairing on Finitely Generated Free R-Modulus

In this section, we will construct trilinear pairing on finitely generated free R-modules with rank 3. At the end of this section, we will also discuss an auxiliary result which will be helpful in the next section.

According to the terminology as in the references [15–18], let R be a commutative ring with unity, P be a finitely generated free R-module with rank 3 and (l, m, n) be a generating pair for P. We consider elements $a = u_1 l + v_1 m + w_1 n$, $b = u_2 l + v_2 m + w_2 n$, $c = u_3 l + v_3 m + w_3 n$ in P, where $u_i, v_i, w_i \in P$ for each $i = 1, 2, 3$.

For some fixed $\alpha, \beta, \gamma \in R$ where all α, β and γ are not zero at the same time, we construct a pairing map

$$f_{\alpha,\beta,\gamma} : P \times P \times P \rightarrow P \tag{2}$$

defined by

$$f_{\alpha,\beta,\gamma}(a,b,c) = [u_1(v_2 w_3 - v_3 w_2) + v_1(u_3 w_2 - u_2 w_3) + w_1(u_2 v_3 - u_3 v_2)].(\alpha l + \beta m + \gamma n) \tag{3}$$

It can be easily seen that the pairing map (2) defined by (3) is a nontrivial and well-defined map.

For this, if $a = a'$, $b = b'$ and $c = c'$ then we have $u_i = u_i'$, $v_i = v_i'$ and $w_i = w_i'$ for each $i = 1, 2, 3$ by independency of (l, m, n). This implies $f_{\alpha,\beta,\gamma}(a,b,c) = f_{\alpha,\beta,\gamma}(a',b',c')$. Therefore, the map is well defined.

2.1 Proposition

The pairing $f_{\alpha,\beta,\gamma}(a,b,c)$ has the following properties:

2.1.1 Identity

$f_{\alpha,\beta,\gamma}(a,a,a) = 0$ for all $a \in P$.

Proof Let $a \in P$. Then we have

$$\begin{aligned}
f_{\alpha,\beta,\gamma}(a,a,a) &= [u_1(v_1 w_1 - w_1 v_1) + v_1(w_1 u_1 - u_1 w_1) + w_1(u_1 v_1 \\
&\quad - v_1 u_1)](\alpha l + \beta m + \gamma n) \\
&= 0.
\end{aligned}$$

2.1.2 Bilinearity

If $a, b, c, d \in P$ then we have

$$\begin{aligned}
f_{\alpha,\beta,\gamma}(a+b,c,d) &= f_{\alpha,\beta,\gamma}(a,c,d) + f_{\alpha,\beta,\gamma}(b,c,d), \\
f_{\alpha,\beta,\gamma}(a,b+c,d) &= f_{\alpha,\beta,\gamma}(a,b,d) + f_{\alpha,\beta,\gamma}(a,c,d), \\
f_{\alpha,\beta,\gamma}(a,b,c+d) &= f_{\alpha,\beta,\gamma}(a,b,c) + f_{\alpha,\beta,\gamma}(a,b,d).
\end{aligned}$$

and

$$f_{\alpha,\beta,\gamma}(a,b,c+d) = f_{\alpha,\beta,\gamma}(a,b,c) + f_{\alpha,\beta,\gamma}(a,b,d).$$

Proof Let $a,b,c,d \in P$. Then we have

$$
\begin{aligned}
f_{\alpha,\beta,\gamma}(a+b,c,d) &= [(u_1+u_2)(v_3w_4 - v_4w_3) + (v_1+v_2)(u_4w_3 - u_3w_4) \\
&\quad + (w_1+w_2)(u_3v_4 - u_4v_3)](\alpha l + \beta m + \gamma n) \\
&= [u_1(v_3w_4 - v_4w_3) + v_1(u_4w_3 - u_3w_4) + w_1(u_3v_4 - u_4v_3)](\alpha l + \beta m + \gamma n) \\
&\quad + [u_2(v_3w_4 - v_4w_3) + v_2(u_4w_3 - u_3w_4) + w_2(u_3v_4 - u_4v_3)](\alpha l + \beta m + \gamma n) \\
&= f_{\alpha,\beta,\gamma}(a,c,d) + f_{\alpha,\beta,\gamma}(b,c,d)
\end{aligned}
$$

Similarly, it can be easily verified that

$$f_{\alpha,\beta,\gamma}(a,b+c,d) = f_{\alpha,\beta,\gamma}(a,b,d) + f_{\alpha,\beta,\gamma}(a,c,d)$$

$$f_{\alpha,\beta,\gamma}(a,b,c+d) = f_{\alpha,\beta,\gamma}(a,b,c) + f_{\alpha,\beta,\gamma}(a,b,d)$$

2.1.3 Antisymmetry

$f_{\alpha,\beta,\gamma}(a,b,c) = -f_{\alpha,\beta,\gamma}(b,c,a)$ for all $a,b,c \in P$.

Proof Let $a,b,c \in P$. Then we have

$$
\begin{aligned}
f_{\alpha,\beta,\gamma}(a,b,c) &= [u_1(v_2w_3 - v_3w_2) + v_1(u_3w_2 - u_2w_3) + w_1(u_2v_3 - u_3v_2)].(\alpha l + \beta m + \gamma n) \\
&= -[u_2(v_1w_3 - v_3w_1) + v_2(u_3w_1 - u_1w_3) + w_2(u_1v_3 - u_3v_1)].(\alpha l + \beta m + \gamma n) \\
&= -f_{\alpha,\beta,\gamma}(b,a,c)
\end{aligned}
$$

2.1.4 Non-degeneracy

If $a,b,c \in P$ then $f_{\alpha,\beta,\gamma}(a,b,0) = 0 = f_{\alpha,\beta,\gamma}(a,0,c) = f_{\alpha,\beta,\gamma}(0,b,c)$.

Also, if $f_{\alpha,\beta,\gamma}(a,b,c) = 0$ for all $b,c \in P$, then $a = 0$.

Moreover, if $f_{\alpha,\beta,\gamma}(a,b,c) = 0$ for all $c \in P$ then $a = kb$ for some constant k.

Proof Let $a,b \in P$. Then we have

$$f_{\alpha,\beta,\gamma}(a,b,0) = [u_1(0-0) + v_1(0-0) + w_1(0-0)].(\alpha l + \beta m + \gamma n) = 0.$$

In a similar manner, we can show that $f_{\alpha,\beta,\gamma}(a,0,c) = 0$ and $f_{\alpha,\beta,\gamma}(0,b,c) = 0$ for all $a,b,c \in P$.

If $f_{\alpha,\beta,\gamma}(a,b,c) = 0$ for all $b, c \in P$ then we have

$$[u_1(v_2w_3 - v_3w_2) + v_1(u_3w_2 - u_2w_3) + w_1(u_2v_3 - u_3v_2)].(\alpha l + \beta m + \gamma n)$$
$$= 0 \text{ for all } b, c \in P$$

This implies $u_1 = v_1 = w_1 = 0$. Therefore $a = 0$.
Let $f_{\alpha,\beta,\gamma}(a,b,c) = 0$ for all $c \in P$. Then we have

$$[u_1(v_2w_3 - v_3w_2) + v_1(u_3w_2 - u_2w_3) + w_1(u_2v_3 - u_3v_2)].(\alpha l + \beta m + \gamma n) = 0$$

On rearranging the terms in above expression, we get

$$[u_3(v_1w_2 - v_2w_1) + v_3(u_1w_2 - u_2w_1) + w_3(u_1v_2 - u_2v_1)].(\alpha l + \beta m + \gamma n) = 0$$

This implies that $\frac{u_1}{u_2} = \frac{v_1}{v_2} = \frac{w_1}{w_2} = k$ for some constant k i.e. $a = kb$.

3 Construction of a Trilinear Pairing on Elliptic Curves

In this section, we will extend the trilinear pairing (constructed in the previous section) on an elliptic curve over the finite fields. At the end of this section, we will also discuss an auxiliary result which will be useful in the next section.

3.1 Torsion Points on an Elliptic Curve [10]

Let E be an elliptic curve. Then a point $P \in E$ is said to be a torsion point if there exists a positive integer m such that $mP = O$. The smallest such integer is called the order of P. An n-torsion point is a point $P \in E$ satisfying $nP = O$.

Let K be a field with characteristic zero or a prime p (p is relatively prime to n) and let $E = E(\overline{K})$ be an elliptic curve over $\overline{K}$ where $\overline{K}$ is an algebraic closure of K. Also let $E(K)[n]$ denote the subgroup of n-torsion point in $E(K)$, where $n \neq 0$. For our simplicity, we denote $E(\overline{K})[n]$ by $E[n]$.

Let $\{U, V, W\}$ for some fixed generating pair for $E[n]$. Then the points $P, Q, R \in E[n]$ can be expressed as $P = a_1 U + b_1 V + c_1 W$, $Q = a_2 U + b_2 V + c_2 W$, $R = a_3 U + b_3 V + c_3 W$, where a_i, b_i, c_i for each $i = 1, 2, 3$ are integers in $[0, n-1]$.

Now for some fixed integers $\alpha, \beta, \gamma \in [0, n-1]$, where all α, β, γ are not zero at the same time, we construct a map

$$f^n_{\alpha,\beta,\gamma} : E[n] \times E[n] \times E[n] \ \longrightarrow \ E[n] \tag{4}$$

defined by

$$f_{\alpha,\beta,\gamma}^n(P,Q,R) = [a_1(b_2c_3 - b_3c_2) + b_1(a_3c_2 - a_2c_3) + c_1(a_2b_3 - a_3b_2)].(\alpha U + \beta V + \gamma W) \tag{5}$$

It can be easily checked the map (4) defined by (5) is well defined.

3.2 Proposition

The pairing map $f_{\alpha,\beta,\gamma}^n(P,Q,R)$ constructed as above satisfies the following postulates:

3.2.1 Identity

$f_{\alpha,\beta,\gamma}^n(P,P,P) = O$ for all $P \in E[n]$.

Proof Let $P \in E[n]$. Then we have

$$\begin{aligned} f_{\alpha,\beta,\gamma}^n(P,Q,R) &= [a_1(b_1c_1 - b_1c_1) + b_1(a_1c_1 - a_1c_1) + c_1(a_1b_1 \\ &\quad - a_1b_1)].(\alpha U + \beta V + \gamma W) \\ &= O. \end{aligned}$$

3.2.2 Bilinearity

If $P,Q,R,S \in E[n]$, then we have

$$f_{\alpha,\beta,\gamma}^n(P+Q,R,S) = f_{\alpha,\beta,\gamma}^n(P,R,S) + f_{\alpha,\beta,\gamma}^n(Q,R,S),$$

$$f_{\alpha,\beta,\gamma}^n(P,Q+R,S) = f_{\alpha,\beta,\gamma}^n(P,Q,S) + f_{\alpha,\beta,\gamma}^n(P,R,S),$$

and

$$f_{\alpha,\beta,\gamma}^n(P,Q,R+S) = f_{\alpha,\beta,\gamma}^n(P,Q,R) + f_{\alpha,\beta,\gamma}^n(P,Q,S).$$

Proof Let $P, Q, R, S \in E[n]$. Then we have

$$
\begin{aligned}
f^n_{\alpha,\beta,\gamma}(P+Q,R,S) &= [(a_1+a_2)(b_3c_4 - b_4c_3) + (b_1+b_2)(a_4c_3 - a_3c_4) \\
&\quad + (c_1+c_2)(a_3b_4 - a_4b_3)](\alpha U + \beta V + \gamma W) \\
&= [a_1(b_3c_4 - b_4c_3) + b_1(a_4c_3 - a_3c_4) + c_1(a_3b_4 - a_4b_3)](\alpha U + \beta V + \gamma W) \\
&\quad + [a_2(b_3c_4 - b_4c_3) + b_2(a_4c_3 - a_3c_4) + c_2(a_3b_4 - a_4b_3)](\alpha U + \beta V + \gamma W) \\
&= f^n_{\alpha,\beta,\gamma}(P,R,S) + f^n_{\alpha,\beta,\gamma}(Q,R,S),
\end{aligned}
$$

Similarly, it can be easily verified that

$$
f^n_{\alpha,\beta,\gamma}(P, Q+R, S) = f^n_{\alpha,\beta,\gamma}(P,Q,S) + f^n_{\alpha,\beta,\gamma}(P,R,S),
$$

and

$$
f^n_{\alpha,\beta,\gamma}(P, Q, R+S) = f^n_{\alpha,\beta,\gamma}(P,Q,R) + f^n_{\alpha,\beta,\gamma}(P,Q,S).
$$

3.2.3 Antisymmetry

$f^n_{\alpha,\beta,\gamma}(P, Q, R) = -f^n_{\alpha,\beta,\gamma}(Q, P, R)$ for all $P, Q, R \in E[n]$.

Proof Let $P, Q, R \in E[n]$. Then we have

$$
\begin{aligned}
f^n_{\alpha,\beta,\gamma}(P,Q,R) &= [a_1(b_2c_3 - b_3c_2) + b_1(a_3c_2 - a_2c_3) + c_1(a_2b_3 - a_3b_2)].(\alpha U + \beta V + \gamma W) \\
&= -[a_2(b_1c_3 - b_3c_1) + b_2(a_3c_1 - a_1c_3) + c_2(a_1b_3 - a_3b_1)].(\alpha U + \beta V + \gamma W) \\
&= -f^n_{\alpha,\beta,\gamma}(Q,P,R).
\end{aligned}
$$

3.2.4 Non-degeneracy

If $P, Q, R \in E[n]$ then $f^n_{\alpha,\beta,\gamma}(P, Q, O) = O = f^n_{\alpha,\beta,\gamma}(P, O, R) = f^n_{\alpha,\beta,\gamma}(O, Q, R)$.

Also if $f^n_{\alpha,\beta,\gamma}(P, Q, R) = O$ for all $Q, R \in E[n]$, then $P = O$.

Moreover if $f^n_{\alpha,\beta,\gamma}(P, Q, R) = O$ for all $R \in E[n]$, then $P = kQ$ for some constant k.

Proof Let $P, Q \in E[n]$. Then we have

$$
f^n_{\alpha,\beta,\gamma}(P, Q, O) = [a_1(0-0) + b_1(0-0) + c_1(0-0)].(\alpha U + \beta V + \gamma W) = O
$$

In a similar manner, we can show that $f^n_{\alpha,\beta,\gamma}(P, O, R) = O$ and $f^n_{\alpha,\beta,\gamma}(O, Q, R) = O$ for all $P, Q, R \in E[n]$.

If $f_{\alpha,\beta,\gamma}^n(P, Q, R) = O$ for all $Q, R \in E[n]$ then we write

$$[a_1(b_2c_3 - b_3c_2) + b_1(a_3c_2 - a_2c_3) + c_1(a_2b_3 - a_3b_2)].(\alpha U + \beta V + \gamma W)$$
$$= O \text{ for all } Q, R \in E[n]$$

this implies $a_1 = b_1 = c_1 = 0$. Therefore $P = O$.

3.2.5 Compatibility

If $P \in E[nk]$, $Q \in E[n]$ and $R \in E[n]$ then $f_{\alpha,\beta,\gamma}^n(kP, Q, R) = kf_{\alpha,\beta,\gamma}^n(P, Q, R)$, if $P \in E[n]$, $Q \in E[nk]$ and $R \in E[n]$ then $f_{\alpha,\beta,\gamma}^n(P, kQ, R) = kf_{\alpha,\beta,\gamma}^n(P, Q, R)$, also $P \in E[n]$, $Q \in E[n]$ and $R \in E[nk]$ then $f_{\alpha,\beta,\gamma}^n(P, Q, kR) = kf_{\alpha,\beta,\gamma}^n(P, Q, R)$.

Proof Let $P \in E[nk]$, $Q \in E[n]$ and $R \in E[n]$. Then we have

$$f_{\alpha,\beta,\gamma}^n(kP, Q, R) = [ka_1(b_2c_3 - b_3c_2) + kb_1(a_3c_2 - a_2c_3) + kc_1(a_2b_3 - a_3b_2)].(\alpha U + \beta V + \gamma W),$$
$$f_{\alpha,\beta,\gamma}^n(kP, Q, R) = k[a_1(b_2c_3 - b_3c_2) + b_1(a_3c_2 - a_2c_3) + c_1(a_2b_3 - a_3b_2)].(\alpha U + \beta V + \gamma W),$$
$$f_{\alpha,\beta,\gamma}^n(kP, Q, R) = kf_{\alpha,\beta,\gamma}^n(P, Q, R).$$

Similarly, it can be easily verified that

$$f_{\alpha,\beta,\gamma}^n(P, kQ, R) = kf_{\alpha,\beta,\gamma}^n(P, Q, R)$$

and

$$f_{\alpha,\beta,\gamma}^n(P, Q, kR) = kf_{\alpha,\beta,\gamma}^n(P, Q, R).$$

4 Application of Trilinear Pairing to Cryptography

In this section, we will apply trilinear pairing (constructed in the previous section) to elliptic curve cryptography. A protocol defined by a sequence of steps absolutely specifying the actions required by three or more parties in order to achieve a specified objective. In cryptography, a key agreement protocol is a key establishment technique in which a shared secret is derived by three (or more) parties as a function of information contributed by, or associated with, each of these, such that no party can predetermine the resulting value. It is contributory if each party equally contributes to the key and guarantees its freshness. Key authentication is the property whereby one party is associated that no other party aside from an especially identified second party may gain access to a particular secret key. Key authentication is said to be implicit if each party sharing the key is assured that no other party can learn the secret shared key.

For a prime number p (p is large enough) and a positive integer r, we denote $q = p^r$. Let E be an elliptic curve over a finite field F_q. Then given $P \in E(F_q)$ with order n and $Q \in \langle P \rangle$, to find k such that $Q = kP$, is known as elliptic curve discrete log problem (ECDLP) in $E(F_q)$. Also for a given P, aP, bP to find abP is known as Diffie–Hellman problem for elliptic curves. Actually, it is known as Diffie–Hellman key exchange protocol for elliptic curves.

Now the proposed cryptographic schemes can be described as follow:

(i) We select a large prime s such that $E[s] \subseteq E(F_{q^k})$ for some smallest integer k.
(ii) Next, we select a generating pair $\{U, V, W\}$ in $E[s]$ and integers $\alpha, \beta, \gamma \in [0, l-1]$ which determine the pairing $f^s_{\alpha,\beta,\gamma}$.

Let the parameters $(P, Q, R, f^s_{\alpha,\beta,\gamma})$ be publicly known and let $h : E(F_q) \rightarrow Z/l$ be hash functions.

Now our proposed $f^s_{\alpha,\beta,\gamma}$-pairing can be applied to cryptographic scheme namely authenticated key agreement on elliptic curves. To apply the proposed scheme, we assume that three communication parties Alice, Bob and Carol wish to share a common secret information.

4.1 Authenticated Elliptic Curve Diffie Hellman Key Agreement for Three Parties

It consists of the following phase

Phase-1: Key generation phase

- Alice, Bob, and Carol randomly select secret integers $a, b, c \in (1, s-1)$ respectively.
- They respectively compute aP, bP, cP.
- They broadcast the above-computed values.
  Now the public values of the system are $(P, Q, R, aP, bP, cP, f^s_{\alpha,\beta,\gamma})$

Phase-2: Transmission phase

- Alice computes $S_A = a.bP.cP = abcP$ (because $P \in E[n]$) and $f^s_{\alpha,\beta,\gamma}(aP, Q, R)$. She sends $h(S_A)f^s_{\alpha,\beta,\gamma}(aP, Q, R)$ to Bob and Carol.
- Bob computes $S_B = b.aP.cP = abcP$ and $f^s_{\alpha,\beta,\gamma}(bP, Q, R)$. He sends $h(S_B)f^s_{\alpha,\beta,\gamma}(bP, Q, R)$ to Alice and Carol.
- Carol computes $S_C = c.aP.bP = abcP$ and $f^s_{\alpha,\beta,\gamma}(cP, Q, R)$. He sends $h(S_C)f^s_{\alpha,\beta,\gamma}(cP, Q, R)$ to Alice and Bob.
  It is evident that $S_A = S_B = S_C = abcP = S_{ABC}$ (say).

Phase-3: Authenticated secret share key generation phase

- Alice receives $I_A = h(S_B)f^s_{\alpha,\beta,\gamma}(bP,Q,R) \parallel h(S_C)f^s_{\alpha,\beta,\gamma}(cP,Q,R)$.

  Using trilinearity of pairing $f^s_{\alpha,\beta,\gamma}$, Alice obtain $I_A = h(S_{ABC})bcf^s_{\alpha,\beta,\gamma}(P,Q,R)$.

  Alice computes $h(S_A)^{-1}(\text{mod}s)$ to obtain her secret share key as $K_A = ah(S_A)^{-1}I_A$.

- Next, Bob receives
$$I_B = h(S_A)f^s_{\alpha,\beta,\gamma}(aP,Q,R) \parallel h(S_C)f^s_{\alpha,\beta,\gamma}(cP,Q,R)$$
$$= h(S_{ABC})acf^s_{\alpha,\beta,\gamma}(P,Q,R)$$

  To obtain secret share key, Bob calculates $h(S_B)^{-1}(\text{mod}s)$ and compute his shared secret key as $K_B = bh(S_B)^{-1}I_B$.

- Finally, Carol receives
$$I_C = h(S_A)f^s_{\alpha,\beta,\gamma}(aP,Q,R) \parallel h(S_B)f^s_{\alpha,\beta,\gamma}(bP,Q,R)$$
$$= h(S_{ABC})abf^s_{\alpha,\beta,\gamma}(P,Q,R)$$
To obtain secret share key, Carol calculates $h(S_C)^{-1}(\text{mod}s)$ and compute his shared secret key as $K_C = ch(S_C)^{-1}I_C$.

It can be easily checked that $K_A = K_B = K_C = abcf^s_{\alpha,\beta,\gamma}(P,Q,R) = K$ (say).

Thus there has been established an authenticated common secret key among multiparty Alice, Bob, and Carol.

5 Authenticity of the Proposed Schemes

It is obvious from the proposed authenticated elliptic curve Diffie Hellman protocol that the common secret key $K = abcf^s_{\alpha,\beta,\gamma}(P,Q,R)$ is designed by the contribution of each involved party (Alice, Bob, Carol). This results in the complexity for the attacker. For this suppose an active adversary is capable of reforming, delay or interpose the message. Now possible attacks on Bob and Carol can be described as

If K_B or K_C secret common key calculated by Bob or Carol, then it can be represented as $K_B = bf^s_{\alpha,\beta,\gamma}(d_1P,Q,R)$ or $K_C = cf^s_{\alpha,\beta,\gamma}(d_2P,Q,R)$ where d_1 or d_2 are introduced by the adversary. It means that adversary can alter the first flow of the proposed protocol with $f^s_{\alpha,\beta,\gamma}(d_1P,Q,R)$ or $f^s_{\alpha,\beta,\gamma}(d_2P,Q,R)$. To compute $bf^s_{\alpha,\beta,\gamma}(d_1P,Q,R)$ or $cf^s_{\alpha,\beta,\gamma}(d_2P,Q,R)$ adversary requires to calculate $bf^s_{\alpha,\beta,\gamma}(P,Q,R)$ or $cf^s_{\alpha,\beta,\gamma}(P,Q,R)$ respectively. But in the second flow, the only expression calculating $bf^s_{\alpha,\beta,\gamma}(P,Q,R)$ or $cf^s_{\alpha,\beta,\gamma}(P,Q,R)$ is $h(S_B)f^s_{\alpha,\beta,\gamma}(bP,Q,R)$ or $h(S_C)f^s_{\alpha,\beta,\gamma}(cP,Q,R)$ respectively. This shows that for the adversary to compute $bf^s_{\alpha,\beta,\gamma}(P,Q,R)$ or $cf^s_{\alpha,\beta,\gamma}(P,Q,R)$ respectively from $h(S_B)f^s_{\alpha,\beta,\gamma}(bP,Q,R)$ or $h(S_C)f^s_{\alpha,\beta,\gamma}(cP,Q,R)$ is intractable without the knowledge of K_B or K_C.

Similarly, attack on Alice can be described as:

Suppose key calculated by Alice is $K_A = ah(S_A)^{-1}f^s_{\alpha,\beta,\gamma}(d_3P,Q,R)$ where d_3 is introduced by the adversary. Now if assume that $d_3 = d_4h(S_A)$ where d_4 is known by the adversary and independent of $h(S_A)$, then $K_A = ah(S_A)^{-1} f^s_{\alpha,\beta,\gamma}$ $(d_4 h(S_A) P, Q, R) = a f^s_{\alpha,\beta,\gamma}(d_4 P, Q, R)$. Also to calculate $d_4 h(S_A) f^s_{\alpha,\beta,\gamma}(P, Q, R)$ where d_4 is known to the adversary, is intractable without calculating $h(S_A) f^s_{\alpha,\beta,\gamma}(P, Q, R)$. Further if d_3 is independent of $h(S_A)$, then it is impossible to calculate the key of Alice because K_A depends upon $h(S_A)^{-1}$.

6 Conclusion

Using $f^s_{\alpha,\beta,\gamma}$- pairing in cryptography is based on the difficulty of computing $h(S_B) f^s_{\alpha,\beta,\gamma}(bP, Q, R)$, $h(S_C) f^s_{\alpha,\beta,\gamma}(cP, Q, R)$ and $h(S_A) f^s_{\alpha,\beta,\gamma}(aP, Q, R)$, without knowing the secret values a, b and c(of Alice, Bob, and Carol respectively) in the construction of the self-pairing bilinear map $f^s_{\alpha,\beta,\gamma}$. To compute $b f^s_{\alpha,\beta,\gamma}(d_1 P, Q, R)$ or $c f^s_{\alpha,\beta,\gamma}(d_2 P, Q, R)$ adversary requires to calculate $b f^s_{\alpha,\beta,\gamma}(P, Q, R)$ or $c f^s_{\alpha,\beta,\gamma}(P, Q, R)$ respectively. But in the second flow, the only expression calculating $b f^s_{\alpha,\beta,\gamma}(P, Q, R)$ or $c f^s_{\alpha,\beta,\gamma}(P, Q, R)$ is $h(S_B) f^s_{\alpha,\beta,\gamma}(bP, Q, R)$ or $h(S_C) f^s_{\alpha,\beta,\gamma}(cP, Q, R)$ respectively. This shows that for the adversary to compute $b f^s_{\alpha,\beta,\gamma}(P, Q, R)$ or $c f^s_{\alpha,\beta,\gamma}(P, Q, R)$ respectively from $h(S_B) f^s_{\alpha,\beta,\gamma}(bP, Q, R)$ or $h(S_C) f^s_{\alpha,\beta,\gamma}(cP, Q, R)$ is intractable without the knowledge of K_B or K_C. Furthermore to calculate $d_4 h(S_A) f^s_{\alpha,\beta,\gamma}(P, Q, R)$ where d_4 is known to the adversary, is intractable without calculating $h(S_A) f^s_{\alpha,\beta,\gamma}(P, Q, R)$. In fact, it is impossible to calculate the secret key of Alice because her key K_A depends upon $h(S_A)^{-1}$. Thus $f^s_{\alpha,\beta,\gamma}$ pairing with only public values is as hard as solving the discrete logarithm problem on elliptic curves. Our schemes include only one random secret key per user. This is more efficient and secure than using two random secret keys in the known schemes existing in the literature.

Acknowledgements This research work is supported by University Grant commission (UGC) New Delhi, India under the Junior Research Fellowship student scheme and we would like to thank the referees for their precious comments and suggestions, which are really helpful to improve the quality of this article.

References

1. Kumar, M., Gupta, P.: Cryptographic schemes based on Elliptic Curve over the Ring Zp[i]. Appl. Math. 7(3), 304–312 (2016)
2. Weil, André.: Sur les fonctions algébriques à corps de constantes fini. (French). C. R. Acad. Sci. Paris. 210, 592–594 (1940)
3. Joux, A.: A one round protocol for tripartite Diffie-Hellman. In: Algorithmic Number Theory: 4th International Symposium. ANTS-IV Lecture Notes in Computer Science. vol. 1838, pp. 385–393 (2000)

4. Boneh, D., Franklin, M.: Identity-based encryption from the Weil pairing. Adv. Cryptology—CRYPTOLect. Notes Comput. Sci. **2139**, 213–229 (2001)
5. Boneh, D., Lynn, B., Shacham, H.: Short signatures from the Weil pairing. Adv. Cryptology—ASIACRYPT. Lect. Notes Comput. Sci. **2248**, 514–532 (2001)
6. Miller, V.: Use of elliptic curves in cryptography. Adv. Cryptology-CRYPTO. **85**(LNCS 218), 417–426 (1985)
7. Koblitz, N.: Elliptic curve cryptosystem. J. Math. Comput. **48**(177), 203–209 (1987)
8. Silverman, J.: The Arithmetic of Elliptic Curves. Springer, New York (1986)
9. Stinson, D.R.: Cryptography Theory and Practice. Chapman and Hall/CRC, UK (2006)
10. Washington, L.C.: Elliptic Curves Number Theory and Cryptography. Chapman and Hall/CRC, United Kingdom (2008)
11. Hardy, G. H., Wright E. M.: An introduction to the theory of numbers. Oxford University Press. UK (1938)
12. Sklavos, N., Zhang, X.: Wireless Security and Cryptography Specifications and Implementations. Chapman and Hall/CRC, UK (2007)
13. Nemati, H.: Information security and Ethics: Concept, Methodologies, Tools, and Applications. Information Science Reference, New York (2007)
14. Hankerson, D., Menezes, J. A.: Vanstone S., Guide to Elliptic Curve Cryptography. Springer, Germany (2004)
15. Bhattacharya, P.B., Jain, S.K., Nagpaul, S.R.: Basic Abstract Algebra. Cambridge University Press, United Kingdom (1995)
16. Gilbert, W.J.: Modern Algebra with Application. Willey, New York (2004)
17. Gallian, J.A.: Contemporary Abstract Algebra. Narosa Publishing House, New Delhi (1998)
18. Lee, H.S.: A self pairing map and its application to cryptography. Appl. Math. Comput. **152**, 671–678 (2004)

A Comparative Analysis of Various Vulnerabilities Occur in Google Chrome

Saloni Manhas and Swapnesh Taterh

Abstract Nowadays, security has become an integral and essential part of any web browser, to employ security measures in the web browser, first we need to check the vulnerabilities and flaws that are occurring in a particular web browser. In this paper, we will discuss the vulnerabilities that come under Google Chrome web browser and we have described their overall impact factor and reason behind the occurrence of these vulnerabilities. Also, we have discussed their level of damage from low to severe, that is, how much damage these vulnerabilities are causing. We have taken the data of 4 years from 2012 to 2015 and mentioned various vulnerabilities that came across during these years.

Keywords Web browser · Security · Vulnerabilities · Denial of service
Google Chrome

1 Introduction

In the present era, web browsers are used so frequently that it is essential to configure them safely. It is important to understand the overall facilities and characteristics provided by the web browser. Sometimes we install extra features in the web browser in order to improve the computing experience but it can compromise the security of the browser. Thus attackers can make use of certain vulnerabilities and can cause serious damage to the computer system. Attackers can penetrate the computer system by making use of certain vulnerabilities and can intentionally exploit the functioning of the computer. There are five security

S. Manhas (✉) · S. Taterh
AIIT, Amity University, Jaipur, India
e-mail: salonithakur786@gmail.com

S. Taterh
e-mail: swapnesh@hotmail.com

© Springer Nature Singapore Pte Ltd. 2018   51
M. Pant et al. (eds.), *Soft Computing: Theories and Applications*,
Advances in Intelligent Systems and Computing 583,
https://doi.org/10.1007/978-981-10-5687-1_5

parameters that we have taken in order to define the vulnerability impact [1]. These parameters are as follows:

a) Confedientiality
b) Integrity
c) Availability
d) Access Complexity
e) Authentication

These are five parameters of security through which the authenticity of a browser can be understood. Data collection of 4 years from 2012 to 2015 has been done, through which various vulnerabilities are revealed that came across during these years. Therefore, year wise vulnerabilities are described as follows.

2 Analysis of Vulnerabilities

In all the vulnerabilities listed below, the above-written security parameters have been exploited according to the impact of the attack performed by attackers. Out of all the parameters, confidentiality, integrity, and availability are abused severely than rest of the two, as little skills are needed to misuse them [2].

2.1 Vulnerabilities in 2015

Different tools have been discussed to remove distributed denial of service attacks [3]. Denial of service attack is a prominent attack that is found in the vulnerabilities. After discussing vulnerabilities that have arisen in google chrome in the year 2015, a diagram is drawn containing the vulnerabilities score, with the help of a diagram a person will get to know that which vulnerabilities have higher penetration rate as shown as in Fig. 1.

Fig. 1 Vulnerabilities score in 2015 [2]

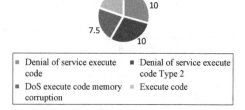

Vulnerabilities score in 2015

6.8
10
7.5
10

▪ Denial of service execute code   ▪ Denial of service execute code Type 2
▪ DoS execute code memory corruption   ▪ Execute code

2.1.1 Denial of Service Execute Code

The midi system in Google Chrome before version 47.0.2526.106 was not capable of handling the data that is being sent, due to which intruders are executing random code, causing a shutdown of service through unnamed path. MIDI supports multiple instruments, volume control, and different options.

2.1.2 Denial of Service Execute Code Type 2

This vulnerability in content or browser occurs in Google Chrome before version 47.0.2526.73, which results in allowing the distant attackers to cause service repudiation by executing inconsistent code. These codes are executed by an attacker to target process and the attacker can influence the miss applying of app cache update jobs.

2.1.3 DoS Execute Code Memory Corruption

This vulnerability occurs in the opj_j2k_copy_default_tcp_and_create_tcd function in j2k.c in OpenJPEG before r3002, as used in PDFium. An open-source project has been started by Google for a PDF software library, where developers will be able to perform into applications designed for a variety of platforms. Google Chrome before 45.0.2454.85, allows attackers to perform inconsistent code and cause a service shutdown by causing a storage allocation failure.

2.1.4 Execute Code

Google Chrome before version 44.0.2403.89 does not make sure that open auto list excludes entire types of harmful files, due to which remote attackers find it easy to perform random code by making available a forged file and influencing a user's previous ways to open files of this type" choice, related to download_commands.cc and download_prefs.cc. As these files contain harmful contents which can leverage the credibility of a user's data.

2.2 Vulnerabilities in 2014

When someone downloads any add-on feature in the browser, there is a possibility that it can consist of certain entry points through which an attacker can peep into the system by executing malicious codes. XSS filter can be used to avoid insecure authentication [4]. The reason behind this could be that while developing a certain application or software some part of it may be left insecure due to poor coding or due to other shortcomings. Following are some vulnerabilities consisting of different attack patterns. The overall vulnerabilities score diagram of top four vulnerabilities occuring in google chrome in 2014 has framed in Fig. 2.

2.2.1 Execute Code

Google Chrome before version 38.0.2125.101 and chrome OS before 38.0.2125.101 does not safely handle the interconnections of extensions Google V8 and IPC, allowing distant attackers to perform arbitrary code through vectors involving JSON data that is related to inappropriate parsing of an escaped index by Parse Json Object in json-parser.h.

2.2.2 Execute Code Type 2

Google Chrome before version 37.0.2062.94 does not appropriately control the interconnection of extensions, Google V8, IPC, and the sync API, permitting remote attackers to perform arbitrary code through undefined vectors.

2.2.3 DoS Execute Code

There are multiple vulnerabilities is found in the layout application in Blink, as used in Google Chrome before version 33.0.1750.117 that provides a way to attackers to cause a service repudiation or possibly have undetailed consequences with the help of vectors which involves

- Running JavaScript code during operation of the update widget positions function
- Make a call into a plugin while performing the update vision positions function.

Fig. 2 Vulnerabilities score
in 2014 [2]

Vulnerabilities score in 2014

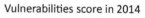

2.2.4 Execute Code Type 3

Google Chrome before version 37.0.2062.94 does not properly handle the interaction of extensions, IPC and Google V8, which allows attackers to execute random code through unspecified vectors.

2.3 Vulnerabilities in 2013

There are certain security flaws in versions of Google Chrome resulting in causing security flaws, due to which browser's security is compromised. Although a safe stack system has been introduced to patch vulnerabilities [5]. Following are some vulnerabilities providing an overview of the scenario. The diagram containing the score of top four vulnerabilities in the year 2013 has created in Fig. 3.

2.3.1 DoS Execute Code Overflow Memory Corruption

Integer overflow in Google Chrome before version 31.0.1650.57 permits attackers to perform random code, causing a denial of service, through un-described vectors, as described during a Mobile Pwn2Own competition at PacSec 2013.

2.3.2 Execute Code

Different un-described vulnerabilities in Google Chrome before version 31.0.1650.48 permits intruders to execute inconsistent code or might be having other effects through unnamed vectors.

Fig. 3 Vulnerabilities score in 2013 [2]

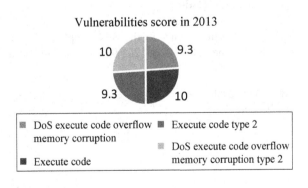

Vulnerabilities score in 2013

2.3.3   Execute Code Type 2

This is a vulnerability in Google Chrome before 28.0.1500.71 which makes a path
for remote users to perform inconsistent code through forged response traffic after a
URL request.

2.3.4   DoS Execute Code Overflow Memory Corruption Type 2

Google Chrome before version 27.0.1453.110 does not accurately manage SSL
sockets, allowing remote attackers to perform random code, causing a repudiation
of service through un-described vectors.

2.4   Vulnerabilities in 2012

Attackers can attack your system through unspecified vectors and violate the
security of the browser, causing denial of service. This can occur due to various
reasons, for example, due to malicious software, sql injections [6] etc. The vulnerabilities score diagram of top four vulnerabilities that affected google chrome in
the year 2012 is designed in Fig. 4.

2.4.1   DoS Execute Code

Google Chrome before 23.0.1271.97 is not capable of handling history navigation
that influences attackers to cause a random code and cause a service shutdown with
the help of undefined vectors.

2.4.2   Execute Code

This vulnerability occurs in the SVG implementation web kit which is used in
Google Chrome before 22.0.1229.94 and it allows intruders to perform inconsistent

Fig. 4 Vulnerabilities score
in 2012 [2]

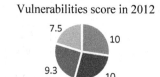

Vulnerabilities score in 2012

code with the help undefined vectors. SVG is an image format for vector graphics. It means Scalable Vector Graphics. Basically, what you work with in Adobe Illustrator. Adobe Illustrator is a program used by artists and graphic designers for the creation of images.

2.4.3 Execute Code Type 2

Race condition in Google Chrome before 22.0.1229.92 permits remote attackers to perform arbitrary code through vectors which are associated with audio devices.

2.4.4 DoS Execute Code Overflow

Google Chrome before 21.0.1180.57 on Linux is not able to hold tabs appropriately, which allows distant intruders to perform random code and causing a service shutdown through undetailed vectors.

3 Overall Evaluation of Vulnerabilities

In the above-given data, we have mentioned various vulnerabilities of Google chrome and its rate of damage on different parameters, so here we are concluding the vulnerabilities that came across during four years, here it is shown that which vulnerability is causing a higher rate of damage [2]. So, a year wise damage potential rate of these vulnerabilities is described below:

In the year 2015, vulnerabilities DoS execute code and DoS execute code type 2 are causing more damage to Google chrome browser as compared to others. In the year 2014, vulnerability execute code and its other three types have caused more damage by following different attack patterns than others. In the year 2013, vulnerabilities DoS execute code memory corruption type 2 and execute code are damaging the Google Chrome browser highly. In the year 2012, vulnerabilities DoS execute code and execute code are causing more damage. Therefore, from the above-mentioned data, we have observed that the two vulnerabilities, DoS execute code and execute code are raised in almost every year, whereas execute code vulnerability is repeating itself in all the four years with different versions of Google Chrome by various attack vectors. The impact factor of Execute code vulnerability continues to be the highest from 2012 to 2014, that is 10 and decreases to 6.8 in 2015. Also, the same category of vulnerability, i.e., Execute code type 2 continues to penetrate the Google Chrome browser with similar impact factor of 9.3 from 2012 to 2013 and with the increases its impact factor 10 in year 2014 by following diverse attack paths in different versions of Google Chrome in and further becomes null in the year 2015.

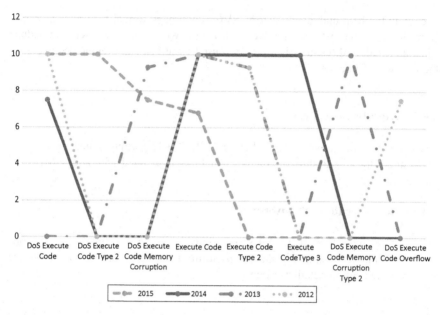

Fig. 5 Overall vulnerabilities diagram

The impact factor of DoS execute code vulnerability continues to vary from 7.5 to 10, which becomes null in the year 2013, whereas rest of the vulnerabilities are also harming the browser but they are not repeating themselves in all the consecutive years as portrayed in Fig. 5.

4 Conclusion

With this relevant information, we have concluded that Execute code vulnerability is abusing the browser enormously by making use of different attack passages. Thus in near future, the vulnerability with the highest impact can be eliminated or its impact can be reduced so that the end user can safely use Google Chrome as a web browser. Researchers can further try to mitigate the vulnerabilities by taking this data into consideration.

References

1. Khan, S.A., Khan, R.A.: A Framework to quantify security: complexity perspective. Int. J. Inf. Educ. Technol. **2**(5), 439–441 (2012)
2. https://www.cvedetails.com/vulnerability-list/vendor_id-1224/product_id-15031/opec-1/ Google-Chrome.html

3. Kaur, H., Behal, S., Kumar, K.: Characterization and comparison of distributed denial of service attack tools. In: Proeedings 2015 International Conference on Green Computing And Internet Things, ICGCIoT 2015, pp. 1139–1145 (2016)
4. Mewara, B., Bairwa, S., Gajrani, J.: Browser's defenses against reflected cross-site scripting attacks. In: 2014 International Conference Signal Propagation Computer Technology, ICSPCT 2014, pp. 662–667 (2014)
5. Chen, G., Jin, H., Zou, D., Zhou, B.B., Liang, Z., Zheng, W., Shi, X.: Safestack: Automatically patching stack-based buffer overflow vulnerabilities. IEEE Trans. Depend. Secur. Comput. **10** (6), 368–379 (2013)
6. K. C. G.: It, Analysis of Security Vulnerabilities for Web Based, pp. 233–236 (2012)

Secure Framework for Data Security in Cloud Computing

Nishit Mishra, Tarun Kumar Sharma, Varun Sharma and Vrince Vimal

Abstract Cloud computing is the new bending curve in the line of Information Technology and computing paradigm. Cloud computing has simultaneously revolutionized business and government sectors by stretching its arms in every possible field, from not only to Information communication technology to medical advancements to a more diverse agriculture field which was untouched by technology for a decade. Cloud computing enlightens path for another dimension of computing which is expected to rise even more than mobile computing, i.e., Internet of Things (IOT) or internet of everything. As new dimensions are being added up to the cloud domain every day this gives a window to hackers and intruders to breach the doors. The security problem is amplified under cloud computing as it introduces new problem domains. One of the major problem domain identified is data security, where security of user's data is the utmost priority. Providing privacy to the user and the data stored by the user is to be ensured by the cloud service provider. This paper aims at the security model for cloud computing which ensures the data security and integrity of user's data stored in the cloud using cryptography.

Keywords Cloud computing · Cryptography advanced encryption standard Message digest

N. Mishra · T. K. Sharma · V. Sharma (✉)
Amity University, Rajasthan, India
e-mail: vsharma@jpr.amity.edu

N. Mishra
e-mail: nishit.mishra@student.amity.com

T. K. Sharma
e-mail: tksharma@jpr.amity.edu

V. Vimal
Indian Institute of Technology Roorkee, Roorkee, India
e-mail: vrince.vimal@gmail.com

© Springer Nature Singapore Pte Ltd. 2018
M. Pant et al. (eds.), *Soft Computing: Theories and Applications*,
Advances in Intelligent Systems and Computing 583,
https://doi.org/10.1007/978-981-10-5687-1_6

61

1 Introduction

Cloud computing is a service-based model which aims at delivering everything as a service. Cloud computing is the pool of resources or services which are provided by the cloud service providers. Cloud computing emphasizes on service model which is based on user demand and a business model of pay per use basis. Cloud computing elevates its users by enabling them to reduce infrastructure cost and helps small business to scale faster without investing much in infrastructure and services. Cloud computing has three service delivery and deployment models which provide its users for selecting services from a varied set of models.

1. Deployment models of cloud computing are described as follows:
a. Public cloud
b. Private cloud
c. Hybrid cloud
d. Community cloud

  - **Public clouds**: Public clouds are based on standard cloud computing model in which a cloud service provider makes resources such as servers, data storage, and applications delivered to the consumers via the Internet.
  - **Private cloud**: Private clouds are also based on the same computing model like a public cloud which provides resources like servers, data storage, and applications but these services are proprietary and mostly tailored for one organization while public cloud aims at delivering it to multiple organizations.
  - **Hybrid cloud**: Hybrid cloud is an alignment of two or more clouds (public, private, and community) that works separately but are bound together to provide a suffice a common purpose.
  - **Community cloud**: Community cloud is an infrastructure service that is shared by several organizations which serve a specific community that has common concerns like medical, military, etc (Fig. 1).

2. Delivery Model of cloud computing is described as follows:

  - Infrastructure-as-a-service (Iaas).
  - Platform-as-a-service (Paas).
  - Software-as-a-service (Saas).

    - **Infrastructure-as-a-service (Iaas)**: Iaas is a form of cloud computing which provides virtualized computing resources over the internet. In Iaas, a third-party service provider hosts services such as hardware, application, network, data, etc., on behalf of consumer and charges on a usage basis. It also liberates its users from time-consuming and side lined tasks like system maintenance, backups, and system recovery planning.
    - **Platform-as-a-service (Paas)**: Paas is a form of cloud computing which enables the consumer to develop and deploy their applications on the service provider's infrastructure. It does not completely replace a business's entire infrastructure but relies on paas service provider's services

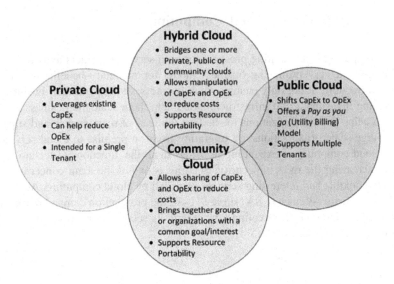

Fig. 1 Cloud deployment model

like software, a database which are essential to develop and deploy an application.

- **Software-as-a-service (Saas)**: Saas is a software distribution model in cloud computing which enables its developers to publish their applications to consumers via Internet by hosting their application on third-party service provider. It liberates an organization or developer to install and run application on their own systems or data centers (Fig. 2).

Fig. 2 Cloud delivery model

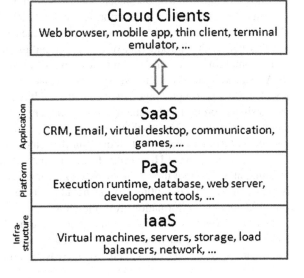

2 Security Issues in Cloud Computing

Cloud computing is an exciting prospect for both service providers and consumers as this gives a very promising business model for monetary benefits as well as technical freedom. It not only enables small businesses to expand their business by eliminating investment into infrastructure and help them scale the business rapidly but also helps large organizations to shed off the load of maintenance and support. Many large-scale organizations are migrating from on premise to cloud services.

As cloud computing is expanding its wings in all the directions of technology it is also increasing the risks of security. Security is one of the main concerns rising above the flexible and interesting services provided by cloud computing. According to Cloud Security Alliance (CSA) survey [1–3], the growing on demand market cap of cloud computing is also creating a buzz with its security concerns and some the security flaws are listed below.

1. Data Breaches and Loss
2. Account or Service Hijacking
3. Denial of Service
4. Malicious Insiders
5. Hypervisor Vulnerabilities

Above are only some of the threats and security issues in cloud computing among them data security is one of the most important ones. All the data is stored on the service provider's cloud which might include confidential financial details of a company or personal data of an individual or banking and social accounts details of users. If cloud service is breached, then all the sensitive data can be compromised and can cost a lot to the stakeholders (Fig. 3).

Data Breaches: Every organization's biggest nightmare is data breach, it escalates issue from an L1 engineer to a CxO of the organization, it keeps the head honchos on their toes as data breach could result in a huge loss to the organization in terms of money and sensitive information. And If this happens in a case of a Data Center then the scenario is even worse. Data breach was even one of the popular attack's before the introduction of cloud computing and after the introduction of cloud computing cases of data breaches have increased exponentially.

An attacker can extract the sensitive information in many ways and one of the popular ways is to determine whether if the multitenant cloud service database is not designed properly, the attack can be triggered if there is flaw in one of the client's application, through that attacker can not only access client's data but can get access to other client's data stored on that cloud.

Account or Service Hijacking: In this kind of attack the attacker can make a phishing attack to fool the client into giving his/her credentials. This kind of attack is very common in computing world; the attacker can get access to cloud and then can steal sensitive information from the cloud.

Denial of Service: Denial of service attack is nothing special but the simple attack in which the user cannot access cloud services due to service not responding. The attacker overloads the system by increasing utilization of resources (disks,

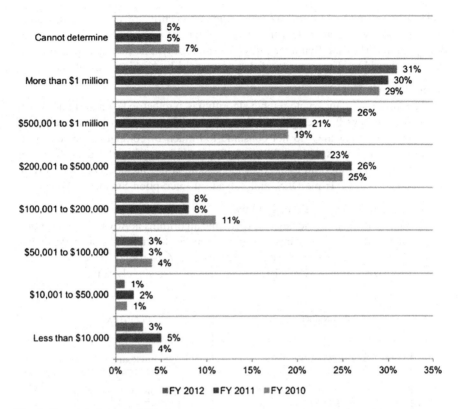

Fig. 3 Economic impact of data loss

processors, and network). This down of service can cause cloud service provider a huge loss In terms of money and can give attackers time to prepare for another attack while the service is down.

Malicious Insiders: These kinds of attacks are implemented by attackers who are part of the organizations and can plan an attack from inside of the organization, through this the attacker can retrieve all the information to the cloud as it has direct access to the data.

Hypervisor Vulnerabilities: Hypervisor has changed the computing worlds by implementing multiple services on a single system. Virtualization not only plays an important role in cloud computing, but also leads to certain vulnerabilities due to which there could be a loss of data and sensitive information.

3 Cryptography

Cryptographic techniques are not only helping in keeping data in the computing world but before that also, from the time of Second World War cryptography techniques have evolved and integrated into computing world. It has expanded its roots in every field and now one cannot ignore security of data without implementation of cryptography algorithm. Cryptography is a combination of three types of algorithms. They are Symmetric key algorithm, Asymmetric key algorithm, and Hashing algorithm. The integrity of data is ensured by hashing algorithms.

A. **Symmetric Key Algorithm**:

The most important cryptographic algorithm is symmetric key algorithm, which uses same keys for encryption and decryption The advantage of these types of algorithm is that they do not consume too much of computing power and it works with high speed while encrypting and decrypting the data. Some of symmetric key algorithms used in cloud computing include Data Encryption Standard (DES), Triple-DES, and Advanced Encryption Standard (AES).

- **Advanced Encryption Standard**

One of the symmetric key cryptographic algorithms is Advanced Encryption Standard. It is a block cipher and it is non-Feistel cipher. Data encryption in AES is of 128-bit block size. It uses 10, 12, or 14 rounds depending on the key size varies, 128 bit, 192 bit, or 256 bit respectively (Fig. 4).

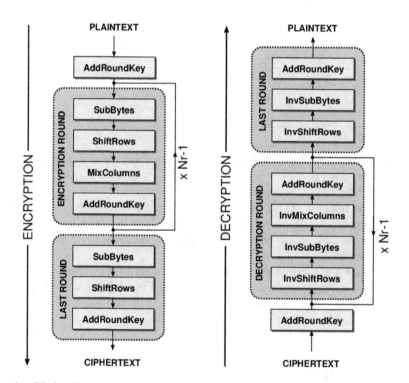

Fig. 4 AES algorithm

B. **Asymmetric Key Algorithm**:
Asymmetric key algorithms are those algorithms that use different keys for encryption and decryption. Private keys are kept private and are used for the decryption of data by the receiver. In cloud computing, asymmetric key algorithms are used to generate keys for encryption. The most common asymmetric key algorithm used in cloud computing is RSA and Diffie–Hellman key exchange.

4 Security Concerns

One of the main concerns in cloud security is data security. It is very important for service providers to maintain confidentiality and privacy of customer's data. Integrity of consumer's data should be intact as there might be a chance that information has been tempered. There can be many situations when an attacker can plan an attack on the cloud and leak out all the information stored in the cloud. One of the simplest ways to secure information on the cloud is to encrypt the data, but sometimes simple encryption is not enough. For example, Google cloud service is very famous among these days and we all use Google Drive and others services to store our important information like documents, photos and videos on cloud, not just documents but Google cloud also stores our activity information like Google timeline, places, and others. All these services are linked to a single user account, i.e., Google account, if someone got hands on to someone's Google account credentials then the attacker can access all the information using a single account, although Google uses a very rigid and redundant way to trace the attack, it could not stop the attack.

Many cloud service providers and aggregators are using the same approach to secure users data in the cloud using a single sign-in account. There are many ways in which an attacker can breach the security and it would become difficult to trace that what information is been leaked out. There have been many cases in which data has been leaked out. One of the famous attacks which was held in 2014, made a big scream in the tech world. That was an attack at Sony Pictures Entertainment which made Sony business come to a halt. The attack was made by a group called The Guardians of peace. The group managed to breach Sony servers revealing sensitive information of 100 Tb like social security numbers, salaries, movies, and other important information. The attack could be stopped or the implications of that attack could be minimized if proper security measures were taken like encryption and controlled use of administrative passwords.

Figure 5 shows possible points of intrusion and data breach. From User's personal database to cloud database of the service provider, an attacker can steal the information. Below mentioned are some of the scenarios where attacks are possible.

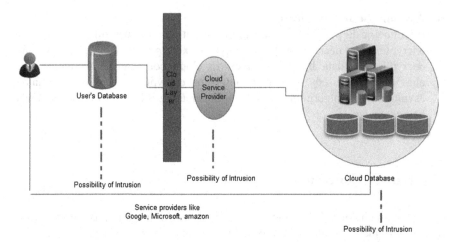

Fig. 5 Possible attacks in cloud computing

- At User level: An attacker can plan an attack at user level while directly attacking at user's database. In this way, there would be no liability of the service provider and there would be no way to stop or minimize the effect of attack until and unless strict security mechanisms were employed at the user level.
- At Service Provider level: Inside attack could be easily planned at service provider level. This is called malicious insiders where someone from within the organization could plan an attack, this could be traced by audit logs. But meanwhile, the attackers are being traced it would be very late for the service provider to take necessary defensive steps.
- At Cloud Database: Cloud database are like large data centers where data of multiple organizations are stored on multiple tape libraries.

These kinds of attacks can be difficult to trace and could keep the organization on its toes.

This paper aims at introducing a methodology using which the of storing information in the cloud could be more agile and difficult to break in. This will help in securing cloud databases and servers by encrypting the information and credentials.

5 Methodology

In today's growing technology, vast infrastructure, high clock speed processors, high memory and large disk spaces performing complex calculations and storing large amount of data is not a big problem in cloud computing, the distributed array of computation resources are capable enough to perform heavy and complex tasks.

This Methodology uses very complex calculations which are difficult for an attacker to solve in a given time and hence the attack can be stopped and prevented in future. The combination of symmetric algorithm and XOR operation with message digest makes the encryption stronger. There is a central key distribution center (CKDC) which stores all the unique keys generated.

This methodology has two schemes i.e. encryption scheme and decryption scheme.

Encryption Scheme

Step 1: User uploads the data to the cloud, a unique dataID will be generated in correspondence to every data item uploaded.

Step 2: Information will be encrypted using symmetric encryption algorithm i.e. AES Algorithm.

Step 3: Perform XOR operation on dataID and AES Key.

Step 4: XOR of AES Key and dataID will be stored to the central key distribution center.

Step 5: Generate message digest of the original message. Which will be stored in KDC and it will be calculated when the data will be retrieved from cloud database. This will ensure data integrity (Fig. 6).

Decryption Scheme

Step 1: Retrieve data stored in the cloud to begin decryption.

Step 2: Generate message digest of data to ensure the integrity of data.

Step 3: Retrieve the Unique key from Key Distribution center and perform XOR operation on dataID and unique key. This will give AES key.

Step 4: Decrypt the message using AES key (Fig. 7).

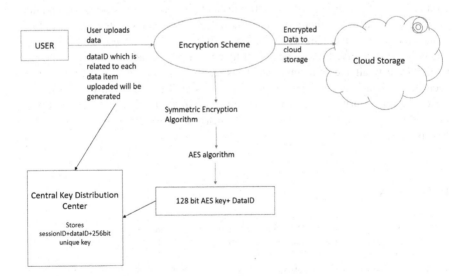

Fig. 6 Encryption scheme

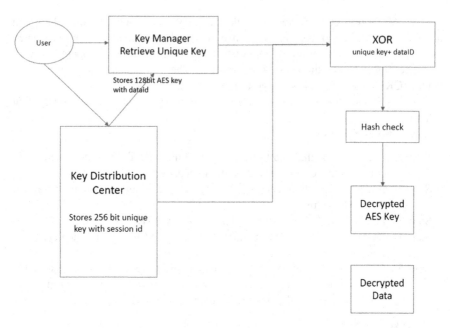

Fig. 7 Decryption scheme

6 Conclusion

Today cloud computing has become the backbone of the society. Most of the users rely on cloud services, not only users but small to big organizations also utilize cloud services for their day to day activities and also uses it for storing information on long term basis. As there are many new technologies emerging which utilizes cloud computing like, Big Data, Internet of things, etc. These new emerging technologies also give birth to new security challenges. To address these challenges Cloud Security alliance puts up an analysis report each year to keep a tab on which security challenges are keeping the administrators on their toes. Many organizations are deploying security mechanisms on their cloud to secure their data, cloud service providers are also upgrading their security measures to ensure that their user's data is kept safe.

References

1. Cloud Security Alliance (2013). The notorious nine: cloud computing threats in 2013. https://downloads.cloudsecurityalliance.org/initiatives/top_threats/The_Notorious_Nine_Cloud_Computing_Top_Threats_in_2013.pdf
2. Behl, A., Behl, K.: An Analysis of cloud computing security issues. In: World Congress on Information and Communication Technologies (2012)

3. Sudha, M., Rao, B.R.K., Monica, M.: A comprehensive approach to ensure secure data communication in cloud environment. Int. J. Comput Appl. (0975–8887), **12**(8) (2010)
4. Infoworld (2008). Gartner: Seven cloud-computing security risks. Accessed Jan 2010
5. Kalpana, P., Singaraju, S.: Data security in cloud computing using RSA algorithm. Int. J. Res. Comput. Commun. Technol. (IJRCCT), **1**(4) (2012), (ISSN 2278-5841)
6. Suresh, K.S., Prasad, K.V.: Security issue and security algorithm in cloud computing. Int. J. Adv. Res. Comput. Sci. Softw. Eng. **2**(10) (2012), (ISSN:2277 128X)
7. Nigoti, R., Jhuria, M., Singh, S.: A survey of cryptographic algorithm for cloud computing. Int. J. Emerg. Technol. Comput. Appl. Sci. (IJETCAS), **4**(2), pp. 13–123 (2013)

An Improved Spider Monkey Optimization Algorithm

Viren Swami, Sandeep Kumar and Sanjay Jain

Abstract Spider Monkey Optimization is the newest member of the Swarm Intelligence-based algorithm, which is motivated by the extraordinary behavior of Spider Monkeys. The SMO algorithm is a population-based stochastic meta-heuristic. The SMO algorithm is well balanced for good exploration and exploitation most of the times. This paper introduces an improved strategy to update the position of solution in Local Leader Phase. The proposed algorithm named as Improved Spider Monkey Optimization (ISMO) algorithm. This method is developed to improve the rate of convergence. The ISMO algorithm tested over the benchmark problems and its superiority established with the help of statistical results.

Keywords Swarm intelligence · Natural-Inspired algorithm · Fission–fusion social structure · Unconstrained optimization problems · Metaheuristic

1 Introduction

Swarm Intelligence refers the natural system that are influenced by colonies of social insects like, fishes, bee, bird flocks, ant, etc. The definition introduced by Bonabeau for the swarm intelligence is "any attempt to design algorithms or distributed problem-solving devices inspired by the collective behavior of social insect colonies and other animal societies" [1]. These social creatures demonstrate some great ability while searching for food, security and mating in complex situations.

V. Swami (✉) · S. Kumar
Faculty of Engineering & Technology, Jagannath University, Jaipur, India
e-mail: swami.viren@gmail.com

S. Kumar
e-mail: sandpoonia@gmail.com

S. Jain
Amity School of Engineering and Technology, Amity University, Jaipur, India
e-mail: jainsanjay17@yahoo.co.in

© Springer Nature Singapore Pte Ltd. 2018
M. Pant et al. (eds.), *Soft Computing: Theories and Applications*,
Advances in Intelligent Systems and Computing 583,
https://doi.org/10.1007/978-981-10-5687-1_7

73

The SMO algorithm is the latest population based strategy that is also stochastic in nature developed by Bansal et al. [2] to solve unconstrained optimization problems. The SMO is motivated by intelligent societal behavior of spider monkeys while searching for rich food sources. The SMO algorithm is based on fission–fusion structure of social living being spider monkey while searching for most suitable food source [2]. It consists of the intrinsic solution of population which denote food source of spider monkeys. The SMO Algorithm tries to keep proper balance between exploration and exploitation while searching for optimal solution. In exploitation it make sure that local optimum solution traversed properly and in exploration it explore global search space in order to avoid problem of trapping in local optimum. It has been observed that SMO is good in exploration of local search.

The recent trend in research is inclined toward algorithms that are inspired by nature in order to solve complex real world problems that are not solvable by classical techniques. The nature inspired algorithms includes algorithms that are inspired by biological process, physical actions and other natural activities. These algorithms show some unconventional approaches that are able to solve optimization problems in field of science, engineering and management. Many researchers have analyzed the behavior and design of the algorithm that can be used to solve nonlinear, non-convex, non-differential, and multi-model problems.

The SMO algorithm is comparatively young algorithm so there is not a large number of a literature. Pal et al. [3] used SMO algorithm in image segmentation and developed a new multi-level thresholding segmentation approach for gray scale images. Gupta et al. [4] carried out a comprehensive study of SMO after incorporating a new operator namely quadratic approximation and solved a large range of scalable and non-scalable benchmark problems and Lennard-Jones problem. Sharma et al. [5] divided the population of spider monkeys into different age groups. It is assumed that younger monkeys are more interacting and frequently change their position in contrast to older monkeys. Gupta and Deep [6] introduced a new probability calculation approach namely tournament selection in SMO algorithm. Gupta and Deep [7] analyzed the behavior of SMO algorithm under different perturbation rate schemes and proposed four editions of SMO are proposed analogous to constant, random, linearly increasing and linearly decreasing perturbation rate variation strategies. Singh et al. [8] developed a binary SMO algorithm and used it for thinning of concentric circular antenna arrays. Singh and Salgotra [9] introduced dual search strategy in SMO. The modified SMO used to synthesize linear antenna array. Sharma et al. [10] developed a new version of SMO with new local search strategy namely power law-based local search. The new strategy was applied to solve model order reduction problem. Al-Azza et al. [11] introduces SMO algorithm for the electromagnetic and antenna community. Agarwal et al. [12] used social spider algorithm in image segmentation and developed a new multi-level thresholding segmentation approach for grayscale images by deploying histogram-based bi-modal and multi-modal thresholding. Kumar et al. proposed three variants of SMO algorithm. Self-Adaptive Spider Monkey Optimization Algorithm for Engineering Optimization Problems [13] that require no manual setting, Fitness Based Position Update in Spider Monkey Optimization Algorithm [14] and Modified position update in spider monkey optimization algorithm [15]. The fitness-based SMO update position of current swarm based on their fitness. It is assumed that highly fitted solution has good solution in their proximity. Almost all variants of SMO are better than other nature-inspired optimization techniques (e.g., ABC, PSO, etc.) in terms of efficiency, accuracy, and robustness.

2   Spider Monkey Optimization

The SMO algorithm is a novel nature-inspired algorithm which is developed by Bansal et al. in 2013 [2]. It is stochastic in nature as it introduces some random component in each step. The SMO strategy mimics the fission–fusion structure of spider monkey. The major characteristics of fission–fusion social structure are described as follow:

Fission–fusion social structure animals survive in group of 40–50 monkeys that divide the member into subgroups for searching food in order to reduce competition.

Global leader (female) is responsible for searching the food source that generally leads in the group. These groups are divided into small subgroups to search for food independently.

Local leader (female) leads the subgroups and responsible for scheduling a well-organized plan for foraging route each day.

These group members search the food sources and modify their position based on the distance from food source.

These group members communicate with all group members to maintain social bond in case of stagnation.

2.1   Phases of SMO Algorithm

The SMO algorithm consists of six major phases followed by initialization phases. These phases suggest that how spider monkey updates their position based on their previous experience and behavior of neighbors.

2.1.1   Initialization of the Population

First, a population of N spider monkey is initialized. Initial population denoted by D-dimensional vector SM_i ($i = 1, 2, ..., N$). Every SMO represents the optimized solution of the problem under consideration. SM_i represents the population of spider monkey. SM_i is initialized as follows:

$$SM_{ij} = SM_{minj} + U[0, 1] \times (SM_{maxj} - SM_{minj}) \tag{1}$$

where SM_{ij} represents the ith food source in the swarm, SM_{minj} and SM_{maxj} are lower and upper bounds of SM_i in jth direction respectively and $U[0, 1]$ is a uniformly distributed random number in the range $[0, 1]$.

2.1.2 Local Leader Phase (LLP)

The second phase is Local Leader Phase. This phase modernizes the location of SMO based on experience of Local and Global group members. These members compare fitness of new location and current location and apply greedy selection. Position updates equation for ith SM of Kth group as follow:

$$SM_{newij} = SM_{ij} + U[0, 1] \times (LL_{kj} - SM_{ij}) + U[-1, 1] \times (SM_{rj} - SM_{ij}) \tag{2}$$

where SM_{ij} represents the ith solution in jth dimension, LL_{kj} denotes the jth dimension of the kth local group leader position. SM_{rj} is the rth solution which is selected randomly from kth group such as $r \neq i$. $U [0, 1]$ is a uniformly distributed random number in the range of 0–1 [2].

2.1.3 Global Leader Phase (GLP)

The GLP phase is just starts after finishing the LLP. Position gets updated according to previous experience of the Global Leader and Local group members with the help of Eq. (3).

$$SM_{newij} = SM_{ij} + U[0, 1] \times (GL_j - SM_{ij}) + U[-1, 1] \times (SM_{rj} - SM_{ij}) \tag{3}$$

where GL_j correspond to the jth dimension of the global leader position and $j \in \{1, 2, ..., D\}$ is randomly selected within the dimension. In this phase, the Spider Monkey (SM_i) updates their position that is based on probabilities ($prob_i$) which are calculated using their fitness [2]. There may be different methods for probability calculation but it must be function of fitness. The fitness of a function indicates about its quality, fitness calculation must include function value.

$$prob_i = 0.9 \times \frac{fitness_i}{fitness_{max}} + 0.1. \tag{4}$$

2.1.4 Global Leader Learning (GLL) Phase

In this phase, SMO modifies position of global leader with help of greedy approaches. Highly fitted solution in current swarm is chosen as global leader. It also performs a check that the position of global leader is modernized or not and modify Global Limit Count accordingly [2].

2.1.5 Local Leader Learning (LLL) Phase

Now in this phase location of local leader is modified with help of greedy approaches. Highly fitted solution in current swarm is chosen as Local Leader. It also performs a check that the location of local leader is modernized or not and modifies Local Limit Count accordingly [2].

2.1.6 Local Leader Decision (LLD) Phase

During LLD phase, decision is taken about the position of Local Leader, if it is not modernized up to a threshold also called as Local Leader Limit (LL_{limit}). In case of no change it randomly initializes position of LL. Position of LL may be decided with the help of Eq. (5).

$$SM_{newij} = SM_{ij} + U[0, 1] \times (GL_j - SM_{ij}) + U[0, 1] \times (SM_{ij} - LL_{kj}) \quad (5)$$

It is clear from the above equation that the updated dimension of this SM is attracted toward global leader and repels from the local leader.

2.1.7 Global Leader Decision (GLD) Phase

This phase takes the decision about position of Global Leader, if it is not modernized up to a threshold is known as Global Leader Limit (GL_{limit}), and then GLD creates subgroups of small size. During this phase, Local Leaders are created for new subgroups using LLL process [2].

3 An Improved Spider Monkey Optimization Algorithm

The Spider Monkey is a latest algorithm in different field of swarm intelligence. In literature there is very little research available on it. The newly proposed Improved Spider Monkey Optimization algorithm improves the performance of basic SMO algorithm. The ISMO suggested some improvement in Local Leader Phase of basic SMO. Position update equation in ISMO takes average of difference of current position and randomly generated positions. It generates a random position in given range for particular problem. This suggested modification accelerates the convergence rate and increase reliability. Here, it is assumed that better fitted solution has optimal solution in their proximity.

$$Y_{ij} = X_{ij} + \phi_{ij} \times (LL_{kj} - ISM_{ij}) + \phi_{ij} \times \left(\frac{SUM}{SN}\right) \tag{6}$$

where

$$SUM = SUM + (X_{ij} - X_{kj})$$

ϕ_{ij} is a uniformly generated random number in range [0,1].

Where ISM_{ij} denotes the jth dimension of the ith ISM, LL_{kj} ensures the jth dimension of the kth local leader group location. The SN represents the food source that is randomly generated by the position for food source. SUM is the average of difference for current position and randomly generated position. This equation updates highly fitted solutions through inspiration from best Swarm Intelligence. This new addition in SMO increases the balance between exploration and exploitation of most feasible solutions.

Algorithm of ISMO

```
Initialization LL_Limit, Sum, SN
For each K ∈ {1,2,.....,MG} do
 For each member SM_i ∈ K^h group do
  For each j ∈ {1,2,...,D} do
   For each M ∈ {1,2,...SN} do
    Sum = Sum + (ISM_ij − ISM_rj)
    If U[0, 1] ≥   pr then
        ISM_newij = ISM_ij + U[0,1] × (LL_kj − ISM_ij) + U[−1,1] × (SUM/SN)
    Else
            ISM_newij = ISM_rj
    End if
   End for
  End for
 End for
End for
```

4 Experimental Analysis

This paper checks the performance of Improved SMO algorithm over some well-known benchmark optimization function f_1 to f_6 (Table 1). The performance of newly proposed algorithm is compared with Basic SMO [2]. The performance comparison is based on standard deviation (SD), mean error (ME), average function evaluation (AFE), and success rate (SR) (Table 2).

Table 1 Test problems

Test problem	Objective function	Search range	Opt. value	D	AE
Six–Hump camel back function	$f_1(x) = (4 - 2.1x_1^2 + \frac{1}{3}x_1^4)x_1^2 + x_1x_2 + (-4 + 4x_2^2)x_2^2$	[−5, 5]	$f(-00898, 0.7126) = -1.0316$	2	1.0E−13
Hosaki problem	$f_2(x) = (1 - 8x_1 + 7x_1^2 - \frac{7}{3}x_1^3 + \frac{1}{4}x_1^4)x_2^2 \times \exp(-x_2)$	$x_1 \in [0, 5]$ $x_2 \in [0, 6]$	−2.3458	2	1.0E−06
Pressure vessel design	$f_3(x) = (1 - 8x_1 + 7x_1^2 - \frac{7}{3}x_1^3 + \frac{1}{4}x_1^4)x_2^2 \times \exp(-x_2)$ Subject to $g_1(x) = 0.0193x_3 - x_1,$ $g_2(x) = 0.00954x_3 - x_2,$ $g_3(x) = 750 * 1728 - \pi x_3^2(x_4 + \frac{4}{3}x_3)$	$1.125 \leq x_1 \leq 12.5,$ $0.625 \leq x_2 \leq 12.5,$ $1.0*10 - 8 \leq x_3 \leq 240$ and $1.0*10 - 8 \leq x_4 \leq 240$	$f(1.125, 0.625, 55.8592, 57.7315) = 7197.729$	30	1.0E−0.5
Rosenbrock	$f_4(x) = \sum_{i=1}^{i=D-1} 100(x_i^2 - x_{i+1})^2 + (1 - x_i)^2$	[−30, 30]	$f(0) = 0$	30	1.0E−01
Salmon problem	$f_5(x) = 1 - \cos(2\pi p) + 0.1 \times p,$ where, $p = \sqrt{\sum_{i=1}^{D} x_i^2}$	[100, 100]	$f(0) = 0$	30	1.0E−01
Pathological	$f_6(x) = \sum_{i=1}^{D-1}\left(\dfrac{\sin^2(\sqrt{100x_{i+1}^2 + x_i^2}) - 0.5}{0.001(x_i - x_{i+1})^4 + 0.50}\right)$	[−100, 100]	$f(0) = 0$	30	1.00E−01

D Dimension, *AE* Acceptable error

Table 2 Comparison of result between SMO and ISMO

Test problem	Algorithm	MFV	SD	ME	AFE	SR
f_1	SMO	−1.03E+00	1.46E–05	1.90E–05	30783.45	41
	ISMO	−1.03E+00	1.52E–05	1.52E–05	22960.92	56
f_2	SMO	−2.35E+00	6.28E–06	6.04E–06	7831.09	85
	ISMO	−2.35E+00	6.06E–06	5.59E–06	3138.71	94
f_3	SMO	7.20E+03	9.49E–04	3.62E–04	28014.45	48
	ISMO	7.20E+03	3.35E–05	2.85E–05	23604.81	62
f_4	SMO	1.65E+00	1.03E+01	1.65E+00	14284.64	96
	ISMO	6.23E–02	5.22E–01	6.23E–02	11936.35	98
f_5	SMO	2.00E–01	7.80E–06	2.00E–01	6418.15	100
	ISMO	2.10E–01	3.00E–02	2.10E–01	15468.15	90
f_6	SMO	1.02E+00	4.47E–01	1.02E+00	50969.76	2
	ISMO	4.54E–01	3.41E–01	4.54E–01	38206.64	33

4.1 Experimental Setting

The proposed ISMO algorithm is compared with Basic SMO technique in order to prove its competence. It is programmed in C programming language with below-mentioned experimental setting.

The size of Swarm $N = 50$

MG = 5 (Maximum group limiting maximum number of spider monkey in a group as MG = $N/10$)

Global Leader Limit = 50

Local Leader Limit = 1500

Pr $\in [0.1, 0.4]$, linearly increasing over iteration.

4.2 Experimental Result Comparison

See Table 2.

5 Conclusion

This paper proposed a coherent and productive variant of SMO that improves the number of function evaluations in comparison to SMO Algorithm. By this algorithm, we can find the feasible solution to understand the swarm intelligence-based algorithm. This process is an extension of the position update in LLP. This algorithm has been tested; it will increase the accuracy and reliability through the average of convergence rate comparison to SMO algorithm. This approach is applied to the 6 benchmarks problems and results prove its superiority over basic SMO algorithm.

References

1. Bonabeau, E., Dorigo, M., Theraulaz, G.: Swarm intelligence: from natural to artificial systems (no.1). Oxford University Press (1999)
2. Bansal, J.C., Sharma, H., Jadon, S.S., Clerc, M.: Spider monkey optimization algorithm for numerical optimization. Memet. comput. 6(1), 31–47 (2014)
3. Pal, S.S., Kumar, S., Kashyap, M., Choudhary, Y., Bhattacharya, M.: Multi-level thresholding segmentation approach based on spider monkey optimization algorithm. In: Proceedings of the Second International Conference on Computer and Communication Technologies, pp. 273–287. Springer, India (2016)
4. Gupta, K., Deep, K., Bansal, J.C.: Improving the local search ability of spider monkey optimization algorithm using quadratic approximation for unconstrained optimization. Comput. Intell. (2016)
5. Sharma, A., Sharma, A., Panigrahi, B.K., Kiran, D., Kumar, R: Ageist spider monkey optimization algorithm. Swarm Evol. Comput. 28, 58–77 (2016)
6. Gupta, K., Deep, K.: Tournament selection based probability scheme in spider monkey optimization algorithm. In: Harmony Search Algorithm, pp. 239–250. Springer, Heidelberg (2016)
7. Gupta, K., Deep, K.: Investigation of suitable perturbation rate scheme for spider monkey optimization algorithm. In: Proceedings of Fifth International Conference on Soft Computing for Problem Solving, pp. 839–850. Springer, Singapore (2016)
8. Singh, U., Salgotra, R., Rattan, M.: A novel binary spider monkey optimization algorithm for thinning of concentric circular antenna arrays. IETE J. Res. 1–9 (2016)
9. Singh, U., Salgotra, R.: Optimal synthesis of linear antenna arrays using modified spider monkey optimization. Arab. J. Sci. Eng. 1–17 (2016)
10. Sharma, A., Sharma, H., Bhargava, A., Sharma, N.: Power law-based local search in spider monkey optimisation for lower order system modelling. Int. J. Syst. Sci.1–11 (2016)
11. Al-Azza, A.A., Al-Jodah, A.A., Harackiewicz, F.J.: Spider monkey optimization (SMO): a novel optimization technique in electromagnetics. In: 2016 IEEE Radio and Wireless Symposium (RWS), pp. 238–240. (2016)
12. Agarwal, P., Singh, R., Kumar, S., Bhattacharya, M.: Social spider algorithm employed multi-level thresholding segmentation approach. In: Proceedings of First International Conference on Information and Communication Technology for Intelligent Systems, Vol. 2, pp. 249–259. Springer International Publishing (2016)
13. Kumar, S., Sharma, V.K., Kumari, R.: Self-adaptive spider monkey optimization algorithm for engineering optimization problems. Int. J. Inf. Commun. Comput. Technol. II, pp. 96–107 (2014)
14. Kumar, S., Kumari, R., Sharma, V.K.: Fitness based position update in spider monkey optimization algorithm. Procedia Comput. Sci. 62, 442–449 (2015). doi:10.1016/j.procs.2015.08.504
15. Kumar, S., Sharma, V.K., Kumari, R.: Modified position update in spider monkey optimization algorithm. Int. J. Emerg. Technol. Comput. Appl. Sci. 2(7), 198–204 (2014)

Neural Network-Based Prediction of Productivity Parameters

Jayant P. Giri, Pallavi J. Giri and Rajkumar Chadge

Abstract This paper emphasized on neural network-based approximation method to predict correct and optimize cycle time for productivity improvement of structural subassembly manufacturing system. Altogether 35 assorted independent variables were considered for the experimentation against cycle time of operation as a productivity measure. 600 experiments were conducted and data were collected over the span of 2 years, which is modeled using Artificial Neural network. Symbolic mathematical model is formulated to reveal black box nature of Neural Network. Scaled conjugate gradient and gradient descent methods are considered as optimization algorithms for ANN, both performs well with diversified topologies. Coefficient of correlation ($R = 0.996$) with the sum of square error in the range of (0.095–0.034) reflecting a better approximation of dependent variable.

Keywords Production cycle time · Transfer function · Dependent variable

1 Introduction

Often it is of interest to study a system to understand the relations between its components or to predict how a system is responsive to changes. Sometimes it is possible to directly experiment with the system. However, this is not always possible, e.g., due to costs when a manufacturing system has to be stopped, changed, or

J.P. Giri (✉) · R. Chadge
Department of Mechanical Engineering, Yeshwantrao Chavan College of Engineering
(Autonomous), Nagpur, Maharashtra, India
e-mail: jayantpgiri@gmail.com

R. Chadge
e-mail: rbchadge@rediffmail.com

J.P. Giri
Laxminarayan Institute of Technology, RTM Nagpur University, Nagpur, Maharashtra, India
e-mail: Pallavijgiri@gmail.com

© Springer Nature Singapore Pte Ltd. 2018
M. Pant et al. (eds.), *Soft Computing: Theories and Applications*,
Advances in Intelligent Systems and Computing 583,
https://doi.org/10.1007/978-981-10-5687-1_8

extended. Often the system even does not yet exist. A model, defined as a representation of the system in order to investigate it, can solve this dilemma. Generally, it is sufficient to abstract the system with a view to analyze the issues under investigation. In terms of modeling and simulation, this abstract is named the simulation model. Simulation enables system analysis with time and space compression, provides a robust validation mechanism under realistic conditions and can reduce the risk of implementing new systems. Validation is achieved using a series of qualitative and quantitative experiments with changes of system variables and structures. The expanding capability of computing systems and the increasing demands of engineers and managers planning, implementing, and maintaining manufacturing systems have been pushing the boundaries of modeling and simulation based research. For many manufacturers, implementing a change in their operation can be risky, so simulation can be used as the test bed for evaluation of new manufacturing methods and strategies. Using engineering discipline, manufacturing systems can be measured through data collection, and processes analysis. Measurement efforts are the first step for better understanding of manufacturing systems. Where processes have been measured and data collected, simulation can be applied as a decision-making tool to enhance system understanding. When systems are not well defined or understood, it is difficult to build accurate models that are worthwhile. Manufacturing and material handling systems can be arbitrarily complex and difficult to understand. The number of possible combinations of input variables can be overwhelming when trying to perform experimentation. After manufacturing system data have been collected and verified, simulation can be used to represent almost any level of detail to provide an accurate representation of a real-world system. From a model of the system, the behavior of the system and its components can be better understood. It is important to note that optimizing on one measure of performance can adversely affect another measure of performance.

Case study of Structural subassembly production which is a batch processing manufacturing system from Sai Industries Pvt. Ltd. MIDC, Nagpur is considered in this paper. Despite some inherent variations which are inbuilt in manufacturing, it specifically comes under the category of deterministic in nature. Present investigation intended toward the formulation of generalized field data-based mathematical model for structural subassembly manufacturing. Decision regarding the disquiet about whether structural subassembly batch processing which is predominantly press working operation can be model using probabilistic or deterministic modeling approach thought out initially. The limitation of stochastic simulation about inability to consider human factor, workplace related factors and environmental factors during formulation of model clearly hints towards the second choice (Deterministic simulation). The approach suggested in the present investigation is to check out the effect of these parameters on performance measures such as production cycle time. Thus, pervasive attempt has been made in this paper to investigate the phenomenon of structural subassembly manufacturing by applying Artificial Neural network (ANN) technique.

Bergmann et al. [1] introduced a novel methodology for approximating dynamic behavior using artificial neural networks, rather than trying to determine exact representations. They suggested using neural networks in conjunction with traditional material flow simulation systems whenever a certain decision cannot be made in the model generation process due to insufficient knowledge about the behavior of the real system. Fowler and Rose [2], there is a need for the invasive use of modeling and simulation for decision support in current and future manufacturing systems, and several challenges need to be addressed by the simulation community to realize this vision. Hosseinpour and Hajihosseini [3] revealed the importance of simulation and according to them, implementing change can be a difficult task for any organization, big or small. For this purpose modeling of complex systems such as manufacturing systems is a strenuous task. Sabuncuoglu and Touhami [4] presented fundamentals of simulation metamodeling using neural networks for job shop manufacturing system. Cimino et al. [5] proposed a methodology for the effectual ergonomic design of workstations contained by industrial plants. The methodology based on multiple design parameters and multiple performance measures prop-up the design and the evaluation of workstations in terms of both ergonomics and work methods. Vainio et al. [6], used neural network for estimation of printed circuit board assembly time. In their study they trained multilayer neural networks to approximate the assembly times of two different types of assembly machines based on several parameter combinations. Abdelwahed et al. [7], presented a performance prediction approach for the product produced after multi-stages of manufacturing processes, as well as the assembly using ANN. Taghi et al. [8], presented designing of a multivariate multistage quality control system using artificial neural networks. Wang et al. [9] thoroughly presented an updated survey on Neural Network applications in intelligent manufacturing. Zain et al. [10], presented regression and ANN models for estimating minimum value of machining performance. Two modeling approaches, regression and Artificial Neural Network (ANN), were applied to predict the minimum Ra value. Dowler [11] in his theses on "Using Neural Networks with limited data to estimate manufacturing cost" concluded that, neural networks are an excellent option for cost estimation even when the amount of data available is limited. Researchers mentioned here have investigated the effect of various parameters such as environmental conditions, ergonomic and anthropometric consideration, and workstation design for various manufacturing scenarios. Unfortunately evidence is not sited regarding the generalized framework which consist of all parameters taken together and checking its effect on productivity. Scene is even more discern for structural subassembly manufacturing type of setups which are predominantly a press shop. In this paper attempt has been made to include other general parameters such as personal factors of operator and specification of the product to model batch manufacturing of structural subassemblies of tractors.

2 Problem Formulation and Design of Experiments

The specific objective of the present investigation is to developed mathematical model for structural subassembly manufacturing process using time honored artificial neural network (ANN). Investigation also provokes the aim of statistical and reliability analysis of any computational problem which can be numerically simulated, mainly the estimation of statistical parameters of response variable. Thus identification of number of influential variables and grouping them logically is essential for the present investigation. This can be achieved because of the ingrained difference in factors which impinge on the production process.

For example, all variable related to machine specification are grouped together. In order to accomplish the intention

The independent variables are grouped and identified as:

1. Anthropometric data of an operator
2. Personal factors of an operator
3. Machine specifications
4. Workplace parameters
5. Specification of the product
6. Environmental conditions

The Dependent variables can be distinguished as:

- Production cycle time

The variable listing with specific symbol for respective groups are depicted in the Table 1a–c.

The quality of results obtained from field research depends on the data gathered in the field. Data collection method is gathering of information to address a research problem. Data that will be subjected to statistical analysis must be gathered in such a way that they can be quantified. For statistical analysis variables must be quantitatively measured. A formal data collection process is necessary as it ensures that data gathered are both defined and accurate and that subsequent decisions based on arguments embodied in the findings are valid. Formal data collection method using calibrated instruments is used for this investigation. Parameters which are constant have been recorded first and then other parameters were recorded. Altogether 600 experiments conducted over the span of 2 years for 35 independent variables and production cycle time as a dependent variable (Fig. 1).

3 Neural Network Modeling

The term neural network applies to a limply related family of models, characterized by a large parameter space and flexible structure, descending from studies of brain functioning. As the family grew, most of the new models were designed for

Table 1 (a) Independent variables and symbol (anthropometric data and personal factors of operator) (b) Independent variables and symbol (machine specification and workplace parameters). (c) Independent variables and symbol (environmental factors and product data)

Anthropometric data of an operator	Symbol	Personal factors of an operator	Symbol
Height of operator	Ht	Age	Ag
Foot breadth	Fb	BMI prime	BMI prime
Arm span	As	Qualification grade	Qgr
Arm reach	Ar	Experience	Exp
Erect sit height	Esh	–	–
Sitting knee height	Skh	–	–
Machine specification	Symbol	Workplace parameters	Symbol
Capacity	C	Height of stool	Hos
Power HP	P	Area of tabletop in Sq.cm.	Areattop
Stroke frequency	Ss	Height of work table in cms	Htw
Stroke speed	Sps	Spatial distance between centroid of stool top and work table	Sd1
Machine age	Aom	Spatial distance between centroid of stool top and WIP table	Sd2
Distance between ram and worktable (cm)	Distrw	–	–
Setting time for machine in min.	stime	–	–
Preventive maintenance time Hrs/week	Mcdtime	–	–
Environmental factors	Symbol	Product data	Symbol
Noise without operation	dB	Length (mm)	L
Noise level with operation	dBstroke	Breadth (mm)	B
Illumination at workstation	Iwt	Thickness (mm)	T
Illumination at sight	Isr	Part Weight (gm)	Wt
Dry bulb temperature	DBT	Machinable length	Mc.len
Wet bulb temperature	WBT	Machine criticality	Mc_criticality

Fig. 1 Working scenario for structural subassembly manufacturing

nonbiological applications, though much of the associated terminology reflects its origin. Specific definitions of neural networks are as varied as the fields in which they are used. While no single definition properly covers the entire family of models, for now, considers the following description

"A neural network is a massively parallel distributed processor that has a natural propensity for storing experiential knowledge and making it available for use." It resembles the brain in two respects:

- Knowledge is acquired by the network through a learning process.
- Interneuron connection strengths known as synaptic weights are used to store the knowledge.

By contrast, the definition above makes minimal demands on model structure and assumptions. Thus, a neural network can approximate a wide range of statistical models without requiring that one hypothesize in advance certain relationships between the dependent and independent variables. Instead, the form of the relationships is determined during the learning process. If a linear relationship between the dependent and independent variables is appropriate, the results of the neural network should closely approximate those of the linear regression model. If a nonlinear relationship is more appropriate, the neural network will automatically approximate the "correct" model structure.

The trade-off for this flexibility is that the synaptic weights of a neural network are not easily interpretable. Thus, if one is trying to explain an underlying process that produces the relationships between the dependent and independent variables, it would be better to use a more traditional statistical model. However, if model interpretability is not important, one can often obtain good model results more quickly using a neural network (Fig. 2).

3.1 Feedforward Neural Network

Although neural networks impose minimal demands on model structure and assumptions, it is useful to understand the general network architecture. The multilayer perceptron (MLP) or radial basis function (RBF) network is a function of

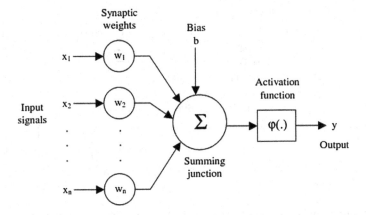

Fig. 2 Model of neuron

predictors (also called inputs or independent variables) that minimize the prediction error of target variables (also called outputs).

The structure depicted in Fig. 3 is known as feedforward architecture, because the connections in the network flow forward from the input layer to the output layer without any feedback loops. In the figure:

- The input layer contains the predictors.
- The hidden layer contains unobservable nodes, or units.

The value of each hidden unit is some function of the predictors; the exact form of the function depends in part upon the network type and in part upon user-controllable specifications.

- The output layer contains the responses.

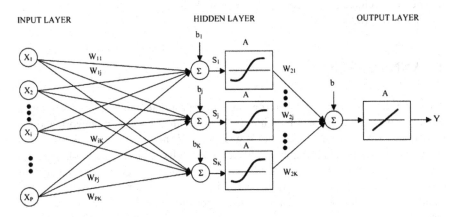

Fig. 3 Feedforward network

Table 2 Different transfer functions in neural network

Function	Definition	Range	Function	Definition	Range
Identity	x	(−inf, +inf)	Unit sum	$\dfrac{x}{\sum_i x_i}$	(0, +1)
Logistic	$\dfrac{1}{1-e^{-1}}$	(0, +1)	Square root	$\sqrt{x}$	(0, +inf)
Hyperbolic	$\dfrac{e^x-e^{-x}}{e^x+e^{-x}}$	(−1, +1)	Sine	$\sin(x)$	[0, +1]
Exponential	e^{-x}	(0, +inf)	Ramp	$\begin{cases} -1 & x \le -1 \\ x-1 & -1 < x < +1 \\ +1 & x \ge +1 \end{cases}$	[−1, +1]
Softmax	$\dfrac{e^x}{\sum_i e^{x_s}}$	(0, +1)	Step	$\begin{cases} 0 & x < 0 \\ +1 & x \ge 0 \end{cases}$	[0, +1]

Each output unit is some function of the hidden units. Again, the exact form of the function depends in part on the network type and in part on user-controllable specifications.

3.2 Activation Function

A function used to transform the activation level of a unit (neuron) into an output signal. Typically, activation functions have a "squashing" effect. Neural Networks supports a wide range of activation functions. Only a few of these are used by default; the others are available for customization (Table 2).

The hyperbolic tangent function (tanh): a sigmoid curve, like the logistic function, except that output lies in the range (−1, +1), often performs better than the logistic function because of its symmetry. Ideal for customization of multilayer perceptron, particularly the hidden layers. Several researchers have indicated that single hidden layer architecture with an arbitrarily large quantity of hidden nodes in the layer, is capable of modeling any categorization mapping. In the present work number of architectures were tried and tested to find out the best one for the prediction of productivity.

4 ANN Modeling

An experimental means for determining an appropriate topology for solving a particular problem involves the training of a larger-than-necessary network and the subsequent removal of unnecessary weights and nodes during training. This approach, called pruning, requires advance knowledge of initial network size, but such upper bounds may not be difficult to estimate. An alternative means for determining appropriate network topology involves algorithms which start with a small network and build it larger; such algorithms are known as constructive algorithms.

4.1 ANN Model with Base Variable as Input and Cycle Time Output

Mathematical Model based on basic 35 independent variables and Cycle time of operation as dependent variable is formulated. Model is based on 600 experiments conducted over the span of two years and considering three bifurcation of entire period (Summer, Winter, Rainy season). Symbolic representation of ANN equation is formulated as follows:

$$X_{1,1} = \left(e^{1*\text{sum(Layer1Cell0)}} - e^{-1*\text{sum(Layer1Cell0)}}\right) / \left(e^{1*\text{sum(Layer1Cell0)}} + e^{-1*\text{sum(Layer1Cell0)}}\right)$$

where sum (Layer1cell0) = $0.24 * X_{0.1} - 0.43 * X_{0.2} - 0.307 * X_{0.3} + 0.17 * X_{0.4} + 0.1888 * X_{0.5} - 0.4 * X_{0.6} + 0.24 * X_{0.7} + 0.4 * X_{0.8} + 0.02 * X_{0.9} - 0.074 * X_{0.10} - 0.023 * X_{0.11} + 0.15 * X_{0.12} + 0.099 * X_{0.13} + 0.02 * X_{0.14} + 0.33 * X_{0.15} - 0.421 * X_{0.16} + 0.29 * X_{0.17} - 0.009 * X_{0.18} + 0.04 * X_{0.19} + 0.077 * X_{0.20} + 0.471 * X_{0.21} - 0.146 * X_{0.22} - 0.363 * X_{0.23} - 0.417 * X_{0.24} + 0.163 * X_{0.25} - 0.237 * X_{0.26} - 0.329 * X_{0.27} + 0.389 * X_{0.28} + 0.196 * X_{0.29} + 0.18 * X_{0.30} - 0.359 * X_{0.31} + 0.04 * X_{0.32} - 0.096 * X_{0.33} - 0.0321 * X_{0.34} - 0.228 * X_{0.35} - 0.421$

$$X_{1,2} = \left(e^{1*\text{sum(Layer1Cell1)}} - e^{-1*\text{sum(Layer1Cell1)}}\right) / \left(e^{1*\text{sum(Layer1Cell1)}} + e^{-1*\text{sum(Layer1Cell1)}}\right)$$

where sum (Layer1cell1) = $0.24 * X_{0.1} - 0.43 * X_{0.2} - 0.307 * X_{0.3} + 0.17 * X_{0.4} + 0.1888 * X_{0.5} - 0.4 * X_{0.6} + 0.24 * X_{0.7} + 0.4 * X_{0.8} + 0.02 * X_{0.9} - 0.074 * X_{0.10} - 0.023 * X_{0.11} + 0.15 * X_{0.12} + 0.099 * X_{0.13} + 0.02 * X_{0.14} + 0.33 * X_{0.15} - 0.421 * X_{0.16} + 0.29 * X_{0.17} - 0.009 * X_{0.18} + 0.04 * X_{0.19} + 0.077 * X_{0.20} + 0.471 * X_{0.21} - 0.146 * X_{0.22} - 0.363 * X_{0.23} - 0.417 * X_{0.24} + 0.163 * X_{0.25} - 0.237 * X_{0.26} - 0.329 * X_{0.27} + 0.389 * X_{0.28} + 0.196 * X_{0.29} + 0.18 * X_{0.30} - 0.359 * X_{0.31} + 0.04 * X_{0.32} - 0.096 * X_{0.33} - 0.0321 * X_{0.34} - 0.228 * X_{0.35} - 0.433$

5 Result and Discussion

Prediction of cycle time as a productivity measure of a small-scale manufacturing unit is considered to established relationship between dependent variable and 35 assorted independent variables.Mathematical Model is developed using fundamentals of Artificial Neural network-based function approximation. Former relationship between output and input was unknown till correct approximation by ANN model. Neural Network is of black box nature, an attempt has been made to showcase mathematical structure of ANN model by using hyperbolic tangential function as a transfer function. Random synapses were generated using training and optimization algorithm and weighted sum of synapses and scaled input data then further constricted through nonlinear transfer function (Table 3).

Table 3 Comparison of various topology of neural network

Topology	Activation function		Training algorithm	Coefficient of correlation (R)	Sum of square errors		Relative error	
	Hidden layer	Output layer			Training	Prediction	Training	Prediction
35-1-1	Tanh	Tanh	Scale conjugate gradient	0.954	3.661	2.182	0.042	0.056
	Sigmoidal	Sigmoidal	SCG	0.948	1.159	0.542	0.050	0.065
	Tanh	Tanh	Gradient descent	0.974	2.265	0.984	0.026	0.025
	Sigmoidal	Sigmoidal	GD	0.976	0.555	0.223	0.026	0.022
35-5-1	Tanh	Tanh	SCG	0.995	0.443	0.211	0.005	0.006
	Sigmoidal	Sigmoidal	SCG	0.993	0.163	0.057	0.007	0.006
	Tanh	Tanh	GD	0.997	0.257	0.163	0.003	0.004
	Sigmoidal	Sigmoidal	GD	0.994	0.109	0.074	0.005	0.008
35-10-1	Tanh	Tanh	SCG	0.997	0.282	0.115	0.003	0.003
	Sigmoidal	Sigmoidal	**SCG**	**0.996**	**0.095**	**0.041**	**0.004**	**0.005**
	Tanh	Tanh	GD	0.996	0.329	0.153	0.004	0.004
	Sigmoidal	Sigmoidal	**GD**	**0.996**	**0.099**	**0.039**	**0.004**	**0.004**
35-15-1	Tanh	Tanh	SCG	0.995	0.357	0.251	0.004	0.006
	Sigmoidal	Sigmoidal	SCG	0.995	0.108	0.065	0.005	0.007
	Tanh	Tanh	GD	0.996	0.388	0.181	0.004	0.005
	Sigmoidal	Sigmoidal	GD	0.994	0.124	0.055	0.006	0.006
35-20-1	Tanh	Tanh	SCG	0.993	0.506	0.370	0.006	0.009
	Sigmoidal	Sigmoidal	SCG	0.979	0.529	0.140	0.024	0.015
	Tanh	Tanh	GD	0.996	0.320	0.149	0.004	0.004
	Sigmoidal	Sigmoidal	**GD**	**0.996**	**0.096**	**0.041**	**0.004**	**0.005**

(continued)

Table 3 (continued)

Topology	Activation function		Training algorithm	Coefficient of correlation (R)	Sum of square errors		Relative error	
	Hidden layer	Output layer			Training	Prediction	Training	Prediction
35-25-1	Tanh	Tanh	SCG	0.991	0.763	0.355	0.009	0.009
	Sigmoidal	Sigmoidal	SCG	0.996	0.087	0.049	0.004	0.005
	Tanh	Tanh	GD	0.995	0.434	0.149	0.005	0.004
	Sigmoidal	Sigmoidal	GD	0.996	0.093	0.039	0.004	0.004
35-30-1	Tanh	Tanh	SCG	0.982	2.001	0.993	0.023	0.027
	Sigmoidal	Sigmoidal	SCG	0.996	0.096	0.034	0.004	0.004
	Tanh	Tanh	GD	0.994	0.636	0.208	0.007	0.006
	Sigmoidal	Sigmoidal	GD	0.994	0.120	0.063	0.006	0.006
35-35-1	Tanh	Tanh	SCG	0.997	0.233	0.091	0.003	0.002
	Sigmoidal	Sigmoidal	**SCG**	**0.996**	**0.097**	**0.034**	**0.004**	**0.002**
	Tanh	Tanh	GD	0.995	0.460	0.174	0.005	0.006
	Sigmoidal	Sigmoidal	GD	0.995	1.104	0.042	0.005	0.006

Topologies with 35-10-1, 35-20-1, 35-35-1 with sigmoidal transfer function in input and output layer appears to be best choice to predict cycle time. Sigmoidal transfer function performs well with both optimization algorithms. Coefficient of correlation in all three cases is closed to 0.996 which is an excellent indication for prediction of correct cycle time after ANN simulation. This ANN simulation will surely helpful for small-scale industries to approximate productivity measures via simulation process instead of cumbersome and exhaustive execution of experiments. It is practically nonviable for small industries to spend time in years to collect and measure huge data based on various influential variables. ANN simulation is rather a better perspective for researchers and small-scale industries for prediction of productivity-related parameters.

Sum of square error and relative error as basic statistical measure to check performance of ANN model is well within stipulated range for the best performing topologies. As relative error does not deviate during training and prediction phase indicating stability and reflected that network is not overtrain.

6   Conclusion

Convincing results are yielded through ANN Model simulation in this investigation. Relationship between 35 independent variables and one dependent variable is captured correctly by ANN. Though ANN modeling is of black box nature and understanding of model is complicated for general Understanding. Enigmatic nature of this hardheaded relationship is tough; but the precision of ANN model must be honored. Prediction of cycle time as a productivity measure using neural network acts as a useful guideline for small-scale industries working on similar line. The paper also elaborates comparative analysis between adoption of two training algorithms and two activation functions. Further, it is also observed that number of hidden nodes in hidden layer also have significant impact on output parameters.

References

1. Bergmann, S., Stelzer, S., Strassburger, S.: On the use of artificial neural net works in simulation-based manufacturing control. J. Simul. **8**(1), 76–90 (2014)
2. Fowler, J.W.: Rose, O.: Grand challenges in modeling and simulation of complex manufacturing systems. Simulation **80**(9), 469–476 (2004)
3. Hosseinpour, F., Hajihosseini, H.: Importance of Simulation in Manufacturing. World Academy of Science, Engineering and Technology, vol. 51 (2009)
4. Sabuncuoglu, I., Touhami, S.: Simulation metamodelling with neural networks: an experimental investigation. Int. J. Prod. Res. **40**(11), 2483–2505 (2002)
5. Cimino, A., Longo, F., Mirabelli, G.: A multimeasure-based methodology for the ergonomic effective design of manufacturing system workstations. Int. J. Ind. Ergon. **39**, 447–455(2009)

6. Vainio, F., Maier, M., Knuutila, T., Alhoniemi, E., Johnsson, M., Nevalainen, O.S.: Estimating printed circuit board assembly times using neural networks. Int. J. Prod. Res. **48** (8), 2201–2218 (2010)
7. Abdelwahed, M.S., El-Baz, M.A., El-Midany, T.T.: A Proposed Performance Prediction Approach for Manufacturing Processes Using ANNs. World Academy of Science, Engineering and Technology, vol. 61 (2012)
8. Taghi, S., Niaki, A., Davoodi, M.: Designing a multivariate–multistage quality control system using artificial neural networks. Int. J. Prod. Res. (2009)
9. Wang, J., et al.: Applications in Intelligent Manufacturing: An Updated Survey. Computational Intelligence in Manufacturing Handbook. CRC Press LLC (2001)
10. Zain, A.M., Harona, H., Qasem, S.N., Sharif, S.: Regression and ANN models for estimating minimum value of machining performance. Appl. Math. Model. **36**, 1477–1492 (2012)
11. Dowler, J.D. Using Neural networks with limited data to estimate manufacturing cost. Ph.D. thesis, Ohio University (2008)

Rigorous Data Analysis and Performance Evaluation of Indian Classical Raga Using RapidMiner

Akhilesh K. Sharma and Prakash Ramani

Abstract In this research work, we propose to classify the Indian classical raga. The research work focuses on preprocessing phase for the audio feature creation and then the music is segmented in various categories according to the raga properties. The extracted feature data set utilized for the raga classification and then the measurement of accuracy has been done. RapidMiner tool is used for the classification purpose and Jaudio is used for the feature extraction. The raga data is used for the north Indian classical music. The classifier chosen for the purpose performed flawlessly and results obtained were above satisfactory level.

Keywords Music information retrieval · Data mining · Classification RapidMiner · Naïve Bayes · J48 · Indian classical music

1 Introduction

The ragas in Indian classical music are very interesting and challenging as well when it needs to be classified in different Ragas. The classification task for the music professionals is very challenging as the data set is increasing day by day and to classify them requires time. In this paper, several approaches have been studied, analyzed and then evaluated the performance of the same in order to get the maximum one. And found significant performance enhancement. The tool utilized in this research work was RapidMiner which provided ease of use.

A.K. Sharma (✉) · P. Ramani
SCIT, Manipal University Jaipur—MUJ, Jaipur, Rajasthan, India
e-mail: akhileshkumar.sharma@jaipur.manipal.edu

P. Ramani
e-mail: prakash.ramani@jaipur.manipal.edu

© Springer Nature Singapore Pte Ltd. 2018
M. Pant et al. (eds.), *Soft Computing: Theories and Applications*,
Advances in Intelligent Systems and Computing 583,
https://doi.org/10.1007/978-981-10-5687-1_9

97

Fig. 1 Indian classical music instrument Harmonium

1.1 Raga Characteristics

The ragas can be identified by analyzing pakad and bandish and getting the most dominant notes of a given raga. Each of the ragas has its own bandish. The varjit swara is very restricted and need to be taken care of and excluded while rendering the raga (It has also been considered as an enemy of raga, and restricted for a given raga).

While performing raga the performer has to remember the structure and raga composition details without which the performer cannot perform well, and the entire composition and performance hamper the emotions attached to it.

Figure 1 depicts the ragas to be created using any instrument, here we took piano and traditional Indian classical instrument harmonium.

Thus, the structural details according to the raga have to be identified, and learned well before performing live. The *raga* composition differs according to the *thaats*, the *thaats* classify the *raga* into various emerging raga from itself.

2 Literature Survey

The thaats have the basic composition derived from the seven keys on the keyboard or musical instruments like harmonium, violin, etc., from which the raga can be emerged and can be further classified in any other category based on the emerging thaat.

Parag et al. (2013) in their research findings used the pitch class [1] and pitch class dyad distributions they have used the HMM techniques to identify the raga but their study not capable enough to perform the statistical evaluations. In the research paper they formulated and proposed a system to identify the raga from the real time environment.

According to Rae and Chordia [2], the tonal classification is based on applying the classifier scheme to the training data and accordingly the results can be achieved on the same.

According to the study of the researcher in machine learning for the computer music, Mr. Bhatkande [3], finds that most modern notation systems are based on Bhatkhande rules, Bhatkhande's notation system consists of symbols for notes indicating their pitch, and symbols for ornaments such as glissandi (meend) and khatka.

Pitch can be represented by a dot that is either placed above or below the note, or may be absent. In these three cases, when the dot is not available above or below the note, it represented the middle octave. In second case, when the dot is placed above the note it indicates the higher octave.

In the third case when the dot is placed below the note, it represents the lower octave. The representation of the notes is done in hindi characters with traditional names of the scale tones. In the Western alphabet, these are Do, Re, Mi, Fa, So, etc. But in Hindi Sa, Re, Ga, Ma, Pa, Dha, and Ni. For the simplicity these can be used simply by the first letter (e.g., S) for the representation purposes only.

Bhatkhande's proposed a system in which, the beats are separated by spaces [3]. Notes or the frequency pitches that occur in the range of the same beat are written next to each other without gap, with a curved symbol underneath to indicate they all occur in the same beat [3].

The many ragas and their notes formation are shown as the notes varied in their structural representation and the formation. The ten thaats are used in order to classify any raga in any particular category. The raga can be framed based on their structures and key combinations. The duration is also plays an important role while creating or performing any raga. The taan, alap, and kan swaras are some of the very important duration related concepts. The kan swaras have no explicit duration. Before each composition the rhythmic cycle (taal), tempo (slow, medium, fast), and part of the bandish (sthayee, antara) are noted [4].

The raga affects maximum emotion or changed behavior of any person. The effect of any raga is maximum when the raga is exercised/performed on the desired time/prahar of the associated raga. If this is true, the raga that needs to perform impact maximum. Emotional qualities include the tempting nature of humans or their mood. The mood is also changed when the raga is changed. It throws a different mood if the raga is of another kind is played or listened. For example, the raga bilawal and yaman produce the bhakti ras-related emotions that creates humans very relaxed and soothing emotions are emerged automatically. But when the different raga can be performed it just put the other effect. Lots of research work have done in this direction and lots more can also be done. The number of ragas has been discussed extensively in NICM [5] music theory [6, 7].

These studies have followed a basic experimental paradigm in which musical excerpts are presented to listeners who are then asked to respond by verbally or by filling up the forms that are option based forms, describing their emotional states, rating emotions on a quantitative scale, or some other measurement. Real musical excerpts rarely vary in any parameter [6, 8].

Most importantly, the difficulty of finding a person's actual emotional state, either through their mental state observation or by eliciting verbal or other responses, makes such research extremely challenging [2].

Parag et al. (2008) in their research paper these kind of research and listener and response study performed, but it has been observed that it is really a very challenging task [2].

2.1 Prahar

The prahar in the ICM (Indian Classical Music) plays an important role for the classical music to take place maximum impact on the minds of the listener. In connection with this type of effect the prahar are identified so as to gain the maximum performance on the performer's choice of the raga to be chosen. This includes the raga-based prahar/time horizons to be remembered by the performer always, (as if he/she chooses different raga that is not relevant with the prahar it will not give maximum impact to the minds of the listener and creates some different emotions that is not related to the raga rendered) [9].

3 Approach

This study includes step wise progress in terms of first phase as feature extraction, second PCP (pitch class profile) construction and then applying model in rapid miner and obtaining the accuracy of each classifier. At last, this study concludes the performance accuracy of each model and results have been shown.

3.1 Classification Problem

The classification of the Indian classical music is found to be difficult as the ragas based on the different structural representations. The 'thaats' are used to classify in different ragas. Thus, it can be classified in different 'thaats' that includes different raga according to their structural detail.

The classification problem is complex when the performer needs to practice at home. Especially, when the performer wants to classify raga details from huge music record data set. At that point of time, the classification techniques help them in order to get the correct raga music in efficient time. This problem is inculcated in this study and it has been observed that by using the feature extraction this problem can be solved in much ease of use and in a less complex way.

3.2 Features Used

This research work focuses on the features that are to be extracted but what kind of features. The music files are signal based files. After preprocessing the music signals are converted into the pitch representations the actual pitches are ranging between the 44.1 kHz and these pitches are combined in the pitch vector files which is nothing but the collections of these pitches. The Spectral roll off, zero cross ratings, spectral centroid, spectral roll off point, Spectral flux, etc. [10], are the other features that are used. These are the features obtained in preprocessing step. This can be used for the classification purpose.

3.3 Feature Extraction

The features extracted [10] for the different Indian classical ragas have been obtained as given in Tables 1 and 2.

The different ragas formed the basis to be classified using their structural forms, so that they have to be divided in term of their structures like raga desh, raga asavari, raga bilawal, etc., and after their structural description obtained as shown in the histogram plots (in Fig. 2), the ragas can be sent to the rapid miner for the detailed classification into different raga classifications.

3.4 Classifier Used

We choose different classifiers as shown and compared the accuracy for different ragas that have been compared.

Naïve Bayes

This classifier is used for the classification task of the Indian classical music ragas. The naïve Bayes classifier chosen because of the Bayesian probability of this

Table 1 Extracted features for the different Indian ragas (a)

Raga/features extracted	Zero cross	Spectral centroid	Spectral rolloff point	Spectral flux
Raga Yaman	40	53	47	46
Raga Bilawal	44	35	53	57
Raga Hamir	50	54	57	26
Raga Bhairav	42	64	64	53
Raga Desh	36	43	24	58
Raga Todi	35	45	54	48
Raga Purvi	49	34	34	36
Raga Asavari	56	54	54	56

Table 2 Feature extracted (pitch vector features) for the different Indian ragas

0 0 0 0 0 0 0 0 0 0 0 0 0 49.2184 48.3304 0 47.7627 48.0442 48.0442 47.7627
48.3304 48.0442 48.0442 48.0442 47.4857 47.4857 47.4857 47.4857 47.4857
47.7627 48.0442 48.0442 47.7627 47.7627 47.7627 47.2131 0 0 0 0 51.4868
51.4868 51.4868 51.4868 51.8366 51.8366 0 0 0 54.9225 54.0979 54.0979
54.0979 54.0979 54.0979 54.0979 54.9225 54.9225 54.9225 54.9225 54.5053
54.5053 54.9225 54.5053 54.5053 54.0979 54.5053 55.35 54.9225 54.9225
54.5053 54.5053 54.5053 54.5053 54.5053 54.5053 54.5053 54.5053 54.5053
54.5053 53.3109 0 0 0 56.238 55.7883 55.7883 56.238 56.6996 56.6996
56.6996 57.174 56.6996 56.6996 56.6996 56.6996 56.6996 56.6996 56.6996
57.174 57.174 57.174 56.6996 56.6996 56.6996 56.6996 0 0 0 59.7627 59.7627
59.7627 59.2131 59.2131 59.2131 58.6804 0 0 0 54.0979 54.0979 54.0979
54.5053 54.5053 54.0979 54.0979 54.5053 54.9225 54.5053 54.5053 54.0979
53.6999 0 0 0 0 0 0 0 0 0 0 0 0 0 0 0 0 0 0 0 0 50.4781 51.4868 51.8366 52.1935
52.558 52.1935 52.558 51.8366 51.8366 52.558 52.9303 52.9303 52.9303
52.9303 52.9303 51.144 0 54.5053 53.3109 52.558 52.1935 52.1935 52.1935
52.558 52.9303 54.0979 54.5053 54.9225 55.7883 56.238 56.238 55.7883
55.35 54.0979 0 0 51.4868 50.1544 50.1544 50.8078 51.144 51.144 51.144
51.144 52.1935 52.1935 51.8366 51.8366 51.8366 51.4868 51.4868 51.4868
51.8366 51.8366 51.4868 51.4868 51.4868 51.4868 51.4868 50.8078 0 52.558

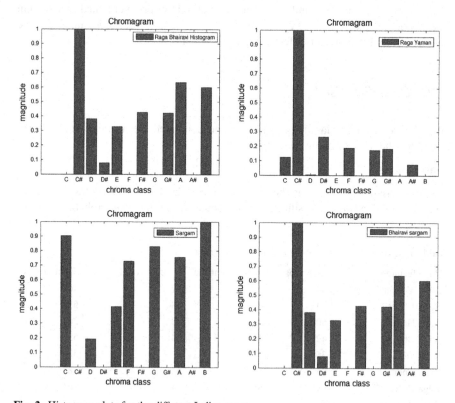

Fig. 2 Histogram plots for the different Indian ragas

classifier fulfills our purpose of the efficient classification. The classifier accuracy has been calculated by observing the ROC curves plotted after the due course of verification and the validation with precision and recall calculation.

KNN

The k-nearest neighbor calculation is also done in order to achieve the increased accuracy level and because of the closest neighbor's availability, so that the correct classification can be obtained. The KNN classifier produced efficient results as verified after due calculation and plotting the ROC curves.

J48 algorithm

This classifier also produces very satisfactory results in the music segmentation. Thus, in this study it has been used to classify the Indian classical music raga details. The j48 algorithm is used to plot the forest trees in order to find the different branches and to prune them when it is required. The j48 also uses the bias function which is very much capable of the effective classification of the different music details. The results are also justified while calculating and drawing the ROC curve and precision and recall.

3.5 Performance Evaluation

We plotted the ROC curve for the performance evaluation to be compared and analyzed with the other methods as well (Fig. 3).

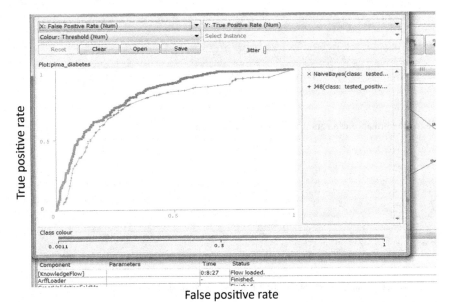

Fig. 3 ROC curve for performance evaluation

4   Results and Discussion

In this study, the results obtained found to be very satisfactory and this was observed that the performance of raga classification improved much and shown using the RapidMiner tool. The results obtained are shown in this paper. The model deployed in the rapid miner for the raga analysis and classification has performed well.

We adopted the performance evaluation model for our research as per the Fig. 4 in the performance evaluation section.

The obtained accuracy was very high (nearly to 84%) for the raga classification. The ragas that we took are preprocessed and their features have also been extracted in order to classify them. The PCP (pitch class profiles) has also been obtained while preprocessing or extracting features. Simultaneously the accuracy and precision also measured for x-validation model 84.40% and also for various other classifier models as well (Figs. 5, 6 and 7).

We compare the approach with some other methods as well (Table 3). The conclusion we found was also satisfactory in terms of the reliable accuracy it provides (Fig. 8).

The results are as shown in Table 3.

The resultant graph is as follows for the accuracy.

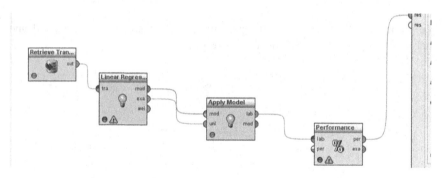

Fig. 4 Performance evaluation in RapidMiner diagram

		true 0	true 1	class precision
accuracy: 84.40% +/- 5.78% (mikro: 84.40%)				
	pred. 0	103	17	85.83%
	pred. 1	22	108	83.09%
	class recall	82.40%	86.40%	

Fig. 5 Accuracy and the precision using RapidMiner x-validation model 84.40%

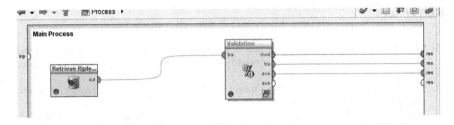

Fig. 6 Classifier x-validation model

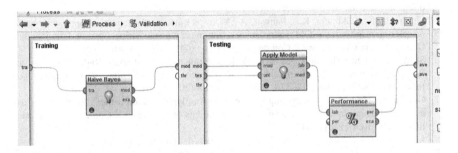

Fig. 7 X-validation model

Table 3 Classifier accuracy on application of models and classifiers on ragas

S. No.	Classifier 1 (NB) (%)	Classifier 2 (KNN) (%)	Classifier 3 (J48) (%)	Classifier 4 (Zero R) (%)
Raga Todi	82	64	52	56
Raga Bilawal	81	73	57	63
Raga Hamir	79	74	68	67
Raga Bhairav	82	67	62	68
Raga Desh	83	75	67	74
Raga Yaman	84	72	69	71

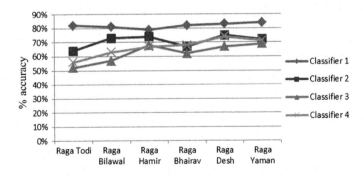

Fig. 8 Accuracy result graph

References

1. Chordia, P.: Automatic raag classification of pitch tracked performances using pitch-class and pitch-class dyad distributions. In: Proceedings of International Computer Music Conference (2006)
2. Rae, A., Chordia, P.: Understanding emotion in raag: an empirical study of listener responses, pp. 110–124. Springer, Berlin (2008)
3. Pandey, G., Mishra, C., Ipe, P.: Tansen: a system for automatic raaga identification. In: Proceeding of the 1st Indian International Conference on Artificial Intelligence, pp. 1350–1363 (2003)
4. Chakraborty, S., Debashis, D.: Pattern classification of Indian classical ragas based on object oriented concepts. Int. J. Adv. Comput. Eng. Arch. 2, 285–294 (2012)
5. Sharma, A.K., Lakhtaria, K.I., Panwar, A., Vishwakarma, S.: An analytical approach based on self organized maps (SOM) in Indian classical music raga clustering. Seventh International Conference on Contemporary Computing (IC3), pp. 449, 453, 7–9 Aug (2014)
6. Shetty, S.A.: Clustering of Ragas Based on Jump Sequence for Automatic Raga Identification. Wireless Networks and Computational Intelligence Communications in Computer and Information Science, pp. 318–328. Springer, Berlin (2012)
7. Gomez, E., Herrera, P.; Estimating the tonality of polyphonic audio files: cognitive versus machine learning modelling strategies. In: Proceedings of International Conference on Music Information Retrieval (2004)
8. Jan, G.T.: Music Information Retrieval Techniques, 1st ed. Georgia (2004)
9. Roy, S., et al.: Modeling high performance music computing using Petri Nets. 2014 International Conference on Control, Instrumentation, Energy and Communication (CIEC), IEEE (2014)
10. Sharma, A.K., Panwar, A., Chakrabarti, P.: Analytical approach on Indian classical raga measures by feature extraction with EM and Naive Bayes. Int. J. Comput. Appl. 107(6), 41–46 (2014)

EPQ Model with Product Stewardship Approach

Pratiksha Saxena, Chaman Singh and Kamna Sharma

Abstract From the past few decades, most of the researchers concentrate on recycling of consumer goods to reduce the impact on environment and health, but now it is time to change our basic approach. In this research paper, we elaborate a nascent issue of product stewardship. Economic production model (EPQ) is a production model in which production occur in a time cycle to fulfill the demand. The proposed model we consider a basic EPQ model with all basic setup and product stewardship. In this paper, we investigate a complete and simple elucidation of product stewardship in a lucid manner which corroborated with many illustrations to convey the intend theme. In this proposed model total cost is calculated with convexity and sensitivity analysis.

Keywords Product stewardship · Green product · EPQ model

1 Introduction

Over the past few decades, almost all the developed and developing countries are taking interest as well as feeling responsibility for the global environmental problems like heat waves, global warming and environmental pollution to resolve these issues development of green product becomes a burning issue for each and every producer. The quality of these products has been a hard core challenge for modern enterprises. Now-a-days most of the people are highly and regularly concern about the raw material of product and recycling of products, when they purchase any of these products. From last few years researcher are concerned about three R's which

P. Saxena · K. Sharma (✉)
Department of Mathematics, Gautam Buddha University, Greater Noida, Uttar Pradesh, India
e-mail: anushka.gautam17@gmail.com

C. Singh
Aacharya Narayan Dev College, Delhi University, Meerut, India

© Springer Nature Singapore Pte Ltd. 2018
M. Pant et al. (eds.), *Soft Computing: Theories and Applications*,
Advances in Intelligent Systems and Computing 583,
https://doi.org/10.1007/978-981-10-5687-1_10

107

are reduce, reuse and recycle, but other than these restricted issue it is also important how can a modern enterprises played a significant role in the manufacturing of green product.

Now, green product may be defined as several ways, but in this paper we concentrate on product steward ship. Over these concerns we emphasis on safety of environment. Product stewardship defined as an important tool to ensure the safety of any industrial product, its raw material, cleaner production of integral component of any chemical, electronic or any hazardous industry. Product stewardship is quite similar to green processing as well as green product but quite different from other green supply chain management. When any enterprises or any well known brand ensure the consumer to serve a fully equipped, well developed, upgraded good quality and useful product, then they also sure that it will be a eco friendly product. But now most of the enterprises are in trouble to satisfy all the criteria's as well as their profit concerns. The solution of that problem may solved by product stewardship. Lot of well-repudiated companies of U.S.A. are following product steward ship pattern. Product stewardship can take the whole responsibility to manage safety, health, and environment aspect of consumer product, raw material etc. Product stewardship helps in reducing the negative impact on environment and maximizes the value of product. It is a share responsibility of producer, supplier, and consumer. They all should concern about it. Product steward ship technically reduces the negative impact on health, environment and consumer product throughout their life cycle. Product stewardship concludes lot of steps in producing green product and technically correlates each other. Technically it may conclude cleaner production of consumer goods, use eco friendly raw material, recycling of goods, reduce scrap value, reduce environmental impact etc. It may achieve by redesigning the products and improving all raw material standard, process should also be redefined. Product stewardship technically reduces the after recycling impact on environmental health and other safety reformation. As we know that every industrial product is recycled then large salvage arises and due to this salvages environmental hazards increasing day by day.

By product stewardship, we can resolve various challenges in these emerging areas of green product with real circumstances and it may also be provide a balance view for those factors which may influence both our decision making factors and consequences of these decisions. So that by using the phenomena of product stewardship manufacture can reduce the risk of supply chain management and profit factor. At earlier stage Weinberg [1] introduced environmentally conscious product stewardship to reduce the impact of product on the environment. Peck and Christy [2] putting the stewardship concept to the practice: commercial moss harvest in northwest USA. Burge [3] proposed environmentally management: Integrated ecological evaluation, remediation, restoration, natural resource damage assessment, and long-term stewardship on the contaminated lands. Dawes [4] proposed integrating science and business model of sustainability for environmentally-challenging industries such as secondary lead smelter. It is a systematic review and analysis of findings. Lun [5] investigate a GMP (green management practice) model with consist of cooperation with supply chain partners, internal management support and environmentally friendly operation. Wong et al. [6] develop green operations and the moderating role of environmental management capability of supplies on manufacturing firm performance. It is a analysis of effect of GO (green operation) on environmental a management capacity of supplier of the firm performance of pollution reduction. In this GO product steward ship and process stewardship is also used. Lawler et al. [7] discover new opportunities for reuse, recycling, and disposal of used reverse osmosis membrane. Plambeck [8] shows how companies can profitably reduce green house emission in their supply chain. On the same side, Dekker et al. [9] develop a model focusing on how good can be design, transport, stored, and consume in an eco-friendly manner. Norlund and Gribkovskaia [10] used speed optimization strategies are introduced in the construction of vessel schedules and the strategies utilize waiting time in vessel schedules to reduce speed. Kristianto and Helo [11] introduced product architecture modularity implications for operations economy of green supply chains. Galeazzo et al. [12] achieve environmentally sound manufacturing process, environmental management, and operation management which need to be implemented in an integrated manner. Wang [13] proposed a comprehensive decision making model for the evaluation of green operations initiatives. Trentesaux and Giret [14] expose an idea to incite researcher to develop sustainability-oriented manufacturing operations control architectures, holonic or multi agent, and to provide a usable generic concept that is easy to appropriate, particularize and implement. Jabbour et al. [15] confirmed that green product practices influence firm's performance and also discuss some results from a survey conducted with Brazilian companies. Curran [16] proposed product steward ship for life cycle analysis and the environment. Jabbour and Jabbour [17] discuss the two linking agendas green human resource management and green supply chain.

In this proposed model Product stewardship is introduced in an EPQ model. Product stewardship is used to resolve the problem of green product. Demand and production rate is constant. Stewardship factor is introduced in the production cost as well as with set up cost. To elucidate the utility of the model total cot function is derived. Sensitivity and Convexity are also shown in the paper.

2 Notations and Assumptions

Notations

P	Production rate
D	Demand rate
T	Total time
PS	Product stewardship cost
p	Production cost
h	Holding cost

A Setup cost
$I_1(t)$ Inventory level in between $0 < t \leq T_1$
$I_2(t)$ Inventory level in between $0 < t \leq T$

Assumptions

1. Demand rate is constant.
2. Production rate is constant.
3. Setup cost is constant.
4. Shortage is not allowed.
5. Deterioration is not considered here.

3   Mathematical Model

In this paper, we explore an EPQ model with product stewardship in constant form. Total time period is divided in two segments. At the time $t = 0$ inventory level is also zero, but when the time cycle starts then inventory level also raise by production and fulfill the demand. In the first one production occurs and demand is fulfilled, but in the next run production is stopped and demand is fulfilled with the stored goods. Inventory level becomes zero at time T.

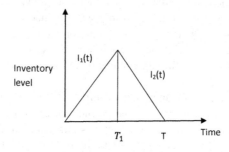

$$I_1(t) = (P - D)t \quad 0 < t \leq T_1$$

$$I_2(t) = D(T - t) \quad T_1 < t \leq T$$

At time T_1

$$T_1 = \frac{DT}{P}$$

$$\text{Set up cost} = \frac{A}{PS}$$

$$\text{Holding cost} = \frac{(h+PS)T^2D}{2}\left(1-\frac{D}{P}\right)$$

$$\text{Production cost} = (p+PS)DT$$

$$TC = \frac{\frac{A}{PS} + \frac{(h+PS)T^2D}{2}\left(1-\frac{D}{P}\right) + (p+PS)DT}{T}$$

3.1 Solution Procedure to Find Out Optimal Solution

In the total cost equation there are two independent variables, T and PS. To optimize the total cost equation we follow the different steps.

Step 1

First, calculate the partial derivatives w.r.t all the independent variables $\frac{\partial f}{\partial T}$ and $\frac{\partial f}{\partial PS}$.

Step 2

Equate the first order partial derivatives to zero and solve for the value of T and PS.

Step 3
Now, to calculate the second order partial derivative w.r.t. all the independent variables.

Step 4

Now, form a Hessian matrix as follows

$$\begin{pmatrix} \dfrac{\partial^2 f}{\partial T^2} & \dfrac{\partial^2 f}{\partial T \partial PS} \\ \dfrac{\partial^2 f}{\partial PS \partial T} & \dfrac{\partial^2 f}{\partial PS^2} \end{pmatrix}$$

Step 5

Find H_1 and H_2, where, H_1 and H_2 denote the first principal minor and second principal minor respectively. If $\det(H_1) > 0$ and $\det(H_2) > 0$, then the matrix is positive definite matrix and f is called convex function.

4 Numerical Example

To resolve the above production model problem following parameters are used in the numerical example. This numerical example also shows some particular results. In this numerical example total time and product stewardship are independent variables. Total cost is found out as a result.

$$A = 500; P = 50; D = 10; p = 0.8; h = 0.2;$$

$$TC = 115.883; T = 11.243; PS = 0.899441.$$

5 Convexity of Total Cost Function

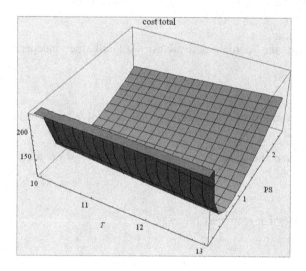

6 Conclusion

In today's world go green is a burning issue for all researchers and scientist. They are not only concern but worried about it. To solve this problem, we approach a nascent but important concept of product stewardship. This concept not only reduce the after effect of product disposal but also reduce the impact on environment in the process of making of any product like carbon emission etc. so that here EPQ model is used to resolve this issue with product stewardship. Product stewardship is a concept, according to this concept product is planned in such a way that it should be recycled properly or without much selvage. Product steward ship is used with setup cost as well as production cost in the proposed model. Convexity and numerical example is used to justify the model.

References

1. Weinberg, L.: Environmentally conscious product stewardship at the Boeing company. Corp. Environ. Strategy **6**(3), 246–257 (1999)
2. Peck, J.L.E., Christy, J.A.: Putting the stewardship concept into practice: commercial moss harvest in northwestern Oregon, USA. For. Ecol. Manag. **225**(1–3), 225–233 (2006)
3. Burger, J.: Environmental management: integrating ecological evaluation, remediation, restoration, natural resources damage assessment and long term stewardship on contaminated lands. Sci. Total Environ. **400**(1–3), 6–19 (2008)
4. Dawes, S.S.: Stewardship and usefulness: policy principles for information-based transparency. Govern. Inf. Q. **27**(4), 377–383 (2010)
5. Lun, Y.H.V.: Green management practices and firm performance: a case of container terminal operations. Resour. Conserv. Recycl. **55**(6), 559–566 (2011)
6. Wong, C.W.Y., Lai, K.H., Shang, K.C., Lu, C.S., Leung, T.K.P.: Green operation and moderating role of environmental management capability of suppliers manufacturing firm performance. Int. J. Prod. Econ. **140**(1), 283–294 (2012)
7. Lawler, W., Bradford-Hartke, Z., Cran, M.J., Duke, M., Leslie, G., Ladewig, B.P., Le-Clech, P.: Towards new opportunities for reuse, recycling and disposal of used reverse osmosis membranes. Desalination **299**(1), 103–112 (2012)
8. Plambeck, E.L.: Reducing green house emission through operations and supply chain management. Energy Econ. **34**(1), S64–S74 (2012)
9. Dekker, R., Bloemhof, J., Mallidis, L.: Operations research for green logistics—an overview of aspect, issues, contributions and challenges. Eur. J. Oper. Res. **219**(3), 671–679 (2012)
10. Norlund, E.K., Gribkovskaia, I.: Reducing emission through speed optimization in supply vessel operation. Transp. Res. D Transp. Environ. **23**, 105–113 (2013)
11. Kristianto, Y., Helo, P.: Product architect modularity implications for operations economy of green supply chain. Transp. Res. E Logistics Transp. Rev. **70**, 128–145 (2014)
12. Galeazzo, A., Furlan, A., Vinelli, A.: Understanding enviormental-operations integration: the case of pollution prevention projects. Int. J. Prod. Econ. **153**, 149–169 (2014)
13. Wang, X.: A comprehensive decision making model for the evaluation of green operations initiatives. Technol. Forecast. Soc. Change **95**, 191–207 (2015)
14. Trentesaux, D., Giret, A.: Go-green manufacturing holons: a step towards sustainable manufacturing operations control. Manuf. Lett. **5**, 29–33 (2015)
15. Jabbour, C.J.C., Jugend, D., de Sousa Jabbour, A.B.L., Gunasekaran, A., Latan, H.: Green product development and performance of Brazilian firms: measuring the role of human and technical aspects. J. Clean. Prod. **87**, 442–451 (2015)
16. Curran, M.A.: Product steward ship: life cycle analysis and the environment. J. Clean. Prod. **112**(1), 1252–1253 (2016)
17. Jabbour, C.J.C., de Sousa Jabbour, A.B.L.: Green human resource management and green supply chain management: linking two emerging agendas. J. Clean. Prod. **112**(3), 1824–1833 (2016)

New Arithmetic Operations in Inverse of Triskaidecagonal Fuzzy Number Using Alpha Cut

A. Rajkumar and D. Helen

Abstract The main target of the paper is to introduce the square root and inverse of new fuzzy number called triskaidecagonal fuzzy number. This new approach deals with lexical scale values and also includes basic arithmetic operations of its inverse by means of alpha cut. It follows with the definitions. The next section proceeds with square root of triskaidecagonal fuzzy number and its linguistic values and arithmetic operation for inverse fuzzy number. This inverse operation can apply in various fields with 13 parameters. Finally, the paper ends with reference and conclusion.

Keywords Fuzzy number · Triskaidecagonal · Triangular fuzzy number
Pentagonal fuzzy number · Nanogonal fuzzy number · Dodecagonal fuzzy number

1 Introduction

The concept of fuzzy number has been extended by many researchers. Nanogonal Fuzzy Number (NFN) and its arithmetic operations based on extension principle of fuzzy sets and α cut are introduced [1]. A good way off Decagonal Fuzzy Number and its operations derived for a pinnacle in this research area [2]. As the complication of unextreme membership function to take trapezoidal, hexagonal, octagonal, decagonal etc., when incertitude arises in 13 different points, researchers try to clear the fuzziness using a new form of fuzzy numbers. Triskaidecagonal, a new form of fuzzy number has been inspected under an uncertain lexical environment which would be easier to illustrate with triskaidecagonal fuzzy lexical scale values. This

A. Rajkumar (✉) · D. Helen
Department of Mathematics, Hindustan Institute of Technology and Science,
Hindustan University, Chennai 603103, India
e-mail: arajkumar@hindustanuniv.ac.in

D. Helen
e-mail: helen.br@outlook.com

© Springer Nature Singapore Pte Ltd. 2018
M. Pant et al. (eds.), *Soft Computing: Theories and Applications*,
Advances in Intelligent Systems and Computing 583,
https://doi.org/10.1007/978-981-10-5687-1_11

paper throws light on how α cut method is used to find square root and inverse fuzzy number by performing arithmetic operation using triskaidecagonal linguistic scale values. Thus, mathematical artistry in this proposed method cannot be denied.

2 Preliminaries

Definition 2.1: The gradual assessment of the membership of elements is permitted by fuzzy set theory in a set which has been described with an aid of membership function valued in the real unit interval [0,1]. A fuzzy number is thus a special case of a convex, normalized fuzzy set of the real line.

Definition 2.2: Arithmetic fuzzy numbers are denoted by $AA_{\overline{DD}}$ and $AB_{\overline{DD}}$. Lingustic variable transformed into dodecagonal fuzzy numbers $[s_i, s_j]$ and $[s_k, s_l]$ is denoted by $LA_{\overline{DD}}$ and $LB_{\overline{DD}}$.

Definition 2.3: If $S = s_0, s_1, \ldots, s_g$ be a finite and totally ordered set with odd linguistic terms where s_i denotes the ith linguistic term, $i \in 0, 1, \ldots, g$ then we call set S the linguistic term set and $(g + 1)$ the cardinality of S. It is usually required that set S has the following properties:

(i)   The set is ordered: $s_i \geq s_j$ if $i > j$
(ii)   There is a negation operator: neg $(s_i) = s_j$ such that $j = g - 1$
(iii) Maximization operator:$\max(s_i, s_j) = s_i$ if $s_i \geq s_j$
       Minimization operator:$\min(s_i, s_j) = s_i$ if $s_i \leq s_j$

Let $\overline{S} = s_l, s_{l+1}, \ldots, s_u$ where $s_l, s_{l+1}, \ldots, s_u \in S$, $s_l \geq s_u$, s_l and s_u are the lower and upper limits, $l, u \in 0, 1, \ldots, g$ respectively, Then we call $\overline{S}$ the uncertain linguistic term.

Triskaidecagonal Fuzzy Number (TDFN)

A fuzzy number $m_{\overline{TD}}$ is a dodecagonal fuzzy number denoted by $m_{\overline{TD}} = (a_1, a_2, a_3, a_4, a_5, a_6, a_7, a_8, a_9, a_{10}, a_{11}, a_{12}, a_{13})$ where are real numbers and its membership function (Table 1).

Table 1 Triskaidecagonal Fuzzy Lexical Scale

Lexical terms	Lexical values
No development	(0, 0, 0, 0, 0, 0, 0, 0.03, 0.06, 0.09, 0.12, 0.15, 0.18)
Very low development	(0.20, 0.22, 0.24, 0.26, 0.28, 0.30, 0.32, 0.34, 0.36, 0.38, 0.40, 0.42, 0.44)
Low development	(0.32, 0.34, 0.36, 0.38, 0.40, 0.42, 0.44, 0.46, 0.48.0.50, 0.52, 0.54, 0.56)
Medium	(0.40, 0.43, 0.46, 0.49, 0.52, 0.55, 0.58, 0.61, 0.64, 0.67, 0.70, 0.73, 0.76)
High development	(0.59, 0.61, 0.63, 0.65, 0.67, 0.69, 0.71, 0.73, 0.75, 0.77, 0.79, 0.81, 0.83)
Very high development	(0.64, 0.68, 0.72, 0.76, 0.80, 0.84, 0.88, 0.92, 1, 1, 1, 1, 1)

$$
\mu_{\overline{\text{TD}}}(x) =
\begin{cases}
\frac{1}{6}\left(\frac{x-a1}{a2-a1}\right) & a1 \leq x \leq a2 \\[4pt]
\frac{1}{6} + \frac{1}{6}\left(\frac{x2-a2}{a3-a2}\right) & a2 \leq x \leq a3 \\[4pt]
\frac{2}{6} + \frac{1}{6}\left(\frac{x-a3}{a4-a3}\right) & a3 \leq x \leq a4 \\[4pt]
\frac{3}{6} + \frac{1}{6}\left(\frac{x-a4}{a5-a4}\right) & a4 \leq x \leq a5 \\[4pt]
\frac{4}{6} + \frac{1}{6}\left(\frac{x-a5}{a6-a5}\right) & a5 \leq x \leq a6 \\[4pt]
\frac{5}{6} + \frac{1}{6}\left(\frac{x-a6}{a7-a6}\right) & a6 \leq x \leq a6 \\[4pt]
1 - \frac{1}{6}\left(\frac{x-a7}{a8-a7}\right) & a7 \leq x \leq a8 \\[4pt]
\frac{5}{6} - \frac{1}{6}\left(\frac{x-a8}{a9-a8}\right) & a8 \leq x \leq a9 \\[4pt]
\frac{4}{6} - \frac{1}{6}\left(\frac{x-a9}{a10-a6}\right) & a9 \leq x \leq a10 \\[4pt]
\frac{3}{6} - \frac{1}{6}\left(\frac{x-al}{a11-a10}\right) & a10 \leq x \leq a11 \\[4pt]
\frac{2}{6} - \frac{1}{6}\left(\frac{x-a11}{a12-a11}\right) & a11 \leq x \leq a12 \\[4pt]
\frac{1}{6} - \frac{1}{6}\left(\frac{a13-x}{a13-a12}\right) & a12 \leq x \leq a13 \\[4pt]
0 & x \geq a13
\end{cases}
$$

Square Root

Let $IM_{\overline{\text{TD}}}$ = (b1*, b2*, b3*, b4*, b5*, b6*, b7*, b8*, b9*, b10*, b11*, b12*, b13*) be a fuzzy number of linguistic scale values. Then, the square root of a fuzzy number is determined by α cut method (Fig. 1).

To justify a new operation with ordinary square root operation with the linguistic scale values

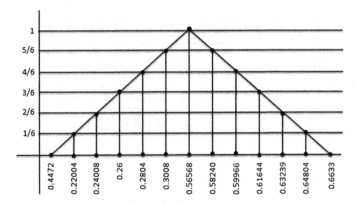

Fig. 1 Square root of arbitrary uncertain linguistic terms

$$IM_{\overline{TD}} = (0.20, 0.22, 0.24, 0.26, 0.28, 0.30, 0.32, 0.34, 0.36, 0.38, 0.40, 0.42, 0.44)$$

$$\alpha IM_{\overline{TD}} = [6\alpha(a2 - a1) + a1, -6\alpha(a13 - a12) + a13]$$

$$\sqrt{\alpha IM_{\overline{TD}}} = \sqrt{[6\alpha(a2 - a1) + a1, -6\alpha(a13 - a12) + a13]}$$

$$= \sqrt{[6\alpha(a2 - a1) + a1}, \sqrt{-6\alpha(a13 - a12) + a13]}$$

$$= \sqrt{0.12\alpha + 0.20}, \sqrt{0.12\alpha + 0.44}$$

When $\alpha = 0 \Rightarrow IM\alpha$ $(/)$ $IN\alpha = \left[\sqrt{0.20}, \sqrt{0.44}\right]$ = [0.4472, 0.6633]

$\alpha = 1/6 \Rightarrow IM\alpha$ $(/)$ $IN\alpha = \left[\sqrt{0.22004}, \sqrt{0.41996}\right]$ = [0.46908, 0.64804]

$\alpha = 2/6 \Rightarrow IM\alpha$ $(/)$ $IN\alpha = \left[\sqrt{0.24008}, 0.39992\right]$ = [0.48997, 0.63239]

$\alpha = 3/6 \Rightarrow IM\alpha$ $(/)$ $IN\alpha = \left[\sqrt{0.26}, \sqrt{0.38}\right]$ = [0.50990, 0.61644]

$\alpha = 4/6 \Rightarrow IM\alpha$ $(/)$ $IN\alpha = \left[\sqrt{0.2804}, \sqrt{0.3596}\right]$ = [0.52952, 0.59966]

$\alpha = 5/6 \Rightarrow IM\alpha$ $(/)$ $IN\alpha = \left[\sqrt{0.3008}, \sqrt{0.3392}\right]$ = [0.54845, 0.58240]

$\alpha = 1 \Rightarrow IM\alpha$ $(/)$ $IN\alpha = \left[\sqrt{0.32}, \sqrt{0.32}\right]$ = [0.56568, 0.56568]

Definition 2.4: Let $IM_{\overline{TD}}$ = ($b1^*$, $b2^*$, $b3^*$, $b4^*$, $b5^*$, $b6^*$, $b7^*$, $b8^*$, $b9^*$, $b10^*$, $b11^*$, $b12^*$, $b13^*$) be a positive number. Then to find the inverse of a fuzzy number by taking α cut of $M_{\overline{TD}}$ using interval arithmetic.

$$\frac{1}{\alpha M_{\overline{TD}}}(x) = \frac{1}{[6\alpha(a2 - a1) + a1, -6\alpha(a13 - a12) + a13]}$$

$$= \left[\frac{1}{-6\alpha(a13 - a12) + a13}, \frac{1}{6\alpha(a2 - a1) + a1}\right]$$

Similarly, for all intervals it is as follows:

$$\frac{1}{\alpha M_{\overline{TD}}}(x) = \left(\begin{array}{l} \left[\frac{1}{-6\alpha(a_{13}-a_{12}) + a_{13}}, \frac{1}{6\alpha(a_2-a_1) + a_1}\right] \\[6pt] \left[\frac{1}{-6\alpha(a_{12}-a_{13})-a_{11} + 2a_{12}}, \frac{1}{6\alpha(a_3-a_2) + 2a_2-a_3}\right] \\[6pt] \left[\frac{1}{-6\alpha(a_{11}-a_{10}) + 3a_{11}-2a_{10}}, \frac{1}{6\alpha(a_4-a_3) + 2a_4-3a_3}\right] \\[6pt] \left[\frac{1}{-6\alpha(a_{10}-a_9) + 4a_{10}-3a9}, \frac{1}{6\alpha(a_5-a_4)-3a_5 + 4a_4}\right] \\[6pt] \left[\frac{1}{-6\alpha(a_9-a_8) + 5a9-4a8}, \frac{1}{6\alpha(a_6-a_5)-4a6 + 5a5}\right] \\[6pt] \left[\frac{1}{-6\alpha(a_8-a_7) + 6a_8-5a7}, \frac{1}{6\alpha(a_7-a_6)-5a_7 + 6a_6}\right] \end{array} \right.$$

3 Inverse of a Fuzzy Number for Addition

3.1 Addition

Let $IM_{\overline{TD}} = (b1^*, b2^*, b3^*, b4^*, b5^*, b6^*, b7^*, b8^*, b9^*, b10^*, b11^*, b12^*, b13^*)$ and $IN_{\overline{TD}} = (d1^*, d2^*, d3^*, d4^*, d5^*, d6^*, d7^*, d8^*, d9^*, d10^*, d11^*, d12^*, d13^*)$ be their corresponding triskaidecagonal fuzzy numbers of linguistic scale values for all $\alpha\varepsilon\ [0,1]$. Then addition operation of α units.

$$\frac{1}{\alpha M_{\overline{TD}}} = \left[\frac{1}{-0.12\alpha + 0.44}, \frac{1}{0.12\alpha + 0.20}\right]$$

$$\frac{1}{\alpha N_{\overline{TD}}} = \left[\frac{1}{-0.18\alpha + 0.76}, \frac{1}{0.18\alpha + 0.40}\right]$$

To certify the inverse fuzzy number for addition, by taking the linguistic fuzzy values

for $\alpha\varepsilon\ [0, 0.166667)$	$\frac{1}{\alpha M_{\overline{TD}}} = \left[\frac{1}{-0.12\alpha+0.44}, \frac{1}{0.12\alpha+0.20}\right]$	$\frac{1}{\alpha M_{\overline{TD}}}(+)\frac{1}{\alpha N_{\overline{TD}}} = \left[\frac{-0.3\alpha+1.2}{0.0216\alpha^2-0.1704\alpha+0.3344}, \frac{-0.3\alpha+1.2}{0.0216\alpha^2-0.084\alpha+0.08}\right]$
for $\alpha\varepsilon\ [0.16667, 0.3333)$	$\frac{1}{\alpha N_{\overline{TD}}} = \left[\frac{1}{-0.18\alpha+0.76}, \frac{1}{0.18\alpha+0.40}\right]$	
for $\alpha\varepsilon\ [0.3333, 0.5)$		
for $\alpha\varepsilon\ [0.5, 0.66666)$		
for $\alpha\varepsilon\ [0.6666, 0.8333)$		
for $\alpha\varepsilon\ [0.83333,1)$		

As for $\alpha\varepsilon\ [0, 0.1666677)$, $\alpha\varepsilon\ [0.16667, 0.3333)$, $\alpha\varepsilon\ [0.3333, 0.5)$, $\alpha\varepsilon\ [0.5, 0.66666)$, $\alpha\varepsilon\ [0.66666, 0.8333)$, and $\alpha\varepsilon\ [0.83333, 1)$ arithmetic intervals are same

$$\frac{1}{\alpha M_{\overline{TD}}}(+)\frac{1}{\alpha N_{\overline{TD}}} = \left[\frac{-0.3\alpha+1.2}{0.0216\alpha^2-0.1704\alpha+0.3344}, \frac{-0.3\alpha+1.2}{0.0216\alpha^2-0.084\alpha+0.08}\right] \text{ for all } \alpha\varepsilon[0,1]$$

When

$$\alpha=0 \Rightarrow \left[\frac{1}{\alpha IM_{\overline{TD}}} + \frac{1}{\alpha IN_{\overline{TD}}}\right] = [0.58851, 7.5]$$

$$\alpha = 1/6 \Rightarrow \left[\frac{1}{\alpha IM_{\overline{TD}}} + \frac{1}{\alpha IN_{\overline{TD}}}\right] = [3.75111, 6.87005]$$

$$\alpha = 2/6 \Rightarrow \left[\frac{1}{\alpha IM_{\overline{TD}}} + \frac{1}{\alpha IN_{\overline{TD}}}\right] = [3.93733, 6.326661]$$

$$\alpha = 3/6 \Rightarrow \left[\frac{1}{\alpha IM_{\overline{TD}}} + \frac{1}{\alpha IN_{\overline{TD}}}\right] = [4.12411, 5.88697]$$

$$\alpha = 4/6 \Rightarrow \left[\frac{1}{\alpha IM_{\overline{TD}}} + \frac{1}{\alpha IN_{\overline{TD}}}\right] = [4.34438, 5.48630]$$

$$\alpha = 5/6 \Rightarrow \left[\frac{1}{\alpha IM_{\overline{TD}}} + \frac{1}{\alpha IN_{\overline{TD}}}\right] = [4.59627, 4.99812]$$

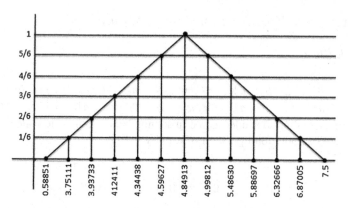

Fig. 2 Addition for inverse two arbitrary uncertain linguistic terms

$$\alpha = 1 \Rightarrow \left[\frac{1}{\alpha IM_{\overline{TD}}} + \frac{1}{\alpha IN_{\overline{TD}}}\right] = [4.84913, 4.84913]$$

Hence, the inverse addition is certified (Fig. 2).

3.2 Multiplication

Let $IM_{\overline{TD}} = (b1^*, b2^*, b3^*, b4^*, b5^*, b6^*, b7^*, b8^*, b9^*, b10^*, b11^*, b12^*, b13^*)$ and $IN_{\overline{TD}} = (d1^*, d2^*, d3^*, d4^*, d5^*, d6^*, d7^*, d8^*, d9^*, d10^*, d11^*, d12^*, d13^*)$ be their corresponding triskaidecagonal fuzzy numbers of linguistic scale values for all $\alpha\epsilon\ [0,1]$. Then addition operation of α units.

$$\frac{1}{\alpha M_{\overline{TD}}} = \left[\frac{1}{-0.12\alpha + 0.44}, \frac{1}{0.12\alpha + 0.20}\right]$$

$$\frac{1}{\alpha N_{\overline{TD}}} = \left[\frac{1}{-0.18\alpha + 0.76}, \frac{1}{0.18\alpha + 0.40}\right]$$

To certify the inverse fuzzy number for multiplication by takes the linguistic fuzzy values

for $\alpha\epsilon\ [0, 0.166667)$	$\frac{1}{\alpha M_{\overline{TD}}} = \left[\frac{1}{-0.12\alpha + 0.44}, \frac{1}{0.12\alpha + 0.20}\right]$	$\frac{1}{\alpha M_{\overline{TD}}}(*)\frac{1}{\alpha N_{\overline{TD}}} = \left[\frac{1}{0.0216\alpha^2 - 0.1704\alpha + 0.3344}, \frac{1}{0.0216\alpha^2 - 0.084\alpha + 0.08}\right]$
for $\alpha\epsilon\ [0.16667, 0.3333)$	$\frac{1}{\alpha N_{\overline{TD}}} = \left[\frac{1}{-0.18\alpha + 0.76}, \frac{1}{0.18\alpha + 0.40}\right]$	
for $\alpha\epsilon\ [0.3333, 0.5)$		
for $\alpha\epsilon\ [0.5, 0.66666)$		
for $\alpha\epsilon\ [0.6666, 0.8333)$		
for $\alpha\epsilon\ [0.83333,1)$		

As for $\alpha\varepsilon$ [0, 0.1666677), $\alpha\varepsilon$ [0.16667, 0.3333), $\alpha\varepsilon$ [0.3333, 0.5), $\alpha\varepsilon$ [0.5, 0.66666), $\alpha\varepsilon$ [0.66666, 0.8333), and $\alpha\varepsilon$ [0.83333, 1) arithmetic intervals are same

$$\frac{1}{\alpha M_{\overline{TD}}}(*)\frac{1}{\alpha N_{\overline{TD}}} = \left[\frac{1}{0.0216\alpha^2 - 0.1704\alpha + 0.3344}\right], \left[\frac{1}{0.0216\alpha^2 - 0.084\alpha + 0.08}\right] \text{ for all } \alpha\varepsilon [0,1]$$

When $\alpha = 0 \Rightarrow \left[\frac{1}{\alpha IM_{\overline{TD}}} * \frac{1}{\alpha IN_{\overline{TD}}}\right] = [2.99043, 12.5]$

$\alpha = 1/6 \Rightarrow \left[\frac{1}{\alpha IM_{\overline{TD}}} * \frac{1}{\alpha IN_{\overline{TD}}}\right] = [3.26158, 10.57093]$

$\alpha = 2/6 \Rightarrow \left[\frac{1}{\alpha IM_{\overline{TD}}} * \frac{1}{\alpha IN_{\overline{TD}}}\right] = [3.613513, 9.191260]$

$\alpha = 3/6 \Rightarrow \left[\frac{1}{\alpha IM_{\overline{TD}}} * \frac{1}{\alpha IN_{\overline{TD}}}\right] = [3.92772, 7.84929]$

$\alpha = 4/6 \Rightarrow \left[\frac{1}{\alpha IM_{\overline{TD}}} * \frac{1}{\alpha IN_{\overline{TD}}}\right] = [4.34873, 6.84931]$

$\alpha = 5/6 \Rightarrow \left[\frac{1}{\alpha IM_{\overline{TD}}} * \frac{1}{\alpha IN_{\overline{TD}}}\right] = [4.848390, 6.040471]$

$\alpha = 1 \Rightarrow \left[\frac{1}{\alpha IM_{\overline{TD}}} * \frac{1}{\alpha IN_{\overline{TD}}}\right] = [5.38793, 5.38793]$

Therefore multiplication is justified (Fig. 3).

3.3 Division

Let $IM_{\overline{TD}} = (b1*, b2*, b3*, b4*, b5*, b6*, b7*, b8*, b9*, b10*, b11*, b12*, b13*)$ and $IN_{\overline{TD}} = (d1*, d2*, d3*, d4*, d5*, d6*, d7*, d8*, d9*, d10*, d11*, d12*, d13*)$ be their corresponding triskaidecagon fuzzy numbers of linguistic scale values for all $\alpha\varepsilon$ [0,1]. Then addition operation of α units.

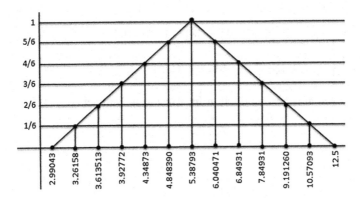

Fig. 3 Multiplication for inverse two arbitrary uncertain linguistic terms

$$\frac{1}{\alpha M_{\overline{TD}}} = \left[\frac{1}{-0.12\alpha + 0.44}, \frac{1}{0.12\alpha + 0.20}\right]$$

$$\frac{1}{\alpha N_{\overline{TD}}} = \left[\frac{1}{-0.18\alpha + 0.76}, \frac{1}{0.18\alpha + 0.40}\right]$$

To justify the inverse fuzzy number for division by taking the linguistic fuzzy values

for $\alpha\varepsilon$ [0, 0.166667)	$\frac{1}{\alpha M_{\overline{TD}}} = \left[\frac{1}{-0.12\alpha + 0.44}, \frac{1}{0.12\alpha + 0.20}\right]$	$\frac{1}{\alpha M_{\overline{TD}}}(/)\frac{1}{\alpha N_{\overline{TD}}} = \left[\frac{-0.18\alpha + .76}{-0.12\alpha - 0.44}, \frac{0.18\alpha + 0.40}{0.12 + 0.20}\right]$
for $\alpha\varepsilon$ [0.16667, 0.3333)	$\frac{1}{\alpha N_{\overline{TD}}} = \left[\frac{1}{-0.18\alpha + 0.76}, \frac{1}{0.18\alpha + 0.40}\right]$	
for $\alpha\varepsilon$ [0.3333, 0.5)		
for $\alpha\varepsilon$ [0.5, 0.66666)		
for $\alpha\varepsilon$ [0.6666, 0.8333)		
for $\alpha\varepsilon$ [0.83333,1)		

As for $\alpha\varepsilon$ [0, 0.1666677), $\alpha\varepsilon$ [0.16667, 0.3333), $\alpha\varepsilon$ [0.3333, 0.5), $\alpha\varepsilon$ [0.5, 0.66666), $\alpha\varepsilon$ [0.66666, 0.8333) and $\alpha\varepsilon$ [0.83333, 1) arithmetic intervals are same $\frac{1}{\alpha M_{\overline{TD}}}(/)\frac{1}{\alpha N_{\overline{TD}}} = \left[\frac{-0.18\alpha + .76}{-0.12\alpha - 0.44}, \frac{0.18\alpha + 0.40}{0.12 + 0.20}\right]$ for all $\alpha\varepsilon[0,1]$

When $\alpha = 0 \Rightarrow \left[\frac{1}{\alpha IM_{\overline{TD}}}(/)\frac{1}{\alpha IN_{\overline{TD}}}\right] = [1.72727, 2]$

$\alpha = 1/6 \Rightarrow \left[\frac{1}{\alpha IM_{\overline{TD}}}(/)\frac{1}{\alpha IN_{\overline{TD}}}\right] = [1.73811, 1.95446]$

$\alpha = 2/6 \Rightarrow \left[\frac{1}{\alpha IM_{\overline{TD}}}(/)\frac{1}{\alpha IN_{\overline{TD}}}\right] = [1.75050, 1.91528]$

$\alpha = 3/6 \Rightarrow \left[\frac{1}{\alpha IM_{\overline{TD}}}(/)\frac{1}{\alpha IN_{\overline{TD}}}\right] = 1.76315, 1.88461]$

$\alpha = 4/6 \Rightarrow \left[\frac{1}{\alpha IM_{\overline{TD}}}(/)\frac{1}{\alpha IN_{\overline{TD}}}\right] = [1.77808, 1.85663]$

$\alpha = 5/6 \Rightarrow \left[\frac{1}{\alpha IM_{\overline{TD}}}(/)\frac{1}{\alpha IN_{\overline{TD}}}\right] = [1.794811, 1.83244]$

$\alpha = 1 \Rightarrow \left[\frac{1}{\alpha IM_{\overline{TD}}}(/)\frac{1}{\alpha IN_{\overline{TD}}}\right] = [1.8125, 1.8125]$

Hence division is also certified in Inverse Fuzzy Number (Fig. 4).

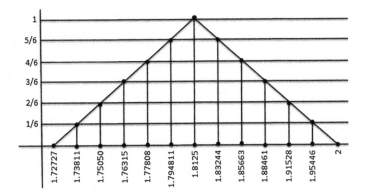

Fig.4 Division for inverse two arbitrary uncertain linguistic terms

4 Conclusion

Thus, triskaidecagonal, a new form of fuzzy number has been observed under linguistic scale values in which arithmetic operation to finding square root and inverse fuzzy number. Hence, an arithmetic operation is derived under inverse triskaidecagonal fuzzy number using elegant and new artistry method. Hence, this new approach makes fuzzy number to get the beauty around this real world.

References

1. Felix, A., Christopher, S., Victor Devadoss, A.: A nanogonal fuzzy number and its arithmetic operation. Int. J. Math. Appl. **3**(2), 185–195 (2015). ISSN: 2347-1557
2. Banerjee, S., Roy, T.K.: Arithmetic operations on generalized trapezoidal fuzzy number and its applications. Turk. J. Fuzzy Syst. **3**(1), 16–44 (2012)
3. Rajkumar, A., Helen, D.: Tree trigger success of door bell using fuzzy number. Int. J. Pure. Appl. Math. **114**(5), 71–77 (2017). ISSN: 1311–8080 (printed version); ISSN: 1314–3395 (online version)

New Arithmetic Operations of Triskaidecagonal Fuzzy Number Using Alpha Cut

A. Rajkumar and D. Helen

Abstract The main target of the paper is to introduce the new fuzzy number called Triskaidecagonal fuzzy number with linguistics values and also includes basic arithmetic operations by means of alpha cut. Thus new operation on Triskaidecagonal under uncertain lexical environment is being projected. It follows with the brief note on Triskaidecagonal. The next section proceeds with Triskaidecagonal fuzzy number and its linguistic values and arithmetic operation. Finally the paper ends with reference and conclusion.

Keywords Fuzzy number · Triskaidecagonal · Triangular fuzzy number
Pentagonal fuzzy number · Nanogonal fuzzy number · Dodecagonal fuzzy number

1 Introduction

As the complexity of restrained membership function to take pentagonal, octagonal [1], dodecagonal [2], etc., when incertitude arises in 13 different points, researchers try to clear the obscurity using modern form of fuzzy numbers. Triskaidecagonal, a new form of fuzzy number has been examined under uncertain linguistic environment which would be simpler to illustrate with Triskaidecagonal fuzzy linguistic scale values. Alpha cut method is the standard method that performs different arithmetic operations including addition, subtraction, multiplication, and division. This paper throws light on how α cut method is used to perform arithmetic operation using Triskaidecagonal linguistic scale values. Thus mathematical artistry in this proposed method cannot be denied.

A. Rajkumar (✉) · D. Helen
Department of Mathematics, Hindustan instutite of technolgy and science,
Hindustan University, Chennai 603103, India
e-mail: arajkumar@hindustanuniv.ac.in

D. Helen
e-mail: helen.br@outlook.com

© Springer Nature Singapore Pte Ltd. 2018
M. Pant et al. (eds.), *Soft Computing: Theories and Applications*,
Advances in Intelligent Systems and Computing 583,
https://doi.org/10.1007/978-981-10-5687-1_12

2 Preliminaries

Definition 2.1: The gradual assessment of the membership of elements is permitted by fuzzy set theory in a set which has been described with an aid of membership function valued in the real unit interval [0,1]. A fuzzy number is thus a special case of a convex, normalized fuzzy set of the real line.

Definition 2.2: Arithmetic fuzzy numbers are denoted by $\mathbf{AA_{\overline{DD}}}$ and $\mathbf{AB_{\overline{DD}}}$. Linguistic variable transformed into dodecagonal fuzzy numbers $[s_i, s_j]$ and $[s_k, s_l]$ is denoted by $\mathbf{LA_{\overline{DD}}}$ and $\mathbf{LB_{\overline{DD}}}$.

Definition 2.3: If $S = s_0, s_1, \ldots, s_g$ is a finite and totally ordered set with odd linguistic terms where s_i denotes the ith linguistic term, $i \in 0, 1, \ldots, g$ then we call set S the linguistic term set and $(g + 1)$ the cardinality of S. It is usually required that set S has the following properties:

(i)   The set is ordered: $s_i \geq s_j$ if $i > j$
(ii) There is a negation operator: neg $(s_i) = s_j$ such that $j = g - 1$
(iii) Maximization operator: max $(s_i, s_j) = s_i$ if $s_i \geq s_j$
      Minimization operator: min $(s_i, s_j) = s_i$ if $s_i \leq s_j$

Let $\overline{S} = s_l, s_{l+1}, \ldots, s_u$ where $s_l, s_{l+1}, \ldots, s_u \in S$, $s_l \geq s_u$, s_l and s_u are the lower and upper limits, $l, u \in 0, 1, \ldots, g$ respectively. Then we call $\overline{S}$ the uncertain linguistic term.

Triskaidecagonal Fuzzy Number (TDFN)

A fuzzy number $m_{\overline{TD}}$ is a dodecagonal fuzzy number denoted by $m_{\overline{TD}} = (a_1, a_2, a_3, a_4, a_5, a_6, a_7, a_8, a_9, a_{10}, a_{11}, a_{12}, a_{13})$ where are real numbers and its membership function (Figs. 1, 2 and Table 1).

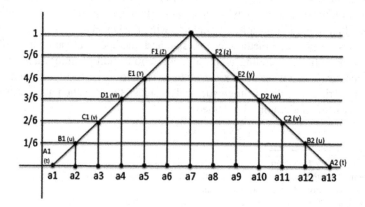

Fig. 1 Triskaidecagonal fuzzy number

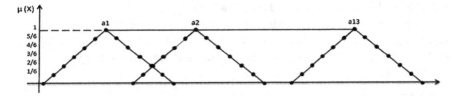

Fig. 2 The Triskaidecagonal fuzzy number from uncertain linguistic term

Table 1 Triskaidecagonal fuzzy linguistic scale

Linguistic terms	Linguistic values
No development	(0, 0, 0, 0, 0, 0, 0, 0.03, 0.06, 0.09, 0.12, 0.15, 0.18)
Very low development	(0.20, 0.22, 0.24, 0.26, 0.28, 0.30, 0.32, 0.34, 0.36, 0.38, 0.40, 0.42, 0.44)
Low development	(0.31, 0.33, 0.36, 0.39, 0.42, 0.45, 0.48, 0.51, 0.54. 0.57, 0.60, 0.63, 0.66)
Medium	(0.40, 0.43, 0.46, 0.49, 0.52, 0.55, 0.58, 0.61, 0.64, 0.67, 0.70, 0.73, 0.76)
High development	(0.60, 0.62, 0.64, 0.66, 0.68, 0.70, 0.72, 0.74, 0.76, 0.78, 0.80, 0.82, 0.84)
Very high development	(0.61, 0.64, 0.68, 0.72, 0.76, 0.80, 0.84, 0.88, 0.92, 0.96, 1, 1, 1)

$$
\mu_{\overline{TD}}(x) =
\begin{cases}
\frac{1}{6}\left(\frac{x-a_1}{a_2-a_1}\right) & a_1 \leq x \leq a_2 \\[4pt]
\frac{1}{6} + \frac{1}{6}\left(\frac{x_2-a_2}{a_3-a_2}\right) & a_2 \leq x \leq a_3 \\[4pt]
\frac{2}{6} + \frac{1}{6}\left(\frac{x-a_3}{a_4-a_3}\right) & a_3 \leq x \leq a_4 \\[4pt]
\frac{3}{6} + \frac{1}{6}\left(\frac{x-a_4}{a_5-a_4}\right) & a_4 \leq x \leq a_5 \\[4pt]
\frac{4}{6} + \frac{1}{6}\left(\frac{x-a_5}{a_6-a_5}\right) & a_5 \leq x \leq a_6 \\[4pt]
\frac{5}{6} + \frac{1}{6}\left(\frac{x-a_6}{a_7-a_6}\right) & a_6 \leq x \leq a_7 \\[4pt]
1 - \frac{1}{6}\left(\frac{x-a_7}{a_8-a_7}\right) & a_7 \leq x \leq a_8 \\[4pt]
\frac{5}{6} - \frac{1}{6}\left(\frac{x-a_8}{a_9-a_8}\right) & a_8 \leq x \leq a_9 \\[4pt]
\frac{4}{6} - \frac{1}{6}\left(\frac{x-a_9}{a_{10}-a_9}\right) & a_9 \leq x \leq a_{10} \\[4pt]
\frac{3}{6} - \frac{1}{6}\left(\frac{x-a_{10}}{a_{11}-a_{10}}\right) & a_{10} \leq x \leq a_{11} \\[4pt]
\frac{2}{6} - \frac{1}{6}\left(\frac{x-a_{11}}{a_{12}-a_{11}}\right) & a_{11} \leq x \leq a_{12} \\[4pt]
\frac{1}{6} - \frac{1}{6}\left(\frac{a_{13}-x}{a_{13}-a_{12}}\right) & a_{12} \leq x \leq a_{13} \\[4pt]
0 & x > a_{13}
\end{cases}
$$

Operation of Triskaidecagonal Fuzzy Numbers

Let $m_{\overline{TD}}$ and $n_{\overline{TD}}$ be two arbitrary uncertain linguistic terms and $m_{\overline{TD}} = (a_1, a_2, a_3, a_4, a_5, a_6, a_7, a_8, a_9, a_{10}, a_{11}, a_{12}, a_{13})$ and $n_{\overline{TD}} = (b_1, b_2, b_3, b_4, b_5, b_6, b_7, b_8, b_9, b_{10}, b_{11}, b_{12}, b_{13})$ be their corresponding Triskaidecagonal fuzzy numbers then

Addition:

$$m_{\overline{TD}}(+)n_{\overline{TD}} = (a_1 + b_1, a_2 + b_2, a_3 + b_3, a_4 + b_4, a_5 + b_5, a_6 + b_6, a_7$$
$$+ b_7, a_8 + b_8, a_9 + b_9, a_{10} + b_{10}, a_{11} + b_{11}, a_{12} + b_{12}, a_{13} + b_{13})$$

Subtraction:

$$m_{\overline{TD}}(-)n_{\overline{TD}} = (a_1 - b_{13}, a_2 - b_{12}, a_3 - b_{11}, a_4 - b_{10}, a_5 - b_9, a_6 - b_8, a_7 - b_7,$$
$$-b_7, a_8 - b_6, a_9 - b_5, a_{10} - b_4, a_{11} - b_3, a_{12} - b_2, a_{13} - b_1)$$

Multiplication:

$$m_{\overline{TD}}(*)n_{\overline{TD}} = (a_1 * b_1, a_2 * b_2, a_3 * b_3, a_4 * b_4, a_5 * b_5, a_6 * b_6, a_7 * b_7, a_8$$
$$*b_8, a_9 * b_9, a_{10} * b_{10}, a_{11} * b_{11}, a_{12} * b_{12}, a_{13} * b_{13})$$

Division:

$$m_{\overline{TD}}(/)n_{\overline{TD}} = \left(\frac{a_1}{b_1}, \frac{a_2}{b_2}, \frac{a_3}{b_3}, \frac{a_4}{b_4}, \frac{a_5}{b_5}, \frac{a_6}{b_6}, \frac{a_7}{b_7}, \frac{a_8}{b_8}, \frac{a_9}{b_9}, \frac{a_{10}}{b_{10}}, \frac{a_{11}}{b_{11}}, \frac{a_{12}}{b_{12}}, \frac{a_{13}}{b_{13}} \right)$$

Numerical Example:

Let $Im_{\overline{TD}}$ (0.20, 0.22, 0.24, 0.26, 0.28, 0.30, 0.32, 0.34, 0.36, 0.38, 0.40, 0.42, 0.44) and $In_{\overline{TD}} = $ (0.40, 0.43, 0.46, 0.49, 0.52, 0.55, 0.58, 0.61, 0.64, 0.67, 0.70, 0.73, 0.76) be two Linguistic variable transform into Triskaidecagonal fuzzy numbers then

Addition:

$Im_{\overline{TD}}(+)In_{\overline{TD}} = $ (0.6, 0.65, 0.7, 0.75, 0.8, 0.85, 0.9, 0.95, 1, 1.05, 1.1, 1.15, 1.2)

Subtraction:

$Im_{\overline{TD}}(-)In_{\overline{TD}} = $ (−0.56, −0.51, −0.46, −0.41, −0.36, −0.31, −0.26, −0.21, − 0.16, −0.11, −0.06, −0.01, 0.04)

Multiplication:

$Im_{\overline{TD}}(*)In_{\overline{TD}} = $ (0.08, 0.0946, 0.1104, 0.1274, 0.1456, 0.165, 0.1856, 0.2074, 0.2304, 0.2546, 0.28, 0.3066, 0.334)

Division:

$Im_{\overline{TD}}(/)In_{\overline{TD}} = $ (0.5, 0.511628, 0.521739, 0.530612, 0.538462, 0.545455, 0.551724, 0.557377, 0.5625, 0.567164, 0.571429, 0.575342, 0.578947)

Definition 2.4: A Triskaidecagonal fuzzy number $\overline{TD}$ can also be defined as $\overline{TD} = A_1(t), B_1(u), C_1(v), D_1(w), E_1(y), F_1(z), A_2(t), B_2(u), C_2(v), D_2(w), E_2(y), F_2(z), t \in [0, 0.16667], u \in [0.16667, 0.3333], v \in [0.3333, 0.5], w \in [0.5, 0.66667] y \in [0.66667, 0.83333]$ and $z \in [0.83333, 1]$ where

$A_1(t) = \frac{1}{6}\left(\frac{x-a_1}{a_2-a_1}\right)$	$A_2(t) = \frac{1}{6}\left(\frac{a_{13}-x}{a_{13}-a_{12}}\right)$
$B_1(u) = \frac{1}{6} + \frac{1}{6}\left(\frac{x-a_2}{a_3-a_2}\right)$	$B_2(u) = \frac{2}{6} - \frac{1}{6}\left(\frac{x-a_{11}}{a_{12}-a_{11}}\right)$
$C_1(v) = \frac{2}{6} + \frac{1}{6}\left(\frac{x-a_3}{a_4-a_3}\right)$	$C_2(v) = \frac{3}{6} - \frac{1}{6}\left(\frac{x-a_9}{a_{11}-a_{10}}\right)$
$D_1(w) = \frac{3}{6} + \frac{1}{6}\left(\frac{x-a_4}{a_5-a_4}\right)$	$D_2(w) = \frac{4}{6} - \frac{1}{6}\left(\frac{x-a_8}{a_9-a_8}\right)$
$E_1(y) = \frac{4}{6} + \frac{1}{6}\left(\frac{x-a_5}{a_6-a_5}\right)$	$E_2(y) = \frac{5}{6} - \frac{1}{6}\left(\frac{x-a_8}{a_9-a_8}\right)$
$F_1(Z) = \frac{5}{6} + \frac{1}{6}\left(\frac{x-a_6}{a_7-a_6}\right)$	$F_2(Z) = 1 - \frac{1}{6}\left(\frac{x-a_7}{a_8-a_7}\right)$

Definition 2.5: The α-cut of the fuzzy set of the universe of discourse X is defined as $\overline{DD} = \{x \in X / \mu_{\overline{A}}(x) \geq \alpha\}$ where $\alpha \in [0, 1]$, $A_1(t)$, $B_1(u)$, $C_1(v)$, $D_1(w)$, $E_1(y)$, $F_1(z)$ is bounded and continuous increasing function over [0, 0.16667], [0.166667, 0.3333], [0.3333, 0.5], [0.5, 0.6666], [0.66667, 0.83333] and [0.83333, 1] respectively. $A_2(t)$, $B_2(u)$, $C_2u(v)$, $D_2(w)$, $E_2(Y)$, $F_2(z)$ is bounded and continuous decreasing function over [0, 0.16667], [0.166667, 0.3333], [0.3333, 0.5], [0.5, 0.6666], [0.66667, 0.83333] and [0.83333, 1] respectively.

$$\overline{TD}_\alpha = \begin{cases} [A_1(\alpha), A_2(\alpha) & \text{for } \alpha \in [0, 0.166667) \\ [B_1(\alpha), B_2(\alpha) & \text{for } \alpha \in [0.166667, 0.3333) \\ [C_1(\alpha), C_2(\alpha) & \text{for } \alpha \in [0.3333, 0.5) \\ [D_1(\alpha), D_2(\alpha) & \text{for } \alpha \in [0.5, 0.66667) \\ [E_1(\alpha), E_2(\alpha) & \text{for } \alpha \in [0.66667, 0.8333) \\ [F_1(\alpha), F_2(\alpha) & \text{for } \alpha \in [0.8333, 1.0) \end{cases}$$

Definition 2.6: If $A_1(x) = \alpha$ and $2(x) = \alpha$, then α-cut operations interval $\overline{TD}_\alpha$ is attained as

$$[6\alpha(a_2 - a_1) + a_1, 6\alpha(a_{13} - a_{12}) + a_{13}] + [6\alpha(b_2 - b_1), -6\alpha(b_{13} - b_{12}) + b_2]$$

Similarly we can obtain α-cut operation interval $\overline{TD}_\alpha$ for $[B_1(\alpha), B_2(\alpha)]$, $[C_1(\alpha), C_2(\alpha)]$, $[D_1(\alpha), D_2(\alpha)]$ and $[E_1(\alpha), E_2(\alpha)], [F_1(\alpha), F_2(\alpha)]$, as follows:

$$[B_1(\alpha), B_2(\alpha)] = [6\alpha(a_3 - a_2) + 2a_2 - a_3, -6\alpha(a_{12} - a_{13}) - a_{11} + 2a_{12}] + 6\alpha(b_3 - b_2) + 2b_2 - b_3, -6\alpha(b_{12} - b_{13}) - b_{11} + 2b_{12}] \text{ for } \alpha\varepsilon [0.16667, 0.3333]$$

$$[C_1(\alpha), C_2(\alpha)] = [6\alpha(a_4 - a_3) - 2a_4 + 3a_3, -6\alpha(a_{11} - a_{10}) + 3a_{11} - 2a_{10}] + [6\alpha(b_4 - b_3) - 2b_4 + 3b_3, -6\alpha(b_{11} - b_{10}) + 3b_{11} - 2b_{10}] \text{ for } \alpha\varepsilon [0.3333, 0.55]$$

$$D_1(\alpha), D_2(\alpha)] = [6\alpha(a_5 - a_4) - 3a_5 + 4a_4, -6\alpha(a_{10} - a_9) + 4a_{10} - 3a_9] + [6\alpha(b_5 - b_4) - 3b_5 + 2b_4 - 6\alpha(b_{10} - b_9) + 4b_{10} - 3b_9] \text{ for } \alpha\varepsilon [0.5, 0.66666]$$

$$[E_1(\alpha), E_2(\alpha)] = [6\alpha(a_6 - a_5) - 4a_6 + 5a_5, -6\alpha(a_9 - a_8) + 5a_9 - 4a_8] + [6\alpha(b_6 - b_5)$$
$$-4b_6 + 5b_5, -6\alpha(b_9 - b_8) + 5b_9 - 4b_8] \text{ for } \alpha\varepsilon\ [0.66666, 0.83333]$$

$$[F_1(\alpha), F_2(\alpha)] = [6\alpha(a_7 - a_6) - 5a_7 + 6a_6, -6\alpha(a_8 - a_7) + 6a_8 - 5a_7] + [6\alpha(b_7 - b_6) - 5b_7 + 6b_6,$$
$$-6\alpha(b_8 - b_7) + 6b_8 - 5b_7] \text{ for } \alpha\varepsilon\ [0.3333, 1].\text{Therefore alpha cut Triskaidecagonal fuzzy number}$$

$$\overline{TD}_\alpha = \begin{cases} [6\alpha(a_2 - a_1) + a_1, 6\alpha(a_{13} - a_{12}) + a_{13}] + [6\alpha(b_2 - b_1), -6\alpha(b_{13} - b_{12}) + b_2] \\ \text{for } \alpha\varepsilon[0, 0.166667] \\ [6\alpha(a_3 - a_2) + 2a_2 - a_3, -6\alpha(a_{12} - a_{13}) - a_{11} + 2a_{12}] + 6\alpha(b_3 - b_2) + 2b_2 - b_3, \\ -6\alpha(b_{11} - b_{10}) - b_{11} + 2b_{12}] \text{ for } \alpha\varepsilon\ [0.16667, 0.3333] \\ [6\alpha(a_4 - a_3) - 2a_4 + 3a_3, -6\alpha(a_{11} - a_{10}) + 3a_{11} - 2a_{10}] + [6\alpha(b_4 - b_3) \\ -2b_4 + 3b_3, -6\alpha(b_{11} - b_{10}) + 3b_{11} - 2b_{10}] \text{ for } \alpha\varepsilon\ [0.3333, 0.55] \\ [6\alpha(a_5 - a_4) - 3a_5 + 4a_4, -6\alpha(a_{10} - a_9) + 4a_{10} - 3a_9] + [6\alpha(b_5 - b_4) - 3b_5 + 2b_4, \\ -6\alpha(b_{10} - b_9) + 4b_{10} - 3b_9] \text{ for } \alpha\varepsilon\ [0.5, 0.66666] \\ [6\alpha(a_6 - a_5) - 4a_6 + 5a_5, -6\alpha(a_9 - a_8) + 5a_9 - 4a_8] + [6\alpha(b_6 - b_5) - 4b_6 + 5b_5, \\ -6\alpha(b_9 - b_8) + 5b_9 - 4b_8] \text{ for } \alpha\varepsilon\ [0.66666, 0.83333] \\ [6\alpha(a_7 - a_6) - 5a_7 + 6a_6, -6\alpha(a_8 - a_7) + 6a_8 - 5a_7] + [6\alpha(b_7 - b_6) - 5b_7 + 6b_6, \\ -6\alpha(b_8 - b_7) + 6b_8 - 5b_7] \text{ for } \alpha\varepsilon\ [0.3333, 1] \end{cases}$$

A new operation for addition, subtraction, multiplication, division on Triskaidecagonal fuzzy number

Addition: Let $\mathbf{IM_{\overline{TD}}} = (b_1, b_2, b_3, b_4, b_5, b_6, b_7, b_8, b_9, b_{10}, b_{11}, b_{12}, b_{13})$ and $\mathbf{IN_{\overline{TD}}} = (d_1, d_2, d_3, d_4, d_5, d_6, d_7, d_8, d_9, d_{10}, d_{11}, d_{12}, d_{13})$ be their corresponding Triskaidecagonal fuzzy numbers for all $\alpha\varepsilon$ [0,1], The addition operation of α-cuts denoted as $\mathbf{IM_{\overline{TD}}} + \mathbf{IN_{\overline{TD}}}$

To certify a new operation with ordinary addition operation with the lexical scale values (Fig. 3).

$$\mathbf{IM_{\overline{TD}}} = (0.20, 0.22, 0.24, 0.26, 0.28, 0.30, 0.32, 0.34, 0.36, 0.38, 0.40, 0.42, 0.44)$$

$$\mathbf{IN_{\overline{TD}}} = (0.40, 0.43, 0.46, 0.49, 0.52, 0.55, 0.58, 0.61, 0.64, 0.67, 0.70, 0.73, 0.76)$$

for $\alpha\varepsilon$ [0, 0.166667)	IM $\alpha = (0.12\alpha + 0.20, -0.12\alpha + 0.44)$	IM α (+)IN $\alpha = (0.30\alpha + 0.60, -0.30\alpha + 1.20)$
for $\alpha\varepsilon$ [0.16667, 0.3333)	IN $\alpha = (0.18\alpha + 0.40, -0.18\alpha + 0.76)$	
for $\alpha\varepsilon$ [0.3333, 0.5)		
for $\alpha\varepsilon$ [0.5, 0.66666)		
for $\alpha\varepsilon$ [0.6666, 0.8333)		
for $\alpha\varepsilon$ [0.83333,1)		

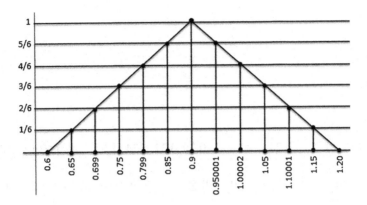

Fig. 3 Addition of two arbitrary uncertain linguistic terms.

As for $\alpha\varepsilon$ [0, 0.1666677), $\alpha\varepsilon$ [0.16667, 0.3333), $\alpha\varepsilon$ [0.3333, 0.5), $\alpha\varepsilon$ [0.5, 0.66666), $\alpha\varepsilon$ [0.66666, 0.8333) and $\alpha\varepsilon$ [0.83333, 1) arithmetic intervals are same, IM $\propto$ (+)IN $\propto$ = (0.30α + 0.60, −0.30α + 1.20) for all $\alpha\varepsilon$ [0,1].

When $\alpha = 0 \Rightarrow$ IMα(+)INα = [0.60, 1.20]; $\alpha = 0.16667 \Rightarrow$ IMα(+)INα = [0.65, 1.15];
$\alpha = 0.3333 \Rightarrow$ IMα(+)INα = [0.699, 1.10001]; $\alpha = 0.5 \Rightarrow$ IMα(+)INα = [0.75, 1.05]
$\alpha = 0.6666 \Rightarrow$ IMα(+)INα = [0.799, 1.00002]; $\alpha = 0.8333 \Rightarrow$ IMα(+) INα = [0.85, 0.950001]
$\alpha = 1 \Rightarrow$ IMα(+)INα = [0.9, 0.9]

Since the addition of two alpha cuts lies within interval. Hence it is verified.

Subtraction
Let IM$_{\overline{TD}}$ = $(b_1, b_2, b_3, b_4, b_5, b_6, b_7, b_8, b_9, b_{10}, b_{11}, b_{12}, b_{13})$ and IN$_{\overline{TD}}$ = $(d_1, d_2, d_3, d_4, d_5, d_6, d_7, d_8, d_9, d_{10}, d_{11}, d_{12}, d_{13})$ be their corresponding Triskaidecagonal fuzzy numbers for all $\alpha\varepsilon$ [0,1], The subtraction operation of α-cuts IM$_{\overline{TD}}$ − IN$_{\overline{TD}}$

To justify a new operation with ordinary subtraction operation with the lexical scale values

for $\alpha\varepsilon$ [0, 0.166667)	IM $\propto$= (0.12α + 0.20, −0.12α + 0.44)	IM $\propto$ (+)IN $\propto$= (0.30α + 0.60, −0.30α + 1.20)
for $\alpha\varepsilon$ [0.16667, 0.3333)	IN $\propto$= (0.18α + 0.40, −0.18α + 0.76)	
for $\alpha\varepsilon$ [0.3333, 0.5)		
for $\alpha\varepsilon$ [0.5, 0.66666)		

(continued)

(continued)

for αε [0.6666, 0.8333)	
for αε [0.83333,1)	

$$IM_{\overline{TD}} = (0.20, 0.22, 0.24, 0.26, 0.28, 0.30, 0.32, 0.34, 0.36, 0.38, 0.40, 0.42, 0.44)$$

$$IN_{\overline{TD}} = (0.40, 0.43, 0.46, 0.49, 0.52, 0.55, 0.58, 0.61, 0.64, 0.67, 0.70, 0.73, 0.76)$$

As for αε [0, 0.1666677), αε [0.16667, 0.3333), αε [0.3333, 0.5), αε [0.5, 0.66666), αε [0.66666, 0.8333) and αε [0.83333, 1) arithmetic intervals are same $IM\alpha(-)IN\alpha = (0.30\alpha - 0.56, -0.30\alpha + 0.04)$ for all αε [0,1].

When $\alpha = 0$ => $IM\alpha(-)IN\alpha = [-0.56, 0.04]$; $\alpha = 1/6$ => $IM\alpha(-)IN\alpha = [-0.51, -0.09]$
$\alpha = 2/6$ => $IM\alpha(-)IN\alpha = [-0.46, -0.6]$; $\alpha = 3/6$ => $IM\alpha(-)IN\alpha = [-0.41, -0.11]$
$\alpha = 4/6$ => $IM\alpha(-)IN\alpha = [-0.36, -0.16]$; $\alpha = 5/6$ => $IM\alpha(-)IN\alpha = [-0.31, 0.21]$
$\alpha = 1$ => $IM\alpha(-)IN\alpha = [-0.26, -0.26]$

Therefore the subtraction of two alpha cuts lies within interval. Hence it is verified (Fig. 4).

Multiplication
Let $IM_{\overline{TD}} = (b_1, b_2, b_3, b_4, b_5, b_6, b_7, b_8, b_9, b_{10}, b_{11}, b_{12}, b_{13})$ and $IN_{\overline{TD}} = (d_1, d_2, d_3, d_4, d_5, d_6, d_7, d_8, d_9, d_{10}, d_{11}, d_{12}, d_{13})$ be their corresponding Triskaidecagonal fuzzy numbers for all αε [0,1], The multiplication operation of α-cuts $IM_{\overline{TD}} * IN_{\overline{TD}}$.

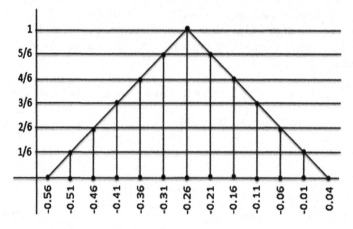

Fig. 4 Subtraction of two arbitrary uncertain linguistic terms

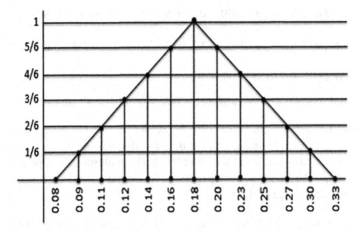

Fig. 5 Multiplication of two arbitrary uncertain linguistic terms.

To verify a new operation with ordinary multiplication operation with the linguistic scale values (Fig. 5).

$$\text{IM}_{\overline{TD}} = (0.20, 0.22, 0.24, 0.26, 0.28, 0.30, 0.32, 0.34, 0.36, 0.38, 0.40, 0.42, 0.44)$$

$$\text{IN}_{\overline{TD}} = (0.40, 0.43, 0.46, 0.49, 0.52, 0.55, 0.58, 0.61, 0.64, 0.67, 0.70, 0.73, 0.76)$$

for $\alpha \varepsilon$ [0, 0.166667)	IM $\alpha = (0.12\alpha + 0.20, -0.12\alpha + 0.44)$	IM α (+)IN $\alpha = (0.30\alpha + 0.60, -0.30\alpha + 1.20)$
for $\alpha \varepsilon$ [0.16667, 0.3333)	IN $\alpha = (0.18\alpha + 0.40, -0.18\alpha + 0.76)$	
for $\alpha \varepsilon$ [0.3333, 0.5)		
for $\alpha \varepsilon$ [0.5, 0.66666)		
for $\alpha \varepsilon$ [0.6666, 0.8333)		
for $\alpha \varepsilon$ [0.83333,1)		

As for $\alpha \varepsilon$ [0, 0.1666677), $\alpha \varepsilon$ [0.16667, 0.3333), $\alpha \varepsilon$ [0.3333, 0.5), $\alpha \varepsilon$ [0.5, 0.66666), $\alpha \varepsilon$ [0.66666, 0.8333) and $\alpha \varepsilon$ [0.83333, 1) arithmetic intervals are same $\text{IM}\alpha(*)\text{IN}\alpha = (0.216\alpha^2 - 0.084\alpha, + 0.08 + 0.0216\alpha - 0.1704\alpha + 0.3344)$ for all $\alpha \varepsilon$ [0,1].

When $\alpha = 0 \Rightarrow \text{IM}\alpha(*)\text{IN}\alpha = [0.5, 0.5789]$; $\alpha = 1/6 \Rightarrow \text{IM}\alpha(*)\text{IN}\alpha = [0.0945999, 0.30659]$

$\alpha = 2/6 \Rightarrow IM\alpha(*)IN\alpha = [0.11039, \quad 0.27999]; \quad \alpha = 3/6 \Rightarrow IM\alpha(*)N\alpha = [0.1274, 0.2546]$

$\alpha = 4/6 \Rightarrow IM\alpha(*)IN\alpha = [0.1456, \quad 0.2304]; \quad \alpha = 5/6 \Rightarrow IM\alpha(*)IN\alpha = [0.1649, 0.2073]$

$\alpha = 1 \Rightarrow IM\alpha(*)IN\alpha = [0.1856, 0.1856]$

Hence multiplication of two alpha cuts lies within interval. Hence it is justified.

Division: Let $IM_{\overline{TD}} = (b_1, b_2, b_3, b_4, b_5, b_6, b_7, b_8, b_9, b_{10}, b_{11}, b_{12}, b_{13})$ and $IN_{\overline{TD}} = (d_1, d_2, d_3, d_4, d_5, d_6, d_7, d_8, d_9, d_{10}, d_{11}, d_{12}, d_{13})$ be their corresponding Triskaidecagonal fuzzy numbers for all $\alpha\varepsilon$ [0,1]. The multiplication operation of α-cuts $IM_{\overline{TD}}/IN_{\overline{TD}}$. To verify a new operation with ordinary division operation with the linguistic scale values (Fig. 6).

$$IM_{\overline{TD}} = (0.20, 0.22, 0.24, 0.26, 0.28, 0.30, 0.32, 0.34, 0.36, 0.38, 0.40, 0.42, 0.44)$$

$$IN_{\overline{TD}} = (0.40, 0.43, 0.46, 0.49, 0.52, 0.55, 0.58, 0.61, 0.64, 0.67, 0.70, 0.73, 0.76)$$

for $\alpha\varepsilon$ [0, 0.166667)	IM $\propto= (0.12\alpha + 0.20, -0.12\alpha + 0.44)$	IM $\propto$ (/)IN $\propto= \left(\frac{0.12\alpha + 0.20}{0.18\alpha + 0.40}, \frac{-0.12\alpha + 0.44}{-0.18\alpha\alpha + 0.76}\right)$
for $\alpha\varepsilon$ [0.16667, 0.3333)	IN $\propto= (0.18\alpha + 0.40, -0.18\alpha + 0.76)$	
for $\alpha\varepsilon$ [0.3333, 0.5)		
for $\alpha\varepsilon$ [0.5, 0.66666)		
for $\alpha\varepsilon$ [0.6666, 0.8333)		
for $\alpha\varepsilon$ [0.83333,1)		

As for $\alpha\varepsilon$ [0, 0.1666677), $\alpha\varepsilon$ [0.16667, 0.3333), $\alpha\varepsilon$ [0.3333, 0.5), $\alpha\varepsilon$ [0.5, 0.66666), $\alpha\varepsilon$ [0.66666, 0.8333) and $\alpha\varepsilon$ [0.83333, 1) arithmetic intervals are same $IM\propto(*)$ $IN\propto = \left(\frac{0.12\alpha + 0.20}{0.18\alpha + 0.40}, \frac{-0.12\alpha + 0.44}{-0.18\alpha\alpha + 0.76}\right)$ for all $\alpha\varepsilon$ [0,1].

When $\alpha = 0 \Rightarrow IM\alpha(/)IN\alpha = [0.5, \quad 0.5789]; \quad \alpha = 1/6 \Rightarrow IM\alpha(/)IN\alpha = [0.5116, 0.57528]$

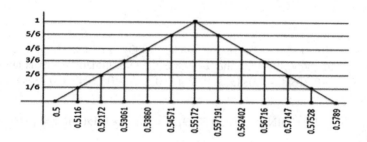

Fig. 6 Division of two arbitrary uncertain linguistic terms.

$\alpha = 2/6 \Rightarrow$ IMα(/)INα = [0.52172, 0.57147]; $\alpha = 3/6 \Rightarrow$ IMα(/)INα = [0.53061, 0.56716]

$\alpha = 4/6 \Rightarrow$ IMα(/)INα = [0.53860, 0.562402]; $\alpha = 5/6 \Rightarrow$ IMα(/)INα = [0.54571, 0.557161]

$\alpha = 1 \Rightarrow$ IMα(/)INα = [0.55172, 0.55172]

Certified division of two alpha cuts lies within interval.

3   Conclusion

Thus Triskaidecagonal, a new form of fuzzy number had been examined under linguistic scale values in which arithmetic operation can be performed by standard alpha cut method with numerical examples. This will be applied in biomedical, numerical analysis, operation research, etc.

References

1. Selvam, P., Rajkumar, A., Sudha Easwari, J.: A new method to find octagonal fuzzy numbers [Ofn]. In: International Conference on Recent Advances in Technology, Engineering and Science' (ICRATES'16), July 2016 (ISBN: 978-93-5258-740-7)
2. Selvam, P., Rajkumar, A., Sudha Easwari, J.: Dodecagonal fuzzy number [DDFN]. In: International Conference on Recent Advances in Technology, Engineering and Science' (ICRATES'16) July 2016 (ISBN: 978-93-5258-740-7)
3. Dutta, P., Boruah, H., Ali, T: Fuzzy Arithmetic with and without using α-cut method. Int. J. Latest Trends Comput. 2(1), 99 (2011) (E-ISSN: 2045-5364)
4. Felix, A., Devadoss, A.V.: A new decagonal fuzzy number under uncertain linguistic environment. Int. J. Math. Appl. 3(1), 89–97 (2015)
5. Banerjee, S.: Arithmetic operations on generalized trapezoidal fuzzy number and its applications. J. Turk. Fuzzy Syst. Assoc. 3(1), 16–44 (2012)
6. Felix, A., Christopher, S., Victor Devadoss, A.: A nanogonal fuzzy number and its arithmetic operation. Int. J. Math. Appl. 3(2), 185–195 (2015) (ISSN: 2347-1557) (Loyola College, Chennai)
7. Zimmermann, H.J: Fuzzy Set Theory and Its Application, 4th edn. Springer (2011)
8. Zadeh, L.A.: Fuzzy sets. Inf. Control 8(3), 338–353 (1965)
9. Zadeh, L.A.: The concept of a Linguistic variable and its application to approximate reasoning (Part II). Inf. Sci. 8, 301–357 (1975)
10. Rajarajeshwari, P., Sudha, A.S., Karthika, R.: A new operation on hexagonal fuzzy number. Int. J. Fuzzy Log. Syst. 3(3), 15–26 (2013)
11. Rezvani, S.: Multiplication operation on trapezoidal fuzzy numbers. J. Phys. Sci. 15, 17–26 (2011)
12. Malini, S.U., Kennedy, F.C.: An approach for solving fuzzy transportation using octagonal fuzzy numbers. Appl. Math. Sci. 54, 2661–2673 (2013)
13. Dubois, D., Prade, H.: Operations on fuzzy numbers. Int. J. Syst. Sci. 9(6), 613–626 (1978)
14. Rajkumar, A., Helen D.: Tree trigger success of door bell using fuzzy number. Int. J. Pure. Appl. Math. 114(5), 71 – 77 (2017). ISSN: 1311-8080 (printed version); ISSN: 1314-3395 (online version)

Loss and Cost Minimization with Enhancement of Voltage Profile of Distribution Systems Using a Modified Differential Evolution Technique

S. Mandal, K.K. Mandal and B. Tudu

Abstract Distribution system is one of the important parts of power transfer systems. Large amount of power losses take place in the distribution system. Many modern heuristic techniques have been successfully applied over the years to solve the problem. Satisfactory performance of evolutionary algorithms like differential evolution is highly dependent on the setting of control parameters. One of the major difficulties for them is the selection of parameters which usually depend on specific problem under consideration. Sometimes, wrongly selected parameters may lead to premature convergence or even stagnation. It has been found that self-adaptation is a viable alternative option in setting control parameters. In this paper, a new hybrid optimization algorithm based on differential evolution and chaos theory called logistic map differential evolution (LMDE) is presented for minimization of loss and cost of distribution networks. Enhancement of voltage profile is also considered. The motivation behind the present work is to get rid of early convergence. To validate and demonstrate the effectiveness of the proposed method, a sample test system is considered. The simulation results clearly show that it can successfully avoid premature convergence. A comparison result is also presented which shows the capability of the proposed technique.

1 Introduction

Power generated in the power generating stations is transmitted through transmission lines and distribution systems to the end users. Because of low voltage level, large amount power loss takes place in the distribution systems. This not only lead to financial loss to the distribution companies (DISCOs), but may also force the voltage profile to go below the specified limits. This is very important for a

S. Mandal
Department of Electrical Engineering, Jadavpur University, Kolkata 700032, India

K.K. Mandal (✉) · B. Tudu
Department of Power Engineering, Jadavpur University, Kolkata 700098, India
e-mail: kkm567@yahoo.co.in

© Springer Nature Singapore Pte Ltd. 2018
M. Pant et al. (eds.), *Soft Computing: Theories and Applications*,
Advances in Intelligent Systems and Computing 583,
https://doi.org/10.1007/978-981-10-5687-1_13

deregulated electricity market where competition is introduced and every attempt is made by each distribution company to maximize the profit as well as improvement of service quality. Several methods including installation of static VAR compensating devices have been adopted to compensate the reactive power and reduction of losses at the distribution level. Installation of shunt capacitors is one the simple and cost-effective methods for loss minimization of distribution networks. Enhancement of voltage profile is another purpose of capacitor placements and thus network reliability is ensured.

Loss reduction and profit maximization in distribution network is a nonlinear optimization problem. Researchers and scientists have applied several methods over the years to address the problem. These include mixed integer nonlinear programming [1], bacteria foraging [2], hybrid honey bee colony algorithm [3], genetic algorithm [4], particle swarm optimization technique [5], etc. A new technique using gravitational search algorithm was proposed by Shuaib et al. [6]. Opposition-Based Differential Evolution Algorithm was suggested by Muthukumar et al. [7] for Capacitor Placement on Radial Distribution System. Rao et al. [8] applied successfully a method using plant growth simulation algorithm for optimal capacitor placement in radial distribution system and presented promising results. A comparative study was presented by Lee et al. [9] using several variants of particle swarm optimization techniques for capacitor placement in distribution systems. Recently, Chiou et al. [10] developed and presented a novel algorithm for optimal capacitor placement in distribution systems. More recently, an improved harmony search algorithm was proposed Ali et al. [11] for optimal location and sizing of capacitors for radial distribution systems.

Differential Evolution (DE) is a powerful stochastic optimization technique and was introduced by Storn and Price in 1995 [12] as a population-based optimizer for nonlinear engineering problems. The present paper introduces a new algorithm combining chaos theory with differential evolution for loss minimization and profit maximization of distribution networks in order to avoid premature convergence. The simulation results are presented and compared with other modern techniques.

2 Problem Formulation

Minimization of loss and cost of distribution network by capacitor placement is formulated in the present section. Minimization of system loss and thereby annual cost is the primary objective. Thus, the objective function can be represented as

$$\text{Minimize} f = \text{Minimize}\left(\text{Cost}_{\text{System}}\right) \tag{1}$$

where $\text{Cost}_{\text{System}}$ consists of cost of power loss and capacitors required. To ensure system security and service quality, the voltage at every bus has to be maintained within acceptable limits and can be is represented as

$$V_{\min}^{\text{bus}} \leq V_i^{\text{bus}} \leq V_{\max}^{\text{bus}} \tag{2}$$

where V_i^{bus} indicates the voltage at bus i, $V_{\min}^{\text{bus}}$ and $V_{\max}^{\text{bus}}$ represents are the minimum and maximum limits of voltage respectively.

Figure 1 shows one-line diagram of a feeder. Power flow can be calculated as follows:

$$P_{i+1} = P_i - P_{Li+1} - R_{i,i+1}\left[\left(P_i^2 + Q_i^2\right)\big/|V_i|^2\right] \tag{3}$$

$$Q_{i+1} = Q_i - Q_{Li+1} - X_{i,i+1}\cdot\left[\left(P_i^2 + Q_i^2\right)\big/|V_i|^2\right] \tag{4}$$

$$|V_{i+1}|^2 = |V_i|^2 - 2\left(R_{i,i+1}\cdot P_i + X_{i,i+1}\cdot Q_i\right) + \left(R_{i,i+1}^2 + X_{i,i+1}^2\right)\frac{\left(P_i^2 + Q_i^2\right)}{|V_i|^2}, \tag{5}$$

where P_i and Q_i indicate the real and reactive powers flowing out of ith bus respectively. The real and reactive load at the ith bus is represented by P_{Li}, Q_{Li} respectively. $R_{i,i+1}$ and $X_{i,i+1}$ represents the resistance and reactance of the line section between buses i and $i + 1$ respectively.

The power loss of the line section connecting buses i and $i + 1$ can be determined as

$$P_{\text{Loss}}(i, i+1) = R_{i,i+1}\cdot\frac{P_i^2 + Q_i^2}{|V_i|^2} \tag{6}$$

Net power loss of the entire network is represented as $P_{N,\text{Loss}}$ and can be determined by summing the losses of all lines. It can be expressed as

$$P_{N,\text{Loss}} = \sum_{i=0}^{n} P_{\text{Loss}}(i, i+1) \tag{7}$$

In practice, a finite number of standard capacitor sizes are available in the market and they are normally integer multiples of the smallest size Q_0^c. In general, capacitors of larger size have lower unit prices. The available capacitor size is usually limited to

Fig. 1 Single-line diagram of a distribution network

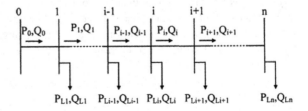

$$Q_c^{\max} = IQ_0^c, \tag{8}$$

where I is an integer. Therefore, for each installation location, there are I capacitor sizes $\{Q_0^c, 2Q_0^c, \ldots, IQ_0^c\}$ available. Let $\{K_1^c, K_2^c, \ldots, K_I^c\}$ be their corresponding equivalent annual cost per kVAr. Thus, the objective function $\text{Cost}_{\text{System}}$ can be expressed as

$$\text{Cost}_{\text{System}} = K_{P,\text{Cost}}\, P_{N,\text{Loss}} + \sum_{i=1}^{n} K_i^c Q_i^c \tag{9}$$

where $K_{P,\text{Cost}}$ is the equivalent annual cost per unit of power loss in \$/(kW − year). The capacitor placed at any bus is limited to

$$Q_i^c \leq \sum_{i=1}^{n} Q_{li} \tag{10}$$

Net saving, i.e., profit can be determined by comparing the annual cost of the compensated systems and the uncompensated system.

3   Differential Evolution (DE)

DE or Differential Evolution is a simple, easy to implement, and yet powerful optimization technique in the field of evolutionary computation. It is capable of solving various complex optimization problems including those with discontinuous, non-convex, and nonlinear solution space. It is comprised of four steps; initialization, mutation, crossover, and selection.

3.1   Initialization

Like all evolutionary computation, DE starts with a population of solutions. Let the population be denoted by P consisting of N_P vectors called individuals that evolve over G generation. Each individual consisting of D decision or problem variables represents a potential candidate solution. It is to be noted that the population size (N_P) remains fixed throughout the entire optimization process. These individuals are initialized randomly in order to search the entire solution space uniformly. The first step in DE is to generate an initial population consisting of all candidate solutions by assigning random values of each decision variable. Thus

$$X_{j,i}^{(0)} = X_j^{min} + rand_j \left(X_j^{max} - X_j^{min} \right)$$
$$i = 1, 2, \ldots, N_p, \; j = 1, 2, \ldots, D$$
(11)

$$P^{(G)} = \left[X_i^{(G)}, \ldots, X_{N_P}^{(G)} \right]$$
(12)

$$X_i^{(G)} = \left[X_{1,i}^{(G)}, \ldots, X_{D,i}^{(G)} \right]^T,$$
$$i = 1, \ldots, N_P$$
(13)

where, $X_{j,i}^{(0)}$ is the initial value of jth decision variable at ith population. X_j^{max} and X_j^{min} are the upper bound and lower bound of the jth decision variable respectively. $rand_j$ is a random number in the range (0, 1) generated anew for each value of j.

3.2 Mutation Operation

After the generation of initial population, new offspring is created through the mutation operator. Mutation operation is responsible for creating mutant vectors (V_i) by using randomly selected parent vector or target vector (X_i) vectors as follows:

$$V_i^{(G)} = X_k^{(G)} + f_m \left(X_l^{(G)} - X_m^{(G)} \right),$$
(14)

where X_k, X_l, X_m are randomly chosen vectors $\in [1, \ldots, N_P]$ and $k \neq l \neq m \neq i$. The mutation factor f_m is a user-defined parameter and is introduced to control perturbation and to improve convergence. Its range is normally between [0, 2].

3.3 Crossover Operation

The mutant vectors ($V_i^{(G)}$) and parent vectors ($X_i^{(G)}$) participate in creating trial vectors ($U_i^{(G)}$). The crossover operation creates trial vectors which were then used in the selection process. A trial vector is generated by combining the mutant vectors and parent or target vectors based on probability distribution. The crossover operation can be described

$$U_{j,i}^{(G)} = \begin{cases} V_{j,i}^{(G)}, & \text{if } \eta_j \leq C_R \text{ or } j = q \\ X_{j,i}^{(G)}, & \text{otherwise} \end{cases}, \tag{15}$$

where and η_j is a uniformly distributed random number within $[0, 1]$ generated anew for each value of j. The crossover factor $C_R \in [0, 1]$ is used to control the diversity of the population while $q \in (1, 2, \ldots, D)$ indicates a randomly selected index. It is to ensure that at least one vector is taken from mutant vector in the formation of trial vectors. It is to be noted that a crossover factor one indicates that all the trial vectors are taken from mutant vectors while a crossover factor zero indicates that all the trial vectors are taken from parent vectors.

3.4 Selection Operation

Selection operator is responsible in selecting vectors that will be used in creating population for the next generation. Selection operator is used to compare the fitness of the parent (target) vectors with the corresponding mutant vectors. The vector with better fitness is selected for the next generation and can be described as

$$X_i^{(G+1)} = \begin{cases} U_i^{(G)}, & \text{if } f\left(U_i^{(G)}\right) \leq f\left(X_i^{(G)}\right) \\ X_i^{(G)}, & \text{otherwise} \end{cases} \tag{16}$$

where f is the objective (fitness) function.

4 Logistic Map Differential Evolution (LMDE)

Chaos is a dynamical system which is characterized with ergodicity, randomicity, and unpredictability. It is very sensitive initial conditions. For systems associated with chaos, a small change in the initial conditions can produce entirely different results. Optimization algorithms using chaos theory are stochastic search methodologies and are different from the existing evolutionary algorithms. Chaotic sequences display an unpredictable long-term behavior due to their sensitiveness to initial conditions [13]. This feature can be utilized to track the chaotic variable as it travels ergodically over the searching space. Crossover ratio and mutation factors are the two important user-defined parameters and successful operation of differential evolution is heavily dependent on these two parameters. This paper utilizes chaotic sequence for automatic adjustment of DE parameters. This helps to escape from local minima and improves global convergence.

In many chaotic search optimization algorithms, logistic chaotic function is utilized to generate a chaotic sequence. This can be implemented [14] as

$$y(k) = ay(k-1)[1 - y(k-1)], \qquad (17)$$

where k is the sample and a is the control parameter, $0 \leq a \leq 4$. The above expression is deterministic displaying chaotic dynamics when $a = 4$ and $y(0) \notin \{0, 0.25, 0.5, 0.75, 1\}$. In this case, $y(k)$ is distributed in the range of $(0,1)$ provided the initial $y(0) \in (0, 1)$.

In the present work, mutation factor and crossover ratio are dynamically controlled using Eq. (17) as follows:

$$f_m(G) = a \cdot f_m(G-1) \cdot [1 - f_m(G-1)] \qquad (18)$$

$$C_R(G) = a \cdot C_R(G-1) \cdot [1 - C_R(G-1)] \qquad (19)$$

where G is the current iteration number.

5 Results and Discussions

A 23 kV, nine-section feeder [7, 10] is chosen as a test system to demonstrate the effectiveness and suitability of the proposed algorithm. Figure 2 shows the 23 kV, nine-section feeder test system. The proposed algorithm is implemented using MATLAB code on 3.0 GHz, 4.0 GB RAM PC.

For a direct comparison, the annual cost per unit of power loss is considered to be same as in [7, 10] and is \$168/(kW-year) [7, 10]. Other data including feeder impedance, three-phase load, available capacitor with annual cost, possible capacitor sizes are taken from [7, 10] and not shown due to space limitation. The limits on bus voltages are as follows: The magnitudes of bus voltages are assumed to be limited as $V_{min}^{bus} = 0.90$ p.u. and $V_{max}^{bus} = 1.10$ p.u.

The population size (N_P), the initial mutation factor (f_m) and the crossover ratio (C_R) are chosen as 40, 0.65, and 0.85 respectively for the present study. The maximum iteration was set at 500. All the buses are considered to be available for compensation. The annual costs, system power loss both before and after compensation, capacitor addition at the desired location are shown in Table 1.

Fig. 2 Single-line diagram of the test system

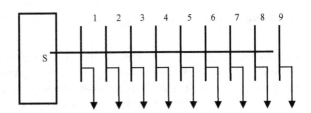

Table 1 Results including voltage profile, annual cost, and capacitor and power loss

Bus No.	Uncompensated voltage (p.u)	Placed (Qc) (kVar)	Compensated voltage (p.u)
0	1	0	1.0000
1	0.9929	0	0.9999
2	0.9874	4050	1.0047
3	0.9634	1200	0.9928
4	0.9619	1500	0.9824
5	0.9480	0	0.9606
6	0.9072	900	0.9546
7	0.8890	450	0.9405
8	0.8587	300	0.9168
9	0.8375	450	0.9001
Total cap. size (Mvar)		8.8500	
Total loss (MW)	0.7837	0.6729	
Annual cost in ($/year)	131,675	114,779	
Net saving ($/year)	16,896		

It is seen from Table 1 that voltage profile for all the buses are improved and well within the system limits. It is also observed from Table 1 that for the uncompensated system, some of the bus voltage limits are violated. For example, voltages limits at bus no. 5, 6, 7, 8, and 9 are below the specified limits. The annual cost is found to be \$114,779 while the system power loss is 0.6729 MW in comparison with uncompensated cases where the annual cost is \$131,675 and power loss is 0.7837 MW. Thus a reduction of 14.13% loss is achieved. The computation time is found to be 100.12 s.

Figure 3 shows the convergence characteristics for optimal annual cost. It also compares the convergence characteristics of the proposed LMDE with classical DE where the annual cost is \$115,078. It seems the algorithm using classical DE converges to local minimum. It is clear from Fig. 3 that the proposed algorithm can successfully avoid premature convergence.

The result is also compared with other methods like fuzzy reasoning [10], plant growth simulation algorithm (PGSA) [8], opposition-based differential evolution (OBDE) [7], hybrid CODEQ [10] and is shown in Table 2. From Table 2, it is clearly seen that proposed method can produce better results.

6 Conclusion

Loss minimization and profit maximization is a challenging and complex problem of distribution network. Loss reduction as well as improvement of voltage profile is the basic objective. It is a complex optimization problem and many optimization techniques and algorithm have been applied to solve it. An algorithm based on

Fig. 3 Covergence
characteristics for annual cost

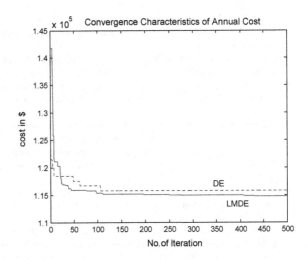

Table 2 Comparison of results with different methods

	OBDE [7]	PGSA [8]	Fuzzy resoning [10]	HCODEQ [10]	Proposed method
Total loss (MW)	0.6949	0.6949	0.7048	0.6756	0.6729
Annual cost in ($/year)	118,340	118,340	119,420	115,395	114,779

differential evolution and chaos theory has been developed and successfully applied on a test system to avoid the premature convergence. The results simulation results shows that it has the capability to produce good quality solutions. The results obtained by the proposed algorithm are also compared with other modern heuristic techniques like fuzzy reasoning, plant growth simulation algorithm, etc.

Acknowledgements The authors express their sincere thank to Jadavpur University, Kolkata, India for providing necessary support to carry out this work. This work is also supported by the program DRS, UPE II, UGC, Govt. of India.

References

1. Nojavan, S., Jalali, M., Zare, K.: Optimal allocation of capacitors in radial/mesh distribution systems using mixed integer nonlinear programming approach. Int. J. Electric Power Syst. Res. **107**, 119–124 (2014)
2. Mohamed Imran, A., Kowsalya, M.: Optimal size and siting of multiple distributed generators in distribution system using bacteria foraging optimization. Swarm Evol. Comput. 58–65 (2014)
3. Taher, S.A., Bagherpour, R.: A new approach for optimal capacitor placement and sizing in unbalanced distorted distribution systems using hybrid honey bee colony algorithm. Int. J. Electr. Power Energy Syst. **49**, 430–448 (2013)

4. Vuletic, J., Todorovski, M.: Optimal capacitor placement in distorted distribution networks with different load models using penalty free genetic algorithm. Electr. Power Energy Syst. **78**, 174–182 (2016)
5. Yu, X., Xiong, X., Wu, Y.: A PSO based approach to optimal capacitor placement with harmonic distortion consideration. Electr. Power System Res. **71**, 27–33 (2004)
6. Shuaib, Y.M., Kalavathi, M.S., Rajan, C.C.A.: Optimal capacitor placement in radial distribution system using gravitational search algorithm. Int. J. Electr. Power Energy Syst. **64** (1), 384–397 (2015)
7. Muthukumar, R., Thanushkodi, K.: Opposition based differential evolution algorithm for capacitor placement on radial distribution system. J. Electr. Eng. Technol. **9**(1), 45–51 (2014)
8. Rao, R.S., Narasimham, S.V.L., Ramalingaraju, M.: Optimal capacitor placement ina radial distribution system using plant growth simulation algorithm. Int. J. Electr. Power Energy Syst. **33**(5), 1133–1139 (2011)
9. Lee, C.S., Ayala, H.V.H., Coelho, L.S.: Capacitor placement of distribution systems using particle swarm optimization approaches. Int. J. Electr. Power Energy Syst. **64**, 839–851 (2015)
10. Chiou, J.P., Chan, C.F.: Development of a novel algorithm for optimal capacitor placement in distribution systems. Int. J. Electr. Power Energy Syst. **73**, 684–690 (2015)
11. Ali, E.S., Abd Elazim, S.M., Abdelaziz, A.Y.: Improved harmony algorithm for optimal locations and sizing of capacitors in radial distribution systems. Int. J. Electr. Power Energy Syst. **79**, 275–284 (2016)
12. Storn, R., Price, K.: Differential evolution: a simple and efficient adaptive scheme for global optimization over continuous spaces, Technical Report TR-95-012. International Computer Science Institute, Berkeley, USA (1995)
13. Hilborn, R.C.: Chaos and nonlinear dynamics: an introduction for scientists and engineers, 2nd edn. Oxford University Press, New York (2004)
14. Li, Y., Deng, S., Xiao, D.: A novel hash algorithm construction based on chaotic neural network. Neural Comput. Appl. **20**(1), 133–141 (2011)

Performance Analysis of Two Multi-antennas Relays Cooperation with Hybrid Relaying Scheme in the Absence of Direct Link for Cooperative Wireless Networks

Sunil Pattepu

Abstract The Multi-antenna Relays Cooperation-Hybrid(MARC-H), this study investigates the impact of MARC-H on two fixed multi-antenna relays network. To improve the network performance and range extension, incremental relaying technique is applied, where the source node communicates with destination through two fixed multi-antenna relays. MARC-H opts decode-and-forward (DF) or amplify-and-forward (AF) at two multi-antenna relays based on received SNR in first and second phase. In particular, this study analyzed the outage performance of MARC-H, when multi-antenna relay2 is performing Switch and Stay Combining (SSC) while destination is performing maximal ration combining (MRC) and selection combining (SC). For relative performance comparison, the proposed scheme MARC-H and conventional MARC-DF and MARC-AF is presented and it is conclude that the proposed scheme is outperforms the other in terms of outage.

1 Introduction

In cooperative wireless network, cooperative relaying is an efficient technique for achieving spatial diversity in distributed wireless networks. The basic idea of cooperation in wireless network is to utilize the coordination among nodes to transmit the source node message signal to destination node through cooperation to achieve the diversity gain, it can be used to improve the performance and range extension of the network. The most frequently used cooperative relaying schemes have been presented in Refs. [1–3], where the relay node helps to other (source) node to forward the message signal, it may simply amplify the received signal and forward (AF) or decode the received signal and then forward (DF) to other (destination) node. The DF-scheme performance will be degraded when the cooperative

S. Pattepu (✉)
Electronics Engineering, KIIT University Bhubaneswar, Bhubaneswar, Odisha, India
e-mail: pattepu.sunilfet@kiit.ac.in

© Springer Nature Singapore Pte Ltd. 2018
M. Pant et al. (eds.), *Soft Computing: Theories and Applications*,
Advances in Intelligent Systems and Computing 583,
https://doi.org/10.1007/978-981-10-5687-1_14

147

relay fails to decode the received signal from the source node correctly and the process of encoding cause error at destination node, to overcome such limitations some of the cooperative schemes have been proposed based on signal-to-noise ratio threshold DF-schemes and hybrid schemes [4–6]. Through cooperation among the relay nodes, a virtual antenna array is formulated so as to achieve the special diversity gain by combining transmissions from all relay at destination and using various combining techniques such as Maximal Ratio Combining (MRC) and Selection Combining (SC) studied in Ref. [7]. In early work, the research mainly focuses on the single-antenna relay network, where relay node carries only a single antenna, so multiple relay could construct a virtual antenna array that offers diversity gain similar to that offered by the traditional MIMO network [8–10]. Benefits of multi-antenna fixed and moving relay networks and performance analysis based on number of antennas installed and combiners like MRC and SC have been demonstrated in Refs. [7, 11]. In this paper, the source and destination nodes are equipped with a single antenna, while two multi-antenna relays (relay1 and relay2) installed with L and N antennas respectively. A simulation has been developed in Monte Carlo to assess the overall outage performance of three-phase hybrid relaying network. This work aim at deriving the closed-form expression of the end-to-end system outage probability and evaluating the impact of system and channel parameters including the number of antennas, the positions of relays on the system performance. Besides, the performance comparison in terms of outage between the proposed system and the conventional relay system. Rest of the paper is organized as follows: Section 2, system model of two multi-antenna relays cooperation-hybrid network and instantaneous SNR calculations, Sect. 3, mathematical model for calculating outage probability at the destination for various diversity combining techniques, in Sect. 4, results and discussion. Finally the conclusion of this paper in Sect. 5.

2 System Models

As shown in Fig. 1, system model consider a three-phase two multi-antenna relays system, consisting of one source node (s) one destination node (d) and two relay nodes (r_1 and r_2) equipped with L and N number of antennas (r_{1_L} and r_{1_N}) respectively. In this system, the communication between any two links is modeled as a Rayleigh fading channel with Additive White Gaussian Noise (AWGN), where the fading coefficient of the channels varies independently from one bit to another.

$h_{sr_{1_L}}$, $h_{sr_{2_N}}$, h_{r_1d}, $h_{r_1r_{2_N}}$ and h_{r_2d} are the fading coefficients of the channels from source to relay1 r_{1_L} ($r_{1_1}, r_{1_2}, \ldots, r_{1_L}$) source to relay2 r_{2_N} ($r_{2_1}, r_{2_2}, \ldots, r_{2_N}$) relay1 to relay2 and destination respectively. Similarly $n_{sr_{1_L}}$, $n_{sr_{2_N}}$, $n_{r_1r_{2_N}}$ and n_{r_2d} are corresponding additive white Gaussian noises. As the system model considers three-phase communication, therefore, one bit transmission consists of three-phases.

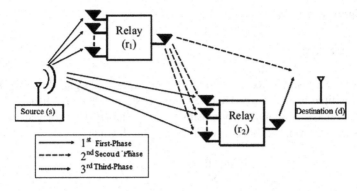

Fig. 1 Three-phase multi multi-antenna relays cooperative hybrid system

Due to half-duplex nature of two multi-antenna relays, the source broadcast the message signal to multi-antenna relay1 (r_{1L}), and multi-antenna relay2 (r_{2N}) in first phase ($s \rightarrow r_{1L}$, and $s \rightarrow r_{2N}$), the two multi-antenna relays r_{1L} and r_{2N} receives signal from source through multiple links $s \rightarrow r_{1L}$ and $s \rightarrow r_{2N}$ respectively, and multi-antenna r_1 and multi-antenna r_2 retransmit to the destination in second and third phase respectively, depending on relaying scheme. The r_2 will perform the Switch and Stay Combining (SSC) for signal received via r_1 and s links, and the Signal received at d via r_1 and r_2 links are combined using MRC and SC.

2.1 First Phase

In the first phase, the s broadcast the modulated message signal to two multi-antenna relays ($s \rightarrow r_{1L}$ and $s \rightarrow r_{2N}$). The received signal at two multi-antenna relay nodes (r_{1L} and r_{2N}) denoted by $y_{sr_{1L}}$ and $y_{sr_{2N}}$ respectively, where

$$y_{sr_{1L}} = h_{sr_{1L}} \sqrt{p}a + N_{sr_{1L}} \tag{1}$$

$$y_{sr_{2N}} = h_{sr_{2N}} \sqrt{p}a + N_{sr_{2N}} \tag{2}$$

Therefore, the corresponding instantaneous SNR at the two multi-antenna relay nodes can be calculated as

$$\gamma_{sr_{1L}} = \frac{\rho}{N_{sr_{1L}}} |h_{sr_{1L}}|^2 = \bar{\gamma} |h_{sr_{1L}}|^2 \tag{3}$$

$$\gamma_{sr_{2N}} = \frac{\rho}{N_{sr_{2N}}} |h_{sr_{2N}}|^2 = \bar{\gamma} |h_{sr_{2N}}|^2 \tag{4}$$

respectively.

2.2 Second Phase

2.2.1 MARC-DF

In the second phase MARC-DF, the r_1 receives message signal from the source through multiple links ($s \rightarrow r_{1L}$) and select the best link ($s \rightarrow r_{1k}$), then r_1 decodes the best link massage signal and retransmit to the r_{2N} and d. The received signal at d in second phase from the r_1 is denoted by $y_{r_1 d}^{\text{MARC-DF}}$ where

$$\gamma_{r_1 d} = \frac{\rho}{N_{r_1 d}} |h_{r_1 d}|^2 = \bar{\gamma} |h_{r_1 d}|^2 \tag{5}$$

The instantaneous SNR at the destination is calculated as

$$y_{r_1 d}^{\text{MARC-DF}} = h_{r_1 d} y_{sr_1} + N_{r_1 d} \tag{6}$$

The received signal at r_{2N} in second phase from r_1 is denoted by $y_{r_1 r_{2M}}^{\text{MARC-DF}}$ where

$$y_{r_1 r_2}^{\text{MARC-DF}} = h_{r_1 r_{2M}} y_{sr_1} + N_{r_1 r_{2M}} \tag{7}$$

instantaneous SNR at r_2 is calculated as

$$\gamma_{r_1 r_{2M}} = \frac{\rho}{N_{r_1 r_{2M}}} \left| h_{r_1 r_{2M}} \right|^2 = \bar{\gamma} \left| h_{r_1 r_{2M}} \right|^2 \tag{8}$$

2.2.2 MARC-AF

In the second phase MARC-AF, the r_1 receives message from the source through multiple links ($s \rightarrow r_{1L}$) and select the best link ($s \rightarrow r_{1k}$), and r_1 amplifies the received best link message signal and retransmit to r_{2N} and d. The received signal at d and r_{2N} in second phase from the r_1 is denoted by $y_{r_1 d}^{\text{MARC-AF}}$ and $y_{r_1 r_2}^{\text{MARC-AF}}$ where

$$y_{r_1 d}^{\text{MARC-AF}} = \eta_1 [h_{r_1 d} y_{sr_{1k}} + N_{r_1 d}], \tag{9}$$

where $\eta_1 = \sqrt{\dfrac{p_r}{p_s |h_{sr_{1k}}|^2 + \sigma^2}}$ represents the amplification factor

Substituting Eq. (1) in Eq. (9)

$$y_{r_1 d}^{\text{MARC-AF}} = h_{r_1 d} h_{sr_{1k}} \sqrt{\frac{p_r p_s}{p_s |h_{sr_{1k}}|^2 + \sigma^2}} a + h_{r_1 d} \sqrt{\frac{p_r}{p_s |h_{sr_{1k}}|^2 + \sigma^2}} N_{r_1 d} + N_{sr_{1k}} \tag{10}$$

The instantaneous SNR at kth received antenna of multi-antenna r_1 can be calculated as

$$\gamma_{r_1d}^{\text{MARC-AF}} = \frac{AB}{C + D + N_{r_1d}},$$

where

$$A = h_{r_1d}h_{sr_{1k}}\sqrt{\frac{P_rP_s}{P_s|h_{sr_{1k}}|^2 + \sigma^2}}, \quad B = h_{sr_{1k}}h_{r_1d}\sqrt{\frac{P_sP_r}{P_s|h_{sr_{1k}}|^2 + \sigma^2}},$$

$$C = h_{r_1d}\sqrt{\frac{P_r}{P_s|h_{sr_{1k}}|^2 + \sigma^2}}, \quad D = N_{sr_{1k}}h_{r_1d}\sqrt{\frac{P_r}{P_s|h_{sr_{1k}}|^2 + \sigma^2}}$$

$$\gamma_{r_1d}^{\text{MARC-AF}} = \frac{\gamma_{sr_{1k}}\gamma_{r_1d}}{\gamma_{sr_{1k}} + \gamma_{r_1d} + 1}, \tag{11}$$

where $\gamma_{sr_{1k}} = \frac{|h_{sr_{1k}}|^2 P_s}{\sigma_1^2}$ and $\gamma_{r_1d} = \frac{|h_{r_1d}|^2 P_r}{\sigma_2^2}$

Similarly, the received signal at multi-antenna r_{2N} in second phase from the r_1 is denoted by $y_{r_1r_2}^{\text{MARC-AF}}$ where

$$y_{r_1r_{2N}}^{\text{MARC-AF}} = \eta_1[h_{r_1r_{2N}}y_{sr_{1k}} + N_{r_1r_{2N}}] \tag{12}$$

Similarly, substituting Eq. (1) in Eq. (12)

$$y_{r_1r_2}^{\text{MARC-AF}} = h_{r_1r_{2N}}h_{sr_{1k}}\sqrt{\frac{P_rP_s}{P_s|h_{sr_{1k}}|^2 + \sigma^2}}ah_{r_1r_{2N}}\sqrt{\frac{P_r}{P_s|h_{sr_{1k}}|^2 + \sigma^2}}N_{r_1r_{2N}} + N_{sr_{1k}} \tag{13}$$

The instantaneous SNR at Mth received antenna of r_{2M} from r_1 in second phase can be calculated as

$$\gamma_{r_1r_2}^{\text{MARC-AF}} = \frac{\gamma_{sr_{1k}}\gamma_{r_1r_{2M}}}{\gamma_{sr_{1k}} + \gamma_{r_1r_{2M}} + 1}, \tag{14}$$

where $\gamma_{sr_{1k}} = \frac{|h_{sr_{1k}}|^2 P_s}{\sigma_1^2}$ and $\gamma_{r_1r_{2M}} = \frac{|h_{r_1r_{2M}}|^2 P_r}{\sigma_2^2}$

2.2.3 MARC-H

In the second phase MARC-H, the multi-antenna relay r_1 receives message from the source through multiple links ($s \to r_{1L}$) and select the best link ($s \to r_{1k}$), then r_1 opt adaptively either MARC-DF or MARC-AF based on the best link received

SNR. The instantaneous SNR at multi-antenna r_2 and d can be calculated as $(\gamma_{r_1 r_2}^{\text{MARC-H}})$ and $(\gamma_{r_1 d}^{\text{MARC-H}})$ respectively

$$\gamma_{r_1 r_2}^{\text{MARC-H}} = \begin{cases} \gamma_{r_1 r_2}^{\text{MARC-DF}}, & \text{for, } \gamma_{sr_{1k}} \geq \xi \\ \gamma_{r_1 r_2}^{\text{MARC-AF}}, & \text{else} \end{cases} \tag{15}$$

$$\gamma_{r_1 d}^{\text{MARC-H}} = \begin{cases} \gamma_{r_1 r_2}^{\text{MARC-DF}}, & \text{for, } \gamma_{sr_{1k}} \geq \xi \\ \gamma_{r_1 r_2}^{\text{MARC-AF}}, & \text{else} \end{cases}, \tag{16}$$

where is ξ SNR threshold

2.3 The Third Phase

2.3.1 MARC-AF

In the third phase MARC-AF, the received message signal at r_2 from the s in first phase through multiple links ($s \rightarrow r_{2N}$) and select the best link ($s \rightarrow r_{2G}$), and in the second phase the received message signal from r_1 through multiple links ($r_1 \rightarrow r_{2N}$), and select the best link ($r_1 \rightarrow r_{2H}$), r_2 performs Switch and Stay Combining (SSC) of signals received in first and second phase, then r_2 amplifies the combined message signal and retransmit to the d. The received signal at d in third phase from the r_2 is denoted by $y_{r_2 d}^{\text{MARC-AF}}$ where

$$y_{r_2 d}^{\text{MARC-AF}} = h_{r_2 d} y_{r_1 r_{2N}} + N_{r_2 d} \tag{17}$$

$$y_{r_2 d}^{\text{MARC-AF}} = h_{r_2 d}[\eta_2(h_{r_1 r_{2N}} y_{sr_{1k}} + N_{r_1 r_2})] y_{r_1 r_{2N}} + N_{r_2 d}, \tag{18}$$

where $\eta_2 = \sqrt{\dfrac{p_r}{p_s |h_{r_1 r_2}|^2 + \sigma^2}}$

$$y_{r_2 d}^{\text{MARC-AF}} = \left[h_{r_2 d} \left[h_{r_1 r_2} h_{sr_{1k}} \sqrt{\frac{p_r p_s}{p_s |h_{sr_{1k}}|^2 + \sigma^2}} a + h_{r_1 r_2} \sqrt{\frac{p_r}{p_s |h_{sr_{1k}}|^2 + \sigma^2}} N_{sr_{1k}} \sqrt{\frac{p_r}{p_s |h_{sr_{1k}}|^2 + \sigma^2}} N_{r_1 r_2} \right] + N_{r_2 d} \right]$$

$$y_{r_2 d}^{\text{MARC-AF}} = \left[h_{r_2 d} h_{r_1 r_2} h_{sr_{1k}} \sqrt{\frac{p_r p_s}{p_s |h_{sr_{1k}}|^2 + \sigma^2}} a h_{r_2 d} h_{r_1 r_2} \sqrt{\frac{p_r}{p_s |h_{sr_{1k}}|^2 + \sigma^2}} N_{sr_{1k}} h_{r_2 d} \sqrt{\frac{p_r}{p_s |h_{sr_{1k}}|^2 + \sigma^2}} N_{r_1 r_2} + N_{r_2 d} \right..$$

$$y_{r_2d}^{MARC-AF} = \eta_2 \left[h_{r_2d} h_{r_1r_2} h_{sr_{1k}} \sqrt{\frac{p_r p_s}{p_s |h_{sr_{1k}}|^2 + \sigma^2}} a h_{r_2d} h_{r_1r_{2N}} \frac{p_r}{\sqrt{\left(p_s |h_{sr_{1k}}|^2 + \sigma^2\right)\left(p_s |h_{r_1r_{2N}}|^2 + \sigma^2\right)}} \right. $$

$$+ h_{r_2d} \frac{p_r}{\sqrt{\left(p_s |h_{sr_{1k}}|^2 + \sigma^2\right)\left(p_s |h_{r_1r_{2N}}|^2 + \sigma^2\right)}} ,$$

$$y_{r_2d}^{MARC-DF} = h_{r_2d} h_{r_1r_{2N}} h_{sr_{1k}} \frac{p_r \sqrt{p_s}}{\sqrt{\left(p_s |h_{sr_{1k}}|^2 + \sigma^2\right)\left(p_s |h_{r_1r_{2N}}|^2 + \sigma^2\right)}}$$

$$+ h_{r_2d} h_{r_1r_{2N}} \frac{p_r}{\sqrt{\left(p_s |h_{sr_{1k}}|^2 + \sigma^2\right)\left(p_s |h_{r_1r_{2N}}|^2 + \sigma^2\right)}} h_{r_2d} \frac{p_r}{\sqrt{\left(p_s |h_{sr_{1k}}|^2 + \sigma^2\right)\left(p_s |h_{r_1r_{2N}}|^2 + \sigma^2\right)}}$$

The instantaneous SNR of end-to-end for third phase can be calculated as

$$\gamma_{r_2d}^{MARC-DF} = \frac{abc}{d+e+f} + N_{r_2d}$$

$$a = h_{r_2d} h_{r_1r_{2N}} h_{sr_{1k}} \frac{p_r \sqrt{p_s}}{\sqrt{\left(p_s |h_{sr_{1k}}|^2 + \sigma^2\right)\left(p_s |h_{r_1r_{2N}}|^2 + \sigma^2\right)}} ,$$

$$b = h_{r_2d} h_{r_1r_{2N}} \frac{p_r}{\sqrt{\left(p_s |h_{sr_{1k}}|^2 + \sigma^2\right)\left(p_s |h_{r_1r_{2N}}|^2 + \sigma^2\right)}} ,$$

$$c = h_{r_2d} \frac{p_r \sqrt{p_s}}{\left(p_s |h_{sr_{1k}}|^2 + \sigma^2\right)\left(p_s |h_{r_1r_{2N}}|^2 + \sigma^2\right)} ,$$

$$d = h_{r_2d} \frac{p_r \sqrt{p_s}}{\left(p_s |h_{sr_{1k}}|^2 + \sigma^2\right)\left(p_s |h_{r_1r_{2N}}|^2 + \sigma^2\right)} ,$$

$$e = N_{sr_{1k}} h_{r_2d} \frac{p_r}{\left(p_s |h_{sr_{1k}}|^2 + \sigma^2\right)\left(p_s |h_{r_1r_{2N}}|^2 + \sigma^2\right)} ,$$

$$f = N_{r_1r_{2N}} h_{r_2d} \frac{p_r}{\sqrt{\left(p_s |h_{sr_{1k}}|^2 + \sigma^2\right)\left(p_s |h_{r_1r_{2N}}|^2 + \sigma^2\right)}} \$$

$$\gamma_{r_2d}^{MARC-DF} = \frac{abc}{E[d+e+f]} + N_{r_2d}$$

$$\gamma_{r_2d}^{MARC-AF} = \frac{|h_{sr_{1k}}||h_{r_1r_{2M}}|^2|h_{r_2d}|^3 p_r^3 \sqrt{p_s}}{|h_{sr_{1k}}|^2 p_s \sigma_2^2 \sigma_3^2 + |h_{r_1r_{2M}}|^2 p_s \sigma_1^2 \sigma_3^2 + |h_{r_2d}|^2 p_r \sigma_1^2 \sigma_2^2 + \sigma_1^2 \sigma_2^2 \sigma_3^2}$$

$$\gamma_{r_2d}^{MARC-AF} = \frac{\gamma_{sr_{1k}} \gamma_{r_1r_{2M}} \gamma_{r_2d}}{\gamma_{sr_{1k}} + \gamma_{r_1r_{2M}} + \gamma_{r_2d} + 1}$$

where $\gamma_{sr_{1k}} = \frac{|h_{sr_{1k}}|p_s}{\sigma_1^2}$, $\gamma_{r_1r_{2M}} = \frac{|h_{r_1r_{2M}}|^2 p_s}{\sigma_2^2}$ and $\gamma_{r_2d} = \frac{|h_{r_2d}|^2 p_r}{\sigma_3^2}$. Similarly, the received signal at d through the link $s \rightarrow r_2 \rightarrow d$ $\left(y_{r_2d}^{MARC-AF}\right)$, similar to the link $s \rightarrow r_1 \rightarrow d$ in second phase, the received signal at d and instantaneous SNR can be calculated as

$$y_{r_2d}^{MARC-AF} = \eta_2[h_{r_2d}y_{sr_{2G}} + N_{r_2d}]$$

$$\gamma_{r_2d}^{MARC-AF} = \frac{|h_{sr_{2G}}|^2|h_{r_2d}|^2 p_s p_r}{|h_{sr_{2G}}|^2 p_s \sigma_1^2 + |h_{r_2d}|^2 p_r \sigma_2^2 + \sigma_1^2 \sigma_2^2}$$

$$\gamma_{r_2d}^{MARC-AF} = \frac{\gamma_{sr_{2G}} \gamma_{r_2d}}{\gamma_{sr_{2G}} + \gamma_{r_2d} + 1},$$

where $\gamma_{sr_{2G}} = \frac{|h_{sr_{2G}}|^2 p_s}{\sigma_1^2}$ and $\gamma_{r_2d} = \frac{|h_{r_2d}|^2 p_r}{\sigma_2^2}$

2.3.2 MARC-DF

In the third phase MARC-AF, the received message signal at r_2 from the in first phase through multiple links ($s \rightarrow r_{2N}$) and select the best link ($s \rightarrow r_{2G}$), and in the second phase the received message signal from through multiple links ($r_1 \rightarrow r_{2N}$), and select the best link ($r_1 \rightarrow r_{2H}$), r_2 performs Switch and Stay Combining (SSC) of signals received in first and second phase, then r_2 decodes the SSC message signal and retransmit to the d. The received signal at d in third phase from the is denoted by $y_{r_2d}^{MARC-DF}$ where

$$y_{r_2d}^{MARC-DF} = h_{r_2d}y_{sr_2} + N_{r_2d} \tag{19}$$

The instantaneous SNR at the destination is calculated as

$$\gamma_{r_2d} = \frac{\rho}{N_{r_2d}}|h_{r_2d}|^2 = \bar{\gamma}|h_{r_2d}|^2 \tag{20}$$

2.3.3 MARC-H

In the third-phase MARC-H, the multi-antenna r_2 received signal in phase one from the source through multiple links ($s \to r_{2N}$) and select the best link ($s \to r_{2G}$), and in the second phase the received message signal from through multiple links ($r_1 \to r_{2N}$), and select the best link ($r_1 \to r_{2H}$), r_2 performs Switch and Stay Combining (SSC) of signals received in first and second phase, then r_2 decodes the SSC message signal and then r_2 opt adaptively either MARC-AF or MARC-DF based on the received SNR retransmit to the d. The instantaneous SNR at d can be calculated as ($\gamma_{r_2d}^{\text{MARC-H}}$) where

$$\gamma_{r_2d}^{\text{MARC-H}} = \begin{cases} \gamma_{r_2d}^{\text{MARC-DF}}, & \text{for, } \gamma_{sr_2} \geq \xi \\ \gamma_{r_2d}^{\text{MARC-AF}}, & \text{else} \end{cases} \tag{21}$$

3 Outage Analysis

In the system model, the PDF of received SNR is assumed as exponentially distributed (as we are using relay fading channel), which can be written as [12]

$$f_{\gamma_{xy}}(\gamma) = \frac{\lambda_{xy}^n \gamma^{n-1}}{(n-1)!} \exp(-\lambda_{xy}\gamma),$$

where $\in \{s, r_1, r_2\}$ and $\in \{r_{1L}, r_{2N}, d\}$ and $\lambda_{xy} = \frac{(2l_{xy}/l_{sd})^\psi}{\omega}$, ψ is path-loss exponent, λ_{xy} is representing the link distance between x and y, which is normalized by reference distance $l_{sd}/2$, ω is SNR at reference point.

3.1 The First Phase

In the first phase source broadcast the message signal to both multi-antenna ($r_1s \to r_{1L}$) and multi-antenna relay2 ($s \to r_{2N}$). r_1 selection of best link ($s \to r_{1k}$) and it use MARC-H to retransmit in second phase. In this condition mutual information (I), transmitted by source is greater than the target data rate, the probability of r_1 transmission and the conditional PDF is given as respectively

$$P[I > R] = \int_\xi^\infty f_{\gamma_{sr_{1k}}}(\gamma)d\gamma \tag{22}$$

$$f_{\gamma_{sr_{1k}|I > R}}(\theta) = \lambda_{r_1d} \exp(-\lambda_{r_1d}) \tag{23}$$

$$f_{\gamma_{sr_{1k}|I>R}}(\theta) = \lambda_{r_1 r_{2N}} \exp(-\lambda_{r_1 r_{2N}}) \tag{24}$$

The PDF of received SNR at r_2 through the link $(s \rightarrow r_{2G})$ is calculated as

$$f_{sr_{2G}}(\theta) = f_{sr_{2G}|I>R}(\theta)P[I>R] \tag{25}$$

The CDF of received SNR at r_2 through the link $(s \rightarrow r_{2G})$ using PDF is calculated as

$$F_{\gamma_{sr_{2G}}}(\theta) = \int_0^\theta f_{\gamma_{sr_{2G}}}(\theta)\mathrm{d}\theta \tag{26}$$

3.2 The Second Phase

The PDF of received SNR at d through the link $(r_1 \rightarrow d)$ is calculated as

$$f_{r_1 d}(\theta) = f_{r_1 d|I>R}(\theta)P[I>R] \tag{27}$$

The PDF of received SNR at Relay2 r_2 is calculated as

$$f_{r_1 r_{2M}}(\theta) = f_{r_1 r_{2M}|I>R}(\theta)P[I>R] \tag{28}$$

The CDF of received SNR at r_2 through the link $(r_1 \rightarrow r_{2M})$ is using above PDF calculated as

$$F_{\gamma_{r_1 r_{2M}}}(\theta) = \int_0^\theta f_{\gamma_{r_1 r_{2M}}}(\theta)\mathrm{d}\theta \tag{29}$$

The total (Final) PDF of received SNR at r_2 through the links $s \rightarrow r_1 \rightarrow r_{2M}$ and $s \rightarrow r_{2G}$, when r_2 is performing SSC can be given as

$$f_\gamma^{SSC}(\gamma) = \begin{cases} \dfrac{F_{\gamma_{sr_{2G}}}(\xi_{SSC})F_{\gamma_{r_1 r_{2M}}}(\xi_{SSC})}{F_{\gamma_{sr_{2G}}}(\xi_{SSC})+F_{\gamma_{r_1 r_{2M}}}(\xi_{SSC})}\{f_{sr_{2G}}(\gamma)+f_{r_1 r_{2M}}(\gamma)\}\gamma<\xi_{SSC} \\[2ex] \dfrac{F_{\gamma_{sr_{2G}}}(\xi_{SSC})F_{\gamma_{r_1 r_{2M}}}(\xi_{SSC})}{F_{\gamma_{sr_{2G}}}(\xi_{SSC})+F_{\gamma_{r_1 r_{2M}}}(\xi_{SSC})}\{f_{sr_{2G}}(\gamma)+f_{r_1 r_{2M}}(\gamma)\} \\[2ex] +\dfrac{f_{\gamma_{sr_{2G}}}(\gamma)F_{\gamma_{r_1 r_{2M}}}(\xi_{SSC})}{F_{\gamma_{sr_{2G}}}(\xi_{SSC})+F_{\gamma_{r_1 r_{2M}}}(\xi_{SSC})}+\dfrac{F_{\gamma_{sr_{2G}}}(\xi_{SSC})f_{\gamma_{r_1 r_{2M}}}(\gamma)}{F_{\gamma_{sr_{2G}}}(\xi_{SSC})+F_{\gamma r_1 r_{2M}}(\xi_{SSC})}\gamma\geq\xi_{SSC} \end{cases}$$

$$\tag{30}$$

3.3 The Third Phase

The PDF of received SNR at d through the link $r_2 \to d$ is calculated as

$$f_{r_2 d}(\theta) = f_{r_2 d | I > R}(\theta) P[I > R] \tag{31}$$

The CDF of received SNR at destination through the link $r_2 \to d$ is using with the help of calculated as

$$F_{\gamma_{r_2 d}}(\theta) = \int_0^\theta f_{\gamma_{r_2 d}}(\theta) d\theta \tag{32}$$

The total end-to-end PDF of received SNR at destination in third phase when destination is performing SC and MRC respectively

$$f_\gamma^{SC}(\gamma) = \frac{d}{d\gamma} \left\{ F_{\gamma_{r_1 d}^{MARC-H}}(\gamma) F_{\gamma_{r_2 d}^{MARC-H}}(\gamma) \right\} \tag{33}$$

$$f_\gamma^{MRC}(\gamma) = \left\{ f_{\gamma_{r_1 d}^{MARC-H}}(\gamma) \oplus f_{\gamma_{r_2 d}^{MARC-H}}(\gamma) \right\}, \tag{34}$$

where $\oplus$ convolution

Outage probability (P_{out}): Outage occurs when the received SNR falls below a specified threshold. Here, for a given threshold ξ, we can calculate outage probability as follows [11, Eq. (14)]

$$P_{out} = \int_0^\xi f_\gamma(\gamma) d\gamma, \tag{35}$$

where $f_\gamma(\gamma) \in \left\{ f_\gamma^{SC}(\gamma), f_\gamma^{MRC}(\gamma) \right\}$

4 Results and Discussion

The Multi-antenna Relays Cooperation-Hybrid(MARC-H), to analyze the performance of this network, s located at (0,0 m), d located at (0,30 m), between s and d LOS, r_1 located at (0,10) and r_2 located at (0,20). In this paper the path-loss exponent assumed as 3 and the value of R to be unity. For simulation of MARC-H, random variables of received SNR (exponentially distributed) have been generated for three phases (first phase, second phase, and third phase). With the help of

Fig. 2 Comparative outage performance of various cooperative relaying schemes for diversity combining technique MRC, for L and $N = 4$

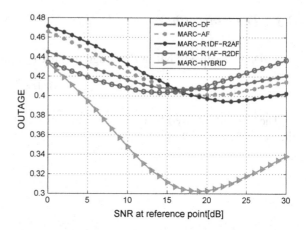

generated random variable. r_1 compares the received SNR with a particular threshold, i.e., for decode-and-forward, if received SNR is greater than required threshold, r_2 performs SSC of signals received through s and through r_1, for amplify-and-forward simply amplifies the signal received from s and r_1 and forward to r_2 and d. r_1 performs SC of signal received from s and r_2 performs SSC of the signals received from s and r_1. From the random variables of received SNR at, outage probability has been calculated. For comparing relative outage performance of MARC-H, four similar cases are simulated using Monte Carlo, case1: MARC-AF, case2: MARC-DF, case3: MARC-R1DF-R2AF and case4: MARC-R1AF-R2DF. Comparative outage probability for L and $N = 4$ have been plotted in Figs. 2 and 3 when the d using two diversity combining techniques MRC and SC, it is observed that, the proposed scheme improves network performance.

Fig. 3 Comparative outage performance of various cooperative relaying schemes for diversity combining technique SC, for L and $N = 4$

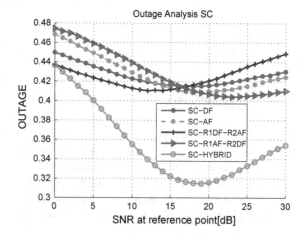

5 Conclusion

This study investigated the impact of MARC-H network of two fixed receiving multi-antenna relays between source and destination nodes in Rayleigh fading channel for BPSK modulation. The outage performance of MARC-H at has been studied when r_1 and r_2 using hybrid relaying schemes. SSC combining is used at the multi-antenna r_2 while SC and MRC are used at destination, to analyze the end-to-end outage performance of the network. MARC-AF and MARC-DF are used for relative comparison. Comparison show that MARC-H improves the performance for proposed network compared to others relaying schemes in terms of outage probability.

References

1. Karmakar, S., Varanasi, M.K.: The diversity-multiplexing tradeoff of the dynamic decode-and-forward protocol on a MIMO half-duplex relay channel. IEEE Trans. Inf. Theory **57**, 6569–6590 (2011)
2. Gerdes, L., Weiland, L., Utschick, W.: A zero-forcing partial decode-and-forward scheme for the Gaussian MIMO relay channel. In: IEEE International Conference on Communications (ICC), Budapest, pp. 3349–3354 (2013)
3. Onat, F.A., Fan, Y., Yanikomeroglu, H., Thompson, J.S.: Asymptotic BER analysis of threshold digital relaying schemes in cooperative wireless systems. In: IEEE Wireless Communications and Networking Conference, pp. 488–493, Las Vegas (2008)
4. Laneman, J.N., Tse, D.N.C., Wornell, G.W.: Cooperative diversity in wireless networks: efficient protocols and outage behavior. IEEE Trans. Inf. Theory **50**(12), 3062–3080 (2004)
5. Bao, V.N.Q., Kong, H.Y.: Distributed switch and stay combining for selection relay networks. IEEE Commun. Lett. **13**(12), 914–916 (2009)
6. Onat, F.A., Adinoyi, A., Fan, Y., Yanikomeroglu, H., Thompson, J.S., Marsland, I.D.: Threshold selection for SNR-based selective digital relaying in cooperative wireless networks. IEEE Trans. Wireless Commun. **7**(11), 4226–4237 (2008)
7. Hu, J., Beaulieu, N.C.: Performance analysis of decode-and-forward relaying with selection combining. IEEE Commun. Lett. **11**, 489–491 (2007)
8. Onat, F.A., Adinoyi, A., Fan, Y., Yanikomeroglu, H., Thompson, J.S.: Optimum threshold for SNR-based selective digital relaying schemes in cooperative wireless networks. In: Wireless Communications and Networking Conference, WCNC 2007, pp. 969–974. IEEE, Kowloon (2007)
9. Katiyar, H., Bhattacharjee, R.: Performance analysis of twohop regenerative relay network with generalized selection combining at multi-antenna relay (2010)
10. Michalopoulos, D.S., Karagiannidis, G.K.: Distributed switch and stay combining (dssc) with a single decode and forward relay. IEEE Commun. Lett. **11**, 408–410 (2007)
11. Simon, M.K., Alouin, M.-S.: Digital communication over fading channels, 2nd edn. Wiley (2005)
12. Wang, C., Zhu, G., Wang, Y., Chang, Y., Yang, D.: Outage probability and the BER of amplifyand- forward relay network. In: IEEE, 978-1-4244-1645-5/08 (2008)
13. Som, P., Chockalingam, A.: Performance analysis of space-shift keying in decode-and-forward multihop MIMO networks. IEEE Trans. Veh. Technol. **64**, 132–146 (2015)
14. Bao, V.N.Q., Kong, H.Y.: Performance analysis of multi-hop decode-and-forward relaying with selection combining. J. Commun. Netw. **12**, 616–623 (2010)

15. Adinoyi, A., Yanikomeroglu, H.: Cooperative relaying in multi-antenna fixed relay networks. IEEE Trans. Wirel. Commun. **6**, 533–540 (2007)
16. Katiyar, H., Bhattacharjee, R.: Outage performance of dualhop opportunistic relaying compared to a single-hop SIMO under rayleigh fading. In: Proceedings of NCC 2010, Chennai, India, 29–31 Jan 2010
17. Wang, B., Zhang, J., Host-Madsen, A.: On the capacity of mimo relay channels. IEEE Trans. Inf. Theory **51**, 29–43 (2005)

The Use of Soft Computing Technique of Decision Tree in Selection of Appropriate Statistical Test For Hypothesis Testing

Munish Sabharwal

Abstract This study is an interdisciplinary application of soft computing through use of decision tree for solving the complex and intricate problem of determining the appropriate statistical method to analyze the collected data for hypothesis testing and describes how to select the appropriate statistical test for hypothesis testing in a research project. The methodology adopted is that the most commonly used statistical tests were identified through a selective literature review on the methodology of research publications of diverse and wide-ranging disciplines as well as statistical handbooks, dictionaries and based on them the basic factors to select the appropriate statistical test for hypothesis testing are formulated. Finally the formulated basic factors and observations are mapped using a decision tree for determining the appropriate statistical method to analyze the collected data or hypothesis testing. The study provides a decision tree for determining the appropriate statistical method for hypothesis testing.

Keywords Decision tree · Data analysis · Statistical test selection
Hypothesis testing · Soft computing · Test of normality

1 Introduction

Statistics is the mathematical science that has evolved over the years and in the current age statistical theory and methods are being to applied to various disciplines like Social Sciences (Psychology, Demography, etc.), Management (like Quality control, Operations research, Business Analytics, etc.), medical Sciences (like Biostatistics, Epidemiology, etc.), Computer Science (Like Business analytics, Machine Learning etc.), Sciences (Physics, Biology, Astronomy), Engineering (like mechanics, chemical processes etc.), economics, environmental science, geography, Actuarial science etc.

M. Sabharwal (✉)
School of Engineering and Technology, KITE Group of Institutions, Meerut, UP, India
e-mail: mscheckmail@yahoo.com

© Springer Nature Singapore Pte Ltd. 2018
M. Pant et al. (eds.), *Soft Computing: Theories and Applications*,
Advances in Intelligent Systems and Computing 583,
https://doi.org/10.1007/978-981-10-5687-1_15

Statistics helps in obtaining more valid and reliable interpretation of the results by the data analysis using latest tools of the field. Statistical methods are one of the most prominent tools used to facilitate the transformation of the data obtained in scientific research into knowledge. However, due to technical approaches, language, and methodology options, difficulties are often experienced in the selection of the correct statistical tool and depending on these issues, incorrect applications of statistical methods are frequently seen in the data analysis phase of research [1–4].

The researcher puts together a few observations from his experiences which are the key motivations behind this study. First, most undergraduate and graduate students are familiar to some degree with descriptive statistical measures such as those of central tendency and those of dispersion but falter at inferential statistics. Second, with reference to numerous undergraduate, graduate, and doctoral students as well as research professional for different disciplines toil hard but still struggle to complete their research project timely and adeptly due to their inability to determine the appropriate statistical method they can use to analyze the data collected by them, depending upon the type of data. Third, the statistical training at both the undergraduate and postgraduate levels puts entire emphasis on learning how to perform statistical calculations rather than on learning how to determine the appropriateness of the statistical test. Finally, there is over reliance on parametric analyses and lack of familiarity with nonparametric tests and distribution-free tests even when they are the more appropriate choice.

The most frequent error made by researchers is the selection of an inappropriate test for a particular data type and consequently conduct data analysis that leads to incorrect interpretation of research findings and false conclusions resulting in wastage precious time and resources. There are many different statistical methods that can be used to analyze the data that have been collected but for accurate reporting of results it is important that the tests used are appropriate for the type of data that have been collected and knowing which statistical test to apply in which situation is an essential skill.

Computers are facilitators and not surrogates for researchers [5]; today the use of computer in contemporary research is so extensive that it is difficult to conceive a research project without the support of IT and it is an arduous task to timely conclude research studies, particularly those involving complex computations, data analysis, and modeling without the use of computers [6].

The research gap is that the existing studies mostly give fractional flowcharts or tables for selection of very few appropriate statistical tests based on type of data and whether it is independent or dependent and also these fractional flowcharts or tables do not include Post Hoc Test selection.

The novelty of the current study is that it formulates the basic factors for statistical test selection and then uses an interdisciplinary application of soft computing through use of decision tree for solving complex and intricate problem of determining the appropriate statistical test used to analyze the collected data for hypothesis testing by putting forward a single decision tree covering more than 40 tests from over a 100 that are currently in use, which makes this decision tree good enough to tackle majority of studies and it also includes Post Hoc Test selection.

2 Methodology

The most commonly used statistical tests were identified through a selective literature review on the methodology of research publications of diverse and wide-ranging disciplines as well as statistical handbooks, dictionaries.

Then basic factors to select the appropriate statistical test for hypothesis testing are formulated. Finally the formulated basic factors and observations are mapped using a decision tree for determining the appropriate statistical method used to analyze the collected data for hypothesis testing.

3 Discussion

The most commonly used statistical tests were identified through a selective literature review on the methodology of research publications of diverse and wide-ranging disciplines as well as statistical handbooks, dictionaries, etc., and based on them the following basic factors to select the appropriate statistical test for hypothesis testing are formulated:

1. The prerequisites for good data analysis are having a well-defined hypothesis as it helps to distinguish the outcome variable and the exposure variable.
2. The next step is data collection (if the data is being collected using survey method then it is a must to validate your questionnaire with the use of factor analysis and check for multicollinearity in your data) followed by consolidation, cleansing, summarizing, organizing, and transformation of collected data.
3. Find out the number of samples, and an optimum sample size should be selected because the lesser the sample size, the greater the chances of a non-normal distribution. The sample size relies on the confidence interval and confidence level chosen as well as the population size.
4. The choice of statistical test depends upon the scale of measurement or data type [continuous (interval/ratio) or categorical (nominal/ordinal)] and what we want to do with the data [Descriptive statistical analysis or Inferential statistical analysis (Comparison, Association, Prediction)].
5. If Descriptive Statistical Analysis is to be done, it quantitatively describes or summarizes features of a collection of information and interprets measures of central tendency and dispersion using graphical tools like Pie chart, Bar Graph, Histogram, etc.
6. If Inferential Statistical Analysis is to be done, it makes inferences about populations using data drawn from the population and infers properties about a population. For inferential statistics one must ascertain whether the sample follows a normal distribution curve and this can be done using any of the following methods:

- Testing for Normality by comparing the mean, median and mode, if they are approximately numerically equal, the data can be assumed to follow normal distribution.
- Testing for Normality by plotting a histogram of the data, if it looks like a symmetrical bell-shaped curve, it may be assumed to be normally distributed.
- Testing for Normality by plotting a Box plot, check that there are very few "outliers" and those "outliers" that are close to the box plot can be used to test for symmetry, which on most occasions is an adequate replacement for normality.
- Testing for Normality by Plotting a Q–Q (Quantile–Quantile) Plot, if the points in the QQ-normal plot lie on a roughly straight diagonal line, it may be assumed that the data is normally distributed.
- Testing for Normality by measuring the skewness and kurtosis by using standard statistical test of normality like Wilk–Shapiro Test, Kolmogorov–Smirnov Test, Anderson–Darling test, D'Agostino's K-squared Test, Lilliefors Test, Cramérvon Mises Criterion, Jarque–Bera Test, etc. These entire tests are based on null hypotheses that data are taken from the population that follows the normal distribution. P value is determined to see alpha error. If significant P value < 0.05, data is not following the normal distribution.
- If the sample is following a normal distribution then parametric tests are used, otherwise nonparametric tests should be used. The additional pre-requirements for parametric tests are that samples have been randomly drawn from the same population and observations within a group are independent, which implies that the samples have the same variance.

7. Determine whether the samples are 'independent' ('unpaired'/'unrelated') or 'dependent' ('paired'/'related'). Groups or data sets are considered as independent samples if there is no probability of the values in one group or data set being related to or being influenced by the values in the other groups or data sets whereas groups or data sets are regarded as dependent samples if the values in one data set are being related to or being influenced by the values in the other data sets. For example, independent sample (grades of males versus female students) or dependent sample (grades of male students after a remedial class).

8. If comparison between groups is to be done, that is to find out if there are differences between groups:

- When the groups are unpaired, the data is quantitative and follows a normal distribution curve then in such cases parametric tests such as One Sample T-Test, Unpaired T Test, One-Way ANNOVA (F-Test), etc., are applied depending on the number of groups being compared.
- When the groups are unpaired, the data is quantitative and does not follow a normal distribution curve then in such cases nonparametric tests such as Wilcoxon Test, Mann–Whittney Test, Kruskal–Wallis Test, etc., are applied depending on the number of groups being compared.

- When the groups are unpaired, the data is categorical then nonparametric tests such as Chi-Square Test, Binominal Test, Fisher Test, Wilcoxon Test, Mann–Whittney Test, Kruskal–Wallis Test, Crosstabs Test, etc., are applied depending on the data being of ordinal or nominal type and the number of groups being compared.
- When comparing more than two sets of unpaired data by applying equivalent multiple group comparison parametric or nonparametric tests and if these tests return a statistically significant P value ($P < 0.05$) then only they should be followed by a post hoc test to find out that the difference lies between exactly which two data sets. Post hoc (Multiple Group Comparison) tests for parametric data are Turkey–Kramer Test, Newman–Keuls Test, Bonferroni Test, Dunnett's Test, Scheffe's Test, etc., and for nonparametric data are Dunn's Test.
- When the groups are paired, data is quantitative and follows a normal distribution curve then in such cases parametric tests such as Paired T Test, Repeated measures of ANOVA, etc., are applied depending on the number of groups being compared.
- When the groups are paired, data is quantitative and does not follow a normal distribution curve then in such cases nonparametric tests such as Wilcoxon Test, Friedman Test, etc., are applied depending on the number of groups being compared.
- When the groups are paired, data is categorical data then nonparametric tests such as McNemar Test, Cross Tabs, Wilcoxon Test, Crochrane Q Test, Friedman Test, etc., are applied depending on the data being of ordinal or nominal type and the number of groups being compared.
- When one is comparing more than two sets of paired data by applying equivalent multiple group comparison parametric or nonparametric tests then if these tests return a statistically significant P value ($P < 0.05$) then only they should be followed by a post hoc test to find out that the difference lies between exactly which two data sets. A post hoc (Multiple Group Comparison) test for parametric data is Wilcoxon's Matched Pair Signed Rank Test and for nonparametric data is Dunn's test.

9. If association between two variables is to be checked, that is relationship between two variables without a true independent variable, these are correlation tests and they suggest the power of the association between two variables as a correlation coefficient which may vary in degree from 0 (i.e., having no correlation at all) to 1 (i.e., having a perfect correlation). In a perfect correlation knowing the value of one variable, we may exactly predict the value of the other variable but it still does not imply causality.

   When association between two quantitative variables is to be checked and if data follows a normal distribution curve, then in such cases Pearson Correlation Coefficient and Scatter Diagrams should be applied and if data does not follow a normal distribution curve, then in such cases Spearman Coefficient and Scatter Diagrams should be applied.

When association between two qualitative variables is to be checked, if the data is ordinal then in such cases Spearman Coefficient and Scatter Diagrams should be applied and if the data is nominal, then in such cases Contingency Coefficient should be applied.

10. When value of a dependent variable is to be predicted based on independent variable, that is relationship between dependent and a true independent variable, if the two variables (Dependent and Independent) are quantitative and linearly related, in such case linear regression analysis is applied as it can formulate a mathematical equation which can be utilized to predict the value of dependent variable for a particular value of the independent variable.

When two quantitative variables are linearly related and data follows a normal distribution curve, then in such cases Simple Linear Regression should be applied and if data does not follow a normal distribution curve, then in such cases nonparametric Regression should be applied. When variables are more than two then for quantitative data Multiple Linear Regression or Multiple Nonlinear Regression are applied.

When two qualitative variables are linearly related and the data is ordinal then in such cases nonparametric Regression should be applied and if data in nominal then in such cases Simple Logistic Regression should be applied.

4 Conclusion

With the assumption that the user has understanding of basic statistical terms, the decision tree for determining the suitable appropriate test for hypothesis testing for novice users is formulated as given in Fig. 1.

The decision tree as shown in Fig. 1 will be useful for especially those users who are not well versed with statistics. The decision tree in Fig. 1 covers more than 40 tests from over a 100 that are currently in use, which makes this decision tree good enough to tackle majority of studies and it also includes Post Hoc Test selection.

Post selection for appropriate statistical test for data analysis, that data can be analyzed using software like SPSS, STATISTICA, R, SAS, etc.

The results proposed in the decision tree have been authenticated by applying on five different data sets and the projected results were found to be correct.

5 Practical Application

A practical application of the formulated decision tree can be done by developing a web-based open source application based on it where researchers are asked questions based on the formulated basic factors to select the appropriate statistical test for hypothesis testing. This would help the researchers decide which statistical

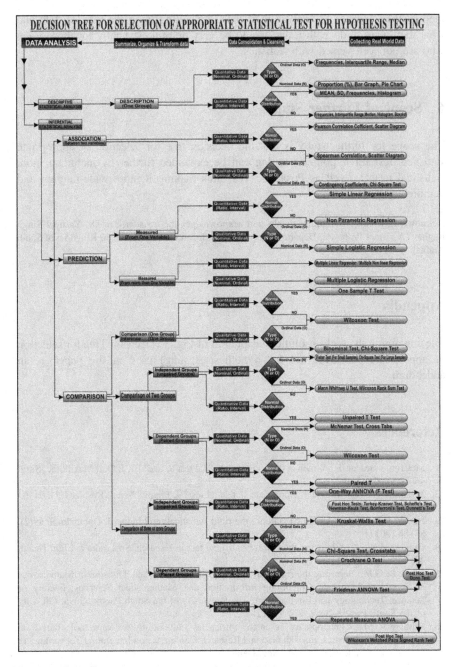

Fig. 1 Self-adaption

method they can use to analyze their data, depending upon the type of data and what they want to do with it and would be very helpful especially for those users who are not well versed with statistics.

6 Scope of Further Study

The scope for further study is that the decision tree for selection of appropriate statistical test for hypothesis testing can be expanded further by including more statistical tests as well as Post Hoc Test thus making it more wide-ranging and comprehensive.

Acknowledgements The author thanks and acknowledges the assistance of Dr. Yashpal Singh Raghav (Assistant Professor, Department of Statistics, Jazan University, Jazan, Kingdom of Saudi Arabia).

Appendix

File containing the high-resolution adaptation of Fig. 1 (Decision Tree for selection of appropriate statistical test for hypothesis testing) used in the paper as an addendum.

References

1. McCrum-Gardner, E.: Which is the correct statistical test to use? Br. J. Oral Maxillofac. Surg. **46**(1), 38–41 (2008)
2. Jaykaran, C.: How to select appropriate statistical test? J. Pharm. Negat. Results **1**(2), 61–63 (2010)
3. Nayak, B.K., Hazra, A.: How to choose the right statistical test? Indian J. Ophthalmol. **59**(2), 85–86 (2011)
4. Gunawardena, N.: Choosing the correct statistical test in research. Sri Lanka J. Child Health **40**, 149–153 (2011)
5. Khan, M.G.M.: Selecting the right way of analyzing research data. Division of Mathematics & Statistics, School of Computing Information and Mathematical Sciences, Faculty of Science, Technology and Environment, The University of the South Pacific, Suva, Fiji, City of the South Pacific, Suva, Fiji. (2012)
6. Sabharwal M.: Contemporary research: intricacies and aiding software tools based on expected characteristics. In: 10th National Research Conference on Integrating Technology in Management Education, AIMA, New Delhi, 28–29 Mar 2016
7. Scales, A.Y., Petlick, J.H.: Selecting an appropriate statistical test for research, conducted in engineering/graphics education: a process. In: Proceedings of the 2004 American Society for Engineering Education Annual Conference & Exposition Copyright © 2004, American Society for Engineering Education (2004)

8. Nayak, B.K.; Why learn research methodology? Indian J. Ophthalmol. **57**(3), 173–174 (2009)
9. Bettany-Saltikov, J., Whittaker, V.J.: Selecting the most appropriate inferential statistical test for your quantitative research study. J. Clin. Nurs. **23**, 1520–1531 (2014)
10. Yuen, B.: Choosing the appropriate descriptive statistics, graphs and statistical tests, University of Southampton (2013). Alacaci, C.: Inferential statistics: understanding expert knowledge and its implications for statistics education. J Stat. Educ. **12**(2), (2004)
11. Cramer, D., Howitt, D.L.: The Sage dictionary of statistics: a practical resource for students in the social sciences. Sage, London (2004)
12. du Prel, J.B., Röhrig, B., Hommel, G., Blettner, M.: Choosing Statistical Tests. Dtsch Arztebl Int. **107**(19), 343–348 (2010)
13. Everitt, B.S., Skrondal, A.: The Cambridge dictionary of statistics, 4th edn. Cambridge University Press, Cambridge (2010)
14. Fakuda, H., Ohashi, Y.: A Guideline for Reporting Results of Statistical Analysis. Jpn. J. Clin. Oncol. **27**(3), 121–127 (1997)
15. Heiman, G.W.: Basic statistics for the behavioral sciences. Houghton Mifflin Co, New York (2006)
16. Herman, A., Notzer, N., Libman, Z., Braunstein, R., Steinberg, D.M.: Statistical education for medical students-concepts are what remain when the details are forgotten. Stat. Med. **26**(23), 4344–4351 (2007)
17. Howell, D.C.: Fundamental statistics for the behavioral sciences. Brooks/Cole, California (2004)
18. Karan, J., Goyal, J.P., Bhardwaj, P., Yadav, P.: Statistical reporting in Indian pediatrics. Indian Pediatr. **46**, 811–812 (2009)
19. Karan, J., Kantharia, N.D., Yadav, P., Bhardwaj, P.: Reporting statistics in clinical trials published in Indian journals: a survey. Pak. J. Med. Sci. **26**, 212–216 (2010)
20. Johnson, L.R., Karunakaran, U.D.: How to choose the appropriate statistical test using the free program "statistics open for all" (SOFA). Ann. Commun. Health **2**(2) 54–66 (2014)
21. Marusteri, M., Bacarea, V.: Comparing groups for statistical differences: how to choose the right statistical test? Biochem. Medica. **20**(1), 15–32 (2010)
22. McDonald, J.H.: Handbook of biological statistics, 3rd edn., pp. 293–296. Sparky House Publishing, Baltimore, Maryland (2014)
23. Parab, S., Bhalerao, S.: Choosing statistical test. Int. J. Ayurveda Res. **1**(3), 187–191 (2010)
24. Parikh, M.N., Hazra, A., Mukherjee, J., Gogtay, N.: Hypothesis testing and choice of statistical tests. In: Research Methodology Simplified: Every Clinician A Researcher, pp. 121–128. New Delhi, Jaypee Brothers (2010)
25. Park, H.M.: Hypothesis testing and statistical power of a test. Test **4724**(812), 1–41 (2010)
26. Petlick, J., Scales, A.: Selecting an appropriate statistical test for research. In: Engineering/Graphics Education: A Useful Procedure Paper presented at 2004 Annual Conference, Salt Lake City, Utah (2004) https://peer.asee.org/1337
27. Vowler, S.L., Analyzing data-choosing appropriate statistical methods. Hosp. Pharm. **14**, pp. 39–44 (2007)
28. Spriestersbach, A., Röhrig, B., du Prel, JB., Gerholday, A., Blettner, M.: Descriptive statistics: the specification of statistical measures and their presentation in tables and graphs part—part 7 of a series on evaluation of scientific publications. Dtsch Arztebl Int. **106**(36), 578–583 (2009)

Integration of GIS, Spatial Data Mining, and Fuzzy Logic for Agricultural Intelligence

Mainaz Faridi, Seema Verma and Saurabh Mukherjee

Abstract With increasing population and decreasing crop production, there is an enormous need to increase land under cultivation. This paper attempts to explore the applicability of spatial data mining integrated with Geographic Information System (GIS) and fuzzy logic for Agricultural Intelligence. The research uses thematic agricultural data of Jodhpur District of Rajasthan state and mines spatial association rules between groundwater, wastelands, and soils of Jodhpur District which are then used to create Mamdani fuzzy inference system for determining the utilization of wastelands. A taluk-wise map of Jodhpur district is created from the fuzzy values showing the utilization of wastelands. Analysis of results showed that out of 36,063 hectares of mined pattern, Phalodi taluk of Jodhpur district contains the largest wasteland area and the area under the medium type of utilization is the largest. It could be suggested that wastelands having a substantial groundwater underneath can be irrigated for agriculture and/or producing fodder and firewood.

Keywords Spatial association rule mining · GIS · Fuzzy inference system
Agriculture intelligence

M. Faridi (✉) · S. Mukherjee
Department of Computer Science, Banasthali University, Vanasthali, Rajasthan, India
e-mail: mainaz.faridi@gmail.com

S. Mukherjee
e-mail: mukherjee.saurabh@rediffmail.com

S. Verma
Department of Electronics, Banasthali University, Vanasthali, Rajasthan, India
e-mail: seemaverma3@yahoo.com

© Springer Nature Singapore Pte Ltd. 2018
M. Pant et al. (eds.), *Soft Computing: Theories and Applications*,
Advances in Intelligent Systems and Computing 583,
https://doi.org/10.1007/978-981-10-5687-1_16

171

1 Introduction

The improvements in GPS, GIS, high-resolution remote sensing, and location sensitive services, provide us the ability to store geo-referenced, location-sensitive information, or spatial characteristics along with the transactional data [1] resulting in the collection of a large amount of spatial and spatial temporal data. Spatial data mining is used to understand spatial data, establish the interrelationship between spatial and non-spatial data, expose the spatial distribution patterns, build spatial knowledge-base and envisage the future trends [2]. Till recently, GIS alone was used not only to capture, store, retrieve geospatial data but also to query and analyze them and visualize the results in form of maps [3]. The functions such as spatial and non-spatial query and selection, map creation and overlays, classification, etc., provide GIS with the capability to be used as spatial data analysis tool. But even the modern GIS provides the limited tools for complex spatial data analysis and knowledge discovery. Furthermore, it has an inadequate ability to incorporate the decision maker's preferences, experiences, intuitions, and judgments into the problem-solving process [4]. GIS alone does not have the capability of finding relationships between spatial and non-spatial attributes.

Similarly, spatial data mining and GIS alone provides limited methods to deal with the uncertainty in spatial data. Spatial association rules work on crisp data and suffer from sharp boundary problem. GIS has a limitation that spatial attributes, their relationship, etc., must be defined in advance in a precise and accurate manner which cannot be assured every time [5]. Hence, fuzzy logic could be used to solve the above issues. Thus the massive outburst of geo-referenced and location-sensitive data, advancements in technology, remote sensing, digital cartography along with the need of addressing the uncertainty in spatial data and use of linguistic terms accentuates the significance of integration of spatial data mining, GIS, and fuzzy logic [6]. Integrated use of GIS and spatial data mining can have applications in remote sensing, cartography, weather monitoring and forecast agriculture, urban planning, wildlife monitoring, environmental studies, demography, health services, etc.

2 Agriculture Intelligence

The term "Agriculture Intelligence" proposed by Ghadiyali et al. [7], is a method where a collection of information, events, and issues from various agricultural activities is done, and analyzed to be used in making well-informed decisions. Any organization related to agriculture (like those involved in agriculture production, animal husbandry, government agencies, agricultural investment, insurance and consultancy agencies, etc.) can provide its input to Agriculture Intelligence.

The information and data provided by the contributing organizations are analyzed and turned into intelligence which is then utilized to assist them in taking efficient and timely decisions. Agriculture Intelligence can be used especially where accurate and timely forecasts of the future tendencies or agricultural conditions can be obtained like in predicting crop production at regional, country, and even at global levels. Agriculture Intelligence plays a major role in predicting crop production at regional, country, and even at global levels. Agriculture Intelligence allows for better trend and market analysis, prediction of crop production, forecasting cost of agricultural products thereby helping in making proactive business decisions [8–10].

2.1 Agriculture in India

Although the country has witnessed an enormous growth in agriculture after its independence its share in country's GDP has lowered to 13–15%. Besides other factors like insufficient use of modern agricultural practices and advancement in technology, delayed or less monsoon, etc. [11], utilization of agricultural lands into industrial lands or other non-agriculture lands is of significant concern.

Indian National Policy for Farmers of 2007 stated that "prime farmland must be conserved for agriculture except under exceptional circumstances, provided that the agencies that are provided with agricultural land for non-agricultural projects should compensate for treatment and full development of equivalent degraded or wastelands elsewhere" [12].

Thus, one of the ways of increasing agriculture production is to bring more and more wastelands under cultivation. For this, one has to first filter out those wastelands which when provided with minimum irrigation water, could be utilized for production of crops or else fodder for animals [13, 14]. In this research, the authors apply Spatial Association Rule Mining to strain those wastelands of Jodhpur District which have a good amount of groundwater underneath. Thus, SARM can be used to extract intelligence in agriculture.

3 Spatial Association Rule Mining

Association Rule Mining (ARM) [15] is one of the most eminent and widely used data mining techniques for extracting interesting correlations, frequent patterns, regularities, or associations in the large transaction databases [16]. Similarly, spatial association rules can be mined in spatial and geo-referenced databases by considering spatial predicates and properties. The spatial predicates may represent topological relationships between spatial objects, such as disjoint, intersects, inside/outside, adjacent to, covers/covered by, equal, etc., or spatial orientation or ordering, such as left, right, north, east, etc., or contain some distance information, such as close to, far away, etc. [17, 18]. The strength and reliability of an association rule are measured by two factors: support and confidence. A spatial association rule can be stated as

is_$a(x,$ city) $\wedge$ close _to($x,$ equator) $\rightarrow$ is_hot(x) (98%)

4 Fuzzy Logic

Fuzzy logic was introduced by a Computer Science Professor, Lotfi A. Zadeh in 1965 for modeling the uncertainty and vagueness of natural language [19]. Crisp set defines a sharp, unambiguous boundary between the members and non-members of the class by assigning 0 or 1 to distinguish between them. But in the real world, there are many cases, when assigning a 0 or 1 is not sufficient enough to represent subjective or vague ideas. Fuzzy set assigns a membership value to an element with a real number from 0 to 1 representing the degree to which an element belongs to a given set and removes sharp boundary problem. Fuzzy Inference System (FIS) is a computer model that performs the reasoning based on fuzzy set theory and fuzzy rules of the "IF-Then" form [20]. A Fuzzy Inference System consists of a fuzzifier, fuzzy rule base, fuzzy inference engine and defuzzifier.

5 Dataset Description

Jodhpur District of Rajasthan state of India is chosen as the study area for this research. The district is a semi-arid region of Rajasthan state and lies between 250 51′ 08″ and 270 37′ 09″ North latitude and 710 48′ 09″ and 730 52′ 06″ East longitude encompassing a geographical area of 2,256,405 hectares. The datasets for land use, groundwater, and soil in GIS format were collected from ISRO center of Jodhpur.

Due to the extraneous information present in the original datasets, useful data are extracted and cleaned during data preprocessing step. Table 1 shows the selected attributes of the datasets while Fig. 1a–c show their corresponding maps. Finally, an intersection is done for all the three datasets which creates a new map layer showing all those areas of the district which are either salt affected wastelands or wastelands without scrubs, have cambid/fluvent/orthent soil and have substantial ground water beneath Fig. 1d.

Table 1 Dataset description

Dataset	Selected attributes
Groundwater	Very good to good, Good, Good but saline, Good but moderate, Moderate
Wastelands	Salt-affected, Land without scrub
Soil suborder	Cambids, Orthents, Fluvents

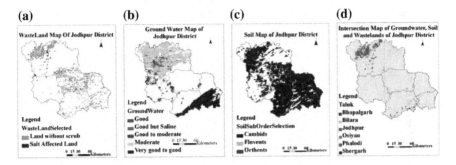

Fig. 1 Jodhpur District map for (a) waste-land, (b) groundwater, (c) soil and (d) intersection of all three datasets

6 Integration of GIS, Spatial Data Mining, and Fuzzy Logic

In this paper, a unified framework for integration of GIS, spatial data mining, and fuzzy logic for Agriculture Intelligence is proposed (Fig. 2). The framework contains all the steps necessary for the complete data mining process as proposed by [21]. At the bottom lies the data repository containing data which may be stored in different locations and format. In the data preprocessing step, the data in GIS format

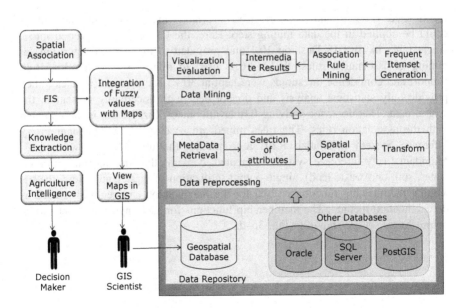

Fig. 2 Framework for integration of GIS, spatial data mining, and fuzzy logic

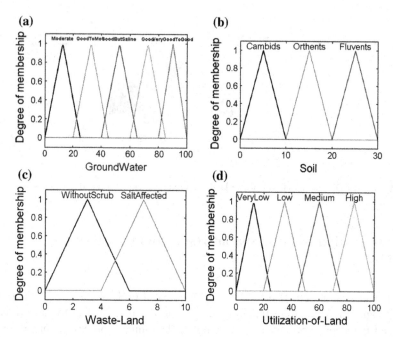

Fig. 3 Membership functions for **a** groundwater, **b** soil, **c** wasteland and **d** utilization of wastelands

is extracted, cleaned, and transformed to a format where data mining algorithms could be applied in the data mining step. Here, the GIS data is converted into a single table in ArcGIS Desktop 10.1 and provided to the Apriori algorithm for mining of association rules. Weka 3.6 is used to mine association rules. Both support and confidence thresholds were kept at 30%. The association rules mined provide only statistical analysis and do not deal with uncertainty present in the spatial data. Therefore, for deriving further inferences, fuzzy logic is incorporated with the association rules. The rules mined are used to create Mamdani Fuzzy Inference System in Matlab R2013a to predict the utilization of wastelands. Mamdani fuzzy inference system so created, contains three-inputs (viz., groundwater, soil, and waste-land), single-output (utilization of wasteland). Here, the triangular membership function is used for the fuzzification and centroid is used as defuzzification method. The membership functions for each input fuzzy set and an output fuzzy set are shown in Fig. 3a–d.

Fig. 4 Map showing
utilization of wasteland

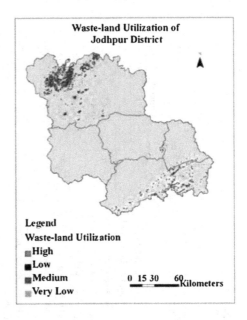

For more analysis, the rules are categorized on the basis of taluks of Jodhpur District. Thus, a file containing the fuzzy values of the output variable, utilization of wasteland for the study area was produced using the Mamdani FIS. These fuzzy values were then exported into a GIS environment, and a final map was created showing the taluk-wise utilization of wastelands (Fig. 4). At the last, this map is studied for the statistical interpretation. Thus the above framework provides the map to the environmentalist, government agencies, land use planning committees, agricultural researchers, etc., with the knowledge they are interested in, which otherwise was simply a topographic representation.

7   Results and Discussion

The association rules mined for the Jodhpur District are shown in Table 2. These rules are then used to create Mamdani FIS. The rule viewer for FIS is shown in Fig. 5. The first three columns represent the selected feature type viz. groundwater, soil, and waste-land while the fourth column represents the output variable-utilization of land. The surface viewer is a three-dimensional curve representing the mapping between the input and output variables. Since the curve has the limitation of representing a two-input and one output variables, the Fig. 6 shows a surface viewer with groundwater and soil as input variables and utilization of land as the output variable.

Table 2 Spatial association rules for groundwater, soil and wastelands of Jodhpur district

S. No.	Body	Implies	Head	Support	Confidence
1	Groundwater = very good to good	→	Waste-land = land without scrub	0.31	1
2	Groundwater = very good to good and soil suborder = cambids	→	Waste-land = land without scrub	0.30	1
3	Soil suborder = cambids	→	Waste-land = land without scrub	0.93	0.98
4	Groundwater = moderate and soil suborder = cambids	→	Waste-land = land without scrub	0.53	0.98
5	Groundwater = moderate	→	Waste-land = land without scrub	0.57	0.98
6	Groundwater = very good to good	→	Soil suborder = cambids	0.31	0.95
7	Groundwater = very good to good and waste-land = land without scrub	→	Soil suborder = Cambids	0.31	0.95
8	Groundwater = very good to good	→	Soil suborder = cambids and waste-land = land without scrub	0.31	0.95
9	Waste-land = land without scrub	→	Soil suborder = cambids	0.98	0.93
10	Groundwater = moderate and waste-land = land without scrub	→	Soil suborder = cambids	0.60	0.92
11	Groundwater = moderate	→	Soil suborder = cambids	0.57	0.92
12	Groundwater = moderate	→	Soil suborder = cambids and waste-land = land without scrub	0.57	0.9
13	Waste-land = land without scrub	→	Groundwater = moderate	0.98	0.57
14	Soil suborder = cambids	→	Groundwater = moderate	0.93	0.56
15	Waste-land = land without scrub	→	Groundwater = moderate and soil suborder = cambids	0.98	0.52

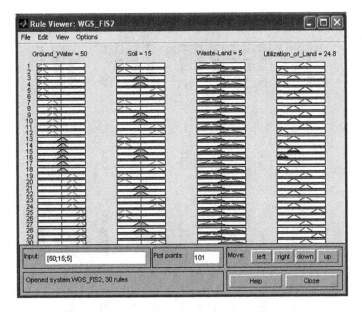

Fig. 5 The rule viewer representing the FIS for groundwater, soil and waste-land

Fig. 6 The surface viewer, showing mapping between input and output parameters

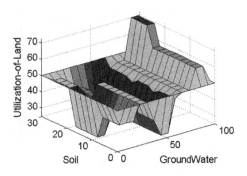

The area under the mined pattern produced from the output map is shown in Table 3, covering a total of 36,062.71 hectares of land. The analysis reveals that Phalodi taluk contains the largest wasteland area while Shergarh taluk contains the smallest area of wasteland. The distribution of area with respect to groundwater, soil, and wasteland is showed in the graph (Fig. 7). Utilization of wastelands among all the taluks of Jodhpur district is shown in Table 4 and a graph depicting the same is shown in Fig. 8.

Table 3 Taluk wise distribution of area for groundwater, soil, and waste-land

	Taluk						Area (Hectares)
	Bhopalgarh	Bilara	Jodhpur	Osiyan	Phalodi	Shergarh	Total
Groundwater							
Moderate	0.00	0.00	0.00	0.00	26,963.10	0.00	**26,963.10**
Good to moderate	0.00	75.49	66.40	0.00	129.94	0.00	**271.82**
Good but saline	0.00	0.00	0.00	215.99	91.16	0.00	**307.15**
Good	23.05	88.87	0.00	135.16	42.13	5.84	**295.05**
Very good	262.49	4691.80	3271.29	0.00	0.00	0.00	**8225.58**
Total	**285.54**	**4856.15**	**3337.69**	**351.15**	**27,226.33**	**5.84**	**36,062.71**
Soil suborder							
Cambids	285.55	4856.15	2974.16	350.15	25,370.34	5.27	**33841.62**
Orthents	0.00	0.00	0.00	1.01	1855.99	0.57	**1857.57**
Fluvents	0.00	0.00	363.53	0.00	0.00	0.00	**363.53**
Total	**285.55**	**4856.15**	**3337.69**	**351.15**	**27,226.33**	**5.84**	**36,062.71**
Waste-land							
Land without scrub	285.55	4856.15	3337.69	135.16	27,028.17	5.84	**35,648.57**
Salt-affected land	0	0.00	0.00	215.99	198.16	0.00	**414.15**
Total	**285.55**	**4856.15**	**3337.69**	**351.15**	**27,226.33**	**5.84**	**36,062.71**

Table 4 Utilization of wastelands

Utilization of waste-land	Area (Hectares)
High	6032.35
Medium	22,976.44
Low	3929.35
Very Low	3124.57

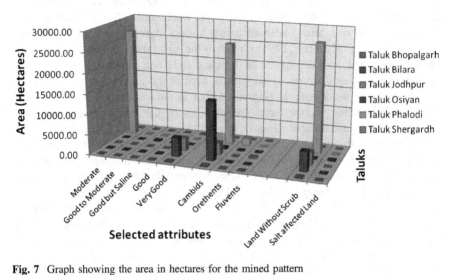

Fig. 7 Graph showing the area in hectares for the mined pattern

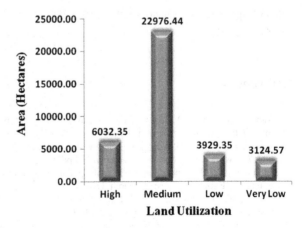

Fig. 8 Area for utilization of wastelands

8 Conclusion

The advancement in collection, storage, and retrieval techniques of digital data has resulted in accumulation of huge amount of data, which is beyond the human ability to comprehend and handled by single field (like statistics, spatial data mining, GIS, artificial intelligence, etc.), accentuates the need to develop multidisciplinary approach to extract useful information. In this paper, a framework for the integration of spatial data mining and GIS with fuzzy logic is presented for knowledge extraction to be used for Agriculture Intelligence. The association rules for groundwater, soil, and wasteland using Apriori algorithm are extracted, and used for creating Mamdani FIS to predict the utilization of wastelands. The knowledge gained in this way could be used by environmentalist, government agencies, farmers, agriculture advisory firms, agricultural researchers, and analysts, etc., to strain the wastelands having substantial groundwater underneath and use them for reclamation. Furthermore, the analysis of the mined area shows that majority of wastelands has medium utilization. The majority of wastelands is the lands without scrubs and has considerable groundwater underneath; these lands can be used for cultivation of crops. The wastelands which have moderate groundwater can be used to produce firewood and fodder for animals. Plant species like *Acacia jacquemontii, Acacia senegal, Albizia lebbeck, Anogeissus rotundifolia, Prosopis cineraria, Salvadora oleoides, Tecomella undulata, Leucaena leucocephala, Tephrosia purpurea, and Crotalaria medicaginea* can be grown. Thus, the research suggests that more and more of wastelands could be reclaimed and brought under cultivable lands, thereby increasing the crop production, along with providing an additional source of income to the rural people, helping in maintaining ecological balance and providing timber and fodder for local use.

Acknowledgements The authors will like to thank ISRO, Jodhpur Centre for providing necessary data for the research.

References

1. Mennis, J., Guo, D.: Spatial data mining and geographic knowledge discovery—an introduction. Comput. Environ. Urban Syst. **33**(6), 403–408 (2009)
2. www.iasri.res.in/ebook/win_school_aa/notes/spatial_data_mining.pdf
3. http://www.manage.gov.in/studymaterial/GIS.pdf
4. Eldrandaly, K.: Expert systems, GIS, and spatial decision making: current practices and new trends. In: Expert Systems: Research Trends, pp. 207–22 (2007)
5. Ladner, R., Petry, F.E., Cobb, M.A.: Fuzzy set approaches to spatial data mining of association rules. Trans. GIS **7**(1), 123–138 (2003)
6. Tang, H., McDonald, S.: Integrating GIS and spatial data mining technique for target marketing of university courses. In: ISPRS Commission IV, Symposium, pp. 9–12 (2002)
7. Ghadiyali, T., Lad, K., Patel, B.: Agriculture intelligence: an emerging tool for farmer community. In: Proceedings of Second International Conference on Emerging Application of Information technology (IEEE), (2011) doi:10.1109/EAIT.2011.36

8. Cojocariu, A., Stanciu, C.O.: Data warehouse and agricultural intelligence. Agric. Manag./ Lucrari Stiintifice Seria I, Manag. Agric. **11**(2). (2009)
9. Munroe, F.A.: Integrated agricultural intelligence: a proposed framework. Vet. Ital. **43**(2), 215–223 (2006)
10. http://www.gdacorp.com/agricultural-intelligence
11. Dwivedy, N.: Challenges faced by the agriculture sector in developing countries with special reference to India. Int. J. Rural Stud. **18**(2), (2011)
12. National Policy for Farmers: (2007). http://agricoop.nic.in/imagedefault1/npff2007.pdf
13. Faridi, M., Verma, S., Mukherjee, S.: Impact of ground water level and its quality on fertility of land using GIS and agriculture business intelligence. In: Proceedings of Geomatrix'12: An International Conference on Geospatial Technologies and Applications, IIT Bombay (2012)
14. Faridi, M., Verma, S., Mukherjee, S.: Association rule mining for ground water and wastelands using apriori algorithm: case study of Jodhpur district. Int. J. Adv. Res. Comput. Sci. Softw. Eng. **5**(6), 751–758 (2015)
15. Agrawal, R., Imieliński, T., Swami, A.: Mining association rules between sets of items in large databases. ACM Sigmod Rec. **22**(2), 207–216 (1993)
16. Kotsiantis, S., Kanellopoulos, D.: Association rules mining: a recent overview. GESTS Int. Trans. Comput. Sci. Eng. **32**(1), 71–82 (2006)
17. Han, J., Kamber, M.: Data mining: concepts and techniques. Morgan Kaufmann Publishers (2001)
18. Koperski, K., Han, J.: Discovery of spatial association rules in geographic information databases. In: International Symposium on Spatial Databases, pp. 47–66. Springer Berlin Heidelberg (1995)
19. Zadeh, L.A.: Fuzzy sets. Inf. control **8**(3), 338–353 (1965)
20. Sowan, B.I.: Enhancing fuzzy associative rule mining approaches for improving prediction accuracy. Integration of fuzzy clustering, apriori and multiple support approaches to develop an associative classification rule base. Doctoral dissertation, University of Bradford (2012)
21. Fayyad, U., Piatetsky-Shapiro, G., Smyth, P.: From data mining to knowledge discovery in databases. AI Mag. **17**(3), (1996)

Hybrid Micro-Generation Scheme Suitable for Wide Speed Range and Grid Isolated Remote Applications

Radhey Shyam Meena, Mukesh Kumar Lodha, A.S. Parira and Nitin Gupta

Abstract The research work mainly focuse on grid isolated operation suitable for low wind speed areas for micro-generation purpose and to meet the requirements of local household and critical loads. A hybrid generation system is designed where electricity is generated by wind turbine coupled Induction Generator (IG) in conjunction with a solar-PhotoVoltaic (PV) panel. The generation system designed in this work is capable of harnessing power for wide ranges of wind speeds. The system uses a three-phase inverter which provides the necessary reactive power support to the IG in addition to the fixed capacitors. The three-phase inverter is capable to fault-tolerant operation which enables more reliability on the generation. A suitable storage battery is used which stores the excess energy during low loading or high generation conditions. The main strategy is to mount the solar system in the same tower of the wind turbine so that the space requirement is minimal. The results of proposed system show better quantitative performance.

Keywords Micro-generation · Hybrid solar-wind system · Induction generator
Battery storage system · Energy conversion techniques

R.S. Meena (✉) · A.S. Parira
National Solar Mission Division, Ministry of New and Renewable Energy,
New Delhi 110003, India
e-mail: rshyam.mnre@gov.in

A.S. Parira
e-mail: anindya.parira@nic.in

M.K. Lodha
Department of Electrical Engineering, SBCET, Jaipur 302013, India
e-mail: mukesh_lodha85@yahoo.co.in

N. Gupta
Department of Electrical Engineering, MNIT, Jaipur 302017, India
e-mail: nitineed@gmail.com

© Springer Nature Singapore Pte Ltd. 2018
M. Pant et al. (eds.), *Soft Computing: Theories and Applications*,
Advances in Intelligent Systems and Computing 583,
https://doi.org/10.1007/978-981-10-5687-1_17

1 Introduction

Over the last few decades, a plentiful amount of research has been done in generation of electrical energy from renewable source like wind, solar, biomass, biofuel. As conventional sources of energy are becoming scarce and costly day by day, wind and solar energy has evolved into an attractive energy source for electric utilities.

Induction generators (IG) are invariably used for generation of electricity from wind due to their ruggedness and low cost. However, wide ranges of wind speed could not be harnessed for generation. Also due to intermittent nature of wind, problems of voltage and frequency regulation are quite common in the available schemes. The structure of wind turbines, as well as the fact that the wind energy rate is uncontrollable, necessitates the problem of regulating the process of power generation. This engineering challenge has been alleviated by the construction of variable speed wind turbines, which are designed to regulate the power captured over a range of operating speeds. IG are extensively used for harnessing electricity from wind via wind turbines [1–3]. Three-phase cage rotor IGs are widely used for wind energy conversion systems having advantages of ruggedness and low cost [4–7]. The induction generators have great potential for small-scale installations in fractional kW ranges besides large-scale generation [8]. The scheme of [8] is shown in Fig. 1a. Single-phase IGs are also widely used for isolated load applications [9–12]. These IGs can be of self-excited type with balanced or unbalanced terminal loads [9, 10], energized from a variable excitation source like an inverter [11, 12]. A similar single-phase IG is shown in Fig. 1b. However, problems are faced in determining excitation requirements of these IGs.

The self-excited induction generators are provided with reactive power using capacitor banks connected across the induction machine stator terminals to maintain output voltage when the rotor is driven above its synchronous speed to run as a generator [13]. Grid-isolated single-phase IGs are also available in relevant literatures for feeding domestic loads [13, 14].

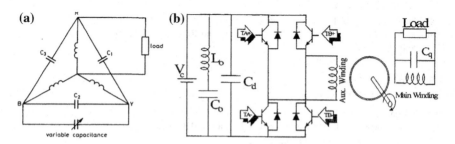

Fig. 1 **a** Small low-cost generator [8], **b** Single-phase inverter assisted IG [11]

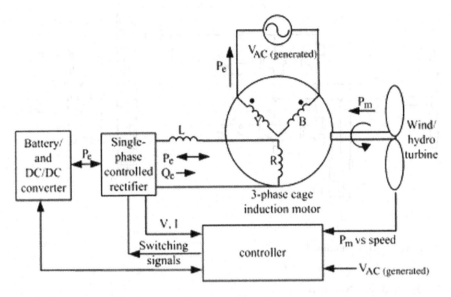

Fig. 2 Scheme proposed in [22]

Three-phase induction IGs can be operated as stand-alone, single-phase generator with fixed excitation source [15–17]. Analytical approaches are also presented to determine the excitation requirement for such machine [18]. Attempts are also made using different schemes of series-parallel capacitor connections with three-phase machines for power generation in single-phase mode [19, 20]. All these schemes are highly dependent on control of bulk capacitors chosen for excitation. Furthermore, in [21], better voltage regulation is achieved using additional bulk load for desired outcome of steady single-phase power generation. In [22], modeling and analysis of single-phase induction generator is proposed as shown in Fig. 2. Paper [23] uses PV sourced inverter for voltage regulation and providing variable excitation for such a scheme. Hybrid schemes involving wind turbine coupled IGs and solar-PhotoVoltaic (PV) [24, 25] or wind- and diesel-based generations [26] for supplying isolated critical loads are also becoming popular [27, 28]. These schemes involving wind and PV sources increase the reliability of the generation schemes especially for grid isolated and critical loads. A PV-excited IG is shown in Fig. 3.

While several attempts are made for grid connected large-scale generation, not major work is done for small-scale domestic applications operational for low wind speeds in isolation from grid mode. A PV assisted and inverter excited induction generator based wind energy conversion system capable of generation in wide speed range is proposed in herewith for application of the same.

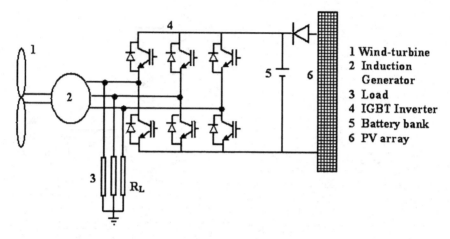

1 **Wind-turbine**
2 **Induction Generator**
3 **Load**
4 **IGBT Inverter**
5 **Battery bank**
6 **PV array**

Fig. 3 Integrated PV-excited IG [24]

2 Proposed System

A PV assisted and inverter excited induction generator based wind energy conversion system capable of generation in wide speed range is design in this work. In this work, three-phase a 1 kW, 415 V, 50 Hz, eight-pole, induction machine is used as IG. The proposed scheme is shown in Fig. 4.

The variable excitation required by the IG is provided by a three-phase variable frequency inverter. The inverter will be capable of fault-tolerant operation for more reliability on the generation scheme especially hilly Himalayan region or northeastern region where frequent maintenance facility can be difficult. The inverter DC

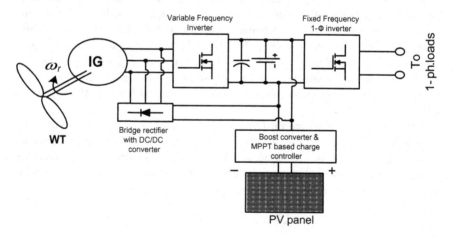

Fig. 4 Block diagram of the proposed scheme

bus is connected to a solar-PV panel along with a storage battery. The generated voltage is rectified to DC using a three-phase rectifier. The rectified voltage, along with the voltage obtained from the PV panel and boost converter is used to supply the domestic and critical loads. For this purpose, the voltage is converted to AC using a fixed frequency single-phase inverter. A storage battery may be used across the DC bus of the inverter to store the surplus energy. This is helpful especially when the generation is low from the PV during low insolation periods or during night time.

3 Control Technique for the System

3.1 Utilization of Wide Speed Ranges

The system designed to utilize wide ranges of wind speeds typically from 0.5 to 2.0 p.u. range. For, this purpose, the variable frequency inverter is switched strategically to provide required excitation according to wind speed available. For this purpose, a variable reference voltage as a function of wind speed is used.

3.2 Improvement of Power Quality

Since the system use adjustable excitation from an inverter and the load is fed from a fixed frequency inverter, non-sinusoidal voltage can be generated at the load terminals when the inverter output voltage contains harmonics. For this reason, selective harmonic elimination based switching scheme is used. An optimization technique is used to adjust and minimize the unwanted harmonics from the output.

3.3 Control for the PV Panel

The PV panel is used, since the PV panel will remain connected with the generator throughout the day, a Maximum Power Point Tracking (MPPT) based control used. To maintain continuity to the load during low insolation/night time or when the generation from the IG is not sufficient, a suitable storage battery is used in the proposed system.

4 Methodology

Wind-PV hybrid energy is one of the most popular energy sources where both sources are renewable with complementary characteristics. The present system can harness wider ranges of wind speed. The system can support load even in the absence of wind using the PV panel battery support. No cost of fuel, routine maintenance is far less than conventional plants and no fuel transportation cost is necessary.

An Energy Management Scheme (EMS) for controlling the multiple energy sources in the proposed configuration can be derived for efficient delivery of power to the load during varying physical conditions of solar insolation, wind speed, battery backup with varying loads. The EMS algorithm can be represented in eight possible physical conditions as

Condition 1 (When Solar Insolation (SI), Wind Speed (WS) and Load (L) are high): Load is supplied majorly by the wind turbine with PV share; Inverter supplies the reactive power. Storage battery (SB) is in discharging mode if load is very high.

Condition 2 (When SI and WS high, L is low): Load is supplied majorly by the wind turbine with PV share with inverter supply the reactive power, SB is in charging mode.

Condition 3 (When WS and L are high, SI is low): Active power supplied by wind turbine, PV panels with SB supplies reactive power.

Condition 4 (When WS is high, SI and L are low): Active power supplied by wind turbine, only reactive power supplied by inverter panels, SB is discharging mode.

Condition 5 (When WS is low, SI and L are high): Only active power supplied by PV panels, reactive power is supplied by inverter, SB is in discharging mode if PV panel falls insufficient.

Condition 6 (When WS and L are low, SI is high): Only active power supplied by PV panels, reactive power is supplied by inverter, SB is in charging mode.

Condition 7 (When WS and SI are low, L is high): Active power supplied by turbine with support from power supplied by PV panels, SB is in discharging mode until battery voltage is permissible.

Condition 8 (When WS, SI and L are low): Active power demanded by load is supplied by turbine with PV panels, SB is in charging mode if load is met successfully and generation is surplus.

5 Result

In this work, simulation result shows the better performance than the previous results. In Fig. 5a hybrid of wind-PV configurations show an output voltage which has fluctuation with time but its better than old system. Figure 5b presents three-phase output voltage variation with time in hybrid system.

In Fig. 6 output of wind turbine is presented in which variation in rotor speed with time, variation in torque with time, and stator current of all phases with time define the performance of wind turbine.

(a)   **(b)**

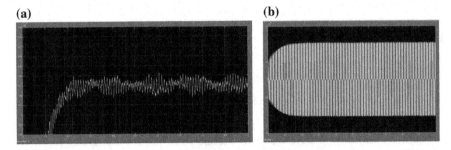

Fig. 5 **a** Hybrid output voltage 5, **b** three-phase output voltage V_{abc}

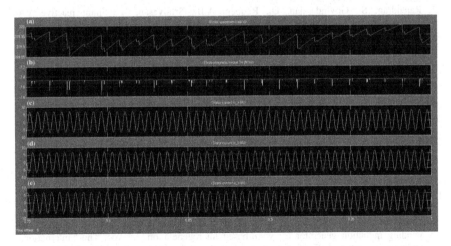

Fig. 6 **a** Rotor speed (rad/s), **b** Torque N/m, **c** Stator current phase a (A), **d** Stator current phase a (A), **e** Stator current phase b (A), **f** Stator Current Phase c (A)

6 Conclusion

In this research, a proper design basis and it is thus important to develop a system theory especially the overall coordinated operation and control are developed. Different simulation environments like MATLAB/Simulink, PSIM or PSCAD software are used to develop the primary system. Based on the simulation results, the desired system can be optimized for generation. The technical feasibility and performance also analyzed using HOMER for cost prediction which shows quantitative results. After more advanced in power electronics, equipments' power quality also improved.

References

1. Patel, M.R.: Wind and solar power systems: Design, analysis and operation, (CRC Press, 2006, 2nd edn.)
2. Bassett, E.D., Potter, F.M.: Capacitive excitation for induction generators, Trans. Am. Inst. Electr. Eng. **54**(5), pp. 540–545 (1935)
3. Grantham, C., Sutanto, D., Mismail, B.: Steady-state and transient analysis of self-excited induction generators, IEE Proceedings B, Electric Power Applications. **136**(2), pp. 61–68 (1989)
4. Wang, Y.J., Huang, Y.S.: Analysis of a stand-alone three-phase self-excited induction generator with unbalanced loads using a two-port network model. IET Electr. Power Appl. **3**(5), 445–452 (2009)
5. Rajendran, S., Govindarajan, U., Reuben, A.B., Srinivasan, A.: Shunt reactive VAr compensator for grid-connected induction generator in wind energy conversion systems. IET Power Electronics **6**(9), 1872–1883 (2013)
6. Ahmed, T., Nishida, K., Nakaoka, M.: Advanced control for PWM converter with variable-speed induction generator. IEEE Trans. Ind. Appl. **42**(4), 934–945 (2006)
7. Basic, M., Vukadinovic, D.: Online efficiency optimization of a vector controlled self-excited induction generator. IEEE Trans. Energy Convers. **31**(1), 373–380 (2016)
8. Elder, J.M., Boys, J.T., Woodward, J.L.: Self-excited induction machine as a small low-cost generator, IEE Proc. C, Generation Transmission and Distribution **131**(2), 33–41 (1984)
9. Murthy, S.S.: A novel self-induced self-regulated single phase induction generator. I. Basic system and theory. IEEE Trans. Energy Convers. **8**(3), 377–382 (1993)
10. Murthy, S.S., Singh, B., Sandeep, V.: A novel and comprehensive performance analysis of a single-phase two-winding self-excited induction generator. IEEE Trans. Energy Convers. **27**(1), 112–127 (2012)
11. Ojo, O., Omozusi, O., Jimoh, A.A.: The operation of an inverter-assisted single-phase induction generator. IEEE Trans. Energy Convers. **47**(3), 632–640 (2000)
12. Chatterjee, A., Chatterjee, D.: Photovoltaic assisted excitation control of 1-phase dual winding induction generator for wind-based microgeneration, in Proc. Third international conference Computer, Communication, Control and Information Technology, IEEE, India, pp. 1–5, Feb. 7–8 (2015)
13. Bansal, R.C.: Three-phase self-excited induction generators: an overview. IEEE Trans. Energy Convers. **20**(2), 292–299 (2005)
14. Benghanem, M., Bouzid, A.M., Bouhamida, M., Draou, A.: Voltage control of an isolated self-excited induction generator using static synchronous compensator. AIP J. Renew. Sustain. Energ. **5**, 043118 (2013)

15. Smith, O.J.M.: Three-phase induction generator for single-phase line. IEEE Trans. Energy Convers. **2**(3), 382–387 (1987)
16. Fukami, T., Kaburaki, Y., Kawahara, S., Miyamoto, T.: Performance analysis of a self-regulated self-excited single-phase induction generator using a three-phase machine. IEEE Trans. Energy Convers. **14**(3), 622–627 (1999)
17. Chan, T.F., Lai, L.L.: Single-phase operation of a three-phase induction generator with the Smith connection. IEEE Trans. Energy Convers. **17**(1), 47–54 (2002)
18. Chan, T.F., Lai, L.L.: Steady-state analysis and performance of a stand-alone three-phase induction generator with asymmetrically connected load impedances and excitation capacitances. IEEE Trans. Energy Convers. **16**(4), 327–333 (2001)
19. Mahato, S.N., Singh, S.P., Sharma, M.P.: Capacitors required for maximum power of a self-excited single-phase induction generator using a three-phase machine. IEEE Trans. Energy Convers. **23**(2), 372–381 (2008)
20. Chan, T.F., Lai, L.L.: Single-phase operation of a three phase induction generator with the Smith connection, IEEE Trans. Energy Convers. **17** (2014)
21. Gao, S., Bhuvaneswari, G., Murthy, S.S., Kalla, U.: Efficient voltage regulation scheme for three-phase self-excited induction generator feeding single-phase load in remote locations. IET Renew. Power Gener. **8**(2), 100–108 (2014)
22. Madawala, U.K., Geyer, T., Bradshaw, J.B., Vilathgamuwa, D.M.: Modeling and analysis of a novel variable-speed cage induction generator. IEEE Trans. Industr. Electron. **59**(2), 1020–1028 (2012)
23. Chatterjee, A., Roy, K., Chatterjee, D.: A Gravitational Search Algorithm (GSA) based Photo-Voltaic (PV) excitation control strategy for single phase operation of three phase wind-turbine coupled induction generator, Elsevier Energy. **74**, pp. 707–718 (2014)
24. Daniel, S.A., AmmasaiGounden, N.: A novel hybrid isolated generating system based on PV fed inverter-assisted wind-driven induction generators. IEEE Trans. Energy Convers. **19**(2), 416–422 (2004)
25. Nehrir, M.H., LaMeres, B.J., Venkataramanan, G., Gerez, V., Alvarado, L.A.: An approach to evaluate the general performance of stand-alone wind/photovoltaic generating systems. IEEE Trans. Energy Convers. **15**(4), 433–439 (2000)
26. Hirose, T., Matsuo, H.: Standalone hybrid wind-solar power generation system applying dump power control without dump load. IEEE Trans. Industr. Electron. **59**(2), 988–997 (2012)
27. Meena, R.S., Sharma, D., Birla, D.K.: PV-Wind Hybrid System with Fuel Cell & Electrolyzer, Int. J. Eng. Res. ISSN: 2347–5013 **4**(12), pp. 673-679 (2015)
28. Meena, R.S., Lodha, M.K.: Analysis of Integrated Hybrid VSC Based Multi Terminal DC System Using Control Strategy, Int. J. Adv. Res. Electr. Electron. Instrum. Eng. ISSN (Print): 2320 – 3765 ISSN (Online): 2278 – 8875 **4**(12), pp. 9823–9830 (2015)
29 Meena, R.S., Dubey, B., Gupta, N., Sambariya, D.K., Parira, A.S., Lodha, M.K.: Performance and feasibility analysis of integrated hybrid system for remote isolated communities. IEEE Publication, 2016 IEEE International Conference on Electrical Power and Energy Systems, ISBN 978-1-5090-2476- 6/16/$31.00 ©2016 IEEE (2016)
30 Meena, R.S., Nigam, D., Gupta, N., Lodha, M.K.: Control strategy of a stand-alone hybrid renewable energy system for rural home application. IEEE Publication, 2016 IEEE Seventh India International Conference on Power Electronics (IICPE 2016), ISBN 978-1-5090-4530-3/16/$31.00 ©2016 IEEE (2016)
31 Meena, R.S.: Sustainable development of remote isolated communities using integrated hybrid system, Akshaya Urja, Mag. Ministry. N. Renew. Energy. **10**(2), pp. 42–45 (2016)

A Model Framework for Enhancing Document Clustering Through Side Information

Kajal R. Motwani and B.D. Jitkar

Abstract Mining textual data is the need of this era. In most text mining applications, side information is associated with the text documents. This side information consists of document origin information, links present in document,
user-access behavior which can be retrieved from web logs, and different kinds of
non-textual attributes. Such side information may include explanatory data which
can be beneficial for text clustering. Conventional text clustering methods are
available that perform quite well but do not consider such attributes. As side
information consists of meaningful data, it can assist in enriching the quality of
clusters by incorporating such side information into clustering process.
Nevertheless, not all side information is important. Hence it should be incorporated
carefully. Herein, an effective clustering technique is proposed which adds only
important side information in the clustering process and determines if incorporating
side information improves the cluster quality. The clustering technique is extended
to classification problem.

Keywords Data mining · Document clustering · Side information

1 Introduction

Document clustering has been perceived as an important field in knowledge discovery. Document clustering has numerous applications in automatic document
organization, customer segmentation, social networks, web applications, digital
collections, information retrieval, etc. It is also referred to as text clustering. Text
clustering groups similar documents together. Most often the data available
nowadays is in the form of text. Textual data is complex data to process. It is

K.R. Motwani (✉) · B.D. Jitkar
D. Y. Patil College of Engineering & Technology, Kolhapur, Maharashtra, India
e-mail: kajal.motwani06@gmail.com

B.D. Jitkar
e-mail: bjitkar@rediffmail.com

© Springer Nature Singapore Pte Ltd. 2018
M. Pant et al. (eds.), *Soft Computing: Theories and Applications*,
Advances in Intelligent Systems and Computing 583,
https://doi.org/10.1007/978-981-10-5687-1_18

represented carefully. To handle textual data, many text specific algorithms are available.

In certain text mining applications, side information [1] is available along with text documents. This side information can be effectively used for purposes of text clustering. Such side information consists of huge amount of information that can assist in efficient clustering of documents. Some examples of such side information include

- Web documents contain information about the authors, when and where the documents were created, location, links having important information, user-tags, etc.
- Web logs contain browsing history of users. These logs show correlations and hence can be used to improve the quality of mining process.

Also, only some side information is important, hence only important side information must be included in the mining process or else it can result in degrading of cluster quality. Traditional document clustering algorithms are available. But these do not consider such side information while mining.

The objective here is to design a clustering technique that is based on use of both textual attributes and side information. The system proposed here identifies important side information by using Shannon Information Gain and Gaussian distribution and incorporates this side information into clusters by estimating probability to enhance the quality of document clustering. Also, the clustering technique is extended to design classification technique. In classification technique, labels are generated considering the side information and enriched clusters obtained from the clustering technique. Based on these labels, large documents can be clustered in lesser time.

The paper is organized as follows: Sect. 2 gives the related work on the topic. Section 3 elaborates the proposed methodology. Section 4 presents experimental results. Section 5 gives the conclusion and future directions.

2 Literature Survey

In the database and information retrieval communities, text clustering has been studied widely [2]. Document clustering algorithms are designed for search result clustering wherein effective information is presented to the user [3]. A famous document clustering algorithm that has been used for ages is the Scatter/Gather technique [4]. This technique combines K-means and agglomerative hierarchical clustering. Another technique is the collection clustering [5]. Google news is an example of this wherein the cluster must return to user the latest breaking news every time. Text clustering has been also used for language modeling [6]. These models have been employed for topic detection. Yet, these methods for text clustering are designed only for genuine clustering. These do not make use of side information while clustering.

Charu C. Aggarwal, Yuchen Zhao [1], and Philip S. Yu have proposed a technique that uses side information for clustering text documents. Herein noisy side information is eliminated using Gini index and probability is used for adding side information into mining process. They have also shown how to extend the clustering technique to classification problem.

TF-IDF is one of the simplest techniques that assign weights to the words based upon their relevance across documents. Results in [7] show that TF-IDF (Term Frequency-Inverse document frequency) categorizes relevant words which can enhance the query retrieval by well representation of data. K-means algorithm is a well known partitional clustering algorithm [2]. It partitions the data into clusters based upon the distances between data points and the mean data point. M. Steinbach, G. Karypis, V. Kumar [8] have proposed a variant of K-means, called bisecting K-means, that outperforms hierarchical clustering.

Yiming Yang, Jan O. Perdersen [9] have shown that IG (Information gain) and CHI (χ^2) perform better than mutual information and term strength. Information gain (IG) identifies unique features and eliminates unimportant features. Christopher M. Bishop [10] has given the details about the different ways in which the Gaussian distribution can be used for machine learning. A probabilistic matrix factorization model is proposed in [11]. The model uses Gaussian process priors for including side information. Venue and date of game are used as side information to determine scores of basketball game.

Charu C. Aggarwal, Yuchen Zhao, Philip S. Yu [1] have used the interest ratio to determine the importance of side information with respect to clusters. For finding out interest ratio, they make a naïve Bayes approximation.

Liqiang Geng, Howard J. Hamilton [12] have given survey on various interestingness measures which are useful for find out only meaningful patterns and these reduce time and space complexity.

3 Proposed Methodology

The proposed system consists of two techniques. The first is the clustering technique and the second is the classification technique. The side information is referred herein as side attributes. Figure 1 shows the system architecture diagram. The entire system works as follows: The input given to the system is a collection of documents. Preprocessing is done to remove stopwords and perform stemming on text content. Next, initial text-based clusters are formed using K-means algorithm. By computing Shannon information gain and Gaussian distribution, unwanted side attributes are discarded. Next using interest ratio, coherence between side attributes and text attributes is determined and accordingly clusters are enhanced. This technique is extended further to generate labels.

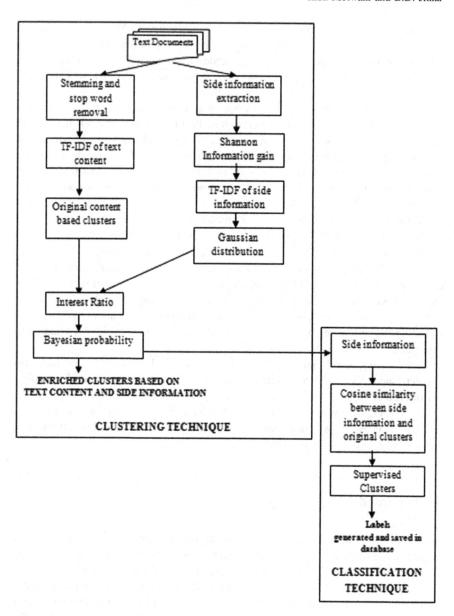

Fig. 1 System architecture diagram

3.1 Clustering Technique

As a first step, basic data mining tasks are performed, i.e., preprocessing of textual data is carried out. Stemming is performed to reduce words to their base forms by removing suffixes like 'ly',' 'ing', 'ed', etc. For example, the word 'playing' is reduced to 'play'. Porter Stemmer algorithm is used for performing stemming [13]. Stopwords like 'the, for, is, as, an', etc., are discarded. For this an English stopwords list is used which comprises of 500 stopwords. Special symbols if present are also eliminated.

After the text content is preprocessed, initial text-based clusters are formed based entirely on the text content (abstract of the text document). For this, first the TF-IDF of words in the document is computed. TF-IDF assigns weights to each word. TF-IDF of each word in the document is computed [7] as follows:

$$w_a = f_{w,a} * \log(|A|/f_{w,A}) \tag{1}$$

where

$f_{w,a}$ number of times word w appears in text document a,
$|A|$ size of text documents in collection,
$f_{w,A}$ number of text documents in which word w appears in A.

Based on these TF-IDF values, documents get clustered. K-means algorithm [2] is used for clustering documents. K-means algorithm partitions the documents into clusters where each document belongs to cluster with nearest mean.

- Identification of important side information

Side attributes associated with the documents are extracted and processed. Many side attributes are not important for clustering as they can degrade its quality. So such unimportant side attributes are discarded. For eliminating such side attributes, Shannon Information gain [9] is used. It is a feature selection measure and it indicates impurity and is computed [15] as follows:

$$\mathbf{Info}(A) = -\sum_{i=1}^{m} p_i \log_2(p_i) \tag{2}$$

where

p_i is the is the non-zero probability that an arbitrary tuple in A belongs to class X_i and is estimated by, $|X_{i,A}|/|A|$.

A predetermined threshold is set to 0.4, based on which side attributes above 0.4 are considered as useful and important side attributes and hence extracted. Side attributes below this threshold are noisy side attributes and so are eliminated.

The TF-IDF of side attributes obtained from Shannon information gain is computed using (1). Next in order to ensure that only informative side attributes get added, Gaussian distribution is used. The Gaussian distribution fetches side attributes that are highly distributed. Gaussian distribution is known as normal distribution. Gaussian function is computed [10] as follows:

$$f(x|\mu, \sigma^2) = \frac{1}{\sigma\sqrt{2\pi}} e^{-\frac{(x-\mu)^2}{2\sigma^2}} \tag{3}$$

where

x TF-IDF of side attributes,
μ mean of distribution,
σ standard deviation of distribution,
σ^2 variance of distribution.

Based on the value obtained from Gaussian function, a range is computed as follows:

$$\begin{aligned} \text{minrange} &= m - \text{p} \\ \text{maxrange} &= m + p \end{aligned} \tag{4}$$

where

m mean,
p Gaussian distribution.

The side attributes having TF-IDF values in minrange to maxrange are considered. At this stage only important side attributes are extracted.

- Enhanced clusters formation

For formation of enhanced clusters based on both text attributes and side attributes, interest ratio is computed. The interest ratio defines the importance of side attributes with respect to the clusters. It is defined as the ratio of the probability of a document belonging to cluster C_j, when the rth side attribute is present, to the ratio of the same probability unconditionally [1]. It is computed [1] as follows:

$$I(r,j) = \frac{P^a(x_{ir} = 1|T_i \in C_i)}{P^a(x_{ir} = 1)} \tag{5}$$

where

value of $P^a(x_{ir} = 1)$ fraction of the documents in which the side attribute is present.

value of $P^a(x_{ir} = 1|T_i \in C_i)$ fraction of the documents in cluster C_j in which side attribute is present.

In order to find out interest ratio, the difference between side attribute TF-IDF and mean TF-IDF of each document present in the initial text based clusters is determined. Then the average of these differences is computed. Next, we check whether individual difference obtained for each document is less than the average difference. If the difference is less, then such documents are considered as they show interest of adding side attributes. Then, the maximum number clusters that show interest of adding side attributes are determined. Based upon this, a condition is set which checks in how many clusters out of the total clusters that side attributes are correlated. If the probability comes out to be more than a 0.5 then the side attributes and clusters are considered. Once side attributes having correlation with clusters are determined, they are used for forming clusters. For this, the document with smallest difference is taken and TF-IDF of side attribute is added to this difference. This is done for all the documents having smallest difference in each initial text based cluster. This results in generation of new TF-IDF values for documents. Based on the new TF-IDF values, K-means algorithm is used to form new clusters. Hence, we obtain enhanced clusters which are based on both the side attributes and text attributes.

3.2 Classification Technique

When there are large number of documents to be clustered, classification labels can be of great use. When documents are given as input initially, after the enhanced clusters are formed, the labels are generated. These labels are stored in the database. Next time when we input documents, rather than going through the entire procedure of clustering technique, the system compares the TF-IDF values of side attributes with the labels. If there is match then the documents get clustered there itself and this saves a lot of time. For example, when 40 documents are given as input to the system, it will initially go through entire clustering and classification system and form enriched clusters and further form labels and store them in database. Next time when 60 documents are given as input, the system will directly cluster the documents if match is found based on the labels. If no match is found it will again go through entire system and generate new labels. This can be seen in Fig. 2.

- Label generation

For generating labels, centroid and Atkinson index value is computed by taking into account the side attributes obtained from Interest ratio. Atkinson index [12] is a diversity measure of inequality. It is used to find out the deviation of TF-IDF value of each side attribute from mean TF-IDF value of side attributes. Next similarity is found between side attributes and documents in enhanced clusters. For this, cosine similarity is used. Here, for each side attribute a revised TF-IDF value is generated as follows:

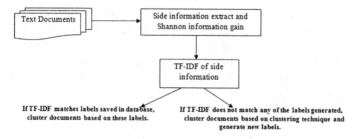

Fig. 2 Document clustering based on labels

$$\text{revised\_TF-IDF} = \text{TF-IDF (side information)} + \text{centroid} + \text{Atkinson Index} \tag{6}$$

Cosine similarity is computed between the revised_TF-IDF of side attribute and mean TF-IDF value of each document in enhanced clusters. Cosine similarity [14] is computed as follows:

$$\mathbf{SIM}(X_i, Y_j) = \frac{\sum_{k=1}^{t}\left(TERM_{ik}.TERM_{jk}\right)}{\sqrt{\sum_{k=1}^{t}\left(TERM_{ik}\right)^2.\sum_{k=1}^{t}\left(TERM_{jk}\right)^2}} \tag{7}$$

where

$\sum_{k=1}^{t}\left(TERM_{ik}.TERM_{jk}\right)$ component-by-component vector product.

$\sqrt{\sum_{k=1}^{t}\left(TERM_{ik}\right)^2}$ length of the term vector for X_i.

$\sqrt{\sum_{k=1}^{t}\left(TERM_{jk}\right)^2}$ length of the term vector for Y_j.

Once cosine similarity is computed, supervised clusters are formed. These are formed by adding the cosine similarity to mean TF-IDF value of documents in only those clusters to which the side attribute is similar. Based on these supervised clusters, labels are generated for each cluster. Labels are in the form of a range. This range is the minimum and maximum value that is associated with each document in supervised clusters that is obtained by adding cosine similarity. These labels are saved in database. Whenever label checking if performed, the TF-IDF of side attributes is checked as to whether it lies in this range. If side attributes lies in this range, then the documents lying in this TF-IDF range are picked up and clustered accordingly. If no match is found, new labels are formed.

4 Results

The proposed system is tested on two datasets. The first dataset is Cora dataset. Cora dataset is obtained from "http://www.cs.umass.edu/~mccallum/code-data. html". The Cora data set consists of 19,396 scientific publications in the computer science field. Side attributes extracted from the Cora data set are title, url, keyword, author, address, and affiliation. The other dataset used is the Reuters-21578 dataset and it is obtained from "http://www.daviddlewis.com/resources/testcollections/ reuters21578/". Reuters dataset consists of 21578 Reuters news documents since 1987. Side attributes extracted from Reuters-21578 dataset are title, places, dateline and author.

- For testing the performance of clustering system; Precision, Recall, F-measure are computed Figs. 3 and 4, Tables 1 and 2.
- Accuracy measure is used to determine the performance of classification technique Fig. 5, Table 3.
- The running time needed in order to cluster documents using labels and without using labels is computed. Average percentage improvement is also computed. Results below reflect that labels can be of great use to cluster large number of documents Figs. 6 and 7, Tables 4 and 5.

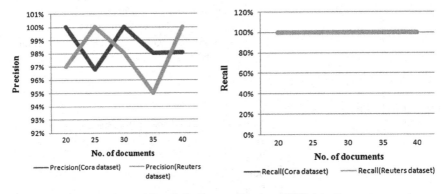

Fig. 3 Graph of Precision and Recall for Cora and Reuters-21578 datasets

Table 1 Precision, Recall, F-measure for Cora dataset	No. of documents	Precision (%)	Recall (%)	F-measure (%)
	20	100	100	100
	25	96.77	100	98.36
	30	100	100	100
	35	98.03	100	99
	40	98.07	100	99.02

Table 2 Precision, Recall, F-measure for Reuters-21578 dataset

No. of documents	Precision (%)	Recall (%)	F-measure (%)
20	97	100	98
25	100	100	100
30	98	100	97.4
35	95	100	96
40	100	100	100

Fig. 4 Graph of F-Measure for Cora and Reuters-21578 datasets

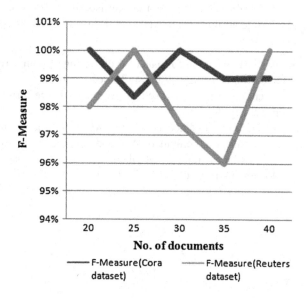

Fig. 5 Graph displaying accuracy

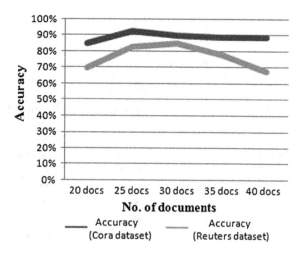

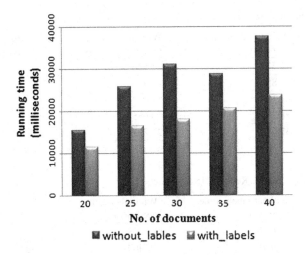

Fig. 6 Graph displaying running time with labels and without labels for Cora dataset

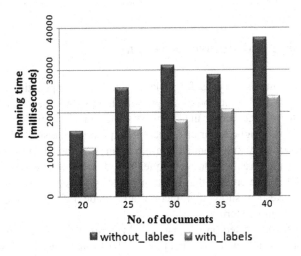

Fig. 7 Graph displaying running time with labels and without labels for Reuters-21578 dataset

Table 3 Accuracy

No. of documents	Accuracy (Cora dataset) (%)	Accuracy (Reuters-21578 dataset) (%)
20	85	70
25	92.78	82.67
30	90	85
35	89	78
40	88.70	67.78

Table 4 Running time for Cora dataset

No. of documents	Without labels, time needed (ms)	With labels, time needed (ms)	Percentage improvement (%)
20	15,571	11,520	26
25	25,850	16,472	30
30	31,115	18,022	42
35	28,799	20,665	28
40	37,642	23,861	36

Table 5 Running time for Reuters-21578 dataset

No. of documents	Without labels, time needed (ms)	With labels, time needed (ms)	Percent improvement (%)
20	16,634	13,678	17
25	27,980	19,400	30
30	32,897	26,089	20
35	33,980	25,965	23
40	39,034	31,890	18

5 Conclusion and Future Directions

Document clustering is very useful in large number of applications. Side information, i.e., other kinds of attributes are present along with most of text documents. Herein, a clustering technique that considers the side information and the text content together is designed. For effectively incorporating the side information into clustering process, only important side information is extracted and used. Also it is extended to design classification technique. Based upon the clustering technique, classification labels are created that assist in clustering large number of documents in lesser time. Results show that incorporating side information enhances the quality of document clustering.

Here, the system is tested on two datasets. In future, the system can be tested on different datasets. Another future direction can be to make use of clustering technique for optimizing search. Web applications can be designed in such a way that they fetch only relevant documents for user when user enters side information as a query. Also, the clustering technique can be designed efficiently to make use of browsing history of user, i.e., web logs.

References

1. Aggarwal, C.C., Zhao, Y., Philip, S.Y.: On the Use of Side Information for Mining Text Data. IEEE Trans. Knowl. Data Eng. (2014). doi:10.1109/TKDE.2012.148
2. Aggarwal, C.C., Zhai, C.-X.: Mining Text Data. Springer, New York, USA (2012)
3. Zamir, O., Etzioni, O.: Grouper: a dynamic clustering interface to Web search results. Elsevier Journal of Computer Networks(1999)
4. Cutting, D., Karger, D., Pedersen, J., Tukey, J.: Scatter/Gather: A cluster-based approach to browsing large document collections. Proc. ACM SIGIR Conf. New York, USA, pp. 318–329 (1992)
5. Radev, D.R., Hovy, E., McKeown, K.: Introduction to Special Issue on Summarization. MIT Press Journals, pp. 399–408 (2002)
6. Liu, X., Croft, W.B.: Cluster-based retrieval using language models. Proc. ACM SIGIR conference on Research and development in information retrieval, NY, USA, pp. 186–193 (2004)
7. Ramos, J.: Using TF-IDF to determine word Relevance in document Queries. Department of Computer Science, Rutgers University (2003)
8. Steinbach, M., Karypis, G., Kumar, V.: A comparison of document clustering techniques. Proc. Text Mining Workshop KDD, pp. 109–110 (2000)
9. Yang, Y., Perdersen, J.O.: A comparative study of feature selection In text categorization. Proc. ICML Conf, CA, USA, pp. 412–420 (1997)
10. Bishop, C.M.: Pattern Recognition and Machine Learning. Springer Verlag (2006)
11. Adams, R.P., Dahl, G.E., Murray, I.: Incorporating Side Information in Probabilistic Matrix Factorization with Gaussian Processes. arXiv preprint (2010)
12. Geng, L., Hamilton, H.J.: Interestingness Measures for Data Mining: A Survey. ACM Computing Surveys (2006)
13. Porter, M.F.: An algorithm for suffix stripping. Emerald Insight, pp. 211–218 (1996)
14. Salton, G.: An Introduction to Modern Information Retrieval. U.K., McGraw Hill, London (1983)
15. Han, J., Kamber, M., Pei, J.: Data Mining: Concepts and Techniques, Third Edition. Elsevier (2012)

Patient-Specific Modeling for Pattern Identification in Clinical Events Using Data Mining Models and Evolutionary Computing

Sourabh Yadav and Anil Saroliya

Abstract Nowadays due to the technological advancement in collecting, processing and producing the information available, it is quite challenging to obtain the relevant information out from various sources of data. According to earlier research in homeopathy, a pattern system of treatment is being followed, which ultimately helps in finding the cure for the disease. One of the main goals of this system of treatment is to individualize the model in a manner that patient-specific prediction can be fixated with the help of pattern identification. By adopting the context of homeopathy system, the aim of this paper focuses on identifying appropriate parameters from a large quantity of raw clinical data, which are responsible for developing celiac disease in the human body. The result is based on the extensive quantitative research of a clinical database, shows that one parameter is correlating with another and such kind of pattern identification will be useful in the curing of celiac disease.

Keywords Machine learning (ML) · Homeopathy · Parameters
Celiac disease (CD) · Patient-specific model

1 Introduction

As the role of information technology in a clinical environment has taken place in such a manner that, the whole medical system is apparently relying on this. So moving towards the goal of achieving an identification and prediction methodology, reasons those are responsible for developing a disease in human can be acquired, in the context of being supportive to cure an associated disease. Over the previous decades, main advances in the area of molecular biology, united with developments in genomic technologies, have headed towards a volatile growth in the natural

S. Yadav (✉) · A. Saroliya
Amity School of Engineering and Technology, Amity University, Jaipur, Rajasthan, India
e-mail: sourabhyadav9200@gmail.com

A. Saroliya
e-mail: asaroliya@jpr.amity.edu

© Springer Nature Singapore Pte Ltd. 2018
M. Pant et al. (eds.), *Soft Computing: Theories and Applications*,
Advances in Intelligent Systems and Computing 583,
https://doi.org/10.1007/978-981-10-5687-1_19

information generated closely by the community of science [1]. Initially, this work had gone through an intense data mining process in the data warehouse of patient records analyzed in a homeopathy clinic. It is found that the role of Machine Learning becomes popular in such cases of medical domain, as digitization trends took an important role. The models of system biology contribute in a healthcare perspective that individualizes models with a goal of computing predictions, specific to patients for the advancement of species of interest associated with time [2]. Making a system trained so that it is capable of producing results based on a prediction by considering the similar specific parametric values, so that a relation can be mapped. By training a system to make it able for producing prediction-based result on the strategy that if the specific parameters are having values in common to other patients, they can be related to each other. The holistic medicine of traditional and current social requirements of getting the patient treated in a hasty and appreciable way, homeopathy supported by computer deals with multiple ways that can be adopted in a fashion which are suggested as finest implicit in the context of a process delivering de-emphasizing gender (de-gendering) [3]. There can be a model which gives the idea about the occurrence of the disease in humans; also it can identify the common factors required for having that disease. In the present scenario, the hospitals and clinics are getting the information recorded in such an increasing amount that gradually overwhelming decision maker of human, this issue causes an important reason behind Machine Learning (ML), hence ML has gained so much attention in the domain of medical science, with the help of ML algorithms, events those are likely to come in future can be predicted by using available information for individual patients [4]. Due to the multipart interaction of many disease- and patient-level issues, positive management of the disease is hardly a "solitary size fits all" state and is often further effective when tailored to specific patient requirements, local philosophy, and open resources [5].

The subsequent sections of this paper are: Sect. 1 introduces the problem domain, Sect. 2 describes the logical flow of work carried out in this research. In Sect. 3, the disease which is selected for this research work is explained, Sect. 4 states the methodology used. In Sect. 5, the simulation and results are included. Finally, Sect. 6 includes the conclusion and future scope of this research.

2 Related Work

It is also worth observing that data sets in the medical field are becoming primarily longer (i.e., we have more samples collected through time) and wider (i.e., we store more variables) [4]. However, such evolution has not relieved machine learning specialists from getting to identify their datasets and problematic domains [6]. In the context of predicting the clinical events, a process of spending hundreds of hours in a clinic observing the patients and their characteristics by interacting with them was initiated. There is a designated time slot of up to 2 h per patient; so that a clear image of patient's mental generals can be drawn by a thorough questioning.

By the means of mental generals, it is found that the disease is associated with the situations in which the patient lived in or is living, while physical generals are the physiological parameters like thirst, appetite, etc. The surroundings of the patient are the main cause for the occurrence of disease, as the system of homeopathy describes. By going through a deep analysis of identifying the reasons behind having a disease, it is found that whenever the hormonal level is disturbed, the disease is initiated. It is well known that with every mental or emotional annoyance a hormone is allied, so the level of annoyance is directly affecting the amount of hormone to be secreted. So the introduced prediction is mainly based on parametric values of patients as their mental and physical generals appear, after examining hundreds of records of the patients having celiac disease (allergy from wheat, rye, and barley), it is found that the prediction can be done by correlating variables responsible for the development of CD. The search process for suitable parameter settings of algorithms itself is a problem of optimizing, optimization algorithms, known as meta-algorithms, can be well-defined to complete this task [7]. At first, the records found in a clinical database were investigated thoroughly, then a management of the records and finding the parameters (i.e., thirst, appetite, will power, etc.) of a patient's mental and physical generals, took place. It is noticed that most of the patients are having a timid mindset. Then the parameters are managed in an excel sheet by minimizing the characteristics and assigning a numeric value which is ranged from 1 to 10. All of the data in this excel sheet is created by surveying a sample size of 200 patients having CD. The age is further divided (from a range of 1–100) in 10 intervals (i.e., Age Interval 1 = 1 year to 10 years, Age Interval 2 = 11 years to 20 years etc.) and this work clarifies that in earlier intervals the occurrence of disease completely relies on sensitivity level of patient's mother and if CD occurs in intervals later on then it relies on the patient's sensitivity level. The base of homeopathy is a simple principle of nature; it is entirely constructed on the indications of patients, and homeopathy rests on a basic natural principle [8]. As the sensitivity taken into two parts, one is of mother's and another is of patient's, shows that if the patient lies in the first three age intervals then the sensitivity of mother is higher than the sensitivity level of the patient. It clarifies that if the mother is having some issues or complexities which are the reasons of emotionally annoying her. In surroundings of the mother, during the first trimester of pregnancy are complex or disturbed, then the child is prone to have the disease. It is also found that if the disease has occurred in later intervals then the sensitivity value of the mother is lesser and here the patient is having the higher level of sensitivity as compared to the mother. By this study, it is being noticed that the parameter sensitivity is correlated with the age of a patient in the context of CD occurrence. By correlating the parameters, i.e., sensitivity (of both patient and mother) with age gives the result that the disease is associated with sensitivity level. So age and sensitivity are the important parameters to be mapped for identification of CD occurrence. Hence, by having a clear image of the sensitivity factor the related remedies can be given to the patient to support a successful treatment. As CD is incurable but by opting this approach of treatment when the wheat was introduced to the patients there was a development of tolerance found in their body, so that they were able to have normal food. Conventionally the job of training a subject is done through an introduction of training and tuning phase during which certain feedback is offered to the subject [9]. In direction of programmed diagnostics systems to execute in stages of sensitivity and specificity related to that of human specialists, it is crucial that a healthy framework on problem-specific strategy parameters is framed [10].

3 Celiac Disease End Points

Celiac disease (CD) is an autoimmune disease, in which the patient suffers from a situation called gluten sensitivity. The patient of CD faces physiological symptoms when the gluten (a type of protein found in wheat, rye, and barley) introduced to them orally. The moment at which body is introduced with gluten, the antibody (IgA from all five antibodies) of immune system marks the substances of gluten at the wall of small intestine, means the way body reacts to the gluten. The next phase is a war-like situation at the small intestine where the antibody starts to destruct the marked protein resulting as a damaged inner wall of the small intestine. This wall is responsible for food consumption so with a damaged wall the patient faces Inflammation such as diarrhea (Lack of H_2O), steatorrhea (Lack of essential fats) and anemia (lack of Fe and Vitamin B12). As the disease is concerned the patient faces malnutrition which directly affects the appetite, activity and energy level, body weight and growth of the patient. There are two main endpoints in a CD patient associated with gluten tolerance, i.e., no tolerance and nominal tolerance. No tolerance means that the body is rejecting the gluten completely and the patient gets severe physical symptoms like stomach ache, motions, vomiting, rashes, etc., while in another case the patient can tolerate a nominal amount of gluten.

The goal of this work focuses on predicting which of these end points can occur and how to avoid them by providing the best treatment, in which it is very complex as well as difficult to make a decision in the practice of treating CD patients.

4 Proposed Methodology

The clinical database of patients having CD is used to collect the data for this study. First, the case studies of the patients were analyzed and understood thoroughly. Then this data is organized into the excel files where variables were assigned a numeric value (absolute) in a range from 1 to 10. By this experiment, it is found that if the disease occurs in an early interval of age then the sensitivity of the mother is higher and if the disease occurs in a later interval then the sensitivity of patient is higher. By creating a data model the flow of work can be understood as shown in Fig. 1.

Moreover, the sample table of only nine patients for indication purpose is presented below, as the actual sample size is 200, in this table, only the important

Fig. 1 Flow of work

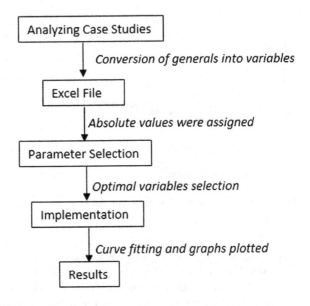

Table 1 Patient-specific records

Patients	Age	Sensitivity of patient	Sensitivity of mother
P1	1	5	9
P2	3	8	6
P3	4	9	5
P4	2	6	8
P5	1	4	10
P6	3	8	6
P7	5	10	4
P8	4	9	5
P9	2	6	8

parameters are shown; as they were selected out of several parameters from the excel file. The table shown here is clarifying the dependency of sensitivity on age as the earlier intervals are having higher values of mother's sensitivity level and later on intervals are having higher values of patient's sensitivity level (Table 1).

5   Results

5.1   *Quantitative Analysis Between Subjects*

The experiment totally acquired 4600 cells of featured parametric values. At first, the parameters were assigned absolute values ranged from 1 to 10. Then by a thorough research, some of them were selected as they fit best with each other by

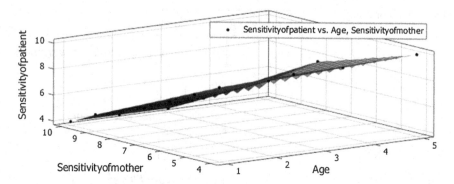

Fig. 2 Three-dimensional view of the correlation between sensitivity (i.e., of patient and mother) and age

correlating them in terms of occurrence of the disease. From the strategy applied on these selected parameters, it is clear that, while the suffering from a situational or emotional (good or bad) cause the subjects are tended to develop the disease as their hormonal level increases or decreases. But in the aspect of the quantitative analysis, it is mandatory to keep in knowledge that the parameters, which directly causes the emotional discomfort resulting in a hormonal imbalance, are to be selected carefully. The MATLAB tool is used for the computation of this study showing the results in Figs. 3 and 4 respectively.

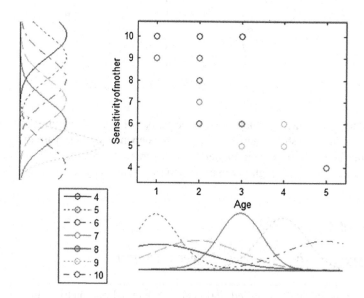

Fig. 3 Scatter Histogram showing the sensitivity level of mother is higher in earlier age intervals

Linear interpolant:

$f(x, y)$ = piecewise linear surface computed from p

where x is normalized by mean 2.33 and std 1.249

and where y is normalized by mean 7.365 and std 1.881

Coefficients:

p = coefficient structure

As Figs. 2 and 3 give an insight of the correlation between age and sensitivity, it can be stated that the two parameters are relating with each other in such a way that a pattern identification is achieved.

5.2 Pattern Classification Results

Pattern found here was performed throughout each patient, and the minimum sensitivity value is 4 and the maximum value is 10 for all of the experimented patients (a total of 200). Further identification of patterns and success rate of this model depends on the relevant parameters. Results achieved, clarifies the interdependency of the efficient parameters, providing the accuracy of pattern identification in the occurrence of the disease in a human. The tool for plotting a graph shown in Fig. 4 is SCILAB. In this figure, the dependency of sensitivity level on age can be clearly identified.

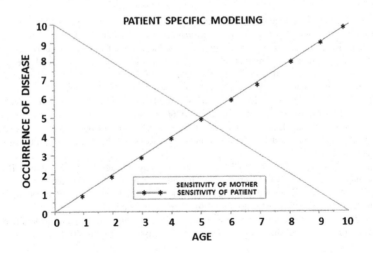

Fig. 4 Age versus Sensitivity

6 Conclusion and Future Scope

It is experimented and identified the dependencies in the data, behind occurrence of a disease based on the symptoms appeared in patients, by which it is easier to find the irrelevancy or disturbance resulting in psychosomatic disorder, so the remedy which fits best can be given in order to treat the patient and promising a healthy life. As this work provides insight, regarding the development of CD in the human body that is supportive of giving the patient best treatment, as this disease is incurable, there are some patients having the tolerance of Gluten by using this approach. This work is focused only on Celiac Disease, such type of work can be useful for other incurable diseases also.

References

1. Pal, S.K., Bandyopadhyay, S., Ray, S.S.: Evolutionary computation in bioinformatics: A review. IEEE Trans. Syst. Man Cybern. Part C (Applications and Reviews) **36**(5), 601–615 (2006)
2. Tronci, E. et al.: Patient-specific models from inter-patient biological models and clinical records. In: Proceedings of the 14th Conference on Formal Methods in Computer-Aided Design. FMCAD Inc, 207–214 (2014)
3. Degele, N.: Gender, computers, and holistic knowledge: the case of homeopathy. In: Technology and Society, 1999. Women and Technology: Historical, Societal, and Professional Perspectives. Proceedings. 1999 International Symposium on. IEEE, 153–161 (1999)
4. Esteban, C. et al.: Predicting clinical events by combining static and dynamic information using recurrent neural networks. *arXiv preprint* arXiv:1602.02685 (2016)
5. Thomas, K.L., Shah, B.R., Elliot-Bynum, S., Thomas, K.D., Damon, K., LaPointe, N.M.A., Calhoun, S., Thomas, L., Breathett, K., Mathews, R., Anderson, M., Califf, R.M., Peterson, E.D.: Check it, Change it a community-based, multifaceted intervention to improve blood pressure control. Circ. Cardiovasc. Qual. Outcomes **7**(6), 828–834 (2014)
6. Dieleman, S., De Fauw, J., Kavukcuoglu, K.: Exploiting cyclic symmetry in convolutional neural networks. *arXiv preprint* arXiv:1602.02660 (2016)
7. Dwivedi, S.K., Sukhadeve, P.P.: Comparative structure of Homoeopathy language with other medical languages in Machine Translation System, ICACCI 2013. Adv. Comput. Commun. Inform. 775–778 (2013)
8. Almeida, R.M.V.R.: Reports on Homeopathic therapy efficacy: a survey of two Brazilian newspapers. In: IEEE Pan American Health Care Exchange, 297–300 (2011)
9. Atyabi, A. et al.: Reducing training requirements through evolutionary based dimension reduction and subject transfer. *arXiv preprint* arXiv:1602.02237 (2016)
10. Georgiou, H.: Algorithms for image analysis and combination of pattern classifiers with application to medical diagnosis. *arXiv preprint* arXiv:0910.3348 (2009)

Centric-Oriented Novel Image Watermarking Technique Based on DWT and SVD

Somanshi Ojha, Abhay Sharma and Rekha Chaturvedi

Abstract Due to the rapid growth in the field of digital media and communication there is a great demand for the copyright protection of multimedia data over internet. In order to address this issue, a robust and efficient watermarking scheme must be used, so as to obtain the ownership of real creator. The proposed watermarking algorithm uses a novel image watermarking technique which is based on Discrete Wavelength Transform (DWT) and Singular Value Decomposition (SVD). The algorithm consists of Region of interest (ROI), which is the extracted central part of particular dimensions of the cover image. DWT has been applied to the ROI, extracted from the cover image and then SVD is taken for embedding watermark information.The proposed method withstands the various geometric and filtering attacks such as—Cropping, Rotation, Poisson noise, Gaussian noise, etc. Experimental results show the proposed method leads over the existing ones.

Keywords Digital image watermarking · DWT · SVD · ROI · PSNR

1 Introduction

Today internet has become a major part of human life. It gave us the opportunity to connect, share, and communicate with different people across the globe. Communication and information sharing over it is done in the most effective and efficient manner. It includes exchange of bulk data which may be text, image, video

S. Ojha (✉)
Department of Computer Science and Engineering, Amity University,
Jaipur, Rajasthan, India
e-mail: somanshi.ojha@gmail.com

A. Sharma · R. Chaturvedi
Amity University, Jaipur, Rajasthan, India
e-mail: asharma2@jpr.amity.edu

R. Chaturvedi
e-mail: rchaturvedi@jpr.amity.edu

© Springer Nature Singapore Pte Ltd. 2018
M. Pant et al. (eds.), *Soft Computing: Theories and Applications*,
Advances in Intelligent Systems and Computing 583,
https://doi.org/10.1007/978-981-10-5687-1_20

or audio. These data may be confidential and may represent intellectual properties of an individual [1]. Hence, to protect the intellectual property of creators, distributors or simple owners of such confidential data watermarking came as a way. Watermarking is a process of embedding data and hiding information [2]. Basically it is the technology of concealing or hiding additional data into the multimedia data such as video, image, audio, and text.Watermark data, which holds copyright information is embedded in a cover image. This watermarked image or data prevents illegal copying, distribution and modification of underlying data over internet [3].

This paper focuses on a novel image watermarking technique which protects copyrighted multimedia data from various watermarking attacks. Watermarking has been classified under various types [4]. According to types of documents, it has been classified as image, video, audio or text watermarking. Based on human perception it can be visible, which is translucent overlaid and is visible to users and can be invisible, in which modified pixels are used as evidence of ownership and can also be dual, which is a combination of visible and invisible watermarking. Based on working domain, digital image watermarking is classified into spatial domain and transform domain. In Spatial domain watermarking, watermark information is embedded directly on pixels, whereas in Frequency domain image is firstly converted into transform domain then watermark information is embedded. Some of the transform based watermarking techniques used are discrete wavelet transform (DWT), discrete cosine transform (DCT) [5, 6]. Digital image watermarking can be blind, non-blind and semi-blind based on techniques used for extracting watermark. In Blind watermarking only secret key is needed, in non-blind watermarking both host image and secret key is needed and in semi-blind secret key and some part of watermark is needed.

2 Method Used

The proposed watermarking scheme includes concepts like YCbCr color space, DWT and SVD model. Detailed description of each concept is in the following section.

2.1 YCbCr Color Space

YCbCr is a well know color model which was defined in the ITU-R BT.601 standards of International Telecommunication Union [7]. It is widely used for video and digital photography systems. It represents three colors similarly as RGB color space. The Y component stores the luminance information, the chrominance information is stored by two different chroma components Cb and Cr, where Cb is blue difference chroma component and Cr is red difference chroma component. YCbCr is derived from RGB using following Eq. (1).

$$[Y\ Cb\ Cr] = [R\ G\ B] \begin{bmatrix} 0.299 & -0.168935 & 0.499813 \\ 0.587 & -0.331665 & -0.418531 \\ 0.114 & 0.50059 & -0.081282 \end{bmatrix} \qquad (1)$$

Similarly, RGB can be recovered using Eq. (2)

$$[R\ G\ B] = [Y\ Cb\ Cr] \begin{bmatrix} 1.0 & 1.0 & 1.0 \\ 0.0 & -0.344 & 1.77 \\ 1.403 & -0.714 & 0.0 \end{bmatrix} \qquad (2)$$

2.2 Singular Value Decomposition

Singular values are defined on rectangular matrices and is approached through the singular value decomposition [8, 9]. It works for both singular and numerically near-singular matrices. For a rectangular real or complex matrix, it came out to be the powerful liner algebra factorization technique. The SVD theorem determines:

Suppose A is $m \times n$ matrix then factorization of A is given using following Eq. (3)

$$A = U \times S \times V^T \qquad (3)$$

where, U and V are orthogonal matrices of size $m \times m$ and $n \times n$ respectively, $UU^T = I$ and $VV^T = I$. The columns of U are the left singular vectors, columns of V^T are termed as right singular vectors and S is a diagonal matrix of $m \times n$ has singular values. SVD have wide variety of application in the field of pattern identification, processing of noisy signals.

2.3 Discrete Wavelet Transform

DWT breaks off the signal into mutually orthogonal set of wavelets [7, 10]. It is very useful in analysis, de-noising and compression of signals and images. It is similar to HVS (human visual system) therefore it has been widely used for digital watermarking of images and is more accurate model aspect of HVS than any other transform domain. DWT when applied to the image divides it into four components namely LL, HL, LH, and HH. LL is approximation image component, HL is horizontal detailed component, LH is vertical detailed component and HH is diagonal detailed component. These four components are non-overlapping and multi-resolution sub-bands. It provides multi-scale decomposition of function (Fig. 1).

Fig. 1 DWT decomposition

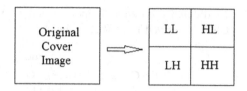

3 Proposed Methodology

In this paper we have proposed a centric based novel watermarking technique using DWT and SVD. For this we have used YCbCr color scheme in which a colored RGB image is converted into YCbCr and Y component is extracted for watermarking. Using Y-part of cover image, Region of interest (ROI) is extracted which is central part of image in our scheme. Then this ROI is transformed using DWT method (Fig. 2).

3.1 Watermark Embedding

1. Load the color RGB cover image of size 512 × 512.
2. Convert the RGB image into YCbCr color space and extract Y-part.
3. Extract the center portion X from Y-part of the cover image of size 256 × 256.
4. Perform single level DWT on the extracted image X.
5. Apply SVD to HL sub-band of cover image and Its USV components are U_img, S_img and V_img.
6. Read colored watermark image and convert it to grayscale image and resize it to 256 × 256.
7. Perform single level DWT on the watermark image w.

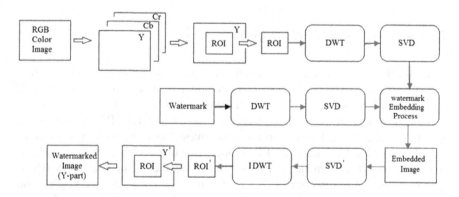

Fig. 2 Watermark embedding process

8. Apply SVD to watermark image and computed *USV* components are named as: U_Wimg, S_Wimg, V_Wimg.
9. Take strength factor $\alpha = 0.10$
10. Proceed watermarking by modifying the singular values additively

$$S\_wimg = S\_img + \alpha * S\_Wimg.$$

11. Obtain the value of h_HL1 as

$$h\_HL1 = U\_img * S\_wimg * V\_img'.$$

12. Apply inverse DWT

$$\text{Watermarked image} = \text{IDWT2}(h\_LL, h\_LH, h\_HL1, h\_HH, \text{'haar'}).$$

13. Paste this watermarked image back to the cover image in the intervals same as from where it has been extracted.
14. Concatenate Y, Cb, Cr components and get RGB watermarked image.
15. End.

3.2 Watermark Extraction

1. Read watermarked image.
2. Apply DWT on watermarked image.
3. Apply SVD to watermarked image. Its *USV* components are U_wmimg, S_wmimg and V_wmimg.
4. Read original watermark image.
5. Apply DWT on watermark image.
6. Apply SVD to watermark image and Its computed *USV* components are U_wimg, S_wimg and V_wimg.
7. Read the Cover image.
8. Apply DWT on cover image.
9. Apply SVD to cover image. It's computed *USV* components are U_img, S_img and V_img.
10. Set

$$S\_wimg1 = (S\_wmimg - S\_img)/\alpha$$

11. Finally compute w_HL1 as

$$w\_HL1 = U\_wimg * S\_wimg1 * V\_wimg.$$

12. Apply inverse DWT and calculate Extracted watermark

Watermark $= \text{IDWT2}(w\_LL, w\_LH, w\_HL1, w\_HH, \text{'haar'})$.

13. End.

4 Performance Evaluation

4.1 PSNR

PSNR stand for peak signal to noise ratio. PSNR uses a term mean square error in denominator so, there is inverse relationship between MSE and PSNR which shows that low error rate produce higher PSNR value. It is used to compare the quality of original image and watermarked image. Formula for the calculation of PSNR value of watermarked image is obtained by following Eq. (4)

$$\text{PSNR} = 10 \log_{10}(R^2/\text{MSE}) \tag{4}$$

4.2 NCC

NCC stands for normalized cross-correlation. It checks the measure of similarity between the original watermark and the extracted watermark after application of various attacks on it. Formulation of NCC is given in Eq. (5).

$$\text{NCC} = \frac{\sum_{i=1}^{M} \sum_{j=1}^{N} [w(i,j).w'(i,j)]}{\sum_{i=1}^{M} \sum_{j=1}^{N} [w(i,j)]^2} \tag{5}$$

NCC values varies from 0 to 1, if NCC values are close to 0, this means that the attack has altered the watermark to large extend and if its near 1, then its mean that the applied attacks could not altered the watermark much and it is more similar to original watermark.

5 Experimental Results

To check the performance and robustness of the method four mostly used colored cover images Lena, mandrill, Pepper and cameraman each of size 512×512 are taken. Amity logo and aeroplane are used as watermark images each of size 256×256.

A quantitative index, Peek Signal to Noise Ration (PSNR) has been measured to compare the original host image and watermarked image. The proposed method shows less distortion as PSNR (>35 dB) for all the standard test cover images namely Lena, Mandrill, Pepper and cameraman and watermark images Amity logo and Aeroplane image. To check the robustness of proposed method against various attacks—Salt and pepper noise, Poisson noise, Gaussian noise (0.006), Median filtering (3 × 3), Mean filtering (3 × 3), Wiener Filtering (3 × 3), Rotation (1°), Scaling (0.9), Scaling (1.2), Cropping (10%), JPEG Compression(10%), JPEG Compression (90%), normalized cross-correlation (NCC) metric values are evaluated. Table 1 shows the PSNR values of different host images and Table 2

Table 1 PSNR values of different host images

S. No.	Used watermark	Cover image	PSNR
1	Amity logo	Lena	58.48
		Mandrill	58.49
		Cameraman	58.59
		Pepper	58.62
2	Aeroplane	Lena	69.41
		Mandrill	68.84
		Cameraman	68.14
		Pepper	69.26

Table 2 Comparison of robustness in terms of NC values obtained after applying attacks on the watermarked image

Amity logo				Aeroplane image	
Lena				Mandrill	
S. No.	Types of attacks	Manish et al. [11]	Proposed	Manish et al. [11]	Proposed
1	Salt and pepper	0.97	0.9859	0.94	0.9914
2	Poisson noise	0.94	0.9947	0.90	0.9927
3	Gaussian noise	0.91	0.9591	0.87	0.9696
4	Median	0.79	0.9962	0.74	0.9976
5	Mean	0.98	0.9917	0.97	0.9955
6	Cropping (10%)	0.97	0.9993	0.94	0.9997
7	Wiener	0.93	0.9980	0.91	0.9964
8	Scale (0.9)	0.98	0.9994	0.97	0.9999
9	Scale (1.2)	0.98	0.9979	0.97	0.9996
10	Rotate (1°)	0.97	0.9993	0.94	0.9988
11	JPEG (10%)	0.72	0.7401	0.70	0.7261
12	JPEG (20%)	0.88	0.9022	0.86	0.8999

compares the robustness of the proposed scheme with the scheme in [11] after application of various attacks on watermarked image.

a.   Cover Images

b.   Watermark Images

6   Conclusions

In this paper centric-oriented novel watermarking technique based on DWT and SVD in YCbCr color space has been used. The main focus of the paper is to embed the watermark image into the central part of the cover image as it's much difficult to extract something from the center. The Proposed scheme withstands the various geometric and filtering attacks to much greater extend. The experimental result demonstrates that the proposed method leads the existing ones.

References

1. Podilchuk, C.I., Delp, E.J.: Digital watermarking: algorithms and applications. IEEE Signal Process. Mag., 33–45 (2001)
2. Gonzales, F.P., Hernandez, F.R.: A tutorial on digital watermarking. In: Proceedings of IEEE 33rd Annual 1999 International Carnahan Conference on Security Technology, pp. 286–292 (1999)
3. Potdar, V.M., Han, S., Chang, E.: A survey of digital image watermarking techniques. In: International Conference on Industrial Informatics, pp. 709–716 (2005). doi:10.1109/INDIN. 2005.1560462
4. Ganic, E., Eskicioglu, A.M.: Robust DWT-SVD domain image watermarking: embedding data in all frequencies, pp. 166–174. Magdeburg, ACM Multimedia and Security Workshop (2004)

5. Bei, Y., Yang, D., Liu, M., Zhu, L.: A multi-channel watermarking scheme based on HSV and DCT-DWT. In: Proceedings of the IEEE International Conference on Computer Science and Automation Engineering, pp. 305–308 (2011)
6. Lee, D., Kim, T., Lee, S., Parik, J.: Genetic algorithm-based watermarking in DWT domain. LNCS **4113**, 709–716 (2006)
7. Dharwadkar, N.V., et al.: The image watermarking scheme using edge information in YCbCr color space. In: International Proceedings of Computer Science and Information Technology, 56.127 (2012)
8. Ding, W., Yan, W., Qi, D.J.: Comput. Sci. Technol. 17, 129 (2002). doi: 10.1007/BF029662205
9. Nguyen, T.H., Duong, D.M., Duong, D.A.: Robust and high capacity watermarking for image based on DWT-SVD. In: IEEE RIVF International Conference on Computing Communication Technologies Research, Innovation, and Vision for Future (RIVF) (2015)
10. Abu-Errub, A., Al-Haj, A.: Optimized DWT based image watermarking. In: First International Conference on Application of Digital Information and Web Technologies, IEEE, pp. 4–6 (2008)
11. Gupta, M., Gupta, R., Parmar, G., Saraswat, M.: Digital image watermarking using steerable pyramid transform and uncorrelated color space. In: Proceedings of 9th International Conference on Industrial and Information System (ICIIS), Gwalior, pp. 1–5, doi:10.1109/ICIINFS.2014

Is Open-Source Software Valuable for Software Defect Prediction of Proprietary Software and Vice Versa?

Misha Kakkar, Sarika Jain, Abhay Bansal and P.S. Grover

Abstract Software Defect Prediction (SDP) models are used to identify the defect prone artifacts in a project to assist testing experts and better resource utilization. It is simple to build a SDP model for a project having historical data available. The problem arises when we want to build SDP model for a project, which has limited or no historical data available. In this paper, we have tried to find out whether data from Open-Source Software (OSS) projects is helpful in building SDP model for proprietary software having no or limited historical data. For collection of data for training SDP model a tool is developed which extract metric data from open-source software projects source codes. Three-benchmarked dataset from NASA projects are used as proprietary software dataset for which software defects are predicted. machine learning algorithms: LR, kNN, Naïve Bayes, Neural Network, SVM, and Random Forest are used to build SDP models. Using popular performance indicators such as precision, recall, F-measure, etc., the performances of these six SDP models are compared. The study concluded that when SDP models are trained using data from OSS projects then they are able to predict software defects for proprietary software with greater accuracy in comparison to SDP models predicting defects for OSS projects when trained using proprietary software data.

Keywords Software defect prediction · Open source software · Proprietary software · Machine learning algorithm

M. Kakkar (✉) · A. Bansal
Department of Computer Science & Engineering, Amity University Uttar Pradesh, Noida, Uttar Pradesh, India
e-mail: mkakkar@amity.edu

S. Jain
Amity Institute of Information Technology, Amity University Uttar Pradesh, Noida, Uttar Pradesh, India
e-mail: ashusarika@gmail.com

P.S. Grover
KIIT Group of Colleges, Gurugram, India
e-mail: drpsgrover@gmail.com

© Springer Nature Singapore Pte Ltd. 2018
M. Pant et al. (eds.), *Soft Computing: Theories and Applications*,
Advances in Intelligent Systems and Computing 583,
https://doi.org/10.1007/978-981-10-5687-1_21

1 Introduction

Software defect indicates an error in the software code, which affects the behavior of the program [1]. Software defect prediction enables quality support staff to predict defect prone modules/ files in advance, which in-turn helps in effective resource allocation and utilization. This effective resource allocation, apart from other tools, supports greatly in meeting project time-lines, as generally exhaustive testing is not possible because of stringent release schedules [2, 3]. A plethora of work has been done in the field of SDP [2–5]. To prepare data for software defect prediction model, metric data is collected from software archives consisting of various configuration and change control management systems, which is then mapped with various bugs tracking system to label instances into buggy and clean. Two techniques (1) keyword searching [6]; (2) bug report reference searching [6] can be used to perform this mapping. In keyword searching technique keywords like "bug"; "fault", "fixed" are searched to identify the defect-prone artifacts, and in reference searching technique reference to bug reports are searched [1, 4]. The main contributions of this paper are: it gives a dedicated metrics extraction tool—StaticCodeAnalyzer to extract relevant software metrics from source code of open-source software, then this data is used for carrying out a performance analysis for SDP model build for within project and cross-project defect prediction. This study tries to analyze how SDP model built from open-source software works for NASA project benchmarked data set and vice versa.

We present a background work and data collection method in Sects. 2 and 3 whereas model building is given in Sect. 4. Finally, result analysis and conclusion is presented in Sects. 5 and 6.

2 Related Work

Software defect prediction is an important research area nowadays [1, 4]. He et al. [7] in their study concluded that the SDP model's accuracy is affected by the training data for defect prediction. They selected simplified set of metric for training SDP model and the performance of their model improves by using this metric set. This kind of metric set selection is very useful where there is limited resources supply. Basic classifiers like Naïve Bayes also perform well with this simplified set of metric for Within Project Defect Prediction (WPDP). The approach is not tested for Cross-Project Defect Prediction (CPDP). Feng et al. [8] proposed a context-aware rank transformation for predictors using open source dataset. Open-source data sets from SourceForge and GoogleCode were used in the study for testing the approach for CPDP models. Addition of context factors information in the training dataset improves the model performance. He et al. [9] focuses on building an improved CPDP model by considering distribution characteristic of the training instance values. Three publicly available datasets were used—PROMISE,

ReLink, and AEEEM in the study for building CPDP model whose performance was slightly better but not comparable to WPDP model performance. Eleven open-source projects were used by Fukushima et al. [10] to propose JIT cross-project models. These models were trained using other projects historical data as a feasible solution for projects with limited or no historical data. It was concluded that the performance of JIT cross-project models depends on the careful selection of training dataset. Verma et al. [11] built based on WPDP using a two-stage data preprocessing to increase accuracy. Feature selection techniques were used by Khoshgoftaar et al. [12]. They concluded that data quality affects the model performance. This study uses a set of 28 metrics (mostly CK metric suit) for building software defect prediction models, which are further tested in WPDP setup as well as CPDP setup. This metric suit is widely used for building SDP model [13, 14].

3   Data Collection Method

To collect data from open-source software repositories, a tool naming StaticCodeAnalyzer is developed. Figure 1 shows the interface of the tool which analyses the sources codes of projects written in Java Programming language to extract metric information of the projects. StaticCodeAnalyzer is written using Java Language and uses majority of Java in-built function to calculate the metrics. The tool used equations given in Aggarwal et al. [15] for computing metrics for which built-in functions were not available. Project source codes are taken from two most

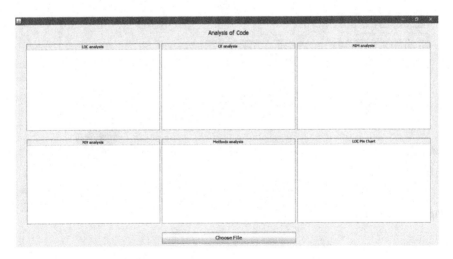

Fig. 1 Interface for StaticCodeAnalyze

popular open source repositories SourceForge and GoogleCode. Both SourceForge and GoogleCode have abundant projects to limit the number of projects to be review, we constraint our search based on three parameters: first is the lifespan of the projects, second—number of commits throughout their lifespan, and third—use of bug tracking system. For our study, we have included only those projects, which have a lifespan of more than one year. Then we performed a statistical analysis based on the number of commits on the projects having more than one year of lifespan to decide the threshold value of number of commits. Based on box plot analysis of number of commits for the project the threshold value came out to be 30 (25% quartile value). Thus only those projects are included in the study, which have a lifespan of more than one year and having more than 30 number of commits.

In a total of 580 projects, 58,100 source code files were analyzed by StaticCodeAnalyzer, which collectively gave us a dataset of 58 100 instances. A total of 28 metrics were calculated for each instance by our tool (Fig. 2). All these 580 projects used Bug Tracking system, information from which was used to classify these 58,100 instances into buggy and clean. Table 1 provides the list of all the calculated metrics along with their description and level of granularity. Metrics are calculated for all granular levels, i.e., from highest project level to the lowest method level. As for the proprietary software dataset, we have used benchmarked dataset of three NASA project—PC2, PC3, and PC4, which again are widely used in software defect prediction.

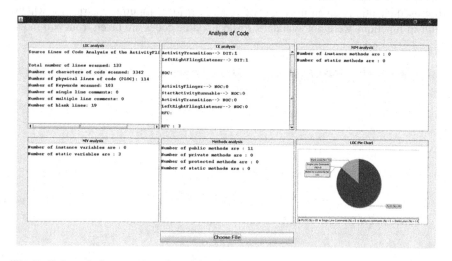

Fig. 2 Code analysis

Table 1 List of collected metrics

Metrics name	Description	Granularity
TLOC	Total lines of code for project	Project
TNF	Total no. of files	
TND	Total no. of classes	
LOC	Lines of code	File
CL	Comment lines	
N_Stmt	No. of statements	
N_func	No. of functions	
RCC	Ratio of comments to code	
MNL	Max. level of nesting	
NREV	No. of revisions	
NFIX	No. of times file is involved in a bug fix	
addedloc_total	Total of lines added	
deletedloc_total	Total of lines deleted	
modifiedloc_total	Total of lines deleted	
WMC_total	Weighted method count	Class
DIT_total	Inheritance Tree Depth	
REC_total	Response for class	
NOC_total	No. of children	
CBO_total	Coupling between objects	
NIV_total	Total no. of instance variables	
NIM total	Total no. of instance method	
NOM_total	Total no. of local methods	
NPRM_total	Total no. of public methods	
NPM_total	Total no. of protected methods	
NPRM_total	Total no. of private methods	
CC	Mcbabe cyclomatic complexity	Method
FANIN_total	No. of input data	
FANOUT_total	No. of output	

4 Model Building

We have designed our experiment to find answer for the question, is proprietary software valuable for software defect prediction of open-source software and vice versa. The proposed methodology is explained in Fig. 3. The study is divided in two setups: WPDP setup and CPDP setup. In WPDP setup, training as well as test data is from one project only. As programming language and development methodology of all 580 open source projects are same, they are considered together as one project by combining together their data in one file. In CPDP setup, data for training and testing are from different projects. In this data from all the proprietary projects are also combined together as one.

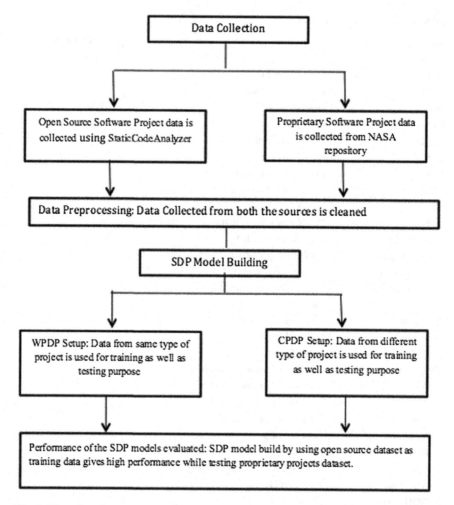

Fig. 3 Flow chart for proposed methodology

Thus, SDP models were built once by using proprietary projects data as training data (denoted by NASA) and open-source projects data as testing data (denoted by OSS). SDP models were also built by revering both training and test data.

5   Results and Analysis

5.1   WPDP Setup

Six machine learning algorithms—LR, Naïve Bayes, kNN, Neural Network, SVM, Random Forest are used to build SDP models. These six classifiers belong to different families of classifiers as Logistic Regression and Naïve Bayes belongs to linear classification family. Naïve Bayes follows the Bayesian rule for its classification, i.e., classification is done based on the probabilistic measure of the instances. KNN is based on the distance of the instance form of its nearest neighbor. For our problem we have used euclidean distance to measure the distance between two instances. Neural Network and SVM are nonlinear problem solving algorithms. Neural Networks try to simulate the behavior of human brain and SVM uses sequential minimal optimization techniques for the purpose of classification. Random forest is an ensemble-based decision tree algorithm. Random forest averages the decision of multiple trees to reduce the variance. For each of the classifier, we have developed SDP models using both training and test data from same project.

Classification results for this setup are shown in Table 2. Five most popular performance measures (Classifier Accuracy (CA), Area Under Curve (AUC), F-measure, Precision and Recall) are used to compare the performances of these classifiers.

Table 2 WPDP Result

WPDP		Naïve Bayes	Logistic Regression	Neural Network	kNN	SVM	Random forest
PC2	CA	0.94	0.97	0.97	0.96	0.98	0.98
	AUC	0.75	0.63	0.71	0.49	0.53	0.74
	F	0.95	0.96	0.98	0.98	0.99	0.99
	Prec	0.97	0.97	0.97	0.98	0.98	0.98
	Recall	0.95	0.97	0.98	0.99	1	1
PC3	CA	0.86	0.89	0.89	0.86	0.89	0.89
	AUC	0.75	0.81	0.83	0.71	0.69	0.77
	F	0.92	0.94	0.94	0.92	0.94	0.94
	Prec	0.91	0.9	0.9	0.91	0.89	0.89
	Recall	0.94	0.98	0.97	0.93	0.99	0.99
PC4	CA	0.86	0.87	0.87	0.81	0.86	0.86
	AUC	0.74	0.84	0.85	0.72	0.77	0.79
	F	0.92	0.92	0.92	0.89	0.92	0.92
	Prec	0.86	0.88	0.88	0.89	0.86	0.86
	Recall	0.99	0.97	0.97	0.89	0.99	0.99
OSS	CA	0.67	0.68	0.7	0.73	0.69	0.64
	AUC	0.66	0.63	0.71	0.78	0.69	0.55
	F	0.79	0.79	0.8	0.79	0.77	0.79
	Prec	0.68	0.7	0.71	0.78	0.67	0.66
	Recall	0.93	0.91	0.9	0.8	0.93	0.99
NASA	CA	0.89	0.91	0.90	0.89	0.89	0.92
	AUC	0.75	0.80	0.84	0.74	0.70	0.76
	F	0.94	0.93	0.72	0.95	0.97	0.94
	Prec	0.91	0.90	0.76	0.91	0.91	0.90
	Recall	0.95	0.99	0.80	0.90	0.99	0.98

5.2 CPDP Setup

To compare the performance of SDP models in CPDP setup, same six machine learning classifiers were used in CPDP setup. In this setup, SDP models were built first by using data from proprietary NASA software as training data and data from open-source software as testing data (NASA => OSS). Then SDP models were built by reversing training and test data (OSS => NASA). Table 3 shows the result obtained for this setup.

The performance of these SDP models was compared using popular performance indicators—CA, AUC, F-measure, Precision and Recall. CA is defined as the total number of instances that are correctly predicted (either true or false) divided by the total number of instances used for testing. F-measure is the harmonic mean of precision and recall. Precision is the proportion between corrected true instances and the total true instances whereas recall is the proportion between true instances and the total number of instances. AUC is area under ROC curve that is plotted between TPR and FPR (TPR = true correctly classified/total true; FPR = false incorrectly classified/total false). All the performance indicators are measured under 10-cross validation to remove any type of biasness for classification. It can be observed that recall in almost all cases are very good and sometimes it has reached to a perfect score also.

The results obtained under two setups are compared in Table 4. When SDP model uses open-source software projects data for its training purpose, then it predicts proprietary software project defect with high accuracy almost same as that achieved in WPDP. SDP model build by using open-source dataset as training data gives high performance while testing proprietary projects dataset. These results are even comparable (or sometimes) better than their WPDP counterpart. Although it is not the same in the vice versa case as the performance of SDP model has decreased if we take proprietary projects dataset as training data and open-source project dataset as testing data.

Table 3 CPDP Results

CPDP		Naïve Bayes	Logistic Regression	Neural Network	kNN	SVM	Random forest
NASA => OSS	CA	0.68	0.69	0.71	0.73	0.69	0.66
	AUC	0.67	0.70	0.73	0.79	0.69	0.55
	F	0.80	0.80	0.80	0.80	0.80	0.80
	Prec	0.69	0.70	0.72	0.79	0.69	0.66
	Recall	0.94	0.93	0.90	0.81	0.94	0.99
OSS => NASA	CA	0.90	0.91	0.91	0.88	0.91	0.92
	AUC	0.77	0.82	0.86	0.71	0.66	0.79
	F	0.94	0.95	0.95	0.93	0.95	0.96
	Prec	0.92	0.91	0.92	0.92	0.91	0.92
	Recall	0.96	0.99	0.98	0.94	0.99	1

Table 4 Comparison

	WPDP	CPDP	
		NASA => OSS	OSS => NASA
CA	0.86	0.69	0.91
AUC	0.72	0.69	0.77
F	0.9	0.80	0.95
Prec	0.86	0.71	0.92
Recall	0.95	0.92	0.98

6 Conclusion

This study was steered to inspect the influence of open source project dataset in building the SDP model. It can be concluded that SDP model build by using open-source dataset as training data gives high performance while testing proprietary projects dataset. Threats to validity for the conclusion include that a lesser number of proprietary software data sets are used in the study. It also includes the method of gathering the open-source software dataset. In future, we will explore more proprietary software dataset to further generalize this conclusion.

References

1. ANSI/IEEE, Standard Glossary of Software Engineering Terminology, STD-729–991, ANSI/IEEE (1991)
2. Wahono, R.S.: A systematic literature review of software defect prediction: Research trends, datasets, methods and frameworks. J. Softw. Eng. 1, 1–16 (2015)
3. Hall, T., Beecham, S., Bowes, D., Gray, D., Counsell, S.: A systematic literature review on fault prediction performance in software engineering. IEEE Trans. Software Eng. (2012)
4. Shepperd, M., MacDonell, S.: Evaluating prediction systems in software project estimation. Inf. Softw. Technol. (2012)
5. Radjenović, D., Heričko, M., Torkar, R., Živkovič, A.: Software fault prediction metrics: A systematic literature review. Inf. Softw. Technol. (2013)
6. Schröter, A., Zimmermann, T.: If your bug database could talk. In: Proc. 5th Int. Symp. Empir. Softw. Eng. Vol. II Short Pap. Posters, pp. 3–5 (2006)
7. He, P., Li, B., Liu, X., Chen, J., Ma, Y.:An empirical study on software defect prediction with a simplified metric set. Inf. Softw. Technol. (2015)
8. Zhang, F., Mockus, A., Keivanloo, I., Zou, Y.: Towards building a universal defect prediction model. In: Proc. 11th Work. Conf. Min. Softw. Repos. - MSR 2014, pp. 182–191 (2014)
9. He, P., Li, B., Ma, Y.: Towards cross-project defect prediction with imbalanced feature sets
10. Fukushima, T., Kamei, Y., Mcintosh, S., Yamashita, K., Ubayashi, N., Fukushima, T., Mcintosh, S., Yamashita, K., Ubayashi, N., Hassan, A.E.: Studying just-in-time defect prediction using cross-project models. Empir. Softw. Eng. pp. 172–181 (2014)
11. Singh, P., Verma, S., Vyas, O.P.: Cross company and within company fault prediction using object oriented metrics. Int. J. Comput. Appl. 74(8), 975–8887 (2013)
12. Khoshgoftaar, T.M., Gao, K., Napolitano, A., Wald, R.: A comparative study of iterative and non-iterative feature selection techniques for software defect prediction. Inf. Syst. Front. (2014)

13. Gyimothy, T., Ferenc, R., Siket, I.: Empirical validation of object-oriented metrics on open source software for fault prediction. IEEE Trans. Softw. Eng. **31**(10), 897–910 (2005)
14. Xu, J., Ho, D., Fernando Capretz, L.: An empirical validation of object-oriented design metrics for fault prediction. J. Comput. Sci. **4**(7), 571–577 (2008)
15. Aggarwal, K.K., Singh, Y., Kaur, A., Malhotra, R.: Empirical study of object-oriented metrics. J. Object Technol. **5**(8), 149–173 (2006)

ANN-Based Modeling for Flood Routing Through Gated Spillways

Pallavi J. Giri and A.R. Gajbhiye

Abstract The significance of dependent and independent variables relationship of hydrological data of reservoir is presented in this paper. It highlights on checking the significance of nonlinear fitting to establish a relationship. Multilayered perception neural network model is developed with diverse topologies and tested with different optimization algorithms. Relative error and mean square error is considered as performance parameters for training neural network. Overall, ANN model performs well with coefficient of correlation within the range of 0.94–0.99.

Keywords Mean square error · Hydrological data · Nonlinear transfer function

1 Introduction

Reservoir operation is encountered with a specific problem regarding the operation of spillway gates during flooding condition. Unprecedented opening of spillway gates results in a large amount of flow release which in turn results in indemnity downstream of the reservoir. On the contrary, improper release through gates or inadequate release may hint towards threat factor for the safety of the dam. Therefore, engineering practices and rulings are used for the operation of spillway gates during an incoming flood in systems without flood forecast [10]. Prediction of flow discharge in the natural stream flows is the key parameter in the flood management problems. The compound channel notion is the correct method for the simulation of the river problems. Predicting the flow discharge in compound open channels by classical formulas such as Manning and Chezy leads to the advent of implausible error in comparison to the measured data, so researchers revised these

P.J. Giri (✉)
Laxminarayan Institute of Technology, RTM Nagpur University, Nagpur, Maharashtra, India
e-mail: Pallavijgiri@gmail.com

A.R. Gajbhiye
Department of Civil Engineering, Yeshwantrao Chavan College of Engineering, Nagpur, Maharashtra, India

© Springer Nature Singapore Pte Ltd. 2018
M. Pant et al. (eds.), *Soft Computing: Theories and Applications*,
Advances in Intelligent Systems and Computing 583,
https://doi.org/10.1007/978-981-10-5687-1_22

approaches and proposed more advanced analytical approaches such as divided channel method, etc. Although the revised approaches have increased the accuracy of flow discharge calculation, flood management needs to implement more accurate approaches in the river engineering problems. Recently, by advancing the AI models in water engineering problems, the artificial neural network models have been used widely for flow discharge prediction. Bahram Malekmohammadi et al. [11] presented methodology based upon combining a Genetic Algorithm (GA) reservoir operation optimization model for a cascade of two reservoirs, a hydraulic-based flood routing simulation model in downstream river system.

Morteza Zargar et al. [22] presented overview of various researchers worked in the domain of reservoir operation. Various researchers revealed that flood inflow hydrograph which is single known based on real-time data or flood forecasting system is modeled and corresponding control strategy policy for the flood is formulated using various optimization techniques. Decision variables for the optimization model were releases from reservoir at specific time step. Precise objective function, corresponding practical constraints and independent variables are important to derive mathematical expression of the flood control problem. Various objective functions for different reservoirs have been considered by researchers so far flood control and optimal reservoir operation. Unver and Mays (1990) presented the general notion of an optimization–simulation model for the operation of a Nat Hazards river–reservoir system under flooding conditions. In various studies [1, 3, 7, 9, 15, 18, 20], flood's outflow crest and the water level in downstream area minimization were the main objectives. Giri et al. [14] presented overview of different optimization methods used for reservoir operation and its control.

Methods used to achieve reservoir control have been presented in [2, 6, 12, 13, 16, 19, 21]. Deterministic operating procedure for the reservoir control is presented by Beard. Dynamic programming technique for water reservoir management presented by Ozelkan et al. [13], and Oshimaa and Kosudaa [12] describes a distribution reservoir control approach with demand prediction using the deterministic-chaos method. Revealed that it is difficult or almost impossible to build an intelligent reservoir system without information of hydrological pattern. Haktanir and Kisi [6] presented a model with a set of operation rules with 10 stages for controlling the spillway gate opening.

2 Study Area for Case Study

See Figs. 1 and 2.

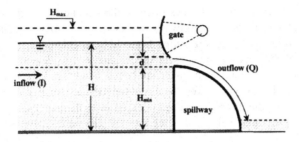

Fig. 1 Basic components of a reservoir during a flood operation

Fig. 2 Actual reservoir site

2.1 Information About Kamthikhairy (Pench) Irrigation Project

Name in the National Register of Dams is Kamthikhairy Dam and, Official Designation of Kamthikhairy (Pench) Dam Irrigation Project is "Kamthikhairy (Pench) Dam, D-01100". However, local and popular name is "Pench Dam, Pench Lake". Pench Irrigation Project is constructed as part of irrigation project by the Government of Maharashtra. It was completed in 1976. It is built on Pench River near city of Parseoni in Nagpur District of Maharashtra. The dam is an Earth fill dam. The length of dam is 2248 m (7375.33 ft), while the height of the dam above lowest foundation is 32 m (105.0 ft). The project has Ogee type of spillway. The length of the spillway is 237 m (777.56 ft). The dam has 16 Radial type of spill gates. The dam's catchment area is 466.1 Thousand Hectors. Maximum/Gross storage capacity is 230 MCM. Live storage capacity is 180.00 MCM. Almost all the water bodies make for good picnic spots and so does this place. The location has famous Tiger Reserve which is an added attraction.

Altogether 16 independent parameters have been considered against total discharge a dependent parameter. Independent parameters are live storage of reservoir,

reservoir level, time, number of gates opened, the distance through which gate is opened, rainfall, discharge over spillways, and evaporation losses. Table 1 shows a snapshot of various parameters recorded from the actual site.

The specific objective of the present investigation is to develop mathematical model for flood routing through gated spillways using artificial neural network (ANN). Investigation also provokes the aim of function approximation (discharge in this case) to predict the discharge values released from the gated spillways. Further comparisons with actual values of discharge are compared with mathematically simulated values.

3   Neural Network Modeling

The term neural network applies to a limply related family of models, characterized by a large parameter space and flexible structure, descending from studies of brain functioning. As the family grew, most of the new models were designed for non-biological applications, though much of the associated terminology reflects its origin. Specific definitions of neural networks are as varied as the fields in which they are used.

The trade-off for the neural network flexibility is that the synaptic weights of a neural network are not easily interpretable. Thus, if one is trying to explain an underlying process that produces the relationships between the dependent and independent variables, it would be better to use a more traditional statistical model. However, if model interpretability is not important, one can often obtain good model results more quickly using a neural network (Fig. 3).

3.1   Feedforward Neural Network

Neural networks are not specifically imposed upon model structure and associated assumption; it is predominantly useful to understand general network architecture.

The multilayer perceptron (MLP) or radial basis function (RBF) network is a function approximation technique to establish relationship between independent variables and dependent variables. The basic performance is imposed upon mini-mization of the prediction error of dependent variables.

The structure depicted in Fig. 4 is known as feedforward architecture because the independent variables nodes in the network from input layer are connected in forward direction with hidden layer nodes and further connected to the output layer without any feedback loops. In this figure:

Table 1 Snapshot of recorded readings

Reservoir level 1	Reservoir level 2	Drainage losses	Live storage	Day's out flow	Total out flow	Day's inflow	Total inflow	Day's rainfall	Total rainfall
324.1300	208.8220	0.0080	158.8220	2.8390	28.6410	0.0000	21.6290	14.0000	43.0000
324.1000	207.9020	0.0080	157.9020	2.3620	31.2730	1.7120	23.3410	0.0000	43.0000
324.0400	206.4670	0.0080	156.4670	2.3750	33.6480	0.9400	24.2810	66.0000	109.0000
324.0600	206.9450	0.0080	156.9450	2.3910	36.0390	2.8690	27.1500	10.0000	119.0000
324.1000	207.9020	0.0080	157.9020	2.5030	38.5420	3.4600	30.6100	1.0000	120.0000
324.1600	209.3430	0.0080	159.8400	2.5140	41.0560	3.9550	34.5650	4.0000	124.0000
324.1200	208.3820	0.0080	158.3820	2.5330	43.5890	1.5720	36.1370	0.0000	124.0000
324.0100	205.7520	0.0080	155.7520	2.5690	46.1580	0.0610	36.1980	0.0000	124.0000
324.0400	206.4670	0.0080	156.4670	0.7450	46.9030	1.4600	37.6580	0.0000	124.0000
324.0400	206.4670	0.0080	156.4670	0.7450	47.6480	0.7450	38.3250	0.0000	124.0000
324.1300	208.8220	0.0080	158.8220	0.7450	48.3940	3.1010	41.4260	0.0000	124.0000
324.3300	213.4800	0.0080	163.4800	0.9930	49.3870	5.6510	47.0770	0.0000	124.0000
324.4900	217.3520	0.0080	167.3520	0.9760	50.3630	4.8480	51.9250	6.0000	130.0000
324.6100	220.3010	0.0080	170.3010	0.9760	51.3300	3.9160	55.8410	0.0000	130.0000
324.8200	225.5080	0.0080	175.5080	0.9700	52.3000	6.1770	62.0180	0.0000	130.0000
324.8900	227.2580	0.0070	177.2580	0.5840	52.8840	2.3340	64.3520	0.0000	130.0000
324.8300	225.7580	0.0070	175.7580	1.0080	53.9200	0.0000	64.3520	0.0000	130.0000
324.7500	223.7860	0.0070	173.7860	2.1050	55.9970	0.1330	64.4850	0.0000	130.0000
324.7000	222.5250	0.0070	172.5250	2.6640	58.6610	1.4030	65.8880	0.0000	130.0000
324.6400	221.0410	0.0070	171.0410	3.1630	61.8240	1.6790	67.5670	0.0000	130.0000
Evaporation losses1	Evaporation losses 2		Discharge over spillway	Gate opening time		No. of gates opened	Distance through which GATE is opened		Total discharge through GATEs
8.8000	0.2114		21.1230	13.0000		2	0.2500		120.7360
10.4000	0.2492		21.1230	13.0000		2	0.2500		120.7360

(continued)

Table 1 (continued)

Evaporation losses1	Evaporation losses 2	Discharge over spillway	Gate opening time	No. of gates opened	Distance through which GATE is opened	Total discharge through GATEs
4.8000	0.1146	21.1230	13.0000	2	0.1500	120.7360
8.0000	0.1910	21.1230	13.0000	3	0.0500	120.7360
8.0000	0.1917	21.1230	13.0000	3	0.0500	120.7360
8.0000	0.1924	21.1230	13.0000	3	0.1500	120.7360
8.8000	0.2110	21.1230	13.0000	2	0.2500	120.7360
10.4000	0.2479	21.1230	13.0000	2	0.2500	120.7360
10.4000	0.2484	19.3200	13.0000	2	0.2500	120.7360
10.4000	0.2484	18.2560	13.0000	2	0.2500	120.7360
10.4000	0.2498	17.1920	13.0000	2	0.2500	120.7360
10.4000	0.2525	16.1280	13.0000	2	0.3000	121.5000
9.6000	0.2352	15.0640	13.0000	2	0.3000	121.5000
9.2000	0.2269	14.0000	13.0000	4	0.3000	121.5000
9.2000	0.2296	12.9360	13.0000	4	0.3000	121.5000
8.4000	0.2104	11.8720	13.0000	4	0.3000	126.7360
9.1000	0.2272	126.7360	13.0000	16	0.3000	506.9440
9.8000	0.2436	488.0770	13.0000	4	0.3000	425.6500
9.8000	0.2429	349.1800	13.0000	4	0.3000	400.3200
9.8000	0.2421	0.0000	13.0000	4	0.3000	374.9900

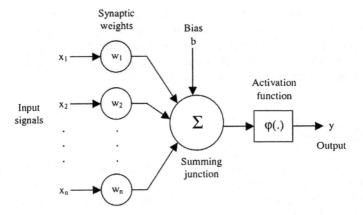

Fig. 3 Model of neuron

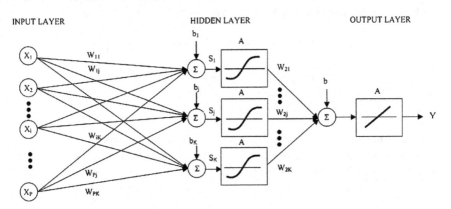

Fig. 4 Feedforward network

- The input layer contains the independent variables.
- The hidden layer contains unobservable nodes or units. The value of each hidden unit is weighted sum of input and synapses constricted through nonlinear transfer function; the scrupulous form of the function is contingent upon the network type and upon user-controllable stipulations.
- The output layer comprises the dependent variables. Each output unit is some function of the hidden units.

4 ANN Modeling

Experimental ANN modeling largely depends upon specific topology (representation of number of input units, number of hidden nodes, and output in tandem) for solving a problem involves training of the network and handling and removal of unnecessary weights and nodes during training. This approach is called pruning, requires advance knowledge of initial network size, but such upper bounds may not be difficult to estimate. An alternative method for determining appropriate network topology includes algorithms which start with a small network and extended to larger; such algorithms term as constructive algorithms.

4.1 ANN Model

Mathematical model based on basic 16 independent variables and total discharge as dependent variable is formulated. Model is based on more than 1800 data points collected over the span of 1 year. Symbolic representation of ANN equation is formulated as follows (Fig. 5):

ANN model is based on hyperbolic tangential transfer function for 16 independent variables and discharge through gate as a dependent variable.

$$X_{1,1} = \left(e^{1*sum(Layer1Cell26)} - e^{-1*sum(Layer1Cell26)} \right) / $$
$$\left(e^{1*sum(Layer1Cell26)} + e^{-1*sum(Layer1Cell26)} \right),$$

where

$sum(Layer1cell1) = 0.24 * X_{0,1} - 0.43 * X_{0,2} - 0.307 * X_{0,3} + 0.17 * X_{0,4} + 0.1888 * X_{0,5}$
$- 0.4 * X_{0,6} + 0.24 * X_{0,7} + 0.4 * X_{0,8} + 0.02 * X_{0,9} - 0.074 * X_{0,10} - 0.023 * X_{0,11} + 0.15 * X_{0,12} + 0.099*$
$X_{0,13} + 0.02 * X_{0,14} + 0.33 * X_{0,15} - 0.421 * X_{0,16},$

Learning process: training data , weight matrix

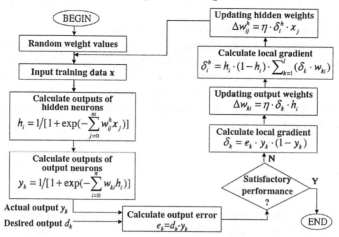

Fig. 5 Algorithm flowchart for neural network

$$X_{1,2} = \left(e^{1*\text{sum(Layer1Cell27)}} - e^{-1*\text{sum(Layer1Cell27)}}\right) /$$
$$\left(e^{1*\text{sum(Layer1Cell27)}} + e^{-1*\text{sum(Layer1Cell27)}}\right)$$

where

$$\text{sum(Layer1cell2)} = 0.24 * X_{0,1} - 0.43 * X_{0,2} - 0.307 * X_{0,3} + 0.17 * X_{0,4} + 0.1888 * X_{0,5}$$
$$- 0.4 * X_{0,6} + 0.24 * X_{0,7} + 0.4 * X_{0,8} + 0.02 * X_{0,9} - 0.074 * X_{0,10} - 0.023 * X_{0,11} + 0.15 * X_{0,12} + 0.099 *$$
$$X_{0,13} + 0.02 * X_{0,14} + 0.33 * X_{0,15} - 0.421 * X_{0,16}$$

$$X_{1,3} = \left(e^{1*\text{sum(Layer1Cell28)}} - e^{-1*\text{sum(Layer1Cell28)}}\right) /$$
$$\left(e^{1*\text{sum(Layer1Cell28)}} + e^{-1*\text{sum(Layer1Cell28)}}\right)$$

where

$$\text{sum(Layer1cell3)} = 0.24 * X_{0,1} - 0.43 * X_{0,2} - 0.307 * X_{0,3} + 0.17 * X_{0,4} + 0.1888 * X_{0,5}$$
$$- 0.4 * X_{0,6} + 0.24 * X_{0,7} + 0.4 * X_{0,8} + 0.02 * X_{0,9} - 0.074 * X_{0,10} - 0.023 * X_{0,11} + 0.15 * X_{0,12} + 0.099 *$$
$$X_{0,13} + 0.02 * X_{0,14} + 0.33 * X_{0,15} - 0.421 * X_{0,16}$$

$$X_{1,4} = \left(e^{1*\text{sum(Layer1Cell29)}} - e^{-1*\text{sum(Layer1Cell29)}}\right) /$$
$$\left(e^{1*\text{sum(Layer1Cell29)}} + e^{-1*\text{sum(Layer1Cell29)}}\right),$$

where

$$\text{sum(Layer1cell4)} = 0.24 * X_{0,1} - 0.43 * X_{0,2} - 0.307 * X_{0,3} + 0.17 * X_{0,4} + 0.1888 *$$
$$X_{0,5} - 0.4 * X_{0,6} + 0.24 * X_{0,7} + 0.4 * X_{0,8} + 0.02 * X_{0,9} - 0.074 * X_{0,10} - 0.023 * X_{0,11} + 0.15 * X_{0,12}$$
$$+ 0.099 * X_{0,13} + 0.02 * X_{0,14} + 0.33 * X_{0,15} - 0.421 * X_{0,16}$$

$$\text{Opi7} = \left(e^{1*\text{sum(Layer2Cell0)}} - e^{-1*\text{sum(Layer2Cell0)}}\right) /$$
$$\left(e^{1*\text{sum(Layer2Cell0)}} + e^{-1*\text{sum(Layer2Cell0)}}\right),$$

where

$$\text{sum(Layer2cell00)} = -0.466 * X_{1,1} + 0.132 * X_{1,2} - 0.009 * X_{1,3} - 0.331 * X_{1,4}$$
$$+ 0.047 * X_{1,5} + 0.118 * X_{1,6} + 0.329 * X_{1,7} + 0.228 * X_{1,8} + 0.069 * X_{1,9} - 0.129 * X_{1,10} - 0.521 * X_{1,11}$$
$$- 0.077 * X_{1,12} + 0.248 * X_{1,13} + 0.141 * X_{1,14} + 0.288 * X_{1,15} + 0.033 * X_{1,16}$$

5 Result and Discussion

There are many supported training algorithm available to train neural network which are as follows: Trainb (Batch training with weight and bias learning rules), Trainbfg (BFGS quasi-Newton BP), Trainbr (Bayesian regularization), Trainc (Cyclical order incremental update), Traincgb (Powell–Beale conjugate gradient BP), Traincgf (Fletcher–Powell conjugate gradient BP), Traincgp (Polak–Ribiere conjugate gradient BP), Traingd (Gradient descent BP), Traingda (Gradient descent with adaptive learning rate BP), Traingdm (Gradient descent with momentum BP), Traingdx (Gradient descent with momentum & adaptive linear BP), Trainlm (Levenberg–Marquardt BP), Trainoss (One-step secant BP), Trainr (Random order incremental update), Trainrp (Resilient backpropagation (Rprop), Trains (Sequential order incremental update), Trainscg (Scaled conjugate gradient BP). Tables 2, 3, and 4 represents the performance of neural network for three different topologies such as 16-5-1 (16 input, 5 hidden nodes, 1 output), 16-10-1 and 16-10-1. Finally, coefficient of correlation for function approximation and network performance is represented in terms R and mean square error. Figures 6 and 7 represent regression plot and network performance plot for 16-5-1 topology with Trainscg as a training and optimization algorithm.

Table 2 Comparison of various optimization algorithms

Iteration–I topology 16-5-1				
Training algorithm	Transfer function		R	MSE
	Hidden layer	Output layer		
Trainlm	Logsig	Logsig	0.964	0.18577
	Tansig	Tansig	0.976	0.1789
Traingd	Logsig	Logsig	0.903	1.623
	Tansig	Tansig	0.806	8.623
Traingda	Logsig	Logsig	0.709	5.689
	Tansig	Tansig	0.698	6.235
Traingdx	Logsig	Logsig	0.823	2.36
	Tansig	Tansig	0.752	3.256
Trainoss	Logsig	Logsig	0.693	25.693
	Tansig	Tansig	0.526	29.365
Trainr	Logsig	Logsig	0.693	12.562
	Tansig	Tansig	0.596	13.698
Trainrp	Logsig	Logsig	0.698	8.963
	Tansig	Tansig	0.615	9.236
Trainscg	Logsig	Logsig	0.922	1.2567
	Tansig	Tansig	0.833	495.127
Trainbfg	Logsig	Logsig	0.789	123.365
	Tansig	Tansig	0.692	224.632
Trainbr	Logsig	Logsig	0.635	185.362
	Tansig	Tansig	0.623	190.236

Table 3 Comparison of various optimization algorithms

Iteration–II topology 16-10-1				
Training algorithm	Transfer function		R	MSE
	Hidden layer	Output layer		
Trainlm	Logsig	Logsig	0.984	0.1687
	Tansig	Tansig	0.979	0.1889
Traingd	Logsig	Logsig	0.911	1.256
	Tansig	Tansig	0.825	7.223
Traingda	Logsig	Logsig	0.698	6.92
	Tansig	Tansig	0.627	7.025
Traingdx	Logsig	Logsig	0.793	3.86
	Tansig	Tansig	0.702	4.284
Trainoss	Logsig	Logsig	0.623	32.003
	Tansig	Tansig	0.629	32.365
Trainr	Logsig	Logsig	0.694	12.562
	Tansig	Tansig	0.667	13.698
Trainrp	Logsig	Logsig	0.788	8.963
	Tansig	Tansig	0.630	9.236
Trainscg	Logsig	Logsig	0.936	1.2567
	Tansig	Tansig	0.856	495.127
Trainbfg	Logsig	Logsig	0.809	123.365
	Tansig	Tansig	0.682	212.632
Trainbr	Logsig	Logsig	0.605	145.022
	Tansig	Tansig	0.613	156.006

Table 4 Comparison of various optimization algorithms

Iteration–III topology 16-15-1				
Training algorithm	Transfer function		R	MSE
	Hidden layer	Output layer		
Trainlm	Logsig	Logsig	0.994	0.1287
	Tansig	Tansig	0.989	0.1689
Traingd	Logsig	Logsig	0.923	1.146
	Tansig	Tansig	0.865	6.223
Traingda	Logsig	Logsig	0.712	5.92
	Tansig	Tansig	0.685	6.025
Traingdx	Logsig	Logsig	0.823	2.16
	Tansig	Tansig	0.812	2.284
Trainoss	Logsig	Logsig	0.693	12.003
	Tansig	Tansig	0.669	13.365
Trainr	Logsig	Logsig	0.714	10.562
	Tansig	Tansig	0.697	13.698

(continued)

Table 4 (continued)

Iteration–III topology 16-15-1				
Training algorithm	Transfer function		R	MSE
	Hidden layer	Output layer		
Trainrp	Logsig	Logsig	0.793	7.663
	Tansig	Tansig	0.680	9.236
Trainscg	Logsig	Logsig	0.951	1.1267
	Tansig	Tansig	0.906	300.127
Trainbfg	Logsig	Logsig	0.789	111.365
	Tansig	Tansig	0.711	96.632
Trainbr	Logsig	Logsig	0.611	185.022
	Tansig	Tansig	0.609	196.006

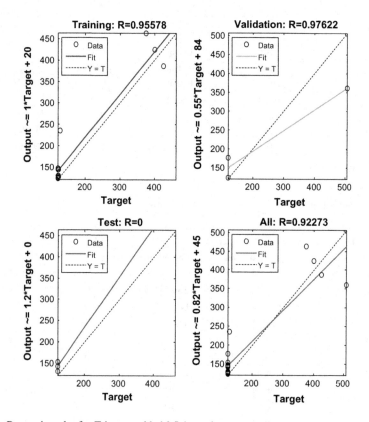

Fig. 6 Regression plot for Trianscg with 16-5-1 topology

Fig. 7 Performance plot for Trianscg with 16-5-1 topology

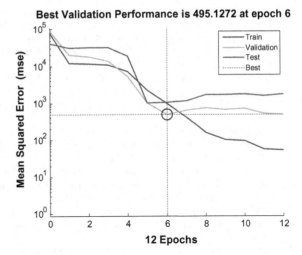

6   **Conclusion**

Persuasive results are yielded through ANN Model simulation in this investigation. Relationship between 16 independent variables and 1 dependent variable is captured correctly by ANN. It is observed that sigmoidal transfer function with Levenberg–Marquardt as a training and optimization algorithm is performing the best for all topologies. The corresponding coefficient of regression and mean square error for 16-10-1 topology with Trainlm yielded best results for present investigation. The paper also elaborates comparative analysis between adoption of different training algorithm and two activation function. Further, it is also observed that number of hidden nodes in hidden layer also have significant impact on output parameters. This investigation is ultimately helpful for finalization of reservoir operation schedule based on total discharge prediction.

References

1. Ahmed, S.M.S., Mays, L.W.: Model for determining real-time optimal dam releases during flooding conditions. J. Nat. Hazards **65**(3), 1849–1861 (2013)
2. Beard, L.R: Flood control operation of reservoirs. J. Hydr. Div. (ASCE), **89**(1), 1–23 (1963)
3. Chang, L.C., Chang, F.J., Hsu, H.C.: Real-time reservoir operation for flood control using artificial intelligent techniques. Int. J. Nonlinear Sci. Numer. Simul. **11**(11), 887–902 (2010)
4. Cheng, C.T., Chau, K.W.: Fuzzy iteration methodology for reservoir flood control operation. J. Am. Water Resour. Assoc. **37**(5), 1381–1388 (2001)
5. Cheng, C.T., Chau, K.W.: Flood control management system for reservoirs. Environ. Model Softw. **19**(12), 1141–1150 (2004)
6. Haktanir, T., Kisi, O.: Ten-stage discrete flood routing for dams having gated spillway. J. Hydrol. Eng. (ASCE), **6**(1), 86–90 (2001)

7. Karaboga, D., Bagis, A., Haktanir, T.: Controlling spillway gates of dams by using fuzzy logic controller with optimum rule number. Appl. Soft Comput. **8**(1), 232–238 (2008)
8. Kumar, D.N., Baliarsingh, F., Raju, K.S.: Optimal reservoir operation for flood control using folded dynamic programming. Water Resour. Manag. **24**(6), 1045–1064 (2010)
9. Li, Y., et al.: Novel multi objective shuffled frog leaping algorithm with application to reservoir flood control operation. J. Water Resour. Plan Manag. (ASCE), **136**(2), 217–226 (2010)
10. Linsley, R.K., et al.: Water Resources Engineering, 4th edn. Mc-Graw-Hill Book Company, New York (1992)
11. Malekmohammadi, B., Zahraie, B., Kerachian, R.: A real-time operation optimization model for flood management in river-reservoir systems. Nat. Hazards **53**, 459–482 (2010)
12. Oshimaa, N., Kosudaa, T.: Distribution reservoir control with demand pre-diction using deterministic-chaos method. Water Sci. Technol. **37**(12), 389–395 (1998)
13. Ozelkan, E.C., et al.: Linear quadratic dynamic programming for water reservoir management. App. Math. Model. **21**(9), 591–598 (1997)
14. Giri, P.J., Gajbhiye, A.R.: Modeling and Simulation of gate operations of dam for flood management: an approach, Pratibha: Int. J. Sci. Spirituality, Bus. Technol. (IJSSBT), **3**(1), 83–97 (2014)
15. Qin, H., Zhou, J., Lu, Y.: Multi-objective cultured differential evolution for generating optimal trade-offs in reservoir flood control operation. Water Resour. Manag. **24**(11), 2611–2632 (2010)
16. Simonovic, S.P.: Reservoir systems analysis: closing gap between theory and practice. J. Water Resour. Plan Manag. (ASCE), **118**(3), 262–280 (1992)
17. Turgeon, A.: Daily operation of reservoir subject to yearly probabilistic constraints. J. Water Resour. Plan Manag. **131**(5), 342–350 (2005)
18. Valerino, O.C.S., Koike, T., Yang, K., Yang, D.: Optimal dam operation during flood season using a distributed hydrological model and a heuristic algorithm. J Hydrol. Eng. (ASCE), **15**(7), 580–586 (2010)
19. Wurbs, R.A.: Reservoir-system Simulation and Optimization Models. J. Water Resour. Plan Manag. (ASCE), **119**(4), 445–472 (1993)
20. Wei, C.C., Hsu, N.S.: Optimal tree-based release rules for real-time flood control operations on a multipurpose multi reservoir system. J. Hydrol. **365**(3), 213–224 (2009)
21. Yeh, W.: Reservoir management and optimization models: a state of the art review. Water Resour. Res. **21**(12), 1797–1818 (1985)
22. Zargar, M., Samani, H.M.V., Haghighi, A.: Optimization of gated spillways operation for flood risk management in multi-reservoir systems. Int. J. Nat. Hazards **82**(1), 299–320 (2016)

Teaching–Learning-Based Optimization on Hadoop

S.M. Jagdeo, A.J. Umbarkar and P.D. Sheth

Abstract The challenges of evolutionary algorithms (EA) are high dimensionality of problems, curse of dimensionality, premature convergence, and controlling parameters. There are various advancements that happen to the EAs over a period of time. Various kinds of EAs are the GA, PSO, TLBO, GSA, etc. Teaching–learning-based algorithm (TLBO) is the recent algorithm introduced in 2011. Since its introduction, many researchers have modified TLBO and applied in various real-world problems. Various modifications in TLBO have been done to enhance and get the original version better. There is no clear information about how much the TLBO algorithm has evolved and the research has been done. Hence, this paper presents modifications of TLBO done in chronological way. Advancements in technologies also happened based on processing of data. Various new technologies like Hadoop based on distributed computing are introduced to process data quickly when compared to sequential code. TLBO can be fused with Hadoop for enhancement in processing speed thus a novel model of Hybrid TLBO is presented. Lastly, new approaches towards EAs are discusses.

Keywords Evolutionary algorithm (EA) · Metaheuristic · Hadoop
MapReduce · Teaching–learning-based optimization (TLBO) · Function optimization

S.M. Jagdeo (✉) · A.J. Umbarkar
Department of Information Technology, Walchand College of Engineering, Sangli, MS, India
e-mail: shashankjagdeo15@gmail.com

A.J. Umbarkar
e-mail: anantumbarkar@rediffmail.com

P.D. Sheth
Department of MCA, Government College of Engineering, Karad, MS, India
e-mail: pranalisheth@gmail.com

© Springer Nature Singapore Pte Ltd. 2018
M. Pant et al. (eds.), *Soft Computing: Theories and Applications*,
Advances in Intelligent Systems and Computing 583,
https://doi.org/10.1007/978-981-10-5687-1_23

1 Introduction

A metaheuristic algorithm is a population-based probabilistic algorithm which has two important groups, i.e., Swarm Intelligence (SI) and evolutionary algorithms (EA). Swarm Intelligence based algorithms are Particle Swarm Optimization (PSO), Artificial Immune Algorithm (AIA), Ant Colony Optimization (ACO), Shuffled Frog Leaping (SFL), Fire Fly (FF) algorithm, Differential Evolution (DE), Artificial Bee Colony (ABC), etc. EAs are Evolutionary Programming (EP), Differential Evolution (DE), Evolution Strategy (ES), Genetic Algorithm (GA), etc. Such algorithms require common controlling parameters and their own algorithm-specific controlling parameters. For example, PSO uses inertia weight, cognitive, and social parameters; GA uses operators like crossover probability, mutation probability, and selection; ABC uses a number of onlooker bees, scout bees, employed bees, etc. The impact of improper tuning of the algorithm-specific parameters may degrade the performance and increase the computational time required for the algorithm, and thus may lead to an improper solution. Nevertheless, another important aspect is the converging rate for solving an optimization problem over the quality of solutions [1, 2].

Teaching–learning-based optimization (TLBO) algorithm was introduced by Rao in 2011 and is an optimization technique used in finding exact or approximate solutions to the problems. TLBO does not required algorithm-specific parameters and requires only common controlling parameters for its execution, such as the number of generations and population size. Less number of controlling parameters does not infer it as a best algorithm as compared to other EAs.

This paper shows the various modifications and hybridization of TLBO algorithm and recent new approaches towards EA. The rest of this paper is organized as follows: Sect. 2 gives a brief description of TLBO. Section 3 describes the literature review of TLBO algorithm. Section 4 gives the distributed EA models and suitable distributed computing platforms. Section 5 gives description of TLBO on MapReduce. Section 6 gives conclusion on this work.

2 Brief Description of TLBO

Teaching–learning-based optimization Algorithm are computer algorithms that search for good solutions to a problem from among a large number of possible solutions. Teaching–learning-based optimization (TLBO), which works on the philosophy of teaching and learning, is first proposed by R.V. Rao et al. in 2011. This procedure resolves a many practical problems also computational problems from different fields. Some applications of TLBO are optimization, manufacturing, machine learning, economics, and population genetic and social system [1, 3]. TLBO is a population-based optimization method inspired by teacher–student teaching–learning mechanism. Like all other population-based methods, it also uses a population of solutions into direct the search to get a global solution. Here, population indicates a group of learners and different design variable indicate the different subjects offered to learners. Based on the above data, the result is calculated which is known as fitness, same as in other population-based optimization techniques. The best solution of the whole population is treated as a teacher since the most learned people in the society are considered as a teacher. The TLBO algorithm process is divided into parts. The first part is considered as teacher phase, whereas the second part is considered as the learner phase. The first phase is teacher phase, in which learner learns through teacher while in the second phase, i.e., learner phase learner does the interaction between themselves in order to increase their knowledge. We discuss the complete implementation of TLBO in the further subsection.

2.1 Initialization

Following are the notations used for describing the TLBO:

m: number of subjects, i.e., "design variables or dimension".
n: number of learners, i.e., "population size".
NOI: Termination Criteria.

The population X is randomly initialized by a search space bounded by a matrix of n-rows and m-columns, where random distribution is in the range of search space.

2.2 Teacher Phase

The most experienced and knowledgeable person in the whole class is teacher. So the best learner among the entire population is treated as teacher, i.e., best fitness value among the entire population. The result of the best learner is calculated by considering all the subjects.

Since population size is n, i.e., ($k = 1, 2, 3, ..., n$) and number of subjects is m, i.e., ($j = 1, 2, 3, ..., m$) at any sequential teaching cycle say 'i'. Let M_{ij} is mean result of the learners in particular subject 'j'. $X_{\text{total-}k\text{best},I}$ is best learner among population by considering all the subjects.

The best learner in whole class is treated as a teacher for a particular cycle. The main job of the teacher is to improve the knowledge level of entire class. It depends on two factors, i.e., quality of teaching by the teacher and quality of learners present in class. Based on the above factor, the mean difference is calculated between the teacher and mean result of the learners in each subject is expressed as Eq. 1,

$$\text{Difference\_Mean}_{j,i} = r_i \left(X_{j,kbest,i} - T_f \cdot M_{j,i} \right) \tag{1}$$

T_f is the teaching factor used to change the mean value given in Eq. 2. The value of T_f is randomly decided by the algorithm and not given as input to the algorithm. Its value may be either 1 or 2. r_i is random number whose range is between 0 and 1.

$$T_f = \text{round}[1 + \text{rand}(0,1)\{2 - 1\}] \tag{2}$$

Based on the Difference_Mean$_{j,k,i}$, the initial solution is updated in the teacher phase by using Eq. 3.

$$X'_{j,k,i} = X_{j,k,i} + \text{Difference\_Mean}_{j,k,i} \tag{3}$$

Here, $X'_{j,k,i}$ is the updated value of $X_{j,k,i}$. $X'_{j,k,i}$ is accepted if it gives better fitness value then $X_{j,k,i}$. Finally, the updated populations turn into the input to the learner phase.

2.3 Learner Phase

In learner phase, the learners interact among each other in pairs to enhance their knowledge. Interaction is helpful for learner to enhance their knowledge if another learner is having more knowledge as compared to other. The learning trend is shown below: Let P and Q be the two learners, chosen randomly. The selection should be such that $X'_{\text{total}-P,i} \neq X'_{\text{total}-Q,i}$., where $X'_{\text{total}-P,i}$ and $X'_{\text{total}-Q,i}$ are the input values form teacher phase.

$$X''_{j,P,i} = X'_{j,P,i} + r_i \left(X'_{j,P,i} - X'_{j,Q,i} \right), \quad \text{If } X'_{\text{total}-P,i} > X'_{\text{total}-Q,i} \tag{4}$$

$$X''_{j,P,i} = X'_{j,P,i} + r_i \left(X'_{j,Q,i} - X'_{j,P,i} \right), \quad \text{If } X'_{\text{total}-Q,i} > X'_{\text{total}-P,i} \tag{5}$$

Above equations are for maximization problem, the reverse is for minimization. $X''_{j,P,i}$ is accepted if its gives better fitness value. Continue all the phases until the stopping criteria is met.

3 Literature Review

Table 1 shows that TLBO has evolved with various modifications and hybridization done till date.

Table 1 Literature review of TLBO algorithm

Sr. No	Authors	Description
1	Rao et al. (2012)	TLBO algorithm proposed for solving the constrained optimization problems of mechanical design
2	Satapathy and Naik (2012)	Proposed a modified TLBO algorithm by incorporating a random weighted differential vector. Comparisons were made with those of OEA, HPSO-TVAC, CLPSO, APSO and variants of DE such as JADE, jDE and SaDE and different variants of ABC
3	Rao and Patel (2012)	Introduced the notion of elitism in the TLBO algorithm and investigated its effect on the performance of the algorithm for the 35 constrained optimization problems [3]
4	Biswas (2012)	Cooperative coevolutionary TLBO algorithm with a modified exploration strategy for large scale global optimization was proposed [11]
5	Gonzalez-Alvarez (2012)	The multiobjective teaching–learning-based optimization (MOTLBO) was used to solve one of the most important optimization problems in Bioinformatics, the Motif Discovery Problem (MDP) [12]
6	Jiang and Zhou (2013)	Proposed a hybrid differential evolution and TLBO (hDE-TLBO) algorithm to solve the multiobjective short-term optimal hydrothermal scheduling [13]
7	Niknam (2013)	Introduced a modified phase based on the self-adaptive mechanism in the TLBO algorithm in order to achieve a balance between the exploration and exploitation capabilities of the TLBO algorithm [14]
8	Rao and Patel (2013)	Introduced the concept of number of teachers and adaptive teaching factor. The presented modifications speeded up the convergence rate of the basic TLBO algorithm. This work applied successfully to the multiobjective optimization of heat exchangers [15]
9	Tuo (2013)	Proposed an Improved Harmony Search Based Teaching Learning (HSTL) optimization algorithm in order to maintain a balance between the convergence speed and the population diversity [16]
10	Camp and Farshchin (2014)	Applied a modified TLBO algorithm to fixed geometry space trusses with discrete and continuous design variables [17]
11	Ghasemi (2014)	Applied a modified version of TLBO algorithm and Double Differential Evolution (DDE) algorithm for solving Optimal Reactive Power Dispatch (ORPD) problem [18]

(continued)

Table 1 (continued)

Sr. No	Authors	Description
12	Krishnasamy and Nanjundappan (2014)	Integrated the TLBO algorithm with Sequential Quadratic Programming (SQP) to well tune the better solutions obtained by the TLBO algorithms
13	Lim and Isa (2014a)	Proposed to combine the PSO algorithm with the TLBO algorithm in order to offer an alternative learning strategy if a particle fails to improve its fitness [19]
14	Lim and Isa (2014b)	Proposed an improved TLBO algorithm adapted to the enhanced framework of PSO known as Bidirectional Teaching and Peer Learning Particle Swarm Optimization (BTPLPSO)
15	Pholdee (2014)	Integrated the differential operators into the TLBO with a Latin hypercube sampling technique for making of an initial population in order to improve the flatness of a strip during strip coiling process
16	Rao and Patel (2014)	Proposed a multiobjective improved TLBO algorithm for unconstrained and constrained multiobjective function optimization. The algorithm used a grid-based approach to adaptively evaluate the non-dominated solutions maintained in an external archive. The performance evaluated by implementing it on the unconstrained and constrained test problems of CEC 2009 competition [20]
17	Shabanpour-Haghighi (2014)	Proposed a modified TLBO algorithm with self-adaptive wavelet mutation strategy and fuzzy clustering technique to control the size of the repository and selection of smart population the next iteration [21]
18	Xing and Gao (2014)	Proposed hybrid TLBO combines for solution evolution and a variable neighborhood search (VNS) for fast solution enhancement for Permutation flow shop scheduling [22]
19	Yang (2014)	Proposed a new compact TLBO algorithm to combine the strength of the original TLBO and to reduce the memory requirement through a compact structure that utilized an adaptive statistic description to replace the process of a population of solutions [23]
20	Zou (2014a)	Proposed a modified TLBO algorithm with Dynamic Group Strategy. Experiments were conducted on 18 benchmark functions with 50 dimensions [24]
21	Zou (2014b)	A new TLBO variant called bare-bones teaching–learning-based optimization was presented to solve the global optimization problems [25]

(continued)

Table 1 (continued)

Sr. No	Authors	Description
22	Patel and Savsani (2014)	The MO-ITLBO algorithm uses a grid-based approach in order to keep diversity in the external archive. To allow the MO-ITLBO algorithm to handle problems with several objective functions, Pareto dominance is included
23	Wang (2015)	Proposed a hybrid TLBO-DE algorithm for chaotic time series prediction. To show the effectiveness of TLBO-DE approach, it was applied to three typical chaotic nonlinear time series prediction problems [26]
24	Chen (2015)	In order to decrease the computation cost and improve the global performance of the original TLBO algorithm, the area copying operator of the producer–scrounger (PS) model was introduced into TLBO for global optimization problems [27]
25	Dokeroglu (2015)	Proposed a set of TLBO-based hybrid algorithms to solve the challenging combinatorial optimization problem of quadratic assignment
26	Lahari (2015)	Partition based clustering using GA and TLBO algorithms was carried out [28]
27	Umbarkar (2015)	Applied OpenMP TLBO algorithm over multicore system. parallelization of TLBO [29]
28	Yang (2015)	A hybrid bird mating optimizer algorithm with TLBO for global numerical optimization was proposed
29	Rao (2015)	A non-dominant sorting TLBO (NSTLBO) algorithm is introduces to solve the multiobjective optimization problem of the fused decomposition process (FDM)
30	Zhao (2015)	To improving the performance of TLBO, a new Teaching–Learning-Based Optimization with Crossover (TLBOC) is introduced. The TLBOC incorporated the conventional crossover operation of differential evolution (DE) algorithm into teaching phases for balancing local and global searching effectively. The estimation of distribution operation is used to predict a learning elitist which helps to boost learning efficiency of each student in learning phase
31	Gong, Chen, Zhan, Zhang, Li, Zhang and Li (2015)	Distributed EAs and their models in a complete survey [4]

4 Distributed Evolutionary Algorithm Models and Suitable Distributed Computing Platforms

Many real-world optimization problems such as TSP, N-Queens, and graph partitioning are complex in nature and hard to solve as it required large computational power as this kind of problem cannot be solved within polynomial time and thus required exponential time. In order to minimize the computational efforts required by the algorithm, most of the EA is heading towards distributed computing. The main purpose of distributing computing is to distribute the task across multiple nodes. This entire node will perform similar kinds of task with different data items. Parallelization of the evolutionary task may be done at dimensions, population, operational levels, etc. Figure 1 shows distributed approaches of EA [4].

4.1 Population-Based Distribution

In population-based distribution, the entire population is divided into multiple subpopulations and each subpopulation is distributed to multiple processor or salve. Each slave will perform the individual processing on subpopulation and fine result is given to the master. Figure 2 shows the population-based distribution approach and Fig. 3 shows various models to implement population-based distribution.

4.2 Dimension-Based Distribution

In dimension-based distribution, the partitioning is based on dimension or subspace. Figure 4 shows the dimension-based distribution approach and Fig. 5 shows various models to implement dimension-based distribution [4].

Fig. 1 Distribution approaches

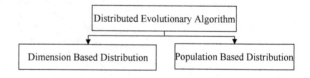

Fig. 2 Population-based distribution

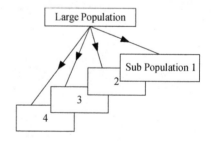

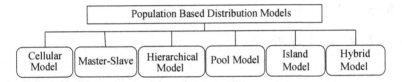

Fig. 3 Population-based distribution model

Fig. 4 Dimension based
distribution

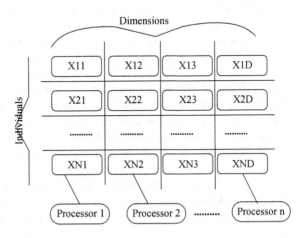

4.3 Distributed Computing Platforms

Various programming platforms such as MPI, Parallel Python, MapReduce, etc., are used to implement distributed EA. Figure 6 shows distributed computing platforms for the distributed EAs.

Fig. 5 Dimension based
distribution models

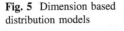

Fig. 6 Distributed
computing platforms

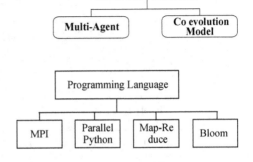

5 Recent Approaches

5.1 TLBO on MapReduce

MapReduce is a framework programming that enables programmers to write programs that process large data sets in parallel. The components of MapReduce are jobTracker, taskTracker, and jobHistoryServer [5]. The execution is carried out in fault-tolerant manner on large clusters of commodity hardware. The model was developed by Jeffrey Dean and Sanjay Ghemawat at Google in 2004. MapReduce is a combination of two functions namely map and reduce function. Mainly MapReduce programming language is used in Hadoop cluster. Figure 7 exhibits TLBO over Hadoop framework using MapReduce. Input data set is split into various chunks by MapReduce which are then processed by map tasks parallely. Output of the maps is sorted by the framework which is then given to the reduce task. File system is used to store the inputs and outputs. MapReduce framework handles monitoring, scheduling, and re-execution of the tasks. MapReduce has gained a considerable amount of attention for implementing DEA [6, 7].

Figure 5.1 shows the implementation of TLBO over Hadoop cluster. Hadoop cluster is master–slave architecture with one master and multiple slaves. Hadoop uses the divide and conquer mechanism. In master–slave architecture, master will assign task to slaves and slaves have to perform that task and the result is given

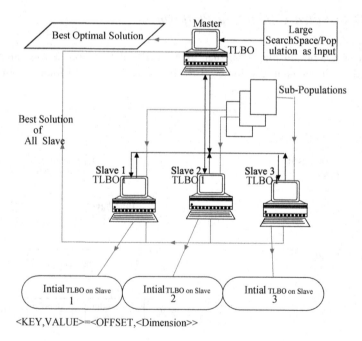

Fig. 7 TLBO on Hadoop

Table 2 Distributed evolutionary algorithms using MapReduce

Sr. No.	Authors	Carried out work
1	McNabb, Monson and Seppi (2007)	MapReduce Particle Swarm Optimization (MRPSO) is a parallel implementation solution of Particle Swarm Optimization for computationally intensive functions. Radial basis function (RBF) network is treated as optimization problems which use large amounts of data
2	Tagawa and Ishimizu (2010)	Concurrent differential evolution based on MapReduce
3	Wu, Wu and Yang (2012)	A MapReduce based Ant Colony Optimization approach to combinatorial optimization problems
4	Hans, Mahajan and Omkar (2015)	Big data clustering using genetic algorithm on Hadoop MapReduce

back to the master. Hadoop framework is used for processing large scale data. In EA, search space or population can be treated as large data. The entire large population is divided into subpopulation by the Hadoop [8, 9].

Each subpopulation is processed by the individual slaves and sent to the master in key value format. Here, key is offset whereas value is dimensions and fitness value computed for each individual. All the combine results of slaves are given to master and further TLBO is executed on the combine result of all the slaves. After completion of job, master will return the best among best solution. This approach can solve the problem of high dimensionality and large computational time taken by the EAs. Some recent papers are shown in Table 2 [10].

5.2 Distributed Computing for Evolutionary Algorithm

The advantages of adopting distributed computing for EA are: Finding substitute solution for same problem, parallel search from multiple points in the space, easy parallelization, more efficient search without parallel hardware, higher efficiency than sequential EAs, and speedup due to use of multiple CPUs.

The disadvantages of adopting distributed computing for EA are: Not suitable for small input, single node failure in case of master–slave architecture, troubleshooting and diagnosing problems in a distributed system are difficult.

6 Conclusion

TLBO is more effective and efficient optimization method for the mechanical design optimization problems. It is a parameter-less optimization method. This new optimization method can be easily used in other engineering design optimization problems.

Recent work shows that most of the EAs take more computation time when the dimension of the problem increases. Another issue associated with the evolutionary algorithm that is it has slow convergence rate, premature convergence, and large time complexity for optimizing the computational intensive objective functions. From the above review analysis, it is been noticed that the traditional sequential algorithm does not provide a satisfactory result with an increase in dimensionality in a reasonable amount of time. This problem puts new challenge in the concept of an evolutionary algorithm. To overcome on such issue, most of evolutionary algorithms are heading towards the Distributed Evolutionary Algorithm (DEA) using Distributed Computing Platforms. This approach solves above-mentioned problem. It gives an opportunity to solve complex and high-dimensional problem using divide and conquer mechanism by distributing task on different computing nodes.

Various distributed models, programming environment, and physical platform would be used to implement the EAs and as well the TLBO algorithm. The distributed EA models would be divided into two groups, namely population-distributed and dimension-distributed. In population distribution, different models such as master–slave, island, cellular, hierarchical, etc., can be used. In dimension distribution, coevolution and multi-agent models can be used. Different models may have different characters and features. This approach in the field of EA is likely to solve the above problems as well as refine the quality of the solution with respect to dimensionality and size of the population.

References

1. Rao, R., Waghmare, G.: A comparative study of a teaching–learning-based optimization algorithm on multi-objective unconstrained and constrained functions. J. King Saud. Univ. Comput. Inf. Sci. **26**, 332–346 (2014)
2. Satapathy, S., Naik, A.: Improved teaching learning based optimization for global function optimization. Decis. Sci. Lett. **2**, 23–34 (2012)
3. Rao, R., Patel, V.: An elitist teaching learning based optimization algorithm for solving complex constrained optimization problems. Int. J. Ind. Eng. Comput. **3**, 535–560 (2012)
4. JiaoGong, Y., Chen, W., Zhan, Z., Zhang, J., Li, Y., Zhang, Q., Li, J.: Distributed EAs and their models: a survey of the state-of-the-art. Appl. Soft Comput. **34**, 286–300 (2015)
5. http://www.tutorialspoint.com/map_reduce/. Accessed 05 July 2016
6. White, T.: Hadoop: the definitive guide: O'Reilly Media, Inc. (2012). ISBN: 1449311520
7. Jiang, W.: A Map-Reduce-Like System for Programming and Optimizing Data-Intensive Computations on Emerging Parallel Architectures. The Ohio State University (2012)
8. McNabb, A.W, Monson, C., Seppi, K.: Parallel PSO using mapreduce. IEEE Congr. Evolut. Comput. (CEC), 7–14 (2007)
9. http://www.ntu.edu.sg/home/epnsugan. Accessed 05 July 2016
10. Tagawa, K., Ishimizu, T.: Concurrent differential evolution based on MapReduce. Int. J. Comput. **4**, 161–168 (2010)
11. Biswas, S., Kundu, S., Bose, D., Das, S.: Cooperative co-evolutionary teaching-learning based algorithm with a modified exploration strategy for large scale global optimization. In:

Swarm, Evolutionary, and Memetic Computing, pp. 467–475. Lecture Notes in Computer Science (2012)

12. Gonzalez-Alvarez, D., Rodriguez, V., Gomez-Pulido, M., Sanchez-Perez, J.M.: Multiobjective teaching-learning-based optimization (MO-TLBO) for motiffinding. In: Proceedings of 13th IEEE International Symposium on Computational Intelligence and Informatics, Budapest, Hungary. doi:10.1109/cinti.2012.6496749

13. Jiang, X., Zhou, J.: Hybrid DE-TLBO algorithm for solving short term hydro-thermal optimal scheduling with incommensurable Objectives, pp. 2474–2479. In: Proceedings of IEEE 32nd Chinese Control Conference (2013)

14. Niknam, T., Rasoul, A., Aghaei, J.: A new modified teaching-learning algorithm for reserve constrained dynamic economic dispatch. IEEE Trans. Power Syst. **28**, 749–763 (2013)

15. Rao, R., Patel, V.: Multiobjective optimization of heat exchangers using a modified teaching-learning-based optimization algorithm. Appl. Math. Model. **37**, 1147–1162 (2013)

16. Tuo, S., Yong, L., Zhou, T.: An improved harmony search based on teaching-learning strategy for unconstrained optimization problems. Math. Probl. Eng. doi:10.1155/2013/413565

17. Camp, C., Farshchin, M.: Design of space trusses using modified teaching-learning-based optimization. Eng. Struct., 87–97 (2014)

18. Ghasemi, M., Ghanbarian, M., Ghavidel, S., Rahmani, S., Moghaddam, E.: Modified teaching learning algorithm and double differential evolution algorithm for optimal reactive power dispatch problem: a comparative study. Inf. Sci. **278**, 231–249 (2014)

19. Lim, W., Isa, N.: Teaching and peer-learning particle swarm optimization. Appl. Soft Comput. **18**, 39–58 (2014)

20. Rao, R., Patel, V.: A multiobjective improved teaching-learning based optimization algorithm for unconstrained and constrained optimization problems. Int. J. Ind. Eng. Comput. **5**(1), 1–22 (2014)

21. Shabanpour, A., Seifi, A., Niknam, T.: A modified teaching learning based optimization for multiobjective optimal power flow problem. Energy Convers. Manag. **77**, 597–607 (2014)

22. Xing, B., Gao, W.: Teaching-learning-based optimization algorithm. In: Innovative Computational Intelligence: A Rough Guide to 134 Clever Algorithms. Intelligent Systems Reference Library, vol. 62, pp. 211–216 (2014)

23. Yang, Z., Li, K., Guo, Y.: A new compact teaching-learning-based optimization method. In: Intelligent Computing Methodologies. Lecture Notes in Computer Science, vol. 8589, pp. 717–726 (2014)

24. Zou, F., Wang, L., Hei, X., Chen, D., Yang, D.: Teaching-learning-based optimization with dynamic group strategy for global optimization. Inf. Sci. **273**, 112–131 (2014)

25. Zou, F., Wang, L., Hei, X., Chen, D., Jiang, Q., Li, H.: Bare-bones teaching-learning-based optimization. Sci. W. J. doi:10.1155/2014/136920

26. Wang, L., Zou, F., Hei, X., Yang, D., Chen, D., Jiang, Q., Cao, Z.: A hybridization of teaching–learning-based optimization and differential evolution for chaotic time series prediction. Neural Comput. Appl. **25**(6), 1407–1422 (2015)

27. Chen, D., Zou, F., Wang, J., Yuan, W.: A teaching–learning-based optimization algorithm with producer–scrounger model for global optimization. Soft Comput. **19**, 745–762 (2015)

28. Lahari, K., Murty, M., Satapathy, S.: Partition based clustering using genetic algorithm and teaching learning based optimization: performance analysis. In: Proceedings of the 49th Annual Convention of the Computer Society of India, vol. 338, pp. 191–200 (2015)

29. Umbarkar, A., Rothe, A., Sathe, A.: OpenMP teaching-learning based optimization algorithm over multi-core system. Intell. Syst. Appl. **7**, 57–65 (2015)

30. Hans, N., Mahajan, S., Omkar, S.: Big data clustering using genetic algorithm on Hadoop mapreduce. Int. J. Sci. Technol. Res. **4** (2015)

Design and Analysis of Broadband Microstrip Antenna Using LTCC for Wireless Applications

Parul Pathak (Rawat) and P.K. Singhal

Abstract This paper reports a new geometry with defected ground for broadband applications. The operating bandwidth of the proposed geometry is 2.04 GHz suitable for modern communication systems. Asymmetric slots are etched into the ground and rectangular patch is optimized for bandwidth enhancement. The proposed antenna by simulation achieved the bandwidth of 71.14% with uniform gain over the entire band. A reasonably high dielectric constant LTCC substrate had been assimilated and it is simulated using CST microwave design tool.

Keywords Microstrip antenna · Defective ground · Broadband
Radiating patch

1 Introduction

The antenna is a one of the most imperative components in communication systems. Antennas that are low profile are customarily used in the process of most of the applications. LTCC (low temperature co-fired ceramic) has several benefits, such as its high thermal conductivity, high temperature resistance, low dielectric loss, the admirable features for high-frequency and high-Q, it also bids high-speed and functionality for handy electronic devices, so it is preferably appropriate material of a little antenna [1].

Microstrip patch antennas can take different shapes, excited in different ways but all have a rudimentary element that comprises of a solo patch of conductor on the upper surface of beached dielectric substrate [2–5]. Microstrip antennas are commonly used as of their countless advantages, such as the light weight, low profile,

P.P. (Rawat) (✉) · P.K. Singhal
Electronics and Communication Department, MITS, Gwalior, India
e-mail: parul.pathak5@gmail.com

P.K. Singhal
e-mail: pks_65@yahoo.com

© Springer Nature Singapore Pte Ltd. 2018
M. Pant et al. (eds.), *Soft Computing: Theories and Applications*,
Advances in Intelligent Systems and Computing 583,
https://doi.org/10.1007/978-981-10-5687-1_24

and conformity. A conventional microstrip printed antenna has a very narrow band that disqualifies its use in modern communication systems, researchers have ended numerous labors to conquest this problem and catch stretched bandwidth. There are quite a lot of approaches to upsurge the bandwidth of microstrip antenna such as amending the shape of a common radiator patch by fit in slots and stubs, swelling the substrate thickness, and acquaint with parasitic element either in coplanar or stack configuration, etc. The bandwidth of microstrip antenna can be augmented by means of air as substrate, but this may clue to hulking antennas and hence inopportune to use [6–12].

The use of defected ground is one of the techniques which is very commonly used nowadays because generally it provides excellent bandwidth enhancement with compact size. In this work, a rectangular patch antenna with defected ground having operating bandwidth of 71.14% in communication band is reported. The simulation is conceded out by means of a finite integration method (CST Microwave). The antenna geometry is presented in Sect. 2, the simulation outcomes are depicted in Sect. 3. Lastly, concluding interpretations are given in Sect. 4.

2 Antenna Geometry

The initial microstrip patch antenna entails of three layers that are ground, substrate, and single patch which are simpler than any conventional wideband microstrip patch antenna. The material castoff for substrate is LTCC with dimension 30 mm × 30 mm × 2.54 mm. The finite ground plane and patch is considered to be perfect electric conducting material (PEC), where dimension of ground plane is 30 mm × 30 mm × 0.018 mm and height of patch is also 0.018 mm. The microstrip patch antenna is a fundamentally low bandwidth antenna so for augmenting the bandwidth slots of different shape and size are cut in the ground.

The initial geometry consists of a rectangular patch having dimensions 13 mm × 12 mm as shown in Fig. 1. The ground plane has no defects. The radiation performance of initial geometry is shown in Figs. 2 and 3.

As per results obtained in Fig. 2, the initial geometry is resonating at 5.92 GHz. Although, the feedline has good matching with the antenna, as visible from Fig. 3, the operating bandwidth in both the bands is very low and some modifications need to be done in the geometry for improved radiation performance.

In direction to spread bandwidth and additional radiation characteristics slots are etched in the ground. The size and shape of slots in the patch and ground are optimized for best results and final geometry is shown in Fig. 4.

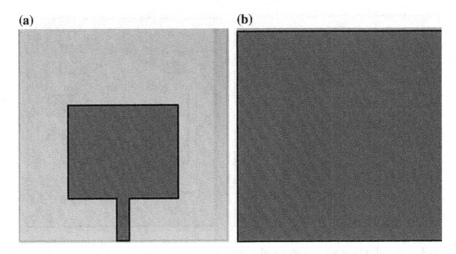

Fig. 1 **a** Patch. **b** Ground structure

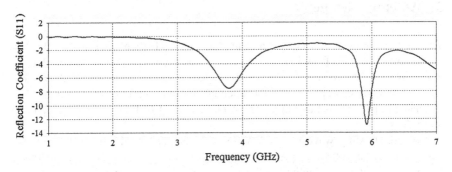

Fig. 2 Variation of return loss with frequency for initial geometry

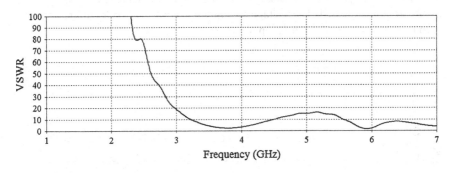

Fig. 3 Variation of VSWR with frequency for initial geometry

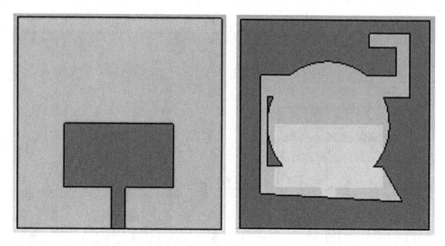

Fig. 4 Top and bottom view of proposed geometry

3   Antenna Geometry

The characteristics of this antenna are simulated by CST microwave studio. By this tool return loss, VSWR and radiation pattern are obtained. The simulated −10 dB return loss of the projected design is displayed in Fig. 5. Return loss shows the range of frequency of operation of the antenna with minimum loss of the signal. The simulated impedance bandwidth of about 71.14% (2.04 GHz) is reached at −10 dB return loss.

Voltage standing wave ratio (VSWR) is demarcated as the proportion of the maximum signal voltage to the minimum signal voltage attained by the standing wave. The greater the amplitude of the standing wave, the greater the impedance mismatch. The proportion of the extreme voltage to the smallest voltage would be 1 (1:1) in case of perfect impedance matching which is achieved when standing waves are not generated. The voltage standing wave ratio (VSWR) falling below 2 is considered to be under the acceptable limits.

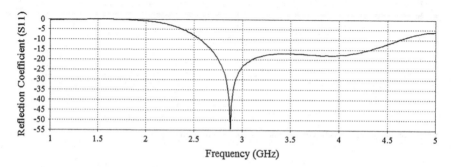

Fig. 5 Variation of return loss with frequency for proposed geometry

The simulated VSWR values at resonating frequency are approaching to unity (1.02) which connotes that still antenna geometry has brilliant cup tie with the microstrip feed line as shown in Fig. 6.

The elevation radiation patterns of projected design at resonant frequency contained by the impedance bandwidth region are shown in Fig. 7.

It might be experiential that the radiation patterns at resonant frequency are unwavering and thorough going radiations are focused normal to the patch geometry. Both the patterns are omnidirectional in nature.

The distinction of gain with frequency is displayed in Fig. 8, the gain is found to be uniform in nature with average value of nearly 2.5 dBi over the entire bandwidth.

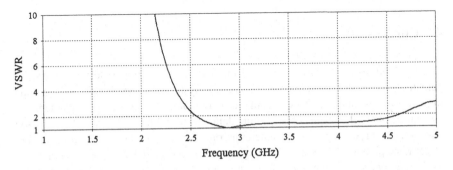

Fig. 6 Variation of VSWR with frequency for proposed geometry

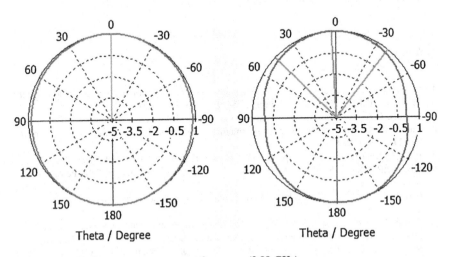

Fig. 7 E and H plane pattern at resonant frequency (2.88 GHz)

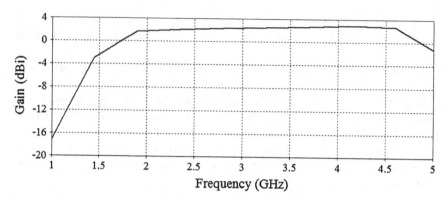

Fig. 8 Variation of directive gain as a function of frequency (GHz)

4 Conclusion

The reported work displays a fresh geometry of rectangular patch antenna with defected ground for modern communication systems. Even though the antenna improvement is under process, its simulation outcomes are inspiring. The antenna is resonating at single frequency with impedance bandwidth of 71.14%. The radiation pattern at desired range of frequencies is observed to be stable and more or less alike in shape and direction of thorough going intensity in each instance is normal to the geometry of patch and will be appropriate for antenna will be suitable for medical imaging, Wireless Broadband (WiBro), vehicular radar, ground penetrating radar, and transportable wireless uses.

References

1. Lamminen, A.E.T., Saily, J.: 60-GHz patch antennas and arrays on ltcc with embedded-cavity substrates. IEEE Trans. Antenna Propag. **56**(9), 2865–2874 (2008)
2. Balanis, C.A.: Antenna Theory, Analysis and Design, John Wiley & Sons
3. Carver, K.R., Mink, J.W.: Microstrip antenna technology. IEEE Trans. Antennas Propag **AP-29**, 2–24 (1981)
4. Bahl, J., Bhartia, P.: Microstrip Antennas. Artech House, Dedham, MA (1980)
5. James, J.R., Hall, P.S., Wood, C.: Microstrip antenna theory and design. Peter Peregrinus, U. K., London (1981)
6. Huynh, T., Lee, K.F.: Single-layer single-patch wideband microstrip antenna, Electron. Lett. 1310–1312 (1995)
7. Rawat, S., Sharma, K.K.: A compact broadband microstrip patch antenna with defected ground structure for C-band applications, Cent. Eur. J. Eng. Springer, 287–292 (2014)
8. Sharma, M.D., Katariya, A., Pandit, A.K.: Design and simulation of broad band aperture coupled microstrip antenna. In: IEEE Fourth International Conference on Computational Intelligence and Communication Networks (CICN), 32–36 (2012)

9. Rawat, S., Sharma, K.K.: Stacked configuration of rectangular and hexagonal patches with shorting pin for circularly polarized wideband performance. Cent. Eur. J. Eng. Springer **4**, 20–26 (2014)

10. Thakare, V.V., Singhal, P.K.: Bandwidth analysis by introducing slots in microstrip antenna design using ann. Prog. Electromagnet. Res. M **9**, 107–122 (2009)

11. Rawat, S., Sharma, K.K.: Annular ring microstrip patch antenna with finite ground plane for ultra-wideband applications. Int. J. Microwave Wireless Technol. 179–184 (2015)

12. Agrawal, A., Singhal, P.K., Jain, A.: Design and optimization of a microstrip patch antenna for increased bandwidth. Int. J. Microwave Wireless Technol. **5**, 529–535 (2013)

An OTSU-Discrete Wavelet Transform Based Secure Watermark Scheme

Devendra Somwanshi, Mahima Asthana and Archika Agarwal

Abstract In today's world, Internet is becoming essential for every human. There are various issues and challenges in data security in Internet system like the authenticity of content or matter, data validation, etc. These factors play crucial role in problems like unauthorized use, modification, etc., of the intellectual properties. Watermarking can resolve the stealing problem of intellectual properties. This technique is used to hide data or identifying information within images, audio, video, and documents which cannot be easily extracted by an unauthorized person. This paper shows an OTSU-DWT (Discrete Wavelet Transform) Algorithm which considers security along with quality for watermarking. The experimental result shows that the peak signal-to-noise ratio (PSNR) value and normalized correlation (NC) value are different for different tagged image file format (TIFF) image files. It could be seen that the PSNR value of original image does not decrease even after embedding the watermark on the contrary, it increases indicating quality improvement. Further, the NC values between original watermark and extracted watermark image showed that there is degradation of quality in extracted watermark.

Keywords Watermarking · Discrete wavelet transform (DWT)
OTSU method · Peak Signal-to-Noise ratio (PSNR) · Normalized correlation (NC)

D. Somwanshi (✉) · M. Asthana · A. Agarwal
School of Engineering and Technology, Poornima University, Jaipur, India
e-mail: imdev.som@gmail.com

M. Asthana
e-mail: mahimaasthana@gmail.com

A. Agarwal
e-mail: archikaagarwal@gmail.com

© Springer Nature Singapore Pte Ltd. 2018
M. Pant et al. (eds.), *Soft Computing: Theories and Applications*,
Advances in Intelligent Systems and Computing 583,
https://doi.org/10.1007/978-981-10-5687-1_25

1 Introduction

Date hiding and image security are some of the emerging issues in today's era of digitalization. Watermarking is one of the most popular techniques used for identifying information, date hiding within digital multimedia. In this technique information is hidden inside a signal, which cannot be extracted by unauthorized person. Watermarks are of various types, few watermarks are visible, but mostly invisible watermarks are preferred. Digital watermarks are generally embedded inside the information, so that proprietorship of the information cannot be prerogative by third parties. In digital watermarking embedding process, quality is generally not degraded [1].

In this paper, discrete wavelet transform technique has been applied on original and watermark images with the combination of OTSU Method [2]. One of the main contributions of the proposed technique is PSNR value of original image which does not decrease even after embedding the watermark; after all it improves the quality. OTSU method performs the image thresholding on clustering basis and it reduces the gray level image into a binary image. This method also provides the embedded images with improved PSNR values and extracted watermark images with improved NC values. In this we shall focus on the binary watermark scheme.

This paper comprises of nine sections. Section 2 gives brief idea about the available literature; Sect. 3 presents the proposed work. Section 4 gave the details of data used that includes the input images and properties of input images. Section 5 presents the performance parameters. Performance comparison of Basic DWT algorithm and OTSU-DWT algorithms is presented in Sect. 6. Section 7 presents the comparison with other contemporary works. Results and conclusion are given in the Sect. 8, followed by the future work in Sect. 9. And finally, references.

2 Literature Review

Literature is full of watermarking schemes based on discrete cosine transform (DCT) [3], lifting wavelet transform [4], discrete wavelet transform (DWT), and singular value decomposition (SVD) [5], combination of DWT-DCT [6], Haar wavelet transform [7], etc., but in all, these quality of images were degraded. Both security and quality parameters were rarely considered together.

OTSU method was introduced in 2004 and later on various researchers used these methods [2, 8–10]. This method has advantages like simplicity of calculation of the threshold. In case, the noise removal and character recognition implementations were good, this method can be used for thresholding. But OTSU has too much of noise in the form of the background being detected as foreground [11] and also this does not work well with variable illumination [11].

3 Proposed Work

OTSU method is named after scientist Nobuyuki OTSU. This method is used to reduce gray level image to binary image and also automatic clustering based image thresholding can be performed using this method. This algorithm is based on bi-modal histogram which states that image contains two classes of pixels foreground pixel and background pixels. In this, two classes are separated to calculate the optimum threshold. Process is applied to obtain minimal intraclass variance, and maximal interclass variance, as the sum of pair-wise squared distance is constant [2]. The architecture of proposed work is shown in Fig. 1.

3.1 DWT Algorithm

The steps of DWT process are shown below [12]:

Step 1: Read original image in any format and resize it into 512 × 512 pixels.
Step 2: Read watermark image in any format and resize it into 64 × 128 pixels.
Step 3: Divide the original image into 4 × 4 sub-blocks.
Step 4: Apply 1-DWT to each of sub-blocks of original image and save.
Step 5: Get embedded image.
Step 6: Divide the watermark image into 4 × 4 sub-blocks.
Step 7: Apply 1-DWT to each sub-block of watermarked image.
Step 8: Get extracted watermarked image.
Step 9: Calculate PSNR value of embedded image and NC value of extracted watermark image.

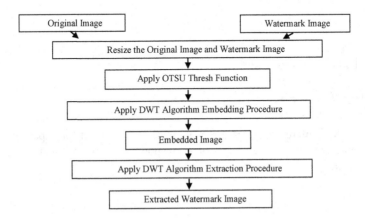

Fig. 1 Architectural diagram of proposed work

3.2 OTSU-DWT Algorithm

The Steps of OTSU-DWT Algorithm are shown below:

Step 1: Read original image in any format and resize it into 512×512 pixels.

Step 2: Read watermark image in any format and resize it into 64×128 pixels.

Step 3: Call OTSU thresh function and calculate maximum thresh value, i.e., 120.

Step 4: Divide the original image into number of 4×4 sub-blocks.

Step 5: Apply 1 level DWT to each of sub-blocks of original image. After that the high-frequency components (High Low (HL) and Low High (LH)) subband are divided into nonoverlapping blocks of 2×2 size. For embedding operation of watermarking process, even columns of HL and odd columns of LH are selected.

Step 6: Block $B(p, q)$ and watermark bit W_b is selected, Mean value $M(p, q)$ of four coefficients in each selected block $B(p, q)$ is calculated.

Step 7: Embedded the watermark bit W_b by calculating integer value R.

Step 8: Get embedded image.

Step 9: Divide the watermarked image into number of 4×4 sub-blocks.

Step 10: Apply 1 level DWT to each of sub-blocks of watermarked image. After that the high frequency components (HL and LH) subband are divided into nonoverlapping blocks of 2×2 size. For extraction of watermarked image even columns of HL and odd columns of LH are selected.

Step 11: Block $B(p, q)$ is selected, Mean value $M(p, q)$ of four coefficients in each selected block $B(p, q)$ is calculated.

Step 12: Extract the watermark bit W_b by calculating the integer value R.

Step 13: Get extracted watermark image.

Step 14: Calculate PSNR value of embedded image and NC value extracted watermark image.

4 Details of Data Used

4.1 Input Images

Table 1 shows different original and watermark images used for experimentation. Group represents the combination of corresponding input original images and watermark images.

Table 1 Original and watermark images used

S. No.	Group No.	Original image	Watermark image
1	Group 1	house.tiff	gaadi.tiff
2	Group 2	bander.tiff	pirate.tiff
3	Group 3	buddhi.tiff	ana.tiff
4	Group 4	mirch.tiff	ghadi.tiff
5	Group 5	tree.tiff	boat.tiff
6	Group 6	girl.tiff	city.tiff
7	Group 7	ghaar.tiff	numbers.tiff
8	Group 8	leena.tiff	room.tiff
9	Group 9	crown.tiff	truck.tiff

(continued)

Table 1 (continued)

S. No.	Group No.	Original image	Watermark image
10	Group 10	marbles.tiff	desert.tiff
11	Group 11	chini.tiff	ruler.tiff

4.2 Properties of Input Images

Original image is the image which is secured through watermarking technique and watermark image is the image which is embedded into original image. Table 2 shows properties of original images and watermark images.

Table 2 Properties of input images

S. No.	Original image	Properties		PSNR value	Watermark image	Properties	
		Size (in pixels)	Size (in KB)			Size (in pixels)	Size (in KB)
1	house.tiff	676 × 598	1177.6	45.50	tank.tiff	292 × 150	128
2	bander.tiff	512 × 512	768	20.40	pirate.tiff	1024 × 1024	1024
3	buddhi.tiff	256 × 256	192	79.20	ana.tiff	512 × 512	256
4	mirch.tiff	512 × 512	768	85.35	ghadi.tiff	256 × 256	64.1
5	tree.tiff	512 × 512	768	69.89	boat.tiff	512 × 512	256
6	girl.tiff	256 × 256	192	22.79	city.tiff	256 × 256	64.1
7	ghaar.tiff	256 × 256	192	35.00	numbers.tiff	256 × 256	64.1
8	leena.tiff	512 × 512	768	42.15	room.tiff	512 × 512	256
9	crown.tiff	512 × 512	768	33.25	truck.tiff	512 × 512	256
10	marbles.tiff	256 × 256	192	35.00	desert.tiff	512 × 512	256
11	chini.tiff	256 × 256	192	24.56	ruler.tiff	512 × 512	256

5 Performance Measures Used

5.1 Peak Signal-to-Noise Ratio (PSNR)

Measure of peak error is represented by PSNR. Quality measurements of compressed or reconstructed images are done using this ratio. Higher value of PSNR indicates improvement in quality. Mean squared error (MSE) is calculated for computation of PSNR values. PSNR is calculated using the following equations.

$$PSNR = 10 \log_{10}(MAX^2/MSE) \tag{1}$$

$$= 20 \log_{10}(MAX/\sqrt{MSE}) \tag{2}$$

where MAX is the maximum possible pixel value of the image.

And MSE represents the cumulative squared error between the compressed and the original image. MSE is calculated using the following equation.

$$MSE = 1/M * N \sum_{i=0}^{M-1} \sum_{j=0}^{N-1} [I(i,j) - K(i,j)]^2 \tag{3}$$

where

M is number of rows
I is original image
N is number of columns
K is reconstructed image

5.2 Normalized Correlation (NC)

The NC represents the normalized correlation factor. Similarity between original watermark and extracted watermark is measured using this factor. NC is calculated using the following equation.

$$NC = 1/N * M \sum_{i=0}^{N-1} \sum_{j=0}^{M-1} W(i,j) \otimes W(i,j)' \tag{4}$$

where

W is original image
W' is watermarked image

6 Performance Comparison of Basic DWT Algorithm and OTSU-DWT Algorithms

The comparative analysis of performance of Basic DWT and OTSU-DWT using image quality parameters like PSNR and NC values is as follows Table 3.

By analyzing Fig. 2, it is seen that improvement in PSNR value between original image and embedded image after applying Basic DWT algorithm is highest for bander.tiff and lowest for mirch.tiff. It is also seen that improvement in PSNR value between original image and embedded image after applying OTSU-DWT algorithms are highest for bander.tiff and lowest for mirch.tiff Table 4.

By analyzing Fig. 3, it is seen that NC value between watermark image and extracted watermark image after applying Basic DWT algorithm is highest for city.tiff and lowest for numbers.tiff. It is also seen that NC value between watermark image and extracted watermark image after applying OTSU-DWT algorithm is highest for city.tiff and lowest for numbers.tiff.

Table 3 Performance analysis of Basic DWT and OTSU-DWT algorithms

Images taken for experiment (Original)	Original images PSNR values.	PSNR values of embedded Images after applying Basic DWT algorithm	PSNR values of embedded images after applying OTSU-DWT algorithm	PSNR improvement after embedding watermark with Basic DWT (%)	PSNR improvement after embedding watermark with OTSU-DWT (%)
house.tiff	45.50	46.70	47.23	2.63	3.80
bander.tiff	20.40	22.46	26.65	10.09	30.63
buddhi.tiff	79.20	80.23	87.03	1.300	9.88
mirch.tiff	85.35	85.45	87.03	0.11	1.96
tree.tiff	69.89	70.00	71.28	0.15	1.98
girl.tiff	22.79	24.70	26.78	8.38	17.50
ghaar.tiff	35.00	35.18	37.18	0.51	6.22
leena.tiff	42.15	43.04	45.91	2.11	8.92
crown.tiff	33.25	34.79	39.47	4.63	18.70
marbles.tiff	35.00	37.01	37.14	5.74	6.11
chini.tiff	24.56	25.33	27.31	3.13	11.19

	house.tiff	bander.tiff	buddhi.tiff	mirch.tiff	tree.tiff	girl.tiff	ghaar.tiff	leena.tiff	crown.tiff	marbles.tiff	chini.tiff
PSNR Improvement after Embedding Watermark with Basic DWT	2.63%	10.09%	1.30%	0.11%	0.15%	8.38%	0.51%	2.11%	4.63%	5.74%	3.13%
PSNR Improvement after Embedding Watermark with OTSU-DWT	3.80%	30.63%	9.88%	1.96%	1.98%	17.50%	6.22%	8.92%	18.70%	6.11%	11.19%

Fig. 2 Performance improvement of PSNR values for Basic DWT and OTSU-DWT

Table 4 Performance comparison of Basic DWT and OTSU-DWT algorithms for NC values

Watermark Images (WI)	NC value (Extracted WI after applying Basic DWT algorithm)	NC values (Extracted WI after applying OTSU-DWT algorithm)	Percentage decrement of NC values between WI and Extracted WI after applying Basic DWT algorithm (%)	Percentage decrement of NC values between WI and Extracted WI after applying OTSU-DWT algorithm (%)
tank.tiff	0.78	0.80	22	20
pirate.tiff	0.42	0.56	58	44
ana.tiff	0.43	0.47	57	53
ghadi.tiff	0.72	0.73	28	27
boat.tiff	0.79	0.79	21	21
city.tiff	0.38	0.41	62	59
numbers.tiff	0.88	0.88	12	12
room.tiff	0.64	0.67	36	33
truck.tiff	0.64	0.65	36	35
desert.tiff	0.63	0.64	37	36
ruler.tiff	0.52	0.56	48	44

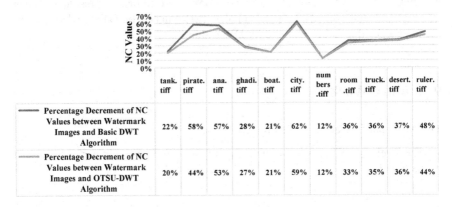

	tank.tiff	pirate.tiff	ana.tiff	ghadi.tiff	boat.tiff	city.tiff	numbers.tiff	room.tiff	truck.tiff	desert.tiff	ruler.tiff
Percentage Decrement of NC Values between Watermark Images and Basic DWT Algorithm	22%	58%	57%	28%	21%	62%	12%	36%	36%	37%	48%
Percentage Decrement of NC Values between Watermark Images and OTSU-DWT Algorithm	20%	44%	53%	27%	21%	59%	12%	33%	35%	36%	44%

Fig. 3 Comparative performance analysis of NC values for Basic DWT and OTSU-DWT

Table 5 Common comparable contemporized work

Ref.	Embedding rule	Parameters taken by researchers	Parameters considered in work	Reviewed research's results	Results obtained with OTSU-DWT
[5]	DWT and SVD	Size: OI = 512*512	Size: OI = 512*512	PSNR = 26.5	PSNR = 87.03 NC = 0.88
		Size: WI = 256*256	Size: WI = 256*256		
[13]	1-DWT	Size: OI = 512*512	Size: OI = 512*512	PSNR = 56.20 SR = 0.99	PSNR = 87.03 NC = 0.88
		Size: WI = 256*256	Size: WI = 256*256		

7 Comparison with Other Contemporary Works

In the past, many authors have done the experiment on watermarking using discrete wavelet transform. They have used similar performance parameters such as PSNR, etc. The results of PSNR cannot be compared as it differs from image to image. The researchers have considered different images for their experimentation than the used in this work. However, a comparison table is prepared as shown in Table 5. These details are for those works that included DWT as watermarking algorithm.

8 Result and Conclusion

Two algorithms Basic DWT and OTSU-DWT were implemented and used for watermarking on different TIFF image file formats. The performance of these algorithms was checked through PSNR and NC values after embedding the watermark into original image. The experimental analysis could fetch the best PSNR value 87.03 for mirch.tiff as embedded image after applying An OTSU-DWT algorithms and best NC value 0.88 for numbers.tiff as extracted watermark image after applying An OTSU-DWT algorithms. The results obtained from experiments validated the claim that performance of OTSU-DWT watermarking algorithm is better that only DWT algorithm.

9 Future Work

The work can further be carried out as follows:

- The level of DWT used may be varied and the improvement in performance can be tested.

- Other watermarking algorithms can also be used on varied images to check their performance.
- The proposed approach can be tested against security attacks.
- Few other parameters can also be considered to make an efficient analysis.

References

1. Chaturvedi, N.: Various digital image watermarking techniques and wavelet transforms. Int. J. Emerg. Technol. Adv. Eng. IJETAE **2**, 363–366 (2012)
2. Vala, H.J., Baxi, A.: A review on Otsu image segmentation algorithm. Int. J. Adv. Res. Comput. Eng. Technol. (IJARCET) **2**(2), pp. 387 (2013)
3. Aniyan, A., Deepa, J.: Hardware implementation of a robust watermarking technique for digital images. IEEE Recent Adv. Intell. Comput. Syst. (RAICS), pp. 293–298 (2013)
4. Arya, M.S., Siddavatam, R., Ghrera, S.P.: A hybrid semi-blind digital image watermarking technique using lifting wavelet transform—Singular value decomposition. In: IEEE International Conference on Electro/Information Technology (EIT), pp. 1–6 (2011)
5. Ayangar, V.R., Talbar, S.N.: A novel DWT-SVD based watermarking scheme. In: International Conference on Multimedia Computing and Information Technology (MCIT), pp. 105–108 (2010)
6. Deb, K., Al-Seraj, M.S., Hoque, M.M., Sarkar, M.I.H.: Combined DWT- DCT based digital image watermarking technique for copyright protection. In: 7th International Conference on Electrical & Computer Engineering (ICECE), vol. 20–22, pp. 458–461 (2012)
7. Huffmire T., Sherwood T.: Wavelet-Based phase classification. In: 15th International Conference on Parallel Architectures and Compilation Techniques (PACT), pp. 95–104 (2006)
8. Kumar, S., Pant, M., Ray, A.: Differential evolution embedded Otsu's method for optimized image thresholding. In: Information and Communication Technologies (WICT), World Congress on, pp. 325–329 (2011) Doi:10.1109/WICT.2011.6141266
9. Chen, Y., Chen, D.R., Yang, X., Li, Y.: Fast Two-Dimensional Otsu's thresholding method based on integral image. In: International Conference on Multimedia Technology (ICMT), Ningbo pp. 1–4 (2010) Doi:10.1109/ICMULT.2010.5630923
10. Sezgin, M., Sankur, B.: Survey over image thresholding techniques and quantitative performance evaluation. J. Electron. Imaging **3**(1), 14–165 (2013). doi:10.1117/1.1631315
11. Devi, H.K.A.: Thresholding: A Pixel-Level image processing methodology preprocessing technique for an OCR system for the brahmi script. Anc. Asia. **1**, 161–165 (2006). doi:10.5334/aa.06113
12. Sheth, R.K., Nath, V.V.: Secured digital image watermarking with discrete cosine transform and discrete wavelet transform method. In: International Conference on Advances in Computing, Communication, & Automation (ICACCA) (Spring), Dehradun, India pp. 1–5 (2016) Doi:10.1109/ICACCA.2016.7578861
13. Sujatha, S.S.: Feature based blind approach for robust watermarking. In: IEEE International Conference on Communication Control and Computing Technologies (ICCCCT), pp. 608–611 (2010)

Analysis and Parameter Estimation of Microstrip Circular Patch Antennas Using Artificial Neural Networks

Monika Srivastava, Savita Saini and Anita Thakur

Abstract Microstrip circular patch antenna play a vital role in the mobile communication area. In round shape patch antenna, radius of round patch is important for generating the required resonant frequency. So proper functioning and high accuracy is required for design specifications of microstrip antenna. This paper discusses the design of round shape microstrip patch antenna using ANN. The mathematically calculated results using ANN model are compared with the Computer Simulation Technology (CST) to verify the results. The calculated resonant frequency is retained at S-band (2–4 GHz). S-band is used for RF power and high data rate application.

Keywords Microstrip patch antenna · Artificial neural Network (ANN)
Resonant frequency · Back propagation algorithm

1 Introduction

Microstrip patch antennas are utilized in large number of areas like mobile antenna system, satellite communication, aircraft system, and biomedical system. It has some effective properties like more economical, light weight, simple in fabrication, and more reliable [1]. Nowadays, these attractive aspects make microstrip antenna more in demand. Many configurations such as rectangular, elliptical, circular, ring, and square are used as patch in microstrip antenna. Circular patch is study in this paper because it is quite simple to fabricate [2]. The resonant frequency is essential to determine because overall antenna performance confides on it. Resonant frequency

M. Srivastava (✉) · S. Saini · A. Thakur
Electronics and Communication Department, Amity University, Noida, India
e-mail: monikasrvstv@gmail.com

S. Saini
e-mail: savitasaini91@gmail.com

A. Thakur
e-mail: athakur@amity.edu

© Springer Nature Singapore Pte Ltd. 2018
M. Pant et al. (eds.), *Soft Computing: Theories and Applications*,
Advances in Intelligent Systems and Computing 583,
https://doi.org/10.1007/978-981-10-5687-1_26

285

also depends on the radius of the round patch which is calculated by formula [3]. This range of resonant frequency is used for radio astronomy, microwave devices and communication, wirelesses LAN, modern radars, etc. So fast, efficient method is required to compute the design parameter of antenna. ANN has inherent property to estimate the result with missing data or inappropriate data. Many prediction-based solutions are obtained using ANN [4]. It is also time reducing and fast method to handle the big data bank [5]. In this paper, proposed ANN model is generating the resonant frequency using different radius for circular patch antenna with dielectric constant of material is used for antenna. For learning of ANN model, back propagation algorithm is used. In this algorithm, inputs are trained to the corresponding target outputs. Calculated result of ANN model is compared with the CST Software.

The rest paper is arranged as follows. Section 2 defines the antenna design analysis, Sect. 3 discusses the synthesis of design of circular microstrip antenna by using NN (neural network), Sect. 4 shows the results and Sect. 5 defines the conclusion.

2 Antenna Design Analysis

Circular patch antenna operates at 3.96 GHz. A FR-4 substrate is used to print the circular patch antenna. The dielectric constant of FR-4 is 4.7 [6]. A circular patch antenna has been printed on dielectric substrate. A large ground plane is formed on the bottom side of the substrate. There are two main feeding techniques to evaluate the circular patch antenna via microstrip method and coaxial feed method [7]. The circular patch has been impressed on the dielectric substrate. The dominant mode is TM_{110} of microstrip circular antenna.

$$(f_r)_{mno} = \frac{1}{2\pi\sqrt{\mu\varepsilon}}\left(\frac{X'_{mn}}{a}\right) \tag{1}$$

To calculate the f_r (resonant frequency) Eq. (1) is used.

$$a = \frac{F}{\left\{1 + \frac{2h}{\pi\varepsilon_r F}\left[ln\left(\frac{\pi F}{2h} + 1.7726\right)\right]\right\} \wedge 1/2} \tag{2}$$

where

$$F = \frac{8.791 \times 10^9}{f_r\sqrt{\varepsilon_r}} \tag{3}$$

$$(f_r) = \frac{1.8412v_0}{2\pi a\sqrt{\varepsilon_r}} \tag{4}$$

ε_r Dielectric constant of the substrate
h Height of the substrate in mm

a Radius of the patch
υ_o Speed of light

The structure of the circular patch antenna is printed on a FR-4 substrate with thickness of $h = 1.5$ mm and relative dielectric constant $\varepsilon_r = 4.7$ as shown in Fig. 1 side view of circular patch antenna. The dimension of the dielectric substrate is taken 50×42 mm, where width (W_s) and length (L_s) of the substrate show in Fig. 2 as front view of microstrip round shape antenna. The radius of the microstrip round shape antenna has been calculated 10 mm at desired resonating frequency 3.8 GHz using Eq. (1) and thickness of the patch was considered 0.01 mm. The patch antenna has been printed on one side of a dielectric substrate and a metallic ground plane of size 50×42 mm is formed on the other side of the substrate.

The antenna structure is fed by a coaxial feed of characteristic impedance of 50 Ω through ground plane. Front view of the microstrip circular antenna is shown

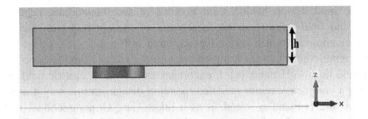

Fig. 1 Side view of microstrip circular antenna

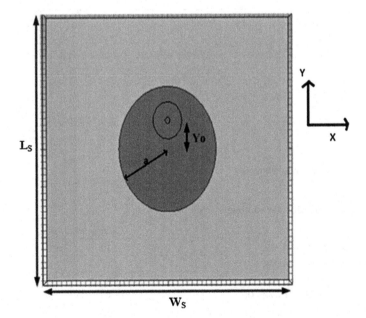

Fig. 2 Front view of microstrip circular antenna

in Fig. 3. To design a coaxial probe, the dielectric constant is fixed 4.7, the inner (D_{in}) and outer diameter (D_{out}) of coaxial probe is chosen 1 and 6 mm respectively. The feed location for the patch antenna is found to be of the distance ($y_0 = 2.7$ mm) away from the center of the patch along the y-axis, where the appropriate impedance matching of 50 Ω has been achieved. Table 1 shows the parameter of proposed microstrip circular antenna.

Figure 3 shows the S_{11} magnitude in dB plot of the circular patch antenna using coaxial feed method at 3.8 GHz. Because in coaxial feeding method it can be easily to set the impedance matching by adjusting the feed position [8]. It has been observed that the circular patch antenna radiates at single resonating frequency. Figure 3 shows the S parameter graph with return loss value is −12.49 dB at gain of 5.6 dB for circular patch antenna.

The E-plane radiation pattern is directional since the rotating electric current enters the radio wire through the feed line circular domain intersection and leaves the receiving wire through the transmitting edge of the circular plot [8]. They frame an electric field design having field maxima at the transmitting edges toward radiation and field minima at the focal point of the patch. So an arch molded radiation example is framed which has appeared in Fig 4.

The magnetic field that is prompted because the electric field is opposite to the electric field lines in H-plane [8]. It encompasses the whole electric field lines and producing a complete circular modeled radiation design. In this way, the radiation example of the H-Plane is round fit as a fiddle (i.e., Omnidirectional in nature) which is appeared in Fig. 5. It is observed in Figs. 4 and 5 that the gain is to be 5.6

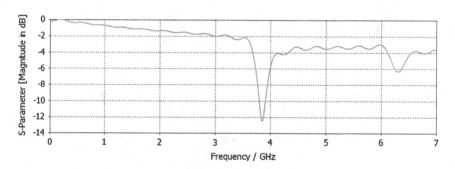

Fig. 3 S parameter versus frequency

Table 1 Circular patch Antenna design parameter

Parameters	Dimension (mm)
Radius of the round patch (a)	10
Thickness of the substrate (h)	1.5
Outer diameter of coaxial feed (D_{out})	6
Inner diameter of coaxial feed (D_{in})	1
Feed location from the center of the patch (y_0)	2.7

Fig. 4 E-plane radiation pattern of circular patch antenna

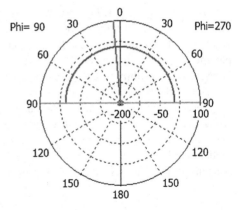

Fig. 5 H-plane radiation pattern of circular patch antenna

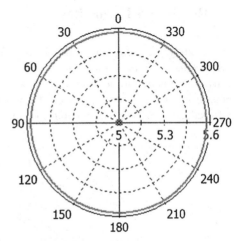

Table 2 Performance parameters of round shape patch antenna

Frequency (GHz)	Return loss magnitude (dB)	BandWidth (MHz) at −10 dB reference level	Gain (dBi)
3.82	−12.49	105	5.6

dBi at 3.8 GHz. Table 2 shows the performance parameter of circular patch antenna on the basis of H and E plane radiation pattern.

3 Design of Microstrip Round Shaped Antenna Using ANN (Artificial Neural Network)

In proposed neural network model, three-layer architecture is used where one hidden layer is between input and output layer. ANN architecture for microstrip circular antenna is shown in Fig. 6. As input radius and dielectric constant is the parameter to input layer. For training to the model, 400 data are used. In 400 data, 260 data are used for training, 60 data are used for testing, and rest 60 data used for validation. Equation (1) is used for generating the input parameter. The output parameter is the resonant frequency. Sigmoid activation function is used in hidden layer. The algorithm is used for learning to network is Levenberg–Marquardt algorithm which has rapid convergence.

These training algorithms of neural network are utilized to train the samples which have only aim to minimize the error to acquire the resonant frequency at defined dimensions [4]. By varying the number of neurons in hidden layer, ANN model is trained. Best model is selected on the basis of minimum mean square that is achieved at given number of neuron to the hidden layer [9].

4 Result and Discussion

In this study, we used nftool fitting tool of MATLAB to determine the best possible test results. In fitting nftool, we declare the input layer, hidden layer, and the activation function to be used to get a desired output. By changing the number of neuron in hidden layer, we trained the network. Best performance ANN model is selected on the basis of minimum mean square error. Table 3 shows the comparison

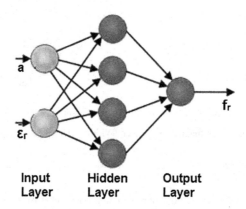

Fig. 6 Architecture of ANN for circular patch antenna

Table 3 Shows the MSE versus number of iterations with varying the number of neuron in hidden layer

S. No.	No. of hidden neurons	Mean square error	No. of iterations
01	5	5.7426e−10	85
02	10	2.9786e−11	259
03	15	1.7594e−12	568
04	20	1.9127e−08	56

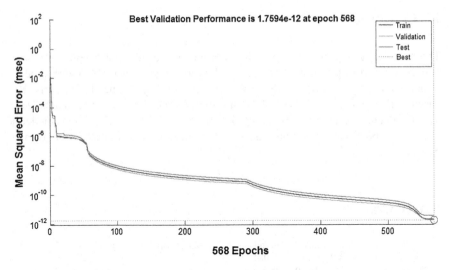

Fig. 7 Graph of performance of training, validation, and test data sets with respect to epochs for circular patch antenna parameter

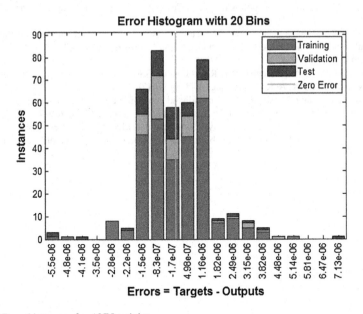

Fig. 8 Error histogram for ANN training

of mean square error with respect to number of iteration while changing the number of neuron in hidden layer.

Figure 7 shows the performance of training, validation, and test data sets with respect to epochs for circular patch antenna. It is seen that the best training performance was meet at epoch 568 and the mean square error is 1.7594×10^{-12}. Figure 8 shows the error histogram of proposed neural network model for circular patch antenna.

5 Conclusion

The target of this paper is to calculate accurately the resonant frequency of the circular microstrip patch antenna with the help of artificial neural network model. This neural network model possesses the high accuracy during the analysis. By varying the number of neuron in hidden minimum mean square error is achieved. Mostly, antenna design process is time consuming but with neural network model it reduced the process time and increased the accuracy to synthesis process of antenna parameter design.

References

1. Mailloux, R.J., McIlvenna, F., Kernweis, N.P.: Microstrip array technology. IEEE Trans. Antennas Propag. **AP-29**, 25–37 (1981)
2. Pozar, D.M., Gardiol, D.H.: Broadhand Patch Antennas the Analysis and Design of Microstrip Antennas and Arrays. IEEE Press, New York (1995)
3. Balanis, C.A.: Antenna Theory, Analysis and Design, 3rd edn., pp. 844–849. Wiley (2005)
4. Harston, M.C., Pap, R.: Handbook of Neural Computing Application. Academic Press, London (1990)
5. Pattnaik, S.S., Panda, D.C., Devi, S.: Tunnel-based artificial neural network technique to calculate the resonant frequency of a thicksubstrate microstrip antenna. Microw. Opt. Technol. Lett. **34**, 460–462 (2002)
6. Guo, Y.J., Paez, A., Sadeghzadeh, R.A., Barton, S.K.: A circular patch antenna for radio LAN's. IEEE Trans. Antennas Propag. **45**(1), 177–178 (1997)
7. Asem, A.-Z., Yang, F., Kishk, A.: A broadband center-fed circular patch-ring antenna with a monopole like radiation pattern. IEEE Trans. Antennas Propag. **57**(3), 789–792 (2009)
8. Cheston, T.C., Frank, J.: Phased Array Radar Antenna. McGraw Hill, New York (1990)
9. Maren, A., Harston, C., Pap, R.: Handbook of Neural Computing Application. Academic Press, London (1990)

Identification of Top-Ranked Features Using Consensus Affinity of State-of-the-Art Methods

Barnali Sahu, Satchidananda Dehuri and Alok Kumar Jagadev

Abstract Feature selection is one of the vital preprocessing tasks of biological data mining process. Although plentiful feature selection techniques have been developed for processing small- to large-dimensional dataset in general but most of them are classifier bound rather than selection of a set of generic features. Hence to address this issue, this work is making an effort to reveal generic features with consensus affinity using state-of-the-art methods based on which various classification models produce unbiased and stable results. In particular, here, we have focused on set of significant features from four benchmark microarray datasets selected by state-of-the-art filter methods such as signal-to-noise ratio, significant analysis of microarray, and t-test and then compared with the features selected by preferred evolutionary wrappers such as Particle Swarm Optimization (PASO), Ant Colony Optimization (ACE), and Genetic Algorithms (Gas). The classification accuracies of several classifiers are measured based on the selected features by the above-mentioned filter techniques and evolutionary wrapper techniques.

Keywords Classification · Feature selection · Filter · Wrapper

B. Sahu (✉)
Department of Computer Science and Engineering, Siksha O Anusandhan University,
Bhubaneswar 751030, Odisha, India
e-mail: barnalisahu@soauniversity.ac.in

S. Dehuri
Department of Information and Communication Technology, Fakir Mohan University,
Vyasa Vihar, Balasore 756019, Odisha, India
e-mail: satchi.lapa@gmail.com

A.K. Jagadev
School of Computer Engineering, KIIT University, Bhubaneswar 751024, Odisha, India
e-mail: alok.jagadev@gmail.com

© Springer Nature Singapore Pte Ltd. 2018
M. Pant et al. (eds.), *Soft Computing: Theories and Applications*,
Advances in Intelligent Systems and Computing 583,
https://doi.org/10.1007/978-981-10-5687-1_27

293

1 Introduction

In biological data mining particularly microarray data, gene selection is a challenging task for researchers since it contains hundreds of thousands of genes in comparison to number of samples [1]. Microarray data can be expressed as an expression matrix of $M \times N$, where M and N represent number of genes and samples, respectively and e_{ij} represents the expression value of it gene and th sample. Selecting S genes from M is known as gene (feature) selection, where $S \subset M$. Many feature (gene) selection methods have been developed under the umbrella of filter and wrapper approaches [2]. Filter method evaluates the features (genes) without the support of any classifier and selects the features with best statistical scores and arbitrary thresholds [3]. On the other hand, wrapper approach is classifier dependent. Filter approach is simple with low computational cost in comparison to wrapper approach. Wrapper approach uses either greedy or stochastic search strategy to find the relevant feature subset. Evolutionary algorithms such as GA [4], PASO [5], and ACE [6] have proved to be better in solving feature selection problem. Generally, different criteria may lead to different optimal feature subsets provided that every criterion tries to measure the discriminating ability of a feature or a subset of features to distinguish different class label [7]. The objective of this work is to identify the generic feature subset based on which the predictive ability of all the classifier will strengthen with no bias. The classification is only used to evaluate the feature subset selected for consideration. The features selected by filter methods and wrapper methods are compared based on the classification accuracies.

The paper is organized as follows: Section 2 discusses the basic background of this work. Section 3 provides the overall framework of the proposed work. Section 4 provides experimental results and statistical analysis of the classifiers based on Friedman's test and Niminy test. Section 5 provides the conclusion.

2 Basic Materials

This section is providing the background details of the proposed work.

2.1 Signal-to-Noise Ratio (SNR)

It is defined as the ratio of signal to the noise power. Mathematically, defined as follows:

$$\text{SNR} = \frac{\mu_1 - \mu_2}{\sigma_1 + \sigma_2}, \tag{1}$$

where μ_1, μ_2, σ_1 and σ_2 are the mean and standard deviations of positive and negative classes, respectively.

2.2 t-statistics

This is a statistical test to identify differentially expressed genes between two groups. It is the distance between the positive and negative class based on standard deviation.

$$t = \frac{\mu_1 - \mu_2}{\sigma}, \tag{2}$$

$$\sigma = \sqrt{\frac{\sigma_1^2}{n_1} + \frac{\sigma_2^2}{n_2}}, \tag{3}$$

Here, μ_1, μ_2, σ_1 and σ_2 are the mean and standard deviations of class 1 and class 2, respectively. Here, $n1$ and $n2$ are total number of positive and negative samples.

2.3 Significance Analysis of Microarrays (SAM)

SAM is used to choose significant genes based on differential expression between sets of samples.

$$d = \frac{\mu_1 - \mu_2}{\sigma + \sigma_0}, \tag{4}$$

$$\sigma = \sqrt{\left(\frac{1}{n_1} + \frac{1}{n_2}\right)\frac{\sigma_1 + \sigma_2}{n_1 + n_{2-2}}}, \tag{5}$$

where σ is standard deviation and the σ_0 term is here to deal with cases when the variance gets too close to zero and $n1$ and $n2$ are total number of positive and negative samples, respectively.

2.4 Binary Particle Swarm Optimization (BPSO)

The salient feature of PSO is that knowledge is optimized by social interaction. In Binary PSO, the position of the particle is represented as a binary string form generated randomly. In the process of feature selection, the fitness of the particle is the classification accuracy of 1-NN classifier. The fitness value of each particle is represented as $pbest$ and the best fitness value among a group of $pbest$ is the global best and is represented as $gbest$. Tracking the features of $pbest$ and $gbest$ with regard to the position and velocity, we can finally get the optimal feature subset. The particle in the population is updated according to the following equations:

$$v_{pd}^{new} = w * v_{pd}^{old} + c_1 * \text{rand}_1 * (pbest_{pd} - x_{pd}^{old}) + c_2 \text{ rand}_2(gbest_d - x_{pd}^{old}), \quad (6)$$

if

$$v_{pd}^{new} \notin (V_{min}, V_{max}) \text{ then } v_{pd}^{new} = \max(\min(V_{max}, v_{pd}^{new}), V_{min}) \quad (7)$$

$$S(v_{pd}^{new}) = \frac{1}{1 + e^{-v_{pd}^{new}}}, \quad (8)$$

If $(\text{rand} < S(v_{pd}^{new}))$ then

$$x_{pd}^{new} = 1, \text{ else } x_{pd}^{new} = 0 \quad (9)$$

In the above equation, w is the inertia weight, c1 and c2 are acceleration factors and rand, rand1 and rand2 are random numbers. The old and new velocities of the particles are v_{pd}^{old} and v_{pd}^{new}, respectively.

2.5 Ant Colony Based Feature Selection

In ant colonies, pheromone secreted by ant is used as a communication medium to trace the food source. This concept can be used to select features from a high-dimensional data set. The approach is described bellow in terms of pseudocode.

ACO-NN Wrapper Method

Step 1. Initialization

> Set $T_b = cc$ where cc is a constant
> Set maximum number of iteration
> Set k, where k-best subsets will influence the subsets of next iteration.
> Set p, where m-p is the number of features each ant will start with in the second and following iteration.

Step 2. Randomly assign a subset of m features to S_j and go to step 4.
Step 3. Select the remaining features for each ant

> For $mm = m - p+1$ to m
> Given subset S_j, choose feature f_i that maximizes the selection
> $S_j = S_j \cup \{f_i\}$.

Replace the duplicated subsets
End for

Step 4. Evaluate the $F = fitness\ (S_j)$.
Step 5. Update *BL*
Step 6. For each feature f_i, update the pheromone trail
Step 7. Using the feature subset of the best k ant

For $j = 1\ to\ na$
Randomly produce $m\text{-}p$ features for ant j, to be used in the next iteration and store it in S_j.
End for

Step 8: Continue till maximum number of iteration

The symbols used in the above pseudocode are defined as follows.

Features of the original set, $F = \{f_1,\ f_2...f_n\}$, *na*: number of artificial ants to search the feature space, T_b: the intensity of pheromone associated with feature f_b, for each ant j, a list of feature subset selected is $S = \{s_1,\ s_2,\ ....,\ s_m\}$, *L*: list of previously tested subsets, *k*: is the best k subsets which will be used to influence the feature subsets of the next iteration, *BL*: list of the best k subsets

2.6 Genetic Algorithm Based Feature Selection

The objective of GA-based feature selection is to find a population of best sequence for every feature vector, which maximizes the classification accuracy. A fitness function is generated to evaluate the fitness of each chromosome in the population. In our proposal (i.e., Sect. 3), we have taken k-nearest neighbor classifier, where the classification accuracy is the fitness value of each chromosome. The detail of GA and pseudocode to carry out the GA-based feature selection is given in [8].

3 Proposed Work

Our work is a three phase method (see Fig. 1). In phase 1, the microarray datasets are normalized using z-score normalization, in phase 2 the gene selection methodologies are applied in two different approaches, i.e., filter approach and evolutionary wrapper approach. The filter technique includes gene selection based on signal-to-noise ratio (SNR), significant analysis of microarray (SAM), t-test and evolutionary wrapper approach includes Binary Particle Swarm Optimization (BPSO), Ant Colony Optimization (ACO), and Genetic Algorithm (GA). All the evolutionary wrappers use the predictive accuracy of k-NN classifier to determine the quality of the selected features. Twenty in specific significant genes are selected

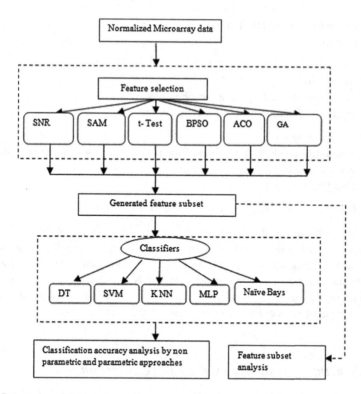

Fig. 1 Proposed model

by each above-mentioned approaches from the given data set. In phase 3, the classifiers used in the work are decision tree, Support Vector Machine (SVM), k-Nearest Neighbor (k-NN), Multilayer Perceptron (MLP), and Naïve Bayes (NB) for measuring the accuracy of our framework. We use 10-fold cross validation to validate the performance of the classifiers for all the data set. As multiple classifiers and multiple data sets are used in the current work, the classification accuracies are analyzed based on hypothetical testing such as Fried man's approach (non parametric) and Neimany's approach (parametric). Identification of generic features for which all the classifiers gives a stable result is identified for different data set based on filter approach.

4 Experiments and Comparative Analysis

The experiment is done in MATLAB2010a. Table 1 describes the data set and parameter setting for BPSO, ACO, and GA are given in Table 2. Tables 3, 4, 5, and 6 show the 20 significant selected features by different selection approaches for

Table 1 Data set description

Data set	# Genes	# Samples	Positive classes	Negative classes	Classes
DLBCL [9]	2648	77	58	19	2
Prostate [10]	2136	102	52	50	2
Lymphoma [11]	5000	38	27	11	2
Breast cancer [12]	47294	128	84	44	2

Table 2 Parameter setting

PSO		ACO		GA	
C1	2	# Features	20	Population	70
C2	2	MaxIt	30	Max generation	30
Vmax	6	K	6	Selection	Roulette wheel
Vmin	−6	Cc	1	Crossover	Single point
Wmax	0.9	P	6	Mutation type	Flip bit
Wmin	0.4			Distance measure in k-NN	Euclidian
MaxIt	30			K	5
# Features	20				

Table 3 20 Significant genes selected from DLBCL dataset

	Gene accession number
SNR	'D55716_at', 'D79987_at', 'D80008_at', 'HG1980HT2023_at', 'HG4074HT4344_at', 'HG417HT417_s_at', 'L17131_rna1_at', 'L36720_at', 'M14328_s_at', 'M57710_at', 'M63138_at', 'U63743_at', 'U73379_at', 'V00594_at', 'V00594_s_at', 'X02152_at', 'X13293_at', 'X56494_at', 'X62078_at, 'X74801_at'
SAM	'X02152_at', 'U63743_at', 'V00594_s_at', 'HG1980HT2023_at', 'L19686_rna1_at', 'AB002409_at', 'M14328_s_at', 'L17131_rna1_at', 'X56494_at', 'X62078_at', 'HG2279HT2375_a', 'X74801_at', 'X67951_at', 'D79987_at', 'U73379_at', 'M63138_at', 'M23323_s_at''D55716_at', 'U05340_at', 'S73591_at'
t-test	'U63743_at', 'X02152_at', 'V00594_s_at', 'HG1980-HT2023_at', 'L19686_rna1_at', 'AB002409_at', 'D79987_at' 'M14328_s_at', 'Z50115_s_at', 'X74801_at', 'L17131_rna1_at', 'U05340_at', 'X62078_at', 'HG4272-HT4542_at', 'M23323_s_at', 'X56494_at', 'L36720_at', 'U73379_at', 'U12595_at', 'HG2279-HT2375_at'
BPSO	'X69910_at', 'X70476_at', 'HG1980HT2023_a', "HG4662HT5075_at', 'L17131_rna1_at', 'M64571_at', 'X67951_at', 'X56494_at', 'M63138_at', 'HG2279-HT2375_a', 'S73591_at', 'HG417-HT417_s_a', 'V00594_at', 'M57710_at', 'J04988_at', 'J03909_at', 'M63379_at', 'M12529_at', 'X52851_rna1_at', 'M55998_s_at'
ACO	'U88629_at', 'V00572_at', 'X02317_at', 'L02326_f_at', 'M23323_s_at', 'HG4272-HT4542_a', 'S73591_at', 'U49835_s_at', 'V00594_at', 'X91911_s_at', 'M63379_at', 'M13792_at', 'X52773_at', 'AFFX-HUMGAPDH/M', 'X05360_at', 'X06985_at', 'M96684_at', 'HG2380-HT2476_s' 'U14187_at', 'X91868_at'
GA	'X81625_at', 'M20867_s_at', 'HG3117-HT3293_at', 'U33921_at', 'L34673_at', 'D87116_at', 'U44754_at' 'S76638_at', 'U33053_at', 'X51688_at', 'L19183_at', 'X05855_s_at', 'U47414_at', 'X67235_s_at', 'S74017_at', 'U03851_at', 'D50310_at', 'J02645_at', 'U90878_at' 'L16842_at'

Table 4 20 Significant genes selected from prostate dataset

	Gene accession number
SNR	'1251_g_at', '1740_g_at', '32598_at', '34840_at', '35276_at', '36174_at', '36491_at', '37366_at', '37639_at' '38028_at', '38087_s_at', '38291_at', '38634_at', '38827_at', '39756_g_at', '40282_s_at', '41468_at', '41706_at', '575_s_at', '914_g_at'
SAM	'37639_at', '41706_at', '41468_at', '32598_at', '37366_at', '575_s_at', '36174_at', '34840_at', '38827_at', '36491_at', '38634_at', '1740_g_at', '40282_s_at', '35276_at', '38087_s_at', '36495_at', '33904_at', '39073_at', '1251_g_at', '914_g_at'
t-test	'37639_at', '41706_at', '32598_at', '41468_at', '37366_at', '575_s_at', '36174_at', '34840_at', '38827_at', '36491_at', '38634_at', '1740_g_at', '35276_at', '40282_s_at', '38087_s_at', '36495_at', '33904_at', '1251_g_at', '39073_at', '914_g_at'
BPSO	'33683_at', '33705_at', '33818_at', '31791_at', '33859_at', '41203_at', '617_at', '41300_s_at', '41435_at' '39054_at', '556_s_at', '38634_at', '31906_at', '41106_at', '837_s_at', '39657_at', '1389_at', '424_s_at' '39135_at', 'AFFX-M27830_M_a'
ACO	'39030_at', '39120_at', '41268_g_at', '553_g_at', '444_g_at', '34775_at', '685_f_at', '757_at', '38827_at', '291_s_at', '41610_at', '39329_at', '1890_at', '36203_at', '33904_at', '38604_at', '35823_at', '1661_i_at', '1514_g_at', '1662_r_at'
GA	'35828_at', '39472_s_at', '1635_at', '37399_at', '36617_at', '36945_at', '725_i_at', '2089_s_at', '39332_at' '39331_at', '40643_at', '36100_at', '34777_at', '631_g_at', '35017_f_at', '39423_f_at', '34255_at', '31527_at', '41338_at', '1052_s_at'

Table 5 20 Significant genes selected from lymphoma dataset

	Gene accession number
SNR	'M23197_at', 'X59417_at', 'X17042_at', 'M16038_at', 'U05259_rna1_at', 'M84526_at', 'D49950_at', 'M55150_at', 'M81933_at', 'U82759_at', 'Y12670_at', 'M92287_at', 'U46751_at', 'U22376_cds2_s_at', 'M28170_at', 'X95735_at', 'X52142_at', 'U50136_rna1_at', 'M80254_at', 'M27891_at'
SAM	'M86933_s_at', 'M28170_at', 'M16038_at', 'U05259_rna1_at', 'M84526_at', 'AF007551_at', 'D42073_at', 'U46751_at', 'U82759_at', 'M80254_at', 'Y12670_at', 'M23197_at', 'M92287_at', 'U22376_cds2_s_at 'X17042_at', 'X95735_at', 'M55150_at', 'X52142_at', U50136_rna1_at', 'M27891_at'
t-test	'U50136_rna1_at', 'X95735_at', 'M55150_at', 'M16038_at', 'Y12670_at', 'M23197_at', 'M81933_at', 'U82759_at', 'D49950_at', 'X17042_at', 'M84526_at', 'L08246_at', 'Y00787_s_at', 'M80254_at', 'U4651_at', 'M62762_at', 'M27891_at', 'M63138_at', 'M28130_rna1_s_at', 'M81695_s_at'
BPSO	'X15722_at', 'U79254_at', 'L07615_at', 'M58378_cds1_at', 'M15353_at', 'X91117_rna1_at', 'D25274_at', 'AB000896_at', 'D10522_at', 'U71087_at', 'U20998_at', 'X15722_at', 'U79254_at', 'L07615_at', 'M58378_cds1_at', 'M15353_at', 'X91117_rna1_at', 'D25274_at', 'AB000896_at', 'D10522_at'
ACO	'L36847_at', 'X86098_at', 'X02874_at', 'L41067_at', 'U89336_cds4_at', 'U16720_rna1_s_at', 'M14758_at', 'M54995_at', 'M58459_at', 'U25975_at', 'D83657_at', 'D26155_s_at', 'M63438_s_at', 'U72621_at', 'X57522_at', 'D86970_at', 'S80335_at', 'D12763_at', 'S76067_at', 'HG25930HT26386_at'
GA	'M74096_at', 'M92432_at', 'J03600_at', 'Z46967_at', 'M30269_at', 'U46751_at', 'U39412_at', 'L10413_at', 'X59871_at', 'X58022_at', 'X98743_at', 'S82185_at', 'M63438_s_at', 'X66358_at', 'X17576_at', 'J00148_cds2_f_at', 'M94055_at', 'M19045_f_at', 'X12451_at', 'D86479_at'

Table 6 20 Significant genes selected from breast cancer dataset

	Gene accession number
SNR	'GI_14249703-S', 'GI_16507967-S', 'GI_18152766-S', 'GI_22748948-S', 'GI_22779933-S' 'GI_29126237-S', 'GI_29728071-S', 'GI_29738585-S', 'GI_34304343-A', 'GI_34452698-S', 'GI_37551139-S', 'GI_37552339-S', 'GI_38146007-A', 'GI_38455428-S', 'GI_40788002-S', 'GI_42660596-S', 'GI_4503602-S', 'GI_4503928-S', 'GI_7706686-S', 'GI_9951924-S'
SAM	GI_4503602-S, GI_14249703-S, GI_9951924-S, GI_29738585-S, GI_38455428-S, GI_22779933-S, GI_37552339-S, GI_38146007-A, GI_29728071-S, GI_4503928-S, GI_22748948-S, GI_37551139-S, GI_34452698-S, GI_40788002-S, GI_42660596-S, GI_7706686-S, GI_4885496-S, GI_21614543-S, GI_34304343-A, GI_18584419-S
t-test	'GI_4503602-S', 'GI_14249703-S', 'GI_22779933-S', 'GI_9951924-S', 'GI_38146007-A', 'GI_34452698-S', 'GI_29738585-S', 'GI_38455428-S', 'GI_37552339-S', 'GI_37551139-S', 'GI_40788002-S', 'GI_22748948-S', 'GI_4503928-S', 'GI_29728071-S', 'GI_34304343-A', 'GI_42660596-S', 'GI_29126237-S', 'GI_42657473-S', 'GI_7706686-S', 'GI_34147362-S'
BPSO	'GI_28882054-S', 'GI_21426826-S', 'GI_21389370-S', 'Hs.344829-S', 'GI_40018632-S, GI_37541410-S, GI_7108334-S, GI_24497576-A, 'hmm21458-S', GI_4757883-S, GI_30102943-S, GI_13376079-S, GI_14589926-S, GI_42661231-S, GI_42794759-I, Hs.520142-S, hmm22444-S, hmm19540-S, GI_28559079-I, hmm24555-S
ACO	'GI_38257154-I', 'GI_38261961-S', 'GI_9951924-S', 'GI_19920322-A', 'GI_21614543-S, 'Hs.507307-S', GI_42659459-S, GI_31341936-S, GI_24497500-S, GI_22035691-A, GI_29728071-S, GI_12751474-S, GI_42741670-S, GI_34222365-S, GI_31542212-S, GI_4507450-S, GI_38327636-S, GI_4505168-S, GI_10947028-S, GI_17158004-S 'U14187_at', 'X91868_at'
GA	'GI_27734964-S', 'GI_34447230-S', 'GI_31542640-S', 'GI_28872791-A', 'GI_6715588-S', 'GI_40788019-S', 'GI_38261972-S', 'GI_23510383-A', 'GI_34330189-S', 'GI_5031970-S', 'GI_21071004-S', 'GI_4504098-S', 'GI_20070134-S', 'GI_25777737-S', 'GI_4505878-S', 'GI_22748650-S', 'GI_4557286-S', 'GI_19913405-S', 'GI_45433498-S', 'Hs.117020-S'

DLBCL, prostate, Lymphoma and Breast cancer data set, respectively. In the above table, the highlighted gene accession numbers are the common genes selected by the filter approaches, on the other hand, feature selected by the evolutionary wrappers does not contain any common features. Likewise different classifiers are giving different accuracies for the same set of genes in a data set. The difference is due to the parametric variations among learning algorithms. For all the four data sets, the k-NN classifier with $k = 3$ gives better result with Euclidian distance as another parameter. Classification and regression tree is showing significant result for Lymphoma data set. Support vector machine classifier with linear kernel is used. In MLP classifier since the selected genes are 20 and the numbers of classes are 2, a 20-L-2 MLP, i.e., a MLP with 20 inputs, L hidden neurons and 2 outputs, was trained with the samples on the selected gene subset. The training set is used to train the MLP for p epochs and the test set error was recorded for each of the epochs and the average is taken. The number of hidden layer is 1 with 2 nodes. With the increase in the hidden layer the error rate increases hence all the experiments are done with 1 hidden layer. Friedman's test (non parametric) and Nemenyi test (parametric) approach has been chosen for further analysis of classifier. For both the test, the null hypothesis is "all the classifiers perform equally" and rejection of the null hypothesis means that "there exist one pair of classifiers with significantly different performance". Tables 7, 8, 9, 10, 11, and 12 represent the comparative performance of various classifiers and their corresponding ranks for multiple data set with common genes selected from various feature selection technique. Tables 13 and 14 represent the rejection or acceptance of null hypothesis and pair-wise hypothesis test of the classifiers, respectively.

Table 7 The comparative performance of multiple classifiers and their subsequent ranks for various data set with common genes chosen by SNR

Domain with selected features index	DT	SVM	k-NN	MLP	NB
DLBCL: [286, 327, 450, 759, 920, 1157, 1186, 1807, 1856, 1988, 2000, 2157, 2208, 2318]	73.39 (1)	91.10 (5)	88.23 (3)	76.08 (2)	90.91 (4)
Prostate: [39, 134, 432, 728, 783, 939, 987, 1181, 1226, 1303, 1469, 1642, 1737, 1945, 1970, 2035, 2101]	78.04 (1)	92.16 (4)	91.18 (3)	78.34 (2)	93.14 (5)
Lymphoma: [26, 526, 851, 1374, 1555, 1602, 1645, 1818, 2145, 2247, 2259, 2349, 2591, 2646, 268, 3408, 4393, 4828]	72.03 (1)	90.05 (4)	91.74 (5)	82.32 (2)	87.67 (3)
Breast cancer: [1801, 1801, 2519, 3069, 6484, 6672, 9815, 10065, 13738, 13906, 16094, 16259, 16850, 17276, 18430, 21136, 22090, 22150, 25365, 26034]	80.87 (2)	89.06 (4.5)	88.28 (3)	80.56 (1)	89.06 (4.5)
Ri	5	17.5	14	7	16.5

Table 8 The comparative performance of various classifiers and their subsequent ranks for multiple data set with common genes selected by SAM

Domain with selected features index	DT	SVM	k-NN	MLP	N B
DLBCL: [2000, 1807, 1988, 450, 786, 14, 920, 759, 2157, 2208, 460, 2318, 327, 1856, 1186, 286]	87 (1.5)	93.51(4)	98.12 (5)	87 (1.5)	90.91 (3)
Prostate: [2000, 1807, 1988, 450, 786, 14, 920, 759, 2157, 2208, 460, 2318, 327, 1856, 1186, 286]	89.04 (3)	88.24(2)	94.18 (5)	82.12 (1)	93.14 (4)
Lymphoma: [39, 134, 432, 728, 783, 939, 987, 1181, 1226, 1303, 1469, 1737, 1945, 1970, 2035, 2101]	83.46 (1)	90.37(3)	94.74 (5)	86.03 (2)	91.67 (4)
Breast cancer: [26, 526, 851, 1555, 1645, 1818, 2247, 2259, 2591, 2646, 2687, 3408, 4393, 4828]	77.66 (1)	91.41(5)	88.28 (2.5)	89.22 (4)	88.28 (2.5)
Ri	6.5	14	17.5	8.5	13.5

Table 9 The comparative performance of various classifiers and their subsequent ranks for multiple data set with common genes selected by *t*-test

Domain with selected features index	DT	SVM	k-NN	MLP	NB
DLBCL: [1807, 2000, 1988, 450, 786, 14, 327, 920, 2318, 1856, 460]	87(1)	93.51 (4)	100 (5)	93.06 (3)	90.91 (2)
Prostate: [1807, 2000, 1988, 450, 786, 14, 327, 920, 2318, 1856, 460]	98.04 (4)	98.20 (5)	96.16 (2)	97.04 (3)	93.14 (1)
Lymphoma: [39, 134, 432, 728, 939, 987, 1226, 1303, 1469, 1737, 1945, 1970, 2035, 2101]	100 (5)	97.37 (2)	94.74 (1)	99.32 (4)	98.67 (3)
Breast cancer [26, 526, 851, 1555, 1602, 1818, 2145, 2259, 2646, 3408, 4393, 4828]	97.66 (4)	95.84 (1)	97.53 (3)	98.01 (5)	96.56 (2)
Rj	14	12	11	15	8

Table 10 The comparative performance of multiple classifiers and their subsequent ranks for various data set with common features selected by BPSO

Domain with selected features index	DT	SVM	k-NN	MLP	NB
DLBCL: [1272, 652, 193, 2281, 2284, 554, 612, 645, 759, 903, 1154, 1157, 1186, 1191, 1204, 1345, 1987, 2127, 2157, 2262]	72.04 (1)	98.91 (4)	100 (5)	78.04 (2)	80.12 (3)
Prostate: [67, 154, 292, 318, 580, 583, 594, 605, 1409, 1511, 1538, 1624, 1871, 1898, 1926, 1941, 2011, 2031, 2047, 2083]	73.28 (1)	92.16 (5)	91.18 (4)	83.06 (3)	74.16 (2)
Lymphoma: [2494, 245, 2396, 2197, 498, 499, 2100, 2501, 3502, 3700, 3124, 3618, 3923, 4112, 4267, 4278, 4319, 4434, 4512, 4863]	78(1)	97.37 (5)	94.74 (4)	89.32 (3)	88.67 (2)
Breast cancer: [907, 1889, 5308, 5394, 7704, 9445, 9548, 10479, 1482517845, 21275, 21684, 24647, 29042, 37265, 38082, 39584, 40780]	67.14 (1)	100 (5)	94.53 (4)	88.06 (3)	84.89 (2)
Ri	4	19	17	11	9

Table 11 The comparative performance of various classifiers and their equivalent ranks for different data sets with common genes selected by ACO

Domain with selected feature index	DT	SVM	k-NN	MLP	NB
DLBCL: [674, 914, 998, 1191, 1309, 1345, 1467, 1719, 1935, 1986, 1987, 2001, 2034, 2049, 2126, 2445, 2446, 2541, 2562]	96.10 (4)	87.40 (1)	88(2)	90.02 (3)	98.70 (5)
Prostate: [89, 120, 121, 161, 230, 616, 711, 873, 951, 1400, 1469, 1503, 1532, 1567, 1917, 1962, 2017, 2030, 2060, 2074]	96.08 (5)	81.37 (3)	78.43 (2)	90.67 (4)	77.45 (1)
Lymphoma: [483, 931, 1243, 2315, 2463, 2485, 2549, 2641, 3134, 3278, 3491, 3476, 3512, 3601, 4026, 4239, 4362, 4613, 4489, 4806]	97.37 (5)	76.32 (1)	84.21 (3)	92.54 (4)	82.34 (2)
Breast cancer: [204, 649, 2745, 5645, 5972, 7662, 9815, 11214, 11467, 13661, 17013, 17016, 20874, 21649, 22347, 36510]	97.66 (4)	85.16 (2)	84.38 (1)	93.34 (3)	98.03 (5)
Rj	18	7	8	14	13

Table 12 The comparative performance of various classifiers and their subsequent ranks for different data set with common genes selected by GA

Domain	DT	SVM	k-NN	MLP	NB
DLBCL: [266, 393, 496, 584, 756, 764, 830, 964, 1347, 1352, 1396, 1617, 1630, 1686, 1703, 1954, 2038, 2115, 2256, 2383]	77.12 (2)	97.01 (5)	88.23 (4)	73.06 (1)	78.31 (3)
Prostate: [10, 114, 203, 256, 648, 712, 750, 874, 909, 1019, 1085, 1190, 1569, 1570, 1593, 1602, 1793, 1931, 2048, 2071]	76.08 (3)	93.53 (5)	63.73 (1.5)	79.32 (4)	63.73 (1.5)
Lymphoma: [438, 499, 1332, 1757, 1903, 2034, 2209, 2259, 2342, 2480, 2885, 3157, 3882, 3902, 3967, 4071, 4306, 4370, 4794, 4814]	77.37 (1)	94.21 (5)	86.84 (3)	87.54 (4)	78.07 (2)
Breast cancer: [3714, 4015, 4638, 6391, 7067, 7943, 8974, 9519, 11520, 13810, 13892, 17023, 18439, 22190, 22491, 22894, 23030, 23849, 24464, 26226]	74.53 (1)	98.28 (5)	89.06 (4)	82.12 (3)	79.06 (2)
Ri	7	20	12.5	12	8.5

Table 13 Acceptance/rejection of null hypothesis using Friedman's test

Methods	Rejection/acceptance of hypothesis
SNR	$Q = 12.85 > 9.48773$, hypothesis rejected
SAM	$Q = 7.9 < 9.48773$, hypothesis accepted
t-test	$Q = 3 < 9.48773$, hypothesis accepted
BPSO	$Q = 12.8 > 9.48773$, hypothesis rejected
ACO	$Q = 8.3 < 9.48773$, hypothesis accepted
GA	$Q = 10.15 > 9.48773$, hypothesis rejected

Table 14 Nemenyi test chart for the performance analysis of different classifiers

Methods	$q_{DT,SVM}$	$q_{SVM,kNN}$	$q_{kNN,MLP}$	$q_{MLP,naiveBays}$	$q_{naiveBays,DT}$
SNR	-11.26	3.15	6.30	-8.55	10.36
BPSO	-13.51	1.80	5.40	1.80	4.50
GA	-11.71	6.75	0.45	3.15	1.35

5 Conclusions

This paper suggests the generic feature subset selection based on filter approach and wrapper approach along with their validation with a few forefronts classifiers. From the experimental analysis, it is evident that the feature subset selected by the filter approaches is common, whereas the feature subsets selected by different wrapper approaches are different. All the classifiers considered the selected features stable because the null hypothesis is accepted using the hypothetical analysis for SAM, *t*-test, and ACO. Also, the classifiers performed differently as the hypothesis is rejected using SNR, BPSO, and GA. The features selected using SNR provide a stable result for almost all the classifiers. Then, it has been observed from the experimental analysis that the features selected by the filter methods are generic because they provide stable classification accuracy in comparison to evolutionary wrapper methods.

References

1. Mollaee, M., Moattar, M.H.: A novel feature extraction approach based on ensemble feature selection and modified discriminant independent component analysis for microarray data classification. Biocybern. Biomed. Eng. **36**(3), 521–529 (2016)
2. Cai, H., Ruan, P., Michael, N., Akutsu, T.: Feature weight estimation for gene selection: a local hyper linear learning approach. BMC Bioinform. **15**(70), 1–13 (2014)
3. Apolloni, J., Leguizamóna, G., Alba, E.: Two hybrid wrapper-filter feature selection algorithms applied to high-dimensional microarray experiments. Appl. Soft Comput. **38**, 922–932 (2016)
4. Zhao, M., Ren, J., Ji, L., Zhou, M.: Parameter selection of support vector machines and genetic algorithm based on change area search. Neural Comput. Appl. **21**(1), 1–8 (2012)
5. Nieto, J.G., Alba, E.: Parallel multi-swarm optimizer for gene selection in DNA microarrays. Appl. Intell. **37**(2), 255–266 (2012)
6. Ke, L., Feng, Z., Zhigang, R.: An efficient ant colony optimization approach to attribute reduction in rough set theory. Pattern Recognit. Lett. **29**(9), 1351–1357 (2008)
7. Majhi, P., Paul, S.: Rough set based maximum relevance-maximum significance criterion and Gene selection from microarray data. Int. J. Approx. **52**(3), 408–442 (2011)
8. Dehuri, S.: Application of parallel genetic algorithms for attribute selection in knowledge discovery in databases. ANVESA: Interdiscip. Res. J. Fakir Mohan Univ. **9**(1), 16–29 (2014)
9. http://www.broadinstitute.org/cgi-bin/cancer/datasets.cgi
10. http://datam.i2r.astar.edu.sg/datasets/krbd/ProstateCancer/ProstateCancer.html
11. http://csse.szu.edu.cn/staff/zhuzx/Datasets.html
12. http://ico2s.org/datasets/microarray.html

Soil Temperature Influencing Evaporation Estimation Using Data-Driven Tools

Narhari D. Chaudhari, Neha N. Chaudhari, Ankush R. Pendhari
and Milind M. Thombre

Abstract Evaporation of water is an important player in hydrological cycle. Planning and designing water resources projects need evaporation. Pan evaporation data is used by hydrologist and water planners for irrigation scheduling and water balance studies. Empirical equations need huge data. Class A pan readings are popular for measuring evaporation but it is difficult to provide pan at every irrigation project. Evaporation is highly nonlinear in nature, estimation of evaporation at pan less station is important for irrigation department as well as researchers and hydrologists which initiates the need of an alternative tool with accuracy in results and lesser data requirement. The data-driven models which work on the data rather than physics of the process can play their role here. Purpose of the present study is to note the response of these data-driven models to soil temperature along with other climatic factors as input for Nagpur climatic station in Maharashtra, India. It is noticed that model tree (MT model, $r = 0.88$) with all nine climatic factors as inputs performed well compared to Artificial Neural Networks (ANN) and Genetic Programming (GP). Model tree (MT model, $r = 0.83$) with only influential four inputs also performed good. These models can be of great use to farm managers and hydrologists.

Keywords Soil · ANN · GP · MT · Nagpur

N.D. Chaudhari (✉) · A.R. Pendhari · M.M. Thombre
Department of Civil Engineering, Gokhale Education Society's, R. H. Sapat College
of Engineering, Management Studies and Research, Nasik, Maharashtra, India
e-mail: chaudhari_nd@rediffmail.com

A.R. Pendhari
e-mail: ankushpendhari1@yahoo.com

M.M. Thombre
e-mail: thombremilind21@gmail.com

N.N. Chaudhari
Civil Engineering Graduate, Nashik, India
e-mail: ndc.civil@gmail.com

© Springer Nature Singapore Pte Ltd. 2018
M. Pant et al. (eds.), *Soft Computing: Theories and Applications*,
Advances in Intelligent Systems and Computing 583,
https://doi.org/10.1007/978-981-10-5687-1_28

1 Introduction

Evaporation is a key element of hydrological process and its estimation is important in planning and design of many water resources projects. Soil scientists, climate engineers, and irrigation engineers use this data of pan evaporation for various reasons like irrigation scheduling and water balance studies.

Conventionally, determination of evaporation is done by energy budget, water budget, empirical, mass transfer, Penman method, and measurement [8, 9, 12, 26, 32, 41]. The process of evaporation is complex and nonlinear in nature as observed in many of the estimation methods mentioned above. Various scientists contributed with empirical methods to estimate evaporation and Evapotranspiration (ET).

Empirical methods need lot of data like heat entering in and out of the water, maximum sunshine, actual duration of sunshine in hours, actual vapor pressure, average air temperature, and saturation vapor pressure at mean air temperature [10].

The methods of water budget, energy budget, and mass transfer are accurate but need information of few parameters which cannot be measured easily [20]. The National Weather Service class A pan evaporation is used extensively as an index of evapotranspiration and for estimating lake evaporation. Lastly, pan evaporation which is widely used and popular for measurement is employed but it is not simple to install and maintain Class A pan at every place of proposed or existing reservoir and agricultural field. Estimation of pan evaporation practically at pan less station is very important to soil scientist and hydrologists.

Complex and non-linear nature of evaporation demands alterative data driven technique which can estimate pan evaporation, assess the influence of climatic variables, specially, the influence of soil temperature with reasonable accuracy eliminating excessive data measurement.

The data-driven models which work on the data rather than physics of the process can play their role here. Characterizing the system under study using data analysis is the basis of Data-driven Modeling (DDM). Neural networks, fuzzy systems, genetic programming, model tree, and support vector regressions can be grouped as data-driven modeling techniques which focus on constructing models that will serve the purpose served by "knowledge-driven" models explaining physical behavior [37].

Recently, data-driven techniques have been successfully employed by many hydrologists for hydrological modeling [3, 16, 17, 24, 25, 31] employed data-driven techniques in hydrology. Genetic Programming (GP) though new seems to have found its foot hold in hydrology. Within the last decade, several studies have reported the use of an M5 model tree, a decision-tree-based regression approach, for water resource applications [6, 13, 15, 19, 20, 22, 28, 29, 34–36, 38] employed data-driven techniques for estimating evapotranspiration.

Particularly in the field of evaporation modeling, data-driven techniques have gained momentum, since last 10 years [1, 2, 4, 5, 7, 11, 14, 18, 23, 39] estimated evaporation employing data-driven techniques.

Objective of the study is to assess response of data-driven models to various climatic variables with soil temperature and without soil temperature. We next present information on the study area and data followed by brief details of data-driven techniques. The model formulation and assessment is then discussed, followed by the results and a discussion. Finally, concluding remarks are given.

2 Study Area and Data

City of Nagpur lies at 21.15°N and 79.24°E, at an elevation of 310 m from sea level situated in northeastern parts of the state of Maharashtra India. May is the hottest month with the daily maximum temperature of 46.4 °C. December is the coldest month with the daily minimum temperature of 3.2 °C. The average daily value of evaporation is 6.11 mm. Figure 1 shows study area [42]. Data used for Nagpur station is for period of 2000–2001 (1.5 years).

Fig. 1 Study area and data (http://www.mapsofindia.com/maps/maharashtra/)

3   Data-Driven Techniques

An ANN is a nonlinear model that makes use of a parallel programming structure capable of representing arbitrarily complex nonlinear processes that relate the inputs and outputs of any system [21, 30]. It provides better solutions than traditional statistical methods when applied to poorly defined and poorly understood complex systems that involve pattern recognition. The basic unit in the artificial neural network is the node. Nodes are connected to each other by links known as synapses, associated with each synapse there is a weight factor. Usually, neural networks are trained so that a particular set of inputs produces as nearly as possible a specific set of target outputs.

GP which derives inspiration from nature evolves the best individual (program) through combination of crossover, mutation, and reproduction processes. It works on the Darwinian principle of "survival of the fittest" [27]. The commercial software Discipulus was used to develop GP model.

Model tree generalizes the concepts of regression tree, which have constant values at their leaves [40]. Therefore, they are analogous to piece-wise linear functions (and hence nonlinear). The M5 model tree [33] is a binary decision tree having linear regression functions at the terminal (leaf) nodes, which can predict continuous numerical attributes. Tree-based models are constructed by a divide-and-conquer method. Weka software is employed to build MT models

4   Model Formulation and Assessment

The present models were tried with data set of total daily records 530 values out of which first 400 (75%) values were used for training the model and remaining 130 (25%) were withheld for testing purpose, nine input climatic variables were maximum and minimum humidity (MHS and Mhs), maximum and minimum air temperature (MTX and MTN), wind speed (MWS), maximum and minimum soil temperatures at two depths (5.75 and 10.15 cm). Daily pan evaporation (MEP) was set as output. Data used was for the period 1.5 years. Furthermore, four inputs having higher correlation with output were used to develop three data-driven models (ANN, GP and MT). Four inputs used for developing models were maximum air temperature (MTX), maximum and minimum soil temperature at 5.75 cm depth, and maximum humidity (MHS).

The selection of inputs was done by calculating correlation of every input with the output as done by [13] in their work on evapotranspiration estimation. Though correlation coefficient represents a linear relationship between two variables still, it gives at least some idea for choice of inputs though the process is highly nonlinear as in this case.

Climatic factors were set as inputs and the output layer consisted of 1 neuron pertaining to the value of measured pan evaporation. Neurons in the hidden layer

Table 1 Error statistic for data-driven models

Data-Driven Models	MSRE	RMSE	MAE	CE	d	r
ANN 4-7-1 (Influential inputs)	0.28	2.52	1.76	0.52	0.82	0.80
MT (Influential inputs)	0.16	2.18	1.49	0.65	0.88	0.83
GP (Influential inputs)	0.14	2.09	1.53	0.67	0.88	0.76
ANN 9-2-1 (All inputs)	0.17	2.80	2.06	0.42	0.80	0.74
MT (All inputs)	**0.09**	**2.03**	**1.35**	**0.70**	**0.90**	**0.88**
GP (All inputs)	0.17	2.55	1.71	0.51	0.81	0.75

Note MSRE is Mean Square Relative Error, *RMSE* is Root Mean Squared Error, *MAE* is Mean Absolute Error, *CE* is Coefficient of Efficiency and *d* is Index of Agreement

were fixed by trial and error process. For present work, a three-layered Feed Forward Neural Net with Levenberg Marquardt (LM) as the training algorithm was used. The transfer functions used were "logsig" between the input and hidden layer and "purelin" between the hidden and the output layer. The data was normalized between -1 and 1. Performance function was mean squared error (mse). The networks were trained till a very low value of mean squared error was achieved in each case. ANN 9-2-1 was finalized for first case of all 9 inputs and ANN4-7-1 was finalized for second case of four influential inputs. The trained weights and biases were then used to test the unseen data. Performance of the networks was checked with respect to testing sets by calculating the correlation coefficient between the observed and estimated pan evaporation values and by plotting the time series plots between the same. Additionally five errors measures (MSRE, RMSE, MAE, CE and d) were used for model assessment. Refer Table 1 for error measures.

The genetic programming models were developed with similar input configuration as mentioned above. The initial control parameters were initial population size of 2048, mutation frequency 95%, and the crossover frequency, 53% with a fitness function of mean squared error

M5 model was developed for the best model with similar input configuration as explained in modeling strategy and data division as used in ANN and GP models.

5 Results and Discussion

It is noticed that the MT model ($r = 0.88$) with all nine inputs along with soil temperature works well than GP and ANN ($r = 0.75$ and $r = 0.74$, respectively). There is marginal difference between GP and ANN for nine inputs combination. However for four influential inputs, ANN model works better than GP. MT model ($r = 0.83$) again works better than GP and ANN for four influential inputs. For influential input models, there is improvement in estimation results of GP and ANN only. Performance of MT model is decreased compared to nine input combination model, however it is working better than ANN and GP in both the cases.

Inclusion of soil temperature does not show satisfactory results if compared with earlier cause effect data-driven models developed by the author using regular inputs like with maximum humidity, minimum humidity, sunshine, and minimum temperature as inputs without soil temperature. Model tree showed excellent result with $r = 0.95$.

Figures 2 and 3 show time series plots for all nine inputs and four influential inputs. Figure 4 shows response of MT model to nine-input model and four-input model which has also been reflected in scatter plots as well (Figs. 5 and 6). The best performing MT model ($r = 0.88$) shows excellent error statics with least MSRE, RMSE and MAE (0.09, 2.03 and 1.35, respectively) and maximum values for coefficient of efficiency and index of agreement (0.70, 0.9 respectively).

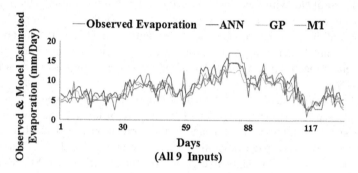

Fig. 2 Time series plot with nine inputs for *ANN*, *GP*, and *MT* model

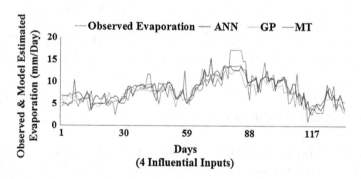

Fig. 3 Time series plot with four influential inputs for *ANN*, *GP*, and *MT* Model

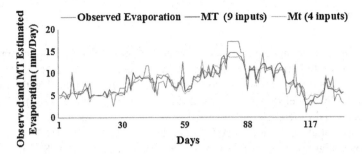

Fig. 4 Time series plot with observed, influential inputs, and all nine for *MT* model

Fig. 5 Scatter plots between
observed and model estimated
evaporation for nine inputs

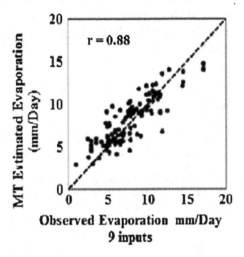

Fig. 6 Scatter plots between
observed and model estimated
evaporation for four inputs

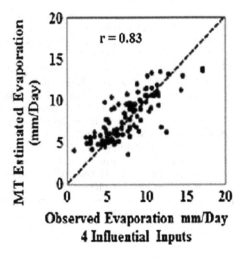

6 Conclusion

Model with all nine inputs including soil temperature as one of the input parameters shows that the MT model ($r = 0.88$) works well than GP and ANN ($r = 0.75$ and $r = 0.74$ respectively). However, for influential four inputs model with soil temperature, ANN model shows improvement by 6% ($r = 0.74$ to $r = 0.80$). GP model shows marginal difference of 1% ($r = 0.76$ to $r = 0.76$) refer Table 1.

For influential input models, MT model declines by 5% ($r = 0.88$ to $r = 0.83$), i.e., shows slightly poor result compared to earlier result ($r = 0.88$) with all nine inputs. ANN shows improvement in results by 6% for four input model ($r = 0.80$) compared to nine input model ($r = 0.74$).

Inclusion of soil temperature does not show satisfactory results if compared with earlier study by the author (MT model, $r = 0.95$) with only four inputs (maximum humidity, minimum humidity, sunshine, and minimum temperature) without soil temperature.

References

1. Abghari, H., Ahmadi, H., Besharat, S., Rezaverdinejad, V.: Prediction of daily pan evaporation using wavelet neural networks. Water Resour. Manage. **26**(12), 3639–3652 (2012)
2. Arunkumar, R., Jothiprakash, V.: Reservoir evaporation prediction using data-driven techniques. J. Hydrol. Eng. **18**(1), 40–49 (2013)
3. Bowden, G.J., Maier, H.R., Dandy, G.C.: Real-time deployment of artificial neural network forecasting models: understanding the range of applicability. Water Resour. Res. **48**(10) (2012). Doi:10.1029/2012WR011984
4. Chang, F.J., Chang, L.C., Kao, H.S., Wu, G.R.: Assessing the effort of meteorological variables for evaporation estimation by self-organizing map neural network. J. Hydrol. **384**(1–2), 118–129 (2010)
5. Chung, C.H., Chiang, Y.M., Chang, F.J.: A spatial neural fuzzy network for estimating pan evaporation at un-gauged sites. Hydrol. Earth Syst. Sci. **16**(1), 255–266 (2012)
6. Ditthakit, P., Chinnarasri, C.: Estimation of pan coefficient using m5 model tree. Am. J. Environ Sci. **8**(2), 95 (2012)
7. Eslamian, S.S., Amiri, M.J.: Estimation of daily pan evaporation using adaptive neural-based fuzzy inference system. Int. J. Hydrol. Sci. Technol. **1**(3/4), 164–175 (2011)
8. Fritschen, L.J.: Energy balance method. Proceedings, American Society Of Agricultural Engineers, Conference on Evapo-Transpiration and its Role in Water Resources Management, December 5–6, Chicago, IL. St. Joseph, MI. 34-37 (1966)
9. Guitjens, J.C.: Models of alfalfa yield and evapotranspiration. J. Irrig. Drain. Div. Proc. Am. Soc. Civ. Eng. **108**(IR3), 212–222 (1982)
10. Gianniou, S.K., Antonopoulos, V.Z.: Comparison of different evaporation estimation methods applied to Lake Vegoritis. Conference at Steven's institute of technology, Kefalonia, Greece. Published in: PROTECTION (2008)
11. Guven, A., Kisi, O.: Daily pan evaporation modeling using linear genetic programming technique. Irrig. Sci. **29**, 135–145 (2011)

12. Harbeck, G.E.: A practical field technique for measuring reservoir evaporation utilizing mass-transfer theory. Geological Survey Professional Paper 272-E, US Government Printing Office., Washington, D.C. 101–105 (1962)

13. Jain, S.K., Nayak, P.C., Sudheer, K.P.: Models for estimating evapotranspiration using anns and their physical interpretation. Hydrol. Processes, Wiley Inter Science **22**(13), 2225–2234 (2008)

14. Jalal, S., Kişi, O.: Application of artificial intelligence to estimate daily pan evaporation using available and estimated climatic data in the Khozestan province (South Western Iran). J. Irrig. Drain Eng. **137**(7), 412–425 (2011)

15. Jalal, S., Nazemi, A., Sadraddini, A., Landeras, G., Kisi, O., Fard, A., Marti, P.: Global cross-station assessment of neuro-fuzzy models for estimating daily reference evapotranspiration. J. Hydrol. **480**, 46–57 (2013)

16. Karimi, S., Kisi, O., Jalal, S., Makarynskyy, O.: Neuro-fuzzy and neural network techniques for forecasting sea level in darwin harbor, australia. Comput. Geosci. **52**, 50–59 (2013)

17. Kasiviswanathan, K.S., Soundhara, R., Pandian, R., Saravanan, S., Agarwal, A.: Genetic programming approach on evaporation losses and its effect on climate change for Vaipar basin. IJCSI Int. J. Comput. Sci. Issues, **8**(5), No 2, pp. 269–274 (2011)

18. Keskin, M.E., Terzi, O.: Artificial neural network models of daily pan evaporation. J. Hydrol. Eng. ASCE **11**(1), 65–70 (2006)

19. Khan, A.S., See, L.: Rainfall-runoff modeling using data driven and statistical methods. Inst. Electr. Electron. Eng. (IEEE) (2006)

20. Kim, S., Kim, H.S.: Neural Networks and Genetic Algorithm Approach for Nonlinear Evaporation and Evapotranspiration Modeling. J. Hydrol. **351**(3–4), 299–317 (2008)

21. Kisi, O.: Modeling evaporation using multi-layer perceptron and radial basis neural networks. Proceedings of 7th International Congress on Advances in Civil Engineering, Istanbul, 341 (2006)

22. Kisi, O.: The potential of different ANN techniques in evapotranspiration modeling. Wiley Inter Sci. Hydro. Process. **22**, 2449–2460 (2008)

23. Kisi, O.: Daily pan evaporation modeling using MLP and RBNN. Hydrol. Process., Wiley Interscience. **23**(2), 213–223 (2009)

24. Kisi, O., Dailr, A., Cimen, M., Jalal, S.: Suspended sediment modeling using genetic programming and soft computing techniques. J. Hydrol. **450**, 48–58 (2012)

25. Kisi, O., Jalal, S., Tombul, M.: Modeling rainfall-runoff process using soft computing techniques. Comput. Geosci. **51**, 108–117 (2013)

26. Kohler, M. A., Nordenson, T. J., and Fox, W. E.: Evaporation from pans and lakes. Weather Bureau Research Paper 38., US Department of Commerce, Washington, D.C. (1955)

27. Koza, J.R.: Genetic programming on the programming of computers by means of natural selection. MIT Press, A Bradford Book (1992)

28. Kumar, M., Raghuwanshi, N.S., Singh, R., Wallender, W.W., Pruitt, W.O.: Estimating evapotranspiration using artificial neural network. J. Irrig. Drainage Eng. ASCE **128**(4), 224–233 (2002)

29. Londhe, S.N., Dixit, P.R.: Forecasting stream flow using model trees. Int. J. Earth Sci. Eng. **4**, 282–285 (2011)

30. Namdar, K.D., Shorafa, M., Omid, M., Fazeli, S.M.: Application of artificial neural networks in modeling soil solution electrical conductivity. Soil Sci. **175**(9), 432–437 (2010)

31. Nourani, V., Parhizkar, M.: Conjunction of SOM-based feature extraction method and hybrid wavelet-ANN approach for rainfall–runoff modeling. J. Hydroinformatics **15**(3), 829–848 (2013)

32. Penman, H.L.: Natural evaporation from open water, bare soil and grass. Proc. R. Soc. Lond. **193**, 120–145 (1948)

33. Quinlan, J.R.: Learning with continuous classes. In: Proc. AI'92 (Fifth Australian Joint Conf. on Artificial Intelligence) (ed. by A. Adams & L. Sterling), 343–348. World Scientific, Singapore (1992)

34. Rahimikhoob, A.: Estimating daily pan evaporation using artificial neural network in a semi-arid environment. Theoret. Appl. Climatol. **98**, 101–105 (2009)
35. Sattari, M.T., Pal, M., Yurekli, K., Unlukara, A.: M5 model trees and neural network based modelling of et0 in Ankara Turkey. Turk. J. Eng. Env. Sci. **37**, 211–219 (2013)
36. Siek, M., and Solomatine D.P.: Tree-like machine learning models in hydrologic forecasting: optimality and expert knowledge. Geophysical Research Abstracts, 9 (2007)
37. Solomatine, D.P., Ostfeld, A.: Data-driven modelling: some past experiences and new approaches. J. Hydroinformatics, IWA Publishing **10**(1), 3–22 (2008)
38. Stravs, L., Brilly, M.: Development of a low flow forecasting model using the m5 machine learning method. Hydrol. Sci. **52**, 466–477 (2007)
39. Sudheer, K.P., Gosain, A.K., Rangan, D.M., Saheb, S.M.: Modeling evaporation using artificial neural network algorithm. Hydrol. Process. **16**(16), 3189–3202 (2002)
40. Witten, I.H., Frank, E.: Data Mining: Practical machine learning tools and techniques with java implementations. Morgan Kaufmann, San Francisco (2005)
41. Young, A. A.: Evaporation from water surface in California: summary of pan records and coefficients. Bulletin 54.Public Works Department, Sacramento, CA, 1881–1946. (1947)
42. http://www.mapsofindia.com/maps/maharashtra/

Vibrational Frequencies of Parallelogram Plate with Circular Variations in Thickness

Amit Sharma

Abstract The present study emphasizes on the effect of two-dimensional circular variations in thickness on non-homogeneous parallelogram plate. The temperature variation on the plate is considered bi-parabolic in nature. It is assumed that density varies linearly due to non-homogeneous viscoelastic material of the plate. All the differential equations of motions are solved by using Rayleigh–Ritz technique. To find the vibrational frequencies of parallelogram plate under variation of parameters such as thickness, temperature, skew angle, and non-homogeneity, frequency equation is calculated using MAPLE software. Results are displayed in tabular form.

Keywords Parallelogram plate · Vibrational frequencies · Circular variation

1 Introduction

Non-homogeneous plates with thermal or temperature effect have wide variety of applications in the engineering structures (mechanical and electrical). To make better design in enginerring like turbines, jet engine, and wings of an aircraft, it is mandatory to study vibrations because these structures have high intenstiy heat. Under high intensity flux, significant changes in the property of material occur. Plates with temperature play an essential role in moderen engineering structures. Also plates with variable varying thickness or tappered plates are oftenly used because of less economic and lighten weight. Therefore to study the vibrations completely, it is required to study plate's characherstics under the nonuniformity and temperature enviornment.

A. Sharma (✉)
Amity University Haryana, Gurgaon 122413, Haryana, India
e-mail: dba.amitsharma@gmail.com

© Springer Nature Singapore Pte Ltd. 2018
M. Pant et al. (eds.), *Soft Computing: Theories and Applications*,
Advances in Intelligent Systems and Computing 583,
https://doi.org/10.1007/978-981-10-5687-1_29

The main object of scientists and researchers is to optimize vibrations and get desired frequency. Various studies based on effect of temperature (one dimensional), non-homogeneity (linear or parabolic) as well as variation in thickness (linear, parabolic, and sinusoidal) have been done. However, very less work is done on two-dimensional temperatures and no work has been reported on circular variation in thickness.

Gupta and Khanna [1] have studied the effect of temperature, linear variation in thickness on vibrational frequencies of parallelogram plate. Gupta et al. [2] have discussed vibration of viscoelastic parallelogram plate by taking parabolic variation in thickness. Gupta and Kumar [3] have provided the vibrations of parallelogram plate with thermal, two-dimensional linear variations in thickness. Transverse vibration of skew plates with variable thickness has been discussed by Singh and Sexena [4]. Khanna and Arora [5] have studied the effect of sinusoidal thickness variation on vibrations of non-homogeneous parallelogram plate with bilinearly temperature variation. Singh and Chakarverty [6] have discussed flexural vibration of skew plates using boundary characteristic orthogonal polynomial in two variables. Sharma et al. [7, 8] have studied the vibration of orthotropic rectangular plate with two-dimensional temperatures and isotropic non-homogeneous parallelogram plate with linear temperature along with thickness variation. Khanna and Kaur [9–11] have studied the effect of two-dimensional parabolic temperatures on vibrational modes with linear variation in thickness and exponential variation in Poisson ratio. Khanna and Singhal [12, 13] have discussed vibration of rectangular plate in two different scenarios. In the first scenario, they studied the effect of plate's parameter with different boundary conditions with 2D parabolic variation in thickness and 2D linear variation in temperature. In second scenario, they studied free vibration of clamped viscoelastic plate with 2D linear variation in thickness and 2D parabolic variation in temperature. Leissa [14] has provided a fruitful monograph on vibrations which contained different sizes and shapes.

In this work, the authors have attempted to study the effect of two-dimensional circular variations in thickness to frequency modes of vibrations on parallelogram plate. The authors have considered linear variation in density and bi-parabolic temperature variation in the present study. First two modes of vibrations are obtained using Rayleigh–Ritz technique as numerical data which is represented in the form of tables for different combinations of value of tapered parameters, thermal gradient, non-homogeneous constant, and skew angle.

2 Modeling and Solution of the Problem

2.1 Material Analysis and Corresponding Equations

A thin, isotropic parallelogram plate with skew angle θ denoted by $P(a, \zeta, b)$ is shown in Fig. 1. The skew coordinate and corresponding boundary conditions of

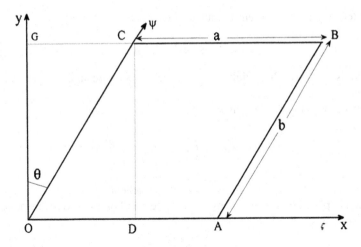

Fig. 1 Thin isotropic parallelogram plate with sides a and b

the plate are given by $\zeta = x - y \tan \theta$, $\psi = y \sec \theta$ and $\zeta = 0, a$, $\psi = 0, b$, respectively.

For free vibration of the plate, the displacement is assumed as [14]

$$\phi(\zeta, \psi, t) = \Phi(\zeta, \psi) \sin \omega t \qquad (1)$$

where $\Phi(\zeta, \psi)$, ω is known as the highest displacement at time t and angular moment respectively.

As material of the plate is assumed non-uniform, therefore thickness varies circularly in both the directions

$$j = j_0 \left(1 + \beta_1 \sqrt{1 - \frac{\zeta}{a}}\right)\left(1 + \beta_2 \sqrt{1 - \frac{\psi}{b}}\right) \qquad (2)$$

where β_1, β_2 are known as tapered constants along ζ and ψ axis, thickness becomes constant at $\beta_1 = \beta_2 = 0$.

It is also assumed that the temperature variation on plate is subjected to be bi-parabolic in nature as

$$T = T_0 \left(1 - \frac{\zeta^2}{a^2}\right)\left(1 - \frac{\psi^2}{b^2}\right) \qquad (3)$$

Here, T represents the temperature excess above the reference on the plate and T_0 is the temperature at $\zeta = a$, $\psi = b$.

The temperature dependent modulus of elasticity is taken as

$$K = K_0(1 - \gamma T) \tag{4}$$

where K_0 is the Young's modulus at reference temperature and γ is known as slope variation.

From Eq. (3), Eq. (4) becomes

$$K = K_0\left[1 - \alpha\left(1 - \frac{\zeta^2}{a^2}\right)\left(1 - \frac{\psi^2}{b^2}\right)\right] \tag{5}$$

and $\alpha = \gamma T_0$, $(0 \le \alpha < 1)$ is known as thermal gradient.

Since the plate's material is considered to be non-homogeneous, linear variation is taken in density as

$$\rho = \rho_0\left(1 - \alpha_1 \frac{\zeta}{a}\right) \tag{6}$$

where α_1 $(0 \le \alpha_1 \le 1)$ are known as constants of non-homogeneity.

For obtaining vibrational frequency of parallelogram plate, two term deflection function along with boundary condition is given by

$$\Phi = \left[\left(\frac{\zeta}{a}\right)\left(\frac{\psi}{b}\right)\left(1 - \frac{\zeta}{a}\right)\left(1 - \frac{\psi}{b}\right)\right]^2\left[C_1 + C_2\left(\frac{\zeta}{a}\right)\left(\frac{\psi}{b}\right)\left(1 - \frac{\zeta}{a}\right)\left(1 - \frac{\psi}{b}\right)\right] \tag{7}$$

$$\Phi = \Phi_{,\zeta} = 0, \ \zeta = 0, \ a \text{ and } \Phi = \Phi_{,\psi} = 0, \ \psi = 0, \ b \tag{8}$$

2.2 Solution to Find Vibrational Frequencies

The expression for maximum kinetic energy (K) and maximum strain energy (P) at parallelogram plate for transverse vibration modes shape $\Phi(\zeta, \psi)$ as in [14] are

$$K = 1/2\,\omega^2\cos\theta \int\int j\rho\,\Phi^2 d\zeta d\psi \tag{9}$$

and

$$P = 1/2\cos^3\theta \int\int D\left[\begin{array}{l}(\Phi_{,\zeta\zeta})^2 - 4\sin\theta\,\Phi_{,\zeta\zeta}\Phi_{,\zeta\psi} + 2(\sin^2\theta + v\cos^2\theta)\Phi_{,\zeta\zeta}\Phi_{,\psi\psi} \\ + 2(1 + \sin^2\theta - v\cos^2\theta)(\Phi_{,\zeta\psi})^2 - 4\sin\theta\,\Phi_{,\zeta\psi}\Phi_{,\psi\psi} + (\Phi_{,\psi\psi})^2\end{array}\right]d\zeta d\psi \tag{10}$$

A comma in Eq. (10) represents the partial derivative of Φ with respected to corresponding independent variables and $D = Kj^3/12(1 - v^2)$ is known as rigidity of flux with Poisson ratio $v\,(=0.345)$. From Eqs. (2) and (5), rigidity of flux becomes

$$D = \frac{j_0^3 K_0 \left[1 - \alpha\left(1 - \frac{\zeta^2}{a^2}\right)\left(1 - \frac{\psi^2}{b^2}\right)\right]\left[\left(1 + \beta_1\sqrt{1 - \frac{\zeta}{a}}\right)\left(1 + \beta_2\sqrt{1 - \frac{\psi}{b}}\right)\right]^3}{12(1 - v^2)} \quad (11)$$

Using Eqs. (2), (6) and (11) in Eqs. (9) and (10), we get

$$K = 1/2\rho_0 j_o \omega^2 \cos\theta \int_{\zeta=0}^{\zeta=a} \int_{\psi=0}^{\psi=b} \left[\left(1 + \beta_1\sqrt{1 - \frac{\zeta}{a}}\right)\left(1 + \beta_2\sqrt{1 - \frac{\psi}{b}}\right)\left(1 - \alpha_1\frac{\zeta}{a}\right)\Phi^2\right] d\psi\, d\zeta$$

$$(12)$$

$$P = \frac{K_0 j_0^3}{24(1 - v^2)\cos^3\theta} \int_{\zeta=0}^{\zeta=a} \int_{\psi=0}^{\psi=b} \left[\begin{array}{l}\left[1 - \alpha\left(1 - \frac{\zeta^2}{a^2}\right)\left(1 - \frac{\psi^2}{b^2}\right)\right]\left[\left(1 + \beta_1\sqrt{1 - \frac{\zeta}{a}}\right)\left(1 + \beta_2\sqrt{1 - \frac{\psi}{b}}\right)\right]^3 \\ \left[(\Phi,_{\zeta\zeta})^2 - 4\sin\theta\,\Phi,_{\zeta\zeta}\Phi,_{\zeta\psi} + 2(\sin^2\theta + v\cos^2\theta)\Phi,_{\zeta\zeta}\Phi,_{\psi\psi}\right. \\ \left.+ 2(1 + \sin^2\theta - v\cos^2\theta)(\Phi,_{\zeta\psi})^2 - 4\sin\theta\,\Phi,_{\zeta\psi}\Phi,_{\psi\psi} + (\Phi,_{\psi\psi})^2\right]\end{array}\right] d\psi\, d\zeta$$

$$(13)$$

To find the frequency equation, Rayleigh–Ritz technique is applied. The technique emphasis on the phenomena that maximum strain energy (P) must equal to maximum kinetic energy (K). Therefore,

$$\delta(P - K) = 0 \quad (14)$$

Using Eqs. (12) and (13), Eq. (14) becomes

$$\delta\left(P_{\max} - \lambda^2 K_{\max}\right) = 0 \quad (15)$$

where

$$P_{\max} = \frac{1}{\cos^4\theta} \int_{\zeta=0}^{\zeta=a} \int_{\psi=0}^{\psi=b} \left[\begin{array}{l}\left[1 - \alpha\left(1 - \frac{\zeta^2}{a^2}\right)\left(1 - \frac{\psi^2}{b^2}\right)\right]\left[\left(1 + \beta_1\sqrt{1 - \frac{\zeta}{a}}\right)\left(1 + \beta_2\sqrt{1 - \frac{\psi}{b}}\right)\right]^3 \\ \left[(\Phi,_{\zeta\zeta})^2 - 4\sin\theta\,\Phi,_{\zeta\zeta}\Phi,_{\zeta\psi} + 2(\sin^2\theta + v\cos^2\theta)\Phi,_{\zeta\zeta}\Phi,_{\psi\psi}\right. \\ \left.+ 2(1 + \sin^2\theta - v\cos^2\theta)(\Phi,_{\zeta\psi})^2 - 4\sin\theta\,\Phi,_{\zeta\psi}\Phi,_{\psi\psi} + (\Phi,_{\psi\psi})^2\right]\end{array}\right] d\psi\, d\zeta$$

$$K_{\max} = \int_{\zeta=0}^{\zeta=a} \int_{\psi=0}^{\psi=b} \left[\left(1 + \beta_1\sqrt{1 - \frac{\zeta}{a}}\right)\left(1 + \beta_2\sqrt{1 - \frac{\psi}{b}}\right)\left(1 - \alpha_1\frac{\zeta}{a}\right)\Phi^2\right] d\psi\, d\zeta$$

and

$$\lambda^2 = \frac{12\rho_0\omega^2 a^4(1-v^2)}{K_0 j_o^2} \text{ is known as frequency parameter.}$$

Equation (15) consists of two unknowns, i.e., C_1, C_2 because of Eq. (7) which can evaluated by

$$\frac{\partial(P_{max} - \lambda^2 K_{max})}{\partial C_i} = 0, \ i = 1, 2 \tag{16}$$

On simplifying Eq. (16), we get

$$q_{i1}C_1 + q_{i2}C_2 = 0, \ i = 1, 2 \tag{17}$$

where q_{i1}, q_{i2} $(i = 1, 2)$ comprises with frequency λ^2.

To obtained vibrational frequencies, it is necessary that determinant of coefficient matrix from Eq. (17) must be zero

$$\begin{vmatrix} q_{11} & q_{12} \\ q_{21} & q_{22} \end{vmatrix} = 0 \tag{18}$$

Equation (18) gives vibrational frequency modes for parallelogram plate as λ_1 and λ_2 for different combinations of value of skew angle θ, tapering constants β_1, β_2, thermal gradient α and non-homogeneity constant α_1. All the results are displayed in the form tables.

3 Results Discussion and Comparisons

The effect of parameters to vibrational frequencies of parallelogram plate like thermal gradient (α), skew angle (θ), non-homogeneity constant (α_1), and especially tapered constants $(\beta_1$ and $\beta_2)$ (circular variation) is studied. All the results are displayed in the form of tables.

Tables 1 and 2 provide the vibrational frequencies for both modes corresponding to tapered constants $(\beta_1$ and $\beta_2)$ which vary circularly in both the directions for fixed value of aspect ratio $(a/b = 1.5)$ and skew angle $(\theta = 30°)$. From these two tables, authors conclude that frequencies for both modes increase when tapered constants $(\beta_1$ and $\beta_2)$ increase from 0 to 1 for the following cases.

Table 1 Taper constant (β_1) versus frequency (λ) for $a/b = 1.5$, $\theta = 30°$

β_1	$\alpha = 0.0$, $\alpha_1 = 0.0$, $\beta_2 = 0.4$		$\alpha = 0.4$, $\alpha_1 = 0.4$, $\beta_2 = 0.6$	
	λ_1	λ_2	λ_1	λ_2
0.0	96.22	383.42	107.20	427.35
0.2	112.33	448.88	124.83	498.91
0.4	128.55	514.68	142.58	570.87
0.6	144.88	580.86	160.41	643.09
0.8	161.15	646.84	178.29	715.48
1.0	177.51	713.11	196.21	788.00

Table 2 Taper constant (β_2) versus frequency (λ) for $a/b = 1.5$, $\theta = 30°$

β_2	$\alpha = 0.0$, $\alpha_1 = 0.0$, $\beta_1 = 0.4$		$\alpha = 0.4$, $\alpha_1 = 0.4$, $\beta_1 = 0.6$	
	λ_1	λ_2	λ_1	λ_2
0.0	98.97	395.85	112.34	450.12
0.2	113.54	454.27	127.90	512.22
0.4	128.55	514.68	143.76	576.83
0.6	143.86	576.38	160.41	643.09
0.8	159.38	638.95	177.11	710.48
1.0	175.05	702.15	194.00	778.67

Case 1 $\alpha = 0.0$, $\alpha_1 = 0.0$, $\beta_2 = 0.4$, Case 2 $\alpha = 0.4$, $\alpha_1 = 0.4$, $\beta_2 = 0.6$ (in Table 1). Case 3 $\alpha = 0.0$, $\alpha_1 = 0.0$, $\beta_1 = 0.4$, Case 4 $\alpha = 0.4$, $\alpha_1 = 0.4$, $\beta_1 = 0.6$ (in Table 2).

After analyzing both tables simultaneously, authors also conclude that the values of frequencies are slightly higher in Table 2 for case 3 as compared to case 1 in Table 1 but when the value of tapered constants becomes $(\beta_1 = \beta_2 = 0.4)$ the value of frequencies coincides and after that frequencies decreases for case 3 in Table 2 as compared to case 1 in Table 1. Also case 2 in Table 1 has less frequencies than case 4 in Table 2 and frequencies coincide when tapered constants become $(\beta_1 = \beta_2 = 0.6)$ after that frequencies for case 4 in Table 2 are slightly less then as compared to case 2 in Table 1.

Table 3 contains both vibrational frequency modes corresponding to thermal gradient (α) for fixed value of aspect ratio $(a/b = 1.5)$ and skew angle $(\theta = 30°)$ and displays the facts that frequency vibration decreases with the continuous increment in thermal gradient for two cases.

Case 5 $\alpha_1 = 0.0$, $\beta_1 = 0.4$, $\beta_2 = 0.6$ and Case 6 $\alpha_1 = 0.4$, $\beta_1 = 0.6$, $\beta_2 = 0.4$.

On the other hand, authors conclude that when the value of non-hommogenity constant (α_1) increases from 0 to 0.4 and value of tapered constant is $(\beta_1$ and $\beta_2)$ interchanged, the frequencies for both modes increase.

Table 4 contains the vibrational frequncies corresponding to skew angle (θ) for fixed value of aspect ratio $(a/b = 1.5)$. From Table 4, authors conclude the fact that

Table 3 Thermal gradient (α) versus frequency (λ) for $a/b = 1.5$, $\theta = 30°$

α	$\alpha_1 = 0.0, \beta_1 = 0.4, \beta_2 = 0.6$		$\alpha_1 = 0.4, \beta_1 = 0.6, \beta_2 = 0.4$	
	λ_1	λ_2	λ_1	λ_2
0.0	143.86	576.38	161.70	647.31
0.2	136.00	545.02	153.09	613.08
0.4	127.66	511.74	143.97	576.83
0.6	118.73	476.14	134.23	538.14
0.8	109.07	437.66	123.71	496.45

Table 4 Skew angle (θ) versus frequency (λ) for $a/b = 1.5$

θ	$\alpha = \alpha_1 = \beta_1 = \beta_2 = 0.0$		$\alpha = \alpha_1 = \beta_1 = \beta_2 = 0.2$		$\alpha = \alpha_1 = \beta_1 = \beta_2 = 0.4$		$\alpha = \alpha_1 = \beta_1 = \beta_2 = 0.6$	
	λ_1	λ_2	λ_1	λ_2	λ_1	λ_2	λ_1	λ_2
0°	54.15	217.54	72.64	292.55	93.54	377.26	116.27	469.20
30°	74.00	294.60	99.31	396.54	127.95	511.98	159.22	637.99
60°	232.24	910.49	311.74	1226.69	401.78	1585.39	500.30	1978.49

Table 5 Non-homogeneity (α_1) versus frequency (λ) for $a/b = 1.5$, $\theta = 30^0$

α_1	$\alpha = 0.2, \beta_1 = 0.4, \beta_2 = 0.6$		$\alpha = 0.4, \beta_1 = 0.6, \beta_2 = 0.8$	
	λ_1	λ_2	λ_1	λ_2
0.0	136.00	545.02	169.40	680.34
0.2	143.29	573.93	178.46	716.20
0.4	151.90	608.00	189.14	758.40
0.6	162.26	648.95	202.00	809.06
0.8	175.09	699.48	217.89	871.46

with the increase in skew angle (θ) form 0° to 60°, the vibrational frequencies increase for both modes of vibrations horizontally and vertically for all the cases.

Case 7 $\alpha = \alpha_1 = \beta_1 = \beta_2 = 0.0$, Case 8 $\alpha = \alpha_1 = \beta_1 = \beta_2 = 0.2$, Case 9 $\alpha = \alpha_1 = \beta_1 = \beta_2 = 0.4$, Case 10 $\alpha = \alpha_1 = \beta_1 = \beta_2 = 0.6$.

Table 5 accommodates the vibrational frequencies corresponding to non-homogeneity constant (α_1) for fixed value of aspect ratio ($a/b = 1.5$) and skew angle ($\theta = 30°$). From Table 5, one can easily conclude that frequencies increase with the increase in non-homogeneity constant (α_1) from 0 to 0.8 for both the cases.

Case 11 $\alpha = 0.2, \beta_1 = 0.4, \beta_2 = 0.6$, Case 12 $\alpha = 0.4, \beta_1 = 0.6, \beta_2 = 0.8$

It is interesting to note the fact that when the increment in values of thermal gradient (α) and tapered constants (β_1 and β_2) are of 0.2 as from case 11 to case 12, frequencies for both modes also increase. Also with the increment of 0.2 in the above-said parameters, rate of increase in vibrations is much higher in case 12 as compared to rate of increase corresponding to case 11.

4 Conclusion

From above results and discussions, authors conclude the following points:

1. On the simultaneous evaluation of Tables 1 and 2, authors conclude that the optimum value of frequencies is $\lambda_1 = 128.55$, $\lambda_2 = 514.68$ when the value of parameters are $\alpha = \alpha_1 = 0.0$, $\beta_1 = \beta_2 = 0.4$, $a/b = 1.5$, $\theta = 30°$ and $\lambda_1 = 160.41$, $\lambda_2 = 643.09$ when the value of parameters are $\alpha = \alpha_1 = 0.4$, $\beta_1 = \beta_2 = 0.6$, $a/b = 1.5$, $\theta = 30°$.
2. Due to circular variations in thickness, frequencies and variations in frequencies are less as compared to parabolic, exponential, or sinusoidal variation in thickness as reported in literature by others researchers.
3. All the parameters involved in the study directly effect the vibrational frequencies.
4. By choosing appropriate variation in appropriate parameter along with appropriate value, the desired frequencies can be obtained.

References

1. Gupta, A.K., Kumar, M., Khanna, A., Kumar, S.: Thermal effect on vibrations of parallelogram plate of linearly varying thickness. Adv. Stud. Theor. Phys. **4**(17), 817–826 (2010)
2. Gupta, A.K., Kumar, A., Gupta, Y.K.: Vibration of visco-elastic parallelogram plate with parabolic thickness variation. Appl. Math. **1**, 128–136 (2010)
3. Gupta, A.K., Kumar, M., Kumar, S., Khanna, A.: Thermal effect of vibration of a parallelogram plate of bi-direction linearly varying thickness. Appl. Math. **2**, 33–38 (2011)
4. Singh, B., Sexena, V.: Transverse vibration of skew plates with variable thickness. J. Sound Vib. **206**(1), 1–13 (1997)
5. Khanna, A., Arora, P.: Effect of sinusoidal thickness variation on vibrations of non-homogeneous parallelogram plate with bi-linearly temperature variation. Indian J. Sci. Technol. **6**(9), 5228–5234 (2013)
6. Singh, B., Chakraverty, S.: Flexural vibration of skew plates using boundary characteristic orthogonal polynomials in two variables. J. Sound Vib. **173**(2), 157–178 (1994)
7. Sharma, A., Sharma, A.K., Raghav, A.K., Kumar, V.: Effect of vibration on orthotropic visco-elastic rectangular plate with two dimensional temperature and thickness variation. Indian J. Sci. Technol. **9**(2), 7 pages (2016)
8. Sharma, A., Raghav, A.K., Kumar, V.: Mathematical study of vibration on non-homogeneous parallelogram plate with thermal gradient. Int. J. Math. Sci. **36**(1), 1801–1809 (2016)
9. Khanna, A., Kaur, N.: Effect of thermal gradient on natural frequencies of tapered rectangular plate. Int. J. Math. Anal. **7**(16), 755–761 (2013)
10. Khanna, A., Kaur, N.: Vibration of non-homogeneous plate subject to thermal gradient. J. Low Freq. Noise Vib. Act. Control **33**(1), 13–26 (2014)

11. Khanna, A., Kaur, N.: A study on vibration of tapered rectangular plate under non-uniform temperature field. Mechanika **20**(4), 376–381 (2014)
12. Khanna, A., Singhal, A.: Effect of plate's parameter on vibration of isotropic tapered rectangular plate with different boundary conditions. J. Low Freq. Noise Vib. Act. Control **35** (2), 139–151 (2016).
13. Khanna, A., Singhal, A.: A study on free vibration of visco-elastic tapered plate with clamped ends. Romanian J. Acoust. Vibr. **12**(1), 43–48 (2015)
14. Leissa, A.W.: Vibration of Plate. NASA SP-160, (1969)

Impact of Dense Network in MANET Routing Protocols AODV and DSDV Comparative Analysis Through NS3

Girish Paliwal and Swapnesh Taterh

Abstract Today's mobile communication networking technology is people-centric revolution that means the people put in the center and build the devices network to fulfill the people requirements. This era is shifting towards the wireless technology sharing mobile network with low cost, high performance, and low energy consumption. The paradigm shifting to mobile communication, the MANET communication network provides three kinds of networking: Wireless Mesh Network (WMN), Wireless Sensor Network (WSN), and Vehicular Ad hoc Networking (VANET). The MANET networking is popular because there is no requirement for any fixed infrastructure. The mobile ad hoc networking is having many challenges like: security, topology, energy, routing, etc., the researcher wants to find out the most efficient solution for reducing these challenges, in this paper, we are focusing on one of the most important challenges to maintaining the link between mobile nodes in MANET, using through AODV and DSDV routing protocols analysis. These protocols are analyzed and find out the impact on networking performance when increasing the density of the MANET. AODV and DSDV performance analyzes through NS3 network simulator.

Keywords MANET · VANET · AODV · DSDV · NS3

1 Introduction

The wireless network communication is the prominent area of research for the last few years and it is growing very fast. The wireless networks are of two type based on the physical structure. One kind of network that has the fixed physical network structure and another type which does not have the fixed physical structure, is also

G. Paliwal (✉) · S. Taterh
Amity University, Jaipur, Rajasthan, India
e-mail: gpaliwal@jpr.amity.edu

S. Taterh
e-mail: staterh@jpr.amity.edu

© Springer Nature Singapore Pte Ltd. 2018
M. Pant et al. (eds.), *Soft Computing: Theories and Applications*,
Advances in Intelligent Systems and Computing 583,
https://doi.org/10.1007/978-981-10-5687-1_30

327

Fig. 1 Mobile ad hoc network architecture using IEEE 802.11 [2]

known as Mobile Ad hoc Network. Mobile ad hoc network is not having the fixed physical structure. The mobile ad hoc networks (MANETs) are the type of wireless mesh networking. MANET communication devices are simply called nodes. The nodes are freely moving from one location to another, either within the network group or outside the network group, due to this the MANET topology is not fixed, it changes very infrequently and unpredictably. The researcher tries to find the efficient protocol for MANET to maintain their links between nodes, automatically healing route, self-motivated, self-organized, and maintenance. These properties of MANET make it so popular and attract the focus of researcher [1], (Fig 1).

MANET routing protocols face the challenges related to the routing path and numerous kinds of algorithms have been implemented. Mobile ad hoc network has too much potential to handle in emergency and rescue situations.

MANET is growing and requires communication protocols which allow nodes to communicate over the wireless link without fail. To allow such on the fly communication networks, different kinds of ad hoc routing protocols are developed but here we include only two protocols for the simulation study: AODV and DSDV. The Ad hoc On-demand Distance Vector (AODV) routing protocol maintains the routing link to other nodes in the network whenever such routes are required [3].

The protocols testing for a communication network in the real-world environment has different kinds of problems and the real-world testing is too much costly and very time consuming process. Testing is on the basis of different scenarios. Every time when scenario is changed, it is implemented throughout the network having the numbers of nodes. Because of these problems, simulation tools are used to create scenario and evaluate the performance of the network. In this article, we are using the network simulation tool NS3. In NS3 simulator, AODC and DSDV routing protocols are implemented and performance matrix is calculated and analysis is done in terms of impact on MANET performance when density of network increased [4].

2   MANET Overview and Related Work

In MANET, a communication device is establishing the link through communication medium and it has working sender node as well as receiver node. Therefore, MANET communication is based on the nodes cooperation and coordination of each other. There is no centralized controlling system and it is formed in two forms: single hop and multiple hop. The MANET network topology is continuously changed without any prediction. Because of this, dynamic property of MANET required more prominent routing protocol that maintains the path between nodes. On the basis of routing methodology, the MANET routing protocols are classified in three categories: proactive, reactive, and hybrid.

2.1   Proactive Routing Protocol (Table Driven)

The proactive routing protocol or table-driven protocol maintains the up-to-date and uniform routing information proactively. Each node is maintaining the routing table of the recent topology structure in their own routing tables. A periodic update message is required to send by each node to maintain the routing table consistent after each significant change in the existing network topology. Among the other proactive routing protocol DSDV, one of the most effective is table-driven routing protocol [2].

2.1.1   Destination Sequenced Distance Vector (DSDV) Routing Protocol

DSDV is a proactive MANET routing protocol. It finds the shortest path from the available list of paths using the Bellman–Ford algorithm. The DSDV source to destination path is loop free. In DSDV routing protocol, node periodically sends their routing information with their neighbor nodes. It also uses the path sequence number to identify the best routing path. The routing table is updated when the sending node has the higher sequence number than the existing similar path. The DSDV does not support multipath routing and consumed much bandwidth.

2.2   Reactive Routing Protocol (On Demand)

In MANET, reactive routing protocols are not maintaining the prior routing information for all nodes the whole time. When nodes are interested to share data or transmit data to a specific node, then the routing protocol gets the information to a specific node which is having the path or not. This is indicating the reactive routing protocols on on-demand basis concept. Reactive routing protocol identifies the route information through distance vectored routing algorithm. The reactive routing protocols are reducing the bandwidth consumption. In reactive routing protocol, the AODV protocol is the most efficient routing protocol [5].

2.2.1 Ad Hoc On-Demand Distance Vector (AODV) Routing Protocol

AODV is a reactive MANET routing protocol. AODV allowed multihop routing protocol. In MANET, two or more nodes are willing to communicate then AODV routing protocol search and establish routing path between the nodes. AODV routing protocol is based upon the distance vector algorithm and it is periodically updated. It provides best and secure transmission path to send data packets to destination. It reduced the network's band width consumption, because it searches path when nodes are required. AODV broadcasts four kinds of messages as like: RREQ, RREP, Hello, and RERR. RREQ broadcasts when establishing a between source node and destination node. When destination node is received the RREQ message, it sends the RREP message to the source node. To check the recent connection whether it is established successfully or not, the sender sends the Hello message to the receiver. RERR message is sent when the recent path testing failed. The major drawback is delay during route discovery [6–8].

3 Proposed Work

Most of the research work has been pointed in MANET routing protocol performance using the different existing conditions, such as node density in a specific area, energy consumption, type of traffic, speed of nodes, size of network, quality of services, etc. This paper is focusing on the routing protocol performance on the basis of MANET network density. Based on the simulation result, it analyzes the scaling relations and also discusses the throughput and network overhead when density changes. Here, we chose the one proactive (DSDV) and one reactive (AODV) routing protocol for analysis. AODV and DSDV routing protocols are evaluated with various kinds of performance matrices. Here, we simulated the DSDV and AODV routing protocols with NS3 network simulator and determined which protocol performed best under the simulation conditions.

4 Methodology

To perform analysis of AODV and DSDV using the NS3 network simulation tool, the number of parameters are applied. Those parameters are used to calculate and analyze the performance of the network.

4.1 Simulator

The NS3 Network Simulator used here analyzes the performance of AODV and DSDV. NS3 is an open source network simulator and it is an emulator as well. NS3 is a discrete event simulator. NS3 simulator core and different types of models are implemented in object-oriented programming language C++. NS3 has the different framework and built in libraries. These libraries can be linked to (C++/python) simulator program, either statically or dynamically. NS3 is not a NS2 extension. It does not support NS2 files. NS3 API is implemented through python programming language. NS3 produced simulation result in two formats, first the packet capturing file extension .pcap and another trace file extension .tr. It provides the flexibility of simulation result capturing different types [1, 9].

4.2 Metrics Value

After simulation of proactive routing protocol (DSDV) and reactive routing protocol (AODV), evaluation of the following metrics values is done [10]:

Throughput
It is measured as the ratio of the total amount of received data and the total simulation run time.

Delay (End to End)
The average end-to-end delay measured by total time taken by sent packets, divided by total time taken by received packets [10].

Packet Delivery Ratio
It is a ratio of packet generated by source and the number of packets delivered to the destination [10].

Packet Loss
Number of dropped packed when the buffered becomes full or stored packed time to live exceeds

Routing Overhead
Broadcasting the number of control packet to maintaining the route.

4.3 Simulation Setup

In this paper, we analyze the DSDV and AODV routing protocol and test the network overhead and throughput. The simulation self-created scenario which has the following parameter values. Table 1 describe the each parameter values that are used during the network simulation.

Table 1 Simulation parameters

Parameters	Values
Simulator	NS3
Number of nodes	50, 70, 90, 110, 130, 150
Simulation time	100 s
Simulation area	1000 × 1000 m
Packet size	512 Bytes
Packet rate	5 Packet/S
Mobility model	Random way point
Node speed	20 m/s
Transmission power	7.5 dBm
Startup time	50 s
Channel type	Wireless
MAC type	802.11
Protocols	DSDV, AODV

5 Simulation Results

The simulation results after simulation completed of each routing protocol in the NS3 environment have observed as per the following data table.

According to the simulation result shown in Table 2 there is clear indication that when the network density increased both the routing protocols AODV and DSDV are affected but the effect of the density implies in the same ratio in specified scenario. The Fig. 2 clears the picture in one sight and what is the impact of the density [11].

Simulation results under specific parameters of the network density shown in the Table 3, the table having the average throughput value of the network.

Comparative analysis of AODV and DSDV routing protocols cleared that when density increases, throughput increased in DSDV routing protocol whereas in the AODV routing protocol throughput decreases when density increased. The Fig. 3 shows the throughput of AODV and DSDV routing protocol comparative analysis [10].

Table 2 Simulation result of routing overhead

Nodes	AODV	DSDV
50	92.71	87.52
70	96.68	90.03
90	97.12	90.89
110	97.26	91.04
130	97.65	91.77
150	97.68	92.11

Fig. 2 Node density versus AODV and DSDV Routing Overhead

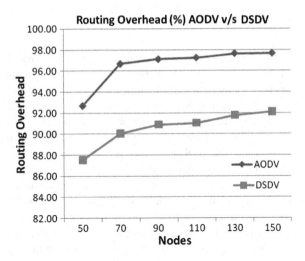

Table 3 Simulation result of throughput

Nodes	AODV	DSDV
50	616.52	439.02
70	552.23	461.97
90	548.26	514.00
110	520.16	556.83
130	439.92	597.45
150	448.57	649.32

Fig. 3 Node density versus AODV and DSDV throughput analysis

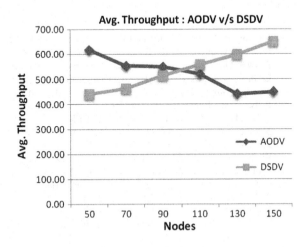

6 Conclusion

After comparative analysis of AODV and DSDV routing protocols, we find out the impact of the network density on the routing protocols performance. It is also analyzed that both routing protocols AODV and DSDV have different impacts. In AODV routing protocol analysis, it is observed that after increasing the density of networking nodes, the network routing overheard increases. Because it broadcasts the large number of control message packets for maintaining the route. Due to this reason, the AODV routing protocol performance gets affected. It is analyzed that when density increases in the network, AODV routing protocol throughput decreases. Whenever, the DSDV routing protocol has effected the density increasing of the network. It is also analyzed that DSDV has the same effect of routing overhead due to density of network. But its performance is far better due to the routing table information storage. When the network density increases, the DSDV routing protocol throughput increases. In this paper, it is identified that when network density increases the DSDV works more efficiently rather than AODV for the specified network scenario.

Future Work: Mobile ad hoc network is used very wildly by the group of networks but in future it is more usable when AODV routing performance increases because this kind of network node has the limited resources so that the AODV routing protocol is more suitable for MANET but it required some improvement in its efficiency, security, scalability, and quality of services. After that it is analyzes with other existing routing protocols.

References

1. Upadhyay, A., Phatak, R.: Performance Evaluation of AODV DSDV and OLSR Routing Protocols with Varying FTP Connections in MANET. IJRCCT 2, 531–535 (2013)
2. Paliwal, G., Mudgal, A. P., Taterh, S.: A Study on Various Attacks of TCP/IP and Security Challenges in MANET Layer Architecture. Proceedings of Fourth International Conference on Soft Computing for Problem Solving 195–207 (2015)
3. Paliwal, G., Taterh, S.: A Topology Based Routing Protocols Comparative Analysis for MANETs. Int. J. Adv. Eng. Res. Sci. (IJAERS) 3, 161–166 (2016)
4. Maurya, P.K., et al.: An overview of AODV routing protocol. Int. J. Mod. Eng. Res. (IJMER) 2, 728–732 (2012)
5. Walia, H., Singh, E.M., Malhotra, R.: A Review: Mobile AdQ Hoc Routing Protocols. Int. J. Future Gener. Commun. Networking 9, 193–198 (2016)
6. Bruno, R., Conti, M., Gregori, E.: Mesh networks: commodity multihop ad hoc networks. IEEE Commun. Mag. 43, 123–131 (2005)
7. Royer, E.M., Perkins, C.E.: An implementation study of the AODV routing protocol. Wireless Communications and Networking Confernce, 2000. WCNC. 2000 IEEE 3, 1003–1008 (2000)
8. Royer, E.M., Toh, C.-K.: A review of current routing protocols for ad hoc mobile wireless networks. IEEE Pers. Commun. 6, 46–55 (1999)

9. Singh, A., Dhaka, V., Singh, G.: Comparative Analysis of Dynamic Path Maintenance Routing Protocols for Mobile Ad-Hoc Networks. Indian J. Sci. Technol. **9** (2016)
10. Mehmood, Z., Iqbal, M., Wang, X.: Comprehensive experimental performance analysis of DSR, AODV and DSDV routing protocol for different metrics values with predefined constraints. Int. J. Inform. Technol. Comput. Sci. (IJITCS) **6**, p. 24 (2014)
11. Mohsin, R.J., Woods, J., Shawkat, M.Q.: Density and mobility impact on MANET routing protocols in a maritime environment. Science and Information Conference (SAI), 2015 1046–1051 (2015)

Design and Development of Intelligent AGV Using Computer Vision and Artificial Intelligence

Saurin Sheth, Anand Ajmera, Arpit Sharma, Shivang Patel and Chintan Kathrecha

Abstract The main aim of this paper is to develop a smart material handling system using an AGV (automated guided vehicle). The task is to transport a container of a fixed size from a defined start point to a defined end point. There is an overhead camera located at the boundary of the arena, in such a way that complete arena can be seen in a single frame. The camera will be capturing real-time images of the vehicle to determine its position and orientation using OpenCV library. The computer will also perform the task of path planning by using various artificial intelligence algorithms like RRT (rapidly random exploring tree) and A* (A Star). The outcome of this process will be the shortest path from beginning point to finish point while avoiding the obstacles. The commands should be enough for the robot to understand where it should go next, i.e., the next pose for the robot. This process continues until the goal is reached. To achieve this, few algorithms are developed for shape detection and edge detection. They help in determining the obstacles and the free area/ path where robot can traverse. The image from overhead camera is used to make the shortest global path from start to end using image processing. The computer will do this using various packages in ROS (robot operating system). This global path will generate waypoints for robot to traverse and the image will also

S. Sheth (✉) · A. Ajmera · A. Sharma · S. Patel · C. Kathrecha
Mechatronics Department, G H Patel College of Engineering and Technology,
Gujarat, India
e-mail: saurinsheth@gcet.ac.in

A. Ajmera
e-mail: anand_10594@yahoo.com

A. Sharma
e-mail: arpit07091993@gmail.com

S. Patel
e-mail: shivaang14@gmail.com

C. Kathrecha
e-mail: chintan.kathrecha@gmail.com

© Springer Nature Singapore Pte Ltd. 2018
M. Pant et al. (eds.), *Soft Computing: Theories and Applications*,
Advances in Intelligent Systems and Computing 583,
https://doi.org/10.1007/978-981-10-5687-1_31

337

provide current pose for the robot. Though the orientation of the obstacles varies the path of AGV and will always follow the shortest path. Thus, AGV shows the artificial intelligent.

Keywords AGV · OpenCV · RRT · Artificial intelligence · A star
Robot operating system (ROS)

1 Introduction

1.1 What Is AGV?

AGV is a system consisting of an unmanned, battery-powered vehicle, a guidance system, and other associated components. It is a programmable system which can be programmed according to the need of the organization's shop floor requirements, it can take material/product through a variety of different paths based on the traffic along the paths using programming to get to the desired location. AGV's can work throughout the year, round the clock increasing the productivity of organization. There are four main components of an AGV system. They are as follows:

- The Vehicle, which is used to traverse through the shop floor without human operator.
- The Guide Path, which guides the vehicle to move along the specified path of travel avoiding obstacles.
- The Control Unit is consisting of various controllers, sensors, encoders, etc., for proper monitoring of the system.
- The Computer Interface: Several AGVs are connected to the main host computer and other systems such as Automated Storage and Retrieval Systems (AS/RS) to perform the functions efficiently as in flexible manufacturing systems.

1.2 Role of AGV

The use of material handling system within the organization helps improve the productivity within the organization and reduces the dependency on labor for the organization, which is a key problem for any organization. There is scope for reducing labor in material handling systems by using technology in a competitive global environment. It can perform tasks even in hazardous environments. Its application involves: (a) Handling of raw materials in an organization, (b) Movement of the work in process (WIP) inventory, (c) Handling of pallets using automotive technology.

1.3   Existing AGV

Table 1 shows the advantages and disadvantages of various existing AGVs on the basis of the guidance systems used.

2   Literature Review

Saurin Sheth et al. have developed a color-guided vehicle system and special purpose machine using mechatronics system, which includes machine vision, pneumatics, Arduino, and Zigbee wireless module [1, 2, 3, 4]. V. Chauhan et al. have developed sophisticated approach for detection of colored object detection using computer vision [5] and also further helped in improvement in the color

Table 1 Advantages and disadvantages of existing AGV systems

Type of guidance system	Advantages	Disadvantages
Wire guidance [23]	Dirty environment can be handled since the wire is set in concrete floor	Changing to a new route by an AGV requires laying of new guide wires throughout the shop floor, so it should not be used for a frequently changing system
Paint and chemical guidance [23]	These type of systems can be easily moved and are much better than the wire guidance system. A new path can be painted with low cost and is also easy to repair when the path is damaged	It requires a clean environment. The path must remain clear of all types of obstructions including a piece of paper
Laser guidance [23]	New targets can be defined without much difficulty within the existing system and it allows for easy expansion of the system. The reflective targets present in the system can be easily moved without much disruption to the process	It requires additional setup time as compared to other systems discussed above. Amount of programming required is also more as the AGV must be taught where the reflective targets are with respect to pick up and drop off points as well as obstacles that may be permanent fixtures on the factory/warehouse floor
Dead-reckoning guidance [23]	They add flexibility compared to other systems discussed above if smoother surfaces are available to run. The system is less expensive compared to laser guidance system as the controller used is not a high speed controller	Wheel slippage is the most serious problem while using dead-reckoning guidance system. It requires smooth surfaces to run. Since the optical encoder acts as an odometer, when the wheels are turning, the controller assumes that the AGV is moving. Heavy loads on AGVS can also pose a problem

detection algorithm [6]. Hark Hwang et al. have developed a new automated guided vehicle (AGV) dispatching algorithm which is based on the concept of bidding. The given algorithm utilizes the information of WIP and travel time of the AGV by making use of mathematical functions [7]. Ottjes et al. proposed a free range which is dynamic and has benefits over other models. A simulation model compares various layouts of the fixed type with the approach to connect through the shortest route. Explicitly the avoidance of collision will play an important role. It can thus be concluded that this approach has very high potential in terms of transport capacity [8]. K.H. Patel et al. have designed a closed-loop image processing system in a mechatronics system. Closed-loop system aids the robotic arm to align in a required sophisticated arrangement [9]. J.T. Udding et al. have developed an AGV system using zone control methodology. A number of layouts are designed for the workspace of the AGVs. A traffic control strategy is formulated which decouples the traffic congestion problems of the AGV traveling on congested networks consisting of several AGVs. An algorithm is formulated for tackling the problem of traffic congestion using zone control. The goal is to minimize the travel time of the AGV as well as the transportation time [10]. Lothar Schulze et al. have provided an overview on AGVS technology, giving information about recent technological advances made in the field of material handling systems and the results of research through use of various methods such as statistical analyses, characteristics curves, and new approaches [11]. David Portugal et al. have presented the full integration of compact educational mobile robotic platforms built around an Arduino controller board in the robotic operating system (ROS) [12]. ROS serves as a common platform for interfacing of various hardware and software such as controllers, sensors, and image processing software such as OpenCV. By reviewing this, it seems that new school of thought to secure intelligent interaction between environment and robot vehicle is required.

3   Scope, Objective, and Need of Proposed System

Block diagram of the proposed system in Fig. 1a shows the setup of the proposed system. It shows the predefined arena with start point and end point, along with the various static obstacles that are lying around. The obstacles are of various sizes and shapes, which have to be identified by various algorithms like shape detection and edge detection [13]. The image that comes from the center camera located at the top is used to obtain the shortest global path from start to end using image processing with help of the off board laptop having various packages in ROS. This global path will generate waypoints for the robot to traverse. The image will also provide current pose for the robot which helps in estimating the next pose for the robot. Thus, the proposed system uses image processing and artificial intelligence to decide the shortest path in various applications like material handling systems, defense industries, agriculture, etc.

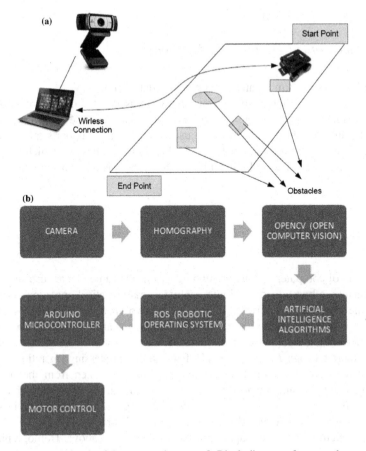

Fig. 1 **a** Conceptual sketch of the proposed system. **b** Block diagram of proposed system

Figure 1b shows the block diagram and concepts used in designing of this intelligent system. First, take the image which is captured by the camera, which is then homographed using the homography function in OpenCV so that a desired image is available for image processing operations in OpenCV. After homography that image is used to find the contours, corners, and centroids of the obstacles in the image and their distances from the start point. The obtained points are given to the artificial intelligence algorithms which will form a desired shortest path of travel between the start point and target point. Now, these points are given to ROS which serves as a common platform for the interfacing of various hardware, software, and sensors. As per the instructions given by ROS, the Arduino controls motor and thus reaching the target point via the shortest path avoiding obstacles.

4 Design Methodology

4.1 Mechanical Design and Its Model

Figure 2 shows the simple and conceptual model of the AGV.

Figure 2 shows different components that were used in the making of automated guided vehicle. The system uses two wheel drives to traverse in workplace. The frame for unit is a hollow frame, i.e., just a frame which supports motor support clamps. Onto this frame, a flat plate is put which acts as a base on which material is loaded or unloaded. This cylindrical block shown here is the conceptual design for motor body.

4.2 Homography

In the field of computer vision, any two images of the same planar surface in space are related by a **homography**. Homography is a mathematical operation performed using the concept of matrices. It consists of two matrices P and Q such that $P = HQ$; where P = output image, Q = input image, and H = homography matrix, Thus after finding the homographic matrix H, we apply it onto the input image Q to get the output image P on which all of the image processing operations can be performed to get the desired results. As the image is taken from the overhead camera placed at a suitable height above the working arena, the image obtained is trapezoidal in shape but in order to process the image, it is needed to be in square form, which can be achieved through the process of homography. Figures 3 and 4 show the example of homographic process as mentioned above. Homography can

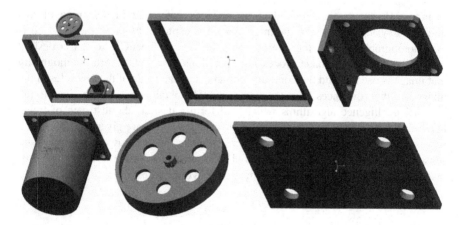

Fig. 2 Conceptual sketch of AGV

Fig. 3 Original image before homography

Fig. 4 Result of homography

also be of use in recognizing alphanumeric through image segmentation when viewing from an angled camera [14].

Figure 5 shows the experimental output achieved by generated codes of image processing. Here, the homography is performed on real-time situation using the live video stream of the camera.

4.3 Image Segmentation, Counter Detection, and Centroid Detection

Now, the image which is obtained after homography is used to find the contours of the obstacles placed in the path of the vehicle. After detecting the contours, the

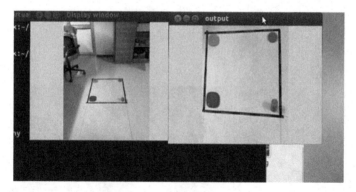

Fig. 5 Experiment's real time result of the homography

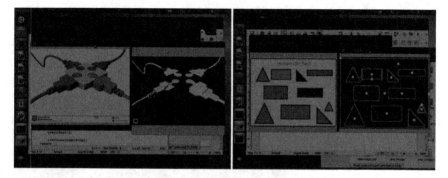

Fig. 6 Image segmentation and contour and centroid detection

centroids of the obstacles and the distances of the centroids from the start point of the arena can be obtained and the distances can be mapped into a 10 ft × 10 ft area. Figure 6 shows the centroid and contour detection.

4.4 Artificial Intelligence

Rapid Exploring Random Tree (RRT)

RRT as shown in Fig. 7, is an AI algorithm which can be used for path planning of AGVS in the proposed system. It works by finding a shortest path between the start and target point. Similar to a RRT is an algorithm called A* which can be used to program the system efficiently to achieve the shortest path. The tree preferentially expands towards large unsearched areas. It tries to connect the samples in random space available in the search space. If the result of searching is feasible then a new state is added to the tree and the tree starts expanding towards the target point taking the shortest possible path among several global paths. Then, check the collision of

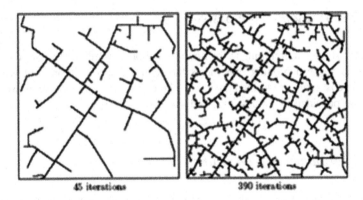

45 iterations   390 iterations

Fig. 7 RRT example

robot with the environment objects and do it until the robot is not in collision and has not reached the random target point. If robot hits an object in environment at a grid point say, for example [6, 6], then take the previous point [5, 5] which is not in collision. Insert all points till the new point example [5, 5] in the dictionary with parent as key and new point as value. Repeat it once again. Put random value == goal position after every 3–4 random values, so that the tree is directed towards the goal point. Experiment with this value. Repeat the whole thing again till last point in the dictionary (Nodes) is not the goal point in other words, the goal has not been reached [15].

A* Search Algorithm

A* search algorithm is used to find the path and graph traversal. It employs a process of finding efficient traversable points in the path called the nodes. It is known for its accuracy and performance. However, practically in travel-routing systems, it is outperformed by some algorithms which can preprocess the graphs to attain better accuracy and performance. However, much of the work done shows that A* outperforms many algorithms and is considered superior to other approaches.

In this algorithm, staring node and ending node is defined or at least stored on the system. Loop of the algorithm starts by the evaluating or expanding first node and determines the neighboring nodes. These new nodes are kept in the "open" nodes while the analyzed nodes are shifted into "closed" one. $f(x) = g(x) + h(x)$: where $f(x)$ is the total weight or total cost while $g(x)$ is called path cost and $h(x)$ is called the heuristic cost, shortest distance between current node and the final node. Algorithm eliminates the high cost paths and finally calculates and provide the sequence of nodes for the shortest path [16, 17, 18].

4.5 Specification of the System

Efficient system is a prerequisite for running any process efficiently. So, an attempt is made to make a product that meets the required specifications but within an affordable cost structure to the consumer. The capabilities of the system use during testing are provided in Tables 2, 3, and 4.

4.6 Environmental Constrains

As this intelligent system cannot be used in just any environment, much of constrains are defined below for successfully implementation of the proposed system.

- The obstacles should remain static throughout the run.
- No moving object should come in the path of AGV.
- The surface on which vehicle is moving needs to be flat.
- The vehicle is restricted to move in a fixed region only.
- The given area should be completely visible in a single frame of camera.

Table 2 Arduino specification

Feature	Value
Flash memory	4 kB
Chip/processor	ATmega328P
Opt voltage	5/7–12 V
CPU speed communication type	16 MHz UART

Table 3 Camera specification

Feature	Value
Frame size	1080 × 720
Resolution	1.3 megapixels
Frame view	30 FPS
Field vew	149°

Table 4 Motor specification

Feature	Value
Type	Regular D.C.
Opt voltage	9–12 V
Opt current	1–6 A

5 Experimentation and Testing

Figure 8 shows the experimental environment for real-time testing. The image in this figure has been taken from the key camera which is being used to guide the AGV. Here, the prototype of AGV is in the left hand side corner which is a starting point and the final position in the right-hand side corner. Figure 9 shows that sample test was carried out. Here, a cardboard of various shapes is used as an obstacle. With the help of the central camera, the images are taken. Then, the homography of the image has been carried out and final images were used to define contours, obstacles, and their distance from the start point. With the help of the artificial intelligence, it will follow the shortest path without touching the obstacles, with help of the algorithms, which commands the motors through ROS interface. A number of tasks are performed in the environment as shown in Fig. 8. Accuracy in positioning of the AGV, while testing, was almost 80% with an error of 5–10 cm. Here, main frame computer was the laptop and also, camera and Arduino Uno were connected through cables with it.

Fig. 8 Experimentation setup environment

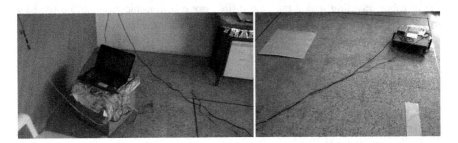

Fig. 9 Images of final testing in the experimental environment

6 Conclusion and Future Scope

This intelligent system proposed works entirely on the open source platform using open source software like ROS, OpenCV, OpenRave, etc., thus, minimizing the overall cost of the system and being productive. It can be used in various industrial applications like (a) manufacturing, (b) Warehousing, (c) Pharmaceuticals, (d) Automotive, etc.

Above system has been developed by using various open source platforms and is thus cost-effective. The productivity of the organizations can increase to a great extent due to optimized path travel. The same concepts can be used for variety of industries.

The proposed system has its scope of working in defense where it can carry ammunitions to remote places of war which are far from the base camp of the army. This application would require advancement in the system which could include accurate positioning and giving direction to the vehicle through the advanced vision and navigation system by use of satellites.

Above proposed system can work only for static obstacles during the run but it can be further extended to work with a dynamic environment using dynamic obstacles and improved navigation technology. For dynamic environment A* will not be of any use, instead many options are available such as D* [19], D* LITE, Anytime A* [20], etc. These algorithms are widely used for unknown or partially known dynamic environment for path planning of the AGV.

Localization of AGV can be further improved using triangulation method for positioning and for orientation of robot, wheel encoders will be an effective choice [21]. This whole system will need ROS-based environment and for that large communication network must be created which can be accessed across the whole environment. Line following technique is also one of the options for navigation of an AGV and in fact, it is one of the popular navigation approaches in industries [22]. Swarm robotics in AGV is also much in popular in modern industries and can also further be improved by improving the swarm communication among them [13].

This intelligent system can also be used in mining application. Additional sensors are required for navigation of the AGV. Mining robots are of great choice when exploring dangerous mines, in terms of poisonous gases and uneven mine caves. Here, satellite navigation or normal wireless navigation may not be an option due to robot traveling inside the caves; fiber optic communication can be used for positioning and transfer of data in this type of environments.

References

1. Parikh, P.A., Joshi, K.D., Sheth, S.: Color guided vehicle—an intelligent material handling mechatronic system. In: 1st International and 16th National Conference on Machines and Mechanisms. 628–635 (2013)
2. Gajjar, B.R., Sheth, S.: Design and automation in back plug press fitting process of ball pen assembly. Appl. Mech. Mat. 2596–2600 (2014)

3. Gajjar, B.R., Sheth, S.: Investigation of automation strategy and its effect on assembly cost: a case study on ball pen assembly line. Int. J. Curr. Eng. Technol. **3**, 89–92 (2014)
4. Virani, M., Vekariya, J., Sheth, S., Tamboli, K.: Design and development of automatic stirrup bending mechanism. In: Proceedings of the 1st International and 16th National Conference on Machines and Mechanisms (iNaCoMM2013), IIT Roorkee, India, Dec 18–20 (2013)
5. Chauhan, V., Sheth, S., Hindocha, B.R., Shah, R., Dudhat, P., Jani, P.: Design and development of a machine vision system for part color detection and sorting. In: Proceedings of Second International Conference on Signals, Systems & Automation (ICSSA), 91–93 (2011)
6. Sheth, S., Kher, R., Shah, R., Dudhat, P., Jani, P.: Automatic sorting system using machine vision. In: Multi-Disciplinary International Symposium on Control (2010)
7. Hwang, H., Kim, S.H.: Development of dispatching rules for automated guided vehicle systems. J. Manuf. Syst. 137–143 (1995)
8. Duinkerken, M.B., Ottjes, J.A., Lodewijks, G.: Comparison of routing strategies for AGV systems using simulation. In: Winter Simulation Conference, 1523–1529 (2006)
9. Sheth, S., Patel, K.H., Patel, H.: Design of automatic fuel filling system using a mechatronics approach. In: CAD/CAM Robotics and Factories of the Future, pp. 785–795. Springer, New Delhi (2016)
10. Li, Q., Adriaansen, A.C., Udding, J.T., Pogromsky, A.Y.: Design and control of automated guided vehicle system: a case study
11. Schulze, L., Behling, S., Buhrs, S.: Automated guided vehicle systems: a driver for increased business performance. In: International Multi Conference of Engineers and Computer Scientists, 19–20 (2008)
12. Araújo, A., Portugal, D., Couceiro, M.S., Rocha, R.P.: Integrating arduino based educational mobile robots in ROS. In: 13th International Conference on Mobile Robots and Competitions, 8–13 (2013)
13. Parikh, P., Shah H.B., Sheth, S.: Development of a multi-channel wireless data acquisition system for swarm robots. Int. J. Eng. Dev. Res. 717–725 (2014)
14. Gupta, M., Patel, A., Dave, N., Goradia, R., Sheth, S.: Text-based image segmentation methodology. Procedia Technol. 465–472 (2014)
15. Zhao, Y.J., Konh, B., Honarvar, B., Joseph, F.O., Podder, T.K., Hutapea, P., Dicker, A.P., Yu, Y.: 3D motion planning for robot-assisted active flexible needle based on rapidly-exploring random trees. J. Autom. Control Eng. 3(5), 360–367 (2015)
16. Flinn, J., Ortiz, H.S., Yuan, S.: A secure routing scheme for networks with unknown or dynamic topology using A-star algorithm. In: International Conference Security and Management (2016)
17. Jiang, B., Wang, Y., Zhao, Z.: Path planning for terrestrial platform based on A-star algorithm. In: 3rd International Conference on Mechatronics, Robotics and Automation. 409–414 (2015)
18. Ducho, F., Babinec, A., Kajan, M., Beo, P., Florek, M., Fico, T., Jurišica, L.: Path planning with modified A star algorithm for a mobile robot. Model. Mech. Mechatron. Syst. 59–69 (2015)
19. Kim, J.G., Kim, D.H., Jeong, S.K., Kim, H.K., Kim, S.B.: Development of navigation control algorithm for AGV using D* search algorithm. Int. J. Sci. Eng. 34–38 (2013)
20. Sharma, D., Dubey, S.K.: Anytime A*algorithm—an extension to A*algorithm. Int. J. Sci. Eng. 1–4 (2013)
21. Jing, C., Wushan, C.: Study on the path tracking and positioning method of wheeled mobile robot. Int. J. Comput. Sci. Eng. Surv. (IJCSES). 1–8 (2015)
22. Parikh, P.A., Shah, H.B., Sheth, S.: A mechatronics design of a line tracker robot using Ziegler Nichols control technique for P, PI and PID controllers. In: International Mechanical Engineering Congress (IMEC), 13–15 (2014)
23. Chandak, A., Bodhale, R., Burad, R.: Optimal shortest path using HAS, A star and Dijkstra algorithm. Imperial J. Interdisc. **2**(4), 978–980 (2016)

Study and Comparative Analysis of Various Image Spamming Techniques

Jasvinder Kumar, Swapnesh Taterh and Deepali Kamnthania

Abstract Image spamming is a recently emerging technique that has gained attention because of the filtering problems faced in multimedia data. To compete with the developments in the area of spam filters, spammers started using image spam which resulted in ineffective text analysis of email body. The spam message is embedded into the attached image that is mostly modified randomly to elude signature-based mechanism of detection, and it is made obscure from OCR (Optical Character Recognition) text recognition tools. In this paper, an attempt has been made to study image spam and various image spam filtration techniques. Further, we have discussed different image spam detection techniques applied by researchers along with their pros and cons. Survey suggests that image spam has increased the complexity of spamming. Deployment of spam filtering techniques is mandatory for productivity and integrity of business system. Several approaches followed by research give positive aspect in spamming but there are certain issues with these techniques. Hence, there is a need to have hybrid approach that is cost-effective and feasible.

Keywords Image spamming · Spam filtering · Spam · Filter techniques, Optical Character Recognition

J. Kumar (✉) · S. Taterh
AIIT, Amity University, Jaipur, Rajasthan, India
e-mail: jasskumar2008@yahoo.com

S. Taterh
e-mail: staterh@jpr.amity.edu

D. Kamnthania
VIPS, Affiliated to IP University, Delhi, India
e-mail: deepali102@gmail.com

© Springer Nature Singapore Pte Ltd. 2018
M. Pant et al. (eds.), *Soft Computing: Theories and Applications*,
Advances in Intelligent Systems and Computing 583,
https://doi.org/10.1007/978-981-10-5687-1_32

1   Introduction

Spamming refers to sending unwanted or irrelevant content to users that are not interested, associated or related to that particular content. The most widespread spamming is done through emails where the messages are mostly for advertising purpose. Another purpose might be to carry out some malicious activity with emails, SMS, calls or any such digital data oriented communication process. The intention of the sender is to make the spam content available to maximum people. Webopedia defines spam as, "Internet junk mail". In simple words, it is also known as flooding of data [1]. Spamming has been creating a problem with both servers and general users. The increased volume of data causes delay in services or can reduce the reliability and credibility of services provided by server, occupying bandwidth of the channel and consuming additional space for content storage [2, 3]. With respect to these problems, spamming is a great concern for the Internet community as it has been continuously carried out in digital, web and mobile space [4].

According to a source [5], in the second quarter of 2015, there was 53.4% of spam recorded in the email traffic only. In the second quarter of 2015, the top three countries contributing their share in spam sources were: 14.59% (USA), 7.82% (Russia) and 7.14% (China). Not only text but even the images and videos are nowadays used for spamming leading to more usage and complete wastage of bandwidth and storage memory. Section 2 explains image spamming, spam detection and related background study, Sect. 3 discusses various spam detection techniques followed by Sect. 4 with detailed study of proposed techniques, Sect. 5 discusses the issues with existing techniques and finally the conclusion.

2   Background Study

Earlier, the nature of spamming was textual only (i.e. Text spam), an endless repetition of worthless text [3]. Text spam is known as unsolicited bulk email (UBE), junk mails or unsolicited commercial email (UCE) which is generated as email text messages. Text spam can be sent without prior consent by organizations which have obtained any user's email id through marketing activity or any trans-actional activity. This can be avoided if the genuine email id is not passed on to the unknown persons or sites. Various researchers worked on text-based anti-spam filtering techniques [6] and used machine learning algorithms to overcome the problem [7].

On the other hand, image spam is one of the most modern tricks introduced by spammers. It contains the text message embedded into an image which is sent as email attachments. These emails target different advertisement websites like adult, financial, products, leisure, health political, education, spiritual, etc. Image spam-ming started in late 2004, wherein spam messages were being embedded in images rather than text body of message [8]. The methods like White list/Blacklist, keyword checking which were used in text-based spam are also used in image spam, but these methods are not much effective to solve this problem. Another well-known machine learning classifiers like Support Vector Machine (SVM), Naive Bayesian, Decision tree, Neural Network [4], etc., are used to filter spam. The above-mentioned techniques are used by anti-spam researchers to overcome the tricks employed by spammers and restore efficiency. In this paper, an attempt has been made to carry out detailed survey on various image spamming and anti-spam techniques.

2.1 Spam Filters and Image Spamming

With a global increase of spamming, a spam filter is a technical feature designed to block spam messages from user's inbox. It helps to clean and maintain spam free inboxes. Although the filters cannot resolve spamming problem completely but they do help to countermeasure the effect of spamming. These tools contain several modules which analyse the features of email message (header format, text content, sender and receiver address, etc.) and run on ISP's, mail servers or client mail. The combined output of different modules helps to categorize whether the mail is spam or legitimate. To detect spam, various spam detection tools are available, e.g. [4] sendsafe (2003), recator mailer released in 2007 based on DNS, IP address and message type. Spam assassin is an open source filtering software package that can be used in both client and server sides. It filters incoming as well as outgoing mail messages by assigning value, based on some set of features like header set rules for body, automatic and manual white/black list of address, blocked domain list, time of delivery and Bayesian filtering. A message is considered as spam if calculated value of message is greater than threshold. Over a decade, various authors have given different machine learning algorithms to solve the problem of text spamming [9].

2.2 Image Spamming and Detection

Implementation of spam filters gave a competitive edge to spammers as it has increased the race between spammers and anti-spammers. Now spammers have employed another technique to deceive the anti-spammers, instead of using text spam, spammers are using image spam technique to achieve their target. Various authors define image spam in their own ways, one such definition is: A spam message which includes images in the body of mail message or a kind of email spam where spam text is presented in the form of an image in the mail message [6–8]. In case of text spamming, spammers used to add text in a message whereas in case of image spam [10], they try to entrap images in the following ways:

- Sending image file as an attachment in mail body.
- Sending image via hyperlink in a message.
- Embedding image in the body of a mail.

The main impulse of a spammer is to deceive spam filters which detect text only. For image spam, spammers can fabricate message by employing some of the following tricks:

- Obfuscation: A technique used for blurring, pixilation, low resolution, misspelling words, hand written fonts and adding noise to images. These make it hard for spam filters to detect the spam.
- Animated sketch: This trick includes cartoons, animations and drawing that are presented in various styles, colours and shapes which fascinate user but hard for filters to detect as spam.
- Natural Images: The natural images are different from modified images. Patchy font: Spammers are using uneven or sketchy characters that are hard to detect.
- Certain special properties of natural images make it difficult for spam filters to detect spam.

So, there is a great need to find solution for this cybercrime as it leads to complete wastage of resources like bandwidth. The header part of an email message (header based information) and some threshold value are used to decide spam or ham [11]. The properties of image or files are used by researchers to stop spam [2]. There are techniques that use OCR method to extract text of images into editable electronic document and then traditional filters are used to analyse image spam by finding special keywords from text [12]. To avoid spamming, the task of spam detection has broader scope than the spam filtering. There are various algorithms and techniques that are used for the detection of image spam and can be optimized for better usage. In the current study, the detailed analysis of various algorithms has been carried out.

Recently, spammers are using various image processing methods so that the properties of every single message are different. The font type, background and foreground colours or also by adding altered or rotated image in the email. So, this is a great threat for traditional spam filtering techniques which cannot recognize obfuscated message. The anti-spam filtering techniques which analyse email content are not capable of detecting the spam text hidden in images [13].

Figure 1 shows a normal image seen in the email which you receive normally as an attachment that is not a spam and could be from a regular website that you have subscribed to. Whereas Fig. 2 shows a spam image. It has a hidden purpose and provokes the user to click on some link or visit a particular website. A very normal looking image could also have some hidden spam text behind it which is difficult to differentiate. So, this has caused a decline in rate of spam capture throughout the industry of email security and also increased the income for spam generators [14].

Fig. 1 Normal image [40]

Fig. 2 Spam image [41]

3 Image Spamming Filter Techniques

The image spamming filter techniques can be divided into various categories as described below:

3.1 OCR-based Techniques

The first step towards image spamming was OCR (Optical Character Reader) technique. Also known as intelligent word/character recognition system targets handwritten, print script or cursive text or character at a time. Junk text embedded in image by spammer will be extracted by OCR and then analysed by text based filters that could be used to identify spam or ham. System can easily recognize the shape of still characters while using OCR (partial). Earlier OCR technique can extract only specific font style and text but with recent development efficient OCR system has been proposed which can identify complex font styles and size [15], commercial and open source tools like spam assassin use this technique. However, text extracted from image or email subject and body part are equivalent [16], so existing technique of spam filters like keyword detection and text categorization are used.

Keyword detection is a technique in spam filtering where a list of common words is prepared like sale, Viagra, etc. A library of keywords is maintained and edited as combinations of words are added. Those listed keywords which are frequently and repeatedly used in email message are considered as spam and blocked [16]. Where as in text categorization technique, the text extracted from image using OCR is tested against the text coming from email body. A text classifier trained and tested on image text can also be implemented on text body of email message [17].

These techniques are successful only when no content obfuscation has been used. Content obfuscation is a tricky trick used by spammers to circumvent the working of OCR and make system inefficient [17]. The content obscuring did not exist before 2005. The OCR technique can be fully utilized if the image is not abused by content obscuring. In spite of updating the OCR to recognize more obfuscation, the problem did not resolve because every time new update leads to higher computational cost [15]. On the other hand, some of the authors worked in contrast way instead of recognizing characters or word they made an effort on presence of obfuscation means image spam. Work has been done on the existence of noisy text and text region rather than recognition of text obfuscation. It has been found that this approach is better than earlier one. To reduce the high computational cost, stratified framework has also been used [18]. Framework worked on text extraction by lowering the number of modules in OCR which are not much reliable, this has reduced the module not the cost of running of OCR.

To overcome obfuscation error and block spamming, CAPTCHA emerged as image filtering technique. CAPTCHA (Completely Automated Public Turning Test to tell Computers and Humans Apart) is used to differentiate between human and a robot (i.e. machine). The ability of a human to read any distorted or noisy text which is hard for machine was the main feature of CAPTCHA. To keep spammers away, it efficiently checks whether you are human or machine by the use of images of languages. The images of language take randomly generated text and change them to image file, known as image CAPTCHA. CAPTCHA works on inverse theory of Alan turning test, rather than preventing spamming activities directly hamper spammer [19]. Various CAPTCHAS have been designed like the one given by named image CAPTCHA in which user will be identified as human if the common term of the image which is provided to the user is typed correctly. Others are differentiating images and unusual images. In differentiating images, two images groups will be tested as same or not, to find the odd one out from images provided to user unusual image is used. To circumvent CAPTCHA, few of the methods are adopted by spammers like hiring low cost manpower to identify them; employing bugs in the implementation so that they can easily bypass CAPTCHA, machine learning based attacks [16]. While designing CAPTCHA important characteristics like huge variety in shapes of letters (invariant recognition), segregate compact letters efficiently (segmentation) and correctly identify letters (parsing) should be there for greater efficiency. Modern CAPTCHA known as re-CAPTCHA acts like a security guard is designed to secure and resist against spamming. Use of API in re-CAPTCHA helps user not to solve whole distorted CAPTCHA text, just on single click they can find human or robot, which also saves time [20]. In a recent implementation of CAPTCHA, two-step verification is conducted to ensure the user is a human and not a robot. It is based on the browser's behavioural analysis by interacting with CAPTCHA or also referred to as No CAPTCHA re-CAPTCHA, which includes image identification CAPTCHA. The image identification could be a collection of similar looking images and the user might be asked to differentiate them based on their understanding. Only after it is verified that the user is not a robot, a re-CAPTCHA is generated [20]. It has turned out to be a very effective approach.

3.2 Header-based Technique

To combat with spamming, anti-spammers are using significant part of a message, i.e. header. Just as postal address, an email message has an address which contains information about sender and receiver and successfully delivers the message to recipient. RFC822 and 2822 define header fields also known as trace fields. As shown in Fig. 3, a standard message header contains filed like Return Path: Delivery-date: Date: Message-ID: X-Mailer: From: To: Subject, etc.

Return-path: <sender@senderdomain.tld>
Delivery-date: Wed, 13 Apr 2011 00:31:13 +0200
(3)Received: from mailexchanger.recipientdomain.tld([ccc.ccc.ccc.ccc])
by mailserver.recipientdomain.tld running ExIM with esmtp
id xxxxxx-xxxxxx-xxx; Wed, 13 Apr 2011 01:39:23 +0200
(2)Received: from mailserver.senderdomain.tld ([bbb.bbb.bbb.bbb]
helo=mailserver.senderdomain.tld)
by mailexchanger.recipientdomain.tld with esmtp id xxxxxx-xxxxxx-xx
for recipient@recipientdomain.tld; Wed, 13 Apr 2011 01:39:23 +0200
(1)Received: from senderhostname [aaa.aaa.aaa.aaa]
(helo=[senderhostname])
by mailserver.senderdomain.tld with esmtpa (Exim x.xx)
(envelope-from <sender@senderdomain.tld>)
id xxxxx-xxxxxx-xxxx
for recipient@recipientdomain.tld; Tue, 12 Apr 2011 20:36:08 -0100

Message-ID: <xxxxxxxx.xxxxxxxx@senderdomain.tld>
Date: Tue, 12 Apr 2011 20:36:01 -0100
X-Mailer: Mail Clients
From: Sender Name <sender@senderdomain.tld>
To: Recipient Name <recipient@recipientdomain.tld>
Subject: Message Subject

Fig. 3 A sample message [21]

Fields like FROM, TO, Subject, Date are standard fields used by every user. Other fields are additional information fields which contain routing information from sender mail box to the receiver mail server [21].

In everyday work, it is not necessary to look all the fields of header, but sometime when you feel like phishing attempt or spoof by intruder or want to know routing information, you can easily check mail header and find out sender masquerading. Thus, finding spam by filtering header information is an easy and effective approach. Various researchers used machine learning algorithm while using this technique.

3.3 Image Feature-based Technique

Unlike text content an image also contain some content in itself that is used to identify spam or ham image. These contents are known as features of an image. Based on feature of an image, an image classification technique is performed. Classification is based on high- or low-level features of an image. High-level features are general properties of an image like image size, image width and height, aspect ratio, file size, file format, image metadata, filename, bit depth, etc. [22]. Work has been done on image size, width and height. Low-level feature extraction method includes colour saturation, texture, edge, layout, shape and display of an image, etc. It is based on visual content analysis of an image which is widely used in the field of computer vision and pattern recognition. Based on visual features, the method is broadly categorized into two spam detection techniques.

3.3.1 Image Classification Method

A selective approach is used to extract visual features of an image and is based on these features classifier trained to identify spam or ham mails. Initially, a framework for image classification technique was given [23]. The method extracted visual features based on Embedded text (i.e. number of text region detected, text area), Banner and graphic (geometric characteristics), Image Location Feature which are based on assumption of attached images to personal mail in contrast with images behind web server and referenced in email.

3.3.2 Near-Duplicate Detection

These anti-spam methods are in correspondence with CBIR (Content-based Image Retrieval) techniques with an objective of finding images which look similar to the query image. The logic behind these techniques is similar to image classification methods. Spam images send via email basically originate from a single template, which is further randomized in order to elude signature-based recognition, and are then sent to a large number of users in batches. So, images abstracted and created from a single template should be similar visually (or "near-duplicate"), these images can be identified by comparing them with an existing database of all known image spam. It is asserted that the large volume of image spam incapacitates the spammers to send out any more images generated from the primary parent template [12]; therefore, near-duplicate recognition systems are supposed to stay efficient over time.

4 Techniques Used for Image Spam Detection

Some published image spam filtration techniques are surveyed in this section. Initially, 1654 mail messages were taken from author's own mail box and saved as text file [1]. Based on assumptions that every substring within the subject header and in the message body which is enclosed by white spaces is considered as token. Seventeen hand crafted features were used for evaluation. A multilayer artificial neural network using back propagation algorithm was used to train data. Various indicators like SP (Spam precision) percentage of messages, SR (Spam Recall) the proportion of the number of correctly classified spam messages to the number of messages originally categorized as spam, LP (Legitimate precision) percentage of messages classified as legitimate have been used to classify spam or ham.

Image spam poses a great threat to email communications due to high volumes, bigger bandwidth requirements and higher processing requirements for filtering. A feature extraction and classification framework that operates on features that can be extracted from image files in a very fast fashion has been proposed [24]. The features considered are thoroughly analysed regarding their information gain. The classification performance results for C4.5 decision tree and support vector machine classifiers have been presented. Lastly, the performance has been compared with the one that can be achieved using these fast features to a more complex image classifier operating on morphological features extracted from fully decoded images. The proposed classifier is able to detect a large amount of malicious images while being computationally inexpensive.

But this technique does not provide the required high accuracy so, another classification technique was introduced which compares existing technique of content filtering with new introduced technique and achieved spam identification accuracy from 90 to 99% [25]. For better result, study adopted two methods to automatically classify a mail as spam or ham. First, they used new algorithm for modifying feature extraction and second use JIT (Just-in-time) classifier for fast classification. The technique use wide variety of features which are simpler and easy to compute like image metadata, file size, width and height also some advance features like edge, random pixels, colour, etc. Three classifiers Maximum Entropy (MaxEnt), Naive Bayes (NB) and an ID3 Decision Tree (DT) are used to test on three data sets Personal Ham (PHam) with Personal Spam (PSpam), Personal Ham with Spam Archive (SpamArc) and all data together (All). The study concluded that combinational classification approach reduces time of filtering by 2–3 ms per image.

The use of image features is an effective way to detect spam, so there is another technique that uses four visual features of an image namely colour moment, colour heterogeneity, conspicuousness and self-similarity to propose a selective image modelling classifier framework for spam identification [26]. The framework uses multi-class characterization and a newly proposed MFoM (Maximal Figure of Merit) based learning. To use multi-class characterization, images are grouped into several sub classes and decision rules are designed to identify spam images. Three decision rules are defined as: 1. maximum score, 2. taking an arithmetic average and 3. geometric average. MFoM learning algorithm is used for identification of spam. On the basis of maximum score rule experiment result show that this approach is efficient than single class characterization, output 81.5% spam detection and only 5.6% false-positive rate.

Gao et al. propose a machine learning based image spam hunter to segregate spam images from natural images on texture statistics [27]. Global aid tad colour histogram and gradient orientation histogram features of image are used. Using global features, dissimilar spam images were converted into similar image group by using K-means algorithm than on the trained group of spam images probabilistic boosting tree (PBT) is applied to identify spam or natural images. As a result, system interpreted 89.44% true positive and 0.86 and 0.86% false-positive rate fivefold cross-validation on a database with 928 spam images and 810 normal images.

Apart from image features, researchers [28] have also worked on textual information present in a spam image. Study proposed an edge classification algorithm to extract non text edges on the basis of some edge features. Corner detection algorithm has been used to detect corner points, edge and user text region. Study interpreted 97.6% precision on spam archive dataset with a detection of 96% text in spam images.

Sheu and Chu proposed another text categorization technique based on a two phase filtering method [29]. First, the mail is categorized into four sections: sexual, finance, job hunting, marketing and advertising. Decision tree algorithm on every category fields of header part is extracted, like email subject, date and size of email. Author used this filter on Chinese email and attained an accuracy, precision and recall of 96.5 and 97%. Hui Yin et al. also prove that a single spam filtering technique is not as effective as a multi-heuristic method. So, a combination of ACO (Ant Colony Optimization) and LDA (Linear Discrimination Analysis) algorithms is used [30].

With advancement in technology, now IP address could also be analysed to detect if the email in your inbox is spam or not. Y. Hu et al. have proposed a framework which uses email header fields (host field, destination field, x-mailer field, sender server IP address and email subject) [31]. Study uses Random Forest (RF), C4.5 Decision Tree (DT), Naive Bayes (NB), Bayesian Network (BN), and

Support Vector Machine (SVM) classifiers to obtain accuracy, precision and recall of 96.7, 92.99 and 92.99%, respectively.

An innovative two stage classification framework known as BFMLC (Binary Filtering with Multi-Label Classification) was proposed [32], it is composed of multiple spam detection techniques. Grey scale images which are constantly creating trouble for binary spam filtering process decline the efficiency of filtering system. In order to reduce the problem, BFMLC used public personal data set to identify and classify spam images on different predefined topics. Study shows that BFMLC framework achieves the accuracy of 96.39% in identifying spam mails and a precision of 89.42% on text or mixed mails.

Spam or unsolicited email has now become a great threat which is negatively affecting the electronic messages and their usability. This approach explains Group Method of Data Handling (GMDH) which is based on inductive knowledge acquiring method for detection of spam by automatic content identifying features which can efficiently differentiate between legitimate and spam emails. The performance of complexities of different network models are studied using a public benchmark spam base dataset. According to the results, a 91.7% accuracy of classification can be attained utilizing merely 10 out of 57 available characteristics, which are selected via inductive learning. The most effectual subset of features is selected that is data reduction by 82.5%. Some ways to improve categorization performances by utilizing abductive network committees (ensembles) are prepared with varying subsets from the data of training [33].

Comparing this method with others like Naive Bayesian classification and neural networks depicts that this new Data Handling dependent learning method is capable of providing better accuracy of spam detection and false-positive rate of 4.3% only, requiring short training time. Jitendra et al. prove that when a generic algorithm (GA) is combined with other spam filtering methods then better results are obtained [34]. A collection of spam mail is encoded within a chromosome class and then they undergo genetic operations such as fitness function, mutation and crossover, etc. The GA itself is used to develop rule sets to filter spam.

A two-phase hierarchical spam filtering technique is studied [35]. In first phase, email message is classified as spam or ham on some of required header fields. Then in second phase, content base spam filtering is used to indeed classify mail is spam or not, not confirmed mails are sent to ham folder and again text-based filter or image-based filter based on message will be used. C4.5 Decision Tree (DT), Support Vector Machine (SVM), Multilayer Perception (MP), Nave Bays (NB), Bayesian Network (BN), and Random Forest (RF) classifiers are compared. The study concluded that in phase-I, RF classifier and in Phase-II RF and SVM classifier performed better, efficiency of the system is increased with a precision of 99.99% and 100% with very low FP rate were obtained.

Artificial Neural Network (ANN) was developed and instantly its implementation in spam filtering was recognized. ANN algorithm was used as a multi-level approach as it uses text-based and image spam detection, to improve filtration process, email text is pre-processed before it goes through the ANN [36]. A single-stage simple process is not equipped enough to catch all spam emails.

Another issue to combat in spam filtering techniques is that certain legitimate emails are also sent to the spam, attempts are being made to reduce such false positives to zero [37]. A perfect email spam filtration system that is required today, needs to overcome all or most of the issues that are discussed in the next section.

5 Open Issues with Existing Techniques

In the current scenario, spammers are directly proportional to anti-spammers, researchers are constantly focusing on finding efficient and novel methodologies to handle image spamming. As explained in Sect. 4, image spam is a type of classification problem, so classifier algorithms like neural network, support vector machine (SVM) and naive Bayesian classifier are used to create spam filtering methods. SVM distinguishes objects within discussion boundaries based on decision planes, whereas Bayesian Classifier identifies the most commonly used words in spam.

Since, there is no empirically complete or common evaluation of the image spam filtering techniques so all the pros and cons exhibited by the techniques need to be discussed on theoretical/detailed basis [16].

OCR-based methods are capable of analyzing the high-level images for hidden textual message, another advantage is that this method generates minimal to almost negligible false positives (genuine emails marked as spam), considered that the images in any legitimate messages do not contain any listed spam words. There are certain OCR errors which hamper their effectiveness; the errors depend on features like the OCR characteristics, image resolution and background or font size and face. Some better results were observed if the text to be extracted uses same classifiers as the body of email [18]. Another major drawback is the computational cost that causes increased processing time of the servers. The OCR methods can be obfuscated and disabled easily by the spammers, whereas these tricks do not work on the low-level images. So, the low-level techniques can be used as a part some multi-level spam filtration process where the initial stages will have some other cheaper techniques and just leave the OCR as last option to be certain.

The image classification technique cannot be affected by text obfuscation and another advantage is the capability of generalization. They use machine learning algorithm which allows this technique to scan all hidden patterns in any legitimate or spam email and also capable to detect some new image spam. The proper working of this technique does depend on the correct feature selection. The choice of features is always a guess depending upon the observation of various spam images. Another problem with this filtering technique is its adversarial nature [38]. The operation phase processing time in low-level features will depend upon the feature values and their computation. As processing times go, some authors say that algorithms of classification have a small impact as compared to others [39].

Another spam detection technique is near-duplicate, this method has very low rate of false positives. But, this technique does not adapt with learning algorithm or have generalization capabilities and hence cannot detect any new image that surfaces as a spam. One possible disadvantage of the near-duplicate technique of spam detection is that the distance for each template image needs to be calculated, although distance computation among two images in this case will be faster than a classification algorithm (such as nonlinear kernel or SVM). The distance for each template image will be calculated, so if the template size is big, it will increase the processing time by a big margin.

Finally, which image spam filtering technique is the best, among the ones mentioned here, will depend on a number of factors such as visual properties of the image, textual obfuscation, computational cost, number of false positives generated and generalization capabilities, the features selected, and the method adopted to use them.

6 Conclusion

The response of spammers to each situation is crucial along with complexity of spam images and related data. Also, end users and system administrators need to keep in mind that the spam-filtering techniques should not only be designed, developed and created, but should also be deployed as a priority.

It can be concluded that a combination of selective machine learning algorithms and low-level or high-level image detection techniques can be created so that the advantages of both methods can be utilized for better image spam filtrations. Machine learning algorithms that filter spam take a batch of labelled emails as their input and give correct labels as output for data testing, for labelling the data text categorization techniques are used. Machine learning combined with meta-heuristic or rule-based approach will highlight the positive aspects of both filtering techniques at different levels by arranging them into a new hybrid approach. Techniques that are not based on machine learning use regular expression rules for detecting common spam characteristics or phrases. This combined technique needs to be low in cost to be effective, adaptive and efficient.

References

1. Stuart, I.J., Cha, S., Tappert, C.: A neural network classifier for junk E-mail. Notes Comput. Sci. **3163**, 442–450 (2004)
2. Uemura, M., Tabata, T.: Design and evaluation of a Bayesian-filter-based Image Spam Filtering Method. IEEE (2008)
3. Patidar, V., Singh, D.: A survey on machine learning methods in spam filtering. Int. J. Res. Comput. Eng. Electron. **2**(5) (2013)
4. Khanum, M.A., Ketari, L.M.: Trends in Combating Image Spam E-mails. Information Technology Department, College of Computer and Information Sciences, King Saud University, Kingdom of Saudi Arabia

5. Securelist.com: Spam and phishing in Q2 2015. https://securelist.com/analysis/quarterly-spam-reports/71759/spam-and-phishing-in-q2-of-2015/ (2016)

6. Wang, Z., Josephson, W., Lv, Q., Charikar, M., Li, K.: Filtering image spam with near-duplicate detection. In: 4th Conference Email Anti-spam, CEAS (2007)

7. Soranamageswari, M., Meena, C.: An efficient feature extraction method for classification of image spam using artificial neural networks. 169–172 (2010)

8. Duan, Z.H. et al.: Detecting spam zombies by monitoring outgoing messages. IEEE INFOCOM, Rio de Janeiro, 1764–1772 (2009)

9. Chopra, N.D., Gaikwad, K.P.: Image and text spam mail filtering. Int. J. Comput. Technol. Electron. Eng (IJCTEE). **5**(3), Department of Computer Engineering, SND COE and RC Yeola, Nashik, India (2015)

10. Attar, A., Rad, R.M., Atani, R.E.: A Survey of Image Spamming and Filtering Techniques, pp. 71–105. Springer, Iran (2011)

11. Liu, Q., Qin, Z., Cheng, H., Wan, M.: Efficient modeling of spam images. In: Third International Symposium on Intelligent Information Technology and Security Informatics (IITSI), pp. 663–666 (2010)

12. Mehta, B., Nangia, S., Gupta, M., Nejdl, W.: Detecting image spam using visual features and near duplicate detection. In: Proceedings of the 17th International Conference World Wide Web. ACM, pp. 497–506 (2008)

13. Win, Z.M., Aye, N.: Detecting image spam based on file properties, histogram and hough transform. J. Adv. Comput. Netw. **2**(4), 287 (2014)

14. Kamble, M., Dule, C.: Detecting image spam based on image features using maximum likelihood technique. IJCST, March 2012

15. Youn, S., Cho. H.: Improved spam filter via handling of text embedded image e-mail. J. Electr. Eng. Technol. **9**, 742–748 (2014)

16. Biggio, B., Fumera, G., Pillai, I., Roli, F.: A survey and experimental evaluation of image spam filtering techniques. Pattern Recognit. Lett. **32**, 1436–1446 (2011)

17. Gupta, A., Singhal, C., Aggarwal, S.: Identification of image spam by using low-level and metadata features. Int. J. Netw. Secur. Appl. **4**(2) (2012)

18. Fumera, G., Pillai, I., Roli, F.: Spam filtering based on the analysis of text information embedded into images. Special issue on Machine Learning in Computer Security, Mach. Learn. Res. **7**, 2699–2720 (2006)

19. Kamble, M., Malik, L.G.: Detecting image spam using principal component analysis and SVM classifier. Int. J. Comput. Sci. Inf. Technol. Secur. **2**(6), 1217–1220 (2012)

20. Security.googleblog.com: Are you a robot? Introducing No CAPTCHA re-CAPTCHA. https://security.googleblog.com/2014/12/are-you-robot-introducing-no-captcha.html (2014)

21. Arclab.com: Email header: How to read and analyze the email header fields and information about SPF, DKIM, SpamAssassin. https://www.arclab.com/en/kb/email/how-to-read-and-analyze-the-email-header-fields-spf-dkim.html (2011)

22. Dakhare, B.S., Gaikwad, U.V., Gaikwad, V.B.: Spam detection using email abstraction. Int. J. Innov. Eng. Technol. **3**(2), 33–39 (2013)

23. Wu, C.T., Cheng, K.T., Zhu, Q., Wu, Y.L.: Using visual features for anti-spam filtering. In: Proceedings of the IEEE International Conference on Image Processing, vol. III, pp. 501–504 (2005)

24. Aradhya, H.B., Myers, G.K., Herson, J.A.: Image analysis for efficient categorization of image-based spam e-mail. In: Proceedings of the Eighth International Conference on Document Analysis and Recognition (ICDAR'2005)

25. Dredze, M., Gevaryahu, R., Elias-Bachrach, A.: Learning fast classifiers for image spam. CEAS (2007)

26. Byun, B., Lee, C.H., Webb, S., Pu, C.: A discriminative classifier learning approach to image modeling and spam image identification. In: Proceedings of the Fourth Conference on Email and Anti-Spam (CEAS) (2007)

27. Gao, Y., Sun, S.B., Qin, K.Y.: A new k-means data clustering approach. J. Comput. Inf. Syst. **4**(2), 565–570 (2008)

28. Wan, M., Zhang, F., Cheng, H., Liu, Q.: Text localization in spam image using edge features. In: Proceedings of International Conference on Communications, Circuits and Systems (ICCCAS), pp. 838–842 (2008)

29. Sheu, J.J., Chu, K.T.: An efficient spam filtering method by analyzing e-mail's header session only. Int. J. Innov. Comput. Inf. Control 5(11), 3717–3731 (SCIE), 371129 (2009)

30. Yin, H., Cheng, F., Zhang, D.: Using LDA and ant colony algorithm for spam mail filtering. In: International Symposium on Information Science and Engineering, IEEE. doi:10.1109/ISISE.2009.82, 368–371 (2009)

31. Hu, Y., et al.: A scalable intelligent non-content-based spam-filtering framework. Expert Syst. Appl. 37, 8557–8565 (2010)

32. Cheng, H., Qin, Z., Fu, C., Wang, Y.: A novel spam image filtering framework with multi-label classification. In: ICCCASn, pp. 282–285 (2010)

33. El-Alfy, E.-S.M., Abdel-Aal, R.E.: Using GMDH-based networks for improved spam detection and email feature analysis. Appl. Soft Comput. 11, 477–488 (2011)

34. Jitendra, N.S., Bindu, M.H.: E-mail spam filtering using adaptive generic algorithm. Mod. Educ. Comput. Sci. Int. J. Intell. Syst. Appl. 2, 54–60 (2014). doi:10.5815/ijisa.2014.02.07

35. Khater, I.M., Al-Jarrah, O.M., Al-Duwairi, B.: Hierarchical email spam filtering. Int. J. Comput. Inf. Technol. 5(3), 290–299 (2016)

36. Deshmukh, H. et al: Spam mail detection using artificial neural networks. Imp. J. Interdiscip. Res. 2(5) (2016). ISSN: 2454-1362, http://www.onlinejournal.in

37. Khater, I.M., Al-Jarrah, O.M., Al-Duwairi, B.: Hierarchical email spam filtering. Int. of Comput. Inf. Technol. 5(3) (2016). ISSN: 2279-0764

38. Ghosh, P.: A Framework of email cleansing and mining with case study on image spamming. Inter. J. Adv. Comput. Res. 4(17), 961–965. ISSN (Print): 2249-7277, ISSN (Online): 2277-7970 (2014)

39. Sheu, J.J.: An efficient two-phase spam filtering method based on e-mails categorization. Int. J. Netw. Secur. 9, 34–43 (2009)

40. Boscovs.com: 50% off sale. http://www.boscovs.com/shop/dept/50-off-sale/25000000.htm (2016)

41. Healthymomscircle.com: Work from home jobs for moms. http://healthymomscircle.com/topics/isagenix/work-from-home/ (2014)

42. Youn, S., Cho, H.: Improved spam filter via handling of text embedded image e-mail. J. Electr. Eng Technol. 9, 742–748 (2014)

DTN-Enabled Routing Protocols and Their Potential Influence on Vehicular Ad Hoc Networks

Vijander Singh and G.L. Saini

Abstract Delay-tolerant networks are the particular class of networks which use the concept of store, carry, and forward. In delay-tolerant network, the path is not established until and unless data is transmitted. The vehicles store data, carry it, and forward it to appropriate node whenever there is an opportunity to transmit data, i.e., a vehicle in the desired direction is available. This is also introduced as message passing mechanism.

Keywords VANET · MANET · DTN · Routing protocols

1 Introduction

The main characteristics of VANET provide more customization to develop new services. However, safety and comfort are two most important and relevant areas [1] of the device as shown in Fig. 1 [2]:

(1) **Comfort Applications:** These applications provide traveler more comfort and traffic efficiency during traveling. For example, a comfort application informs you about *traffic information system, nearest or next petrol/gas station, GPRS location tracking, weather information as well as enjoying internet, music, or movies* during the journey. This particular application is directly related to DTNs.

V. Singh (✉) · G.L. Saini
Amity Institute of Information Technology, Amity University Rajasthan,
NH-11C, Jaipur-Delhi Highway, Jaipur, India
e-mail: vijan2005@gmail.com

G.L. Saini
e-mail: glsaini86@gmail.com

© Springer Nature Singapore Pte Ltd. 2018
M. Pant et al. (eds.), *Soft Computing: Theories and Applications*,
Advances in Intelligent Systems and Computing 583,
https://doi.org/10.1007/978-981-10-5687-1_33

367

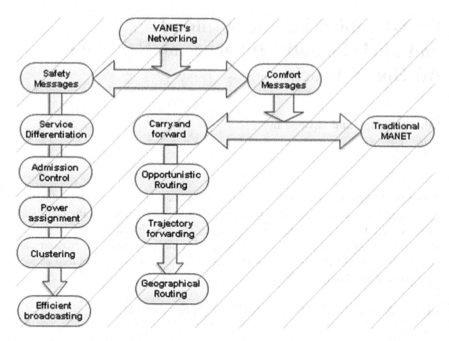

Fig. 1 Categories of VANETs

(2) **Safety Applications:** This type of applications works to enhance the safety of passengers. They work in a different manner such as exchanging safety-relevant information via IVC. Drivers are motivated to use these applications to get maximum safety during travel. Few examples of safety applications are *emergency warning system, traffic sign/signal violation warning, lane-changing assistant, and interaction coordination*. These applications are required for direct vehicle-to-vehicle communication due to stringent delay necessities.

To sort out this major error, nodes keep the packets with themselves and carry until they came into the range of another node, and now the carrying node forward the packets following some protocol based on characteristics of neighbor's node (called carry-and-forward strategy). This strategy is known as DTN technology and improves the packet delivery by store carry-and-forward.

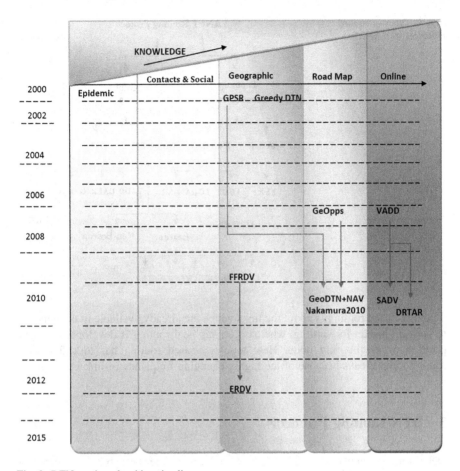

Fig. 2 DTN routing algorithm timeline

2 VANETs Routing Protocols

In Fig. 2, a timeline of the presently available routing protocols [3] for vehicular ad hoc network is given. These protocols are analyzed on the basis of the following specifications: requirement, designing, and objectives. Few of them target building vehicle-to-vehicle network; others are focused on vehicle toward roadside communication. Likewise some other protocols are designed to create communication in delay-tolerant networks. Simply there is a quality of service-oriented protocols, some of which connects vehicles to Internet. Routing protocols are categorized into two streams: topology-based and geographic routing. The information used by the

Fig. 3 Types of routing
protocols in VANETs

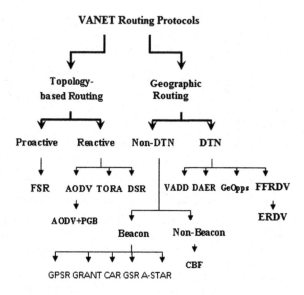

Fig. 3 Types of routing protocols in VANETs

topology-based routing is about the links which are already available in the network to perform packet forwarding, whereas geographical routing is developed to apply position of neighboring node. Here topology-based routing has to suffer from routing route breaks because of the highly changing frequency of link information.

3 DTN Routing Protocols

Delay-tolerant network is developed to maximize the probability of message release. Besides this, DTN minimizes resources (*i.e., network bandwidth, buffer size, and battery energy*) consumption. Moreover, while DTNs applications are designed to tolerate the delay, this does not benefit DTNs utilities to get benefit from the decreased delay but helps to reduce the delivery latency. The prominent delay-tolerant routing vehicular routing protocols [3, 4] are vehicle-assisted data delivery, GeOpps (Geographical Opportunistic Routing), DAER, and FFRDV. These protocols are shown in Fig. 3 and discussed as follows.

3.1 VADD: Vehicle-Assisted Data Delivery

It is processed according to the carry-and-forward [5]. Using the mobility, VADD differs from existing carry-and-forward applications to avoid dependency on traffic model and roadside layouts. For example, a vehicle's speed of driving is controlled using the traffic density of the road as well as the speed limit of the vehicle. Its driving direction can be predicated easily observing roadside model and the increase of velocity of the engine velocity. In this scenario, a vehicle decides at the nearest junctions to select the subsequent path to minimize delay of packet delivery. It supposed that vehicles are ready with pre-overloaded digital plans including steel level map with traffic information, i.e., vehicle speed, traffic density, and other data.

VADD had developed for sparse connected network to that vehicle can try to utilize wireless communications channel effortlessly. It allows vehicles to race with faster speed; else VADD model is disciplined by the following principles:

- It should choose the road with lower traffic density and higher speed if the packet carried through selected roads, It should transmit through wireless channel as much as possible.
- As the characteristics of VANET are unpredictable, it cannot be expected that the packet to be routed successfully on the same path is identified previously as an optimal path. To improve performance, the dynamic algorithm for the selection of path should be performed throughout the packet-ahead method continuously.

VADD has three types of modes for packet forwarding: *StraightWay, Intersection, and Destination*. These modes are completely dependent on the position of packet carrier. The packet shipper selects the best packets forwarding path by switching between different packet modes. Moreover, intersection mode is the mainly critical, efficient, critical, and problematical. It authorizes vehicles additional choices at the junction.

3.2 GeOpps: Geographical Opportunistic Routing

GeOpps is a new DTN routing algorithm which exploits the accessibility of information from the direction-finding system [6] to optimize route a data packet to a sure geographical location. It uses advantages of the vehicle's direction-finding system optional routes to select path and vehicle that can keep the information closer to the targeted destination of the packets. It uses suggested routes of vehicle's navigation system to choose vehicle to get nearest the final destination of a packet. GeOpps consumes information from the navigation system to proficiently route packets. It supposes that vehicles are prepared with navigation systems having informations of the geographical locations of the nearby information stations (*access points to the Backbone*). A navigation system may have different types of sensors including acceleration sensor and speed sensor. It contains maps also to evaluate the present traffic conditions of a special road segment. This information is communicated to the nearest infostations. However, it does not provide constant connectivity between info stations (*mainly in remote areas*) and vehicle; other vehicles are assumed to perform as data mules from the sensing vehicle to get accurate geographical location of the nearest info station

A centralized system combines all the information gathered from various sources to estimate the current traffic conditions on the road. Later, it generates traffic warning for the concerned specific road segment and suggests alternative paths to vehicles as per their requirements. Afterward, traffic management center dispatches this information to the vehicle to warn the drivers. Very first, a warning is sent to the nearest information, from where they are channelized to the affected road segments using the maps of the vehicle networks. It spreads the information close to vehicles; the information reaches to the affected area, local message dissemination techniques (*such as localized epidemic or constrained flooding*). After receiving this information, the navigation system of vehicles evaluates the information in order to recalculate a route to avoid road segments with highly dense traffic and lower speed.

3.3   DAER: Distance Aware Epidemic Routing

In distance aware epidemic routing the packets which need to be forwarded from n_i to n_j are categorized according to the distance from the neighbor node n_j and the destination of packet [7]. $d(\Psi(n_j, t), \Psi(\text{dest}(bk), t))$ is given that it is smaller than the current distance $(\text{dest}(bk), t))$, $d(\Psi(n_i, t)$, where $d(\Psi_1, \Psi_2)$ has been defined as the geographical distance between Ψ_1 and Ψ_2. A package bk with the maximum d $(\Psi(n_i, t), \Psi(\text{dest}(bk), t)) - d(\Psi(n_j, t), (\text{dest}(bk), t))$ is chosen forst for forward. This is called greedy distance forwarding. Once a bundle bk at node ni is forwarded to its neighbor node nj, if bk still remains in buffer at ni, it contains the same probability with other bundles in the buffer to be forward again while the nodes meet with another neighbor. Using this practice, the duplication tree expands immediately. In order to improve the effectiveness of duplication, every time when a bundle is newly forwarded, a result is made whether the package will be trimmed off from node ni. The result is based on the package geographical inclination of approaching to its objective. It compares the current distance $d(\Psi(n_i, i), \Psi(\text{dest}(bk), t))$ to the distance $d(\Psi(n_i, t'), \Psi(\text{dest}(bk), t'))$ traced at the time bk arrived at n_i. If package bk scatters away from its destination along the node ni, it will be trimmed, that is, removed from the buffer at node n_i. This process is named as anti-diffusion pruning in the algorithm. Similarly, for the buffer replacement, a bundle compares its current distance to its destination and the targeted value. It is imminent to or diffusing from its target when it first arrived at current node to conclude. If the bundle diffuses away from its set objective, it contains higher priority to be replaced by a newly arrived bundle. The whole process is called anti-diffusion buffer replacement. This mechanism is introduced due to two reasons. First, a bundle must travel on a node which is moving toward its destination. Else it must be dropped to avoid unnecessary consumption of the limited connection time. Second, when bundle is near to the destination, it should make more copies to expedite the delivery.

Table 1 Characteristics of DTN routing protocols

Protocol	Year	VN-specific	Application	Group	Routing metrics	Optimization		
						Reliability	Redundancy	Message priority
GeOpps	2007	Yes	P2P/V2I	Road map	Nearest point ETA	No	No	No
DAER	2008	Yes	P2P/V2I	Zero knowledge	Distance	No	Multicopy	Distance
VADD	2008	Yes	P2P/V2I	Online	Loc + Density + Speed	No	No	No
FFRDV	2009	Yes	Dissemination	Geographic Loc	Speed	HopAck	No	No

3.4 FFRDV: Fastest Ferry Routing in DTN VANETs

The FFRDV utilizes the concept of transmitting data (*from one node vehicle to another node vehicle on highways*) for applications which can be able to tolerate delay, over fastest ferry [8] where ferry is a reliable source for communication and transmission of messages between nodes. The "FFRDV" considers that the road is divided into lanes and further lanes are divided into logical blocks of some fixed size (between two consecutive milestones). While entering the block the ferries, which carry the data, broadcast a HELLO message. All ferries within the block reply to that HELLO with their speed and coordinates. The ferry which holds data is called Current Ferry (CF). After receiving HELLO messages from all in-block ferries, CF compares its own speed from the speed received in reply HELLO messages. Now, CF elects the fastest ferry (*the ferry having highest speed*) among all ferries running within the block. This elected fastest ferry becomes Designated Ferry (DF). The CF sends bundle having with it to the elected DF and discards the bundle after receiving acknowledgment from DF. Now the CF after discarding bundle becomes normal ferry.

4   Comparison of DTN Routing Protocols

The above-discussed DTN-enabled vehicular ad hoc network routing protocols can be summarized [3] in Table 1. The VN-specific field indicates whether the protocol is for vehicular network or not.

5   Conclusion

Geographic routing and topology-based routing are the categories of "VANET" routing protocols. The previous routing is also known as "DTN"-based routing protocols, which uses adjacent site information for packet forwarding. Later on, other researchers used the information about routes that exist in the network for packet forwarding. The "FFRDV" , "DAER", "VADD", "GeOpps", etc. are the important examples of "DTN"-based VANET routing protocols.

Due to high mobility of vehicular nodes in "VANETs" there are high chances of partitions in the network. In such situation, the protocols developed for "VANETs" cannot work well and an alternative network known as "DTN" (Delay-Tolerant Network) is capable enough to deal with "VANET" characteristics.

References

1. Yousefi, S., Mousavi, M.S., Fathy, M.: Vehicular Ad Hoc Networks (VANETs): challenges and perspectives, In: 6th International Conference on ITS Telecommunications Proceedings, Chengdu, China, pp. 761–766 (2006)
2. Schiller, J.: Mobile Communications, 2nd edn. Addison-Wesley, pp. 375–376 (2003)
3. Tornell, S., Calafate, C., Cano, J.-C., Manzoni, P.: DTN protocols for vehicular networks: an application oriented overview, IEEE Commun. Surv. Tutor. 2nd Quart **17**(2), 868–887, (2015)
4. Nagaraj, U., Kharat, M.U., Dhamal, P.: Study of various routing protocols in VANET. IJCST **2** (4), 45–52 (2011)
5. Zhao, J., Cao, G.: VADD: Vehicle-assisted data delivery in vehicular Ad Hoc networks, IEEE Trans. Veh. Technol. **57**(3), 1910–1922, May (2008)
6. Leontiadis, I., Mascolo, C.: GeOpps: Geographical Opportunistic routing for vehicular networks. In: IEEE International Symposium on World of Wireless, A Mobile and Multimedia Networks, (WoWMoM 2007)
7. Luo, P., Huang, H., et al.: Performance evaluation of vehicular DTN routing under realistic mobility models. In: Proceedings of Wireless Communications and Networking Conference, IEEE Communication Society, pp. 2206–2211 (2008)
8. Yu, D., Ko, Y.-B.: FFRDV: Fastest-Ferry Routing in DTN-enabled Vehicular Ad-hoc networks. In: Proceedings of ICACT. pp. 1410–1414 (2009)

Evaluation of a Thermal Image for Pedestrian Modeling Using the Least-Action Principle

S. Mejia, J.E. Lugo, R. Doti and J. Faubert

Abstract Living labs provide the possibility of doing real-time research in an ecological context corresponding to normal daily activities. In particular, it is important to know how humans respond to environmental changes and different scenarios. The appropriate characterization of individual human displacement dynamics within a crowd remains illusive and for this reason there is a great interest in exploring behaviors with general physical models. In this work, we present a theoretical and experimental study of the natural movement of pedestrians when passing through a limited and known area of a shopping center. The modeling problem for the motion of a single pedestrian is complex and extensive; therefore, we focus on the need to design models taking into account mechanistic aspects of human locomotion. The theoretical study used mean values of pedestrian characteristics, e.g., density, velocity, and number of obstacles. We propose a human pedestrian trajectory model by using the least-action principle, and we compared it with experimental results. The experimental study is conducted in a Living Lab inside a shopping center using infrared cameras. For this experiment, we collected highly accurate trajectories allowing us to quantify pedestrian crowd dynamics. The experiments included 20 runs distributed over 5 days with up to 25 test persons.

Keywords Infrared cameras · Crowd dynamics · Least-action principle
Pedestrian simulation

S. Mejia (✉) · J.E. Lugo · R. Doti · J. Faubert
Visual Psychophysics and Perception Laboratory École d'optométrie,
Université de Montreal, UdeM, Montreal, Canada
e-mail: sergio.mejia.romero@umontreal.ca

S. Mejia · J. Faubert
Center for Interdisciplinary Research in Rehabilitation of Greater Montreal (CRIR),
Greater Montreal, Canada

© Springer Nature Singapore Pte Ltd. 2018   377
M. Pant et al. (eds.), *Soft Computing: Theories and Applications*,
Advances in Intelligent Systems and Computing 583,
https://doi.org/10.1007/978-981-10-5687-1_34

1 Introduction

Pedestrian public areas designated for pedestrian traffic, such as underground corridors, shopping malls, airports among others, are places where multidisciplinary analysis on the interaction and behavior of people is of great interest.

The study of pedestrian flow is not recent; it was developed for helping with the design and planning of buildings. It is also useful in the process of planning evacuations in emergency conditions. To improve service and conditions in pedestrian areas, it is necessary to first understand pedestrian interactions within the facility. This paper explores paths of 25 pedestrians along a known area. After obtaining the trajectory and their points of origin, we evaluated the speed with the objective to calculate the kinetic force of the pedestrian. In the present model, we assume that the principle of least action holds and using this concept we can obtain the potential force. Once the forces of overall pedestrian movement are known, we then calculate the adjustment of the parameters used in the equations of the social force model.

The vector velocity $\bar{v}$, according to Weidmann [1], is different when taking into account three different characteristics: characteristics of the pedestrian (age, weight, sex, etc.), trip characteristics (purpose of the walk, familiarity of the route, length of the path), and the structure properties (floor shape and size of the area).

If we consider all pedestrians to have the same psychological and physical characteristics as a case, the theory of unidimensional flow predicts a homogeneous distribution over the entire walking area with constant velocity [2].

When the flow is heterogeneous (pedestrian with different velocities), by changing the velocity and direction of the pedestrian interactions, then pedestrian flow is distributed with different flux densities within the pedestrian area [3].

Presently, there are several models that can be applied to pedestrian walk simulation. These models were intended to reproduce the pedestrian walking in a known area. Current models describing the trajectory of a pedestrian walking integrate the ideas of pedestrian behavior [4] and social behavior [5]. In this way, it is assumed that the trajectory can be induced for the so-called "Social Force".

2 Social Force Model

The social force model [6] describes the interaction of the pedestrians with their environment (walls, obstacles, another pedestrians, etc.). The social force model is continuous in space and time. The equation of movement for every pedestrian uses Newton's second Law and is given as

$$m_i \frac{dv_i}{dt} = F_i^{\text{mov}} + \xi_i, \qquad (2.1)$$

where $i = 1, 2, \ldots$ indicates the pedestrian i; v_i indicates the velocity of the pedestrian i; ξ_i term represents the alteration of the behavior; F^{mov} defines the pedestrian movement force which contains the sum of the forces, exerted on it, and is defined by

$$F_i^{\text{mov}} = F_i^{\text{kin}} + F_{ij}^{\text{rep}} + F_{iw}^{\text{pot}} + F_{io}^{\text{act}}. \tag{2.2}$$

The *Kinetic Force* of the pedestrian i is calculated as follows: F_i^{kin} is the force for the pedestrian with mass m_i that is moving in $\vec{e}_i$ direction with $\vec{v}_i$ velocity in a τ time:

$$F_i^{\text{kin}} = m_i \frac{\vec{v}_i^0 - \vec{v}_i}{\tau}. \tag{2.3}$$

The *Repulsion Force*. The pedestrian movement is influenced by other pedestrians. Because of this the pedestrian i will react by keeping a distance of value $\vec{R}_{ij}$ from the pedestrian j. It is observed that the closer is the other person the discomfort increases in the direction given by $\vec{e}_{ij}$. In this way the repulsion force exerted by the pedestrian i is represented as

$$F_{ij}^{\text{rep}} = -m_i \frac{v_{ij}^2}{\vec{R}_{ij}} \vec{e}_{ij}. \tag{2.4}$$

The *Potential Force* is the force exerted on the pedestrian due to the scenario. This force is what keeps pedestrians away from the edges of the stage, such as walls, pillars, etc. The force is represented by a least potential gradient; vector $\vec{d}_{iw}$ is the vector distance between the pedestrian i and the wall or obstacle w closest to the individual:

$$\vec{F}_{iw}^{\text{pot}} = m_i \frac{A_w}{B_w} e^{-\frac{\|d_{iw}\|}{B_w}} \left(\frac{d_{iw}}{\|d_{iw}\|} \right). \tag{2.5}$$

The repulsive potential is monotonous decreasing [7], where A_w is the interaction between the edge and the pedestrian i, and B_w is the range of the interaction distance.

The *Attraction Force* is the force that feels the pedestrian i because the place or object g where it meets another pedestrian (Friends, arts, windows, products, dressers, etc.) However, this force usually is determined by the interaction time. So these incentive effects can be modeled by a monotone decreasing potential [8] represented by

$$f_{ig}(\|d_{ag}\|) = -\nabla_{d_{ig}} w_{ag}(\|d_{ig}\|). \tag{2.6}$$

The effects of attraction are responsible for making clusters of pedestrians. The final attractive force is given by

$$F_{ig}(t) = m_i W(\theta_{ig}) f_{ig}(\|d_{ig}\|), \tag{2.7}$$

where $W(\theta_{ij})$ is the weight function, which depends on the angle of view of pedestrian i.

The model equations describe the "social force" pedestrian behavior, but if we want to simulate all the forces that interact with each pedestrian then the computational cost is high. For this reason there is great interest in modeling the trajectory with simple models that have a good approximation for different scenarios.

3 Principle of Least Action

For a particle that moves from an initial position $\vec{r}(t_1)$ to a final position $\vec{r}(t_2)$, without specifying the initial velocity, there exist n possible paths that connect these points. To find out which path the particle takes, we can take any route $\vec{r}(t)$ and call defined action as [9].

$$S[\vec{r}(t)] = \int_{t_1}^{t_2} dt \left(\frac{1}{2} m \dot{\vec{r}}^2 - V(\vec{r}) \right). \tag{3.1}$$

The action is the difference between the kinetic energy (KE) and potential energy (PE) $V(\vec{r})$. The theory predicts that the path chosen by the particle is an extreme value of the action. The difference in the action for two different routes is

$$\delta S = S[\vec{r} + \delta\vec{r}] - S[\vec{r}] = \int_{t_1}^{t_2} \left[-m\ddot{\vec{r}} - \vec{\nabla} V \right] \cdot \delta\vec{r} + \left[m\dot{\vec{r}} \cdot \delta\vec{r} \right]_{t_1}^{t_2}. \tag{3.2}$$

In the above equation if we evaluate at the edges when $\delta\vec{r}(t_1) = \delta\vec{r}(t_2) = 0$, then we have

$$\delta S = \int_{t_1}^{t_2} \left[-m\ddot{\vec{r}} - \vec{\nabla} V \right] \cdot \delta\vec{r}. \tag{3.3}$$

The start condition of the path, when the action is zero, is $\delta S = 0$; this is true for all changes $\delta\vec{r}(t)$ that are made along the route. The only way this can be true is that the expression (3.3) is zero. This is $m\ddot{\vec{r}} = -\vec{\nabla} V$ to require that the action is the extreme value is equivalent to follow the actual path. We can observe that the equation corresponds to Newton's equation.

Fig. 1 **a** Living lab and **b** assemblies infrared cameras

4 Methods

The Living Lab in Alexis Nihon shopping center in Montreal (Fig. 1a) has provided the possibility of placing users in real-time ecological scenarios. This experiment is of great interest because it brings the possibility to include user's behavior in the context of a general physical model.

For the experiments we collect highly accurate trajectories that allow us to study the dynamics of pedestrian crowds. The experiments included 20 runs distributed over 5 days with up to 25 test persons.

Every run was filmed with at least two infrared cameras, which have been mounted at an altitude of 335 cm perpendicular to the floor. The assemblies can be seen in Fig. 1b. For the video analysis we used MATLAB (R2012b). Pedestrian trajectories were recorded from different videos filmed on different days and hours to avoid any trend.

4.1 Tracking Algorithm

For an optimal approximation of the kinetic and potential energy scenario, we obtained 100 videos of pedestrians walking in the study area. From those data we selected 10 pedestrians with a free path and 10 pedestrians confronted to a pedestrian counter flow.

The first step was to identify and label every pedestrian (see Fig. 2a); for every person we localized the centroid and manually registered the pedestrian position for every frame in the video. This was done to avoid any error from the tracking algorithm. With this new pedestrian position data, we used a method to correct the lens distortion of the camera and find the actual position on the scenario (Fig. 2b).

The principle of least action implies that the value of the kinetic force is minimized. That is, theoretically the pedestrian takes the path that represents less effort.

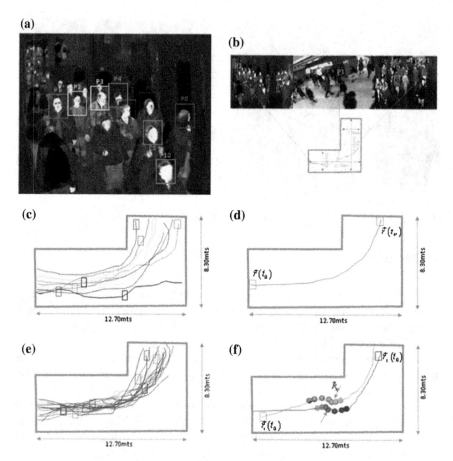

Fig. 2 a This is the frame $t = 0$, from a video in which we labeled pedestrians to know their pathways from frame to frame till $t = N$. **b** We show the trajectory adjustment of the pedestrian starting position. The image is from two cameras we have used. The area in *red* provides the surface where the study has been conducted. **c** We show 10 pedestrian pathways within the scenario. **d** *Blue* is the averaged trajectory; x0 is the subject's path starting point of and xf is the end of the subject's path. **e** We show 20 trajectories and 10 interactions between pedestrians. **f** Finally, in this image we show the interaction of two pedestrians. After the trajectories were obtained, we registered two different velocity groups: the velocity of free walking and the velocity of the pedestrians meeting an obstacle

In the absence of a potential, the trajectory is perfectly described by the action from the average speed of the pedestrian, which is calculated using the ratio of the total distance over the total time. In this way, the pedestrian can walk faster at the beginning and then reduce its speed. But to minimize the action, the pedestrian must go at a uniform speed. In real cases, the pedestrian modifies its velocity along the trajectory, because of the potential energy of the environment and the interaction with the obstacles, as predicted by the social force model. In every case, the action is equal to zero when the trajectory is the real one. Let us consider the real trajectory of the pedestrian that goes at a time t, from an initial point $\vec{r}_1 = \vec{r}(t_0)$ to a final point $\vec{r}_2 = \vec{r}(t_N)$, and is assumed that the action is equal to zero. Knowing the mass of the pedestrian (m_i) you can get the value of the kinetic energy; in theory, it is possible to recover the value of the potential energy for each scenario and every Δt of the trajectory. The next step is to find the values of all the potential force parameters involved to help us gauge the scenario under study and have a better social force model amending pedestrian path. As described before, there are n interaction cases; the main cases for this work are free walking, repulsive force toward an obstacle, and attraction force. We will discuss only the first two and we propose the basis for the experiment in the Living Lab for attractive forces.

5 Results

In order to demonstrate that our theoretical model can get the value of the potential energy that makes the action equal to zero, we took ten trajectories of ten different pedestrians (see method). We also calculated the average of all free walking trajectories for every position and obtained the average trajectory shown in Fig. 2d.

We assumed the mass of every pedestrian as $\bar{m} = 80$ kg.

For trajectory with an obstacle results we obtained the next valued simulation.

5.1 Simulation

The simulation was performed using the social force model described in Sect. 2 and by using the parameters obtained with our analysis. The simulation is performed separately for each scenario described above. Although a width of 45.58 cm shoulders is proposed [10], we considered in this example 80-cm-wide shoulders that gave better results. We obtained the average speed $\bar{v} = 1.34$ m/s with a standard deviation of 0.26 m/s. For simplicity, the attractive forces or fluctuations $\xi(t)$ are not considered. For this situation we have considered Table 1 parameters. We used the method of Euler to calculate the velocity of every pedestrian at the time $t + \Delta t$ as follows as shown in Fig. 3:

$$v_i(t + \Delta t) = v_i(t) + F_i^T(t)\Delta t, \tag{5.1}$$

where Δt is the time interval. After, we calculated the position $\vec{r}_i$ of every pedestrian:

Table 1 Free walking results

Description		Valued
Average velocity	$\bar{v}$	1.34 m/s
Average kinetic force	$\bar{F}_f^{kin}$	54.76 N
Average potential force	∇V	0 N
Range of the interaction distance walls	$\vec{B}_w$	0.8 m
Interaction force parameter of pedestrian walls	$\vec{A}_w$	9.4 m^2/s^2

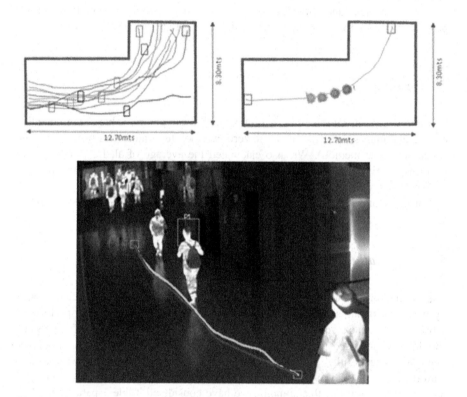

Fig. 3 *Left* Ten trajectories without obstacles for ten pedestrians. *Right* Trajectories' averages in the scenario. In this case the value for the potential energy represents the average that the scenario exerts on the pedestrian. *Bottom* In *blue* we show the real trajectory of a free walking pedestrian. In *red* we show the trajectory that was obtained from the social force model applying the parameters that we obtained with our analysis

$$\vec{r}_i(t + \Delta t) = \vec{r}_i(t) + \vec{r}_i(t + \Delta t). \tag{5.2}$$

For the simulation of the repulsion force between pedestrians we considered two of them walking: one toward the other as can be seen in Fig. 4. We used Eq. (2.4) considering the pedestrian as the center of a radial force (specification circular [11]),

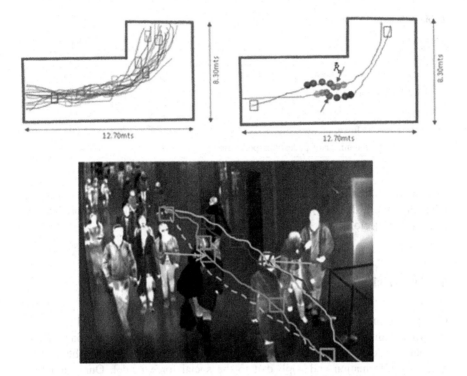

Fig. 4 *Left* Ten trajectories with obstacles of ten pedestrians. *Right* Modified path of a pedestrian that interacts with another pedestrian. *Bottom Blue* and *red lines* show trajectories of the real interaction of two pedestrians walking counter flow. In *yellow* (*solid* and *broken*) we show the simulated trajectories

and with the parameters (Table 2) obtained with the principle of least action. The hall measures are 12.7 m in length and 8.3 m in width. The area has a constant radius of 0.7 m. The initial real distance between subjects is 4 m. When they are close enough at a certain distance determined by $\vec{R}_{ij}$ both pedestrian begin to feel the repulsion force between each other, modifying both trajectories.

To evaluate the validity of the estimations, two types of error index are used in this study, a mean-absolute-relative error (MARE) and a root-mean-square error (RMSE). The formal expressions of MARE and RMSE are presented in the following equations.

$$\text{MARE} = \frac{1}{N}\sum_{ij=1}^{N}\frac{\left|\text{PT}_{ij} - P\hat{T}_{ij}\right|}{\text{PT}_{ij}}, \quad \text{RMSE} = \sqrt{\frac{1}{N}\sum_{ij=1}^{N}\left(\text{PT}_{ij} - P\hat{T}_{ij}\right)^2}, \qquad (5.3)$$

where PT_{ij} is the real position of the pedestrian and $P\tilde{T}_{ij}$ is the position calculated by the simulation. We obtained 25 trajectories of pedestrians. All the trajectories had similar interactions.

Table 2 Trajectory with an obstacle results

Description		Valued
Average velocity	$\bar{v}$	1.17 m/s
Average distance vector between two pedestrian	$\|\vec{R}_{ij}\|$	0.48 m
Average kinetic force	$\bar{F}_i^{\text{kin}}$	62.03 N
Magnitude average social discomfort	$\|\vec{e}_{ij}\|$	0.2 m
Average repulsion force	$\|\bar{F}_i^{\text{rep}}\|$	−8.003 N
Interaction force parameter of pedestrian–pedestrian	$\vec{A}_w$	3.4 m²/s²

The MARE and RMSE are 7.21 and 3.96% in these cases. The errors are relatively small and the results confirmed that the proposed model predicts the pedestrian trajectories.

6 Discussion

Using the principle of least action it is possible to reproduce observed results for real pedestrian movement. In the first scenario, we focussed on a pedestrian walking without obstacles. Using the real trajectories of the experiment we obtained the necessary information and applied it to the social force model. Our simulations were clearly able to reproduce the real observed average trajectories for the free obstacle walking conditions. When a scenario does not represent free walking (obstacles, constraints), the potential energy and the kinetic energy are modified. Note that when the trajectory is real, the action is assumed to equal zero. That is the value of the potential energy changes in each interaction with a new obstacle; however, the value of the action remains. It is shown here that we can clearly reproduce some scenarios and calibrate the model according to different situations. Using different values of potential energy we can obtain the values of the actual pathway. But as an important extension to this model, it would be desirable to stimulate cellular automata that could learn the situation and improve the approximation model to predict the real trajectories with more accuracy. It should be pointed out that the proposed recommendations are based on the assumption that pedestrians are equally distributed on each side of their pathways, which is a situation with extreme bidirectional effects. The purpose for this assumption is to make sure that the recommended pathway parameters satisfy the most demanding situations. For further refined evaluation, since our simulation model can be flexibly expanded, a variety of scenarios can also be evaluated by varying the arrival pattern of pedestrians at each side of their pathways. Furthermore, we should point out that the proposed methodology does not consider pedestrian trip purposes and conditions where, in our case, a significant proportion of pedestrians was elderly and/or young students. The next step is to apply different forces of repulsion and attraction during the pedestrian walk in the Living Lab, to modify its trajectory and ascertain whether this scenario simulation predicts real path.

Acknowledgements This work was supported by an NSERC Discovery operating grant. The author Sergio Mejia is grateful to the Visual Psychophysics and Perception Laboratory and Centre for Interdisciplinary Research in Rehabilitation of Greater Montreal (CRIR) for a postdoctoral fellowship.

References

1. Weidmann, U.: Transporttechnik der Fussganger, vol. 90. ETH-Zurich, Schriftenreihe IVT-Berichte, Zürich, German (1993)
2. Willis, F.N., et al.: Stepping aside: correlates of displacements in pedestrians. J. Commun. **29**(4), 34–39 (1979)
3. Dabbs, J.M., Stokes, N.A.: Beauty is power: the use of space on the sidewalk. Sociometry **38**(4), 551–557 (1975)
4. Helbing, D., Molnár, P.: Social force model for pedestrian dynamics. Phys. Rev. E **51**, 4282–4286 (1995)
5. Lewin, K.: Field Theory in Social Science. Harper and Brothers, New York (1951)
6. Schadschneider, A., Klingsch, W., Klüpfel, H., Kretz, T., Rogsch, C., Seyfried, A.: In: Meyers, R.A. (ed.) Encyclopedia of Complexity and System Science, pp. 3142–3176. Springer, Berlin, Heidelberg (2009)
7. Helbing, D., Johansson, A., Shukla, P.K.: Specification of a microscopic pedestrian model by evolutionary adjustment to video tracking data. Adv. Complex Syst. **10**, 271–288 (2007)
8. Helbing, D., Farkas, I.J., Vicsek, T.: Simulating dynamical features of escape panic. Nature **407**, 487–490 (2000)
9. Feynman, R.P.: Thesis draft, Feynman Papers, Caltech, folder 15.4, f. 32
10. Keith, S.G.: Crowd Dynamics. University of Warwick, UK (2000)
11. Steiner, A., Philipp, M., Schmid, A.: Parameter Estimation for a Pedestrian Simulation Model. Ascona, Suiza, 12–14 Sept 2007

On Comparing Performance of Conventional Fuzzy System with Recurrent Fuzzy System

Anuli Dass and Smriti Srivastava

Abstract Fuzzy controllers are very commonly used as universal approximators. It is seen that while dealing with plants or systems with known order, fuzzy systems prove to be very effective approximators. The concept of fuzzy systems used in the field of identification and control has been extended by introducing a new architecture known as "recurrent fuzzy systems (RFS)". This new architecture enhances the approximation capacity of the conventional fuzzy structure and empowers it to deal with dynamic process of unknown order and structure. Similar to its neural counterpart, fuzzy systems can also be recurrent. This paper presents the identification of two nonlinear systems using the concept of recurrent fuzzy system assuming that their structure is unknown using a single hidden variable. Although the number of hidden variable is kept same for the identification of both the examples but two different structures of RFS (using different numbers of delayed output feedbacks) have been used in this paper. These systems are then also identified using traditional fuzzy systems where the order of the actual plant and the structure of its transfer function are taken into account. A comparative analysis is done between both the identification schemes. A suitable learning algorithm based on backpropagation has also been given to update the parameters associated with the model. The paper highlights that using RFS for approximation of higher order systems reduces the number of rules and thus the complexity to a great extent. Also it shows that even if the exact structure of the plant is unknown, RFS proves to be of great use.

Keywords Recurrent fuzzy systems (RFS) · Hidden variables · Neural network System identification

A. Dass (✉) · S. Srivastava
ICE Division, Netaji Subhas Institute of Technology, New Delhi, India
e-mail: anulidass@gmail.com

S. Srivastava
e-mail: smriti.nsit@gmail.com

© Springer Nature Singapore Pte Ltd. 2018
M. Pant et al. (eds.), *Soft Computing: Theories and Applications*,
Advances in Intelligent Systems and Computing 583,
https://doi.org/10.1007/978-981-10-5687-1_35

1 Introduction

The concept of fuzzy logic has been used in the field of identification and control for quite some time now. Fuzzy systems are used for identification and control of non-linear plants and systems where the traditional conventional techniques fail to give satisfactory results. In [1] a new method based on genetic algorithm has been proposed to extract fuzzy rules for the identification of an unknown system where only input output data is available. In [2] the aim is to extend the fuzzy controller approximation capacity to dynamic processes of unknown order. To achieve this objective a new architecture called recurrent fuzzy system (RFS) has been proposed in [2]. In [3] the effect of evolution of recurrent fuzzy controllers has been investigated by applying the FV representation which provides a set of advantages that would benefit the quality of the knowledge insertion process. Neural networks and fuzzy systems have almost same areas of application. Thus one can very well extend the concept of neural networks to fuzzy logic by co-relating. In [4], the paper demonstrates that neural networks can be used effectively for the identification and control of nonlinear dynamical systems. Models for both identification and control have been depicted. In [5] a new learning method for rule-based feed-forward and recurrent fuzzy systems has been presented. The concept of hidden variables has been elaborated in [6] and it also shows relation between the hidden variables and some bell-shaped inequalities. Gama et al. [7] presents two different RFS models, i.e., TSK-based and linguistic or Mamdani-based. In [8] a probabilistic-based extension of recurrent fuzzy systems has been presented and has also been exemplarily applied to modeling and control of systems in various domains. Reference [9] proposes a recurrent fuzzy system which is designed based on elite-guided continuous ant colony optimization technique. Reference [10] investigates the effect of external feedback in a recurrent fuzzy system and establishes the result that external feedback enhances the recurrent fuzzy system quality leading to better performances of the resulting models. References [11, 12] show some recent work in the field of recurrent fuzzy logic.

Reference [13] shows a technique to represent higher order TSK systems because the conventional technique is cumbersome as the number of rules increases exponentially with increase in the number of the inputs or the order of the system. This problem can also be resolved using the recurrent approach of fuzzy as it also helps in reducing the complexity since it is independent of the order of the respective system.

This paper has attempted the identification of two nonlinear systems assuming that their structure is unknown. The modeling is done by using a recurrent fuzzy system (RFS) which incorporates one hidden variable. Although number of hidden variable is kept constant, two different structures of RFS having a different number of feedback inputs has been used. The overall recurrent fuzzy model proposed is a combination of two fuzzy subsystems which are interconnected via hidden variable. The main idea of the paper is to highlight the importance of RFS when dealing with higher order systems or even when dealing with plants or systems with unknown structure. Section 2 introduces the concept of recurrent fuzzy systems. Section 3 elaborates the learning method and explains how to update the various parameters involved. In Sect. 4 the simulation results depicting successful system identification has been shown. Also the same systems are then identified using the conventional fuzzy systems where the order of these systems is taken into account. Finally, Sect. 5 concludes by doing comparative analysis between the two identification schemes. It is observed that for higher order systems, even if the structure and order are known, the RFS-based identification technique minimizes the number of rules to a great extent and thus simplifies the computation. But in case of unknown structure and order, RFS proves to be the sole solution.

2 Recurrent Fuzzy Systems

The concept of recurrent fuzzy systems crops from the need for identification of those plants and nonlinear systems where the order is unknown or the exact structure is vague. A recurrent fuzzy system is characterized by rules where one or more variables appear in both the premise and consequent parts. A simple rule structure for a first-order recurrent fuzzy system is given as

$$\text{IF } y(t-1) \text{ is } M_1^k \text{ and } u(t-1) \text{ is } M_2^k \text{ THEN } y(t) \text{ is } O^k,$$

where M_i^k represents the membership function for the ith input variable and O^k represents the membership function associated with the output variable.

This is a first-order recurrent fuzzy system because the current output of the above system depends on its single past output and the past input. First-order recurrent systems can be used for the approximation of first-order systems that is

$$y(t) = f(y(t-1), u(t-1)). \tag{1}$$

For higher order systems, a higher order recurrent fuzzy system must be used for the purpose of identification or approximation. Consider the plant

$$y(t) = f(y(t-1), y(t-2), y(t-3), \ldots, y(t-n), u(t-1), u(t-2), \ldots, u(t-m)). \tag{2}$$

Since the plant is nth order, an nth-order recurrent fuzzy system (RFS) must be used for its identification. The rule structure would then be given as

$$\text{IF } y(t-1) \text{ is } M_1^k \text{ and } y(t-2) \text{ is } M_2^k \text{ and } y(t-3) \text{ is } M_3^k, \ldots, \text{ and}$$
$$u(t-1) \text{ is } M_{n+m}^k \text{ THEN } y(t) \text{ is } O^k.$$

Recurrent fuzzy found to be used mainly when dealing with higher order systems, especially when their order as well as their structure is unknown or ambiguous. Basic structure of a recurrent fuzzy model is shown in Fig. 1. The

Fig. 1 General model of
RFS with internal variable h
(t)

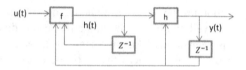

structure consists of two fuzzy subsystems (f and h) interconnected through a variable $h(t)$. This variable is the output of one of the subsystems but is not the ultimate output of the model. It serves only as a connecting link. This internal variable is known as the *"Hidden Variable"*.

When dealing with nth-order RFS, rules may also contain internal variables. The rule shown below has "$h(t)$" as the internal variable:

$$\text{IF } y(t-1) \text{ is } M_1^k \text{ and } h(t) \text{ is } M_2^k \text{ and } u(t-1) \text{ is } M_3^k \text{ THEN } y(t) \text{ is } \beta^k.$$

These internal variables also have their own dynamics and can be described by using recurrent fuzzy rules as

$$\text{IF } y(t-1) \text{ is } M_1^k \text{ and } h(t-1) \text{ is } M_2^k \text{ and } u(t-1) \text{ is } M_3^k \text{ THEN } h(t) \text{ is } \gamma^k.$$

In this paper identification of two nonlinear systems is done by using TS-based fuzzy models. However, two different approaches are used with each plant. In the first scheme a recurrent fuzzy system is used for the identification process. In this case it is assumed that the order of the plants is unknown. In the second case the nonlinear plants are identified using conventional fuzzy systems where their order and structure are taken into consideration. In both the cases the fuzzy equivalent model is realized by using Gaussian membership function and Takagi-Sugeno model is used for the inference and calculation of the output. The Gaussian membership function used is of the following form:

$$\mu = e^{-0.5\left(\frac{x-c}{a}\right)^2}, \tag{3}$$

where "c" denotes the center of the Gaussian function, "a" denotes the spread of the function or its width, and "x" denotes the input variable.

The first example considered is a nonlinear plant given by Eq. (4):

$$y_p(t) = \frac{y_p(t-1)y_p(t-2)y_p(t-3)y_p(t-4)\left[y_p(t-3)-1\right]u(t-2)+u(t-1)}{1+y_p^2(t-2)+y_p^2(t-3)+y_p^2(t-4)}. \tag{4}$$

The structure of the recurrent fuzzy system used for its identification is shown in Fig. 2. The identification is done using a single hidden variable $h(t)$ as shown below.

Fig. 2 RFS with one hidden variable

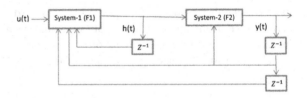

The system is divided into two subsystems: System-1(F1) and System-2(F2). The overall system is a single input single output system with one hidden variable. The output of two subsystems F1 and F2 can be expressed as shown in Eqs. (5) and (6):

$$F1 : h(t) = f(u(t-1), h(t-1), y(t-1), y(t-2)) \tag{5}$$

$$F2 : y(t) = g(u(t-1), h(t), y(t-1), y(t-2)). \tag{6}$$

For each subsystem the individual inputs are fuzzified using Gaussian membership function of Eq. (3). Each input is divided into two fuzzy sets. The objective is to obtain a fuzzy model which would replicate the plant or the nonlinear system of unknown order. This involves optimizing the parameters associated with the membership function for each input as well as the coefficients associated with the output equation. In this paper backpropagation is used for the optimization purpose.

The second example is also a nonlinear plant [4] described by Eqs. (7) and (8):

$$y_p(k) = f(y_p(k-1), y_p(k-2)) + u(k), \tag{7}$$

where

$$f(y_p(k-1), y_p(k-2)) = \frac{y_p(k-1)y_p(k-2)[y_p(k)+2.5]}{1+y_p^2(k-1)+y_p^2(k-2)}. \tag{8}$$

The identification of the plant given by Eqs. (7) and (8) is done using RFS model whose structure is depicted in Fig. 3. Rest of the procedure remains the same.

Observing Figs. 2 and 3, we see that there is a difference in the number of delayed values of the output variable being fed back to the subsystem. The number of delayed output value ($y(t)$) which are fed back to system is solely the choice of the designer and there is no specific rule or boundation. In this paper two different structures of RFS using different numbers of delayed inputs have been used. The

Fig. 3 RFS model

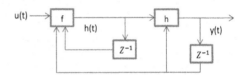

more the number of delayed output value being fed back, the better will be the performance of the approximator. But this would be achieved only at the price of increased complexity as the number of rules would also increase. So a balance has to be created for optimized performance of the overall RFS.

3 Learning in RFS

For the purpose of system identification or modeling using the concept of recurrent fuzzy system, the first step is to initialize all the parameters of the model with random values. The next step is to train the model or optimize the value of its parameters so that the RFS model replicates the actual system. This learning or the training process can be done using various techniques such as backpropagation, genetic algorithm, fish swarm algorithm, ant colony, etc. However, in this paper learning is done using the backpropagation method.

The subsystem F1 has $u(t-1), h(t-1), y(t-1), y(t-2)$ as the input variables and $h(t)$ as the output variable as shown in Fig. 2.

Each of the input variables is fuzzified using Gaussian membership function and two fuzzy sets. Every Gaussian membership function has two parameters associated with it, i.e., "c" (location of the center) and "a" (spread of the function or its width) which are to be optimized. Equations (9)–(16) show the various membership functions:

$$\mu_{11} = e^{-0.5\left(\frac{x-c_{11}}{a_{11}}\right)^2} \tag{9}$$

$$\mu_{12} = e^{-0.5\left(\frac{x-c_{12}}{a_{12}}\right)^2} \tag{10}$$

$$\mu_{21} = e^{-0.5\left(\frac{x-c_{21}}{a_{21}}\right)^2} \tag{11}$$

$$\mu_{22} = e^{-0.5\left(\frac{x-c_{22}}{a_{22}}\right)^2} \tag{12}$$

$$\mu_{31} = e^{-0.5\left(\frac{x-c_{31}}{a_{31}}\right)^2} \tag{13}$$

$$\mu_{32} = e^{-0.5\left(\frac{x-c_{32}}{a_{32}}\right)^2} \tag{14}$$

$$\mu_{41} = e^{-0.5\left(\frac{x-c_{41}}{a_{41}}\right)^2} \tag{15}$$

$$\mu_{42} = e^{-0.5\left(\frac{x-c_{42}}{a_{42}}\right)^2}, \tag{16}$$

where μ_{ij} represents the membership function associated with the ith variable for its jth fuzzy set.

The rules made are of the following form:

If $u(t-1)$ is P_{11} and $h(t-1)$ is P_{21} and $y(t-1)$ is P_{31} and $y(t-2)$ is P_{41}, then
$h(t) = k_0 + k_1 * u(t-1) + k_2 * h(t-1) + k_3 * y(t-1) + k_4 * y(t-2)$.

If $u(t-1)$ is P_{11} and $h(t-1)$ is P_{21} and $y(t-1)$ is P_{31} and $y(t-2)$ is P_{42}, then
$h(t) = k_0 + k_1 * u(t-1) + k_2 * h(t-1) + k_3 * y(t-1) + k_4 * y(t-2)$

If $u(t-1)$ is P_{11} and $h(t-1)$ is P_{21} and $y(t-1)$ is P_{32} and $y(t-2)$ is P_{41}, then
$h(t) = k_0 + k_1 * u(t-1) + k_2 * h(t-1) + k_3 * y(t-1) + k_4 * y(t-2)$

If $u(t-1)$ is P_{11} and $h(t-1)$ is P_{21} and $y(t-1)$ is P_{32} and $y(t-2)$ is P_{42}, then
$h(t) = k_0 + k_1 * u(t-1) + k_2 * h(t-1) + k_3 * y(t-1) + k_4 * y(t-2)$,

where P_{ij} represents the value of the membership function for the ith input variable w.r.t. jth fuzzy set and k_i is the coefficient associated with the mathematical expression for the output.

Similarly, the other 12 rules are also made by taking all possible combinations of the fuzzy sets. Overall, the subsystem F1 has four input variables and each input has two fuzzy sets associated with it, so a total of 16 rules are made taking all the possible combinations.

Using the TS model, the expression for the output of the fuzzy system is given as

$$q_{i\_F1} = k_0 + k_1 * u(t-1) + k_2 * h(t-1) + k_3 * y(t-1) + k_4 * y(t-2), \tag{17}$$

where "q_{i_F1}" denotes the output of the ith rule for the subsystem F1.

The coefficients k_0, k_1, k_2, k_3 actually represent the weights associated with each input in literal terms. These coefficients are also optimized using backpropagation in this paper.

Now consider subsystem F2 shown in Fig. 2. The inputs to the system are $u(t-1), h(t), y(t-1), y(t-2)$ and its output is $y(t)$.

For this subsystem also each of the input variables is fuzzified and is assigned to two fuzzy sets with Gaussian membership function. This fuzzy subsystem also has 16 rules formed in a similar manner as described above for the system F1. The expression for the output of each rule using TS model for this subsystem is given by Eq. (18):

$$q_{iF2} = b_0 + b_1 * u(t-1) + b_2 * h(t) + b_3 * y(t-1) + b_4 * y(t-2), \qquad (18)$$

where "q_{i_F2}" denotes the ith rule's output of the subsystem F2.

Here also the coefficients b_0, b_1, b_2, b_3 are actually the weights associated with each input in literal terms and are optimized using the backpropagation method.

The final output of the recurrent fuzzy system is actually the weighted sum of outputs generated by all the rules. It is given by Eq. (19):

$$y_{\text{crisp}} = \sum_{i=1}^{16} q_i(x_1, x_2, x_3, x_4)\varepsilon_i(x_1, x_2, x_3, x_4), \qquad (19)$$

where

$$\varepsilon_i(x_1, x_2, x_3, x_4) = \mu_{pi}(x_1, x_2, x_3, x_4) / \sum_{j=1}^{16} \mu_{pj}(x_1, x_2, x_3, x_4) \qquad (20)$$

x_1, x_2, x_3, x_4 represent the four inputs to each of the fuzzy subsystems, i.e., F1 and F2 as shown in Fig. 2.

For F1 x_1, x_2, x_3, x_4 represent $u(t-1), h(t-1), y(t-1), y(t-2)$ respectively.
For F2 x_1, x_2, x_3, x_4 represent $u(t-1), h(t), y(t-1), y(t-2)$ respectively.
$\mu_{pi}(x_1, x_2, x_3, x_4)$ is determined assuming the product T-norm.

Let y denotes the model output and y_p denotes the actual plant output. Then error (e) is given by Eq. (21):

$$e = y_p - y \qquad (21)$$

$$\text{Mean square error(MSE)} = E = \frac{1}{2} * e^2. \qquad (22)$$

Based on the error (e) obtained at the end of each iteration, the value of the parameters is updated according to Eq. (23):

$$\text{parameter}_{\text{new}} = \text{parameter}_{\text{old}} - \eta * \frac{\partial E}{\partial \text{parameter}}. \qquad (23)$$

Simple chain rule has been used to calculate the derivative of the mean square error (MSE) w.r.t. each of the parameters for both the subsystems.

First, consider the subsystem F2.

Using backpropagation method the various parameters are updated using Eq. (23). The output expression's coefficients associated with each rule for F2 are updated using the derivative given by Eq. (24):

$$\frac{\partial E}{\partial b_j} = \frac{\partial E}{\partial e} * \frac{\partial e}{\partial y} * \frac{\partial y}{\partial q_{iF2}} * \frac{\partial q_{iF2}}{\partial b_j}, \tag{24}$$

where $j = 1, 2, 3, 4$.

To update the membership function parameters, Eqs. (25) and (26) have been used:

$$\frac{\partial E}{\partial a_{2j}} = \frac{\partial E}{\partial e} * \frac{\partial e}{\partial y} * \frac{\partial y}{\partial \varepsilon_i} * \frac{\partial \varepsilon_i}{\partial \mu p_i} * \frac{\partial \mu p_i}{\partial \mu_{F2}} * \frac{\partial \mu_{F2}}{\partial a_{2j}} \tag{25}$$

$$\frac{\partial E}{\partial c_{2j}} = \frac{\partial E}{\partial e} * \frac{\partial e}{\partial y} * \frac{\partial y}{\partial \varepsilon_i} * \frac{\partial \varepsilon_i}{\partial \mu p_i} * \frac{\partial \mu p_i}{\partial \mu_{F2}} * \frac{\partial \mu_{F2}}{\partial c_{2j}}, \tag{26}$$

where $j = 1, 2$.

Similarly for the subsystem F1, the output expression's coefficients associated with each rule are updated using the derivative given by Eq. (27):

$$\frac{\partial E}{\partial k_j} = \frac{\partial E}{\partial e} * \frac{\partial e}{\partial y} * \frac{\partial y}{\partial q_i} * \frac{\partial q_i}{\partial h(t)} * \frac{\partial h(t)}{\partial q_{iF1}} * \frac{\partial q_{iF1}}{\partial k_j}, \tag{27}$$

where $j = 1, 2, 3, 4$.

Derivatives used to update the membership function parameters are given by Eqs. (28) and (29):

$$\frac{\partial E}{\partial a_{1j}} = \frac{\partial E}{\partial e} * \frac{\partial e}{\partial y} * \frac{\partial y}{\partial q_i} * \frac{\partial q_i}{\partial h(t)} * \frac{\partial h(t)}{\partial \varepsilon_{i\_F1}} * \frac{\partial \varepsilon_{i\_F1}}{\partial \mu p_{i\_F1}} * \frac{\partial \mu p_{i\_F1}}{\partial \mu_{F1}} * \frac{\partial \mu_{F1}}{\partial a_{1j}} \tag{28}$$

$$\frac{\partial E}{\partial c_{1j}} = \frac{\partial E}{\partial e} * \frac{\partial e}{\partial y} * \frac{\partial y}{\partial q_i} * \frac{\partial q_i}{\partial h(t)} * \frac{\partial h(t)}{\partial \varepsilon_{i\_F1}} * \frac{\partial \varepsilon_{i\_F1}}{\partial \mu p_{i\_F1}} * \frac{\partial \mu p_{i\_F1}}{\partial \mu_{F1}} * \frac{\partial \mu_{F1}}{\partial c_{1j}}, \tag{29}$$

where $j = 1, 2$.

4 Simulation and Results

4.1 Example 1

For the first nonlinear plant which is given by Eq. (4), the simulation result of its identification by using RFS is shown in Fig. 4. The input signal used for the both the actual system and the model is given by Eq. (30):

Fig. 4 Identification using
RFS for example 1

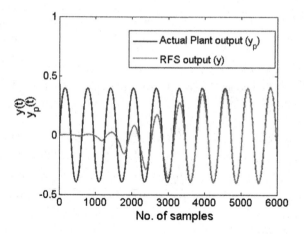

$$u(t) = \sin(0.01t). \tag{30}$$

Figure 5 shows the values of the mean square error and its decreasing pattern
with every instant. The simulation results show successful identification of the
nonlinear plant using recurrent fuzzy system (RFS).

Now, the same plant given by Eq. (4) is identified using conventional fuzzy
system taking its order and structure into consideration. As the structure of the
system given by Eq. (4) is taken into consideration, there will be six inputs to the
fuzzy system, i.e., $u(t-1), u(t-2), y(t-1), y(t-2), y(t-3), y(t-4)$, thus
resulting in a total of 64 rules (assuming each input has two fuzzy sets). When the
same system was modeled using RFS only 32 rules (16 rules for each subsystem)
were sufficient for its successful identification. The simulation results shown in
Figs. 4 and 5 are obtained using only 32 rules.

Fig. 5 Mean square error for
example 1 using RFS

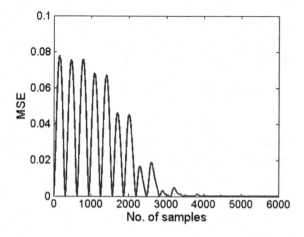

Fig. 6 Identification using conventional fuzzy system for example 1

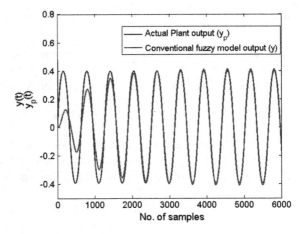

Figure 6 shows the simulation result obtained when the identification is done using conventional fuzzy system taking the order and the structure of the plant into consideration. Figure 7 shows the mean square error obtained in this case.

Observing the simulation results it can be said that although the RFS takes a few more iterations to learn the actual process than the conventional fuzzy system it reduces the computational complexity to a great extent by reducing the number of rules considerably

Fig. 7 Mean square error for example 1 using conventional fuzzy system

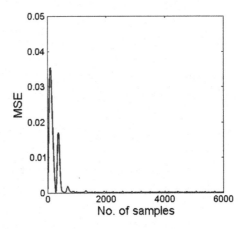

4.2 Example 2

Now take the second nonlinear plant described in Eqs. (7) and (8). This plant is also identified using both the concept of RFS as well as conventional fuzzy system. The input taken for both the recurrent model and the actual plant is given by Eq. (31):

$$u(k) = \sin(2\pi k/25). \tag{31}$$

The simulation result of the identified model using the concept of recurrent fuzzy is shown in Fig. 8. The corresponding mean square error is depicted in Fig. 9.
These results are obtained by taking learning rate (η) = 0.025.

Fig. 8 Identification using RFS for example 2

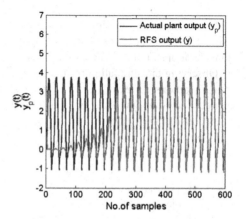

Fig. 9 Mean square error for example 2 using RFS

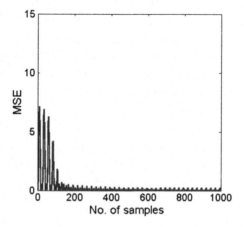

Fig. 10 Identification using conventional fuzzy system for example 2

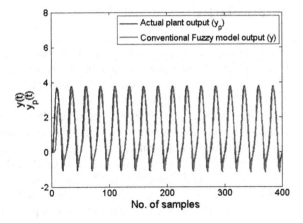

Fig. 11 Mean square error for example 2 using conventional fuzzy system

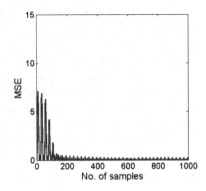

The simulation results clearly show that identification of the plant using recurrent fuzzy system has been successful. Similar to the previous example, this RFS also contains only a single hidden variable.

Now the same plant was identified using conventional fuzzy system taking its order and structure into account. The simulation results are shown in Fig. 10. The mean square error in this case is shown in Fig. 11.

Here, it is clearly seen that although the conventional fuzzy model takes lesser number of iterations for successful identification, the RFS would certainly reduce the computational complexities and would be of great help in case the order of the plant and its structure is unknown or vague.

5   Conclusion

In this paper, modeling of two nonlinear systems is performed using the recurrent fuzzy system (RFS) assuming that the order of these plants is unknown. The concept of recurrent fuzzy mainly finds its application for the identification and control of plants and systems when the order and exact structure are unknown. A recurrent fuzzy system with an order higher than 1 may also involve intermediate variables. These variables are not the final output variables but form a link between the ultimate input and output variables and are given the name "hidden variables". The basic structure of a recurrent fuzzy system (RFS) has been clearly explained in the paper. The paper identifies two different nonlinear systems using the concept of RFS which incorporates only a single hidden variable. The simulation results show that the identification is successful. Also a comparative analysis is done by identifying the same plant using conventional fuzzy system as well where the order of the plant and its actual structure is taken into consideration. It is observed that recurrent fuzzy system is not only useful when the order of the plant is unknown but also reduces the computational complexity as the order of the nonlinear plant increases. It is observed that although the conventional fuzzy system identifies the unknown plant taking lesser number of iterations, RFS also identifies the system while reducing the computational complexity to a great extent. It reduces the number of rules to a great extent when used for modeling of a higher order plant or system. Thus RFS (recurrent fuzzy system) is beneficial not only when the order and structure are unknown but also while dealing with higher order dynamic systems so that the complexity is greatly reduced. These conclusions are clearly proved in the simulation results.

References

1. Wong, C.C., Lin, N.S.: Rule extraction for fuzzy modeling. Fuzzy Sets Syst. **88**, 23–30 (1997)
2. Gorrini, V., Bersini, H.: Recurrent fuzzy systems. In: Proceedings of the Fifth IEEE International Conference on Fuzzy Systems, pp. 193–198. New Orleans (1994)
3. Carlos, K., Roggero, P., Apolloni, J.: Evolution of recurrent fuzzy controllers. In: VI Workshop de Investigadores en Ciencias de la Computación (2004)
4. Narendra, K.S., Parthasarthy, K.: Identification and control of dynamical systems using neural networks. IEEE Trans. Neural Netw. **1**(1) (1990)
5. Surmann, H., Maniadakis, M.: Learning feed-forward and recurrent fuzzy systems: a genetic approach. Elsevier J. Syst. Archit. **47**, 649–662 (2001)
6. Pulmannova, S.: On fuzzy hidden variables. Fuzzy Sets Syst. **155**, 119–137 (2005)
7. Gama, C.A., Evsukoff, A.G., Weber, P., et al.: Parameter identification of recurrent fuzzy systems with fuzzy finite-state automata representation. IEEE Trans. Fuzzy Syst. **16**(1), 213–224 (2008)
8. Diepold, K.J., Lohmann, B.: Transient probabilistic recurrent fuzzy systems. In: IEEE International Conference on Systems Man and Cybernetics (SMC), pp. 3529-3536 (2010)

9. Juang, C.-F., Chang, P.-H.: Recurrent fuzzy system design using elite-guided continuous ant colony optimization. Appl. Soft Comput. **11**(2), 2687–2697 (2011)
10. dos Santos, C.K., Espindola, R.P., Evsufoff, A.G., Development of recurrent fuzzy systems with external feedback. IEEE Latin Am. Tran. **13**(1), 284–290 (2015)
11. Pratama, M., Lughofer, J.L.E., Zhang, G., Er, M-J.: "Increamental learning of concept drift using evolving type-2 recurrent fuzzy neural network. IEEE Trans. Fuzzy Syst. **99** (2016)
12. Chiu, C.-H., Chang, C.-C.: Wheeled human transportation vehicle implementation using output recurrent fuzzy control strategy. IET Control Theory Appl. **8**(17), 1886–1895 (2015)
13. Heydari, G.A., Gharaveisi, A.A., Vali, M.A.: New formulation for representing higher order TSK fuzzy systems. IEEE Trans. Fuzzy Syst. **24**(40), 854–864 (2016)
14. Gorrini, V., Bersini, H.: Recurrent fuzzy systems. In: Proceedings of the Fifth IEEE International Conference on Fuzzy Systems, pp. 193–198. New Orleans (1994)

Analytical Study of Intruder Detection System in Big Data Environment

Ahmed Faud Aldubai, Vikas T. Humbe and Santosh S. Chowhan

Abstract This paper presents analytical study of intruder detection system in big data environment. In recent years, the size of data increases at high speed, through daily increasing in number of people and online applications such as online airline ticket booking, online banking, online payment system, etc., using the Internet and network services, and this leads to a huge amount of data from terabyte to petabytes which are called as the big data. Therefore, predict and analysis of network traffic from a possible intrusion attack through permanent gathering of network traffic data and learning of their characteristics on the fly are the critical aspects. Various experiments are conducted and summed up, to be able to identify different problems in existing network applications and traffics. Analysis of intruder detection over network traffic includes support vector machine (SVM) approach, distance-based classifiers, genetic algorithm, and fuzzy neural network classifiers based on data mining techniques. Similarly, several interesting hybrid techniques are implemented to attain efficient and effective results of intruder detection system analysis over network.

Keywords Big data · Data mining techniques · Intruder detection
Network and traffic analysis

A. F. Aldubai (✉) · S. S. Chowhan
School of Computational Sciences, S.R.T.M University, Nanded, Maharashtra, India
e-mail: Ahmed_Aldubai86@yahoo.com

S. S. Chowhan
e-mail: drschowhan@gmail.com

V. T. Humbe
School of Technology, S.R.T.M University Sub-Campus, Latur, Maharashtra, India
e-mail: vikashumbe@gmail.com

© Springer Nature Singapore Pte Ltd. 2018
M. Pant et al. (eds.), *Soft Computing: Theories and Applications*,
Advances in Intelligent Systems and Computing 583,
https://doi.org/10.1007/978-981-10-5687-1_36

405

1 Introduction

The intrusion detection system is a software application which helps us to protect, monitor our network (NIDS), and host (HIDS) systems, from malicious activity, intruders, attackers, and policy violations when they try to access through network. Increase of internet users leads to increase in network traffic. Self-monitoring software available which work by imposing security policies such as spyware, antivirus, antispam software, and pop-up blocking act at the local client machine side. A firewall can be defined as a device or application that can only able to analyze packet headers and enforce policy at network level. So we need application and tools that can detect and prevent the attacks interchangeably, so here we make use of intrusion detection system. In last a few years, several data mining and machine learning techniques have been applied to network traffic and obtained the good quality detection, accuracy, and performance. In this paper, we have focused on four important techniques that are used for intruder detection over network traffic, which are support vector machine (SVM) technique, distance-based classifiers, genetic algorithm, and fuzzy neural network classifiers. The distinctiveness and restrictions of earlier researches are discussed and typical features of these network traffic analysis are also summarized. The remaining of the paper is planned as follows. A short overview about intruder detection and network traffic analysis is followed by an extensive review of several available network analysis techniques applied; in next section, a comparison of the four network analysis techniques, which are already mentioned above, and the best-recommended technique are presented.

2 Network Traffic Analysis

In present days, network traffic analysis has become more and more important for monitoring the network traffic. In the past, administrators were monitoring only a small number of network devices or less than a thousand computers. The network bandwidth was may be just less or 100 Mbps. At this time, administrators have to monitor higher speed network more than 1 Gbps and various networks such as wireless networks and ATM (Asynchronous Transfer Mode) network, which need modern network traffic analysis technique in order to analysis, manage network, handle the network security, and solve the network problems quickly to avoid network failure. As a result, network traffic analysis has a number of challenges in recent years; intruder detection is the most challenge with network. A variety of approaches are being used by researchers for network traffic analysis. A general framework for network traffic analysis involves some important steps which are as follows collecting the dataset by capturing packages from network traffic, preprocessing followed by actual analysis using data mining techniques, and observations to reveal patterns from the network data. Figure 1 shows the main phases of network traffic analysis.

Fig. 1 General architecture
of intrusion detection system
classification

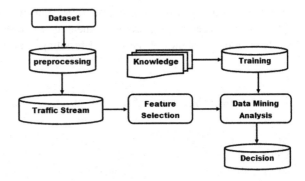

2.1 Datasets

In order to do testing and evaluating in network traffic, all research works should be applied based on common standard dataset.

- **DARPA Dataset**: It was the first dataset of intrusion detection system (IDS) with respect to network traffic analysis. It was released in 1998, which contains two parts: a real-time assessment that was conducted by MIT Lincoln Laboratory and an offline assessment. There are 38 types of attacks in DARPA dataset that fall into the following four categories: user to root (U2R) Probing, denial of service (DoS), and remote to user (R2L).
- **KDD Cup Dataset**: It is a sample dataset which is used most widely for evaluating the network traffic analysis with respect to intrusion detection. It is constructed based on the data captured in DARPA IDS evaluation program. It consists of 4 GB of compressed raw data of 7 weeks of network traffic (i.e., normal and intrusion). It contains 2 million connection records. Moreover, it contains 41 features and 24 training and testing attacks. Using this dataset the data can be classified either as normal or intrusion.
- **NSL-KDD Dataset**: The NSL-KDD dataset is a new updated version of KDD cup dataset [1]. The most important feature in NSL-KDD dataset is that it does not include duplicate records in testing and data redundant instances in training data. For this reason the classifier becomes more accurate. The NSL-KDD is publicly available for researchers and it is an improved version of original KDD cup dataset.

2.2 Preprocessing Method

The most important phase after collecting dataset is the preprocessing method, which used to manipulate real-world data that often contains noisy in specific behavior, dimensionality, inconsistent (outlier values, containing errors), and incomplete, into a comprehensible format. For this reason, the preprocessing methods are necessary before implement data mining techniques, to improve accuracy and the quality of the data of resulting data mining task.

2.3 Feature Selection Method

The next step is feature selection method, which works by selecting only specific features that are important and vital to problem as a subset of the original attributes. To virtualize all classification algorithms, data mining techniques are performed through the removal of irrelevant attributes or redundant. Obviously, the unrelated features may often result in poor modeling, since they are not well related to the class label.

There are two extensive kinds of feature selection methods:

- **Filter Models**: In this model, a crisp criterion on a single feature, or a subset of features, is used to evaluate their suitability for classification. This method is independent of the specific algorithm being used. In order to perform feature selection with this model, a number of different measures are used in order to quantify the relevance of a feature to the classification process (Fig. 2).
- **Wrapper Models**: In this model, the feature selection process is embedded into a classification algorithm, in order to make the feature selection process sensitive to the classification. This approach recognizes the fact that different algorithms may work better with different features (Fig. 3).

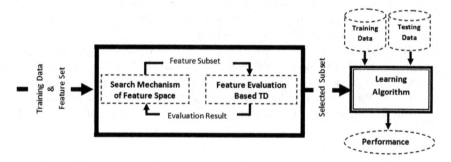

Fig. 2 Filter-based feature selection

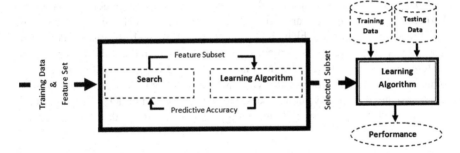

Fig. 3 Wrapper-based feature selection

2.4 Classification Technique

2.4.1 Support Vector Machine (SVM)

Support vector machine (SVM) is a supervised learning method and a binary classifier used for classification and regression. The support vector machine was created by Vapnik 64. By using this technique we can reduce the high dimensions of large data (training and testing). The SVM requires costly time and memory. Several previous studies have reported SVM results of giving higher performance with respect to classification accuracy than the other classification approaches. Figure 4 illustrates the procedure used by support vector machine (SVM).

Koshal and Bag [2] have used two algorithms as hybrid model by combining C4.5 decision tree and support vector machine (SVM) approaches for developing the intrusion detection system. Dataset used was NSL-KDD. A preprocessing of data reduced the dimensionality of entire network traffic dataset using feature selection method. Two algorithms used are correlation-based feature selection (CFS) and consistency-based feature selection (CON); the result is that only 12 features were selected out of 42 features. Later on, they used C4.5 decision tree for their classification, and SVM for categorizing of the attacks. Finally, increase in the accuracy and detection rate and decrease in false alarm rate are found.

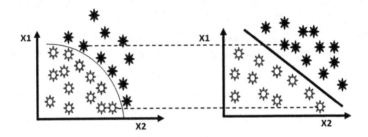

Fig. 4 Illustration of the procedure used by SVM

Kuang et al. [3] adopted a novel support vector machine model by combining kernel principal component analysis (KPCA) with improved chaotic particle swarm optimization (ICPSO), which is proposed to deal with intrusion detection. The KPCA was applied as a preprocessor of SVM to reduce the dimension of feature vectors and shorten training time, and ICPSO is presented to optimize the punishment factor. The proposed hybrid approach is composed of three stages. In the first stage, the principal components are achieved based on KPCA theory which corresponded to the first p biggest eigenvalues, to form the subeigenspace, satisfying

$$\frac{\sum_{i=1}^{p} \lambda i}{\sum_{i=1}^{p} \lambda i} \geq 90\% \tag{1}$$

The second stage is to use this attribute subset as the training dataset, and testing dataset of SVM and ICPSO is used to select the optimal parameter of SVM. The third stage is to use negative mean absolute percentage error (MAPE) as criteria evaluation:

$$\text{MAPE} = \frac{1}{N} \sum_{i=1}^{N} \left| \frac{a_i - b_i}{a_i} \right| \times 100\%. \tag{2}$$

According to authors, the test results indicate that the proposed method shows more excellent detection performance for intrusion detection, higher predictive accuracy, and also saves a lot of training and testing time.

Aslahi-Shahri et al. [4] proposed a method of support vector machine and genetic algorithm (GA) for intrusion detection system. Hybrid algorithm used to reduce the number of features from 45 to 10. The selected feature distribution is done in a way that four features are placed in the first priority, four features in the second, and two features in the third priority. Experiments are conducted using KDD Cup 1999 data; according to authors the results reveal that the proposed hybrid algorithm is capable of achieving a true positive value of 0.973, while the false positive value is 0.017.

Ben Sujitha et al. [5] proposed the layered approach with enhanced fuzzy-based support vector machine algorithm. The feature selection applies the modified multi-objective particle swarm optimization PSO feature selection algorithm; the following two criteria also checked in the modified PSO:

1. If Fitness (xi) = Fitness (Ṕbest) and |xi| < |Ṕbest| then Ṕbest = xi;
2. If any Fitness (Ṕbest) = Fitness (Ḡbest) and |pbest| < |Ḡbest| then Ḡbest = Ṕbest;

The fuzzy rule extraction is done using

$$\left\{ \text{Rule}_i \,\middle|\, \alpha_o^{(i)} \text{ or } w_{i > h_s} \right\}. \tag{3}$$

The fuzzy-based support vector machine algorithm is effectively applicable to detect anomaly attack. The proposed system is tested with the benchmark KDD '99 intrusion dataset as well as real-time captured dataset; the experimental results show the proposed system can effectively identify the signature based and anomaly type of attack with the detection rate up to 99.2% and the false alarm rate is very much reduced.

2.4.2 Distance-Based Classifies

Distance-Based Outlier Detection: It is based on the calculation of distances among objects in the data with clear geometric interpretation. Figure 5 illustrates the procedure used by outlier detection

K-Nearest Neighbors Algorithms: It stores all presented cases and classifies new cases based on a similarity measure (e.g., distance functions); the KNN technique is used for the entire training dataset. Each item distance in the training set must be determined. In the training set particular K closest are considered further. A fresh item is then located in the class that contains nearly all items for this set of K closest items. Figure 6 illustrates the procedure used by KNN. Points in the training set and the three closest items in the training set are shown; it will be placed in the classification where most of these are members.

Jaisankar et al. [6] used a new intelligent agent based on IDS using fuzzy rough set based on outlier detection algorithm to detect outliers and reduced feature selection. Furthermore, it had used fuzzy rough set based on SVM to classify and detect anomalies efficiently. KDD Cup 99 dataset was applied. They have been presented fuzzy rough membership function based on outlier's detection by applying the following definition:

$$W_F^{\{a\}}(F_i) = \frac{\left|[F]_R \cap F\right|}{\left|[F]_R\right|}. \tag{4}$$

Fig. 5 Classification using distance-based outlier detection

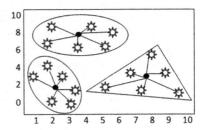

Fig. 6 Illustration of the
procedure used by KNN

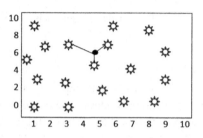

The results of proposed model show that it improves the overall accuracy and reduces the false alarm rate. According to authors it has been proved that the detection accuracy for PROBE and R2L classes of attacks is 99.9%.

Ashfaq et al. [7] proposed a new model for selection algorithm using fuzziness to optimize the training data. The fuzziness of the classifier is given by Eq. (5):

$$F\left(\mu_j\right) = -\frac{1}{C}\sum_{i=1}^{C}\left(\mu_{ij}\log\mu_{ij} + \left(1 - \mu_{ij}\right)\log\left(1 - \mu_{ij}\right)\right). \tag{5}$$

In addition, they used random weight neural network RWNN as a base classifier to obtain a membership vector corresponding to each training sample, and the KDD Cup 99 dataset was used in experiment. The results show that reduced optimized training dataset can effectively increase the accuracy.

Dave et al. [8] designed a model using k-nearest neighbor classification and Dempster theory for performance evaluation of intrusion detection system. After creating the probability density function $f_k(xk)$ between training data dε $D_{\text{train}}(k)$ and rule $rεR_k$, the probability of new connection data dε D_{test} belongs to class k which is represented as

$$P_{K^{(d)}} = \int_{mk(d)}^{1.0} f_k(xk)\mathrm{d}x_{k...} \sum_{kec}f_k(xk)\mathrm{d}x_k \int_{mk(d)}^{1.0} f_{1(x1)\mathrm{d}x1}. \tag{6}$$

The experiment is done based on different datasets such as KDD 99 dataset and DARPA98. The new pattern of intrusion is compared with the existing pattern and generates a new schema with updating a list of pattern and improved the true rate of intrusion detection.

Venkata Lakshmi et al. [9] proposed a subset of records from the KDD Cup 1999 dataset for classification of connection records into normal or attacked data, using k-nearest neighbor algorithms. Moreover, they used k-nearest neighbor algorithm with subset selected feature records proposed by Kok Chin Khor to compare the classifications. Their results show that the feature set (FS5) consisting of seven attributes yields the best results in terms of identifying the maximum number of attacks compared with the whole set of attributes as well as the other feature sets (FS1, FS2, FS3, and FS4).

2.4.3 Fuzzy Neural Network Classifiers

The structures of neural network can deal with ill-defined activities and imprecise data. In addition, fuzzy logic is another tool for modeling uncertainties associated with human perception, cognition, and thinking. Indeed, the neural network approach fuses well with fuzzy logic and some research endeavors have given birth to the field of fuzzy neural networks. Paradigms based upon this integration are believed to have considerable potential in the areas of medical diagnosis, system modeling, pattern recognition, expert systems, and control systems.

Ratnawat et al. [10] developed solution for IDS by combining fuzzy-based rules with neural-based classifier. The fuzzy C-means is a method that allows one piece of data belong to many clusters and minimize the following objective function:

$$J_m = \sum_{i-1}^{N} \sum_{j-1}^{C} u_v^m ||x_i - c_j||^2, \quad 1 \le m < \infty. \tag{7}$$

They proposed a subset of records from the KDD Cup 1999 dataset. Fuzzy rules provide a probability outcome of a signature which is then classified by neuro system. Their results show that proposed system produces 100% accuracy in comparison to neural classifier (99.66%) and fuzzy-based system (70%).

Hassan [11] applied genetic algorithms with fuzzy logic to efficiently detect various types of network intrusions. They carried out an experiment using the standard KDD Cup 99. They used fuzzy confusion matrix to measure the fitness of a chromosome. The proposed model can upload and update new rules to the systems. The experimental results show that the proposed method resulted in good detection rates when using the generated rules to classify the training data itself.

Chandrasekhar [12] had applied a new technique by using neuro fuzzy and radial basis support vector machine. First, it used K-means clustering to the sum of squared error (SSE) using the following equation:

$$J = \sum_{J=1}^{k} \sum_{i=1}^{n} \left\| x_i^{(j)} - c_j^2 \right\|. \tag{8}$$

Then different neuro fuzzy models are trained. Later, they used the radial basis SVM for classification to detect intrusion that has happened or not using the below equation.

The KDD Cup 99 dataset was used. The experimental result shows that they have achieved about 98.80% accuracy in case of DOS attack and in case of R2L; and U2R attacks achieved 97.5%. They reached heights of 97.31% accuracy in case of PROBE attack.

2.4.4 Genetic Algorithms Classifiers

The general procedure of a genetic algorithm begins with a randomly selected population of chromosomes, with representations of the problem to be solved. Several positions of each chromosome are encoded as characters, numbers, or bits. These positions are sometimes referred to as genes and are altered randomly within a range during evolution.

Hoque et al. [13] used genetic algorithm (GA) for intrusion detection system (IDS), and the KDD99 benchmark dataset. They used the standard deviation equation with distance to measure the fitness of a chromosome. They get better detection rate for denial of service and user to root and close detection rate for probe and remote to local.

Dave and Sharma [14] presented a combination of the traditional snort and genetic algorithm together so that the number of rules of snort is decreased. The standard dataset of KDD Cup 1999 was used. In the pre-calculation phase, they have made 23 groups of chromosomes according to algorithm that shows the major steps on training data.

```
Algorithm: Initialize chromosomes for comparison
Input: Network audit data (for training)
Output: A set of chromosomes
1. Range = 0.125
2. For each training data
3. If it has neighboring chromosome within Range
4. Merge it with the nearest chromosome
5. Else
6. Create new chromosome with it
7. End if
8. End for
```

Later they classify all the rules of genetic algorithm and snort based on their functionality. Their results show a decrease in the amount of detection time, CPU utilization, and memory utilization using genetic algorithm and snort IDS.

Sadiq Ali Khan [15] proposed an efficient approach to classify various types of attacks based on rules formulation and on network intrusion detection. Their study shows that GA can be effectively used for the formulation of decision rules in intrusion detection.

The following rule was used to define the initial population. Let

```
DC= don't care
service=µ1
flag=π1
land=Ω1
logged_in=β1
root_shell= µ2
su_attempted= π2
is_hot_login= Ω2
is_guest_login= β2
        If (µ2 or π2 or Ω2 or β2 ) == 0 AND (µ1 or π1
or Ω1 or β1) == DC  Than Categories as "Attacks"
                Else Categories as "Normal"
```

results of their model show that GA is found more efficient in terms of false alarm.

3 Conclusion

The main goal of using intrusion detection with data mining is to improve the detection rate and to reduce the false alarm rate. Intrusion detection in big data environment has become a vital subject of continuous research in various fields of networks. A range of researchers have been implemented an effective network traffic algorithm for the analysis of network traffic. In this study, we analyze previous studies of intrusion detection in network traffic. We enlisted and discussed various approaches and algorithms proposed to predict of network traffic attacks. Moreover, we defined the datasets used, steps of preprocessing and feature selection techniques, classification algorithms, and the metrics used to evaluate the results. For future work, we will implement the SVM and fuzzy k-mean algorithms with KDD dataset using Matlab tool, to get better result than we have analyzed in our study.

References

1. Information Security Centre of Excellence (ISCX), Canada: UNB ISCX NSL-KDD Data set. Retrieved 8 Aug 2016. http://nsl.cs.unb.ca
2. Koshal, J., Bag, M.: Cascading of C4.5 decision tree and support vector machine for rule based intrusion detection system. J. Comput. Netw. Inf. Secur. (2012). doi:10.5815/ijcnis
3. Kuang, F., Zhang, S., Jin, Z., Xu, W.: A novel SVM by combining kernel principal component analysis and improved chaotic particle swarm optimization for intrusion detection, pp. 1–13. Springer, Heidelberg, 22 June 2014

4. Aslahi-Shahri, B.M., Rahmani, R., Chizari, M., Maralani, A., Eslami, M., Golkar, M.J., Ebrahimi, A.: A hybrid method consisting of GA and SVM for intrusion detection system. Nat. Comput. Appl. Forum (Springer), 1–8 (2015)
5. Ben Sujitha, B., Kavitha, V.: Intrusion detection system using F-SVM based layered approach with enhanced MPSO feature selection algorithm. Int. J. Adv. Eng. Technol., 93–99 (2016)
6. Jaisankar, N., Ganapathy, S., Yogesh, P., Kannan, A., Anand. K.: An intelligent agent based intrusion detection system using fuzzy rough set based outlier detection. In: Soft Computing Techniques in Vision Science, SCI 395, pp. 147–153. Springer, Heidelberg (2012)
7. Ashfaq, R.A.R., He1, Y.-l., Chen, D.-g.: Toward an efficient fuzziness based instance selection methodology for intrusion detection system, pp. 1–10. Springer, Heidelberg, 27 June 2016
8. Dave, D., Richhariya, V.: Intrusion detection with KNN classification and DS-theory. Int. J. Comput. Sci. Inf. Technol. Secur. (IJCSITS) 2(2), 274–281 (2012)
9. Venkata Lakshmi, S., Edwin Prabakaran, T.: Application of k-nearest neighbour classification method for intrusion detection in network data. Int. J. Comput. Appl. (0975–8887) 97(7), 34–37 (2014)
10. Ratnawat, N., Jain, A.: A novel intrusion detection system using neural-fuzzy classifier for network security. Int. J. Emerg. Technol. Adv. Eng. 4(6), 900–905 (2014)
11. Hassan, M.Md.M.: Network intrusion detection system using genetic algorithm and fuzzy logic. Int. J. Innov. Res. Comput. Commun. Eng. 1(7), 1435–1445 (2013)
12. Chandrasekhar, A.M., Rahhuveer, K.: Intruder detection technique by using K-means, fuzzy neural network and SVM classifiers, pp. 1–7. In: International Conference on Computer Communication and Informatics, 4–6 Jan 2013
13. Hoque, M.S., Abdul Mukit, Md., Abu Naser Bikas, Md.: An implementation of intrusion detection system using genetic algorithm. Int. J. Netw. Secur. Appl. (IJNSA) 4(2), 109–120 (2012)
14. Dave, M.H., Sharma, S.D.: Improved algorithm for intrusion detection using genetic algorithm and SNORT. Int. J. Emerg. Technol. Adv. Eng. 4(8), 273–276 (2014)
15. Sadiq Ali Khan, M.: Rule based network intrusion detection using genetic algorithm. Int. J. Comput. Appl. (0975–8887) 18(8), 26–29 (2011)

Effect of Intrinsic Parameters on Dynamics of STN Model in Parkinson Disease: A Sensitivity-Based Study

Jyotsna Singh, Phool Singh and Vikas Malik

Abstract Parkinson disease alters the information patterns in moment-related pathways in brain. Experimental results performed on rats (Corinneeurrier et al. in J Neurosci 19(2), 599–609, 1999 [1]) show that the activity patterns change from single spike activity to mixed burst mode in Parkinson disease. However, the cause of this change in activity pattern is not yet completely understood. In this paper, a single-compartment conductance-based model is considered which focuses on subthalamic nucleus and synaptic input from globus pallidus external. Subthalamic nucleus is one of the main nuclei involved in the origin of motor dysfunction in Parkinson disease. This model shows highly nonlinear behavior with respect to various intrinsic parameters. Behavior of model has been represented with the help of activity patterns generated in healthy and Parkinson condition. These patterns have been compared by calculating their cross correlation for different values of intrinsic parameters. Results show that the activity patterns are very sensitive to various parameters and these results also provide insight into the motor dysfunction.

1 Introduction

Brain disorders are most serious health problems and pose a challenge to the society. These disorders are among the most mysterious of all the diseases, and our ignorance of the underlying disease mechanism is a major obstacle to the development of better treatment. One of the diseases caused by brain disorder is Parkinson disease (PD). PD is one of many diseases that are collectively known as movement disorder. PD involves the malfunction and death of vital nerve cells in

J. Singh (✉)
Department of CSE & IT, The NorthCap University, Gurgaon 122017, India
e-mail: singhjyotsna1@gmail.com

V. Malik
Department of Physics, JIIT, Noida, Uttar Pradesh, India

P. Singh
Department of Applied Sciences, The NorthCap University, Gurgaon, India

© Springer Nature Singapore Pte Ltd. 2018
M. Pant et al. (eds.), *Soft Computing: Theories and Applications*,
Advances in Intelligent Systems and Computing 583,
https://doi.org/10.1007/978-981-10-5687-1_37

the brain, called neurons. PD primarily affects neurons in the area of the brain called
the substantia nigra (SNc), one of the nuclei in Basal Ganglia (BG) [2, 3]. Some of
these dying neurons produce dopamine, a chemical that sends messages to the part
of the brain that controls movement and coordination. As PD progresses, the
amount of dopamine produced in the brain decreases, leaving a person normally
unable to control movement. The cause to the depletion of dopamine is not well
known, and hence there is presently no diagnosis and cure at early stage. There are
treatment options at the later stages of diseases, such as medication and surgery, to
manage its symptoms [4].

Apart from SNc, BG has other nuclei which are striatum, subthalamic nucleus
(STN), and globus pallidus (GP). GP is further divided into internal globus pallidus
(GPi) and external globus pallidus (GPe). It has strong connections with the
cerebral cortex, thalamus, and other brain areas. The striatum receives input directly
from cortex and delivers neurotransmitter, which is inhibitory in nature to the GP
through different pathways (1) direct and (2) indirect. STN directly receives inputs
from cortex through (3) hyper-direct pathway [3–5]. The two pathways, direct and
indirect, affect the basal ganglia network in opposite ways and are simultaneously
involved with the control of voluntary movements [2]. It has been acknowledged by
researchers in [6–12] that there is central origin to PD but the localization to the
origin is still not clear. There are few hypotheses of oscillation, bursting, and tremor
generation as proposed in [8, 13]. Some researchers suggest that the circuits in basal
ganglia are itself oscillation and bursting generating circuits [14–17].

To identify the cause to the origin of tremor and bursting generation and further
to narrow down, we have considered a conductance-based model of basal ganglia as
shown in Fig. 1. The connections are disrupted between different nuclei in PD
condition [3]. In normal condition, the dopamine receptor, D1, is excited by
dopamine which produces inhibitory neurotransmitters, γ-aminobutyric acid
(GABA) to the output nuclei, GPi. This suppresses the activity of GPi neurons
resulting in increased activation of the thalamus. The another dopamine receptor,

Fig. 1 Conductance-based
model of basal ganglia

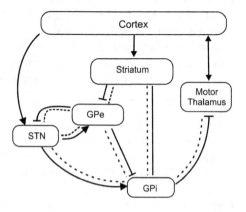

D2, provides GABA to the GPe when it is activated by dopamine. Hence, GPe neurons are inhibited, which results in excited activities in the STN and GPi. As a result, thalamus neurons are less activated.

A possible viewpoint is that the bursting and oscillation generator circuits are basal ganglia thalamocortical circuits [13]. The basal ganglia circuits do not produce tremor oscillations in healthy basal ganglia circuits [18–23]. In contrary the tremor oscillation-related activity is observed in STN in PD condition [24]. However, the reason underlying tremor oscillations generation is still to be understood.

Above studies suggest that STN alone could be the target region to analyze the symptoms and the origin of PD. Therefore, in this paper we have compared the behavior of STN model in healthy primate and in PD condition for exploring the dynamics of the subthalamo–globus loop subjected to various currents. Work has been distributed in various sections. Section 2 explains about the mathematical model and underlying methods. Section 3 is dedicated to the sensitivity-based study of the model for various currents and their results. Section 4 is the conclusion of the work.

2   Methods and Models

Herein, a single-compartment conductance-based model of basal ganglia [3] is considered. STN receives excitatory input from cortex through hyper-direct pathway [2]. GPe receives inhibitory input from the striatum and excitatory input from STN. In addition, there is inhibitory synaptic connection within GPe cells. The STN receives inhibitory input from the GPe and excitatory neurotransmitters through cortex. It increases the frequency of the discharge patterns in GPi neurons [25, 26]. This result leads to the importance of interaction between STN and GP neurons in the direct and hyper-direct pathway in the basal ganglia network [2, 20].

The neuronal model of STN contains the experimentally observed conductance and the timescales of the real phenomenon [27]. Hence, the analysis of the specified model with respect to various currents may provide a potential mechanism for sensitivity of the model in PD.

The models [27] include a leak current (I_l), fast spike-producing potassium (I_k) and sodium currents (I_{Na}), low threshold T-type (I_T) and high-threshold Ca^{2+} currents (I_{Ca}), a Ca^{2+} activated voltage independent after-hyperpolarization K^+ current (I_{AHP}), synaptic current (I_{syn}) and applied current (I_{app}), so that the equation governing the membrane potential V takes the following form:

$$C\frac{dV}{dt} = -I_l - I_k - I_{Na} - I_T - I_{Ca} - I_{AHP} - I_{syn} + I_{app}, \qquad (1)$$

where different membrane currents are given by

$$I_l = g_l.[V - V_l] \tag{2}$$

$$I_k = g_k \cdot n^4 \cdot [V - V_k] \tag{3}$$

$$I_{Na} = g_{Na} \cdot m_\infty^3(V) \cdot h \cdot [V - V_{Na}] \tag{4}$$

$$I_T = g_T \cdot a_\infty^3 \cdot (V) \cdot r \cdot [V - V_{Ca}] \tag{5}$$

$$I_{Ca} = g_{Ca} \cdot s_\infty^2 \cdot (V) \cdot [V - V_{Ca}] \tag{6}$$

$$I_{AHP} = g_{AHP} \cdot \frac{[Ca]}{[Ca] + k_1} \cdot [V - V_k], \tag{7}$$

where k_1 is the dissociation constant of Ca^{2+}-dependent AHP current. I_{app} current is used to adjust the membrane resting potential with the experimental data [28, 29]. The equation for calcium current is described by the equation

$$d\frac{[Ca]}{dt} = \varepsilon[-I_{Ca} - I_T - kCa[Ca]] \tag{8}$$

where ε characterizes the calcium influx and kCa is calcium pump rate. The equation for gating variable m, n, h, s and a is given as

$$\frac{dx}{dt} = \frac{\phi_x[x_\infty(V) - x]}{\tau_x(V)}, \tag{9}$$

where ϕ is the constant in the equation for gating variables. The time constant function τ_x is given by

$$\tau_x(V) = \tau_x^0 + \frac{\tau_x^1}{[1 + \exp[-[V - \theta_x^\tau]/\sigma_x^\tau]}, \tag{10}$$

where θ is the half activation/inactivation variable and τ is the time constant functions. Synaptic current I_{syn} in the STN neuron is computed as a sum of synaptic currents from the GPe and other feedback neurons [27]. Synaptic current I_{syn} is defined by considering the synaptic inputs from GP to STN and from feedback neurons to STN:

$$I_{syn} = g_{gs} \cdot s_g[V - V_{gs}] + g_{fs} \cdot s_f[V - V_{fs}], \tag{11}$$

where g_{gs} and g_{fs} are the synaptic conductance, and s_g and s_f are the synaptic variables. As suggested in above studies, dopamine degeneration is one of the main causes of the Parkinson disease. Synaptic inputs play a major role to show the degeneration of dopamine. To represent this we have considered two synaptic variables, s_g and s_f. These two variables are used to modulate the strength of synaptic input currents. s_g and s_f are varied in [7] range, so that the lower values of s_g and s_f correspond to lower dopamine levels and stronger conductance and higher values represent the opposite. All the synaptic variables in the model circuit are modeled by the following first-order kinetic equation, which describes the fraction of activation channels:

$$\frac{ds}{dt} = \alpha \cdot H_\infty (V_{presyn} - \Theta_g) \cdot [1 - s] - \beta \cdot s, \tag{12}$$

where

$$H_\infty(V) = \frac{1}{1 + \exp^{\frac{-\left[V - \theta_g^H\right]}{\sigma_g^H}}} \tag{13}$$

is a sigmoid function and V_{presyn} is the synaptic potential from neighboring neurons. The values of synaptic variables α, β, θ, and σ are taken from [28]. The values of synaptic strengths in the "normal" state (high dopamine level) are $g_{gs} = 0.695$ and $g_{fs} = 0.215$, and the maximal conductance of the AHP current in STN neuron was set to $g_{AHP} = 4.23$ nS/mm^2. The values of synaptic strengths corresponding to the parkinsonian (low dopamine level) state are $g_{gs} = 1.39$, $g_{fs} = 0.43$, with STN cell's AHP conductance set to $g_{AHP} = 8.46$ nS/mm^2.

3   Analysis of Electrophysiological Properties

This section will analyze the electrophysiological properties of the neurons within STN neuron. STN is an oval-shaped small nucleus which receives inhibitory and excitatory inputs from other neurons within Basal ganglia as shown in Fig. 1 [2]. Often various types of neurons can be found within a certain nucleus. These types can be distinguished on the basis of electrophysiological characteristics. In this section the analysis of analysis of electrophysiological properties of spiking patterns for normal and Parkinson in STN is discussed, and therefore a distinction is made between the two types. Different studies [28–31] reveal that electrophysiological experiment results may differ from case to case. We compute two discharge patterns from the above model, one for STN in normal condition (STN) and one for STN in Parkinson condition (STNP). These discharge patterns are analyzed and compared by computing their cross correlation function (CCF). CCF is a measure of similarity of two series as a function of the lag of one relative to the other. The CCF between A_t and B_{t+i} is called the ith order cross correlation of A and B. The sample estimate of this cross correlation, called r_k, is calculated using the following formula:

$$r_k = \frac{\sum_{j=1}^{n-k}(A_j - \overline{A})(B_{j+k} - \overline{B})}{\sqrt{\sum_1^n (A_j - \overline{A})^2 \sum_1^n (B_j - \overline{B})^2}}. \tag{14}$$

To quantify the sensitivity of STN model in the hyper-direct pathway, we have analyzed time series for STN and STNP for various ionic and synaptic currents. To investigate the model assumptions about physiological properties and connectivity patterns that lead to the best explanation of (i.e., closest fit to) our in vivo electrophysiological data, we studied different activity patterns generated by the mathematical descriptions presented above. All the discharge patterns were generated using Eq. (1). These discharge patterns are then simulated for selected ionic current. The simulations are run for 500 ms. We are displaying the results for parameters which are very sensitive to the model. These parameters are applied current I_{app}, feedback membrane potential V_{fs}, and presynaptic potential V_{presyn}. For the sake of clarity, the specifics of each discharge pattern are described that show the sensitivity trade-off while comparing the discharge patterns obtained from [27]. All the figures have been plotted as a function of time. Detailed sensitivity analysis and results are given in the following sections. While simulating the above model we have used parameters as given in [27].

3.1 Sensitivity Analysis for Applied Current

The discharge patterns that are displayed in basal ganglia thalamocortical network are typically irregular and are correlated with the activity of these cells. Various discharge patterns are generated for I_{app} ranging from 30 to 37. These patterns have shown the sensitivity for I_{app} for both STN and STNP. The original value of applied current was 32, considered by [27]. We have displayed the discharge pattern for $I_{app} = 30$ shown in Fig. 2. Minor change in the value of applied current shows sensitivity trade-off in the cross correlation of STN and STNP computed as shown in Fig. 3.

Fig. 2 Spiking patterns in STN-GPe model in healthy and PD condition at $I_{app} = 30$

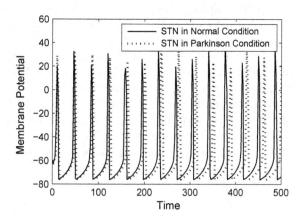

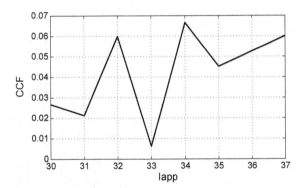

Fig. 3 I_{app} versus cross correlation for healthy and PD spiking patterns

Figure 2 shows that the patterns of variation between two time series of STN and STNP are very similar, except that the time series of STNP is little delayed in nature. It can be easily seen in the trend of the two time series that correlation between these two would improve if there is a linear shift in time for any of the time series. But the correlation computed for I_{app} between 30 and 37 without any linear shift shows a nonlinearity. It can be seen in Fig. 3 that maximum correlation attained is 0.0666 at I_{app} = 34 between I_{app} = 30–37. Variation in I_{app} does not improve the cross correlation between the two time series of STN and STNP.

3.2 Sensitivity Analysis for Presynaptic Membrane Potential

V_{presyn} is the membrane potential of a presynaptic neuron. We have analyzed the effect of V_{presyn} on synaptic variables which ultimately affect STN model in normal and PD condition. The value of V_{presyn} was considered as 31 by [27]. Presynaptic potential may increase or decrease depending upon the connections between neighboring neurons. If the connections are disrupted it may decrease and if the connections are strong, it may increase [30]. Hence, we have analyzed it for disrupted and strong connection. A minor change in the presynaptic value greatly affects the cross correlation function for STN model in normal and PD condition. Activity patterns have been shown for V_{presyn} = 30.5 in Fig. 4. Cross correlation between STN and STNP for analyzed values has been tabulated in Table 1.

Table 1 shows a nonlinearity as compared to the activity patterns showing linear behavior. At lag = 0, CCF is maximum for V_{prsyn} = 31.5. CCF is improved at V_{prsyn} = 31 at lag = 3. It is observed that in STN model there is a time lag in activity patterns in normal and PD condition. This information reveals a linear shift between STN and STNP and the cause to which is not yet identified.

Fig. 4 Discharge patterns generated in STN and STNP at $V_{presyn} = 30.5$

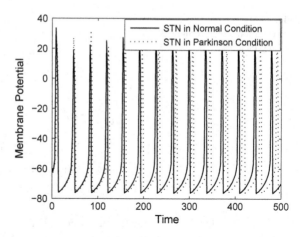

Table 1 Cross correlation function for V_{presyn}

Lag	CCF at $V_{prsyn} = 30.5$	CCF at $V_{prsyn} = 31$	CCF at $V_{prsyn} = 31.5$
0	0.06109	0.05992	0.063652
−1	−0.06389	−0.06501	−0.062725
−2	−0.1434	−0.1456	−0.14553
−3	−0.1613	−0.1631	−0.16363
−4	−0.1693	−0.171	−0.17221
1	0.1968	0.1932	0.19683
2	0.2592	0.2539	0.25803
3	0.2544	0.3037	0.28
4	0.2844	0.2895	0.28479

3.3 Sensitivity Analysis for Feedback Membrane Potential V_{fs}

In this section we have studied the effects on STN model by varying feedback membrane potential V_{fs}. However, the model study demonstrates the general pattern of the change: as the basal ganglia–thalamocortical feedback loop becomes stronger, oscillations are likely to occur. The phenomenon is robust with respect to different kinds of modulation of the dopamine-dependent parameters. The phenomenon is also robust with respect to different values of delays in the feedback loop. While the actual delays are not likely to change in Parkinson disease, they are not well known [27]. The model is somewhat stable for various values of V_{fs}. This is shown in Figs. 5 and 6.

Figures 5 and 6 show that cross correlation is constant for different time ranges and it increases with increase in feedback membrane potential.

Fig. 5 Discharge patterns generated in STN and STNP at $V_{fs} = 1 - 3$

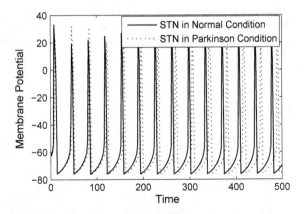

Fig. 6 Discharge patterns generated in STN and STNP at $V_{fs} = 1 - 3$

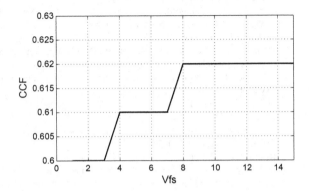

4 Conclusion

Discharge patterns of single-compartment model have been investigated in search of finding the parameters which alter the dynamics in Parkinson Disease. We compared the discharge patterns by analyzing the sensitivity in model due to applied current, presynaptic membrane potential, and feedback synaptic potential. The results of the above study show that STN model in both healthy and Parkinson condition is highly sensitive to various parameters. There is an increase in the discharge rate initially and then the discharge rate decreases irregularly. The two time series, i.e., STN in normal condition and STNP in Parkinson condition, have less cross correlation. Results show that the activity patterns are very sensitive to above-mentioned parameters and they also provide some insights into the disease. Results suggest that there are other intrinsic parameters which can alter the activity patterns in PD and might improve the altered PD patterns which are correlated to healthy primate.

Acknowledgements We thank Department of Science and Technology, Government of India for financial support vide Reference No. SR/CSRI/166/2014(G) under Cognitive Science Research Initiative (CSRI) to carry out this work. We are also thankful to Prof. Karmeshu for providing valuable inputs in this work.

References

1. Corinneeurrier, B., Patrice, C., Bernard, B., Constance, H.: Subthalamic nucleus neurons switch from single-spike activity to burst-firing mode. J. Neurosci. **19**(2), 599–609 (1999)
2. Kang, G., Lowery, M.: Conductance-based model of the basal ganglia in Parkinson's disease. ISSC. University College Dublin. June 10–11 (2009)
3. Santaniello, S., Fiengo, G., Glielmo, L., Grill, W.: Basal ganglia modeling in healthy and Parkinson's disease state. I. Isolated neurons activity. In: Proceedings of the American control conference, USA. July 11–13 (2007)
4. Bergman, H., Feingold, A., Nini, A., Raz, A., Slovin, H., Abeles, M., Vaadia, E.: Physiological aspects of information processing in the basal ganglia of normal and Parkinsonian primates. Trends Neurosci. **21**, 32–38 (1998)
5. Hammond, C., Bergman, H., Brown, P.: Pathological synchronization in Parkinson's disease: networks, models and treatments. Trends Neurosci. **30**, 357–364 (2007)
6. Jankovic, J.: Parkinson's disease: clinical features and diagnosis. J. Neurol. Neurosurg. Psychiatry **79**, 368–376 (2008)
7. Gelb, D.J., Oliver, E., Gilman, S.: Diagnostic criteria for Parkinson disease. Arch. Neurol. **56**, 33–39 (1999)
8. Jellinger, K.A.: Post mortem studies in Parkinson's disease: is it possible to detect brain areas for specific symptoms. J. Neural Transm. Suppl. **56**, 1–29 (1999)
9. Rivlin, M., Marmor, O., Heimer, G., Raz, A., Nini, A.: Basal ganglia oscillations and pathophysiology of movement disorders. Curr. Opin. Neurobiol. **16**, 629–637 (2006)
10. Moran, A., Bergman, H., Israel, Z., Bar, I.: Subthalamic nucleus functional organization revealed by parkinsonian neuronal oscillations and synchrony. Brain **131**, 3395–3409 (2008)
11. Wichmann, T., Kliem, M.A., Soares, J.: Slow oscillatory discharge in the primate basal ganglia. J. Neurophysiol. **87**, 1145–1148 (2002)
12. Weinberger, M., Hutchison, W.D., Lozano, A.M., Hodaie, M., Dostrovsky, J.O.: Increased gamma oscillatory activity in the subthalamic nucleus during tremor in Parkinson's disease patients. J. Neurophysiol. **101**, 789–802 (2009)
13. Deuschl, G., Raethjen, J., Baron, R., Lindemann, M., Wilms, H.: The pathophysiology of Parkinsonian tremor: a review. J. Neurol. **247**(5), V33–V48 (2000)
14. Llinas, R.: Rebound excitation as the physiological basis for tremor: a biophysical study of the oscillatory properties of mammalian central neurons in vitro. Movement Disorders 339–351 (1984)
15. Guehl, D., Pessiglione, M., Francois, C., Yelnik, J., Hirsch, E.C.: Tremorrelated activity of neurons in the 'motor' thalamus: changes in firing rate and pattern in the MPTP vervet model of parkinsonism. Eur. J. Neurosci. **17**, 2388–2400 (2003)
16. Pare, D., Curro, R., Steriade, M.: Neuronal basis of the Parkinsonia resting tremor: a hypothesis and its implications for treatment. Neuroscience **35**, 217–226 (1990)
17. Wichmann, T., DeLong, M.R.: Oscillations in the basal ganglia. Nature **400**, 621–622 (1999)
18. Zirh, T.A., Lenz, F.A., Reich, S.G., Dougherty, P.M.: Patterns of bursting occurring in thalamic cells during Parkinsonian tremor. Neuroscience **83**, 107–121 (1998)
19. Raethjen, J., Govindan, R.B., Muthuraman, M., Kopper, F., Volkmann, J.: Cortical correlates of the basic and first harmonic frequency of Parkinsonian tremor. Clin. Neurophysiol. **120**, 1866–1872 (2009)

20. Lenz, F., Kwan, H., Martin, R., Tasker, R.R., Dostrovsky, J.O.: Single unit analysis of the human ventral thalamic nuclear group. Tremor-related activity in functionally identified cells. Brain **117**, 531–543 (1994)
21. Volkmann, J., Joliot, M., Mogilner, A., Ioannides, A.A., Lado, F.: Central motor loop oscillations in Parkinsonian resting tremor revealed by magnetoencephalography. Neurology **46**, 1359–1370 (1996)
22. Surmeier, D.J., Mercer, J.N., Chan, C.S.: Autonomous pacemakers in the basal ganglia: who needs excitatory synapses anyway. Curr. Opin. Neurobiol. **15**, 312–318 (2005)
23. Bevan, M.D., Atherton, J.F., Baufreton, J.: Cellular principles underlying normal and pathological activity in the subthalamic nucleus. Curr. Opin. Neurobiol. **16**, 621–628 (2006)
24. Levy, R., Hutchison, W.D., Lozano, A.M., Dostrovsky, J.O.: High-frequency synchronization of neuronal activity in the subthalamic nucleus of Parkinsonian patients with limb tremor. J. Neurosci. **20**, 7766–7775 (2000)
25. Kitai, S.T., Deniau, J.M.: Cortical inputs to the subthalamus: intracellular analysis. Brain Res. **214**, 411–415 (1981)
26. Nambu, A., Tokuno, H., Hamada, I., Kita, H., Imanishi, M., Akazawa, T., Ikeuchi, Y., Hasegawa, N.: Excitatory cortical inputs to pallidal neurons via the subthalamic nucleus in the monkey. J. Neurophysiol. **84**, 289–300 (2000)
27. Dovzhenok, A., Rubchinsky, L.L.: On the origin of tremor in Parkinson's disease. PLoS ONE **7**(7), e41598 (2012)
28. Terman, D., Rubin, J.E., Yew, A.C., Wilson, C.J.: Activity patterns in a model for subthalamopallidal network of basal ganglia. J. Neurosci. **22**, 2963–2976 (2002)
29. Rubin, J.E., Terman, D.: High frequency stimulation of the subthalamic nucleus eliminates pathological thalamic rhythmicity in a computational model. J. Comput. Neurosci. **16**(3), 35–211 (2004)
30. Lau, L., Breteler, M.: Epidemiology of Parkinson disease. Neurology. **5**, 1362–1369 (2006)
31. Dorsey, E., Constantinescu, R., Thompson, J., Biglan, K., Holloway, R., Kieburtz, K., Marshall, F., Ravina, B., Schifitto, G., Siderowf, A., Tanner, C.: Projected number of people with Parkinson disease in the most populous nations. Neurology **68**, 384–386 (2007)

Emotion Recognition via EEG Using Neural Network Classifier

Rashima Mahajan

Abstract Automated assessment of human emotions via physiological signals has gained remarkable significance in the development of affective human–machine interfaces for stress detection. However, manual emotion analysis procedure is solely dependent upon the expertise of the analyst. An attempt has been made in this research to characterize and classify emotions from associated human neural responses via electroencephalography (EEG). Emotion-specific multichannel EEG dataset is acquired using 14-channel emotiv EEG neuroheadset. An efficient EEG signal analysis technique for human emotion classification using temporal and morphological features of EEG is presented. It uses power spectral and maximum/minimum peak features extracted from each EEG segment as an outcome to characterize emotion-specific EEG dynamics. A feed-forward neural network classifier is configured using Levenberg–Marquardt training algorithm to classify human emotions in two categories, viz, normal and stress states with classification accuracy of 60%. The experimental results reveal that the methodology adopted can further be explored to develop an automated clinical application to assist patients suffering from stress disorders with more efficient classification rates.

Keywords Emotion · Electroencephalogram (EEG) · Peak features
Neural network · Spectrum · Stress

1 Introduction

Noninvasive neural state recognition parameter electroencephalography (EEG) is being widely explored these days for automated assessment of neural disorders. Neural psychiatric and stress disorders are posing a serious health threat for elderly and medically challenged people worldwide; therefore, early and accurate diagnosis

R. Mahajan (✉)
Department of ECE, School of Engineering, G. D. Goenka University,
Sohna, Gurgaon, India
e-mail: rashima.mahajan@gdgoenka.ac.in

© Springer Nature Singapore Pte Ltd. 2018
M. Pant et al. (eds.), *Soft Computing: Theories and Applications*,
Advances in Intelligent Systems and Computing 583,
https://doi.org/10.1007/978-981-10-5687-1_38

of such disorders is the need of an hour. Physical counseling procedure may cause misdiagnosis and also delay diagnosis and management process. The primary step toward computer-aided recognition and classification of distinct human emotional states is an accurate and efficient analysis of correlated neural responses. An EEG waveform is composed of certain time domain and shape-related parameters. These EEG parameters indicate overall neural activity of brain and thus reflect the human psychological state. Any deviation from these standardized parameters for normal brain is an indication of neural disease. By capturing and characterizing these deviations in acquired EEG signals, it is possible to recognize the correlated neural state. However, these subtle variations in time domain are difficult to analyze visually by a naked eye. Hence, a need is there to develop computer-assisted EEG-based diagnostic tools for neural disorders that can help physicians to monitor subject's brain state accurately and act faster in the case of emergency conditions.

Since decades numerous techniques have been developed and utilized to map captured neural activity with a certain set of EEG features of interest, an extensive literature regarding emerging research in the field of affective computing using human neural responses has been explored. Focus has been paid to explore the ability of EEG signals to portray emotional activity of human subjects. The importance of emotional neuroscience and methods of detection of correlated activities is well discussed in the literature [1–6]. The progressive research in automated emotion analysis techniques provides a strong platform to revolutionize health care using biomedical informatics.

Extensive research has been published in the area of human brain mapping to analyze EEG signals using distinct set of linear and nonlinear features [7]. Wang et al. utilized frequency-domain features of EEG to classify emotions in four categories, viz, joy, relax, sad, and fear with considerably low accuracy [8]. A group of researchers assessed ongoing brain activity from composition of brain oscillations using probability-classification analysis of EEG spectral patterns [9]. The results obtained from assessment of these brain oscillations were twofold. It reflects neuronal synchronization and rhythmic level of neural activation for disease detection. However, this technique based on brain oscillations interaction fails to interpret the degree of severity and abnormality in neural disorders. The application of discrete wavelet transform (DWT) has been widely explored to accurately capture and localize EEG features in effective and disease-based brain mapping [10–12]. However, conventional DWT techniques lack in their performance due to considerably large number of computations. This can be overcome by using lifting-based wavelet transform instead of convolution-based implementations that help in increasing the computational efficiency up to 100% with finite precision [13]. Lot of researchers explored the use of event-related EEGs to investigate the emotional development by capturing corresponding neural variations using event-related potentials (ERPs) [14, 15]. A research has been also done to capture electrical activity of the brain using spectral analysis of acquired EEG signals in four EEG subbands, beta, alpha, theta, and delta [16–20].

Above findings reveal the discrimination ability of EEG, which can be explored further to map and capture neural variations during emotional elicitations. However, various EEG classification techniques are still lacking in their performance due to large feature vector used and thus response time of the system. Considering these facts, power spectral and peak-related features have been used in this research to characterize emotion-specific variations in the captured EEG signals. Once the feature mapping is done in time domain and frequency domain, the signals are needed to be classified. With the growth of computer-aided diagnostics, several decision support systems have been developed. Artificial neural networks (ANN) have shown remarkable classification performance where formulating an algorithmic solution is a cumbersome task [21]. In such scenarios, neural network-based classifiers are configured using extracted feature set by learning the desired mapping between input and output signals of the system.

In this research, an EEG-based BCI is developed to capture and characterize variations in human neural responses via EEG during emotion elicitations. An attempt has been made to classify human emotions using fusion of peak-related and power spectral features of acquired EEG. This hybrid feature vector is used to train a three-layered feed-forward neural network classifier. This configured classifier categorizes the captured human emotions in two states, viz, normal and stressed.

2 Materials and Methods

A block diagram of emotion recognition technique using neural network classifier is presented in Fig. 1. The whole methodology is composed of four stages, *i.e.,* acquisition of EEG signals, preprocessing, feature extraction from each EEG segment, followed by classification using neural network classifier. A preliminary work highlighting the detailed design considerations and power spectral analysis of acquired EEGs for emotion recognition using EEG-based effective brain mapping has been reported earlier [20]. The algorithm has been further extended to utilize power spectral analysis and peak-related features of EEG to classify emotional states using neural network classifier.

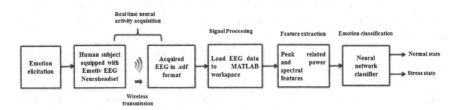

Fig. 1 Block diagram of emotion recognition via EEG using neural network classifier

2.1 EEG Signal Acquisition

Six subjects (3 females, 3 males), aged 40 ± 10 years, with no consumption of any medicine before the test, participated in this experiment to construct the required EEG signal dataset. Emotional elicitations were evoked using external music stimuli to three human subjects and self-recalling of a common depressive incident to the rest of three subjects [20]. The corresponding human EEG signals were captured from each subject using emotiv EEG neuroheadset unit. The EEG dataset is acquired at a sampling frequency of 128 Hz and is saved as .edf (European data format) file.

2.2 Preprocessing

The preprocessing of emotion-specific EEG dataset is done in such a way that so as to obtain a pass band of 0.25–50 Hz. This pass band is obtained since four EEG subbands are delta (δ, 0–4 Hz), theta (θ, 4–8 Hz), alpha (α, 8–12 Hz), and beta, β, with a range between 13 and 30 Hz. Thus, each EEG record is band pass filtered using a zero-phase FIR (finite impulse response) filter.

2.3 Feature Extraction

After preprocessing phase, a significant set of features is extracted which can accurately map the acquired EEG dataset into correlated emotional states. Two sets of features are extracted from each EEG record to characterize the related emotional activity. First set consists of four peak-related features corresponding to variation of peak values of the EEG signal. The four peak-related features extracted from each EEG record are number of positive peaks, number of negative peaks, average of positive peaks, and average of negative peaks. This first involves detection of the peak points in each EEG segment using a noise tolerant and fast peak finding MATLAB function *peakfinder*. It locates local maxima or minima point very accurately even in a noisy input signal. It returns the maxima (positive peaks) of data if input argument *extrema* > 0 and the minima (negative peaks) of data if *extrema* < 0. The peak points are located in each EEG segment of the acquired data and are plotted. Once the positive and negative peak points are located above four peak-related features are computed.

Other four features of the hybrid feature vector based on power spectrum are computed. These include mean spectral power estimation in delta (δ), theta (θ), alpha (α), and beta (β) EEG subbands. The subband frequencies are set for δ (0.5–4 Hz), θ (4–8 Hz), α (8–13 Hz), and β (13–30 Hz) bands, respectively. A power spectral density computation is done based on periodogram technique to extract mean EEG spectral power in each subband. Power spectral density is a measure of respective power strength at each frequency and is expressed as energy per unit frequency [22]. Each EEG is band pass filtered by configuring a FIR filter at 128 Hz sampling frequency using Hamming window function to determine the respective EEG subband power values.

Therefore, by combining peak-related features and mean power spectral features to form a hybrid feature vector, a robust emotion-specific feature extraction methodology is presented here. Thus, information of each EEG segment is stored as an eight-element vector, with first four elements representing peak-related features followed by four elements representing mean spectral power estimation in each EEG subband.

2.4 Classification of Emotional States

A three-layered feed-forward neural network classifier is configured using hybrid feature vector consisting of four peak-related and four mean power spectral-related features. These features are extracted from each EEG segment to classify human emotions into two states, viz, normal and stress states. The whole feature set is divided into training and testing sets. A three-layer feed-forward neural network with one input layer, one hidden layer, and one output layer is created using neural network toolbox peak-related feature vector of length four for each EEG segment followed by mean subband power spectrum feature vector of length four. The number of neurons in the hidden layer of the network is set manually to select the best classification results, whereas the output layer is composed of two neurons each representing the corresponding category of the associated emotional state as shown in Table 1. The training feature vector is fed to train the feed-forward neural network. The network is trained using a fast and reduced memory 'Levenberg Marquardt (trainlm)' back-propagation algorithm [21] in order to minimize the mean squared error (MSE). Once the training procedure is over, the network is capable of classifying the EEG signals efficiently in two emotional states. The neural network function 'Sim' is used to simulate the trained feed-forward neural network using the testing feature set in order to classify the input EEG segments.

The detailed procedure of peak-related feature set and mean power spectrum-based emotion classification using neural network classifier is depicted in the flow chart shown in Fig. 2.

Table 1 Neural network classifier output analysis

Output neurons		Corresponding emotion state
O_1	O_2	
1	1	Normal
0	0	Stressed

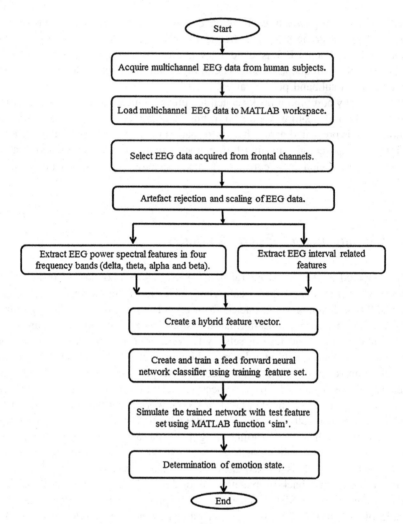

Fig. 2 Detailed flow chart of emotion recognition via EEG

3 Results and Discussion

The various results recorded from an experiment to map an emotional activity of human subjects with respective neural activation via electroencephalography are presented and described in this section. In this experiment, the EEG signals for each subject were recorded while eliciting their emotions using external music stimuli and self-recalling of a common incident. The-real time EEG signals (.edf) acquired through emotiv EEG neuroheadset are imported and analyzed in Matlab workspace.

At first the EEG signals are band pass filtered to obtain a pass band of 0.25–50 Hz using zero-phase FIR filter in Matlab. This is followed by the extraction of two sets of temporal and power spectral features. The temporal feature set is formed by extracting four peak-related features, i.e., number of positive peaks P_+, number of negative peaks P_-, average of positive peaks A_+, and average of negative peaks A_-. It is constituted by first locating the positive and negative peak points in the acquired emotion-specific EEG records. The peak points are located using Matlab function '*peakfinder*'. The inputs sample EEG record and corresponding peaks are plotted in Figs. 3 and 4, respectively. Once the peak points are identified, a total number of positive and negative peaks along with their average are computed. This forms the four-element peak-related feature set.

Another set of power spectral features is also extracted from acquired emotion-specific EEG dataset. Four EEG power spectral features are computed using periodogram technique by setting subband frequencies for δ (0.5–4 Hz), θ (4–8 Hz), α (8–13 Hz), and β (13–30 Hz) bands, respectively. This forms the four-element

Fig. 3 Input EEG record

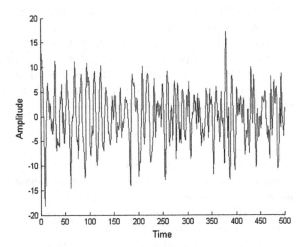

Fig. 4 Positive and negative peaks located in acquired EEG signal

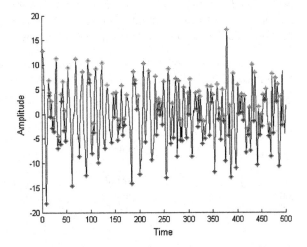

power spectral feature set consisting of mean spectral power estimation in delta (P_δ), theta (P_θ), alpha (P_α), and beta (P_β) EEG subbands. Thus, information of each EEG record is stored as an eight-element vector, with first four elements representing temporal peaking parameters followed by four elements representing power spectral parameters. This leads to a formation of compact eight- dimensional *EEG_hybrid feature vector*/EEG record as [P_+, P_-, A_+, A_-, P_δ, P_θ, P_α, P_β].

The whole extracted feature set from all EEG records is partitioned into training and testing feature set. A three-layer feed-forward neural network classifier is configured with eight input neurons and two output neurons. The 80% of dataset has been used to train the three-layer feed-forward neural network classifier. During testing phase, the classifier is tested with the 20% of the left dataset. The number of hidden layer neurons is finalized after running several simulations with a different number of hidden layer neurons, e.g, 10, 15, 20, 25, 30, 35, etc., during the training phase using '(trainlm)' algorithm. Comparison of various data obtained reveal that the best performance is obtained corresponding to the minimum MSE (mean square error) using 25 neurons.

Finally, the performance and efficiency of the proposed EEG-based emotion classification technique based on hybrid feature set are evaluated in terms of overall accuracy and the error rate. Accuracy is defined as the ratio of the number of correctly classified segments divided by the total number of segments. The corresponding error rate can be calculated as

$$\text{Error rate} = 1 - \text{Accuracy}. \tag{1}$$

The values of accuracy and error rate of the proposed EEG-based emotion classification technique obtained during testing phase using six different values of neurons in the hidden layer are tabulated in Table 2. The best classification accuracy of 60% is achieved in classifying two emotion states using proposed method with 25 neurons in the hidden layer.

Analysis of overall findings reveals that the compact feature set comprising four peak-related and four power spectral features can be utilized to recognize human emotional states from acquired EEG. However, a need is there to improve classification accuracy percentage. Further, the results are obtained from multiple EEG

Table 2 Output classification accuracy and error rate

Number of hidden layer neurons	Training algorithm	Classification accuracy	Error rate
10	trainlm	0.5211	0.4789
15	trainlm	0.5514	0.4486
20	trainlm	0.5730	0.4270
25	**trainlm**	**0.6021**	**0.3979**
30	trainlm	0.5883	0.4117
35	trainlm	0.5691	0.4309

datasets acquired from only six human subjects. Thus, efforts are underway to validate the results on extended EEG dataset with more efficient signal analysis and classification techniques.

4 Conclusion

Automated analysis and classification of human emotions such as stress state using physiological signals like EEG are the need of an hour. Since manual stress analysis is solely dependent upon the behavior analysis of the stressed person by peers and expertize of the psychiatrist, a technique for automated classification of human emotions using neural network classifier has been presented. It uses peak-related and power spectral features of EEG. The simulation results show that the developed technique is able to attain an overall classification accuracy of 60% with a small feature set when 25 neurons are used in the hidden layer of the feed-forward neural network classifier. This method has a potential to be developed as an automated emotion assessment tool for the detection of stress-related emotions, however, with certain modifications. The scope of this method is to validate the results on extended EEG dataset with more efficient signal analysis and classification techniques.

References

1. Sakkalis, V.: Review of advanced techniques for the estimation of brain connectivity measured with EEG/MEG. Comput. Biol. Med. **41**(12), 1110–1117 (2011)
2. Li, X., Hu, B., Zhu, T., Yan, J., Zheng, F.: Towards affective learning with an EEG feedback approach. In: Proceedings of 1st ACM International Workshop on Multimedia Technologies for Distance Learning, pp. 33–38. ACM (2009)
3. Petrantonakis, P., Hadjileontiadis, L.: A novel emotion elicitation index using frontal brain asymmetry for enhanced EEG-based emotion recognition. IEEE Trans. Inf Technol. Biomed. **15**, 737–746 (2011)
4. Kaiser, J.: Self-initiation of EEG-based communication in paralyzed patients. Clin. Neurophysiol. **112**, 551–554 (2001)
5. Subha, D.P., Joseph, P.K., Acharya, R., Lim, C.M.: EEG signal analysis: a survey. J. Med. Syst. **34**(2), 195–212 (2010)
6. Mahajan, R., Bansal, D., Singh, S.: A real time set up for retrieval of emotional states from human neural responses. Int. J. Med. Pharm. Sci. Eng. **8**(3), 16–21 (2014)
7. Schwilden, H.: Concepts of EEG processing: from power spectrum to bispectrum, fractals, entropies and all that. Best Pract. Res. Clin. Anaesthesiol. **20**(1), 31–48 (2006)
8. Wang, X.W., Nie, D., Lu, B.L.: EEG-Based Emotion Recognition Using Frequency Domain Features and Support Vector Machines, ICONIP 2011, Part I, LNCS 7062, pp. 734–743
9. Fingelkurts, A.A., Rytsala, H., Suominen, K., Isometsa, E., Kahkonen, S.: Composition of brain oscillations in ongoing EEG during major depression disorder. Neurosci. Res. **56**(2), 133–144 (2006)

10. Murugappan, M., Nagarajan, R., Yaacob, S.: Comparison of different wavelet features from EEG signals for classifying human emotions. In: Proceedings of IEEE Symposium on Industrial Electronics and Applications. pp. 836–841. Kuala Lumpur (2009)
11. Murugappan, M., Nagarajan, R., Yaacob, S.: Combining spatial filtering and wavelet transform for classifying human emotions using EEG signals. J. Med. Biol. Eng. **31**(1), 45–51 (2010)
12. Ubeyli, E.D.: Combined neural network model employing wavelet coefficients for EEG signal classification. Digit. Signal. Process **19**, 297–308 (2009)
13. Subasi, A., Ercelebi, E.: Classification of EEG signals using neural network and logistic regression. Comput. Methods Programs Biomed. **78**, 87–99 (2005)
14. Cong, F., Phan, A.H., Zhao, Q., Huttunen-Scott, T., Kaartinen, J., Ristaniemi, T.: Benefits of multi-domain feature of mismatch negativity extracted by non-negative tensor factorization from EEG collected by low-density array. Int. J. Neural Syst. **22**(6), 1250025, pp. 1–19 (2012)
15. Zhao, Q., Sun, J., Cong, F., Chen, S., Tang, Y., Tong, S.: Multidomain Feature Analysis for Depression: a Study of N170 in Time, Frequency and Spatial Domains Simultaneously", The 35th Annual International Conference of the IEEE Engineering in Medicine and Biology Society, pp. 5986–5989. Japan, Osaka (2013)
16. Chanel, G., Rebetez, C., Bétrancourt, M., Pun, T.: Emotion assessment from physiological signals for adaptation of game difficulty. IEEE Trans. Syst. Man Cybern. **41**(6), 1052–1063 (2011)
17. Kawasaki, M., Karashima, M., Saito, M.: Effect of emotional changes induced by sounds on human frontal alpha-wave during verbal and non-verbal tasks. Int. J. Biomed. Eng. Technol. **2** (3), 191–202 (2009)
18. Gandhi, T., Kapoor, A., Kharya, C., Aalok, V.V., Santhosh, J., Anand, S.: Enhancement of inter-hemispheric brain waves synchronisation after Pranayama practice. Int. J. Biomed. Eng. Technol. **7**(1), 1–17 (2011)
19. Wang, X.W., Nie, D., Lu, B.L.: EEG-based emotion recognition using frequency domain features and support vector machines. Neural Inf. Process. Berlin, Germany: Springer **7062**, 734–743 (2011)
20. Mahajan, R., Bansal, D.: Depression Diagnosis and management using EEG based affective brain mapping in real time. Int. J. Biomed. Eng. Technol. **18**(2), 115–137 (2015)
21. Hagan, M.T., Menhaj, M.B.: Training feedforward networks with the Marquardt Algorithm. IEEE Trans. Neural Networks **5**, 989–993 (1994)
22. Sanei, S.: Chambers, J.: EEG Signal Processing, p. 55. Wiley (2007)

Trust-Enhanced Multi-criteria Recommender System

Arpit Goswami, Pragya Dwivedi and Vibhor Kant

Abstract Recommender system aims to solve the information overload problem by recommending a set of items that are suitable for users. Recently, the incorporation of multiple criteria into traditional single-criterion recommender system has increased the interest. In this paper, we propose a novel trust-enhanced multi-criteria recommender system using fuzzy rating in collaborative filtering framework. We have also designed a hybrid approach of traditional multi-criteria recommender system and trust-enhanced multi-criteria recommender system to reduce data sparsity problem. The empirical results show that our proposed approach demonstrates efficient recommendation as compared to traditional approach.

Keywords Multi-criteria recommender system · Trust · Collaborative filtering Fuzzy sets

1 Introduction

In the age of the Internet, the amount of information increased exponentially which leads to information overload problem. It is not easy to retrieve information from such a huge collection with traditional information filtering system. Recommender system (RS) provides a solution to the information overload problem by suggesting

A. Goswami (✉) · P. Dwivedi
Department of Computer Science, Motilal Nehru National Institute of Technology,
Allahabad 211004, UP, India
e-mail: goswami.arpit475@gmail.com

P. Dwivedi
e-mail: pragyadwi86@mnnit.ac.in

V. Kant
Department of Computer Science, The LNMIIT, Jaipur 302031, Rajasthan, India
e-mail: vibhor.kant@gmail.com

© Springer Nature Singapore Pte Ltd. 2018
M. Pant et al. (eds.), *Soft Computing: Theories and Applications*,
Advances in Intelligent Systems and Computing 583,
https://doi.org/10.1007/978-981-10-5687-1_39

the items and ensuring that customer will like them. RS are widely used in web applications dealing with news [1], movies [2], books [3], and e-learning [4, 5].

Collaborative filtering (CF) is the best-known technique for generating recommendations to users. Sometimes, it is not easy to present user likeness in terms of single ratings, for example, a user likes a movie based on its 'story' but not on the 'visuals'. It means that overall rating for a movie does not reflect the real situation. Therefore, incorporation of multiple criteria ratings into traditional recommender systems provides a way of reflecting user preferences efficiently. Recommendation generation in multi-criteria recommender system (MCRS) may be difficult through traditional CF because it may have data sparsity problem [6].

To overcome the problem of data sparsity, trust between the users is used to increase the overlapping between the user profiles which in turn will help to provide better recommendation [7]. However, trust computation among users is quite a challenging task based on available ratings. Since trust is a subjective notion, it can be measured efficiently through fuzzy ratings. Therefore, we incorporate trust based on fuzzy ratings into MCRS. In this paper, we proposed a system that utilizes the trust-based CF in MCRS which helps in dealing with trust vagueness among users and data sparsity problem:

- First of all, we have developed traditional MCRS (Tr_MCRS) in collaborative filtering framework.
- Second, trust-enhanced MCRS (FT_MCRS) is developed in which trust is computed on each criterion using fuzzy ratings and we consider trust as a weight on each criterion in prediction task. After that we compute overall rating by aggregation technique.
- Finally, we have developed hybrid trust-enhanced recommender system based on Tr_MCRS and FT_MCRS.

The remainder of the paper is organized as follows: Sect. 2 briefly describes multi-criteria recommender system, collaborative filtering, and trust-based recommender system. In Sect. 3, the proposed scheme is described. Experimental evaluation and result analysis have been done in Sect. 4, and finally, Sect. 5 illustrates the conclusion and future research directions.

2 Background and Related Work

2.1 Trust-Based Recommender System

Recently, researchers have argued that people rely more on the recommendation derived from trusted users directly [8]. In a trust-based recommender system, recommendations are derived from the users either directly trusted or indirectly trusted by an active user. Trust-based recommender system has greater recommendation efficiency than simple CF-based techniques due to more overlapping between the users. Popular method for trust calculation has been defined in [9], where trust is calculated directly from the ratings. The problem with the direct calculation is that it does not take care of the mood of user. In real-life scenario, some people give a rating 5 (on a scale of 1–5) to the movies they like and other tend to give 4 to the movies they like. So in a real sense, there is no difference between ratings 4 and 5. Therefore, trust computation among users may be efficient if we incorporate fuzzy ratings for dealing with subjective uncertainty associated with user ratings.

3 Proposed Hybrid Multi-criteria Recommendation Framework (FT_Tr_MCRS)

In this section, first we discuss traditional multi-criteria approach and then hybrid approach based on traditional MCRS and trust-enhanced MCRS using fuzzy ratings are discussed.

Multi-criteria RS: Let $M = \{u_1, u_2, u_3, ..., u_m\}$ be the set of users and $N = \{i_1, i_2, i_3, ..., i_n\}$ be the set of items and $C = \{c_1, c_2, c_3, ..., c_x\}$ be the set of criteria in which ratings are provided. The multi-criteria rating on item i by user u is defined as a vector $[c_1(i), c_2(i), ..., c_x(i)]$. Table 1 represents a data matrix in the context of MCRS.

The proposed recommendation framework has the following three modules: Tr_MCRS, FT_MCRS, and FT_Tr_MCRS as described below.

3.1 Tr_MCRS in Collaborative Filtering Framework

For developing Tr_MCRS based on CF module, first we will partition the multi-criteria dataset into individual criterion datasets and then CF techniques will be applied to these individual criterion datasets for predicting unknown criterion ratings. After the prediction of rating on each criterion, aggregation function is used for computing overall predicted rating. CF is a three-step process which comprises similarity computation, neighborhood generation, and rating prediction.

Similarity computation: This step will calculate the similarity between the active user a and the other potential users. Similarity, $Wc(x)$ (a, u), will be calculated on each criterion separately using Cosine similarity on the basis of the co-rated items as given in Eq. (1):

Table 1 MCRS ratings

	Item(I_1)	Item(I_2)	Item(I_3)	Item(I_4)
User(U_1)	$5_{4,4,4,3}$	$4_{2,2,4,3}$	$3_{1,2,3,4}$	??
User(U_2)	$5_{3,3,4,4}$	$4_{4,4,4,4}$	$2_{3,3,2,2}$	$4_{4,4,3,3}$
User(U_3)	$4_{2,3,4,2}$	$3_{2,3,2,3}$	$4_{4,3,4,2}$	$5_{5,5,5,5}$
User(U_4)	$4_{4,4,4,3}$	$2_{2,2,4,3}$	$4_{1,2,3,4}$	$5_{5,4,4,5}$

$$W_{r(x)} = \frac{\sum_{i \in S_a \cap S_u} r_{a,i}^{c(x)} \times r_{u,i}^{c(x)}}{\sqrt{\sum_{i \in S_a \cap S_u} \left(r_{a,i}^{c(x)}\right)^2 \sum_{i \in S_a \cap S_u} \left(r_{u,i}^{c(x)}\right)^2}}, \tag{1}$$

where S_a is the set of items rated by user a, $r_{a,i}^{c(x)}$ is the rating provided by the user a on item i for the criteria x, S_u is the set of items rated by user u, and $r_{u,i}^{c(x)}$ is the rating provided by user u on item i for the criteria x

Neighborhood generation: After similarity computing between users, most k similar users for active user a will be kept in the neighborhood set by using either k-nearest neighbor approach or based on some threshold value. In our work, we have estimated the appropriate value of k empirically.

Rating prediction using CF: On the basis of these most similar k users for each criterion, ratings for each criterion will be computed using Resnick's prediction method [10]:

$$P_{a,i}'^{c(x)} = \overline{u_a^{c(x)}} \frac{\sum_{u \in U | i \in S_u} W_{c(x)}^{(a,u)} \times \left(r_{u,i}^{c(x)} - \overline{u_a^{c(x)}}\right)}{\sum_{u \in U | i \in S_u} \left|W_{c(x)}^{(a,u)}\right|}, \tag{2}$$

where $u^{c(x)}$ is the average rating of the user u for item i on criteria $c(x)$.

Aggregation of predicted ratings: Aggregated rating will be calculated as an average of the predicted ratings on each criterion:

$$P_{a,i}^{CF} = \frac{1}{n} \sum_{x=1}^{n} P_{a,i}'^{c(x)}, \tag{3}$$

where n is the total number of criteria.

3.2   FT_MCRS in Collaborative Filtering Framework

Trust can be used to increase the overlap between the user profiles which in turn will help to recommend more accurate items to an active user. Here we assume the ratings provided by users are fuzzy in nature because it depends on the mood of the user. Sometimes, a user can give a rating 5 (on a scale of 5) to the best movie and sometimes user gives it as 4. In reality, we cannot distinguish sharply the ratings 4 and 5 for a good movie [11].

Trust computation: Half triangular fuzzy number is used to represent the degree of user's (u) experience on item i (shown in Fig. 1). The half triangular fuzzy membership function is defined as follows:

Fig. 1 Membership function
for fuzzy set L

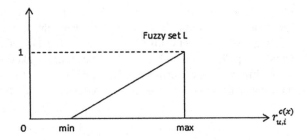

$$\mu_L(i)^{c(x)} = \begin{cases} 0 : & r_{u,i}^{c(x)} = \min \\ \frac{r_{u,i}^{c(x)} - \min}{\max - \min} : & \min < r_{u,i}^{c(x)} < \min \cdot \\ 1 : & r_{u,i}^{c(x)} = \max \end{cases} \quad (4)$$

In next step, the items will be classified into three sets namely preferred items (Pr_Items), non-preferred items (NPr_Items), and indifferent items (In_Items) based on fuzzy ratings calculated in the previous step:

$$\text{Pr\_Items}^{c(x)} = i : \mu_L(i)^{c(x)} > 0.5$$
$$\text{In\_Items}^{c(x)} = i : \mu_L(i)^{c(x)} = 0.5 \quad (5)$$
$$\text{NPr\_Items}^{c(x)} = i : \mu_L(i)^{c(x)} < 0.5$$

Now, trust value between user a and user u for each criterion $c(x)$ can be calculated as

$$\text{Trust}^{c(x)}(a, u) = \frac{1}{2} \left[\frac{\left| \text{Pr\_Items}_a^{c(x)} \cap \text{Pr\_Items}_u^{c(x)} \right|}{\text{Pr\_Items}_a^{c(x)}} + \frac{\left| \text{NPr\_Items}_a^{c(x)} \cap N \ \text{Pr\_Items}_u^{c(x)} \right|}{\text{NPr\_Items}_a^{c(x)}} \right].$$
$$(6)$$

Trust propagation: Trust propagation is used to calculate the indirect trusted between users. It helps to reduce the data sparsity problem. In our work, we have used max–min approach to calculate indirect trust between users as shown in Fig. 2.

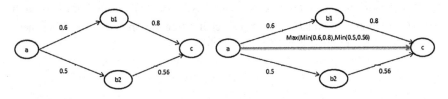

Fig. 2 Trust network

Figure 2 the trust networks in which edges show the trust values between users and these values denote the level of trust between those users.

Neighborhood generation: Neighbors will be selected on the basis of trust values. Top-N users with high trust values will be taken as neighbors to active user.

Rating prediction using trust: In this step, trust is used as a weight to predict the rating of the item which can be calculated as

$$P_{a,i}^{\prime c(x)} = \overline{u_a^{c(x)}} + \frac{\sum_{u \in U | i \in S_u} \text{Trust}_{c(x)}^{(a,u)} \times \left(r_{a,i}^{c(x)} - \overline{u_a^{c(x)}} \right)}{\sum_{u \in U | i \in S_u} \left| \text{Trust}_{c(x)}^{(a,u)} \right|}. \tag{7}$$

Aggregation of predicted ratings: Aggregated rating will be calculated as an average of the ratings on each criterion:

$$P_{a,i}^T = \frac{1}{n} \sum_{x=1}^{n} P_{a,i}^{\prime c(x)}. \tag{8}$$

where n is the total number of criteria.

3.3 Hybrid FT_Tr_MCRS in Collaborative Filtering Framework

Past research has proved that the incorporation of trust in the recommendation task considerably improves the accuracy of the system while reducing the coverage. We propose a hybrid system that enhances accuracy without compromising coverage. Our hybrid system combines two recommendation approaches including Tr_MCRS and FT_MCRS. By considering the possible advantages of both these approaches, the hybrid predicted rating is as follows:

$$P_{a,i} = \begin{cases} 0: & P_{a,i}^{CF} = 0 \ \& \ P_{a,i}^T = 0 \\ P_{a,i}^T: & P_{a,i}^{CF} = 0 \ \& \ P_{a,i}^T \neq 0 \\ P_{a,i}^{CF}: & P_{a,i}^{CF} \neq 0 \ \& \ P_{a,i}^T = 0 \\ \frac{2 \times P_{a,i}^{CF} \times P_{a,i}^T}{P_{a,i}^{CF} + P_{a,i}^T}: & P_{a,i}^{CF} \neq 0 \ \& \ P_{a,i}^T \neq 0 \end{cases} \tag{8}$$

where $P_{a,i}^{CF}$ is the predicted rating for an active user a on item i using Eq. (3) through Tr_MCRS approach, and $P_{a,i}^T$ is the predicted rating for an active user a on item i using Eq. (8) through FT_MCRS approach.

4 Experimental Evaluation

4.1 Experimental Setup

Several experiments were carried out on the Yahoo's movies rating dataset which contains ratings for a movie on four criteria and an overall rating. For experimental purpose the dataset is divided into training (80%) and testing (20%) datasets. The quality of recommendation has been evaluated in terms of evaluation matrices like mean absolute error (MAE) and coverage.

MAE is most widely used performance metric in recommendation task. MAE is calculated as the average absolute deviation between the predicted overall rating $p^{x(0)}$ and the actual overall rating $r^{x(0)}$ assigned by the user:

$$\text{MAE} = \frac{1}{n} \sum_{i=1}^{n} \left| r_{a,i}^{x(0)} - r_{p,i}^{x(0)} \right|. \tag{9}$$

The coverage evaluates the ability of any RS to generate effective recommendation. It is calculated as the percentage of the predicted items to the total items. Let n be the total items and Ip be the set of predicted items:

$$\text{coverage} = \frac{I_p}{n}. \tag{10}$$

4.2 Experiment 1

To demonstrate the effectiveness of proposed FT_Tr_MCRS and FT_MCRS, we compare these approaches with traditional Tr_MCRS via MAE and coverage without using trust propagation mechanism for this experiment. Figures 3 and 4 show the MAE and coverage comparison on different neighborhood sizes (10–50) with an increment of 10. The results presented in Figs. 3 and 4 demonstrate that the proposed FT_Tr_MCRS approach is better than the FT_MCRS approach and traditional Tr_MCRS in terms of coverage. MAE of proposed FT_MCRS is outperformed Tr_MCRS as well as FT_Tr_MCRS. These experiments are performed by taking various neighborhood sizes (10–50).

4.3 Experiment 2

To illustrate the behavior of trust propagation approach in MCRS and for enhancing coverage of fuzzy trust-enhanced MCRS, we compare all approaches via MAE and coverage using propagation mechanism in this experiment. Figures 5 and 6 show

Fig. 3 Mean absolute error (MAE) comparison among Tr_MCRS, FT_MCRS and FT_Tr_MCRS without trust propagation

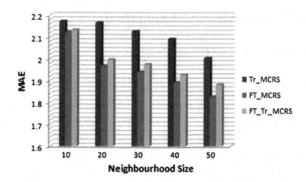

Fig. 4 Coverage comparison among Tr_MCRS, FT_MCRS, and FT_Tr_MCRS without trust propagation

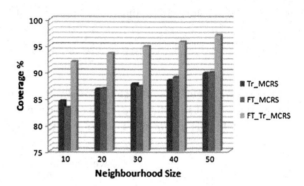

the MAE and coverage comparison on different neighborhood sizes (10–50) with an increment of 10. The results presented in Figs. 5 and 6 demonstrate that the proposed FT_Tr_MCRS approach is better than the FT_MCRS approach and traditional Tr_MCRS in terms of coverage and MAE. These experiments are performed by taking various neighborhood sizes (10–50).

Fig. 5 Mean absolute error (MAE) comparison among Tr_MCRS, FT_MCRS and FT _Tr_MCRS using trust propagation

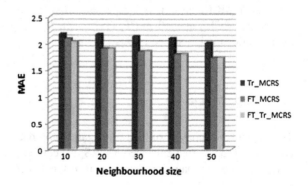

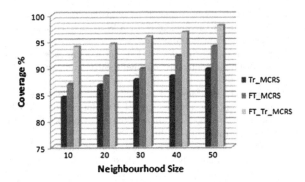

Fig. 6 Coverage comparison among Tr_MCRS, FT_MCRS, and FT_Tr_MCRS using trust propagation

5 Conclusion

This is an attempt toward enhancing the accuracy of MCRS by introducing trust derived from fuzzy ratings into MCRS. In this paper, first we have proposed a trust-enhanced MCRS (FT_MCRS) approach in which trust is used as a weight in the prediction process. By combining traditional MCRS and FT_MCRS, a new hybrid approach FT_Tr_MCRS is proposed. Experimental results reveal that the incorporation of the trust in MCRS has significantly outperformed other traditional MCRS in terms of MAE as well as coverage. Trust propagation in FT_MCRS is used which has increased the coverage and the MAE of the scheme and the hybrid scheme FT_Tr_MCRS also performed better in terms of coverage and MAE after trust propagation. Our proposed work gives an insight into dealing with data sparsity problem by incorporating trust values through trust propagation scheme.

References

1. IJntema, W., Goossen, F., Frasincar, F., Hogenboom, F.: Ontology-based news recommendation. In: Proceedings of the 2010 EDBT/ICDT Workshops, EDBT'10, pp. 16:1–16:6, New York, USA (2010)
2. Kant, V., Bharadwaj, K.K.: Integrating collaborative and reclusive methods for effective recommendations: a fuzzy bayesian approach. Int. J. Intell. Syst. **28**(11), 1099–1123 (2013)
3. Mooney, R.J., Roy, L.: Content-based book recommending using learning for text categorization. In: Proceedings of the Fifth ACM Conference on Digital Libraries, DL '00, pp. 195–204, New York, USA (2000)
4. Dwivedi, P., Bharadwaj, K.K.: Effective trust-aware e-learning recommender system based on learning styles and knowledge levels. Edu. Tech. Soc. **16**(4), 201–216 (2013)
5. Dwivedi, P.: Bharadwaj, K.K.: e-learning recommender system for a group of learners based on the unified learner profile approach. Expert Syst. **32**(2), 264–276 (2015)
6. Adomavicius, G., Manouselis, N., Kwon, Y.: Multi-criteria recommender systems. In: Recommender Systems Handbook, pp. 769–803. Springer US, Boston, MA (2011)
7. Adomavicius, G., Tuzhilin, A.: Toward the next generation of recommender sys- tems: a survey of the state-of-the-art and possible extensions. IEEE Trans. Knowl. Data Eng. **17**(6), 734–749 (2005)

8. Sinha, R.R., Swearingen, K.: Comparing recommendations made by online systems and friends. In: Proceedings of DELOS Workshop: Personalisation and Recommender Systems in Digital Libraries (2001)

9. O'Donovan, J., Smyth, B.: Trust in recommender systems. In: Proceedings of the 10th International Conference on Intelligent User Interfaces, IUI '05, pp. 167–174, New York, USA (2005)

10. Candillier, L., Meyer, F., Boullé, M.: Machine learning and data mining in pattern recognition. In: Proceedings of 5th International Conference, MLDM 2007, Leipzig, Germany. Comparing State-of-the-Art Collaborative Filtering Systems, pp. 548–562. Springer, Berlin, 18–20 July 2007

11. Kant, V., Bharadwaj, K.K.: Fuzzy computational models of trust and distrust for enhanced recommendations. Int. J. Intell. Syst. **28**(4), 332–365 (2013)

Application of PSO Clustering for Selection of Chemical Interface Materials for Sensor Array Electronic Nose

T. Sonamani Singh and R.D.S. Yadava

Abstract In this study PSO has been applied for mining the thermodynamic data on vapor–polymer solvation interactions. The goal of this study is to make polymer selection for the surface acoustic wave (SAW) chemical sensors for electronic nose applications. An electronic nose sensor array is required to generate varying signal patterns corresponding to different vapor types, and the sensor array data is analyzed by pattern recognition methods for extracting specific vapor identities. In this work we considered a specific detection problem, namely, the detection of freshness or spoilage states of fish as food product. Considering the solvation data for 26 potential polymers and 17 likely vapors in the headspace of fish samples, the application of PSO resulted in a set of six polymers for defining the SAW sensor array. The PSO selection was validated by generating simulation data based on a SAW sensor model and analyzing the 6-element sensor array patterns by principal component analysis (PCA).

Keywords Particle swarm optimization · Polymer selection for chemical sensor array · Surface acoustic wave electronic nose · Detection of fish spoilage

1 Introduction

Electronic nose (E-nose) is an odor recognition instrument that mimics the human nose by vapor sniffing, signal processing, and pattern analysis analogous to olfactory receptor neurons, olfactory bulb, and brain, respectively, in human smell sensing organ [1]. E-nose basically comprises three major parts: first, a set of broadly selective chemical vapor sensors (called sensor array) which generates

T. Sonamani Singh (✉) · R.D.S. Yadava
Sensors & Signal Processing Laboratory, Department of Physics,
Institute of Science, Banaras Hindu University, Varanasi 221005, India
e-mail: sonamani.2065@gmail.com

R.D.S. Yadava
e-mail: ardius@gmail.com

© Springer Nature Singapore Pte Ltd. 2018
M. Pant et al. (eds.), *Soft Computing: Theories and Applications*,
Advances in Intelligent Systems and Computing 583,
https://doi.org/10.1007/978-981-10-5687-1_40

signal patterns encoded with odor identities; second, a signal processing unit which applies data processing methods like normalization, scaling, and principal component analysis (PCA) for extraction of identity defining mathematical descriptors (called odor signature or fingerprint); and third, a pattern recognition unit that applies the pattern recognition methods like artificial neural network (ANN), fuzzy inference system (FIS), and/or genetic algorithm (GA) for classification of vapor samples according to their identities. In designing an E-nose each component needs to be optimized individually and cooperatively to achieve efficient E-nose system. An important issue in designing this system is the selection of an optimal set of chemically selective interface material that when deposited on sensors surfaces selectively sorbs different chemical analytes in the vapor samples to generate vapor discriminating sensor output patterns. The objective in this selection process is to search for a minimal set of coating materials that can produce maximally diverse patterns for different target vapor types.

In this paper we have considered polymers as sensor coating material, which are commonly used for chemical functionalization of microelectronic sensors like surface acoustic wave (SAW) sensors and microelectromechanical system (MEMS) sensors. The common approach for selecting polymeric coatings to fabricate a large number of sensors using various prospective polymers, and then evaluate and categorize them by monitoring their performance based on some predefined odor identification task. This approach is both time and cost expensive [2]. To reduce this burden of polymer selection the authors' group has been investigating an alternative approach in which different data mining methods were used to seek an optimal set from a large pool of prospective polymers [3–7]. In this data mining exercise, a database of thermodynamic partition coefficients for various vapor–polymer combinations were calculated by the linear solvation energy relationship (LSER) and vapor and polymer solvation parameters [2, 7]. The partition coefficient (K) quantifies partitioning of a chemical analyte from the vapor phase into the polymer phase in thermodynamic equilibrium. It is defined as $K = C_P/C_V$ where C_P and C_V are concentrations of vapor analyte in polymer phase and in vapor phase, respectively. The partition coefficient data is prepared in the form of a data matrix with vapors in rows and polymers in columns, referred to as K-matrix. The transpose of this matrix (called K^T-matrix with polymers in rows and vapors in columns) is taken as the multivariate data where the polymers are represented as data vectors. This K^T-matrix is then analyzed by applying some data clustering technique which segregates the polymers into clusters according to some measure of similarity among them. Each polymer in a cluster has similar characteristics of interaction with the target vapor analytes. Thus, by selecting a polymer from each cluster we get a set of maximally dissimilar polymers interaction affinities for target vapor.

In earlier studies, the principal component analysis (PCA), the hierarchical clustering, the fuzzy c-means clustering (FCM), the fuzzy subtractive clustering (FSC), and their combination were used for making polymer selection [3–8].

In these studies the discrimination ability of the selected polymers was validated by using virtual SAW sensor arrays for generating synthetic data for some predefined identification tasks. Recently, an investigation validating the FCM-based polymer selection for fish freshness/spoilage detection has also been approached by using virtual MEMS sensor array [9]. In this work, we explore the particle swarm optimization (PSO) as a clustering method for this purpose with an aim to assess its relative suitability in comparison with previously used methods. In order to validate the PSO selection process we generated synthetic data set by employing a SAW sensor array model targeting detection of the freshness and spoilage of fish as a food product.

2 Partitions Coefficient Data (*K*-Matrix)

E-nose as an instrument for food quality monitoring has gained acceptance due to its rapid real-time performance. Many reports on the application of E-nose for fish freshness or spoilage monitoring are available [10, 11]. We also took up this case for the purpose of the present study as the information about volatile species in the headspace of fish samples appears to be well documented, consistent, and reliable. The list of likely volatiles, interferents, and prospective sensing polymers are already presented in detail in [7]. Therefore, for our present analysis we used the K^T-matrix given as Table 5 in [7].

3 Particle Swarm Optimization (PSO) Clustering

Particle swarm optimization (PSO) is population-based swarm intelligence (SI) algorithm which was initially developed for studying the patterns of flocking of birds [12]. In PSO algorithm the potential solutions to the optimization problem are represented by particles, and they are flown in multidimensional search space base on social-psychological tendency to emulate the success (value of the fitness function) of the neighboring individuals (particles) and their own successes, to search for the optimal region (solution of optimization problem) in the search space. The movement or position update of ith particle in N_d-dimensional search space is done by using the relation

$$v_{ij}(t+1) = wv_{ij}(t) + c_1 r_{1j}(t)(y_{ij}(t) - x_{ij}(t)) + c_2 r_{2j}(t)(\hat{y}_j(t) - x_{ij}(t)) \quad (1)$$

$$x_{ij}(t+1) = x_{ij}(t) + v_{ij}(t+1), \quad (2)$$

where v_{ij} is the velocity of ith particle in jth dimension ($j = 1, \ldots, d$); w is the inertia weight (factor by which the present velocity influences the new velocity); c_1 and c_2 are positive acceleration constants used to scale the contribution of the cognitive and social component, respectively; and $r_{1j}(t)$, $r_{2j}(t) \in U(0, 1)$ are random values in the range (0, 1) sampled from a uniform distribution. The second term inside bracket in (1) refers to the cognitive component where the velocity is influenced by the difference between the particle present position $x_{ij}(t)$ and its personal best position till date $y_{ij}(t)$. The personal best is updated by comparing the value of the present fitness function with its previous best. The personal best position of at $(t+1)$-th step is calculated as

$$y_i(t+1) = \begin{cases} y_i(t) & \text{if } f(x_i(t+1)) \geq f(y_i(t+1)) \\ x_i(t+1) & \text{if } f(x_i(t+1)) < f(y_i(t+1)) \end{cases}, \tag{3}$$

where f is the fitness function. The third term inside the bracket in (1) is the social component where the particle new velocity is influenced by the difference in the particle present position and the global best particle position denoted by $\hat{y}$. The global best position at time step t is defined by

$$\hat{y}(t) = \min\{f(y_1(t)), \ldots, f(y_M(t))\}, \tag{4}$$

where M is the total number of swarm particles and $y_1(t), \ldots, y_M(t)$ are the personal best positions of all the swarm particles.

In data clustering by PSO the particle coordinates are specified by a predefined number of cluster centroids. Suppose in a supervised clustering problem one wants to search for N_C number of cluster centers, then for ith particle the coordinates will be designated as

$$x_i = (C_{i1}, \ldots, C_{ik}, \ldots, C_{iN_C}), \tag{5}$$

where C_{ik} is the kth cluster center in the designation of the ith particle. The C_{ik} components of the particle are defined in N_d-dimensional data vector space. Thus, in PSO clustering the particles represent the set of potential cluster centroids. The fitness f of a particle is calculated by using the relation

$$f = \frac{\text{intra}}{\text{inter}}. \tag{6}$$

In this definition the 'intra' term represents the average of all the distances between each data point and its cluster centroid, defined as

$$\text{intra} = \frac{1}{N_D} \sum_{k=1}^{N_C} \sum_{u \in C_k} \|u - C_k\|, \tag{7}$$

where u is the data vector belonging to cluster C_k and N_D is the total number of data vector. This term measures the compactness of the clusters. The 'inter' term is defined as

$$\text{inter} = \min\left\{ \|C_k - C_{kk}\|^2 \right\} \quad \forall k = 1, \ldots, k-1 \quad \text{and} \quad kk = k+1, \ldots, N_C \quad (8)$$

This term measures the average separation between the cluster centroids. The PSO algorithm searches for the set of cluster centroids which have high intracluster compactness and large intercluster separation yielding minimum value for f.

4 Polymer Selection by PSO Clustering

The PSO clustering algorithm was implemented in Matlab environment using the following set of parameters: $N_c = 6, w = 0.72, c_1 = 1.49, c_2 = 1.49$. The set of six polymers selected by this method are PVTD, SXFA, SXPYR, P4 V, PLF, and PBF. Table 1 shows this selection along with the sets of selection made by fuzzy c-means (FCM) and fuzzy subtractive (FS) clustering methods in our previous studies [7, 8]. It can be seen that PSO makes selection closer to FCM as there are four polymers (SXFA, SXPYR, P4V, and PBF) selected commonly. In comparison only two polymers (PLF and P4V) are common with FSC selection. Therefore, in the validation analysis presented in the following section we compare the PSO results with the FCM results.

5 Results and Discussion

The volatiles in the headspace of fish samples that signify the freshness and spoilage conditions are [7] (freshness indicators) 1-pentanol (1PTL), 1-hexanol (1HXL), 1-octanol (1 OCL); (spoilage indicators whose concentration increases after fish

Table 1 The set of polymers selected by PSO, FCM, and FSC clustering methods

PSO	PVTD	PLF	SXFA	SXPYR	P4V	PBF
FCM [7]	OV202	PEI	SXFA	SXPYR	P4V	PBF
FSC [8]	PMPS	PEM	P4V	SXPHB	PMHS	PLF

PVTD poly vinyl tetradecanal, *SXFA* hexafluro-2-propanol substituted polysiloxane, *SXPYR* alkyl aminopyridyl-substituted polysiloxane, *P4 V* poly(4-vinylheduorocumyl alcohol), *PLF* linear functionalized polymer, *PBF* branched functionalized polymer, *OV-202* poly trifluoropropyl methylsiloxane, *PEI* polyethylenimine, *PMPS* polymethyl phenylsiloxane, *PEM* poly ethylenemaleate, *PMHS* polymethyl hydrosiloxane death) trimethylamine (TMA), methyl mercaptan (MM), Ethanol (ET), 1-propanol (1PPL), 1-butanol (1BTL), 2-butanone (2BTN); (spoilage indicators whose concentration fluctuates after fish death) acetone (AC), methyl ethyl ketone (MEK), toluene (TO), ethyl benzene (EB), m-xylene (MX), p-xylene (PX), o-xylene (OX), and styrene (ST).

To examine the efficacy of PSO selected polymer set for making sensor array-based electronic nose, we analyzed responses of a 6-element SAW sensor array model whose sensing elements were assumed to be coated with these polymers. The response calculations were done using the same SAW sensor response model, concentration range for fish headspace volatiles and interferents, and additive frequency noise as detailed in [7]. The synthetic sensor outputs were generated for 100 samples of each vapor by varying the concentration from ppt to ppm level.

The sensor array responses were arranged in the form of 1700 × 6 data matrix with vapor samples in rows and sensors in columns. The data were normalized with respect to vapor concentration and then scaled by logarithmic scaling as established earlier for SAW chemical sensors [2, 7]. The preprocessed data were then subjected to principal component analysis (PCA). The separability of different vapor types (classes) was examined in the principal component (PC) space through score plots for highest eigenvalue components. Figure 1 shows two such plots by taking projections of the PCA transformed data onto PC1-PC2 and PC1-PC3 planes (score plots). From the PC1-PC2 score plot we notice that except 1-butanol all other spoilage volatiles are well separated from the freshness indicator volatiles. Among freshness markers only 1-octanol is clearly demarcated from the rest. However, if we consider all the spoilage volatiles combined into a singles class and all the freshness volatiles combined into another single class, the problem of spoilage versus freshness can be viewed as a two-class problem. The PC1-PC2 score plot clearly projects these two classes well discriminated by the PSO selected polymer set. Qualitatively, similar (or somewhat better) results were obtained with the FCM selected polymer set in [7, see Fig. 3]. In the PC1-PC3 score plot only two vapors methyl mercaptan (spoilage indicator) and 1-octanol (freshness indicator) are distinctly separated from the rest. In this respect the FCM results in [7] are better. The similarity of PSO results with FCM results is understandable in view of the four polymers being common in both selections, Table 1. The PSO as polymer selection method has not been analyzed before. The objective of the present study was to assess its potentiality. The present results indicate that PSO clustering can also be employed for data mining. At present it appears comparable to FCM clustering. However, a detailed comparative analysis is to identify their complementarity and uniqueness.

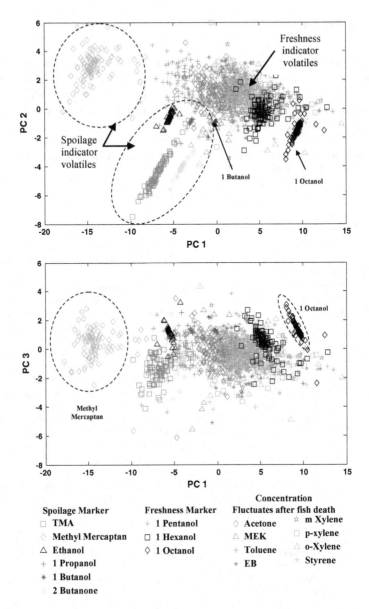

Fig. 1 PSO selection-based PC score plots for fish head space VOCs

6 Conclusion

We conclude that the particle swarm optimization clustering can be employed as data mining tool for electronic nose sensor array designs. It produces polymer selection results comparable to fuzzy c-means clustering method. Further analyses are needed for fully exploring the strength and limitations of this method in comparison to other clustering methods.

Acknowledgements Author T. Sonamani Singh is thankful to UGC, Government of India for providing the BSR fellowship.

References

1. Pearce, T.C., Schiffman, S.S., Nagle, H.T., Gardner, J.W.: Handbook of Machine Olfaction. Wiley, Weinheim (2003)
2. Yadava, R.D.S.: Modeling, simulation, and information processing for development of a polymeric electronic nose system. In: Korotcenkov, G. (ed.) Chemical Sensors—Simulation and Modeling, vol. 3, pp. 411–502. Momentum Press, LLC, New York (2014)
3. Jha, S.K., Yadava, R.D.S.: Designing optimal surface acoustic wave electronic nose for body odor discrimination. Sens. Lett. **9**, 1612–1622 (2011)
4. Jha, S.K., Yadava, R.D.S.: Statistical pattern analysis assisted selection of polymers for odor sensor array. In: 2011 IEEE International Conference Signal Processing, Communication, Computing and Networking Technologies (ICSCCN 2011), pp. 42–47 (2011)
5. Jha, S.K., Yadava, R.D.S.: Data mining approach to polymer selection for making SAW sensor array based electronic nose. Sens. Transducers J. **147**, 108–128 (2012)
6. Verma, P., Yadava, R.D.S.: A data mining procedure for polymer selection for making surface acoustic wave sensor array. Sens. Lett. **11**, 1903–1918 (2013)
7. Verma, P., Yadava, R.D.S.: Polymer selection for SAW sensor array based electronic noses by fuzzy C-means clustering of partition coefficients: model studies on detection of freshness and spoilage of milk and fish. Sens. Actuators, B **209**, 751–769 (2015)
8. Singh, T.S., Verma, P., Yadava, R.D.S.: Fuzzy subtractive clustering for polymer data mining for SAW sensor array based electronic nose. In: Proceeding 6th International Conference on Soft Computing for Problem Solving (SocProS 2016). AISC Series, vol. 546, pp. 245–253 (2017)
9. Singh, T.S., Gupta, A., Yadava, R.D.S.: On development of electronic nose for fish spoilage detection. In: National Conference on Applied Science in Engineering (ASHE 2016), JECRC, Jaipur, India, 9–10 Sep 2016
10. Peris, M., Gilabert, L.E.: A 21st century technique for food control: electronic noses. Anal. Chim. Acta **638**, 1–15 (2009)
11. Horczyczak, E.G., Guzek, D., Moleda, Z., Kalinowska, I.W., Brodowska, M., Wierzbicka, A.: Applications of electronic nose in meat analysis. LWT Food Sci. Technol (Campinas) **36**(3), 389–395 (2016)
12. Kennedy, J., Eberhart, R.: Particle swarm optimization. In: 1995 IEEE International Conference on Neural Networks, 1942–1948 (1995)

Reliability Assessment of Object-Oriented Software System Using Multi-Criteria Analysis Approach

Chahat Sharma and Sanjay Kumar Dubey

Abstract Software reliability is an essential and crucial factor to measure software quality. It is a challenging factor because of its complex nature of software. The complexity of software can be dealt by using object-oriented technology which forms a link between software development and software reliability. This paper assesses the reliability of object-oriented projects. The reliability criterion for object-oriented projects is determined by mapping object-oriented features with the software reliability sub-attributes. On the basis of reliability criterion, selection of a reliable object-oriented project is assessed using Analytic Hierarchy Process (AHP) and Fuzzy Analytic Hierarchy Process (FAHP) for which the results are shown.

Keywords Reliability · Object-oriented software · Analytic hierarchy process
Fuzzy analytic hierarchy process · Assessment

1 Introduction

Software reliability is one of the most necessary and challenging attributes of software quality. ISO/IEC 9126 defines reliability as a set of attributes that bear on the capability of software to maintain its level of performance under stated conditions for a stated period of time [1]. The reliability is important as it provides thrift advantage during development time of software. A major factor which leads to the reliability problems in software is its high complexity. Though users are aware of the reliability of software, they are probably not concerned with reusability of parts

C. Sharma (✉)
Inderprastha Engineering College, 63 Site IV, Sahibabad Industrial Area,
Ghaziabad, UP, India
e-mail: chahat.s.03@gmail.com

S.K. Dubey
Amity University Uttar Pradesh, Sec.-125, Noida, UP, India
e-mail: skdubey1@amity.edu

© Springer Nature Singapore Pte Ltd. 2018
M. Pant et al. (eds.), *Soft Computing: Theories and Applications*,
Advances in Intelligent Systems and Computing 583,
https://doi.org/10.1007/978-981-10-5687-1_41

which make up the source code. Software reliability is hence found as a serviceable measure to plan and control the resources during the development phase to develop high-quality software.

Object-oriented technology can provide a relationship between software development and software reliability [2]. The object-oriented software is made up of attributes such as inheritance, polymorphism, cohesion, etc. that gives a description of the underlying principles of software. Since object-oriented technology provides a link between software development and software reliability, the object-oriented features are mapped to software reliability features to obtain selection criteria. The multi-criteria decision-making approaches such as Analytic Hierarchy Process (AHP) and Fuzzy Analytic Hierarchy Process (FAHP) are used to assess the reliability of object-oriented projects.

The reliability of object-oriented projects can be determined on the basis of following sub-attributes given by ISO/IEC 25010:2011 [3]:

- Maturity: It is the degree to which a system, product, component meets the needs for reliability under normal operation. The factor is mapped with multi-threaded feature of object-oriented programming [4].
- Recoverability: It is the degree to which during a situation of a failure, a system or component can recover the data straightway affected and re-establish the desired state of the product or system. The factor is mapped with high performance feature of object-oriented programming [5].
- Availability: It is the degree to which a software, product, component is operational and accessible when required for use. The factor is mapped with the portability feature of object-oriented programming [6].
- Reliability Compliance: It is the capability of the software product to conform to conventions, standards, or regulations relating to reliability. The factor is mapped to the robust feature of object-oriented programming [7].

The three object-oriented projects Eclipse (P_1), NetBeans (P_2), and Android (P_3) are taken and their reliability is evaluated on the basis of the above-mentioned four selection criteria, i.e., Maturity (C_1), Recoverability (C_2), Availability (C_3), and Reliability Compliance (C_4).

Saaty presented the Analytic Hierarchy Process (AHP) which is based on sectioning a problem into hierarchical pattern [8]. It is one of the widely used and popular Multi-Criteria Decision-Making (MCDM) methods. This approach presents a structured means to organize the decisions which are complex in nature in the order of their effectiveness. However, it suffers certain drawbacks such as inability to cope up with the imprecision and uncertainty. To overcome such kind of problem, fuzzy AHP approach is applied.

Buckley and Laarhoven and Pedrycz extended the Saaty AHP approach using fuzzy set theory to deal with the subjectivity, uncertainty and imprecision [9, 10]. However, their methods were complex and unreliable in nature.

Deng proposed the multi-criteria analysis (MA) approach with fuzzy pairwise comparison which is used to solve multi-criteria analysis problems involving qualitative data [11]. The triangular fuzzy numbers were used to assess the decision-maker's judgments in the pairwise comparison process. The approach is efficient in solving and used to solve practical qualitative MA problems effectively.

The paper is organized as follows: in what follows, the methods and approaches are discussed. In the next section the implementation is described followed by results. Finally conclusion is drawn in context of reliable object-oriented project.

2 Methods and Approaches

A relationship is established between the object-oriented projects (P_1, P_2, and P_3) and the selection criteria's (C_1, C_2, C_3, and C_4) as shown in Fig. 1. The assessment of reliability is done by applying the Analytic Hierarchy Process and Fuzzy Analytic Hierarchy Process.

2.1 The Analytic Hierarchy Process

The Analytic Hierarchy Process (AHP) was proposed by Saaty in 1980. The approach is largely used in the problems requiring decision based on multi-criteria. The decision problem is evaluated by decomposing it into hierarchy of factors. This approach quantifies the importance of factors. The professional developers and experts are involved in the evaluation of each factor [12]. Every judgment is based on a scale and assigned a number corresponding to that scale with the help of Saaty rating scale [13].

The steps of the AHP approach are as follows [14]:

- Reciprocal Matrix: A pairwise comparison matrix is formed in the basis of factors. Each factor is compared with the other factors.
- Eigen Vector: The relative weights of the factors are calculated that is relevant to the problem. The list of weights is called as eigen vector.

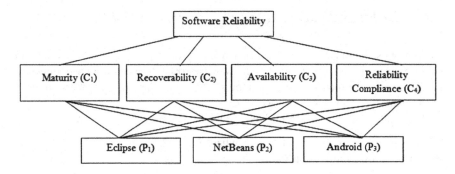

Fig. 1 Proposed relationship hierarchical model

- Consistency Index: After calculating weights, consistency index is computed for the matrix using equation $(\lambda_{max} - n)/(n - 1)$.
- Consistency Ratio: The consistency of judgments is calculated by computing consistency ratio.

Let us consider n number of elements $C_1, C_2, \ldots C_n$ for comparison. The relative weight of element C_i with respect to element C_j is assigned on the basis of Saaty rating scale [8] and denoted as w_{ij} which form a square matrix $A = [w_{ij}]$ of order n. The constraints are that $w_{ij} = 1/w_{ij} \ \forall \ i \neq j$ and $w_{ij} = 1 \ \forall \ i$. The matrix is called as reciprocal matrix represented in Eq. (1):

$$A = \begin{smallmatrix} C_1 \\ C_n \end{smallmatrix} [w_{ij}] = \begin{array}{c} C_1 \\ C_2 \\ \cdot \\ C_n \end{array} \begin{bmatrix} 1 & w_{12} & \cdot & w_{1n} \\ 1/w_{12} & 1 & \cdot & w_{2n} \\ \cdot & \cdot & \cdot & \cdot \\ 1/w_{1n} & 1/w_{2n} & \cdot & 1 \end{bmatrix} \tag{1}$$

Since human judgments are not consistent, so find eigen vector satisfying Eq. (2) as

$$A\omega = \lambda_{max}\omega; \ \lambda_{max} \geq n, \tag{2}$$

where ω is the egen vector and λ_{max} is eigen value. The presence of inconsistency in judgment will be indicated by the difference between λ_{max} and n. Now calculate the consistency index $(C.I.)$ and consistency ratio $(C.R.)$

$$C.I. = (\lambda_{max} - n)/(n - 1) \tag{3}$$

$$C.R. = C.I./R.I. \tag{4}$$

According to Saaty, if the value of $C.R. > 0.1$ it means the judgments are not consistent [15]. In that case, new judgments are taken and the AHP process is repeated untill the value of $C.R. < 0.1$. This way we can apply AHP to get the decision based on the concerned problem.

2.2 The Fuzzy Analytic Hierarchy Process (FAHP)

Deng proposed a multi-criteria analysis (MA) approach to facilitate effectiveness in solving decision-making problems. The MA approach converts the crisp values of professional's judgments to fuzzy numbers. The triangular fuzzy numbers are used in the pairwise comparison matrices [16].

The multi-criteria decision problem generally consists of

- Number of alternatives which is A_i where $i = 1, 2, \ldots n$
- Set of assessment criteria C_j where $j = 1, 2, \ldots m$.

Qualitative as well as quantitative assessment x_{ij} ($i = 1, 2, \dots n; j = 1, 2, \dots m$) called as performance ratings represents the performance of each alternative A_i with respect to criteria C_j to determine decision matrix for the alternatives. A weighting vector $W = (w_1, w_2, \dots w_m)$ is called as criteria weights, which represents the significance of evaluation criteria in relation to the problem to achieve an objective.

The steps of the fuzzy AHP approach are as follows:

1. Structure the decision problem into multi-criteria analysis problem to identify the hierarchy of the problem.
2. Formulate decision matrix X and calculate weighting vector $W \, \forall \, C_i$:

$$
X = \begin{bmatrix}
a_{11} & a_{12} & . & a_{1m} \\
a_{21} & a_{22} & . & a_{2m} \\
. & . & . & . \\
a_{n1} & a_{n2} & . & a_{nm}
\end{bmatrix}
\tag{5}
$$

$$
W = (w_1, w_2, \dots w_m)
\tag{6}
$$

3. Evaluate the fuzzy performance matrix Z by multiplying decision matrix X with weighting vector W:

$$
Z = \begin{bmatrix}
w_1 r_{11} & w_2 r_{12} & . & w_j r_{1j} & \cdots & w_n r_{1n} \\
w_1 r_{i1} & w_2 r_{i2} & . & w_j r_{ij} & \cdots & w_n r_{in} \\
\vdots & \vdots & \vdots & \vdots & \vdots & \vdots \\
w_1 r_{m1} & w_2 r_{m2} & . & w_j r_{mj} & \cdots & w_n r_{mn}
\end{bmatrix}
\tag{7}
$$

4. Determine the interval performance matrix Z_α by applying a-cut operation on the fuzzy performance matrix Z:

$$
Z_\alpha = \begin{bmatrix}
\left[z_{11l}^\alpha, z_{11r}^\alpha\right] & \left[z_{12l}^\alpha, z_{12r}^\alpha\right] & \cdots & \left[z_{1ml}^\alpha, z_{1mr}^\alpha\right] \\
\left[z_{21l}^\alpha, z_{21r}^\alpha\right] & \left[z_{22l}^\alpha, z_{22r}^\alpha\right] & \cdots & \left[z_{2ml}^\alpha, z_{2mr}^\alpha\right] \\
\cdots & \cdots & \cdots & \cdots \\
\left[z_{n1l}^\alpha, z_{n1r}^\alpha\right] & \left[z_{n2l}^\alpha, z_{n2r}^\alpha\right] & \cdots & \left[z_{nml}^\alpha, z_{nmr}^\alpha\right]
\end{bmatrix}
\tag{8}
$$

5. Evaluate the overall crisp performance matrix $Z_\alpha^{\lambda'}$ in regard to the perception toward risk by decision-maker which is denoted by an optimism index λ:

$$
Z_\alpha^{\lambda'} = \begin{bmatrix}
z_{11\alpha}^{\lambda'} & z_{12\alpha}^{\lambda'} & \cdot & z_{1m\alpha}^{\lambda'} \\
z_{21\alpha}^{\lambda'} & z_{22\alpha}^{\lambda'} & \cdot & z_{2m\alpha}^{\lambda'} \\
 & & \cdot & \\
z_{n1\alpha}^{\lambda'} & z_{n2\alpha}^{\lambda'} & \cdot & z_{nm\alpha}^{\lambda'}
\end{bmatrix} \tag{9}
$$

6. Determine the normalized performance matrix Z_α^λ

$$
Z_\alpha^\lambda = \begin{bmatrix}
z_{11\alpha}^{\lambda} & z_{12\alpha}^{\lambda} & \cdot & z_{1m\alpha}^{\lambda} \\
z_{21\alpha}^{\lambda} & z_{22\alpha}^{\lambda} & \cdot & z_{2m\alpha}^{\lambda} \\
\cdot & \cdot & \cdot & \cdot \\
z_{n1\alpha}^{\lambda} & z_{n2\alpha}^{\lambda} & \cdot & z_{nm\alpha}^{\lambda}
\end{bmatrix} \tag{10}
$$

7. Identify the positive ideal solution and negative ideal solution:

$$
A_\alpha^{\lambda+} = \left(Z_{1\alpha}^{\lambda+}, Z_{2\alpha}^{\lambda+} \cdots\cdots Z_{m\alpha}^{\lambda+} \right) A_\alpha^{\lambda-} = \left(Z_{1\alpha}^{\lambda-}, Z_{2\alpha}^{\lambda-} \cdots\cdots Z_{m\alpha}^{\lambda-} \right) \tag{11}
$$

8. Obtain the similarity degree between every alternative and positive ideal solution $S_{i\alpha}^{\lambda+}$ and negative ideal solution $S_{i\alpha}^{\lambda-}$ and then evaluate the overall performance index of every alternative

$$
P_\alpha^\lambda = \frac{S_{i\alpha}^{\lambda+}}{S_{i\alpha}^{\lambda+} + S_{i\alpha}^{\lambda-}} \tag{12}
$$

9. Now order the alternative according to their respective performance index values in the decreasing order.

3   Implementation and Results

In this study, three object-oriented projects are taken which are Eclipse (P_1), NetBeans (P_2), and Android (P_3). The reliability of these projects is determined on the basis of four software reliability evaluation criteria: Maturity (C_1), Recoverability (C_2), Availability (C_3), and Reliability Compliance (C_4).

3.1 Assessment of Object-Oriented Projects Using AHP

The AHP approach is applied on the software reliability relationship model shown in Fig. 1.

3.1.1 AHP of Software Reliability Criterion

First, we determine the significance of each criterion related to software reliability using AHP.

Step 1: Allocation of weights to the criteria

For allocating the weights to the criterion of software reliability, a survey was conducted on professional developers having good experience in object-oriented programming in the form of a questionnaire form. The pairwise relative weight values of criterion C_1 to C_4 are filled as shown in Table 1 using (1) of AHP.

Step 2: Obtaining eigen vector and eigen values

The eigen vector of criterion is obtained by multiplying the values in every row of matrix and then taking its nth ($n = 4$) root. Hence, the eigen vectors (weight) of the criteria's are shown in Table 2. The addition of all Eigen vectors should be 1.0.

Step 3: Calculating the consistency index

Compute the consistency index using Eqs. (2) and (3) to get 0.068.

Step 4: Calculating the consistency ratio

Finally, determine the consistency ratio using Eq. (4). The value of *R.I.* is obtained from Saaty Rating Scale [8] where the corresponding index of consistency is used

Table 1 Pairwise comparison of software reliability criterion

	C_1	C_2	C_3	C_4
C_1	1	1/3	1/2	1/2
C_2	3	1	2	1/2
C_3	2	1/2	1	1
C_4	2	2	1	1

Table 2 Eigen vector values

	C_1	C_2	C_3	C_4	nth root	Eigen vector (ω)
C_1	1	0.333	0.5	0.5	0.537	0.126
C_2	3	1	2	0.5	1.316	0.308
C_3	2	0.5	1	1	1.000	0.234
C_4	2	2	1	1	1.414	0.331
Total					4.267	0.999

for the given order (4 in this case) of matrix: $C.R. = 0.068/0.90 = 0.075$. The value of $C.R.$ is less than 0.1 which shows that the estimation taken is relevant. Further, AHP is applied on object-oriented projects for each criterion to evaluate the most to least reliable project.

3.1.2 AHP of Object-Oriented Projects for Each Software Reliability Criterion

The AHP approach is applied to the object-oriented projects Eclipse (P_1), NetBeans (P_2), and Android (P_3) for each criterion to determine the overall value and rank of the projects.

Step 1: Allocation of weights to the object-oriented projects

In order to allocate the weights to object-oriented projects, a survey is conducted on professional developers using questionnaire form. The form consists of pairwise comparison square matrix of object-oriented projects of each criterion. The pairwise weight values assigned to projects P_1–P_3 of each criterion C_1–C_4 are shown in Tables 3, 4, 5 and 6.

Table 3 Pairwise comparison of object-oriented projects for maturity

	P_1	P_2	P_3
P_1	1	3	1/3
P_2	1/3	1	1/5
P_3	3	5	1

Table 4 Pairwise comparison of object-oriented projects for recoverability

	P_1	P_2	P_3
P_1	1	7	5
P_2	1/7	1	1/3
P_3	1/5	3	1

Table 5 Pairwise comparison of object-oriented projects for availability

	P_1	P_2	P_3
P_1	1	7	3
P_2	1/7	1	1/5
P_3	1/3	5	1

Table 6 Pairwise comparison of object-oriented projects for reliability compliance

	P_1	P_2	P_3
P_1	1	5	3
P_2	1/5	1	1
P_3	1/3	1	1

Step 2: Obtaining eigen vector and eigen values

The eigen vector of all projects of each criterion is obtained following similar process as explained earlier. The values computed from tables are as follows:

Table 3: 0.258, 0.105, and 0.637, Table 4: 0.730, 0.081, and 0.189
Table 5: 0.649, 0.072, and 0.279, Table 6: 0.658, 0.156, and 0.185

Step 3: Calculation of overall values of object-oriented projects

After getting the eigen vectors, the overall values of object-oriented projects P_1–P_3 are obtained using Eq. (13):

$$\text{Object Oriented Project Value} = \sum_{i=1}^{n} \text{Relative value of } P_i * \text{Weight of } C_j \quad (13)$$

0.258 * 0.126 + 0.730 * 0.308 + 0.649 * 0.234 + 0.658 * 0.331 = 0.627.

Similarly, overall values of other projects P_2 and P_3 are computed. The reliable object-oriented project is one which is having the highest value. The ranking of object-oriented projects is done in decreasing order as shown in Table 7. The software reliability order calculated from AHP is **Eclipse > Android > NetBeans**.

3.2 Assessment of Object-Oriented Projects Using Fuzzy AHP

Step 1: The fuzzy AHP approach is applied on the software reliability hierarchy relationship model shown in Fig. 1. On the basis of the brief discussion about software reliability criterion C_1–C_4 and the triangular fuzzy numbers used in the pairwise comparison matrices [11], fuzzy reciprocal judgment matrix of each criterion is formed with respect to object-oriented projects as given in (14)

$$C_1 = \begin{matrix} P_1 \\ P_2 \\ P_3 \end{matrix} \begin{bmatrix} \bar{1} & \bar{3} & \bar{3}^{-1} \\ \bar{3}^{-1} & \bar{1} & \bar{5}^{-1} \\ \bar{3} & \bar{5} & \bar{1} \end{bmatrix} \quad C_2 = \begin{matrix} P_1 \\ P_2 \\ P_3 \end{matrix} \begin{bmatrix} \bar{1} & \bar{7} & \bar{5} \\ \bar{7}^{-1} & \bar{1} & \bar{3}^{-1} \\ \bar{5}^{-1} & \bar{3} & \bar{1} \end{bmatrix}$$

$$C_3 = \begin{matrix} P_1 \\ P_2 \\ P_3 \end{matrix} \begin{bmatrix} \bar{1} & \bar{7} & \bar{3} \\ \bar{7}^{-1} & \bar{1} & \bar{5}^{-1} \\ \bar{3}^{-1} & \bar{5} & \bar{1} \end{bmatrix} \quad C_4 = \begin{matrix} P_1 \\ P_2 \\ P_3 \end{matrix} \begin{bmatrix} \bar{1} & \bar{5} & \bar{3} \\ \bar{5}^{-1} & \bar{1} & \bar{1} \\ \bar{3}^{-1} & \bar{1} & \bar{1} \end{bmatrix} \quad (14)$$

Table 7 Overall project value and ranking of object-oriented projects

	C_1	C_2	C_3	C_4	Overall project value	Rank
Weights	0.126	0.308	0.234	0.331		
P_1	0.258	0.730	0.649	0.658	0.627	1
P_2	0.105	0.081	0.072	0.156	0.107	3
P_3	0.637	0.189	0.279	0.185	0.267	2

Step 2: Obtain the fuzzy decision matrix X shown in (15) considering fuzzy reciprocal judgment matrix (14) and using Eq. (5)

$$X = \begin{bmatrix} 0.08, 0.29, 1.05 & 0.29, 0.66, 1.53 & 0.22, 0.59, 1.37 & 0.18, 0.67, 1.61 \\ 0.05, 0.10, 0.51 & 0.04, 0.08, 0.34 & 0.04, 0.07, 0.28 & 0.08, 0.16, 0.68 \\ 0.19, 0.61, 1.76 & 0.07, 0.23, 0.67 & 0.13, 0.34, 0.88 & 0.08, 0.17, 0.75 \end{bmatrix}$$

$$(15)$$

Step 3: Calculate the weighting vector W, in (17) using Eq. (6) on matrix M in (16) which is pairwise comparison matrix of software reliability criteria C_1, C_2, C_3, and C_4:

$$M = \begin{matrix} C_1 \\ C_2 \\ C_3 \\ C_4 \end{matrix} \begin{bmatrix} \bar{1} & \bar{3}^{-1} & \bar{2}^{-1} & \bar{3}^{-1} \\ \bar{3} & \bar{1} & \bar{2} & \bar{5}^{-1} \\ \bar{2} & \bar{2}^{-1} & \bar{1} & \bar{1} \\ \bar{2} & \bar{2} & \bar{1} & \bar{1} \end{bmatrix}$$

$$(16)$$

$$W = [0.039, 0.121, 0.492 \quad 0.074, 0.036, 1.066 \quad 0.074, 0.233, 0.902 \quad 0.091, 0.310, 1.148]$$

$$(17)$$

Step 4: Based on the matrixes of X and W, obtain the fuzzy performance matrix Z using Eq. (7) as shown in (18):

$$Z = \begin{bmatrix} 0.003, 0.035, 0.517 & 0.021, 0.232, 1.631 & 0.016, 0.137, 1.236 & 0.016, 0.208, 1.848 \\ 0.002, 0.012, 0.251 & 0.003, 0.027, 0.362 & 0.003, 0.016, 0.253 & 0.007, 0.049, 0.781 \\ 0.007, 0.074, 0.866 & 0.005, 0.077, 0.714 & 0.009, 0.079, 0.794 & 0.007, 0.053, 0.861 \end{bmatrix}$$

$$(18)$$

Step 5: The a-cut operation is applied on Z in (18) to obtain the interval performance matrix Z_a using Eq. (8), as shown in (19):

$$Z_\alpha = \begin{bmatrix} 0.019, 0.276 & 0.127, 0.932 & 0.077, 0.687 & 0.112, 1.028 \\ 0.007, 0.132 & 0.015, 0.194 & 0.009, 0.134 & 0.028, 0.415 \\ 0.041, 0.47 & 0.041, 0.395 & 0.044, 0.436 & 0.03, 0.457 \end{bmatrix} \quad (19)$$

Step 6: Calculate the overall crisp performance matrix $Z_\alpha^{\lambda'}$ (20) by using (9) and (19):

$$Z_\alpha^{\lambda'} = \begin{bmatrix} 0.147 & 0.530 & 0.382 & 0.570 \\ 0.069 & 0.105 & 0.072 & 0.222 \\ 0.256 & 0.219 & 0.240 & 0.244 \end{bmatrix} \tag{20}$$

Step 7: Formulate the normalized performance matrix Z_α^{λ} (21) by using (10) and (20):

$$Z_\alpha^{\lambda} = \begin{bmatrix} 0.485 & 0.911 & 0.836 & 0.865 \\ 0.228 & 0.180 & 0.158 & 0.337 \\ 0.845 & 0.376 & 0.525 & 0.370 \end{bmatrix} \tag{21}$$

Step 8: Compute the positive ideal solution and negative ideal solution using Eq. (11):

$$A_\alpha^{\lambda+} = [0.845 \quad 0.911 \quad 0.836 \quad 0.865] A_\alpha^{\lambda-} = [0.228 \quad 0.180 \quad 0.158 \quad 0.337] \tag{22}$$

Step 9: The degree of similarity and the overall performance index of each alternative is computed using Eq. (12). The results are shown in Table 8.

Step 10: According to the performance index of the alternatives, rank them in the decreasing order. The results in Table 8 show that the performance index of P_1 is greater than P_3 and P_2. The software reliability order calculated from fuzzy AHP is **Eclipse > Android > NetBeans**.

Object-oriented Projects	Performance index	Rank
P_1	0.787	1
P_2	0.361	3
P_3	0.404	2

Table 8 Performance index and ranking of object-oriented projects

4 Conclusion

The implementation work assesses the reliability of three object-oriented projects on the basis of reliability criterion, i.e., Maturity, Recoverability, Availability, and Reliability Compliance by pairwise comparison of criteria's and projects. The assessment is done by Analytic Hierarchy Process (AHP) and validated by Fuzzy Analytic Hierarchy Process (FAHP). The results of AHP approach show that eclipse is more reliable object-oriented project which is validated by applying multi-criteria analysis FAHP approach. Eclipse is thereafter found to be the preferred alternative having the highest performance index.

References

1. ISO/IEC 25010:2011. http://en.wikipedia.org/wiki/ISO/IEC_9126. Accessed 24 Aug 2016
2. Antony, J., Dev, H.: Estimating reliability of software system using object oriented metrics. Int. J. Comput. Sci. Eng. Inf. Technol. Res. ISSN 2249-6831, 3(2), 283–294 (2013)
3. ISO/IEC 25010:2011. https://www.iso.org/obp/ui/#iso:std:iso-iec:25010:ed-1:v1:en. Accessed 24 Aug 2016
4. Olsina, L., Covella, G., Rossi, G.: Web Quality. In: Mendes, E., Mosley, N. (eds.) Web Engineering, pp. 109–142. Springer, Heidelberg (2006)
5. Losavio, F., Chirinos, L., Lévy, N., Cherif, A.R.: Quality characteristics for software architecture. J. Object Technol. (Published by ETH Zurich, Chair of Software Engineering) 2 (2), 133–150 (2003)
6. Malaiya, Y.K.: Software Reliability and Security. Encyclopedia of Library and Information Science. Taylor & Francis (2005). doi:10.1081/E-ELIS-120034145
7. Java: http://web.cs.wpi.edu/~kal/elecdoc/java/features.html. Accessed 24 Aug 2016
8. Coyle, G.: The Analytic Hierarchy Process. Practical Strategy. Open Access Material. AHP. Pearson Education Limited (2004)
9. Buckley, J.J.: Ranking alternatives using fuzzy numbers. Fuzzy Sets Syst. **15**, 1–31 (1985)
10. Laarhoven, P.J.M., Pedrycz, W.: A fuzzy extension of Saaty's priority theory. Fuzzy Sets Syst. **11**, 229–241 (1983)
11. Deng, H.: Multicriteria analysis with fuzzy pairwise comparison. Int. J. Approx. Reason. **21**, 215–231 (1999)
12. Mishra, A., Dubey, S.K.: Evaluation of reliability of object oriented software system using fuzzy approach. In: Proceeding of 5th International Conference—The Next Generation Information Technology Summit, Confluence-2014, pp. 806–809 (2014)
13. Saaty, T.L.: The Analytic Hierarchy Process. Mc-Graw Hill, New York (1980)
14. Mishra, A., Dubey, S.K.: Fuzzy qualitative evaluation of reliability of object oriented software system. In: IEEE International Conference on Advances in Engineering and Technology Research (ICAETR), pp. 685–690 (2014)
15. Saaty, T.L.: How to make a decision: the analytic hierarchy process. Eur. J. Oper. Res. **48**, 9–26 (1990)
16. Rouyendegh, B.D., Lesani, S.H.: Object-oriented programming language selection using fuzzy AHP method. Int. J. Anal. Hierarchy Process (Washington, D. C.) 1–17 (2014)

An Extensive Survey on Energy Optimization Techniques Based on Simultaneous Wireless Data and Energy Transfer to Wireless Sensor Network

Ankit Gambhir and Rajeev Arya

Abstract Wireless sensor networks (WSN) have involved many research scholars in recent years. WSN are significantly confined by their limited energy. Energy expenditure is a major concern in the designing and implementation of routing protocols and algorithms for wireless sensor networks because of limitation of power supply. There are many research articles available on optimization of wireless sensor network (WSN), mostly rely on energy-efficient routing protocols such as ant colony optimization, genetic algorithm, etc. However, with the advancement of antenna technology and wireless power transfer through microwaves, the issue of limited energy can be addressed by transmitting power as well as data at the same time. In this paper, authors have presented a wide survey of such optimization techniques that rely on simultaneous transfer of power as well as data to the wireless sensor nodes.

Keywords WSN · Energy · Optimization techniques

1 Introduction

In the recent years, WSN have got consideration from many research communities. A great number of applications, such as medical care, military target tracking, disaster relief, etc., are using this technology. Limited power supply is one of the major concerns as sensor nodes are generally battery-powered devices [1]. The main challenge is to how to decrease the energy utilization of nodes, so that the

A. Gambhir (✉)
GGSIP University, Delhi, India
e-mail: er.ankit.gambhir@gmail.com

R. Arya
Delhi Technical Campus, GGSIP University, Delhi, India
e-mail: rajeev.arya.iit@gmail.com

© Springer Nature Singapore Pte Ltd. 2018
M. Pant et al. (eds.), *Soft Computing: Theories and Applications*,
Advances in Intelligent Systems and Computing 583,
https://doi.org/10.1007/978-981-10-5687-1_42

network life span could be extended. Energy utilization is also a key issue in the designing and implementing algorithms and protocols for WSN. Hence, most research is happening in WSN area which is related to energy optimization. Many routing algorithms are specifically designed for the optimization of energy in WSN and have been proposed. Some algorithms like ant colony optimization (ACO) that take idea from the social behavior of ants [2] and genetic algorithm (GA) which is based on the concept of Darwin's evolution of biological systems that use arbitrary search in the decision space via selection, crossover, and mutation operators in order to reach its destination [2] are designed and used for energy optimization. However, with the advancement of antenna technology and wireless power transfer through microwaves, the issue of limited energy can be addressed by transmitting power as well as data at the same time. Wireless powered communication network eliminates the requirement of frequent replacement of battery or charging thus perk up the performance [3]. In this paper, an extensive survey of energy optimization techniques based on simultaneous transfer of data and power has been presented. Furthermore, recent advances and future research challenges have been discussed.

2 Simultaneous Wireless Data and Energy Transfer (SWDET)

Nikola Tesla first envisage the idea of wireless power transmission (WPT) and demonstrated "the transmission of electrical energy without wires" that rely on electrical conductivity in 1891 [4]. In 1893, lighting of vacuum bulbs had been shown by Tesla without the use of wires for transmission of power, at the World Columbian Exposition [4]. The Wardenclyffe tower that is also called as Tesla tower was constructed by Tesla primarily for electrical power transmission without wires instead of telegraphy [4]. Later in 1961, William C. Brown an American electrical engineer published the foremost paper regarding microwave energy for transmission of power; later on in 1964, he demonstrated a helicopter that received power required for flight from a microwave beam of frequency 2.45 GHz [5]. In 2007 a research group of Massachusetts Institute of Technology (MIT) showed illumination of a light bulb of 60 W using wireless power with efficiency of 40% at a two-meter distance by using a couple of diameter coils of 60 cm [6]. The Same concept has been applied in simultaneous transfer of energy and data to wireless sensor nodes. The recent advances in microwave-based wireless transmission of power open a new way to deal with the issue of limited energy and frequent battery replacements. Any signal that is used for data transmission also bears energy that can be harvested by the node at receiver. This is the principle of simultaneous transfer of energy and data technique. The wireless devices (WD) can use this harvested energy to send out information to or from other nodes. It improves the experience of user and can provide a high throughput in comparison to

Fig. 1 A model for wireless powered communication

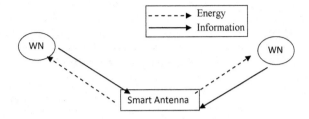

conventional battery-powered wireless sensor networks. Furthermore, with the recent advances in antenna technology and much higher microwave power with an adequate efficiency can be transferred [7] (Fig. 1).

3 Literature Survey

(See Table 1).

Table 1 Literature survey

Authors	Literature	Main contribution
Ioannis krikidis et al.	Simultaneous wireless information and power transfer in modern communication systems [3]	• Discussed techniques such as spatial switching time switching, antenna switching, and power splitting, for SWIPT • Aspects of resource allocation algorithm design for SWIPT systems such as joint power control, user scheduling, energy scheduling, information scheduling, and interference management have been presented • Path loss, energy security, and hardware development have been also discussed
Suzhi Bi et al.	Wireless powered communication: opportunities and challenges [8]	• Outline of high-tech RF-enabled wireless energy transmission (WET) technologies • Comparative analysis of rate as well as energy trade-offs of SWDET receivers • Introduction to harvest-then-transmit protocol and doubly-near-far problem

(continued)

Table 1 (continued)

Authors	Literature	Main contribution
Pulkit Grover et al.	Shannon meets Tesla: wireless information and power transfer [9]	• Considered the case of SWDET across a noisy coupled-inductor circuit with frequency selective channel and AWG noise • Provided solution in discrete and continuous version • Trade-off between information rate and power
Suzhi Bi et al.	Wireless powered communication networks an overview [7]	• Provided an outline on the fundamental model of wireless powered communication model • Presented idea of cognitive WPCN and green WPCN • Performance comparison of various operating WPCN models
Emily Adams et al.	A wireless sensor network powered by microwave energy [10]	• Demonstrated a prototype of a system enables a transfer of wireless energy to nodes • Analysis of node voltage as a function of time has been shown • A complete functional decomposition of the system (prototype) has been presented
Xun Zhou et al.	Wireless information and power transfer: architecture design and rate-energy tradeoff [11]	• Proposed, *dynamic power splitting* (DPS) receiver operation that splits the received signal with adaptable power ratio for energy conservation and data decoding, unconnectedly • Exceptional cases of DPS, on–off power splitting (OPS) power splitting (SPS) and time switching (TS), were examined
Lav R. Varshney	Transporting information and energy simultaneously [12]	• The elementary balance between the rate of energy and information that can be sent over a noisy line has been presented • A coding theorem and a capacity-energy function were also defined
Amin Enayati et al.	3D-antenna-in-package solution for microwave wireless sensor network nodes [13]	• Implemented a 3D antenna-in-package solution for microwave WSN • Designed an antenna in a modular way and its structure and dipole shape pattern have been shown

(continued)

Table 1 (continued)

Authors	Literature	Main contribution
Luca Catarinucci et al.	Switched-Beam Antenna for Wireless Sensor Network Nodes [14]	• Proposed a flexible beam-steering antenna for WSN in ISM band • Demonstrated an inexpensive realization on an FR4 substrate
Giuseppe Anastasi et al.	Energy conservation in wireless sensor networks: a survey [15]	• Presented a systematic and comprehensive nomenclature of the energy-efficient schemes • Different approaches like data-driven and mobility-based schemes for energy conversation have been presented
Seunghyun Lee et al	Cognitive wireless powered network: spectrum sharing models and throughput maximization [16]	• Proposed cognitive radio (CR) WPCN consisting of one single H-AP and distributed wireless powered users, which uses the common spectrum for down-link WET and up-link wireless data transmission (WDT) • Provided analysis as well as simulation results and compared the sum-throughput of the CWPCN (cognitive) with coexisting models
Xiaoming Chen et al.	Enhancing wireless information and power transfer by exploiting multi-antenna techniques [17]	• Performance enhancement by using multi-antenna techniques • Investigated trade-offs based on inadequate feedback multi-antenna techniques for small distance and large-scale MIMO technique for long distance • To improve power transfer efficiency, techniques like energy beamforming by exploiting spatial degree of freedom have been discussed
Yueling Che et al.	Multi-antenna wireless powered communication with co-channel energy and information transfer [18]	• Obtained optimum solutions for the issue of co-channel interference • Achieved best energy-interference trade-off
Ioannis Krikidis et al.	A low complexity antenna switching for joint wireless information and energy transfer in MIMO relay channels [19]	• Investigated a technique for SWDET in MIMO relay channels • The proposed technique was applied to interference in more than one user, where zero-forcing receiver was employed at the relay node

(continued)

Table 1 (continued)

Authors	Literature	Main contribution
Qian Sun et al.	Joint beamforming design and time allocation for wireless powered communication networks [20]	• A multi-input single-output WET model under the protocol of first harvest-then-sent was investigated • Considered integrated time allocation and beamforming designing to make the most of the system sum-output

4 Conclusion

In this paper, we have presented a comprehensive overview of research going on in the domain of wireless powered communication. Recent research articles on energy optimization techniques based on simultaneous transfer of data and power have been considered for the survey. Some authors have worked in the domain of antenna technology and exploited spatial characteristics. Various switching techniques such as time switching, antenna switching, power splitting, etc. have been presented by many authors. Optimization of energy in WSN is a challenging issue and a key area of research. Simulation results and comparative analysis of various parameters have been shown by authors. Trade-off between energy and interference was also discussed by many authors. This paper has provided a summary of recent advancement in the area of transmitting power as well as data to wireless sensor network at the same time. This paper summarizes and organizes recent research results in a way that adds understanding to work in this domain.

References

1. Akyildiz, I.F., Su, W., Sankarasubramaniam, Y., Cayirci, E.: Wireless sensor networks: a survey. Comput. Netw. **38**, 393–422 (2002)
2. Adnan, Md.A., Razzaque, M.A., Ahmed, I., Isnin, I.F.: Bio-mimic optimization strategies in wireless sensor networks: a survey. Sensors (2014). doi:10.3390/s140100299
3. Krikidis, I., Sasaki, S., Timotheou, S., Ding, Z.: Simultaneous information & power transfer in modern communication systems. IEEE Trans. Comm. (2014). doi:10.1109/MCOM.2014. 6957150
4. Nikola, T.: The transmission of electrical energy without wires as a means for furthering peace. Electr. W. Eng. **21** (1905)
5. Brown, W.C., Mims, J.R., Heenan, N.I.: An experimental microwave-powered helicopter. In: IEEE International Convention Record, Vol. 13, Part 5, pp. 225–235 (1965)
6. Goodbye wires: MIT News. http://web.mit.edu/newsoffice/2007/wireless-0607.html. 07 June 2007
7. Bi, S., Zeng, Y., Zhang, R.: Wireless powered communication networks: an overview. IEEE Wirel. Commun. (2016). doi:10.1109/mwc.2016.7462480

8. Bi, S., Ho, C.K., Zhang, R.: Wireless powered communication: opportunities & challenges. IEEE Com. Mag. **53** (2015). doi:10.1109/MCOM.2015.7081084
9. Grover, P., Sahai, A.: Shannon meets Tesla: wireless information & power transfer. In: Proceedings of IEEE International Symposium on Information Theory. Austin, TX, USA, pp. 2363–2367 (2010)
10. Adams, E., Albagshi, A., Alnatar, K., Jacob, G., Mogk, N., Sparrold, A.: A wireless sensor network powered by microwave energy. http://hdl.handle.net/10150/581655
11. Zhou, X., Zhang, R., Ho, C.K.: Wireless.: information & power transfer: architecture design & rate-energy tradeoff (2013). doi: 10.1109/TCOMM.2013.13.120855
12. Varshney, L.R.: Transporting information & energy simultaneously ISIT 2008, Toronto (2008). doi:10.1109/isit.2008.4595260
13. Enayati, A., Brebels, A., de Raedt, W., Vandenbosch, G.A.E.: 3D-antenna-in-package solution for microwave wireless sensor network nodes. IEEE Trans. Antennas Propag. **59**(10) (2011)
14. Luca, C., Sergio, G., Luigi, P., Luciano, T.: Switched-beam antenna for wireless sensor network nodes. Prog. Electromagn. Res. C **39**, 193–207 (2013)
15. Giuseppe, A., Marco, C., Mario, D., Andrea, P.: Energy conservation in wireless sensor networks: a survey. Ad Hoc Netw. **7**, 537–568 (2009)
16. Lee, S., Zhang, R.: Cognitive wireless powered net-work: spectrum sharing models and throughput maximization. IEEE Trans. Cognitive Commun. Netw. (2015). doi:10.1109/tccn.2015.2508028
17. Chen, X., Zhang, Z., Hsiao, H., Zhang, H.: Enhancing wireless information and power transfer by exploiting multi-antenna techniques. IEEE Commun. Mag. (2015). doi:10.1109/mcom.2015.7081086
18. Liu, L., Zhang, R., Chua, K.C.: Multi-antenna wireless powered communication with energy beamforming. IEEE Trans. Commun. (2014). doi:10.1109/TCOMM.2014.2370035
19. Krikidis, I., Sasaki, S., Timotheou, S., Ding, Z.: A low complexity antenna switching for joint wireless information and energy transfer in MIMO relay channels. IEEE Trans. Comm. (2014). doi:10.1109/tcomm.2014.032914.130722
20. Sun, Q., Zhu, G., Shen, C., Li, X., Zhong, Z.: Joint beamforming design and time allocation for wireless powered communication networks. IEEE Commun. Lett. (2014). doi:10.1109/lcomm.2014.2347958

Panorama Generation Using Feature-Based Mosaicing and Modified Graph-Cut Blending

Achala Pandey and Umesh C. Pati

Abstract Panoramic mosaicing has several important areas of applications including computer vision/graphics, virtual reality, and surveillance. Different factors like sensor noise, camera motion, and illumination difference affect the quality of the mosaic due to creation of artificial edges (artifacts) between the images in the resulting panorama. This work presents a technique to overcome such problems for the generation of seamless panoramic images. The proposed technique uses scale-invariant feature transform (SIFT) features for image alignment and modified graph-cut (MGC) for blending the seam in the overlapping region between images. The proposed method is tested on various sets of images to show its effectiveness. Comparison with different mosaicing as well as blending techniques shows that the proposed method achieves improved results for panorama creation.

Keywords Panoramic mosaicing · Alignment · SIFT · Geometric transformations
RANSAC · Seam · Seamless blending

1 Introduction

A panorama contains much more information than an image captured using normal camera as it is the complete surrounding view around a person. In panoramic mosaicing, multiple images are stitched with each other in such a way that a wide angle view of the scene is generated. Panoramic mosaicing has diverse applications in computer graphics and vision [1], object detection [1], virtual tour in real estate, [2], aerial photography [3], and surveillance using videos [4].

A. Pandey (✉) · U.C. Pati
Department of Electronics and Communication Engineering, National Institute
of Technology Rourkela, Rourkela 769008, Odisha, India
e-mail: achala.pandey13@gmail.com

U.C. Pati
e-mail: ucpati@nitrkl.ac.in

© Springer Nature Singapore Pte Ltd. 2018
M. Pant et al. (eds.), *Soft Computing: Theories and Applications*,
Advances in Intelligent Systems and Computing 583,
https://doi.org/10.1007/978-981-10-5687-1_43

Panoramic mosaicing [5] in general has three steps of final mosaic generation, i.e., acquisition of individual images, pre-processing of images, and mosaicing. Mosaicing [3–5] in turn contains registration of images, warping, and blending.

The aim of this work is to generate seamless panoramic mosaic. A feature-based method is used for alignment of images. SIFT features are extracted as these features are robust to scaling, rotation, and to some extent robust to the change in illumination. A modified graph-cut blending is applied to smooth the seam in the overlapping region of input images. The use of these methods makes the algorithm efficient for generation of better quality seamless panorama.

The paper organization is as follows: Sect. 2 gives a brief literature survey on existing mosaicing methods. The fundamental steps of panoramic mosaic generation are reported in Sect. 3. In Sect. 4, the algorithm for the proposed technique has been described. The experimental results using proposed algorithm along with comparative analysis over different sets of images has been demonstrated in Sect. 5. Finally, the paper is concluded in Sect. 6.

2 Related Work

Over the last few decades, the research in the area of image mosaicing has gained a wide popularity. A number of techniques have been introduced to address the problems of image mosaicing. The different approaches for mosaicing are characterized based on the domain of the algorithm used. In correlation-based methods, mosaicing is performed by directly using the image pixels [6]. Fast Fourier transform (FFT) or discrete cosine transform (DCT) is used in frequency domain for mosaicing [7]. Low-level features [7] use edges and corners as distinct features, whereas high-level features use parts of objects for mosaicing [8].

Mainly, all these methods are distinguished in two categories. First, methods that do not use the feature extraction and attempt to use the pixel information directly for minimizing the error function between the two images are called direct methods [9]. Accurate results can be obtained using direct methods as they use all the information present in an image. However, these methods fail in the case of variation in illumination or moving objects in the scene. The second category deals with extraction of distinct features for finding the correspondence between the images [10]. These methods are robust to scaling, rotation, and illumination variation. In addition, these methods can also handle the moving objects in the scene. In case of misalignment and intensity discrepancies, an appropriate blending algorithm is used. Blending using feathering or alpha blending [11] results in loss of edge information due to smoothing of region. Gradient domain blending works well for smoothing the seam, but it leads to color bleeding [12] which changes the color of the objects after blending, because of this gradient based methods are rarely used for blending of panoramic images.

Thus, the aforementioned literature survey reveals that a robust mosaicing scheme is required which can align the images accurately and blend the result using an appropriate blending algorithm.

Therefore, to address these issues the work describes the following: (a) A feature-based technique for accurate alignment of the images, and (b) a blending method that is able to handle intensity and color differences between images for seamless mosaicing.

3 Panoramic Mosaicing

The process of panorama generation has the following three steps: (1) Panoramic image acquisition, (2) pre-processing of acquired images, and (3) mosaicing of images using appropriate algorithm.

A single camera mounted on a tripod can be used to capture images with different rotation angles. An alternative to this is using multiple cameras mounted on tripod to cover the view of different directional angles of the scene. In some cases, omnidirectional image sensors with fisheye lenses or mirrors are used, but their short focal length, high cost, and lens distortion limits their access. The captured images are pre-processed for noise filtration or other sorts of intensity variations before mosaicing. The most common pre-processing step is reduction of noise, lens distortion correction, and camera calibration. These pre-processed images are mosaiced by following the necessary steps of mosaicing. An overview of the framework for panorama generation is shown in Fig. 1.

4 Proposed Technique for Panoramic Mosaicing

The images captured for panorama generation are pre-processed before mosaicing. Once the images are pre-processed the next step is to extract the important features from the images. The following subsections discuss the steps involved in proposed panoramic mosaicing technique.

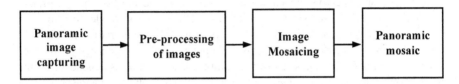

Fig. 1 An overview of the panoramic image generation process

4.1 Feature Extraction and Matching

Scale-invariant feature transform (SIFT) feature [10] is used for feature extraction because of the stability, accuracy, invariance to scaling, and rotation of SIFT features. Therefore, they provide robust matching. In SIFT image data is transformed to scale-invariant coordinates in the following four stages of algorithm:

1. *Scale-space extrema detection*—Differences of successive Gaussian-blurred images called difference of Gaussian (DoG) are taken for candidate point search.
2. *Feature points localization*—Too many candidate points are obtained after scale-space extrema detection, some of which are unstable. Poorly localized or low contrast candidate points are rejected as these are prone to noise.
3. *Orientation assignment*—Every candidate point is assigned one or more orientation based on local gradient directions; it provides invariance to rotation.
4. *Feature point descriptor*—Each candidate point of the image is assigned to a location, scale, and orientation such that the image regions around the candidate points are in 2-D coordinate system and provide invariance to these parameters.

Here, a highly distinctive descriptor is computed for the local image region which is invariant to other variations, e.g., illumination changes or 3-D viewpoint.

4.2 Correspondence Between Features

Feature matching between the images is done after detection of feature points. Similar feature points based on predefined threshold values are selected and rest of the points are rejected. For estimation of homography in order to match the inliers, RANdom SAmple Consensus (RANSAC) [13] has been used. The algorithm works accurately even in the presence of a significant number of outliers.

4.3 Image Transformation and Re-projection

Depending on the geometric distortion of the images, different transformations like affine, rigid, or projective are employed to transform the images. For panoramic mosaicing projective transform is used [5]. This type of model having less parameter provides more accurate and efficient model for panoramic mosaic generation [4]. Such transformation aligns the images on a common compositing or re-projection manifold. The selection of compositing manifold is dependent on the application of the mosaic to be generated. Out of planar, cylindrical and spherical, a planar manifold has been considered for the proposed algorithm.

4.4 Blending of Images

Factors like illumination variation or change in exposure cause difference in intensity values mainly in the overlapping region between images which results in the creation of visible artifacts in the mosaic. Therefore, handling of these regions is different from the rest regions of the mosaic where specific algorithms are used for seam smoothing while avoiding loss of information over that region.

The blending algorithms can be classified as transition smoothing or optimal seam finding. *Transition smoothing* techniques [11] minimize the visible seam by smoothing the region between the images. Feathering or alpha blending approaches fall in this category. In the *optimal seam finding* [12] method, a seam is searched in the region of overlap where intensity difference between the images is minimum. In the present work, a modified graph-cut algorithm (an optimal seam finding method) is used for smooth blending of image regions.

4.5 Modified Graph-Cut

For the region of overlap, a graph is constructed in such a way that each pixel corresponds a graph node and the edge weights indicate the error between the two nodes. A min-cut shows the least error edges over the region of minimum intensity variations. This seam creation is based on the search of minimized energy of a particular energy function. Conventional algorithms (e.g., dynamic programming) were 'memoryless' therefore could not find the optimized seams as in the case of graph-cut.

The graph shown in Fig. 2 is the representation of two input images and the overlapping region between them. Each node in the graph shows a pixel in the overlap region.

Two additional nodes 'S' and 'T' represent the source and sink, i.e., Image 1 and Image 2 in the present case. The arcs connecting the adjacent pixel nodes 'S' and 'T' are labeled with some matching quality cost M. In Fig. 2 red line shows the

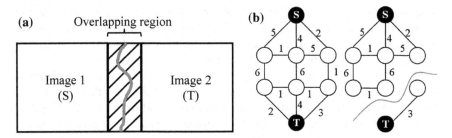

Fig. 2 Example demonstrating minimum cut in a graph. **a** Overlapping region between two images and min-cut. **b** Graph showing min-cut with a total cost of the cut as $2 + 4 + 1 + 1 = 8$

minimum cut which separates the overlap pixels in two parts indicating the pixels to be copied from either of the two images. In the algorithm, the matching quality measure [14] calculates the intensity difference between the pairs of pixels for each color channel L^*, a^*, and b^* of CIE $L^*a^*b^*$ color space. Let 's' and 't' be the positions of two adjacent pixels in the overlapping region, and $A(s)$ and $B(s)$ be the pixel intensities at 's' for old and new patches. Similarly, $A(t)$ and $B(t)$ be the pixel intensities at position 't'. The matching quality cost is defined as

$$M(s,t,A,B) = \|A(s) - B(s)\| + \|A(t) - B(t)\|, \tag{1}$$

where $\|\cdot\|$ denotes the appropriate norm. In the $L^*a^*b^*$ color space, the formula for matching quality cost can be defined as

$$M(s,t) = \sqrt{wt_1(L_A(s) - L_B(s))^2 + wt_2(a_A(s) - a_B(s))^2 + wt_3(b_A(s) - b_B(s))^2}$$
$$+ \sqrt{wt_1(L_A(t) - L_B(t))^2 + wt_2(a_A(t) - a_B(t))^2 + wt_3(b_A(t) - b_B(t))^2}, \tag{2}$$

where $wt_1 + wt_2 + wt_3 = 1$. All the steps of the proposed panoramic mosaicing algorithm have been summarized in Fig. 3.

5   Results and Discussions

The experiments were carried out on an Intel®Core™i7 2600 CPU at 3.40 GHz on MATLAB® under Microsoft® Windows 7 over various image sets. Figure 4 shows the sequence of processing involved. Figure 4a–c show the input images captured with different camera rotations. Initially, SIFT is used for features extraction and RANSAC is used for keeping the inliers while rejecting the outliers. The aligned images are then re-projected on a re-projecting surface (which is planar for this case) for mosaicing. However, the seam or artificial edges in the transition region are generated which can be clearly seen in Fig. 4d. The red boxes show the misalignment and the intensity differences in the transition region. Simple averaging or feathering will not solve the problem as it may lead to over smoothing of the transition region resulting in loss of important information. Graph-cut, on the other hand, results in horizontal artifacts when there is photometric difference between images shown in Fig. 4e and zoomed view is shown in Fig. 4f.

The proposed algorithm employs modified graph-cut for blending. Figure 5a shows the mosaic obtained on applying modified graph-cut with zoomed view (Fig. 5b) of the same region of the image without any artifact. Thus, the proposed method of panoramic mosaicing generates visually appealing seamless mosaics.

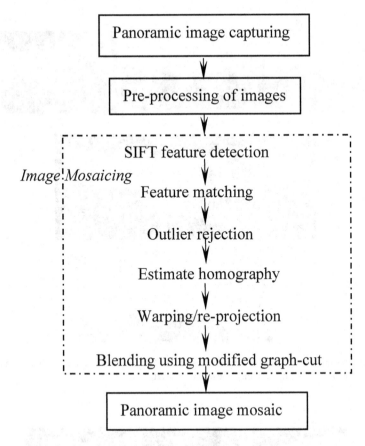

Fig. 3 Proposed algorithm for panoramic image mosaic generation

5.1 Qualitative Analysis

The proposed method of panoramic mosaic generation has also been tested for a set of images consisting of left and right views of a mountain for different mosaicing and blending techniques as it has geometric as well as photometric distortions. Figure 6a, b show the two input images with some overlapping part.

To reduce the photometric difference between the input images, within the mosaic, a number of blending methods have been incorporated. Figure 6a–h shows the results of using different blending methods to smooth the seam. The results clearly show that the averaging, feathering, and minimum blending (Fig. 6d–h) do not overcome the problem of the visible seam. Graph-cut on the other hand [15] finds the optimal seam (Fig. 6g), however, in some cases results in horizontal artifacts when the photometric difference is more. The proposed method is able to reduce such artifacts while producing smooth blend between the images (Fig. 6h).

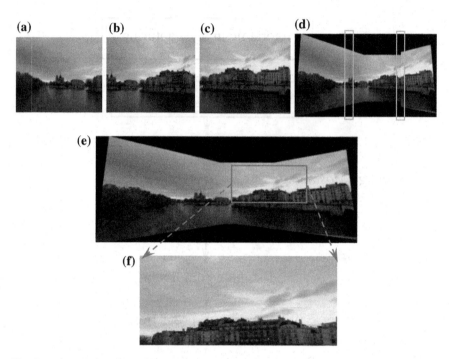

Fig. 4 Panoramic mosaicing: **a–c** input images, **d** misalignment error and intensity discrepancies, **e** blending using graph-cut, **f** zoomed view showing artifacts

Fig. 5 Seam visibility: **a** panoramic mosaic using proposed method, **(b)** zoomed view to show the removed horizontal marks (artifacts) of the mosaic

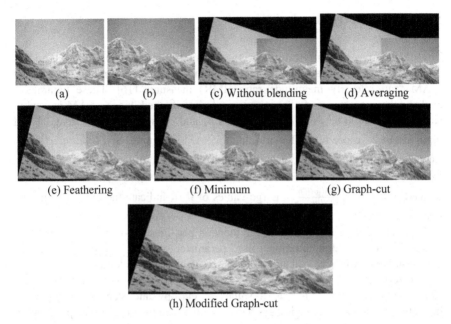

(a)   (b)   (c) Without blending   (d) Averaging

(e) Feathering   (f) Minimum   (g) Graph-cut

(h) Modified Graph-cut

Fig. 6 Blending algorithm comparison

Fig. 7 Panoramic mosaics for five sets of images using proposed technique

The proposed method has also been tested for various sets of images with geometric and photometric differences. The mosaicing results for five sets of images have been shown in Fig. 7.

5.2 Quantitative Analysis

The quantitative analysis results for Fig. 7 are summarized in Tables 1 and 2. Table 1 shows the quality assessment parameter using spectral angle mapper (SAM) and intensity magnitude ratio (IMR) measures [16]. These parameters evaluate the geometric and photometric quality of stitched images. SAM is used to calculate the angle between the two pixels p_1 and p_2 from the input image 1 and input image 2, respectively. The formula for SAM is given as $SAM = \arccos \frac{\langle p_1, p_2 \rangle}{|p_1||p_2|}$. Large values of the angle indicate that pixels/vectors are different from each other. IMR is the ratio of the two 3-D color vectors. For the two pixels p_1 and p_2 it can be defined as $IMR = \frac{\min(|p_1|,|p_2|)}{\max(|p_1|,|p_2|)}$. Large values of IMR indicate the difference in two images, i.e., photometric difference.

Since to measure the quality of the blended part these two parameters may not suffice. Therefore, other commonly used parameters such as entropy and standard deviation have also been calculated for different blending techniques.

The results for quantitative assessment of the five sets of images have been summarized in Table 2. Increased values of entropy and standard deviation is indicative of the fact that the images have been blended over the overlapping region while retaining the important information.

Table 1 Quantitative assessment of mosaics

Images	Average parameters	
	SAM	IMR
Mountain	0.0283	0.4896
Stadium	0.0897	0.7783
Floor	0.0301	0.776
Building	0.0404	0.8378
Scene	0.0222	0.8432

Table 2 Quantitative comparison of different blending approaches

Images	Averaging		Minimum		Feathering		Graph-cut		Modified graph-cut	
	Entropy	Std dev	Entropy	Std dev	Entropy	Std dev	Entropy	Std dev	Entropy	Std dev
Mountain	5.827	0.325	5.931	0.314	5.818	0.327	5.821	0.221	6.529	0.340
Stadium	6.684	0.182	6.630	0.176	6.698	0.183	4.719	0.178	6.787	0.189
Floor	7.573	0.247	7.506	0.254	7.550	0.251	3.967	0.123	7.594	0.262
Building	6.850	0.272	6.821	0.273	6.752	0.272	4.351	0.221	6.986	0.273
Scene	6.966	0.310	6.951	0.310	6.965	0.310	6.978	0.274	7.165	0.311

6　Conclusion

Camera motion, sensor noise, difference in illumination, and parallax affect the process of mosaic generation making the mosaic visually distorted. In this work, an efficient method of mosaicing is presented to mitigate the effect of above-mentioned problems. The algorithm has two steps: first, it uses feature-based technique for image alignment and second, a modified graph-cut method is used for reducing the visual artifacts or seam in the region of overlap. Results have been evaluated by comparing with other mosaicing as well as blending techniques. The qualitative and quantitative comparisons show the superior performance of our method. Presently, the images have photometric and geometric distortion for two or three views of a scene. In future, more images will be considered for panorama generation.

References

1. Hsieh, J.W.: Fast stitching algorithm for moving object detection and mosaic construction. Image Vis. Comput. **22**(4), 291–306 (2004)
2. Chan, S.C., Shum, H.Y., Ng, K.T.: Image based rendering and synthesis. IEEE Signal Process. Mag. **24**(6), 22–33 (2007)
3. Kekec, T., Vildirim, A., Unel, M.: A new approach to real-time mosaicing of aerial images. Robot. Auton. Syst. **62**, 1755–1767 (2014)
4. Marzotto, R., Fusiello, A., Murino, V.: High resolution video mosaicing with global alignment. In: Proceedings of IEEE Computer Society Conference on Computer Vision and Pattern Recognition, Washington, DC., Vol. 1, pp. 692–698 (2004)
5. Gledhill, D., Tian, G.Y., Taylor, D., Clarke, D.: Panoramic imaging—a review. Comput. Graph. **27**, 435–445 (2003)
6. Kuglin, C.D.: The phase correlation image alignment method. In Proceedings of International Conference on Cybernetics Society, pp. 163–165 (1975)
7. Pandey, A., Pati, U.C.: A novel technique for mosaicing of medical images. In: Annual IEEE India Conference (INDICON), Pune, India, pp. 1–5 (2014)
8. Bhosle, U., Chaudhuri, S., Dutta Roy, S.: A fast method for image mosaicing using geometric hashing. IETE J. Res. **48**(3–4), 317–324 (2002)
9. Irani, M., Anandan, P.: About direct methods. In: Vision Algorithms: Theory and Practice, pp. 267–277 (2000)
10. Lowe, D.: Distinctive image features from scale-invariant keypoints. Int. J. Comput. Vision **60**(2), 91–110 (2004)
11. Xiong, Y., Pulli, K.: Fast panorama stitching for high-quality panoramic images on mobile phones. IEEE Trans. Consum. Electron. **56**(2), 298–306 (2010)
12. Perez, P., Gangnet, M., Blake, A.: Poisson image editing. ACM Trans. Graph. (TOG) **22**(3), 313–318 (2003)
13. Fischler, M., Bolles, R.: Random sample consensus: a paradigm for model fitting with application to image analysis and automated cartography. Commun. ACM **24**, 381–395 (1981)

14. Kwatra, V., Schodl, A. Essa, I., Turk, G., Bobick, A.: Graphcut textures: image and video synthesis using graph cuts. ACM Trans. Graph., 277–286 (2003)
15. Pandey, A., Pati, U.C.: Panoramic image mosaicing: an optimized graph-cut approach. In: Proceedings of International Conference on Advanced Computing, Networking, and Informatics ICACNI Bhubaneswar, India, vol. 43, pp. 299–305 (2015)
16. Qureshi, H.S., Khan, M.M., Hafiz, R., Cho, Y., Cha, J.: Quantitative quality assessment of stitched panoramic images. IET Image Proc. 6(9), 1348–1358 (2012)

Design of Semantic Data Model for Agriculture Domain: An Indian Prospective

Poonam Jatwani and Pradeep Tomar

Abstract Agriculture is the backbone of our country's growth and economy. India is the second largest producer of agricultural products. However, due to lack of proper knowledge and information, it lags behind. Existing ontologies in the agricultural domain may lack information about seeds, fertilizers, pesticides, various Govt. Schemes, weather, and soil recommendations along with crop management techniques. So, there is need to convert all agriculture related data into a machine-readable form. In the proposed research work we resort to Semantic Web-based technologies to handle this problem. In particular, we focus on RDF format for publishing and linking data for information sharing so that farmers are provided relevant and contextual information timely and accurately.

Keywords Agriculture · Farmers · Ontology · Semantic web

1 Introduction

India has always been an agrarian society and is known to be a land of farmers. Even centuries ago, Vasco De Gama came to India not for its riches but in search of its agricultural produce like spices. During the British era, in the greed to maximize the commercial benefit, Britishers exploited farmers and forced them to grow cash crops like Indigo, Cotton, etc., and thus disturbed the age-old farming patterns of this county. It also led to uneven distribution of land among farmers creating some with major land banks and rest all with either very small land holdings or as

P. Jatwani (✉)
Department of Computer Science, Government College For Women, Faridabad, India
e-mail: poonam.almadi@gmail.com

P. Tomar
Department of Computer Science and Engineering, School of I.C.T., Gautam Buddha University, Greater Noida,
Uttar Pradesh, India
e-mail: pradeep.tomar@gbu.ac.in

© Springer Nature Singapore Pte Ltd. 2018
M. Pant et al. (eds.), *Soft Computing: Theories and Applications*,
Advances in Intelligent Systems and Computing 583,
https://doi.org/10.1007/978-981-10-5687-1_44

landless laborers. Although it did open avenues like tea and coffee plantation in the hilly region, but the actual farmer did not gain much from it.

After the Green Revolution of 1970 and 1980s major thrust was on to achieve self-sustenance in feeding the population of over 100 billion. Advancement in techniques of production was introduced, use of fertilizers was encouraged so that productivity per acre can be increased but with its benefits, it also brought its own issues. More water was required for assimilation of fertilizers in land. Irrigation is a major source of concern in Indian context. There are states like Jharkhand, Orissa, Chhattisgarh, and MP where farmers get only one produce in a year due to lack of irrigation facilities. Although now we hear reports of people and government initiatives in villages so that round the year water supply can be assured, but it is still far from accomplishing the goal of getting rid from water scarcity.

India is a country where maximum permutations of weather conditions are available. We have Himalayan ranges and long coastal line also. The same crop which can be grown at one place during one season of a year can be grown in another region during another season. Lack of storage, warehousing, and food processing facilities led to a situation of oversupply in the case of a bumper crop thus farmers do not get their desired revenue output. At times, we face situation where agricultural produce especially fruits and vegetables get wasted in India [1].

There is so much of diversity in agriculture produce but the distribution of products and access to services remain untouched in some part of the country. If this diversity is utilized using modern web technology then shortages and excess of perishable food like fruits and vegetables can be controlled, and thus revenue of farmers can be maintained at an optimum level.

So, farmers in India need to be provided with good Agricultural Knowledge Base or an Information Retrieval System using semantic technology to run the farming on a profitable venture in the present competitive world. Lots of data on agriculture have been made but there is no linking of data. Although it is a big challenge to organize data, but this is possible if we have integrated web system and not multiple layers of webs. By incorporating intelligence feature to previous web technology, Semantic web [2] technology has gained immense popularity. Farmer can access the complete information regarding soil, access to mechanical instruments on cooperative basis, information on crop to be grown, quantity of water and fertilizer required, right source for seeds, etc., through Ontology [3].

The paper is organized as follows: In Sect. 2, I will take up the shortcoming of agriculture domain wrt Indian Scenario. These shortcomings motivated me to research further; Sect. 3 describes Work Already Done and its drawbacks. Section 4 describes Design and Development Methodology. Finally, Sect. 5 concludes with future work.

2 Need of Agriculture Ontologies

- Most of the population of India is engaged in farming. India produces variety of agricultural products, but distribution of products and services access remains untouched to some parts of the country. Semantic Web Technology can help in the service and management of resource distribution by utilization of knowledge base.
- Ministry of Agriculture Govt of India itself has a number of web portals for different departments and organizations, directorates where projects related to various departments currently exist. But these websites do not share web services among them, and hence contents are static, non-consistent, and non-integrated. Many times farmers and other stakeholders in agriculture sector have to visit multiple websites to get the desired information.
- Although it is a big challenge to organize and maintain data scattered on the web. Information is spread across various URL sites. But Semantic Web/Ontology Model promise availability of data and information anytime and anywhere in the web.
- India being multilingual country, publish materials in many native languages as per requirement of different states. Retrieval of information from these documents and using it for different information processing task is not possible at the present level of technology. So, sharing and processing information across languages is a major challenge. This varied technology standard results in convenience to user, repeated efforts, multiple source of information, and mismatch of information, hence, confusion.
- The current system is not able to fulfill the farmer query and their day to day problem which the semantic web technology promise to prove better solution in term of cost/benefit.

3 Work Already Done and Its Drawbacks

Several initiatives have been taken by state and central governments to meet the various challenges faced by agriculture sector in the country. The aim is to consolidate the various learning's from the past, integrate all the diverse efforts currently underway to reduce a knowledge deficit. Various organization and institutions are working to improve this knowledge deficit whose ultimate goal is to provide different agriculture related service.

Agropedia [4] is an online agriculture knowledge repository developed by Indian Institute of India-Kanpur (IITK). It facilitates exchange and delivery of information between the agricultural communities through a web portal in multiple languages using knowledge models. Knowledge Models [5] are used to organize, search, and navigate agriculture content to reduce the digital divide between agriculture experts and farmers.

AGROVOC [6] ontology is developed by FAO in 1980, but it still represents Multilingual Thesaurus covering 23 languages and used for searching terms, term definitions, and term relationships. It covers several domains like horticulture, fisheries, food, etc. But user gets limited information about the entered concept. AGROVAC has two models one is based on OWL (Web Ontology Language) and other is based on SKOS (Simple Knowledge Organization System).

Indian Council for Agricultural Research [7] has developed English–Hindi dictionary containing more than one thousand English words pertaining to agriculture domain and their Hindi translations. But the dictionary being in pdf format, retrieval of information from there and using it for different information processing task is not possible at the present level of technology.

INDOAGROVOC [8] Model for Indian Agricultural domain highlights the auto extraction of its web resource using semantic web framework to make uniform structural knowledge base source data that improves reusability and it can be shared among its related sub-domain and other dependent domain.

4   Design and Development Methodology

We are using protégé framework [9] to design and develop agriculture ontology [10] and generate RDF data set to further use in query building and searching techniques. We consider seeds, fertilizers, pesticides, various Govt. Schemes, soil, climate and crop management as a part of agriculture domain. We apply Divide and Conquer algorithm; splitting agriculture domain into further sub-domain. Considering sub-domain individually we create ontologies for these sub-domains separately and finally integrate these ontologies into a single final ontology.

In general, ontology [3] represents a conceptualization of a domain in terms of concepts, attributes, and relations. In a conceptual model, each concept is associated with a set of attributes. Ontology also defines a set of relations among its concepts. Semantic relations defined in ontology allow very complex queries to be answered, which is not possible otherwise [11].

Developed Ontology is used for knowledge representation and helps in organizing scattered data so that farmers can easily extract relevant information by using ontology extraction techniques. Developed ontology will help in Indexing, Retrieving, and organizing data [12] in agriculture domain, thus it reduces the Knowledge deficit in India. Steps are as follows:

1. Development of algorithms to convert existing databases to RDF format.
2. Semantic annotation of texts and identification of semantic roles in the documents.
3. Building ontology from the above sources.
4. Mapping and merging of ontology's created from different resources.
5. Validation of the results.
6. Publishing the final ontology in RDF (Figs. 1,2, and 3).

Fig. 1 Generation of RDF (resource description framework)

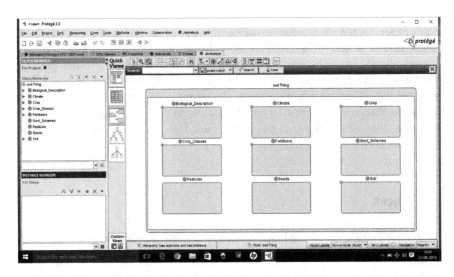

Fig. 2 Creating classes and sub-classes in Protégé ontology editor

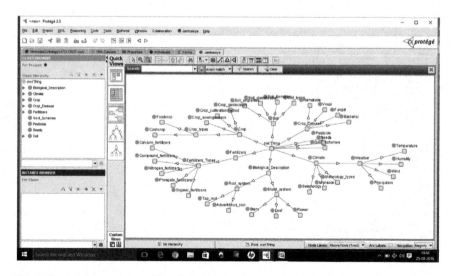

Fig. 3 Protégé framework used for designing and developing ontology

5 Conclusion

In this paper, Protégé framework is used to design our model into RDF/XML/OWL
format dataset which is validated datasets. Its purpose is to convert all the agriculture related data into machine-readable form by constructing ontologies for
sub-domains and integrating them together. However, to achieve this target we have
extracted the information from different documents and tables. In future, we will
use Oracle database to store RDF/OWL datasets to enhance farmer database. We
will also use SPARQL end point for querying the RDF datasets to give
concept-based information retrieval.

References

1. Indian Agriculture—Issues and Reforms with Prof. Gopal Naik [online]. http://spidi2.iimb.
   ernet.in/ ~ tejas/interviews/12.php
2. Berners-Lee, T., Hendler, J., Lassila, O.: The Semantic web. Sci. Am. 34–43(2001)
3. Chandrasekaran, B., Josephson, J.R., Benjamins, R.: What are ontologies, and why do we
   need them? IEEE Intell. Syst. **14**(1), 20–26 (1999)
4. Agropedia [online]: http://www.agropedia.iitk.ac.in
5. Agropedia: The Knowledge and Interaction hub for Indian Agriculture ICRISAT, [online].
   http://vasat.icirisat.org/images/New%20Folder/Agropedia.pdf
6. Caracciolo, C., Stellato, A., Morshed, A., Johannsen, G., Rajbahndari, S., Jaques, Y., Keizer,
   J.: The agrovac linked dataset. Semantic Web **4**(3), 241–248 (2013)
7. http://www.icar.org.in/node/847
8. http://www.cfilt.iitb.ac.in/wordnet/webhwn/wn.php

9. Knublauch, H., Fergerson R.W., Noy, N.F., Musen, M.A.: The protégé OWL plugin: an open development environment for semantic web applications. The Semantic Web-ISWC (2004)
10. Noy, N.F., McGuiness, D.L.: Ontology development 101: a guide to create your first ontology. Technical Report. Stanford Knowledge Systems Laboratory Technical Report KSL-01-05 and Stanford Medical Informatics Technical Report SMI-2001-0880 (2001)
11. Sinha, B., Chandra, S.: Semantic web ontology model for Indian agriculture domain (2013)
12. Jatwani, P., Tomar, P.: Improving Information Retrieval Effectiveness for Web Documents using Domain Ontology. In: Springer International Conference on Recent Developments in Science, Engineering and Technology (REDSET 2016)

Taguchi-Fuzzy Inference System (TFIS) to Optimize Process Parameters for Turning AISI 4340 Steel

Prashant D. Kamble, Atul C. Waghmare, Ramesh D. Askhedkar and Shilpa B. Sahare

Abstract In this paper, an attempt is made to investigate the application of Fuzzy inference system with Taguchi method for Multi-Objective Optimization (MOO) of cutting parameters for turning AISI 4340 steel. The effect of the uncontrollable parameter (spindle vibration) along with the controllable parameters on multiple objectives is studied. Use of uncontrollable factor helped to make the design robust. Coated and uncoated cutting tools and the latest lubrication method (Minimum quantity lubrication) are also considered to match the current scenario of the manufacturing system. Fuzzy logic is used to change multiple objectives to a single objective. The results of ANOVA for MPCI revealed that depth of cut is the most significant machining parameter which affects the multiple performance characteristics followed by feed rate, nose radius, cutting environment and tool type. Based on the response table and the main effect plot of S/N ratio, it is found that the optimal machining parameters are cutting environment = minimum quantity lubrication, nose radius = 1.2 mm, feed rate = 0.35 mm/rev, depth of cut = 1 mm, and tool type = coated (CVD) insert.

Keywords Taguchi-Fuzzy Inference System (TFIS) · Turning process Multi-Objective Optimization (MOO) · ANOVA · Signal to noise ratio

P.D. Kamble (✉) · S.B. Sahare
Department of Mechanical Engineering, Yeshwantrao Chavan College of Engineering, Nagpur, Maharastra, India
e-mail: pdk121180@yahoo.com

S.B. Sahare
e-mail: ssahare83@yahoo.com

A.C. Waghmare
Department of Mechanical Engineering, Umrer College of Engineering, Umrer, District of Nagpur, Maharastra, India
e-mail: dracwaghmare@rediffmail.com

R.D. Askhedkar
Department of Mechanical Engineering, KDK College of Engineering, Nagpur, Maharastra, India
e-mail: rdaskhedkar@rediffmail.com

© Springer Nature Singapore Pte Ltd. 2018
M. Pant et al. (eds.), *Soft Computing: Theories and Applications*,
Advances in Intelligent Systems and Computing 583,
https://doi.org/10.1007/978-981-10-5687-1_45

497

1   Introduction

In the metal cutting process, especially turning, milling process, besides the basic cutting process parameters like cutting speed, feed rate, depth of cut, tool geometry, the environment of cutting and the type of tool plays an important role to decide the performance of quality characteristics. Machining vibration is important in metal cutting operations which may affect the quality characteristics. The machine tool operators always face the problem of chatter in turning process. Over a period of time, the condition of the machine tool gets affected. When the machine is new, there is less chance of producing vibration in the machine. But when machine gets older, due to its continuous usage, vibrations increase. Performance and the efficiency of the machine tool get affected hence, proper quality of products not achieved through the machining operations are performed with the optimal operating condition. Vibration in a machine tool is directly affecting the surface finish of the work material in turning process. So vibration of a machine tool is one of the major factors limiting its performance. In machining, there has been recently and intensive computation focusing on surface roughness at international level. This computation can be observed in turning processes especially in the aviation and automotive industry by increasing the alternative solutions for obtaining more proper surface roughness. Taguchi philosophy is very useful in reducing the number of experiments as compared to other conventional methods like full factorial method which require very much high experimental runs. Saini et al. [1], presents optimization of multi-objective response during CNC Turning using Taguchi-Fuzzy Application. L_{27} Taguchi orthogonal array is used in turning Aluminum alloy 8011with carbide insert with cutting speed, feed, and depth of cut. Material removal rate and surface roughness are output parameters. It is found that feed is the most significant process parameter followed by the depth of cut and cutting speed on the selected response parameters. Vasudevan et al. [2], used grey fuzzy analysis for multi-objective optimization of turning parameters in turning GFRP/Epoxy Composites. Cutting tool nose radius, cutting speed, feed rate, and depth of cut are used as process parameters. Surface roughness parameter, tangential cutting force, and material removal rate are the output performance measures. The parameter combination of tool nose radius of 0.8 mm, cutting speed of 120 m/min, feed rate of 0.05 mm/rev, and depth of cut of 1.6 mm, is evaluated as an optimum combination.Hussain et al. [3], deals with fuzzy rule-based optimization of multiple responses in turning of GFRP Composites. L_{25} OA is used for analyzing surface roughness and cutting force. The results revealed that the optimization technique is greatly helpful for simultaneous optimization of multiple quality characteristics [1]. A grey-fuzzy approach is applied for optimizing machining parameters and the approach angle in turning AISI 1045 steel. L_9 OA is used for experimentation. The optimum conditions are found out by using a hybrid grey-fuzzy algorithm. Ho et al. [4], proposed a method using ANFIS to accurately establish the relationship between the features of the surface image and the actual surface roughness. Kirby et al. [5], discussed the development of a surface roughness prediction system for a turning operation using a fuzzy nets modeling technique.

2 Taguchi-Fuzzy Inference System (TFIS)

Fuzzy logic is a superset of conventional (Boolean) logic that has been extended to handle the concept of partial truth, where the truth value may range between completely true and completely false. A fuzzy inference system (FIS) defines a nonlinear mapping of the input data vector into a scalar output, using fuzzy rules. Fuzzy system is composed of a fuzzifier, an inference engine, a data base, a rule base, and defuzzifier. In the study, the fuzzifier initially uses membership functions to convert crisp inputs into fuzzy sets. Once all crisp input values have been fuzzified into their respective linguistic values, the inference engine will access the fuzzy rule base of the fuzzy expert system to derive linguistic values for the intermediate as well as the output linguistic variables. The fuzzy rule base consists of a group of if-then control rules with the two desirability function values, 1x and 2x one multi-response output y that is:

Rule 1: if x_1 is A_1 and x_2 is B_1 then y is C_1 else
Rule 2: if x_1 is A_2 and x_2 is B_2 then y is C_2 else
.. . .
Rule n: if x_1 is A_n and x_2 is B_n then y is C_n.

A_i and B_i are fuzzy subsets defined by the corresponding membership functions, i.e., μ_{A1} and μ_{B1}. Suppose x_1 and x_2 are the two desirability values, the membership function of the multi-response output y is expressed in Eq. (1).

$$
\mu_{E_0}(y) = \left(\mu_{A1}(x_1) \wedge \mu_{B_1}(x_2) \wedge \mu_{C_1}(y) \right) \ldots \vee \\
\left(\mu_{An}(x_1) \wedge \mu_{B_n}(x_2) \wedge \mu_{C_1}(y) \right) \tag{1}
$$

where $\wedge$ and $\vee$ are the minimum and maximum operation respectively. Equation (1) is illustrated in Fig. 1. Finally, a Centroid Defuzzification method is adopted to transform the fuzzy multi-response output $\mu_{c_0}(y)$ into a non-fuzzy value y_0 Eq. (2).

$$
y_0 = \frac{\sum y \mu_{c_0}(y)}{\sum \mu_{c_0}(y)} \tag{2}
$$

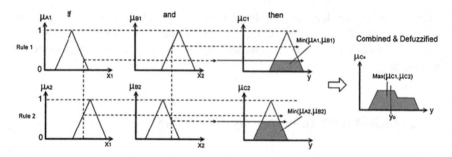

Fig. 1 Mamdani implication methods with fuzzy controller operations

The association of fuzzy inference system with Taguchi method is shown in Fig. 2. Firstly, depending upon number of input factors and their levels the orthogonal array is finalized. Then according to the orthogonal array, experimentation is conducted and outputs are measured. Signal to noise ratio of each output is calculated. Fuzzification is done by converting five outputs into normalized unit less values. Then, by applying the fuzzy rules the defuzzification is done and multiple performance characteristics index (MPCI) is computed. Thus the multiple outputs are converted into a single output. Then by using means of signal to noise ratio and analysis of variance, optimal setting for MPCI is achieved.

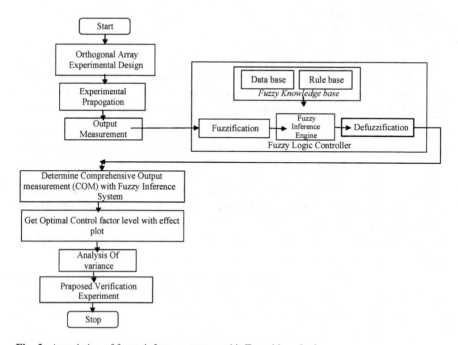

Fig. 2 Association of fuzzy inference system with Taguchi method

Table 1 Process parameters and their levels for experimentation

Process parameters	Abbreviation	Code	Level-1	Level-2	Level-3
Cutting environment	CE	A	DRY	WET	MQL
Nose radius (mm)	NR	B	0.4	0.8	1.2
Feed rate (mm/rev)	FR	C	0.15	0.25	0.35
Depth of cut (mm)	DOC	D	0.5	1	1.5
Tool type	TT	E	Uncoated	Coated (PVD)	Coated (CVD)
Noise factor					
Spindle vibration (m/s^2)	SV	NF	1.7	4.3	6.9

3 Experimental Observation and Analysis

3.1 Experimentation

The experiment is performed on CNC SPINNER15 lathe machine. Test pieces of size 50 mm × 80 mm were cut from AISI 4340 steel bar. The input variables used are cutting environment, nose radius, feed rate, depth of cut and tool type as shown in Table 1 and spindle vibration is selected as a noise factor (uncontrollable parameter). Cutting speed (210 m/min) is kept constant.

Three levels of spindle vibration are selected. First level is recorded by running the spindle at 1337 RPM without a ring. The second level is recorded by adding 45 gm unbalance mass to mild steel ring which is mounted on the spindle (Fig. 3). The third level is achieved by adding 95 g unbalance mass to mild steel ring which is mounted on the spindle. Spindle vibration readings are measured using VM 6360 Vibration meter. The surface roughness is measured by Stylus Profilometer, cutting force is measured by lathe tool dynamometer, tool tip temperature is recorded by non-contact laser type gun and tool wear is measured by image processing method in MATALAB software. MRR is calculated by following formula (Table 2):

Fig. 3 Experimental set up and unbalanced mass attached to mild steel ring

S. No.	Spindle speed (RPM)	Spindle vibration acceleration (m/s^2)		
Table 2 Machine tool's vibration acceleration level readings		Level-1	Level-2	Level-3
1	500	0.7	2.2	4.2
2	750	0.9	2.6	4.7
3	1,000	1.1	3.2	5.3
4	1,250	1.4	3.7	6.1
5[a]	1,337	1.7	4.3	6.9
6	1,500	2.1	4.7	[b]
7	1,750	2.4	5.1	
8	2,000	2.8	5.5	
9	2,250	3.3	6.3	
10	2,500	3.9	[b]	
11	2,750	4.4		
12	3,000	5.2		

[a] Three levels of spindle vibration are taken at spindle RPM 1337 (cutting speed 210 m/min and diameter 50 mm)
[b] Vibration levels at higher spindle speeds were not taken for machine tool safety

$$\text{MRR} = \frac{W_i - W_f}{\rho_s t} \text{ mm}^3/\text{min.}$$

3.2 Data Analysis

Data analysis has been carried out as follows:

For calculating S/N ratio of surface roughness and cutting force a Lower-the-Better (LB) criterion has been selected and for material removal rate Higher-the-Better (HB) criteria has been selected (Table 3). S/N ratios have been normalized based on Higher-the-Better (HB) criterion (Table 4). The Normalized S/N ratios corresponding to individual responses have been fed as inputs to a Fuzzy Inference System (FIS). For each of the input parameters, three Triangular type membership functions (MFs) have been chosen as follows: Low (L), Medium (M) and High (H). The linguistic valuation of COM has been represented by five triangular type membership functions (MFs) have been chosen as follows: Very Low (VL), Low (L), Medium (M), High (H), and Very High (VH). These linguistic values have transformed into crisp values by Defuzzification method. The crisp values (Table 4) have been optimized by using Taguchi philosophy. The predicted optimal setting has been evaluated from Mean Response Plot of MPCIs and it became $A_4 B_2 C_1 D_1 E_1$. FIS combined multiple inputs into a single output (Figs. 4, 5, 6, 7, 8; Tables 5, 6).

Table 3 Observation table with signal to noise ratio

RUN	A	B	C	D	E	SR	MRR	CF	TEMP	TW	S/N_SR	S/N_MRR	S/N_CF	S/N_TTT	S/N_TW
1	1	1	1	1	1	8.3	3.61	415.94	61.14	0.148	−18.3834	11.1542	−52.3805	−32.902	16.5912
2	1	1	2	2	2	6.21	9.99	431.14	66.83	0.103	−15.8636	19.9882	−52.6923	−33.3713	19.7259
3	1	1	3	3	3	4.84	12.47	598.16	66.57	0.155	−13.7044	21.9188	−55.5364	−33.5137	16.2009
4	1	2	1	2	3	7.03	8.14	394.65	62.49	0.121	−16.9452	18.2117	−51.9243	−32.8795	18.3176
5	1	2	2	3	1	5.18	10.05	571.25	73.31	0.167	−14.2887	20.043	−55.1365	−34.1024	15.559
6	1	2	3	1	2	3.14	5.61	413.64	64.15	0.071	−9.941	14.9812	−52.3325	−33.058	22.9192
7	1	3	1	3	2	5.13	8.28	530.96	68.5	0.173	−14.1991	18.3614	−54.5012	−33.7747	15.2281
8	1	3	2	1	3	3.84	4.19	377.16	64.42	0.061	−11.6875	12.4501	−51.5305	−32.9613	24.2507
9	1	3	3	2	1	2.23	10.72	405.63	74.89	0.078	−6.9513	20.606	−52.1626	−34.5106	22.1032
10	2	1	1	1	1	7.16	3.46	412.35	56.4	0.169	−17.0969	10.7866	−52.3054	−32.3119	15.4207
11	2	1	2	2	2	5.61	9.54	431.66	62.8	0.124	−14.978	19.5924	−52.7029	−32.8785	18.1235
12	2	1	3	3	3	4.06	11.85	609.66	64.6	0.175	−12.1633	21.474	−55.7018	−33.2122	15.1231
13	2	2	1	2	3	5.48	7.83	398.14	59.18	0.142	−14.7814	17.8727	−52.0006	−32.4261	16.9645
14	2	2	2	3	1	4.3	9.63	581.9	68.77	0.188	−12.6739	19.6681	−55.297	−33.6077	14.5329
15	2	2	3	1	2	2.48	5.32	418.86	60.12	0.093	−7.8905	14.5231	−52.4415	−32.5644	20.6475
16	2	3	1	3	2	4.67	7.98	546	64.27	0.195	−13.3826	18.0378	−54.7439	−33.3087	14.2181
17	2	3	2	1	3	3.12	4.01	385.34	58.34	0.082	−9.8784	12.0711	−51.7168	−32.2014	21.7015
18	2	3	3	2	1	1.57	10.23	412.8	70.15	0.099	−3.8945	20.2009	−52.3148	−34.0231	20.0913
19	3	1	1	1	1	5.96	3.34	414.46	56.09	0.136	−15.4989	10.4699	−52.3497	−32.2428	17.3603
20	3	1	2	2	2	4.57	9.18	431.34	60.04	0.091	−13.1956	19.2538	−52.6964	−32.5787	20.8517
21	3	1	3	3	3	3.4	11.34	602.79	62.04	0.142	−10.6198	21.0942	−55.6033	−33.0033	16.9347
22	3	2	1	2	3	5.41	7.57	396.08	58.67	0.109	−14.6686	17.5815	−51.9556	−32.3383	19.2653
23	3	2	2	3	1	3.86	9.28	575.45	68.27	0.154	−11.7326	19.3467	−55.2002	−33.5104	16.2385

(continued)

Table 3 (continued)

RUN	A	B	C	D	E	SR	MRR	CF	TEMP	TW	S/N_SR	S/N_MRR	S/N_CF	S/N_TTT	S/N_TW
24	3	2	3	1	2	1.82	5.09	415.72	57.36	0.059	−5.2012	14.1326	−52.3761	−32.2772	24.5961
25	3	3	1	3	2	4.31	7.73	537.04	61.71	0.161	−12.6813	17.7594	−54.6001	−33.0803	15.8569
26	3	3	2	1	3	2.91	3.87	380.4	57.83	0.049	−9.2782	11.7453	−51.6049	−32.1543	26.1409
27	3	3	3	2	1	1.03	9.83	408.49	70.36	0.066	−0.2554	19.8536	−52.2235	−33.9879	23.6159

Table 4 Normalized S/N ratio and comprehensive output measure

RUN	A	B	C	D	E	SR	SR	MRR	CF	TTT	TW	MPCI_FIS	SN_MPCI
1	1	1	1	1	1	8.3	0	0.06	0.8	0.68	0.2	0.42	-7.5873
2	1	1	2	2	2	6.21	0.14	0.83	0.72	0.48	0.46	0.51	-5.8243
3	1	1	3	3	3	4.84	0.26	1	0.04	0.42	0.17	0.41	-7.8014
4	1	2	1	2	3	7.03	0.08	0.68	0.91	0.69	0.34	0.54	-5.2777
5	1	2	2	3	1	5.18	0.23	0.84	0.14	0.17	0.11	0.34	-9.4041
6	1	2	3	1	2	3.14	0.47	0.39	0.81	0.62	0.73	0.54	-5.3469
7	1	3	1	3	2	5.13	0.23	0.69	0.29	0.31	0.08	0.39	-8.0793
8	1	3	2	1	3	3.84	0.37	0.17	1	0.66	0.84	0.55	-5.1584
9	1	3	3	2	1	2.23	0.63	0.89	0.85	0	0.66	0.59	-4.6236
10	2	1	1	1	1	7.16	0.07	0.03	0.81	0.93	0.1	0.4	-7.9814
11	2	1	2	2	2	5.61	0.19	0.8	0.72	0.69	0.33	0.55	-5.213
12	2	1	3	3	3	4.06	0.34	0.96	0	0.55	0.08	0.44	-7.0886
13	2	2	1	2	3	5.48	0.2	0.65	0.89	0.88	0.23	0.54	-5.2951
14	2	2	2	3	1	4.3	0.31	0.8	0.1	0.38	0.03	0.41	-7.8282
15	2	2	3	1	2	2.48	0.58	0.35	0.78	0.83	0.54	0.55	-5.233
16	2	3	1	3	2	4.67	0.28	0.66	0.23	0.51	0	0.44	-7.1989
17	2	3	2	1	3	3.12	0.47	0.14	0.96	0.98	0.63	0.55	-5.1332
18	2	3	3	2	1	1.57	0.8	0.85	0.81	0.21	0.49	0.59	-4.6545
19	3	1	1	1	1	5.96	0.16	0	0.8	0.96	0.26	0.44	-7.1493
20	3	1	2	2	2	4.57	0.29	0.77	0.72	0.82	0.56	0.58	-4.7469
21	3	1	3	3	3	3.4	0.43	0.93	0.02	0.64	0.23	0.47	-6.5193
22	3	2	1	2	3	5.41	0.2	0.62	0.9	0.92	0.42	0.55	-5.2102
23	3	2	2	3	1	3.86	0.37	0.78	0.12	0.42	0.17	0.43	-7.3099

(continued)

Table 4 (continued)

RUN	A	B	C	D	E	SR	SR	MRR	CF	TTT	TW	MPCI_FIS	SN_MPCI
24	3	2	3	1	2	1.82	0.73	0.32	0.8	0.95	0.87	0.64	−3.8866
25	3	3	1	3	2	4.31	0.31	0.64	0.26	0.61	0.14	0.45	−6.969
26	3	3	2	1	3	2.91	0.5	0.11	0.98	1	1	0.65	−3.7372
27	3	3	3	2	1	1.03	1	0.82	0.83	0.22	0.79	0.66	−3.6123

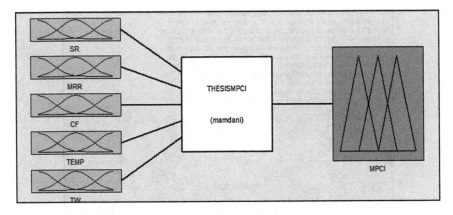

Fig. 4 Proposed fuzzy inference system

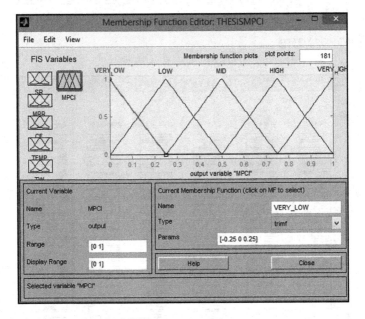

Fig. 5 Membership functions for MPCI

4 Result

The results of ANOVA indicate that depth of cut is the most significant machining parameter in affecting the multiple performance characteristics followed by feed rate, nose radius, cutting environment and tool type. From main effect plot of S/N ratio, the

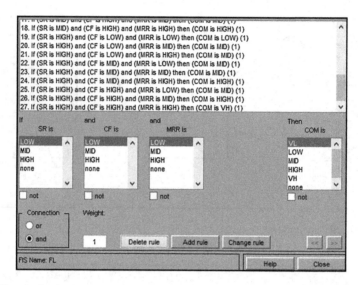

Fig. 6 Fuzzy rule viewers

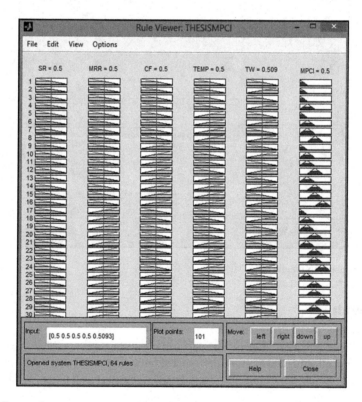

Fig. 7 Fuzzy rules

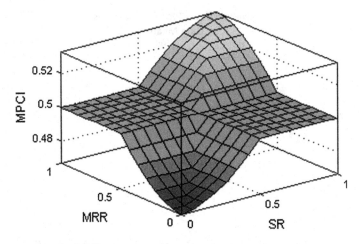

Fig. 8 Surface plot for MPCI

Table 5 Response table for S/N ratio (higher-the-better) of MPCI_FIS

Level	Cutting environment	Nose radius	Feed rate	Depth of cut	Tool type
1	−6.567	−6.657	−6.750	−5.690	−6.683
2	−6.181	−6.088	−6.039	−4.940	−5.833
3	−5.460	−5.463	−5.418	−7.578	−5.691
Delta	1.107	1.194	1.331	2.638	0.992
Rank	4	3	2	1	5

Table 6 Analysis of variance for SN ratio of MPCI_FIS

Process parameters	Dof	SOS	MSOS	F	P	% Contribution
Cutting environment (A)	2	5.6815	2.84075	109.79	0.000	9.31
Nose radius (B)	2	6.4189	3.20945	124.04	0.000	10.52
Feed rate (C)	2	7.9879	3.99395	154.36	0.000	13.09
Depth of cut (D)	2	33.2512	16.6256	642.54	0.000	54.50
Tool type (E)	2	5.1826	2.5913	100.15	0.000	8.49
A * B	4	0.1332	0.0333	1.29	0.406	0.22
A * C	4	0.9047	0.226175	8.74	0.098	1.48
A * D	4	1.3492	0.3373	13.04	0.064	2.21
Residual error	4	0.1035	0.025875			0.17
Total	26	61.0125				100.00

optimal machining parameters obtained are the cutting environment = MQL, nose radius = 1.2 mm, feed rate = 0.35 mm/rev, depth of cut = 1 mm and tool type = coated (CVD) or $A_3B_3C_3D_2E_3$ in short (Figs. 9 and 10).

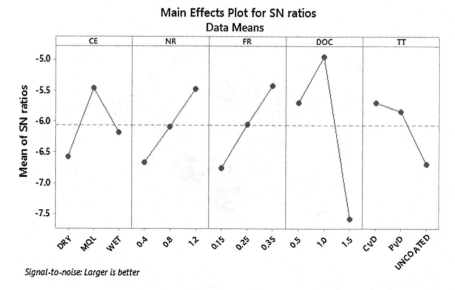

Fig. 9 Main effect plot for S/N ratio of MPCI

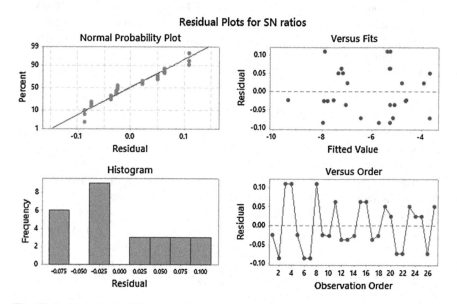

Fig. 10 Residual plot for SN ratio

5   Conclusion

The fuzzy logic with Taguchi method is successfully applied. The multi-objective problem is solved with the application of combined Taguchi-Fuzzy inference system. Multi- responses are converted to a single response by a fuzzy rule base. It is observed that the optimal process condition for lower surface roughness, cutting force, tool tip temperature and tool wear and higher MRR is feed rate 0.35 mm/rev, depth of cut 1 mm with 1.2 mm nose radius of CVD coated insert under Minimum Quantity Lubrication. ANOVA indicates that depth of cut (54.50% contribution) is the most significant machining parameter followed by feed rate (13.09%), nose radius (10.52%), cutting environment (9.31%) and tool type (8.49%).

References

1. Saini, S.K., Pradhan, S.K.: Optimization of multi-objective response during cnc turning using taguchi-fuzzy application. In: 12th Global Congress on Manufacturing and Management, Elsevier, pp. 85–94 (2014)
2. Vasudevan, H., Deshpande, Ramesh, N.C., Rajguru R.: Grey fuzzy multiobjective optimization of process parameters for CNC turning of GFRP/epoxy composites. In: International Conference on Design and Manufacturing, pp. 85–94 (2014)
3. Hussain, S.A., Palani Kumar, K., Gopala Krishna, A.: Fuzzy rule based optimization of multiple responses in turning of GFRP composites. Int. J. Adv. Sci. Technol. **74**, 25–34 (2015)
4. Ho, S.-Y., Lee, K.-C., Chen, S.-S., Ho, S.-J.: Accurate modeling and prediction of surface roughness by computer vision in turning operations using an adaptive neuro-fuzzy inference system. Int. J. Mach. Tools Manuf. **42**(13), 1441–1446 (2002)
5. Daniel Kirby, E., Chen, J.C.: Development of a Fuzzy-Nets-Based Surface Roughness Prediction System in Turning Operations. Elsevier, vol. 53, Issue 1, pp. 30–42 (2007)

Application of Fuzzy Multi-criteria Approach to Assess the Water Quality of River Ganges

R. Srinivas and Ajit Pratap Singh

Abstract The purpose of this study is to develop a fuzzy multi-criteria decision-making framework to evaluate the water quality status of a river basin. The rampant and indiscriminate growth in the urban, agricultural, and industrial sector has directly or indirectly disrupted the water quality of the major rivers by discharging mammoth quantities of wastewaters. Regular and accurate evaluation of water quality of a river has become an important task of water authorities. However, the conventional way of evaluating water quality index has been unsuccessful in incorporating uncertainties and subjectivities associated with water quality analysis. Such limitations can be dealt effectively by using fuzzy logic concepts. The present study proposes an Interactive Fuzzy Water Quality Index (IFWQI) to evaluate the water quality status of river Ganges at Kanpur city, India. Multi-Criteria Decision-Making (MCDM) tool namely Fuzzy Inference System (FIS) of MATLAB has been used to obtain a qualitative and quantitative measure of water quality index at six different sites of Kanpur throughout the year by taking into consideration the six important water quality parameters. The results indicate a significant improvement in the accuracy of the index values and thus providing emphatic information to the planners to decide the remedial measures for sustainable management of river Ganges.

Keywords Multi-criteria Decision-Making · Fuzzy inference · Water quality

R. Srinivas (✉) · A.P. Singh
Civil Engineering Department, Birla Institute of Technology and Science,
Pilani 333031, Rajasthan, India
e-mail: r.srinivas@pilani.bits-pilani.ac.in

A.P. Singh
e-mail: aps@pilani.bits-pilani.ac.in

© Springer Nature Singapore Pte Ltd. 2018
M. Pant et al. (eds.), *Soft Computing: Theories and Applications*,
Advances in Intelligent Systems and Computing 583,
https://doi.org/10.1007/978-981-10-5687-1_46

513

1 Introduction

In general, the problems associated with river water resources management consist of managing the water quality and quantity. Such problems depend on several qualitative and quantitative criteria, and a proper decision-making can be done by integrating all such criteria under a suitable mathematical framework. Effective and efficient evaluation of water quality status of different segments of a river based on critically polluting water quality criteria is one of the most important subject matters of water authorities and planners. Multi-criteria decision-making (MCDM) methods are considered as one of the popular approaches to deal with such problems. Several researchers have presented different MCDM methods in the field of water resource management namely: Elimination Et Choice Translating Reality I (ELECTRE) [1]; Analytic Hierarchy Process (AHP) [2]; Technique for Order Preference by Simulation of Ideal Solution (TOPSIS) [3]; Analytic Network Process (ANP) [4]; and Preference Ranking Organization Method for Enrichment Evaluations (PROMETHEE) [5]. One of the primary limitations of these traditional methods is their inability to incorporate the uncertainties pertaining to complex water resource management problems as they employ the mathematical classic numbers. The uncertainties can be effectively treated in a fuzzy logic system. Therefore, there is a scope to develop a model which combines both MCDM and fuzzy set theory to achieve a fuzzy-based MCDM framework.

Rivers have been playing a very crucial role in sustaining the lives of living beings since age-old by serving them in various ways. Rivers support the majority of the developmental processes of a country like urbanization, agricultural production, power generation, industrial growth, low-cost navigation, and many others. Therefore, proper planning and management of rivers in a sustainable way should be the priority of all the countries. The Ganges river of India is the largest and most sacred river of the country. The river supports more than 450 million people by providing water for drinking, irrigation, and industrial purposes. Out of all the states benefitted by Ganga, Uttar Pradesh derives maximum benefit as the river traverses more than 1100 km in this state. Major cities like Kanpur, Allahabad, and Varanasi are profusely blessed with the holy water of Ganges. However, it is a very unpleasant fact that these cities are expressing their gratitude toward the Ganges by treating it as a drain. Therefore, there is an acute need to compute the water quality status of the river at these sites in order to implement necessary remedial measures to reduce pollution of Ganges. Evaluation of Water Quality Index (WQI) is found to be an effective and efficient tool for assessing the suitability of river water for various beneficial usages [6]. The WQI integrates all water quality parameters by comparison with their respective standard value and thus give a single dimensionless number, which indicates the overall water quality status [7, 8]. However, one of the major limitations of the conventional WQI approach is its inability to incorporate the uncertainty and ambiguity associated with the concentration of the quality parameters [9]. Such limitations and complexities involved in the deterministic and traditional methods for calculating WQI have motivated the development of a more advanced methodology, capable of aggregating and accounting for the vague, inaccurate, and fuzzy information pertaining to water quality. The application of fuzzy logic [9] in modeling water quality indices has shown a good promise [10, 11]. Reference [12] used a Fuzzy Comprehensive Assessment (FCA) to evaluate the soil environmental quality of the Taihu lake watershed. Reference [13] expressed groundwater sustainability in mathematical terms using MATLAB Fuzzy logic toolbox. Though various investigators have also applied the concept of fuzzy multi-criteria in water quality assessment and environmental management [14, 15], there is still enough scope to develop systematic and flexible models, which can be used in assessing water quality by combining multi-criteria approach and fuzzy logic concepts.

The purpose of this study is to propose an interactive fuzzy-based water quality index (IFWQI) which can be computed using a multi-criteria decision-making tool having an artificial intelligence interface known as MATLAB fuzzy inference system. The water quality of river Ganges is analyzed at six different stations of Kanpur city for several beneficial usages.

2 Materials and Methods

As mentioned above, the uncertainties encountered in water quality analysis can be dealt adequately by incorporating concepts of fuzzy logic [16–19]. In this paper, a MATLAB-based fuzzy inference system (FIS) framework has been developed which maps input water quality parameters values to outputs (overall water quality). The methodology involves four steps (a) fuzzification of the crisp input values by mapping them into linguistic variables using membership functions, (b) evaluation of fuzzy inference rules which consists of linguistic rules in the form of IF-THEN statements, (c) aggregation of rule outputs using the fuzzy union of all the individual rule contributions to obtain a single aggregated membership function. (d) defuzzification of aggregated output fuzzy set using centroid method.

2.1 Sampling Sites and Water Quality Parameters

Kanpur, the largest city in the state of Uttar Pradesh is rated as the most polluted city along the Ganges river basin primarily due to more than 700 tanneries indiscriminately discharging wastewater in the river. The length of the Ganges in Kanpur is 38.7 km. In this paper, a total of six sampling stations namely Bithoor Ghat (S1), Rani Ghat (S2), Permat Ghat (S3), Sarsaiya Ghat (S4), Nanarao Ghat (S5), and Siddhnath Ghat (S6) have been chosen from Kanpur city (Fig. 1). The suitability of the water for particular beneficial usage is determined by evaluating FWQI at all the sampling sites. Six critical water quality parameters namely Dissolved Oxygen

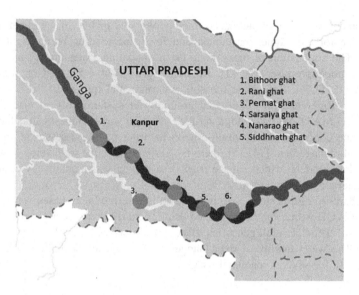

Fig. 1 Study area representing the six sampling stations

(DO), Biochemical Oxygen Demand (BOD), Total Dissolved Solids (TDS), Total Alkalinity (TA), Phosphate (PO_4^{3-}), and Chromium (Cr) have been chosen to evaluate the water quality index.

2.2 Fuzzy Water Quality Index Using MATLAB FIS

In this paper, a fuzzy methodology is proposed to derive water quality index of a particular sampling site. The membership functions of different water quality parameters considered for the study are developed under expert guidance based on the standards given by regulatory bodies [19]. The method used to generate membership functions of the input and output parameters is very simple. Depending upon the nature of the parameter, a particular shape of the membership function is derived. For example, for DO to be considered as "average," its measurements must fall between 2 mg/l to 5 mg/l. Hence, a trapezoidal membership function (μ_{DO}) with ranges [2–5] is chosen to represent the linguistic term 'average' which can be written in equation form "(1)," as given below:

$$\mu_{DO} = \begin{cases} 0, x < 2 \text{ or } x > 5 \\ \frac{(X-2)}{(3-2)}, 2 \le x \le 3 \\ 1, 3 \le x \le 4 \\ \frac{(5-x)}{(5-4)}, 4 \le x \le 5 \end{cases} \tag{1}$$

where x = value of DO (mg/l).

In a similar way, membership functions of other input parameters and output have been derived. In this study, only trapezoidal membership functions have been used, as they are capable of representing the real life situation in a better way. Figures 2 and 3 represent the shapes of the membership functions of input parameter BOD and the output (fuzzy water quality index), respectively, where each color range of the trapezoid represents a linguistic variable as explained further.

'Excellent (E)', 'Good (G)', 'Average (A)', 'Poor (P)', and 'Very Poor (VP)' are linguistic representation of the input water quality parameters having linguistic variables DO, BOD, Cr, TDS, TA and DO and PO_3^{-4}. These linguistic values are assigned to each parameter based on their fuzzified value. 'Very bad (VB)', 'Bad (B)', 'Satisfactory (S)', 'Good (G)' and 'Very good (VG)' are the linguistic representation of the FWQI of a given sampling station along the river. In general, the total number of rules that can be formed is given by R = [number of linguistic variables] $^{(number\ of\ parameters)}$. For example, considering five linguistic representations (i.e., excellent, good, average, poor, and very poor) for 6 parameters corresponding to a given site, there will be 5^6 rules that can be formed under expert guidance to obtain the crisp measure the WQI. Once these rules are formed using the data values of the input criteria, they are aggregated and finally defuzzified to

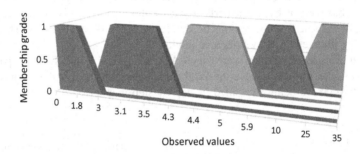

Fig. 2 Membership function for input parameter BOD

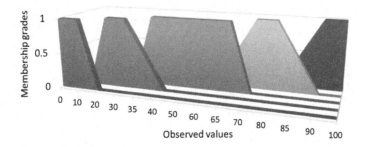

Fig. 3 Membership function for output parameter FQWI

obtain a crisp measurement of WQI. Since the rules and input parameters can be modified, added, or deleted, hence the model developed herein is called an interactive model as the model is flexible to accommodate changes based on the opinion of experts or decision makers. Depending on the observed values of the input parameters, several rules are fired by the system and are aggregated to obtain to fuzzified membership function which upon defuzzification gives the final crisp score of water quality index.

Some of the IF–THEN rules formed under expert guidance for assessing the fuzzy water quality index based on the concentration of DO, BOD, Cr, TDS, TA, and $PO_4{}^{3-}$ are given below:

Rule 1: IF DO is 'excellent' AND BOD is 'poor' AND Cr is 'poor' AND TDS is 'good' AND TA is 'poor' AND $PO_4{}^{3-}$ is 'average' THEN water quality index is 'very bad'.
Rule 2: IF DO is 'very good' AND BOD is 'very poor' AND Cr is 'good' AND TDS is 'good' AND TA is 'very poor' AND $PO_4{}^{3-}$ is 'good' THEN water quality index is 'very bad'.

Table 1 represents the fuzzy trapezoidal membership functions of the input parameters BOD and DO, where all the values are in milligram per liters (mg/l). Similar way, membership functions for other input parameters have been defined. Table 2 represents the fuzzy trapezoidal membership functions of the output parameter FWQI. The categorization of these values into different linguistic variables has been given by the experts based on the standards given by regulatory bodies [19]. Table 3 shows some of the fuzzy rules derived in this study for developing interactive fuzzy inference model.

The entire analysis is performed in MATLAB Fuzzy Logic Toolbox package R2015b (8.6.0). For illustration, the WQI calculations for the month of January for Siddhnath ghat are being explained in step by step manner.

Table 1 Membership function values of BOD and DO

Linguistic representation	BOD [a, b, c, d]	DO [a, b, c, d]
Excellent	[0, 0, 1.8, 3]	[7, 7, 20, 20]
Good	[1.8, 3, 3.5, 5]	[4, 5, 7, 7]
Average	[3.1, 4.3, 5, 10]	[2, 3, 4, 5]
Poor	[4.4, 5.9, 10, 25]	[0, 1, 2, 3]
Very poor	[5.9, 25, 35, 35]	[0, 0, 1, 1]

Table 2 Membership function values of FWQI

Linguistic representation	FWQI [a, b, c, d]
Very good	[80, 90, 100, 100]
Good	[60, 70, 80, 90]
Satisfactory	[30, 35, 65, 70]
Bad	[10, 20, 30, 40]
Very bad	[0, 0, 10, 20]

Table 3 Some fuzzy rules formed using MATLAB FIS

	Input parameters						FWQI Output
Operators	IF	AND	AND	AND	AND	AND	THEN
Parameters	DO	BOD	Cr	TDS	TA	PO_4^{-3}	Resulting integrity
Rule 1	E	E	E	E	P	P	B
Rule 2	E	E	G	E	E	E	G
Rule 3	VP	P	VP	P	P	G	VB
Rule 4	E	A	VP	G	G	G	VB
Rule 5	A	G	A	G	A	VG	S
Rule 6	G	G	G	VP	G	VP	VB
Rule 7	VP	P	P	P	P	P	VB
Rule 8	E	E	VP	VP	E	E	VB
Rule 9	E	E	E	E	E	E	VG
Rule 10	A	E	A	E	A	E	VG
Rule 11	A	E	E	A	A	VG	S
Rule 12	E	G	E	E	E	A	S
Rule 13	A	VP	A	VP	A	VP	VB
Rule 14	G	G	G	G	G	G	G

Fuzzification (**Step 1**): The data values for six parameters [DO, BOD, TDS, TA, PO_4^{3-}, Cr] for the month of January are [6.6; 18.5; 228; 321; 1.01; 2.3] mg/l. These input-data values are fuzzified and represented linguistically using appropriate membership function.

Rule evaluation (**Step 2**): Several rules are fired depending on the input values and their corresponding membership function (s). In this case, one of the rule that is fired is as follows:

IF DO is 'good' AND BOD is 'poor' AND TDS is 'good' AND TA is 'very poor' AND PO_4^{3-} is 'very poor' AND Cr is 'very poor' THEN water quality index is 'very bad'.

As the inference is based on the minimum sub-index, an optimized solution can be obtained by using a disjunction of inputs by means of the "OR" operator: the FWQI is considered "very poor" if any one of the indicators is "very poor." In this study, the Max- Min approach has been used to build the 'FWQI' inference engine. The implication method used is the "min" and the aggregation method is "max." The extension of the union operator (OR) and intersection operator (AND) to fuzzy sets A and B for any value x defined over the same set U is represented in equation "(2)," and "(3)," respectively.

$$\mu_{A \cup B}(x) = \max[\mu_A(x), \mu_B(x)] \tag{2}$$

$$\mu_{A \cap B}(x) = \min[\mu_A(x), \mu_B(x)] \tag{3}$$

Rule aggregation and defuzzification (**Step 3**): All such rules are then aggregated to obtain a single shape of membership function which upon defuzzification gives the WQI value. In this case, the center of gravity (COG) method has been used for defuzzification to determine the output as expressed in equation "(4)." In this case, output value has come equal to 8.22 indicating that water quality during the month of January of at this location can be classified as "very bad."

$$z^* = \frac{\int \mu_c(z).zdz}{\int \mu_c(z)dz}; \text{ where } z \in C, C \text{ is the fuzzy set} \tag{4}$$

3   Results and Discussions

The FWQI values obtained using fuzzy inference system for all the six stations are below 10 indicating that the status of water quality at all the stations is very bad (Fig. 4). A very interesting observation is that the Siddhnath Ghat (S6) has obtained least score of FWQI almost through the year (Fig. 4). This is mainly because of the discharge of mammoth quantities of wastewaters from the tanneries and paper pulp industries consisting high content of BOD and Cr. On the other hand, Rani Ghat (S2) has obtained slightly better score than other sampling stations. Here, the river water is mainly contaminated due to several open drains containing a high content of BOD entering into the river without any treatment. On the basis of the FWQI values obtained, it can be clearly inferred that Ganges river water needs serious treatment before it is used for any beneficial purposes and there is an urgent need to put strict regulations on the industries, establish sewage treatment plants, and create public awareness in the Kanpur region to protect river Ganges. The proposed water quality fuzzy index has also revealed some of the drawbacks of the conventional water quality index approach. The fuzzy-based index is more effective and accurate in determining the overall water quality status of the river based on the standards proposed by the regulatory bodies. Fuzzy logic not just represents water quality linguistically and mathematically but also shows the variation of the water quality parameters in the form of well-defined membership functions. The model

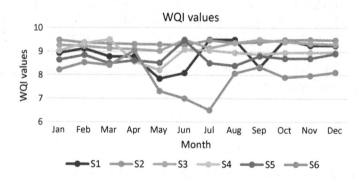

Fig. 4 Final WQI scores of all sampling sites along Ganges

developed also gives the opportunity to the decision makers to assign different membership functions, change the parameters and values of the parameters, and accordingly observe the overall impact on the river. Thus the model is also known as an interactive fuzzy-based model as it gives prerogative to the policy makers to modify it according to their preferences.

4 Conclusions

Application of fuzzy logic to obtain WQI has proved to be an essential informative decision-making tool to obtain both qualitative and quantitative measure of the water quality of river Ganges. The proposed fuzzy inference model can effectively deal with the problems involving uncertainty and linguistic vagueness pertaining to river water quality. In addition, unlike the conventional index method, the FWQI provides a scope for the results to be interpreted both quantitatively and qualitatively with the help of expert defined membership grades. The model provides a scope for better analysis since experts can describe a sampling station quality status as closer to its upper or lower limit. The model developed is flexible and interactive as it allows the decision maker to add, delete or modify the input water quality parameters. Thus, it will assist decision makers in investigating the condition of water quality by incorporating spatial and temporal changes in the river. The ability of the model to provide a framework to the experts to define rules linguistically and thus obtain a crisp measure of water quality makes it an alternate tool for the analysis of river water quality and for sustainable planning in the context of integrated river basin management.

References

1. Raj, A.: Multicriteria methods in river basin planning-a case study. Water Sci. Technol. **31**(8), 261–272 (1995)
2. Minatour, Y., Khazaei, J., Ataei, M.: Earth dam site selection using the analytic hierarchy process (AHP): a case study in the west of Iran. Arab. J. Geosci. (2012). doi:10.1007/s12517-012-0602-x
3. Zarghami, M., Abrishamchi, A., Ardakanian, R.: Multi-criteri decision making for integrated urban water management. Water Resour. Manage **22**(8), 1017–1029 (2008)
4. Razavi Toosi, S.L., Samani, J.M.V.: Evaluating water transfer projects using analytic network process (ANP). Water Resour. Manag. **26**(7), 1999–2014 (1999)
5. Nasiri, H., Darvishi, B.A., Faraji, S.H.A., Jafari, H.R., Hamzeh, M.R.Y.: Determining the most suitable areas for artificial groundwater recharge via an integrated PROMETHEE II-AHP method in GIS environment (case study: Garabaygan Basin, Iran). Environ. Monit. Assess. **185**(1), 707–718 (2013)
6. Miller, W.W., Joung, H.M., Mahannah, C.N., Garrett, J.R.: Identification of water quality differences Nevada through index Application. J. Environ. Quality **15**(3), 265–272 (1986)

7. Cude, C.: Oregon water quality index: a tool for evaluating water quality management effectiveness. J. Am. Water Resour. Assoc. **37**(1), 125–137 (2001)
8. Liou, S., Lo, S., Wang, S.: A generalized water quality index for Taiwan. Environ. Monit. Assess. **96**(1–3), 35–52 (2004)
9. Icaga, Y.: Fuzzy evaluation of water classification. Ecol. Indicators **7**, 710–718 (2007)
10. Lu, R.S., Lo, S.L., Hu, J.Y.: Analysis of reservoir water quality using fuzzy synthetic evaluation. Stoch. Env. Res. Risk **13**(5), 327–336 (1999)
11. Chang, N.B., Chen, H.W., Ning, S.K.: Identification of river water quality using the fuzzy synthetic evaluation approach. J. Environ. Manag. **63**, 293–305 (2001)
12. Shen, G., Lu, Y., Wang, M., Sun, Y.: Status and fuzzy comprehensive assessment of combined heavy metal and organo-chlorine pesticide pollution in the Taihu Lake region of China. J. Environ. Manag. **76**(4), 355–362 (2005)
13. Srinivas, R., Bhakar, P., Singh, A.P.: Groundwater Quality Assessment in some selected area of Rajasthan, India Using Fuzzy Multi-Criteria Decision Making Tool. Elsevier Aquatic Procedia **4**, 1023–1030 (2015)
14. Singh, A.P., Srinivas, R., Kumar, S., Chakrabarti, S.: Water quality assessment of a river basin under Fuzzy Multi-Criteria Framework. Int. J. Water **9**(3), 226–247 (2015)
15. Singh, A.P., Vidyarthi, A.K.: Optimal allocation of landfill disposal site: A fuzzy multi-criteria approach. Iranian J. Environ. Health Sci. Eng. **5**(1), 25–34 (2008)
16. Kommadath, B., Sarkar, R., Rath, B.: A fuzzy logic based approach to access sustainable development of the mining and mineral sector. Sustain. Dev. **20**(6), 386–399 (2011)
17. Singh, A.P., Ghosh, S.K.: Conceptual modeling and management of water quality in a river basin. In: Ramanathan, A.L., Ramesh (eds.) Recent Trends in Hydro Geochemistry. Capital Books, New Delhi, 207–220 (2003)
18. Singh, A.P., Ghosh, S.K.: Uncertainty Analysis in River Basin Water Quality Management. In: Raju, K.S., Sarkar, A.K., Dash, M.L. (eds.) Integrated Water Resources Planning and Management, pp. 260–268. Jain Brothers, New Delhi (2003)
19. Central Pollution Control Board (CPCB), Annual Report: Government of India, New Delhi (2013)

Hodgkin–Huxley Model Revisited to Incorporate the Physical Parameters Affected by Anesthesia

Suman Bhatia, Phool Singh and Prabha Sharma

Abstract Hodgkin–Huxley model describes the action potential phenomenon on the basis of electrochemical properties but does not characterize the anesthetic effects. In this paper, we have proposed a model which reframes Hodgkin–Huxley model to be able to identify the parameters affected by anesthesia. The model comprises of set of partial differential equations that describe how the viscosity of fluid moving along the axon impacts the propagation of action potential. It is observed that with the increase in viscosity of the fluid, there is a reduction in the conduction velocity. The viscosity of the fluid moving along the axon has also been characterized with respect to the temperature, the physical parameter considered in the Hodgkin-Huxley model. The model has been solved using finite difference method and implemented using C++ syntax code in an iterative manner. The results obtained are consistent with the freezing point depression theory for the explanation of anesthesia. The model acts as a framework for drug therapists inducing anesthesia to analyze the target parameters responsible for blocking of action potential propagation and hence for possible therapeutic intervention.

Keywords Hodgkin-Huxley model · Reduced conduction velocity · Anesthesia

Nomenclature

c_m	Membrane capacitance per unit area of membrane
D_{Na}	Diffusivity of sodium ions in fluid
D_K	Diffusivity of potassium ions in fluid
D_L	Diffusivity of chlorine ions in fluid
F	Faraday's constant
$\bar{g}_K$	Conductance of potassium ions per unit area of the membrane
$\bar{g}_L$	Conductance of leakage current per unit area of membrane
$\bar{g}_{Na}$	Conductance of sodium ions per unit area of the membrane

S. Bhatia (✉) · P. Sharma
Department of CSE & IT, The NorthCap University, Gurgaon 122017, India
e-mail: ersuman80@gmail.com

P. Singh
Department of Applied Sciences, The NorthCap University, Gurgaon 122017, India

© Springer Nature Singapore Pte Ltd. 2018
M. Pant et al. (eds.), *Soft Computing: Theories and Applications*,
Advances in Intelligent Systems and Computing 583,
https://doi.org/10.1007/978-981-10-5687-1_47

523

K_B	Boltzman Constant
M_{Na}	Molar mass of sodium ions
M_K	Molar mass of potassium ions
M_L	Molar mass of chlorine ions
r	Axon radius
rad_i	Radius of different ions i
R_a	Resistance per unit axial length
ρ	Density
R_u	Universal gas constant
S	Source term
T	Temperature in kelvin
v_F	Viscosity of fluid inside the axon
v_W	Viscosity of water
V	Membrane voltage
V_{Na}	Equilibrium potential of sodium ions
V_K	Equilibrium potential of potassium ions
V_L	Equilibrium potential of chlorine ions
$\dot{w}_{Na}{}'''$	Rate of addition of mass of sodium ions per unit volume
$\dot{w}_K{}'''$	Rate of addition of mass of potassium ions per unit volume
$\dot{w}_L{}'''$:	Rate of addition of mass of chlorine per unit volume
Y_{Na}	Mass fraction of sodium ions
Y_K	Mass fraction of potassium ions
Y_L	Mass fraction of chlorine ions

Values taken for different parameters are given below [1]

$\bar{g}_{Na}$	$1200 \, S/m^2$
$\bar{g}_K$	$360 \, S/m^2$
$\bar{g}_L$	$3 \, S/m^2$
V_{Na}	$0.050 \, V$
V_K	$-0.077 \, V$
V_{Cl}	$-0.054 \, V$
c_m	$0.01 F/m^2$
R_a	$resistivity/\Pi r^2 \, \Omega/m$
resistivity	$0.354 \, \Omega - m$
rad_{Na}	$0.102 * 10^{-9}$ meters
rad_K	$0.138 * 10^{-9}$ meters
rad_{Cl}	$0.181 * 10^{-9}$ meters

1 Introduction

The generation and propagation of action potential defined by coupling the Hodgkin–Huxley model (further referred to as H-H model) with the Cable theory is given below [2]:

$$\frac{1}{2\Pi r}\frac{\partial}{\partial x}\left(\frac{1}{R_a}\frac{\partial V(x,t)}{\partial x}\right) = \left(c_m\frac{dV}{dt} + \bar{g}_k n^4(V - V_k) + \bar{g}_{Na}m^3 h(V - V_{Na}) + \bar{g}_L(V - V_L)\right)$$

(1)

Above modeling explains the neuronal excitability [3, 4] on the basis of the electrochemical behavior of membrane by taking into consideration the capacitive current and ionic currents. Along with neuronal excitability, action potential propagation through axon of a neuron is also explained clearly. During this propagation through the axon, the action potential actually passes through the fluid moving inside the axon. But there is no complete physiological description given by H-H model as to how this action potential propagation is affected by axoplasmic fluid flow. Also, there is no explanation given for how the anesthesia will affect the action potential conduction through the fluid. Kassahun et al. also highlight the limitation of H-H model in the analysis of thermodynamical properties required for anesthetic effects [5]. Work carried out by Schneider and Pekker [6] shows that the non-propagating action potential effect can be utilized for the description of anesthesia process using Hodgkin-Huxley model. Moreover, research from Meyer Overton Rule [7] till date is focused on the investigation of effect of local and general anesthesia on ligand gated ion channels and voltage gated ion channels [8–10]. Postea and Biel investigate on the possibility of HCN channels as a target for anesthesia, as these channels are known for controlling nerve excitation [11]. In parallel, this phenomenon has also been studied using Soliton theory based on the thermodynamical analysis of membrane [12–15]. Research performed by Graesball et al. is focused on the analysis of membrane heat capacity profiles in the presence of anesthesia [16].

But, still, there is a need to identify the fluid properties that are affected by the anesthesia and hence working in the same direction, a model has been proposed that characterizes the action potential propagation on the basis of viscosity of fluid moving along the axon. This model also investigates the impact of temperature on viscosity and hence on action potential propagation. The impact of temperature is already demonstrated in H-H model [17, 18], but in the proposed model, the coupled effect of viscosity and temperature is analyzed to study action potential phenomenon for a better understanding of anesthetic effects.

2 Methodology

2.1 Model to Incorporate Physical Parameters Affected by Anesthesia

While solving the proposed model to incorporate the impact of viscosity along with temperature on action potential propagation, certain assumptions have been taken into account which are given as follows [19, 20]:

1. Axon is considered to be a perfect cylinder.
2. The model is solved for one-dimensional analysis
3. Axoplasmic fluid is considered to be Newtonian Fluid and continuum
4. Axoplasmic axial fluid velocity is neglected
5. Impact of fluid in cross-sectional area and extracellular area is neglected.

The proposed model comprises of the following four partial differential equations:

Equation for Sodium Ions (Na^+)

$$\frac{\partial(\rho Y_{Na})}{\partial t} = \frac{\partial}{\partial x}\left(\rho D_{Na}\frac{\partial Y_{Na}}{\partial x}\right) + \dot{w}_{Na}''' + \frac{\partial}{\partial x}\left(\frac{\rho D_{Na}\text{Valency}_{Na}FY_{Na}}{R_uT}\frac{\partial V}{\partial x}\right) \quad (2)$$

Equation for Potassium Ions (K^+)

$$\frac{\partial(\rho Y_k)}{\partial t} = \frac{\partial}{\partial x}\left(\rho D_k\frac{\partial Y_k}{\partial x}\right) + \dot{w}_K''' + \frac{\partial}{\partial x}\left(\frac{\rho D_K\text{Valency}_KFY_K}{R_uT}\frac{\partial V}{\partial x}\right) \quad (3)$$

Equation for Leakage Currents, i.e., Chlorine Ions (Cl^-)

$$\frac{\partial(\rho Y_L)}{\partial t} = \frac{\partial}{\partial x}\left(\rho D_L\frac{\partial Y_L}{\partial x}\right) + \dot{w}_L''' + \frac{\partial}{\partial x}\left(\frac{\rho D_L\text{Valency}_LFY_L}{R_uT}\frac{\partial V}{\partial x}\right) \quad (4)$$

$$\frac{\partial(\rho Y_{Na})}{M_{Na}\partial t}\text{Valency}_{Na} + \frac{\partial(\rho Y_K)}{M_K\partial t}\text{Valency}_K + \frac{\partial(\rho Y_L)}{M_L\partial t}\text{Valency}_L = \frac{2}{rF}\times c_m\frac{dV}{dt} \quad (5)$$

The above four Eqs. 2, 3, 4, and 5 have been derived by solving the generalized fluid equation and these equations define transport and conservation of sodium, potassium, and chlorine ions in the fluid moving along the axon [21–25].

Also, the terms $\dot{w}_{Na}'''$, $\dot{w}_K'''$ and $\dot{w}_L'''$ representing the rate of addition of a mass of sodium ions, potassium ions and chlorine ions per unit volume in Eqs. 2, 3, and 4 are given below:

$$\dot{w}_{Na}''' = -\frac{M_{Na}}{\text{Valency}_{Na}F}\frac{2}{r}\bar{g}_{Na}m^3h(V - V_{Na}) \quad (6)$$

$$\dot{w}_K''' = -\frac{M_K}{\text{Valency}_K F} \frac{2}{r} \bar{g}_K n^4 (V - V_K) \tag{7}$$

$$\dot{w}_L''' = -\frac{M_L}{\text{Valency}_L F} \frac{2}{r} \bar{g}_L (V - V_L) \tag{8}$$

The variables m, n, and h are gating variables dependent on temperature [2].

Using Stokes–Einstein equation [26], the equation of diffusivity of ions in the fluid is given as:

$$D_i = \frac{K_B * T}{6 * \Pi * v_F * \text{rad}_i} \tag{9}$$

As per the observation given by Keochlin [27] and Gilbert [28] that only water makes the 87 percent composition of axoplasm, initially we have taken the fluid viscosity equal to the viscosity of water and later on used the following formula for obtaining the actual viscosity of fluid moving along the axon.

$$v_W = 2.414 * 10^{-5} * 10^{\frac{247.8}{T-140}} \tag{10}$$

$$v_F = C * v_W \tag{11}$$

C is the constant that is multiplied with viscosity of water to obtain the viscosity of fluid moving along the axon.

Hence our proposed model which reframes H-H model to incorporate physical parameters consists of a total of sixteen equations. Out of these sixteen equations, ten equations: 2, 3, 4, 5, 6, 7, 8, 9, 10 and 11 are the equations derived by coupling of H-H model Eq. 1 with the generalized fluid equation and rest six equations (not given here) are the equations from H-H model that define the opening and closing rate of gating variables [2].

2.2 Implementation

Finite element method has been used for solving the partial differential equations and for performing discretization, $1m$ length of axon is divided into $10,000$ intervals and radius of axon is taken to be 0.000238. The values of length and radius of axon are taken from [2]. This discretization gives the value of dx as $0.0001m$. Also for time domain, the value of dt is taken to be 0.00001. The external stimulus of $4\,\text{A}/\text{m}^2$ is applied at the very first node of axon for 0.02 milliseconds [29]. The computational code has been written using C++ syntax in an iterative manner. Also for obtaining the stabilized results from unstable and nonlinear equations, the under-relaxation factor is also incorporated.

3   Results

3.1   H-H Model Results Reproduced

As an initial step, the model is validated by reproducing the results given by H-H model. The model is able to describe the action potential phenomenon as described by H-H model and the obtained conduction velocity of 19.5m/s is also in line with the results given by H-H model [2]. Action potential plotted at two different times, depicts the propagation of action potential (Fig.1).

3.2   Temperature Versus Viscosity

The reframed model consisting of physical parameters is checked for its validation by obtaining the inverse relationship between viscosity and temperature. In Figure 2, it is clearly depicted that the viscosity increases with the decrease in temperature and vice versa (Fig. 2).

3.3   Reduced Conduction Velocity

Conduction velocity reduces with the increase in viscosity. This is because when the viscosity is increased, the fluid moving along the axon starts converting into gel state and hence resists the propagation of action potential (Fig. 3).

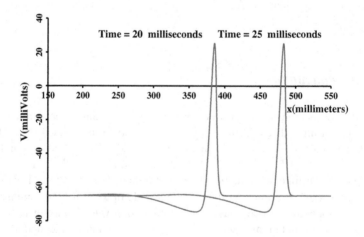

Fig. 1 Action potential at two different times

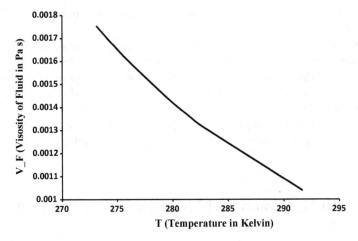

Fig. 2 Representation of inverse relation between temperature and viscosity of fluid inside the axon

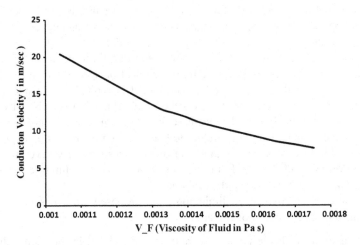

Fig. 3 Conduction velocity reduced due to increase in viscosity and decrease in temperature

4 Discussion and Conclusion

With this model, we are able to demonstrate the coupled effect of temperature and viscosity on the conduction velocity, which was otherwise missing in H-H model. The reduced conduction velocity with the increased viscosity has been utilized to identify the parameters that are affected by the anesthesia and it has been possible to provide the following important results.

1. The action potential conduction velocity is quantified with respect to viscosity at different temperatures.
2. The viscosity of fluid is identified to be one of the physical parameters affected by anesthesia. This conclusion has been drawn based on the following facts:
(a) The conduction velocity of action potential propagation is substantially reduced as viscosity of fluid is increased because increased viscosity converts the fluid to gel state.
(b) As soon as fluid is converted to gel state, it resists the further flow of ions along the axon and hence results in the reduced conduction velocity.

5   Clinical Significance and Limitations of the Proposed Model

The model presented for finding the physical parameters affected by the anesthesia has a significant application in clinical anesthetic studies. Since a neuronal model is extended to define viscosity of fluid along with temperature, it gives a framework for analyzing the neuronal behavior in presence of anesthesia and hence provides an alternative to drug therapists for the preparation of anesthetic drugs that can target the parameters which can result in the conversion of fluid to gel state [30–32].

In spite of finding its applicability from a clinical perspective, there are certain limitations associated with the model. The model is limited by consideration of intracellular fluid only and also neglects the fluid velocity. Also, viscosity of the fluid can be increased with the decrease in temperature and since we cannot decrease the complete body temperature to below zero, this limits the applicability of the model to local anesthesia rather than general anesthesia. However, the model can be further extended to study the impact of pressure for defining anesthesia as done by Graesbaal et al. for lipid membrane [16].

In view of the above-stated limitations, the proposed work has provided the following future directions for research:

1. Investigation of all the physical parameters considered in the model that can resist the propagation of action potential.
2. Sensitivity analysis of all the physical parameters for a better understanding of anesthesia and its impact on action potential propagation for therapeutic intervention [33–35].
3. Reformulation of the model to find its applicability for general anesthesia.

Acknowledgements We thank Prof. Karmeshu, JNU, Delhi and Dr. Pramod Bhatia, The NorthCap University, Gurugram for providing valuable suggestions and guidance in writing this paper.

References

1. Gerstner, W., Kistler, W.M.: Spiking Neuron Models Single Neurons, Populations Plasticity. Cambridge University Press, Cambridge (2002)
2. Hodgkin, A.L., Huxley, A.F.: A quantitative description of membrane current and its application to conduction and excitation in nerve. J. Physiol. **117**, 500–544 (1952)
3. Hille, B.: Ionic Channels of Excitable Membranes. Sinauer (1992)
4. Beilby, M.J.: Action potentials in Charophytes. Int. Rev. Cytol. **257**, 4382 (2007)
5. Kassahun, B.A.T., Murashov, A.K., Bier, M.: Thermodynamic mechanism behind an action potential and behind anesthesia. Biophys. Rev. Lett. **5**(1), 3541 (2010)
6. Shneider, M.N., Pekker, M.: Initiation and blocking of the action potential in an axon in weak ultrasonic or microwave fields. Phys. Rev. E **89**, 052713 (2014)
7. Overton, C.E.: Studies of Narcosis (R. L. Lipnick, trans.). Springer, New York (1991)
8. Scholz, A.: Mechanism of (local) anaesthetics on voltagegated sodium and other ion channels. Br. J. Anaesth. **89**, 52–61 (2002)
9. Yamakura, T., Bertaccini, E., Trudell, J.R., Harris, R.A.: Anesthetics and ion channels: Molecular models and sites of action. Annu. Rev. Pharmacol. Toxicol. **41**, 23–51 (2001)
10. Krasowski, M.D., Harrison, N.L.: General anaesthetic actions on ligand gated ion channels. Cell. Mol. Life Sci. **55**(1), 278–303 (1999)
11. Postea, O., Biel, M.: Exploring HCN channels as novel drug targets. Nat. Rev. Drug Discov. **10**(90), 3–14 (2011)
12. Appali, R., Rienen, U.V., Heimbur, T.: A comparison of the Hodgkin-Huxley model and the soliton theory for the action potential in nerves. In: Iglic, A. (ed.) Adv. Planar Lipid Bilayers Liposomes. vol. 16, pp. 275–299, Academic Press (2012)
13. Barz, H., Schreiber, A., Barz, U.: Impulses and pressure waves cause excitement and conduction in the nervous system. Med. Hypotheses **81**(5), 768–772 (2013)
14. Hardy, A.El., Machta, B.B.: Mechanical surface waves accompany action potential propagation. Nat. Commun. **6**, 6697 (2015)
15. Heimburg, T.: Thermal Biophysics of Membranes. Wiley (2007)
16. Graesbll, K., Sasse-Middelhoff, H., Heimburg, T.: The thermodynamics of general and local anesthesia. Biophys. J. **106**, 2143–2156 (2014)
17. Hodgkin, A.L., Huxley, A.F.: Resting and action potentials in single nerve fibres. J. Physiol. **104**, 176–195 (1945)
18. Hodgkin, A.L., Huxley, A.F., Katz, B.: Measurements of current-voltage relations in the membrane of the giant axon of Loligo. J. Physiol. **116**, 424–448 (1952)
19. Roselli, R.J., Diller, K.R.: Biotransport: Principles and Applications. Springer (2011)
20. Bradley, W.G., William, M.H.: Axoplasmic Flow in Axonal Neuropathies. Brain **96**, 235–246 (1973)
21. Zwillinger, D.: Handbook of Differential Equations. Academic Press, Boston (1977)
22. Landau, L.D., Lifschitz, E.M.: Fluid Mechanics. Pergamon Press, Oxford, England (1987)
23. Cengel, Y.A., Cimbala, J.M.: Fluid Mechanics (Fundamentals and Applications). (McGraw Hill Education (India) Private Limited (2010)
24. Bhatia, S., Singh, P., Sharma, P.: Hodgkin–Huxley model based on ionic transport in axoplasmic fluid. J Integr Neurosci. **16**(4), 401–417 (2017)
25. Roquemore, W.M., Katta, V.R.: Role of flow visualization in the development of UNICORN. J. Visual. **2**(3–4), 257–272 (2000)
26. Salmon, E.D., Saxton, W.M., Leslie, R.J., Karow, M.L., McIntosh, J.R.: Diffusion Coefficient of Fluorescein-labeled Tubulin in the Cytoplasm of Embryonic cells of a sea Urchin: video image analysis of fluorescence redistribution after photobleaching. J. Cell Biology **99**, 2157–2164 (1984)
27. Keochlin, B.A.: On the chemical composition of the axoplasm of squid giant nerve fibers with particular reference to its ion pattern. J. Biophysic. Biochem. Cytol. **1**(6), 511–529 (1955)

28. Gilbert, D.S.: Axoplasm chemical composition in myxicola and solubility properties of its structural proteins. J. Physiol. **253**, 303–319 (1975)
29. Sterratt, D., Graham, B., Gillies, A., Willshaw, D.: Principles of Computational Modelling in Neuroscience. Cambridge University Press (2011)
30. Moreno, J.D., Lewis, T.J., Clancy, C.E.: Parameterization for in-silico modeling of ion channel interactions with drugs. PLoS ONE **11**(3), e0150761 (2016)
31. Catterall, W.A.: Signalling complexes of voltage gated sodium and calcium channels. Neurosci. Lett. **486**, 107–116 (2010)
32. Rajagopal, S., Sangam, S.R., Singh, S.: Differential regulation of volatile anesthetics on ion Channels. Int. J. Nutr. Pharmacol. Neurol. Dis. **5**(4), 128–134 (2015)
33. Heimburg, T., Jackson, A.D.: Thermodynamics of the nervous impulse. In: Nag, K. (ed.) Structure and Dynamics of Membranous Interfaces, pp. 317–339. Wiley, Hoboken (2008)
34. Andersen, S.S.L., Jackson, A.D., Heimburg, T.: Towards a thermodynamic theory of nerve pulse propagation. Prog. Neurobiol. **88**, 104113 (2009)
35. Heimburg, T., Jackson, A.D.: On the action potential as a propagating density pulse and the role of anesthetics. Biophys. Rev. Lett. **2**, 5778 (2007)

Symmetric Wavelet Analysis Toward MRI Image Compression

E. Soni and R. Mahajan

Abstract Every year trillions of medical imaging data are generated through different imaging techniques. Demand for communication of this data through telecommunication networks using the internet is growing rapidly for various applications, hence requires a lot of memory space for storage, and high bandwidth for transmission. Efficient storage and transmission of this data is a problem to be resolved. Compression of this huge medical imaging data would be a suggested solution to this which primarily considers neighboring pixel redundancy. The proposed work includes calculation of compression on a set of ten different real time MRI (512 × 512) images acquired by the neural department of the hospital. The compression of the MRI image includes decomposing the image by a chosen discrete wavelet transform's (DWT) mother wavelet function which is followed by its encoding through set portioning in the hierarchal tree (SPIHT). After applying each of the compression methods their performance is quantified in terms of Compression Ratio (CR), Peak Signal to Noise Ratio (PSNR) and Bits per pixel (BPP). The calculated performance parameters have been compared in order to get a higher compression ratio with considerably high PSNR values from all of the applied mother wavelets. DWT (Discrete Wavelet Transform)-based symmetric mother wavelets like Haar, bior4.4, bior6.8, sym4 and sym20 have been tested and compared in virtue of comparative analysis of them. Among all of the used wavelets, sym4 and sym20 have not been explored yet for compression. MATLAB and wavelet toolbox have been used for implementation of algorithms. Haar wavelet gives a higher compression ratio and a good PSNR value after compres-

E. Soni · R. Mahajan (✉)
Department of ECE, G. D. Goenka University, Sohna, Gurgaon, India
e-mail: rashima.mahajan@gdgoenka.ac.in

E. Soni
e-mail: ekta.soni@gdgoenka.ac.in

© Springer Nature Singapore Pte Ltd. 2018
M. Pant et al. (eds.), *Soft Computing: Theories and Applications*,
Advances in Intelligent Systems and Computing 583,
https://doi.org/10.1007/978-981-10-5687-1_48

sion. The images compressed with these takes less time for uploading and downloading also can be reconstructed without a loss of diagnostic details for utilization of telemedicine services.

Keywords MRI image compression · DWT symmetric wavelets
SPIHT · CR · PSNR

1 Introduction

Medical imaging is defined as the process of acquiring visual representation (images) of the interior body parts for, e.g., brain, chest, etc., and the acquired images are used for clinical analysis. Various medical imaging techniques are mentioned in the literature like Positron emission tomography (PET), Single-photon emission computed tomography (SPECT), Computed Tomography (CT) and Magnetic Resonance Imaging/Functional Magnetic Resonance Imaging (MRI/FMRI). All of these methods are used to detect two-dimensional image signals from the body. Out of which PET and SPECT are used to provide functional information with a low spatial resolution about the brain or any other part of the body. While MRI and CT are high-resolution information sources and produce body pictures in digital form [1]. MRI is a noninvasive technique which is generally used for cancer research, gastroenterology, heart diseases, brain tumors, etc. But it is mainly attempted to classify brain tissues. The accusation of MRI signals can be done by a magnetic field, applied across the body part which is to be detected [2].

MRI and other medical imaging techniques produce plenty of medical data every day, which takes a lot of disk memory space for storage and requires high bandwidth for transmission. As Compression reduces the amount of data required to transmit [3], it may also be implemented on MRI images [4]. For example in the proposed paper 512×512, 8-bit grayscale pixel images have been used. This size of image pixels requires more than 0.2 MB of storage and if this image got compressed by n:1 compression ratio, the capacity of storage increases 'n' times without any distortion [4].

The compression can be divided into two, lossless and lossy. The lossless compression is also called reversible and in this kind of compression the transmitted signal is compressed without any loss of information and at the receiving end, exact signal is received. But it uses higher bandwidth in comparison to lossy compression, whereas lossy compression is the method which compresses the image with some loss of data, but at the same time saves the bandwidth [4–6]. The lossy and lossless compression of the MRI image signal can be achieved in either spatial or transform domain. Spatial domain's techniques directly deal with the image pixels, however, transform domain techniques are applied on the frequency transformed image coefficients. Numbers of image coefficients are less in the transform domain as compared to the spatial domain, hence requires lesser processing time, memory and bandwidth to transmit these signals in the transform domain. The encoding schemes for both spatial and transform domains are different [7]. The Fourier transform is a type of frequency transform which is used to extract the features of an image, but it loses the time information of the signal. Instead of this, short time Fourier transform (STFT) can be used. It adds a window of a particular shape to the signal and provides information about both time and frequency domain. But the precision of the information is limited by the size of the window [8].

Discrete cosine transform (DCT) is a Fourier kind of transform but gives only real terms and results in lossy compression of data. It works on the segmented image and each segment is subject to the transform [3, 5]. DCT has even-symmetric extension properties due to which blocking artifacts reduced to an extent in comparison to DFT, but does not remove fully [9]. Jiansheng et al. [10] have mentioned that blocking effects come when images are broken into blocks and these blocks become visible at the time of higher order compression of the image. DCT also faces band limitation. Wavelet transform was (WT) introduced for compression as it transforms the entire image and compresses it as a single data object, so no more blocking effects present [10]. It is also a windowing technique like STFT but with variable size [8]. These are small waves which preserve both time and frequency information of the signal and generate a timescale view of the signal. Like sine and cosine waves they do not have infinite range (both in negative and positive direction) but these are short waves. The wavelets are generated in the frequency domain by shifting and scaling of their mother wavelets which are also called as basis function [11].

In mathematical form the DWT can be shown as per the following equation [11]:

$$f(X) = \sum_k \alpha_k \emptyset k(x) \tag{1}$$

where

k	= integer index of function
α_k	= coefficients of basic function
$\emptyset k(x)$	= set of functions

and

$$\emptyset r, s(x) = 2^{r/2}\emptyset(2rx - s) \tag{2}$$

Discrete wavelet transform is the implementation of wavelet transform only using the dyadic scales and positions [1]. It is a multi-scale decomposition and multi resolution scheme that reveals data redundancy at several scales and also represents images at high resolutions [5]. It uses only a small number of coefficients to store the information. DWT first decompose an image into four parts by sub sampling horizontal and vertical channels using a combination of low and high pass filters or sub band filters [11]. The image first of all, gets divided into LL, LH, HH, and HL sub-bands. LL is the approximations coordinate of the image while the other sub-bands give detailed information about the image. To get higher resolution, the LL components can be decomposed in further level [8].

According to Talukdar et al. (2007), the DWT works on certain basis function which is also called a mother wavelet. There are a number of different symmetric mother wavelets available for, e.g., Haar, biorthogonal or bior and symlet or sym. Except Haar every mother wavelet has their multiple versions. Among all of the above mentioned mother wavelets the HAAR wavelet is the simplest wavelet Transform and has two coefficients. It is orthogonal and discontinuous in nature. It is used in the proposed work as it is faster, reversible and less memory acquiring than any other wavelet transform. It is symmetric in nature, so exhibit linear phase. It has one vanishing moment unlike others [12]. Biorthogonal (bior) wavelets have been defined by Monika et al. (2013) according to which it posses linear phase properties and are symmetric in nature. They need two wavelet filters as they are Biorthogonal in nature one for decomposition and the other one for reconstruction. Examples of bior filter banks are 1.1, 1.3, 1.5, 2.2, 2.4, 2.6, 2.8, 3.1, 3.3, 3.5, 3.7, 3.9, 4.4, or 5.5, 6.8. In which the first number indicates the order of the decomposition or synthesis filter and the second number indicates the order of the reconstruction or analysis filter. The default is 1.1. In the proposed work properties of 4.4 and 6.8 orders of bior wavelet basis function has been utilized [13]. Symlet is also a type of wavelet which is near to symmetric and has order from 1 to 20, sym4 and sym20 have already been used in this paper.

After transforming the image its encoding should be done in order to compress the image. For DWT various coding schemes are available which are categorized into vector and scalar techniques. Examples of scalar techniques are Joint Photographic Experts Group 2000 (JPEG 2K) and Embedded Block Code for Optimized truncation (EBCOT). JPEG 2K was defined in [9] and had high values of CR 9.12 and PSNR 46.60 dB, but it is not used extensively being a scalar technique. Embedded Zero-tree Wavelet (EZW) and Set partition in hierarchical trees (SPIHT) are vector methods of compression [14, 15]. These methods are based on tree structure formation with the similar data of different sub-bands resulted from the transformation process [5]. These methods come under lossless compression. In EZW the pixels are represented by an embedded bit stream. EZW is a progressive method of transmission and the encoding can be stopped at any point. In [11] authors have implemented EBCOT, SPIHT, and EZW for compression. They achieved highest value of PSNR for EBCOT and comparatively high value of CR for EZW. But the PSNR calculated with all of the schemes is not even in an acceptable range [11].

SPIHT algorithm is an embedded coding method and utilizes zero-tree structures like EZW [2]. Zero-tree structures are the result of similarity across different sub-bands [5, 16]. It is an improved version of EZW algorithm and proposed by Pearlman [17]. It gives higher compression and better performance than EZW. It is a lossless compression method. It is based on ordered bit plane and progressive transmission, Set partitioning sorting algorithm, and spatial orientation trees. The partition is done in the way that can keep the insignificant coefficients together in larger subsets. 'Hierarchical Trees' in SPIHT refers to the quad tree structure and 'Set Partitioning' refers to the way these quad trees are divided. The partitioning of wavelet coefficients should be done on some given threshold value [5, 16].

Different techniques of compression have been explored in the Table 1. By analyzing the results of those techniques it was concluded that they were under-performing than the proposed method in virtue of compression parameters. Compression using bior and symlet mother wavelets have not been experimented extensively in the literature. So this is a new set of techniques to be experimented with for compression. The proposed paper is based on the application of these methods including a well-known wavelet Haar. Although Haar has been used for compression previously also and gave good results, but in this proposed work it was used to compare its results with the new unexplored mother wavelets on real time MRI image signals of 512×512 size.

2 Material and Methods

A block diagram of the proposed MRI image compression procedure is presented in Fig. 1. The procedure of MRI image compression is divided into two parts: encoding and decoding of acquired MRI Images. The encoding process consists of four major steps which include acquisition of MRI images, their format conversion (into a MATLAB compatible format, e.g., .jpg), loading of .jpg images in MATLAB for compression and then applying DWT in conjunction with SPIHT to get a compressed MRI image.

After compression, these images would be either stored or transmitted. Subsequently, the images were decompressed using the inverse of DWT and inverse of SPIHT. On the basis of original and reconstructed images their CR, PSNR, MSE, and BPP would be calculated.

2.1 MRI Image Acquisition

The image acquisition step involves the acquisition of MRI image through real time source, i.e., from the hospital. The acquired image signal was in the DICOM (Digital Imaging and Communications in Medicine) format. This kind of images cannot be directly loaded in MATLAB. In order to load them for further processing 'DICOM converter' software is being used. This software can convert any DICOM image into a MATLAB acceptable format. The provided formats from this software are .png, .jpg, .pnm, .tif and .bmp. After converting the image into a suitable format (.jpg in this proposed work) its size had been computed and which was 512×512 uint8 (8-bit unsigned integer image) pixel images.

Table 1 Comparison of different MRI compression schemes

S. No.	Author and year	MRI image compression technique	Parameters	Image specification	Remarks
1	Korde et al. (2013) [1]	IWT +SPIHT + Run length encoding	PSNR-37.32, MSE-57.50, CR-87.50	256 × 256	ROI method has been implemented for compression
2	Sanchez et al. (2009) [2]	2D IWT + EBCOT + block-based intra band prediction method	See sub-table below	3D medical image	Compression ratio 15% higher than JPEG2000 and H.264/AVC intra-coding and 13% higher than 3D JPEG2000
3	Bairagi et al. (2013) [4]	SPIHT + IWT for non ROI RLE for ROI	MSE = 0.0005, PSNR = 80.85, CR = 59.93 for zero level of distortion	256 × 256 with 8-bit resolution	Both lossy and lossless compression can be achieved in a single image
4	Malý et al. (2003) [5]	DWT(Cdf9/7) + SPIHT	See sub-table below	256 × 256	The results were superior to Haar and comparable to Bior and db
5	Rawat et al. (2015) [11]	1. EZW / 2. SPIHT / 3. EBCOT	See sub-table below	512 × 512	EBCOT gives the superior results in terms of CR, PSNR and execution time than other lossless methods
6	Zhou et al. (2015) [18]	DCT + vector zig-zag order + DPCM + MSVQ + entropy encoding	See sub-table below	512 × 512	Better scheme than JPEG and from VQ and hybrid DCT-VQ
7	Said et al. (1996) [17]	SPIHT	See sub-table below	512 × 512	The proposed scheme SPIHT was compared with EZW and found the proposed scheme better

Parameters sub-table for row 2 (Sanchez et al.):

	H.264/AVC intra-coding (CR)	JPEG2K (CR)	3D JPEG2K (CR)	proposed (CR)
MRI 24:192 × 256:16	2.65:1	2.62:1	2.71:1	3.25:1
30:192 × 256:16	2.58:1	2.51:1	2.68:1	3.21:1

Parameters sub-table for row 4 (Malý et al.):

BPP	0.3	0.5	0.7	0.9
PSNR	29.43	32.07	34.22	36.03

Parameters sub-table for row 5 (Rawat et al.):

	PSNR	MSE	BPP	CR	rate	Execution time
1. EZW	26.9446	31.4063	1.3793	82.758	5.7999	36.5999
2. SPIHT	33.0494	32.221	1	87.5	8	23.4974
3. EBCOT	36.9856	13.0173	1	87.4999	7.999	12.7039

Parameters sub-table for row 6 (Zhou et al.):

PSNR	BPP
34.26	0.49
33.58	0.40

Parameters sub-table for row 7 (Said et al.):

BPP	PSNR
0.5	37.2
0.25	34.1
0.15	31.9

(continued)

Table 1 (continued)

S. No.	Author and year	MRI image compression technique	Parameters	Image specification	Remarks
8	Dodla et al. (2013) [16]	DCT + SPIHT	MSE = 6.72, PSNR = 39.85	512 × 512	SPIHT is a high fidelity and low error method
9	Roy et al. (2016) [9]	1. JPEG 2. JPEG2K	CR / MSE / PSNR: 8.14 / 16.11 / 35.85 9.12 / 11.27 / 46.60	512 × 512	The JPEG2K provides better reconstruction of compressed images
10	Monika et al. (2013) [12]	1. Haar 2. Bior 3. LSHaar 4. LSBior	MSE / PSNR / Execution time: 0.0021 / 74.869 / 0.9048 0.0014 / 76.64 / 0.6708 0.0015 / 76.23 / 0.7488 0.00007 / 79.4779 / 0.4680	64 × 64	Lifting schemes (LS) can be used for speed critical systems like real time image compression
11	Talukdar et al. (2007) [7]	Haar 2DWT	Threshold / CR (hard threshold) / CR (hard threshold): 15 / 15.74 / 14.1 20 / 17.11 / 15.87 25 / 18.47 / 16.95	256 × 256	A threshold value is set to get optimum decomposition level. To achieve better PSNR and CR it is set high but at the cost of coding efficiency
12	Chen 2007 [13]	DCT + CSPIHT	BPP / SPIHT (PSNR) / JPEG2K (PSNR) / DCT + CSPIHT (PSNR): 0.025 / 33.4 / 36.1 / 39.5 0.50 / 49.8 / 50.5 / 52.0	512 × 512	Computational complexity reduced along with storage and transmission bandwidth increase
13	Beladgham et al. (2011) [14]	Cdf9/7 lifting scheme + SPIHT	BPP / PSNR / MSSIM: 0.75 / 35.80 / 0.92	512 × 512	
14	Nagamani et al. (2011) [15]	1. DWT + EZW (lenna) 2. DWT + SPIHT (lenna) 3. DWT + EZW (satellite) 4. DWT + SPIHT (satellite)	CR / PSNR: 3.34 / 27.79 8 / 61.53 1.07 / 13.07 8 / 29.20	512 × 512	This method gives better results with 'lena' images only

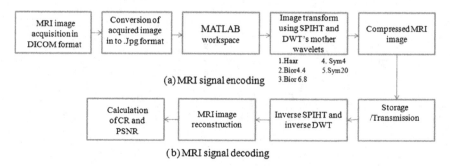

(a) MRI signal encoding

(b) MRI signal decoding

Fig. 1 Block diagram of proposed MRI image compression procedure

2.2 MRI Image Compression

The compression of the acquired MRI 512 × 512 image signal was done by using a combination of DWT (transformation) and SPIHT (encoding). Five different wavelet basis functions had been applied for further comparison purpose. The wavelet basis functions used were Haar, bior4.4, bior6.8, sym4 and sym20. These were so chosen because all of them are symmetric in nature and have not been explored much in the area of medical image compression. For each compression, the number of loops is set to 12 by 'maxloop' function. This function defines the number of steps in which compression of the image should be done. This function should not be set on the lower level because at lower depth levels SPIHT will not give efficient results [5]. After transformation by each of the wavelet basis functions the transformed image was encoded using the SPIHT encoding technique. Results of different compression schemes have been compared to choose an appropriate wavelet.

2.3 Performance Evaluation

After getting the compressed images using different wavelet functions, the comparison had been done on the basis of some performance parameters which are CR, PSNR, MSE, and BPP. These performance parameters are defined below.

CR (Compression ratio)—The compression ratio of an image is defined by the ratio between the original image size and the compressed image size and it can be shown by [11]

$$= \frac{\text{Uncompressed size of signal}}{\text{Compressed size of signal}}$$

The above ratio gives the amount of compression for a particular image. It is a quality measure of a picture. As higher the compression ratio, poorer the quality of

Fig. 2 Detailed flow chart of MRI image acquisition and compression system using MATLAB and wavelet toolbox

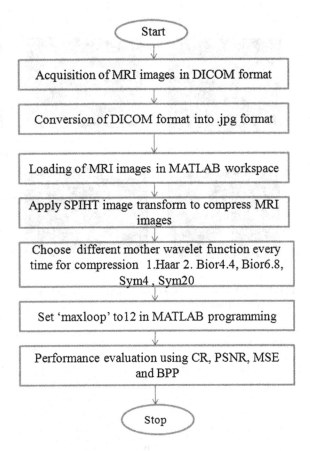

the resulting image. So it's a challenge to make a tradeoff between compression ratio and picture quality in order to save the available memory [14].

MSE (Mean Squared Error) is used to define the distortion rate in the reconstructed image after compression. It is also called as squared error loss. MSE basically measures the average of the square of the error [14]. The mathematical representation of MSE can be shown by the formula given below [12].

$$= \frac{\sum_{M,N}[I_1(m,n) - I_2(m,n)]^2}{M \times N}$$

where 'M' is the number of rows and 'N' is the number of columns in the input image. $I_1(m,n)$ is the input image and $I_2(m,n)$ is the retrieved output image.

PSNR: Peak Signal to noise ratio, this is used to evaluate the quality of the reconstructed images after compression. In simple terminology, it is a ratio between the maximum possible power of the signal and the power of corrupting noise. It is

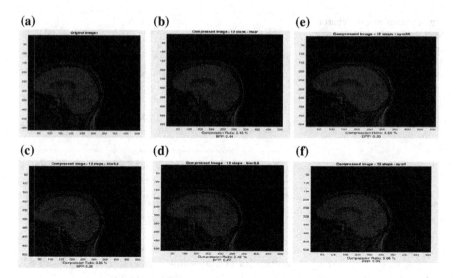

Fig. 3 a–f Original MRI image (**a**), image compressed through Haar (**b**), bior4.4 (**c**), bior6.8 (**d**), sym20 (**e**), and sym4 (**f**)

Table 2 Performance parameters of different images using SPIHT with DWT Haar mother wavelet

S. No.	Image	CR	MSE	PSNR (dB)	BPP
1	Image1	5.46	4.77	41.34	0.44
2	Image2	5.67	4.44	41.66	0.45
3	Image3	4.38	3.81	42.32	0.35
4	Image4	4.71	3.89	42.24	0.38
5	Image5	4.48	4.23	41.86	0.36
6	Image6	5.17	4.53	41.57	0.41
7	Image7	6.28	5.21	40.95	0.50
8	Image8	5.73	4.59	41.51	0.46
9	Image9	5.30	4.23	41.86	0.43
10	Image10	5.04	4.16	41.93	0.40
	Average	5.22	4.39	41.72	0.42

usually expressed in terms of the logarithmic decibel scale [14]. The mathematical representation is given below which shows its inverse relationship with MSE [11].

$$= 10 \log_{10}\left(\frac{R^2}{MSE}\right)(dB)$$

where 'R' is the maximum fluctuation in the input image data type. All images used in this paper are 8-bit unsigned integer data type, for which the value of R is 255. The value of PSNR can be between 30 to 50 dB. It can also be defined as the pixel difference between original and compressed image [19], as CR increases PSNR decreases.

Table 3 Performance parameters of different images using SPIHT with DWT bior4.4 mother wavelet

S. No.	Image	CR	MSE	PSNR (dB)	BPP
1	Image1	3.25	3.43	42.77	0.26
2	Image2	3.33	3.21	43.06	0.27
3	Image3	2.58	2.89	43.57	0.21
4	Image4	2.84	2.87	43.56	0.23
5	Image5	2.60	2.74	43.75	0.21
6	Image6	3.10	3.11	43.20	0.25
7	Image7	3.90	3.68	42.47	0.31
8	Image8	3.42	3.30	42.95	0.27
9	Image9	3.08	3.02	43.33	0.25
10	mage10	2.95	3.01	43.34	0.24
	Average	3.11	3.13	43.2	0.25

Table 4 Performance parameters of different images using SPIHT with DWT bior6.8 mother wavelet

S. No.	Image	CR	MSE	PSNR (dB)	BPP
1	Image1	3.42	2.90	44.35	0.27
2	Image2	3.44	3.00	43.36	0.28
3	Image3	2.74	2.38	44.36	0.22
4	Image4	2.99	2.60	43.98	0.24
5	Image5	2.74	2.44	44.26	0.22
6	Image6	3.30	2.73	43.75	0.26
7	Image7	4.08	3.43	42.78	0.33
8	Image8	3.58	2.97	43.40	0.29
9	Image9	3.21	2.66	43.88	0.25
10	Image10	3.10	2.66	43.89	0.25
	Average	3.26	2.78	43.8	0.26

Table 5 Performance parameters of different images using SPIHT with DWT sym4 mother wavelet

S. No.	Image	CR	MSE	PSNR (dB)	BPP
1	Image1	3.65	3.07	43.25	0.29
2	Image2	3.71	3.11	43.19	0.30
3	Image3	2.84	2.63	43.93	0.23
4	Image4	3.17	2.78	43.69	0.25
5	Image5	2.96	2.56	44.05	0.24
6	Image6	3.49	2.98	43.39	0.28
7	Image7	4.38	3.55	42.75	0.35
8	Image8	3.82	3.08	43.25	0.31
9	Image9	3.41	2.79	43.68	0.29
10	Image10	3.31	3.89	42.24	0.27
	Average	3.47	3.04	43.34	0.28

Table 6 Performance parameters of different images using SPIHT with DWT sym20 mother wavelet

S. No.	Image	CR	MSE	PSNR (dB)	BPP
1	Image1	3.50	2.90	43.50	0.28
2	Image2	3.53	2.77	43.71	0.28
3	Image3	2.79	2.39	44.35	0.22
4	Image4	3.02	2.56	44.04	0.24
5	Image5	2.78	2.50	44.15	0.22
6	Image6	3.35	2.67	43.82	0.27
7	Image7	4.16	3.22	43.05	0.33
8	Image8	3.63	2.99	43.38	0.29
9	Image9	3.30	2.59	44.00	0.26
10	Image10	3.15	2.67	43.87	0.25
	Average	3.32	2.73	43.79	0.26

Table 7 Average value of performance parameters for all the mother wavelets

Name of mother wavelet	Haar	bior4.4	bior6.8	sym4	sym20
CR (avg)	5.22	3.11	3.26	3.47	3.32
MSE (avg)	4.39	3.13	2.78	3.04	2.73
PSNR (avg)	41.72	43.2	43.8	43.34	43.79
BPP (avg)	0.42	0.25	0.26	0.28	0.26

BPP (Bit Per Pixel) is the measure of the number of colors in an image and which is given by 2^{bpp}.

A flow chart of the proposed methodology is shown in the above Fig. 2. It explains the analysis of the detailed flow of processes which starts from the acquisition of images and ends at computing comparative performance of the methods used for compression. MATLAB was used as a tool to compress and to calculate the performance in terms of CR, PSNR, MSE, and BPP, on the basis of these computed performance parameters the best performing wavelet can be found.

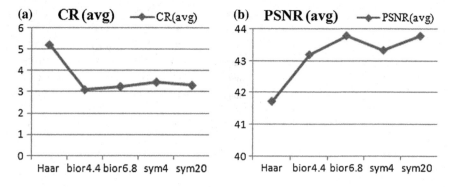

Fig. 4 **a**, **b** Graphical representation of CR and PSNR average values of each wavelet functions

3 Result and Discussion

This section illustrates the different results obtained by applying five DWT basis wavelets: Haar, bior4.4, bior6.8, sym4 and sym20 in alliance with encoding through SPIHT. The performance of the system has been estimated using ten different MRI images which were acquired from the hospital. The algorithm for the proposed system is developed and implemented using MATLAB. The performance parameters obtained from each of the mother wavelets has been compared to figure out the most suitable wavelet for medical images. Figure 3a–f show the original image and its transformed images by using the abovementioned wavelet functions. The image used for compression in the figure is 'Image1'.

The results have been computed by applying each of the five wavelets on all the ten images. These results are shown in the tabular form. Among them, compression results on 'Image1' have been visualized in Fig. 3.

Table 2 is solely related to Haar wavelet applied on ten different images along with their performance parameters. At the end averages of all the parameters have been calculated individually.

Table 3 summarizes the compression results by applying bior4.4 mother wavelet function on each of the ten MRI images with averages of all the parameters calculated at the end. It shows lower values of CR, a higher value of MSE, higher PSNR and lower BPP than the before mentioned Haar wavelet.

Like as Tables 3 and 4 also gives compression results on ten different MRI images but the mother wavelet chosen was bior6.8. The procedure of finding averages of each of the parameters is similar. The result shows a bit higher CR, lower MSE, higher PSNR and higher BPP values than the previous method bior4.4.

The Table 5 is the result of applying sym4 mother wavelet on the MRI images. Its compression parameters were also averaged to reduce the complexity of further computations. It has higher CR, higher MSE, lower PSNR and higher BPP values than bior6.8.

sym20 had been applied on ten different MRI images. The results are shown in Table 6. It also calculates individual compression performance parameters for each of the images and at the end, the result got summarized using averaging of each parameter. It shows lower CR, lower MSE, higher PSNR and lower BPP values than the sym4 wavelet.

Table 7 summarizes all the performance parameters which include their average values for each of the wavelet functions.

The graphical representation of compression parameters CR and PSNR are shown in Fig. 4a, b. By evaluating the given figure comparative analysis can be done for each wavelet. According to that analysis, Haar wavelet attained maximum value for CR and minimum value for PSNR. For the rest of the wavelets, CR values remained almost same, but PSNR values vary in between 43 to 44. The lower versions of the wavelets like sym4 and bior4.4 have almost the same value for PSNR and the higher versions sym20 and bior6.8 share same values of PSNR.

By analyzing the above findings, it is unveiled that the proposed MRI image compression process is comparable and better than the compression methods discussed in the literature for both medical and other images. This paper aims for the comparative analysis of the methods available, which shows that although Haar, bior4.4 and bior 6.8 has been explored in the literature, but not for the medical images and also the used picture size was not same as the image size used in this paper. Moreover, according to literature Table 1 sym4 and sym20 have not been ever applied for compression of images. The analysis shows that sym20 gives a higher value of PSNR and CR without degrading the picture quality of the image. MATLAB was used for compression and decompression of the images along with calculation of their performance parameters.

4 Conclusion

For compressing the medical images a new compression scheme based on discrete wavelet transform using its symmetric mother wavelets in conjunction with zero-tree embedded hierarchal encoding method 'SPIHT' is developed in this paper. This scheme compresses the image with sufficiently high compression ratio without degrading the image quality. The effectiveness of the proposed method has been justified by applying them on data set of ten real MRI images. Haar, bior4.4, bior6.8, sym4 and sym20 are the symmetric wavelets which have been tested for compression individually on each of the images. With all of these wavelets, high values of CR and considerably high PSNR values has been achieved. Haar wavelet gives highest CR (5.22) but lowest PSNR (41.72) values. bior6.8 and sym20 gives a comparable value of PSNR which is 43.8 and 43.79, respectively. But CR of bior6.8 is a bit higher than that of sym20. bior4.4 and sym4 both are comparably low in CR and PSNR. The results obtained prove the postulate according to which PSNR is inversely proportional to the CR and MSE. These compressed MRI images can be used for applications like telemedicine, remote sensing, and satellite communication.

In future IWT (Integer Wavelet Transform) wavelet will be experimented for compression on a similar or a new and richer data set of MRI images.

References

1. Korde, N.S., Gurjar, D.A.: Wavelet based medical image compression for telemedicine application. Am. J. Eng. Res. (AJER). 3(1), 106–111 (2014). e-ISSN 2320-0847
2. Sanchez, V., Abugharbieh, R., Nasiopoulos, P.: Symmetry-based scalable lossless compression of 3D medical image data. IEEE Trans. Med. Imaging 28(7), 1062–1071 (2009)
3. Raid, A.M., Khedr, W.M., El-dosuky, M.A., Ahmed, W.: Jpeg image compression using discrete cosine transform—a survey. Int. J. Comput. Sci. Eng. Surv. (IJCSES) 5(2), 39–47 (2014)

4. Bairagi, V.K., Sapkal, A.M.: ROI-based DICOM image compression for telemedicine. Sadhana **38**(1), 123–131 (2013)
5. Malý, J. Rajmic, P.: Dwt-spiht Image Codec Implementation. Department of Telecommunications, Brno University of Technology, Brno, Czech Republic (2003)
6. Sridevi, M.E.S., Vijayakuymar, V.R., Anuja, R.: A survey on various compression methods for medical images. Int. J. Intell. Syst. Appl. **4**(3) (2012)
7. Talukder, K.H. and Harada K.: Haar wavelet based approach for image compression and quality assessment of compressed image. IAENG Int. J. Appl. Math. **36**(1) (2007)
8. Zhang, Y., Dong, Z., Wu, L., Wang, S.: A hybrid method for MRI brain image classification. Expert Syst. Appl. **38**(8), 10049–10053 (2011)
9. Roy, A., Saikia, L.P.: A comparative study on lossy image compression techniques. Int. J. Curr. Trends Eng. Res. (IJCTER) **2**(6), 16–25 (2016)
10. Jiansheng, M., Sukang L., Xiaomei T.: A digital watermarking algorithm based on DCT and DWT. In: Proceedings of the 2009 International Symposium on Web Information Systems and Applications (WISA'09), pp. 104–107. Nanchang, PR China (2009)
11. Rawat, P., Rawat A., Chamoli S.: Analysis and comparison of EZW, SPIHT and EBCOT coding schemes with reduced execution time. Int. J. Comput. Appl. **130**(2), 24–29 (2015)
12. Monika, C.P., Lalit G.: Lifting scheme using HAAR and biorthogonal wavelets for image compression. Int. J. Eng. Res. Appl. (IJERA) **3**(4), 474–478 (2013)
13. Chen, Y.Y.: Medical image compression using DCT-based subband decomposition and modified SPIHT, data organization. Int. J. Med. Inform. **76**(10), 717–725 (2007)
14. Beladgham, M., Bessaid, A., Abdelmounaim, M.L., Abdelmalik, T.A.: Improving quality of medical image compression using biorthogonal CDF wavelet based on lifting scheme and SPIHT coding. Serb. J. Electr. Eng. **8**(2), 163–179 (2011)
15. Nagamani, K., Ananth, A.G.: EZW and SPIHT image compression techniques for high resolution satellite imageries. Int. J. Adv. Eng. Technol. (IJAET) **2**(2), 82–86 (2011)
16. Dodla, S., Raju, Y.D.S., Mohan, K.M.: Image compression using wavelet and SPIHT encoding scheme. Int. J. Eng. Trends Technol. (IJETT) **4**(9), 3863–3865 (2013)
17. Said, A., Pearlman, W.A.: A new, fast and efficient image codec based on set-partitioning in hierarchical trees. IEEE Trans. Circuits Syst. Video Technol. **6**(3), 243–250 (1996)
18. Zhou, X., Bai, Y., Wang, C.: Image compression based on discrete cosine transform and multistage vector quantization. Int. J. Multimed. Ubiquitous Eng. **10**(06), 347–356 (2015)
19. Vidhya, K., Karthikeyan G., Divakar P., Ezhumalai S.: A Review of lossless and lossy image compression techniques. Int. Res. J. Eng. Technol. (IRJET) **3**(4), 616–617 (2016)

RBFNN Equalizer Using Shuffled Frog-Leaping Algorithm

Padma Charan Sahu, Sunita Panda, Siba Prasada Panigrahi
and K. Rarvathi

Abstract Radial Basis Function Neural Networks (RBFNN) is one of the most popular equalizers to mitigate the channel distortions. The most challenging problem associated with the design of RBFNN Equalizer is the traditional hit and trial method. This paper proposes training of RBFNN equalizer using a recently proposed population-based optimization, Shuffled Frog-Leaping Algorithm (SFLA) and three of its modified forms. It is found from the simulation results that performances of different forms of SFLA for the training of RBFNN equalizers are superior as compared to existing equalizers.

Keywords Radial basis function neural network · Channel equalization
Shuffled Frog-Leaping Algorithm

1 Introduction

Present day research on filter design and channel equalization focuses around the use of swarm and evolutionary algorithms. However, use of the artificial neural network (ANN) is common and popular in a wide range of engineering problems like those of the problem equalization. A detailed review on channel equalization using Multi layer Perceptron (MLP), functional-link artificial NN (FLANN) and neuro-fuzzy systems is provided in [1, 2]. Recent literature on channel equalization shows a pointer for use of neural networks [3, 4]. But ANNs have associated with large complexity and also fail because of over-fitting and local optima. However, in [5] it was demonstrated that the RBFNN has an identical structure to the optimal

P.C. Sahu · S. Panda
Kalam Institute of Technology, Berhampur, Odisha, India

S.P. Panigrahi (✉)
Electrical Engineering, VSSUT, Burla 768018, Odisha, India
e-mail: siba_panigrahy15@rediffmail.com

K. Rarvathi
KIIT University, Bhubaneswar, Odisha, India

© Springer Nature Singapore Pte Ltd. 2018
M. Pant et al. (eds.), *Soft Computing: Theories and Applications*,
Advances in Intelligent Systems and Computing 583,
https://doi.org/10.1007/978-981-10-5687-1_49

Bayesian equalizer, and can be used to implement it RBFNN also finds global minima [6]. In addition to lesser complexity, RBFNN performance is also better than ANN as proved in the literature. These merits of RBFNN in channel equalization became an active area of research [6–9].

Once it is decided to use RBFNN for channel equalization, the problem consists in setting RBFNN parameters (centers and widths) properly, which is the focus of this paper. In work of Chen et al. [5], the RBFNN parameters were determined through simple classical methods as the k-means clustering and the LMS algorithm. To avoid time-consuming process of classical methods, Barreto et al. [10] used Genetic Algorithm (GA) and Feng [11] used Particle Swarm Optimization (PSO). These are used to decide these key and bias parameters. Minimization of the Mean Square Error (MSE) between the desired and actual outputs is actual criteria in the designs formulated in [10, 11]. Still, in PSO there is a limited to a finite search space and hence falls to the local minima [12].

The Shuffled Frog-Leaping Algorithm (SFLA) has some advantages such as simple steps, a few parameters, fast speed, and easy realization [13]. These merits of SFLA and its extensive applications [14, 15] promoted the use of SFLA to train ANN and use in channel equalization [16]. To avoid the limitations of ANN, as discussed above, SFLA has been used to train RBFNN-based equalizer. In addition, exploration of the advancements and improvements in SFLA, this paper introduced three of its modified forms of SFLA for the training of RBFNN equalizer. Merits of this paper can be seen as the use of RBFNN that is advantageous over MLP for the problem and use of improved forms of SFLA, which is also proved in the simulations.

2   SFLA and Modified Forms

2.1   SFLA

The SFLA is gaining importance in engineering optimization problems because of advantages like simplicity in its structure and realization and also because of the necessity of a lesser number of control parameters. Each frog is a solution to the problem in SFLA. A set of frogs with similar structure constitute the population. The population divided into subsets, sub-memeplexes. Motivations that led some of the researchers to propose modified forms of SFLA are discussed in this section. Basic steps in SFLA and need for modifications (for more details on original SFLA, one can refer to [15, 16]) are outlined below:

- Population initialization
  The population is generated by random functions and hence is not uniform. Hence, diversity of the population diversity and ability to search is less.
- Population division
- Local search

Only worst solution updated in SFLA. There is no updating method and principle for the best solution. Hence, convergence is slower. Also, precision is not considered while updating the worst solution.
- Hybrid operation

The frogs with the worst and the best fitness are represented, respectively, as X_W and X_b. If, S is the size of leap step, S_{max} is maximum jump distance allowed, and r is a random number in $(0, 1)$, then the update equations are:

$$S = r(X_b - X_W) \tag{1}$$

$$X'_W = X_W + S, (S < S_{max}) \tag{2}$$

2.2 Modified Forms

To meet the limitations of the SFLA, Jiang et al. [13] proposed a modification. Population initialization is improved by dividing the solution space into p parts consisting of randomly generated individuals. In this way, the population is both random and uniform. In this modified form the population is segregated into searching population and competing population. The best and the second best solution are updated after evolution operation; the simplex method is used to update the second best solution. The convergence speed is improved by adding a Mutation operator to act on the individuals having lower convergence. To improve adaptability, updating rule of Eq. (1) is changed as:

$$S = r(X_b - X_w) * w_i \quad i = 1, 2, \cdots, N \tag{3}$$

Here, N is population size and w_i is the compressibility factor [13] defined as:

$$w_i = w_l + (w_h - w_l) * X_i \tag{4}$$

Here, w_h and w_l, respectively, represent highest and lowest compressibility factor available in the population of frogs. Here, X_i is a factor called relative fitness of a particular frog and defined as:

$$X_i = \frac{X_w - X_{i,n}}{X_w - X_b} \tag{5}$$

$X_{i,n}$ is the fitness of frog 'I' in nth generation.

To improve the convergent velocity, Zhang et al. [17] proposed to introduce the "follow" property of Artificial Fish-Swarm Algorithm (AFSA) in the stage of global information interchange.

Kavousifard et al. in their work [18] proposed to select a modification by updating of the worst frog using a new vector. This new vector considers three factors; the best frog from the population, best individuals in each memeplex and a randomly generated control factor, β.

3 Proposed Training

SFLA and its modified forms find the optimal number of RBFs, centers, etc. This is the basis for training of RBFNN. Each of these parameters is optimized using the proposed training steps as follows:

   i. Initialize a population of memes. Each meme corresponds to a network. Allowed iterations as MaxIteration. Start the first iteration.
  ii. Each meme to be decoded to one ANN, calculate the weights using pseudo–inverse method. Calculate the meme's fitness.
 iii. Position to be updated running SFLA.
 iv. Go to next iteration;
  v. Go back to step ii until reaching the maximum number of iterations.

In this work, RBFNN trained with original SFLA is termed as SRBF, while trained with modified forms proposed by Jiang et al. [13], Zhang et al. [17] and Kavousifard et al. [18] are referred as SRBF-J, SRBF-Z, and SRBF-K, respectively, for convenience of the reader.

4 RBFNN Equalizer

The system considered in this paper is shown in Fig. 1.

A popular linear channel model is FIR model. In this model, transmitted sequence is binary, represented as $x(k)$ at kth time instance and corresponding output at the same instant is, $y_1(k)$ as:

$$y_1(k) = \sum_{i=0}^{N-1} h_i x(k - i) \tag{6}$$

Here, $h_i (i = 0, 1, \cdots N - 1)$ and N, respectively, are the channel taps and N is the channel length. The block 'NL' is nonlinearity inserted in the channel. Most popular form of the nonlinear function is as follows:

$$y(k) = F(y_1(k)) = y_1(k) + b[y_1(k)]^3 \tag{7}$$

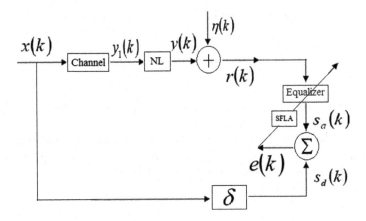

Fig. 1 Communication system with equalizer

Here, b is a constant. The output from this block is:

$$y(k) = \left(\sum_{i=0}^{N-1} h_i x(k-i)\right) + b\left(\sum_{i=0}^{N-1} h_i x(k-i)\right)^3 \tag{8}$$

The channel output $y(k)$ is added with noise, $\eta(k)$ inserted in the channel. The signal at the receiver is $r(k)$ and given by:

$$r(k) = y(k) + \eta(k) \tag{9}$$

Equalizer is used to recover the transmitted symbol, $x(k-\delta)$, from a-priory knowledge on the samples received, 'δ' being the associated transmission delay.

The desired signal can be represented as $d(k)$ and can be defined as follows:

$$d(k) = x(k-\delta) \tag{10}$$

Equalization can be treated as if a problem of classification [6–9], and the equalizer makes a partition in the input space $x(k) = [x(k), x(k-1), \ldots x(k-N+1)]^T$ into two distinct regions.

The Bays theory [16] provides the optimal solution for this where the decision function is as follows:

$$f_{bay}(x(k)) = \sum_{j=1}^{n} \beta_j \exp\left(\frac{-\|x(k) - c_j\|}{2\sigma^2}\right) \tag{11}$$

Since the sequence transmitted is binary, hence the following:

$$\beta_j = \begin{cases} +1 & c_j \in C_d^{(+1)} \\ -1 & c_j \in C_d^{(-1)} \end{cases} \tag{12}$$

Here, $C_d^{(+1)}/C_d^{(-1)}$ and c_j, respectively, represent transmitted symbol, $x(k - \delta) = +1/-1$ and σ^2 is the noise variance.

In Fig. 1, the block "Equalizer" is ANN. SFLA and its modified forms are used to optimize the number of layers and neurons in each layer. In the input layer, there are N number of neurons, same as the number of taps.

The equalizer output is as follows:

$$f_{\text{RBF}}(x(k)) = \sum_{j=1}^{n} w_j \exp\left(\frac{-\|x(k) - t_j\|^2}{\alpha_j}\right) \tag{13}$$

Here, t_j and α_j represent the centers and the spreads of the neurons in hidden layers. The vector w_j contains the connecting weights. The output of the ANN equalizer of Eq. (6) implements the nonlinear function of Eq. (7), For optimal weights, the condition is that t_j is equal to c_j.

The decision at equalizer output is as follows:

$$\hat{x}(k - \delta) = \begin{cases} +1 & f_{\text{ANN}}(x(k)) \geq 0 \\ -1 & \text{elsewhere} \end{cases} \tag{14}$$

Hence, the difference (i.e., $s_a(k) = \hat{x}(k - \delta)$) or (i.e., $s_d(k) = x(k - \delta)$) is the error $e(k)$, and updates the weights.

For l is the number of samples then, Mean Square Error (MSE) is as follows:

$$\text{MSE} = \frac{1}{l} E\left[e^2(k)\right] \tag{15}$$

Bit Error Rate (BER) is the ratio of error bits to transmitted bits: In this paper, MSE and BER are chosen as performance index.

5 SRBF Equalizer

The RBFNN equalizer is made of using channel output states instead of the channel parameter to avoid complexity in modeling the channel. The channel output states set is the data set for the centers. This relation exists on both linear channel and nonlinear channel with one to one mapping between channel input and channel output. On the other words, if the channel output states are known, the RBF equalizer is designed. The objective of the problem can be changed to find the

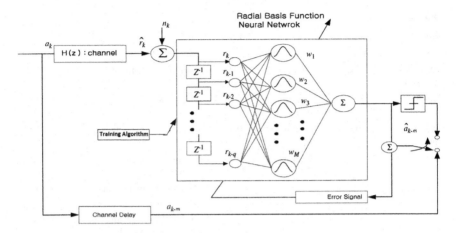

Fig. 2 SRBF equalizer

optimal data set (channel output states). The problem of equalization is now an optimization problem with the objective of reaching Bayesian likelihood. But the relation between the output states of the channel and the Baye's likelihood is a complex or non-realizable formulation, when the structure of nonlinear channel is unknown, we apply a genetic algorithm to solve this complex optimal problem with local minima.

The initial population of N ($m \times n$) memes is the basis for the SRBF equalizer, the channel states are initially chosen from. Each state constitutes the number of memeplexes and each meme represents one symbol that is assigned. As said in the introduction, the objective of this paper is to develop SRBF-based equalizer as shown in Fig. 2. Different forms of SFLA is used here as training algorithm for RBFNN.

6 Simulation Results

For evaluation of the performance of proposed SRBF equalizers, results of contemporary RBFNN GA trained RBFNN (GRBF) [7] and PSO trained RBFNN (PRBF) [8]-based equalizers are reproduced for the purpose of comparison. Parameters selected for the simulations illustrated in Table 1.

Simulations were conducted for the most popular distorted channel with transfer function:

$$H(z) = 0.26 + 0.93z^{-1} + 0.26z^{-2} \tag{17}$$

The equalizer performance is affected by channel nonlinearity. This effect studied in this paper introducing the nonlinearity:

Table 1 Simulation parameters

GA			PSO	SFLA	
Parameter	Value	Parameter	Value	Parameter	Value
Number of iteration	1000	Number of iteration	1000	Number of iterations	1000
Number of individuals	50	Number of particles	50	Population	50
Mutation ratio	0.03	Coefficient C1	0.7	Number of memeplexe	10
Crossover ratio	0.9	Coefficient C2	0.7	Number of memes in memeplexes	10
Mutation type	Uniform			Number of memes in sub-memeplexes	08
Crossover type	Single point			Number of memetic evolutions in each sub-memeplex	10

Fig. 3 MSE of RBFNN equalizers

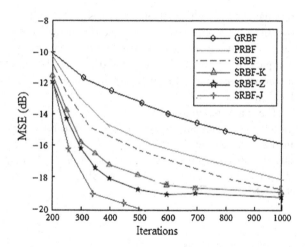

$$y(n) = \tanh[x(n)] \tag{18}$$

MSE was evaluated with a fixed SNR of 10 dB.

Figure 3 depicts the convergence of different equalizers at 10 dB. It is observed from the figure that, proposed SRBF-J outperforms other equalizers. It is also seen that RBFNN trained with SRBF and its modified forms are better than as trained with other nature inspired algorithms like GA and PSO. It is also observed that SRBF-J is a better method for training of RBFNN equalizer as compared to original as well as other modified forms of SFLA. It was observed that SRBF-J requires only 500 iterations to converge while other equalizers fail to converge within 1000 iterations.

BER comparison among RBFNN-based equalizers is depicted in Fig. 4. It is seen from Fig. 4 that the performance of GRBF and PRBF are comparable to each

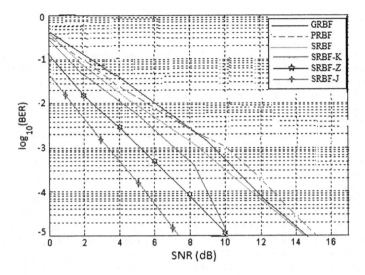

Fig. 4 BER of RBFNN equalizers

other. SRBF equalizers perform better than GRBF and PRBF. Once again SRBF-J performs better than SRBF, SRBF-Z, and SRBF-K equalizers with while SRBF-Z performs better than SRBF-K. It was observed that SRBF-J achieves tolerable error rate for SNR more than 8 dB. The same is achieved by SRBF-Z and SRBF-K with an SNR of 10 dB.

7 Conclusion

This paper introduced novel strategy for RBFNN training using SFLA and its modified forms. This paper also proposed some efficient methods for equalization as proved in simulations. Major contributions by this paper are, RBFNN training using SFLA and its modified forms, use of SRBF in channel equalization and comparison among SFLA and its modified forms while training RBFNN. Significance of the works carried out in this paper as compared to existing RBF-based equalizers is that of a better learning and generalization of the RBF network. Performance of SRBF-based equalizer also better than the existing equalizers as seen from the simulations.

It is also found from the results that SRBF-J performs better than other forms of SFLA. Extension of the work with other kinds of population-based algorithms and also other variants of SFLA may yield better results and a pointer for the future research.

References

1. Burse, K., Yadav, R.N., Shrivastava, S.C.: Channel equalization using neural networks: a review. IEEE Trans. Syst. Man Cybern. Part C: Appl. Rev. **40**(3), 352–357 (2010)
2. Subramanian, K., Savitha, R., Suresh, M.: A complex-valued neuro-fuzzy inference system and its learning mechanism. Neurocomputing **123**, 110–120 (2014)
3. Ruan, X., Zhang, Y.: Blind sequence estimation of MPSK signals using dynamically driven recurrent neural networks. Neurocomputing **129**, 421–427 (2014)
4. Cui, M., Liu, H., Li, Z., Tang, Y., Guan, X.: Identification of Hammerstein model using functional link artificial neural networks. Neurocomputing **142**, 419–428 (2014)
5. Chen, S., Mulgrew, B., Grant, P.M.: A clustering technique for digital communications channel equalization using radial basis function networks. IEEE Trans. Neural Netw. **4**, 570–579 (1993)
6. Gan, M., Peng, H., Chen, L.: Global–local optimization approach to parameter estimation of RBF-type models. Inf. Sci. **197**, 144–160 (2012)
7. Çivicioğlu, P., Alç, M., Beşdok, E.: Using an exact radial basis function artificial neural network for impulsive noise suppression from highly distorted image databases. Lect. Notes Comput. Sci. **3261**, 383–391 (2005)
8. Schilling, R.J., Carroll Jr., J.J., Al-Ajlouni, A.F.: Approximation of nonlinear systems with radial basis function neural networks. IEEE Trans. Neural Netw. **12**(1), 1–15 (2001)
9. Yavuz, O., Yildirim, T.: Design of digital filters with bilinear transform using neural networks. In: 16th IEEE Conference on Signal Processing, Communication and Applications, pp. 1–4 (2008)
10. Barreto, A.M.S., Barbosa, H.J.C., Ebecken, N.F.F.: Growing compact RBF networks using a genetic algorithm. In: Proceedings of the VII Brazilian Symposium on Neural Networks, pp. 61–66 (2002)
11. Feng, H.M.: Self-generating RBFNs using evolutional PSO learning. Neurocomputing **70**, 241–251 (2006)
12. Bergh, V., Engelbrecht, A.P.: A new locally convergent particle swarm optimizer. In: Proceedings of IEEE International Conference on Systems, Man, and Cybernetics, pp. 96–101 (2002)
13. Jiang, J., Su, Q., Li, M., Liu, M., Zhang, L.: An improved shuffled Frog Leaping algorithm. J. Comput. Sci. **10**(14), 4619–4626 (2013)
14. Eusuff, M.M., Lansey, K.E.: Optimization of water distribution network design using the Shuffled Frog Leaping Algorithm. J. Water Resour. Plan. Manage. **129**(3), 210–225 (2003)
15. Kumar, J.V., Kumar, D.M.: Generation bidding strategy in a pool based electricity market using shuffled Frog Leaping Algorithm. Appl. Soft Comput. **21**, 407–414 (2014)
16. Panda, S., Panigrahi, S.P.: A new training strategy for neural network using Shuffled Frog-Leaping Algorithm and application to channel equalization. AEU—Int. J. Electr. Commun. **68**, 1031–1036 (2014)
17. Zhang, X., Zhang, Y., Shi, Y., Zhaoa, L., Cairong Z.: Power control algorithm in cognitive radio system based on modified Shuffled Frog Leaping Algorithm. AEU—Int. J. Electr. Commun. **66**, 448–454 (2012)
18. Kavousifard, A., Samet, H.: A novel method based on modified Shuffled Frog Leaping Algorithm and artificial neural network for power system load prediction. In: Ryzko, et al. (ed.) Emerging Intelligent Technologies in Industry, SCI 369, pp. 35–46.springerlink.com

Efficient Multiprocessor Scheduling Using Water Cycle Algorithm

**Sasmita Kumari Nayak, Sasmita Kumari Padhy
and Chandra Sekhar Panda**

Abstract The multiprocessor scheduling problem consists of a set of tasks to be performed using a finite number of processors. This paper deals with the problem in a heterogeneous processing environment. A nature-inspired meta-heuristic algorithm, Water Cycle Algorithm (WCA) is being used for the purpose. For the purpose of comparison, contemporary strategies using Genetic Algorithm (GA), Bacteria Foraging Optimization (BFO) and Genetic based Bacteria Foraging (GBF) found in the literature also reproduced in this paper. Because of close relationships between the matrixes formed by the problem with those of the WCA, proposed strategy of scheduling outperforms GA and GBF based strategies.

Keywords Water cycle algorithm · Multiprocessor processing · Optimization

1 Introduction

From the issues of load balancing, the problem of multiprocessor scheduling is NP-hard [1]. The job of resource sharing is to allocate the set of task into a set of processors for deterministic execution [1]. Dynamic programming, divide and conquer, branch and bound are the traditional method gives the global optimal value which is regularly prolonged yet don't influence for fathoming the present reality issues. The researchers have determined to minimize the execution time and communication costs using branch and bound method and simulation techniques and they found the complexity of these methods [1, 2]. Traditional methods get stuck on local optima and these are very fast, give accurate answers and

S.K. Nayak
CUTM, Bhubaneswar, Odisha, India

S.K. Padhy (✉)
VSSUT, Burla, Odisha, India
e-mail: chavisiba@rediffmail.com

C.S. Panda
Sambalpur University, Sambalpur, Odisha, India

© Springer Nature Singapore Pte Ltd. 2018
M. Pant et al. (eds.), *Soft Computing: Theories and Applications*,
Advances in Intelligent Systems and Computing 583,
https://doi.org/10.1007/978-981-10-5687-1_50

deterministic [3]. A few focal points of element systems over the static strategies, for example, the dynamic methods don't require any former data or knowledge where as the static techniques must oblige an earlier information to execute every one of the assigned tasks [1].

Recent heuristics methods are [4] broadly useful optimization algorithms, and exactness or relevance or proficiency is not settled to any particular problem area. This scheduling method can be categorized into two heuristics approach such as Meta heuristics and list heuristics [1]. The meta-heuristic method is utilized for taking care of a general computational problem usually heuristics themselves by combining black-box procedures in an efficient way which is given by the user. Whereas in list heuristics the tasks are kept up according to the decreasing order of priority value in a priority queue and then, by using the First in First Out principle, the tasks are allocated to free processors [1]. Works in [1, 2] frame the problem is a multi-objective optimization problem [1, 2] that is time-consuming. Hence, in this work, we study it as single objective function.

In this paper, we use the WCA which a strategy of a global search to get the better solution in multiprocessor scheduling problem. This planned approach proves the higher performance of multiprocessor scheduling problem with experimental results by comparing with other evolutionary and nature based algorithms.

2 The Problem

The objective of the multiprocessor scheduling problem is to reduce the total execution time. This paper considers the Multiprocessor scheduling Problem which comprises of an arrangement of diverse processors (I) having distinctive processing resources including memory, which imply that tasks (J), executed on diverse processor experiences distinctive execution time. Assume that the connected links are indistinguishable. The communication cost between two tasks is experienced, if they executed on diverse processors.

Let fitness_function(I_i) is the fitness function of the processor P_i. The fitness function is utilized to calculate the superiority of the task assignment. Efficient processor utilization is expected to hold the idea of load balancing [5]. In the event that every one of the processors are utilized to their most extreme which are the measures of stillness of processors are successfully diminished then the loads. The fitness function calculates using Eq. (1) [6],

$$\text{fitness\_function}(I_i) = (1/\text{makespan}) \times \max(\text{utilization}) \qquad (1)$$

Taking into account the individual execution of the processor, the average utilization is calculated. The utilization of the individual processor can be calculated by using Eq. (2) [7].

$$\text{utilization}(I_x) = \text{completion\_time}(I_x)/\text{makespan} \tag{2}$$

After that isolating the aggregate of all processors used by the aggregate number of processors gives the average processor utilization. Avoid those processors which are being unused for long period of time, at the point when the average processor usage is optimized [1, 8].

The average of the total execution time of all task are assigned to the processors is called as the objective function of this problem which is calculated as shown in Eq. (3). Let m is the number of processors. The purpose is to minimize Eq. (3). The value clearly indicates the optimum schedule along with the balance in the processor utilization.

$$\text{Objective function} = \min\left\{\frac{\sum_{i=1}^{m} \text{fitness\_function}(I_i)}{m}\right\} \tag{3}$$

The objective is to get the minimum value of total execution time of allocated task.

3 Methodology

The nature-inspired algorithm, WCA, introduced in [9, 10], is derived from the hydrologic cycle process consists of 5 stages such as (a) Transpiration (the evaporation of water is from rivers and lakes at the same time as discharge of water by the plants discharge of water during photosynthesis), (b) Condensation (generation of cloud in atmosphere that condenses and makes it colder), (c) Precipitation (release of water to earth like melting of ice), (d) Percolation (reservation of water in field termed as groundwater) and (e) Evaporation (water from released the underground converted into a stream through evaporation). For a more detailed study on WCA, one can refer to [9, 10]. Steps of WCA are reproduced below for ease of reading.

Step 1: Select the initial parameters known as raindrops. In WCA, the raindrops play the same role as "chromosome" in GA or "particle position" in PSO. The raindrop is defined as [10]:

$$\text{A stream or Raindrop} = [x_1, x_2, x_3, \ldots, x_N] \tag{4}$$

Step 2: Generate the random initial population using Eq. (4) and

$$\text{Population of Raindrops} = \begin{bmatrix} \text{Raindrops}_1 \\ \text{Raindrops}_2 \\ \text{Raindrops}_3 \\ \vdots \\ \text{Raindrops}_{N_{\text{Pop}}} \end{bmatrix} = \begin{bmatrix} x_1^1 & x_2^1 & \cdots & x_{N_{\text{Var}}}^1 \\ x_1^2 & x_2^2 & \cdots & x_{N_{\text{Var}}}^2 \\ \cdots & \cdots & \cdots & \cdots \\ x_1^{N_{\text{Pop}}} & x_2^{N_{\text{Pop}}} & \cdots & x_{N_{\text{Var}}}^{N_{\text{Pop}}} \end{bmatrix} \tag{5}$$

where N_{pop} is the no. of raindrops and N_{Var} defines a number of variables.

Step 3: Compute the fitness of raindrops using [9]:

$$\text{Cost}_i = f\left(x_1^i, x_2^i, \ldots, x_{N_{\text{Var}}}^i\right) \quad i = 1, 2, 3, \ldots, N_{\text{pop}} \tag{6}$$

Step 4: Out of N_{pop} raindrops, N_{rs} best individuals are chosen as rivers and ocean. Find the Number of rivers using as:

$$N_{\text{rs}} = \text{No. of Rivers} + 1 \tag{7}$$

Step 5: The best raindrop is selected as the ocean. The remaining population is considered as streams or raindrops flow to the ocean and river using:

$$N_{\text{Raindrops}} = N_{\text{pop}} - N_{\text{rs}} \tag{8}$$

Step 6: Determining the intensity of the flow of the ocean and rivers as follows:

$$\text{NS}_n = \text{round}\left\{ \left| \frac{\text{Cost}_i}{\sum_{i=1}^{N_{\text{rs}}} \text{Cost}_i} \right| \times N_{\text{Raindrops}} \right\}, \quad n = 1, 2, \ldots, N_{\text{rs}} \tag{9}$$

where, NS_n is the no. of streams which stream to a particular river or ocean.

The flow of a particular stream to a particular river is decided by the distance as:

$$Y \in (0, L \times s), \quad L > 1 \tag{10}$$

where, L is a user defined somewhere around 1 and 2, i.e. $L = 2$ [9].
s is the current distance between stream and river.
Y is a number between 0 and $(L \times s)$ with any distribution.

Step 7: Flowing of the streams into the rivers by using Eq. (11).
The new location for streams and rivers computed as [9]:

$$Y_{Stream}^{i+1} = Y_{Stream}^i + \text{rand} \times L \times \left(Y_{River}^i - Y_{Stream}^i\right) \qquad (11)$$

Step 8: Flowing of the rivers into the ocean (the most downhill location) using:

$$Y_{River}^{i+1} = Y_{River}^i + \text{rand} \times L \times \left(Y_{Sea}^i - Y_{River}^i\right) \qquad (12)$$

where, rand is an equally circulated random number somewhere around 0 and 1 [9, 10].

Step 9: Trading the position of the stream with the river in order to obtain the best solution.

Step 10: Like Step 7, if a river discovers preferred solution over the ocean, the location of the river is traded.

Step 11: Check the conditions of the evaporation using:

$$\text{if} \left|Y_{Sea}^i - Y_{River}^i\right| < s_{max} \quad i = 1, 2, \dots, N_{rs} - 1 \qquad (13)$$

where, s_{max} is a small value near to 0 and controls the search depth, near the ocean.

Evaporation decreases till it rains, hence the value of s_{max} decreases at each of the iterations as:

$$s_{max}^{i+1} = s_{max}^i - \frac{s_{max}^i}{\text{max iteration}} \qquad (14)$$

Step 12: Let, LB and UB are lower and upper bounds characterized by the given problem, respectively. The new arbitrarily produced raindrops form new streams in distinctive areas can be found by using Eqs. (15) and (16) as [10]:

$$Y_{Stream}^{new} = LB + \text{rand} \times (UB - LB) \qquad (15)$$

Let, $\sqrt{\mu}$ says the standard deviation and μ (usually = 0.1 [10]) characterizes the idea of variance which illustrates the scope of seeking area close to the ocean and rand$n$ is a typically distributed random number. The optimum point is determined by the convergence rate [10] and computational execution of the algorithm for a given problems can be found as:

$$Y_{Stream}^{new} = Y_{sea} + \sqrt{\mu} \times \text{rand}n(1, N_{var}) \qquad (16)$$

Step 13: Use Eq. (14) to reduce the value of s_{max} [11].

Table 1 Example of task assignment

Processor	J_1	J_2	J_3	J_4	J_5
I_1	0	1	1	0	0
I_2	0	0	1	0	1
I_3	1	0	1	1	1
I_4	0	0	1	1	0

Step 14: Checking the criteria of convergence. On the off chance that the halting criteria are met, the procedure will stop, else it will go back to Step 7.

The WCA is used for the multiprocessor scheduling is as per the following:

- To begin with, we expect that we have downpour or precipitation [12].
- Initialize the population called as a raindrop.
- The Raindrops are created at random [9] taking into account the no. of processors utilized and the number of tasks that have reached at a specific purpose of time and determine the population size.
- Calculate the good raindrop.
- The good raindrops [9] will go to the river and whatever is left of the raindrops are considered as streams which stream to the rivers and ocean. Here we took sea as the ocean.
- At that point, the best individual or best raindrop is picked as an ocean.
- Repeat the procedure for a greatest number of iterations indicated and gets the optimal value.
- The optimal value will compare with the tasks in the queue to obtain a new schedule to receive another task.

Consequently, the grouping continues changing with time taking into account the coming of new tasks.

Let us consider an illustration of Table 1, where every row stands for the processors which relate to a task that doles out five tasks to four processors, and $[I_1, J_1] = 0$ implies that task J_1 is not allotted to processor I_1 that is 0. $[I_3, J_1] = 1$ implies that the task J_1 is allotted to processor I_3 that is 1.

4 Simulations

In order to evaluate the performance of the proposed strategy using WCA, we have reproduced the performance of GA, BFO, GBF based strategies for the purpose of comparison. Simulations were carried out using MATLAB. Minimization of execution time was taken as the criteria.

Assume the no. of the processor is 4 and the no. of the task is 5. From this combination, we selected 'S_1' number of the task assigned to the processor. In WCA, S_1 is equal to N_{rs} (no. of rivers) i.e. $N_{rs} = S_1$.

The parameters were situated at: $N_{pop} = 100, N_{var} = 10, s_{max} = 1e - 5, LB = -1, L = 2, s$ is any random value, 'g' stands for the no. of streams and the value is set to 20. Arbitrarily picked no. of stream hold in an array 'c' framed utilizing 4 distinct processors and 5 distinct tasks. Once a task is allocated, the value is picked as '1'; else, it is '0'. S_1 is produced utilizing 'c' indicates total no. of tasks allocated. The no. of iterations is picked as 100.

The simulation permitted successively for 100 times each for GBF, BFO, and GA algorithms. Comparing to these entire algorithms with our proposed WCA and getting the feasible solution of multiprocessor scheduling problem. Here we have considered two cases to prove our proposed algorithm WCA finds the best solution according to the number of tasks and processors.

CASE 1: Impact of the number of processors

In this section, we have present the simulation result for different processors with the same number of the task of different execution time. The parameters for this test result are shown below:

No. of Task = 5
No. of Processor = 7, 14, 21, 28, 35
No. of Iteration = 100

The abovementioned parameter are taken into different approaches, which results that our proposed algorithm WCA takes less execution time as depicted in Fig. 1. The minimum execution time with the equal number of tasks assigned to various numbers of processors as shown in Table 2.

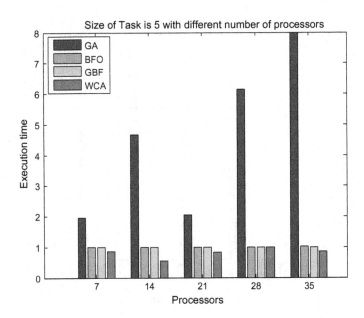

Fig. 1 Impact of number of processors on execution time

No. of processor	GA	BFO	GBF	WCA
7	1.9524	1.0000	1.0000	0.8527
14	4.6667	1.0000	1.0006	0.5564
21	2.0435	1.0000	1.0014	0.8200
28	6.1429	1.0007	1.0014	1.0000
35	8.0000	1.0218	1.0013	0.8462

Table 2 Performance with variation in number of processors (fixed number of task = 5)

CASE 2: Impact of the number of Tasks

In this section, we have present the test result of different execution time for a different number of tasks with the same number of the processor as represented in Fig. 2 and the execution time is shown in Table 3. The parameters for this test result are shown below:

No. of Task = 5, 10, 15, 20, 25
No. of Processor = 4
No. of Iteration = 100

It is seen from the tests that when the quantity of tasks expands, execution time diminishes by using the proposed strategy, WCA.

CASE 3: Impact of the number of tasks and processors

In this section, we have present the simulation of multiprocessor scheduling with different execution time from the different number of tasks and the different number of processors as represented in Fig. 3. The minimum execution time for the

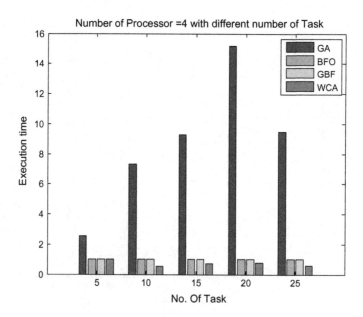

Fig. 2 Impact of number of tasks on execution time

Table 3 Performance with variation in number of tasks (fixed number of processor = 4)

No. of task	GA	BFO	GBF	WCA
7	2.5600	1.0054	1.0000	1.0000
14	7.3182	1.0000	1.0002	0.5259
21	9.2800	1.0000	1.0038	0.7351
28	15.1613	1.0000	1.0010	0.7949
35	9.4545	1.0000	1.0017	0.5720

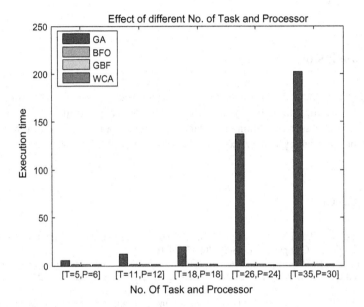

Fig. 3 Impact of change in number of processors and tasks

different number of task on the different number of processors as shown in Table 4. The parameters for this test result are shown below:

No. of Task = 5, 11, 18, 26, 35
No. of Processor = 6, 12, 18, 24, 30
No. of Iteration = 100

Table 4 Performance with variation in number of processors and tasks

No. of task	No. of processor	GA	BFO	GBF	WCA
5	6	5.4000	1.0071	1.0000	0.9278
11	12	12.3636	1.0224	1.0009	1.0000
18	18	19.4375	1.0785	1.0008	0.9114
26	24	137.0000	1.0000	1.0014	0.6478
35	30	202.0000	1.0000	1.0058	1.0000

It is seen from the tests that even if the number of task and processors increases, the proposed strategy, WCA is getting the minimum execution time as compared with the other algorithms.

In all the three cases we have taken 35 maximum numbers of tasks and processors but further, it can increase or decrease the numbers to get the preferred execution time. From the outcomes, we study that the proposed WCA accomplished fundamentally better performance on all values over the GBF, BFO, and GA algorithm.

5   Conclusion

This paper introduced a novel strategy for multiprocessor scheduling using water cycle algorithm. Simulation results prove the better performance of proposed strategy, with lesser execution. The paper also shows a pointer for future work as the proposed strategy can be extended to other scheduling problems.

References

1. Nayak, S.K., Padhy, S.K., Panigrahi, S.P.: A novel algorithm for dynamic task scheduling. Future Gener. Comput. Syst. **28**(5), 709–717 (2012)
2. Peng, D.-T., Shin, K.G., Abdelzaher, T.F.: Assignment and scheduling communicating periodic tasks in distributed real-time systems. IEEE Trans. Software Eng. **23**(12), 745–758 (1997)
3. Omara, F.A., Arafa, M.M.: Genetic algorithms for multiprocessor scheduling problem. J. Parallel Distrib. Comput. **70**(1), 13–22 (2010)
4. Balin, S.: Non-identical parallel machine scheduling using genetic algorithm. Expert Syst. Appl. **38**(6), 6814–6821 (2011)
5. Brucker, P.: Scheduling algorithms, 3rd edn. Springer, Berlin (2001)
6. Heiss, H.-U., Schmitz, M.: Decentralized dynamic load balancing: the particles approach. Inf. Sci. **84**(2), 115–128 (1995)
7. Chiang, T.-C., Chang, P.-Y., Huang, Y.-M.: Multi-processor tasks with resource and timing constraints using particle swarm optimization. Int. J. Comput. Sci. Network. Secur. **6**(4), 71–77 (2006)
8. Abdelmageed, E.A., Earl, W.B.: A heuristic model for task allocation in heterogeneous distributed computing systems. Int. J. Comput. Appl. **6**(1), 1–36 (1999)
9. Chitra, P., Rajaram, R., Venkatesh, P.: Application and comparison of hybrid evolutionary multiobjective optimization algorithms for solving multiprocessor scheduling problem on heterogeneous systems. Appl. Soft Comput. **11**(2), 2725–2734 (2011)
10. Sivanandam, S.N.: Dynamic task scheduling with load balancing using parallel orthogonal particle swarm optimisation. Int. J. Bio-Inspired Comput. (2009)
11. Eskandar, H., Sadollah, A., Bahreininejad, A., Hamdi, M.: Water cycle algorithm—a novel metaheuristic optimization method for solving constrained engineering optimization problems. Comput. Struct. **110–111**, 151–166 (2012)
12. Ashouri, M., Hosseini, S.M.: Application of krill herd and water cycle algorithms on dynamic economic load dispatch problem. IJIEEB **4**, 12–19 (2014)

Equalizer Modeling Using FFA Trained Neural Networks

Pradyumna Mohapatra, Sunita Panda and Siba Prasada Panigrahi

Abstract This work is on the Artificial Neural Network (ANN) training and application in channel equalization. Here, we design a novel training strategy for neural networks. This training uses Firefly Algorithm (FFA) to train ANN. Then, this FFA trained ANN is applied for equalization of nonlinear channels. As proved through simulations, the proposed methodology outperforms the existing ANN-based equalization schemes.

Keywords Equalization · Firefly algorithm · Neural network

1 Introduction

Artificial neural network (ANN) for channel equalization has been used since long [1–4]. The performance of ANNs for nonlinear problems makes them a popular choice in equalization. But, Back Propagation (BP) trained ANNs (1) fall in local minima (2) slow speed of convergence that depend on the selection of parameters like momentum, learning rate, and weights. Hence, evolutionary algorithms like Genetic Algorithm (GA) [5–7], Differential Evolution (DE) [13] and Particle Swarm Optimization (PSO) [8–10] used for ANN equalizer training [11–13].

Poor local search and the premature convergence of GA [14], fall into local minima and limited search space of PSO [15] and sensitivity to the choice of control parameters of DE [16, 17] made this paper for use of FFA [18] for ANN training.

P. Mohapatra
OEC, Bhubaneswar, Odisha, India

S. Panda
ECE, GITAM University, Bangalore, Karnataka, India

S.P. Panigrahi (✉)
EE, VSSUT, Burla 768018, Odisha, India
e-mail: Siba_panigrahy15@rediffmail.com

© Springer Nature Singapore Pte Ltd. 2018
M. Pant et al. (eds.), *Soft Computing: Theories and Applications*,
Advances in Intelligent Systems and Computing 583,
https://doi.org/10.1007/978-981-10-5687-1_51

2   The Problem

Figure 1 depicts a popular digital communication system.

A popular linear channel model is FIR model. In this model, transmitted sequence is binary, represented as $x(k)$ at kth time instance and corresponding output at the same instant is, $y_1(k)$ as follows:

$$y_1(k) = \sum_{i=0}^{N-1} h_i x(k-i) \tag{1}$$

Here, $h_i(i = 0, 1, \ldots, N-1)$ and N, respectively, are the channel taps and the channel length. The block 'NL' denotes the nonlinearity inserted in the channel. One popular form of nonlinear function is as follows:

$$y(k) = F(y_1(k)) = y_1(k) + b[y_1(k)]^3 \tag{2}$$

Here, b is a constant. The block 'NL' output is as follows:

$$y(k) = \left(\sum_{i=0}^{N-1} h_i x(k-i)\right) + b\left(\sum_{i=0}^{N-1} h_i x(k-i)\right)^3 \tag{3}$$

The channel output $y(k)$ is added with noise, $\eta(k)$ inserted in the channel. The signal at the receiver is $r(k)$ and as follows:

$$r(k) = y(k) + \eta(k) \tag{4}$$

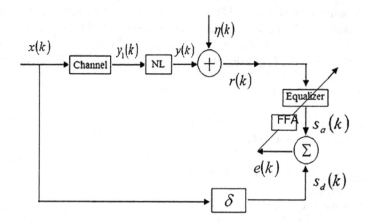

Fig. 1 Baseband model of digital communication system

Equalizer is used to recover the transmitted symbol, $x(k - \delta)$, from a-priory knowledge on the samples received, 'δ' being the associated transmission delay. The desired signal can be represented as $d(k)$ and can be defined as

$$d(k) = x(k - \delta) \tag{5}$$

The problem of equalization is a problem of classification [6–9], and the equalizer makes partition of the input space $x(k) = [x(k), x(k - 1), \ldots x(k - N + 1)]^T$ into two distinct regions.

The Bays theory provides the optimal solution for this where the decision function is as follows:

$$f_{bay}(x(k)) = \sum_{j=1}^{n} \beta_j \exp\left(\frac{-\|x(k) - c_j\|}{2\sigma^2}\right) \tag{6}$$

Since the sequence transmitted is binary, hence:

$$\beta_j = \begin{cases} +1 & c_j \in C_d^{(+1)} \\ -1 & c_j \in C_d^{(-1)} \end{cases} \tag{7}$$

Here, $C_d^{(+1)}/C_d^{(-1)}$ and c_j, respectively, represent transmitted symbol, $x(k - \delta) = +1/-1$ and σ^2 is the noise variance.

In Fig. 1, the block "Equalizer" is ANN. SFLA and its modified forms are used to optimize the number of layers and neurons in each layer. For a number of neurons in the input layer, it is N, same as the number of taps.

The equalizer output is as follows:

$$f_{RBF}(x(k)) = \sum_{j=1}^{n} w_j \exp\left(\frac{-\|x(k) - t_j\|^2}{\alpha_j}\right) \tag{8}$$

Here, t_j and α_j, respectively, represent the centers and the spreads of the neurons in the hidden layer (s). The vector w_j contains the connecting weights. The output from the equalizer of Eq. (6) uses the nonlinear function of Eq. (7). For optimal weights, the condition is t_j is equals to c_j.

The decision at equalizer output is as follows:

$$\hat{x}(k - \delta) = \begin{cases} +1 & f_{ANN}(x(k)) \geq 0 \\ -1 & \text{elsewhere} \end{cases} \tag{9}$$

Also, the difference between the equalizer output (i.e., $s_a(k) = \hat{x}(k - \delta)$) and desired signal (i.e., $s_d(k) = x(k - \delta)$) is termed as error, $e(k)$, and updates the weights.

For l is the number of samples then, Mean Square Error (MSE):

$$\text{MSE} = \frac{1}{l} E\left[e^2(k)\right] \tag{10}$$

Bit error rate (BER) is the ratio of error bits to transmitted bits. In this paper, MSE and BER are chosen as performance index.

3 FFA and ANN Training

3.1 FFA

FFA [18] is a population-based optimization algorithm that mimics firefly. Firefly is known to be unisex and attracted toward each other according to the intensity of lights they produce. Here, the population of FFA is formed by the solution vectors. The parameters of FFA are the weights (w), spread parameters (α), center vector (c), and the bias (β). The mean vector c_i of the ith neuron of hidden layers is represented by $c_i = (c_{i1}, c_{i2}, \ldots, c_{im})$, hence, the parametric vector t_i of each firefly with $IJ + I + MI + J$ parameter is as follows:

$$t_i = \begin{pmatrix} w_{11}^i, w_{12}^i, \ldots, w_{IJ}^i, \alpha_1^i, \alpha_2^i, \ldots, \alpha_I^i, c_{11}^i, c_{12}^i, \\ \ldots c_{1m}^i, \ldots c_{I1}^i, c_{I2}^i, \ldots, c_{Im}^i, \beta_1^i, \ldots \beta_J^i \end{pmatrix} \tag{11}$$

In this work, a firefly represents a specific ANN equalizer. In the training process using FFA, the vector t_i of firefly of corresponding ANN optimizes the fitness function [18]:

$$f(t_i) = \frac{1}{1 + \text{MSE}} = \frac{1}{1 + \frac{1}{Q} \sum\limits_{k=1}^{Q} \|d(k) - y(k)\|^2} \tag{12}$$

Here, $d(k)$ and $y(k)$ are the desired and actual output for training samples x_i of ANN designed by the vector t_i and that of the equalizer. The number of samples used in the training is represented by Q.

3.2 Training

In the proposed training method, ANN formulates guidelines for optimal structure as an administrator. Then FFA plays the role of the teacher to optimize the structure. This teacher once again teaches the ANN that plays the role of student. In this

Advances in Intelligent Systems and Computing

Volume 583

Series editor

Janusz Kacprzyk, Polish Academy of Sciences, Warsaw, Poland
e-mail: kacprzyk@ibspan.waw.pl

The series "Advances in Intelligent Systems and Computing" contains publications on theory, applications, and design methods of Intelligent Systems and Intelligent Computing. Virtually all disciplines such as engineering, natural sciences, computer and information science, ICT, economics, business, e-commerce, environment, healthcare, life science are covered. The list of topics spans all the areas of modern intelligent systems and computing.

The publications within "Advances in Intelligent Systems and Computing" are primarily textbooks and proceedings of important conferences, symposia and congresses. They cover significant recent developments in the field, both of a foundational and applicable character. An important characteristic feature of the series is the short publication time and world-wide distribution. This permits a rapid and broad dissemination of research results.

More information about this series at http://www.springer.com/series/11156

Millie Pant · Kanad Ray
Tarun K. Sharma · Sanyog Rawat
Anirban Bandyopadhyay
Editors

Soft Computing: Theories and Applications

Proceedings of SoCTA 2016, Volume 1

 Springer

Editors
Millie Pant
Department of Applied Science
  and Engineering
IIT Roorkee
Saharanpur
India

Kanad Ray
Department of Physics
Amity School of Applied Sciences, Amity
  University Rajasthan
Jaipur, Rajasthan
India

Tarun K. Sharma
Department of Computer Science
  and Engineering
Amity School of Engineering
  and Technology, Amity University
  Rajasthan
Jaipur, Rajasthan
India

Sanyog Rawat
Department of Electronics
  and Communication Engineering
SEEC, Manipal University Jaipur
Jaipur, Rajasthan
India

Anirban Bandyopadhyay
Surface Characterization Group, NIMS
Nano Characterization Unit, Advanced Key
  Technologies Division
Tsukuba, Ibaraki
Japan

ISSN 2194-5357   ISSN 2194-5365   (electronic)
Advances in Intelligent Systems and Computing
ISBN 978-981-10-5686-4   ISBN 978-981-10-5687-1   (eBook)
https://doi.org/10.1007/978-981-10-5687-1

Library of Congress Control Number: 2017947482

Printed on acid-free paper

This Springer imprint is published by Springer Nature
The registered company is Springer Nature Singapore Pte Ltd.
The registered company address is: 152 Beach Road, #21-01/04 Gateway East, Singapore 189721, Singapore

Preface

It is a matter of pride to introduce the first international conference in the series of "Soft Computing: Theories and Applications (SoCTA)", which is a joint effort of researchers from Machine Intelligence Lab, USA, and the researchers and Faculty members from Indian Institute of Technology, Roorkee; Amity University Rajasthan.

The maiden conference took place in the historic city of Jaipur at the campus of research-driven university, Amity University Rajasthan. The conference stimulated discussions on various emerging trends, innovation, practices, and applications in the field of Soft Computing.

This book that we wish to bring forth with great pleasure is an encapsulation of 149 research papers, presented during the three-day international conference. We hope that the initiative will be found informative and interesting to those who are keen to learn on technologies that address to the challenges of the exponentially growing information in the core and allied fields of Soft Computing.

We are thankful to the authors of the research papers for their valuable contribution in the conference and for bringing forth significant research and literature across the field of Soft Computing.

The editors also express their sincere gratitude to SoCTA 2016 Patron, Plenary Speakers, Keynote Speakers, Reviewers, Programme Committee Members, International Advisory Committee and Local Organizing Committee, Sponsors without whose support the support and quality of the conference could not be maintained.

We would like to express our sincere gratitude to Prof. Sanghamitra Bandyopadhyay, Director, ISI Kolkata, for gracing the occasion as the Chief Guest for the Inaugural Session and delivering a Plenary talk.

We would like to express our sincere gratitude to Dr. Anuj Saxena, Officer on Special Duty, Chief Minister's Advisory Council, Govt. of Rajasthan, for gracing the occasion as the Chief Guest for the Valedictory Session.

We would like to extend our heartfelt gratitude to Prof. Nirupam Chakraborti, Indian Institute of Technology, Kharagpur; Prof. Ujjwal Maulik, Jadavpur University; Prof. Kumkum Garg, Manipal University Jaipur; Dr. Eduardo Lugo,

Université de Montréal; Prof. Lalit Garg, University of Malta for delivering invited lectures.

We express our special thanks to Prof. Ajith Abraham, Director, MIR Labs, USA, for being a General Chair and finding time to come to Jaipur amid his very busy schedule.

We are grateful to Prof. W. Selvamurthy and Ambassador (Retd.) R.M. Aggarwal for their benign cooperation and support.

A special mention of thanks is due to our student volunteers for the spirit and enthusiasm they had shown throughout the duration of the event.

We express special thanks to Springer and its team for the valuable support in the publication of the proceedings.

With great fervor, we wish to bring together researchers and practitioners in the field of Soft Computing year after year to explore new avenues in the field.

Saharanpur, India	Dr. Millie Pant
Jaipur, India	Dr. Kanad Ray
Jaipur, India	Dr. Tarun K. Sharma
Jaipur, India	Dr. Sanyog Rawat
Tsukuba, Japan	Dr. Anirban Bandyopadhyay

Organizing Committee

Patrons-in-Chief

Dr. Ashok K. Chauhan, Founder President, Ritnand Balved Education Foundation (RBEF)
Dr. Aseem Chauhan, Chancellor, Amity University Rajasthan

Patron

Prof. (Dr.) S.K. Dube, Vice Chancellor, AUR

Co-patron

Prof. (Dr.) S.L. Kothari, Pro Vice Chancellor, AUR

General Chair

Prof. (Dr.) Ajith Abraham, Director, MIR Labs, USA
Dr. Millie Pant, Indian Institute of Technology, Roorkee
Prof. (Dr.) Kanad Ray, Amity University Rajasthan

Program Chairs

Dr. Tarun K. Sharma, Amity University Rajasthan
Dr. Sanyog Rawat, Manipal University Jaipur

Organizing Chair

Prof. D.D. Shukla, Amity University Rajasthan
Prof. Jagdish Prasad, Amity University Rajasthan

Finance Chair

Brig. (Retd) S.K. Sareen, Amity University Rajasthan
Mr. Sunil Bhargawa, Amity University Rajasthan

Conference Proceedings and Printing & Publication Chair

Dr. Millie Pant, Indian Institute of Technology, Roorkee
Prof. (Dr.) Kanad Ray, Amity University Rajasthan
Dr. Tarun K. Sharma, Amity University Rajasthan
Dr. Sanyog Rawat, Manipal University Jaipur
Dr. Anirban Bandoypadhyay, NIMS, Japan

Best Paper and Best Ph.D. Thesis Chair

Prof. S.C. Sharma, Indian Institute of Technology, Roorkee
Prof. K.K. Sharma, MNIT, Jaipur
Dr. Anirban Bandoypadhyay, NIMS, Japan

Technical and Special Sessions Chair

Dr. Musrrat Ali, Glocal University Saharanpur
Dr. Sushil Kumar, Amity University Uttar Pradesh

Publicity Chair

Dr. Pravesh Kumar, Jaypee University Noida
Mr. Jitendra Rajpurohit, Amity University Rajasthan
Mr. Anil Saroliya, Amity University Rajasthan
Dr. Divya Prakash, Amity University Rajasthan
Mr. Anurag Tripathi, Arya, Jaipur

Registration Chair

Mr. Amit Hirawat, Amity University Rajasthan
Mr. Jitendra Rajpurohit, Amity University Rajasthan

Outcome Committee

Prof. Kanad Ray, Amity University Rajasthan
Dr. Tarun K. Sharma, Amity University Rajasthan
Prof. (Dr.) P.V.S. Raju, Amity University Rajasthan

Web Administrator

Mr. Chitreshh Banerjee, Amity University Rajasthan

Reporting

Dr. Ratnadeep Roy, Amity University Rajasthan
Ms. Pooja Parnami, Amity University Rajasthan

Hospitality Chair

Mr. Vikas Chauhan, Amity University Rajasthan
Ms. Preeti Gupta, Amity University Rajasthan
Dr. Divya Prakash, Amity University Rajasthan

Mr. Jitendra Rajpurohit, Amity University Rajasthan
Mr. Amit Chaurasia, Amity University Rajasthan
Mr. Deepak Panwar, Amity University Rajasthan

Local Organizing Committee

Prof. Upendra Mishra, Amity University Rajasthan
Dr. Swapnesh Taterth, Amity University Rajasthan
Ms. Parul Pathak, JECRC Jaipur
Mr. Anurag Tripathi, Arya, Jaipur
Mr. Ashwani Yadav, Amity University Rajasthan
Ms. Vaishali Yadav, Amity University Rajasthan
Ms. Bhawana Sharma, Amity University Rajasthan
Ms. Pallavi Sharma, Amity University Rajasthan
Mr. Abhay Sharma, Amity University Rajasthan
Dr. Irshad Ansari, Indian Institute of Technology, Roorkee
Mr. Bilal, Indian Institute of Technology, Roorkee
Mr. Nathan Singh, Indian Institute of Technology, Roorkee
Mr. Sunil K. Jauhar, Indian Institute of Technology, Roorkee
Ms. Meenu Singh, Indian Institute of Technology, Roorkee
Mr. Het, Indian Institute of Technology, Roorkee

International Advisory Board

Aboul Ella Hassanien, University of Cairo, Egypt
Adel Alimi, University of Sfax, Tunisia
Aditya Ghose, University of Wollongong, Australia
André Ponce de Leon F de Carvalho, University of São Paulo, Brazil
Ashley Paupiah, Amity Mauritius
Bruno Apolloni, University of Milano, Italy
Francesco Marcelloni, University of Pisa, Italy
Francisco Herrera, University of Granada, Spain
Imre J. Rudas, Obuda University, Hungary
Javier Montero, Complutense University of Madrid, Spain
Jun Wang, Chinese University of Hong Kong, Hong Kong
Naren Sukurdeep, Amity Mauritius
Mo Jamshidi, University of Texas at San Antonio, USA
Sang-Yong Han, Chung-Ang University, Korea
Sebastián Ventura, University of Cordoba, Spain

way, ANN acts both as an administrator and a student. The methodology follows the following flow chart:

> Initialize ANN_administrator
> for $i = 1, 2, \ldots, N$
>
>> generate FFA_teacher (i)
>> for ANN_student $j = 1, 2, \ldots, N$
>> start ANN_student
>> end
>
> end

when solution is not found

> calculate update
> set maximum number of iterations
> for (FFA_teacher $i = 1, 2, \ldots, N$)
>
>> while (iterations < maximum)
>> for (ANN_student $j = 1, 2, \ldots, N$
>>
>>> test ANN_student (j)
>>
>> end
>> for ANN_student $j = 1, 2, \ldots, M$
>> Update weights of ANN_student (j)
>> end
>
> end
> return global best

end
update global best
end

4 Simulations

For comparison, original GA, PSO, and FFA are considered. Simulation parameters considered are as given in Table 1. Here, Parameters for FFA are as in [18].

For simulations, we have chosen following two popular channels:

$$H(z) = 0.24 + 0.93z^{-1} + 0.26z^{-2} \tag{13}$$

$$H(z) = 0.303 + 0.9029z^{-1} + 0.304z^{-2} \tag{14}$$

Nonlinearity of Eq. (2) introduced.

Table 1 Simulation parameters

GA		PSO		FFA	
Parameter	Value	Parameter	Value	Parameter	Value
Max No. of iterations	1000	Max No. of iterations	1000	Max No. of iterations	1000
Population size	50	Population size	50	Population size	50
Mutation ratio	0.03	Coefficient C1	0.7	Attractiveness	1
Crossover ratio	0.9	Coefficient C2	0.7	Light absorption coefficient	2
Mutation type	Uniform				
Crossover type	Single point				

Fig. 2 MSE performance for channel (13)

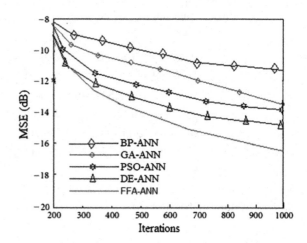

For the purpose of comparison we considered the following:

- ANN-equalizers of [19] represented by BP-ANN
- ANN-equalizers of [11], represented by GA_ANN and PSO_ANN
- ANN-equalizers of [13], represented by DE-ANN

MSE was computed with a fixed Signal to Noise Ratio (SNR) of 10 dB. MSE plots are depicted in Figs. 2 and 3 for channels of Eqs. (13) and (14), respectively. Corresponding BER plots are shown in Figs. 4 and 5

The figures reveal the following:

- Channel of Eq. (13): MSE of proposed FFA-ANN and DE-ANN are comparable till 300 iterations and then FFA-ANN performance is better. But, FFA-ANN performance is better than other equalizers in all conditions. Also, BER of FFA-ANN becomes less than 10^{-5} at SNR of 10 dB outperforms other ANN-equalizers.

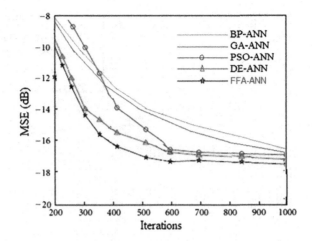

Fig. 3 MSE performance for channel (14)

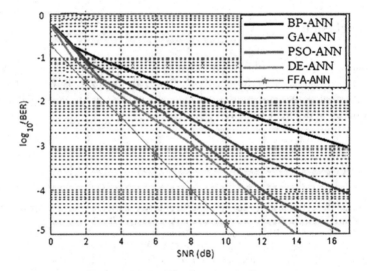

Fig. 4 BER performance for channel (13)

- Channel of Eq. (14): MSE of proposed FFA-ANN and DE-ANN, are comparable up to 600 iterations. In BER after SNR of 2 dB, FFA-ANN outperforms other ANN-based equalizers like those of channel of Eq. (12).

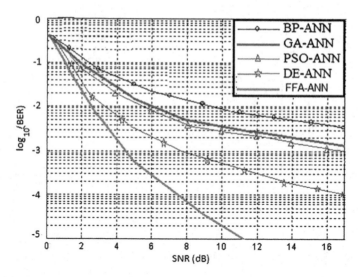

Fig. 5 BER performance for channel (14)

5 Conclusion

This paper proposed a FFA-based training method for neural networks-based equalizer. The superior performance of proposed equalizer as compared to the existing neural networks-based equalizers was evidenced from simulations conducted on three different channels.

References

1. Seyman, M.N., Taşpınar, N.: channel estimation based on neural network in space time block coded MIMO–OFDM system. Digit. Signal Proc. **23**(1), 275–280 (2013)
2. Ruan, X., Zhang, Y.: Blind sequence estimation of MPSK signals using dynamically driven recurrent neural networks. Neurocomputing **129**, 421–427 (2014)
3. Rizaner, A.: Radial basis function network assisted single-user channel estimation by using a linear minimum mean square error detector under impulsive noise. Comput. Electr. Eng. **39** (4), 1288–1299 (2013)
4. Sahoo, H.K., Dash, P.K., Rath, N.P.: NARX model based nonlinear dynamic system identification using low complexity neural networks and robust H_∞ filter. Appl. Soft Comput. **13**(7), 3324–3334 (2013)
5. Montana, D.J., Davis, L.: Training feedforward neural networks using genetic algorithms. Mach. Learn. **12**, 762–767 (2000)
6. Blanco, A., Delgado, M., Pegalajar, M.C.: A real-coded genetic algorithm for training recurrent neural networks. Neural Netw. **14**, 93–105 (2001)
7. Kim, D., Kim, H., Chung, D.: A modified genetic algorithm for fast training neural networks. Lect. Notes Comput. Sci. **3496**, 660–665 (2005)

8. Rakitianskaia, A.S., Engelbrecht, A.P.: Training feedforward neural networks with dynamic particle swarm optimization. Swarm Intell. **6**(3), 233–270 (2012)
9. Vilovic, I., Burum, N., Milic, D.: Using particle swarm optimization in training neural network for indoor field strength prediction. In: ELMAR, International Symposium, pp. 275–278 (2009)
10. Su, R. Kong, L., Song, S., Zhang, P., Zhou, K., Cheng, J.: A new ridgelet neural network training algorithm based on improved particle swarm optimization. In: Third International Conference on Natural Computation, vol. 3, pp. 411–415, 24–27 (2012)
11. Jatoth, R.K., Vaddadi, M.S., Anoop, S.: An intelligent functional link artificial neural network for channel equalization, In: Proceedings of the 8th WSEAS International Conference on Signal processing, ISPRA'09, pp. 240–245 (2009)
12. Das, G., Pattnaik, P.K., Padhy, S.K.: Artificial neural network trained by particle swarm optimization for non-linear channel equalization. Expert Syst. Appl. **41**(7), 3491–3496 (2014)
13. Patra, G.R., Maity, S., Sardar, S., Das, S.: Nonlinear channel equalization for digital communications using DE-trained functional link artificial neural networks. Commun. Comput. Inf. Sci. **168**, 403–414 (2011)
14. Karaboga, N.: Digital IIR filter design using differential evolution algorithm. EURASIP J. Appl. Signal. Process. **8**, 1269–1276 (2005)
15. Bergh, V., Engelbrecht, F.: A new locally convergent particle swarm optimizer. In: Proceedings of IEEE International Conference on Systems, Man, and Cybernetics, pp. 96–101 (2002)
16. Liu, J., Lampinen, J.: On setting the control parameter of the differential evolution method, In: Proceedings of the 8th International Conference of Soft Computing (MENDEL 2002), pp. 11–18 (2002)
17. Das, S., Suganthan, P.N.: Differential evolution: a survey of the state-of-the-art. IEEE Trans. Evol. Comput. **15**(1), 4–31 (2011)
18. Yang, X.S.: Firefly algorithm, stochastic test functions and design optimization. Int. J. Bio-inspired Comput. **2**(2), 78–84 (2010)
19. Zhao, H., Zeng, X., Zhang, J., Li, T., Liu, Y., Ruan, D.: Pipelined functional link artificial recurrent neural network with the decision feedback structure for nonlinear channel equalization. Inf. Sci. **181**, 3677–3692 (2011)

Quantum Particle Swarm Optimization Tuned Artificial Neural Network Equalizer

Gyanesh Das, Sunita Panda and Sasmita Kumari Padhy

Abstract This article uses Artificial Neural Network (ANN) trained with Quantum behaved Particle Swarm Optimization (QPSO) for the problem of equalization. Though the use of PSO in training of ANN finds optimal weights of the network it fails in the design of appropriate topology. But, QPSO is capable of optimizing the network topology. Here, parameters like neurons in each layer, the number of layers etc. are optimized using QPSO. Then, this QPSO tuned ANN then applied to the problem of channel equalization. The superior performance of proposed equalizer is proved through simulations.

Keywords Neural network · Particle swarm optimization · Channel equalization

1 Introduction

Equalization is an established and ever growing field of research. Particle Swarm Optimization (PSO) based equalizers are popular because of easy implementation [1–4]. In QPSO [5, 6], advantages of Quantum mechanics is an addition to PSO. QPSO finds successful use in optimization problems [7–13] also in equalization [14].

On the other hand, because of advantages of ANN in handling nonlinear problems, its use in equalization also goes back to 1990s [15–22]. However, the slow convergence of popular Back Propagation (BP) trained ANN attracted evolutionary algorithms for the training of ANN [23–28].

G. Das
DRIEMS, Cuttack, India

S. Panda
ECE, GITAM University, Bangalore, Karnataka, India

S.K. Padhy (✉)
VSSUT, Burla, Odisha, India
e-mail: chavisiba@rediffmail.com

© Springer Nature Singapore Pte Ltd. 2018
M. Pant et al. (eds.), *Soft Computing: Theories and Applications*,
Advances in Intelligent Systems and Computing 583,
https://doi.org/10.1007/978-981-10-5687-1_52

Use of PSO trained ANN for equalization has been proposed in [29]. In this paper, we have attempted to add the advantages of quantum mechanics to the network swarms generated in [29] and also changed the method of training. Interestingly, proposed approach of training yield much better results

2   The Problem

Communication channel with equalizer model is illustrated in Fig. 1. The impulse response of the channel [30]:

$$H_i(z) = \sum_{j=0}^{p_i-1} a_{i,j} z^{-j} \quad 0 \le i \le n \tag{1}$$

Here $a_{i,j}$ and p_i are weights and length of the channel. For simplicity, transmitted symbol $x_i(n)$, $0 \le i \le n$ was assumed to be binary and from independent, identically distributed (i.i.d) dataset within the bound of $\{\pm 1\}$ satisfying:

$$E[x_i(n)] = 0 \tag{2}$$

$$E\big[x_i(n_1)x_j(n_2)\big] = \delta(i-j)\delta(n_1-n_2) \tag{3}$$

where $E[\cdot]$ is the expectation operator and

$$\delta(n) = \begin{cases} 1 & n = 0 \\ 0 & n \ne 0 \end{cases} \tag{4}$$

The channel output:

$$y(n) = d(n) + d_{co}(n) + \eta(n) \tag{5}$$

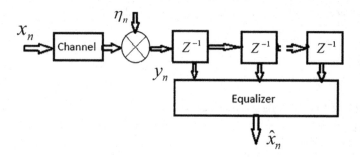

Fig. 1 Model of digital transmission system

Here $d(n)$ desired received signal $d_{co}(n)$ is interfering signal and $\eta(n)$ is noise component assumed to be Gaussian with variance $E[\eta^2(n)] = \sigma_\eta^2$ and uncorrelated with data. The desired and interfering signal:

$$d(n) = \sum_{j=0}^{p_0-1} a_{0,j} x_0(n-j) \tag{6}$$

$$d_{co}(k) = \sum_{i=1}^{n} \sum_{j=0}^{p_i-1} a_{i,j} x_i(n-j) \tag{7}$$

The equalizer recovers the transmitted sequence $x_0(n-k)$ based on channel observation vector, $y(n) = [y(n), y(n-1), \ldots, y(n-m+1)]^T$, where m is the order of equalizer and k is decision delay.

The error $e(n)$ is:

$$e(n) = d(n) - y(n) \tag{8}$$

Since $e^2(n)$ will be positive mean of this square error (MSE) is taken as a performance index. Hence, the problem of equalization now becomes an optimization problem where the objective is to minimize $e^2(n)$ and adapt the optimal weights of the equalizer.

3 QPSO

PSO [31, 32], is inspired by food search behavior of animals and swarm theory. PSO is constituted by some particles representing a solution vector to the problem. Particles are capable of memorizing their past best position and velocity. Each particle in D-dimensional space and denoted by, $X_i = (x_{i1} \ldots x_{id}, \ldots x_{iD})$. and its velocity along each dimension denoted by $V_i = (v_{i1} \ldots v_{id}, \ldots v_{iD})$. Position and velocity are updated in each iteration. The rule to update for a particle i at $(t+1)$th iteration:

$$\begin{aligned} v_{ij}^{t+1} &= w \cdot v_{ij}^t + c_1 \cdot r_{1j}^t \cdot \left(P_{ij}^t - x_{ij}^t\right) + c_2 \cdot r_{2j}^t \cdot \left(P_{gj}^t - x_{ij}^t\right); \quad r_{1j}^t, r_{2j}^t \in U(0,1) \\ x_{ij}^{t+1} &= x_{ij}^t + v_{ij}^{t+1} \end{aligned} \tag{9}$$

Here, c_1 is cognitive coefficient, and c_2 social coefficients. Both c_1 and c_2 are positive and control the movement of the particle. Best previous position for a particle (*pbest*) is denoted by the vector $P_i = (p_{i1} \ldots p_{id}, \ldots p_{iD})$. Similarly, a best previous position for the entire population (*gbest*) is denoted by the vector

$P_g = (p_{g1} \ldots p_{gd}, \ldots p_{gD})$. The inertia weight (w) that optimally reduces from an initial value of 0.9 to 0.4 [32], controls the movement.

In QPSO, MP is mean of *pbest* of all particles in the population. This makes all the particles to depend on their colleagues and determines position distribution in next iteration. Another parameter α termed as contraction-expansion (CE) coefficient is also an addition in QPSO. This CE is responsible for a wider search space in QPSO. For M particles in D-dimensional space, the position of a particle i at $(t+1)$th iteration is updated by:

$$x_{ij}^{t+1} = p_{ij}^t \pm \alpha \cdot \left| x_{ij}^t - \mathrm{MP}_j^t \right| \cdot \ln\left(\frac{1}{u_{ij}^t}\right), \quad u_{ij}^t \in U(0,1) \tag{10}$$

$$p_{ij}^t = \phi_{ij}^t \cdot P_{ij}^t + \left(1 - \phi_{ij}^t\right) \cdot P_{gj}^t, \quad \phi_{ij}^t \in U(0,1) \tag{11}$$

$$\mathrm{MP}^t = \left(\mathrm{MP}_1^t, \ldots \mathrm{MP}_D^t\right) = \left(\frac{1}{M}\sum_{i=1}^{M} P_{i1}^t, \cdots \frac{1}{M}\sum_{i=1}^{M} P_{iD}^t\right) \tag{12}$$

4 ANN Training Using QPSO

The swarm of ANN constructed using the methodology adopted in [29] Then the swarm of ANNs was trained using following steps:

- Initialize a population of particles. Each particle corresponds to a network. Allowed iterations as MaxIteration. Start the first iteration.
- Each particle to be decoded to one ANN, calculate the weights using pseudo–inverse method. Calculate the particle fitness.
- Run QPSO to update the position.
- Go to next iteration;
- Go back to step ii until reaching the maximum number of iterations.

5 Simulations

Simulation parameters had chosen same as in [29] for ease in comparison. Hence, initial values set were $M = 5, N = 25, C_1 = C_2 = 2$ and the velocity factors were set at 0.8. The channel is chosen also same as of [29] with transfer function:

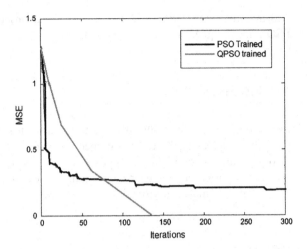

Fig. 2 MSE of the proposed equalizer

$$H(z) = 1 - 0.9z^{-1} + 0.385z^{-2} + 0.771z^{-3} \qquad (13)$$

The same nonlinearity:

$$y(n) = \tanh[x(n)] \qquad (14)$$

In [29], it was proved that PSO trained ANN performs better than existing ANN based equalizers. The objective of this research is to prove that the proposed QPSO trained ANN performs even better than results provided in [29]. MSE and Bit Error rate (BER) were taken as a performance index.

MSE and BER performance of PSO and proposed QPSO trained equalizers are illustrated in Figs. 2 and 3 respectively.

While studying the convergence from Fig. 2, it was observed that proposed equalizer converges after 140 iterations while PSO trained equalizer converges after

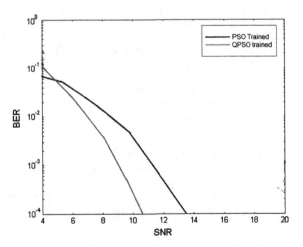

Fig. 3 BER of the proposed equalizer

350 iterations. Similarly, BER curve shows that proposed equalizer outperforms PSO trained equalizer in all noise conditions.

6 Summary

ANN equalizer trained with QPSO is introduced in this paper. Proposed scheme provides an advantage over PSO trained equalizers in that advantage of quantum theory is an addition. Also, proposed training scheme is new and simple to implement. Proposed equalizer outperforms the existing ANN based equalizers as evidenced by simulation results.

References

1. Saha, S., et.al.: Comprehensive learning particle swarm optimization based TSK structure identification and its application in OFDM receiver for nonlinear channel equalization. In: National Conference on Communications (NCC) (2011)
2. Eshwaraiah, H.S., et.al.: Cooperative particle swarm optimization based receiver for large-dimension MIMO-ZPSC systems. In: IEEE Wireless Communications and Networking Conference (WCNC) (2012)
3. Krusienski, D.J.: et.al.: The application of particle swarm optimization to adaptive IIR phase equalization. In: IEEE International Conference on Acoustics, Speech, and Signal Processing (2004)
4. El Morra, H., et.al.: Application of heuristic algorithms for multiuser detection. In: IEEE International Conference on Communications (2006)
5. Sun, J., et.al.: A global search strategy of quantum-behaved particle swarm optimization. In: IEEE Conference on Cybernetics and Intelligent Systems, vol. 1–3, 111–116 (2004)
6. Sun J., et.al.: Adaptive parameter control for quantum-behaved particle swarm optimization on individual level. In: IEEE International Conference on Systems, Man and Cybernetics (2005)
7. Coelho, L.S.: Novel Gaussian quantum-behaved particle swarm optimiser applied to electromagnetic design. IET Sci. Meas. Technol. 1(5), 290–294 (2007)
8. Coelho, L.S., et.al.: Global optimization of electromagnetic devices using an exponential quantum-behaved particle swarm optimizer. IEEE Trans. Magn. 44(6), 1074, 1077 (2008)
9. Coelho, L.S.: A quantum particle swarm optimizer with chaotic mutation operator. Chaos Solitons Fractals 37(5), 1409–1418 (2008)
10. Fei, G., et al.: Parameter estimation online for Lorentz system by a novel quantum behaved particle swarm optimization. Chin. Phys. B 17, 1409–1418 (2008)
11. Li, S., et al.: A new QPSO based bp neural network for face detection. Adv. Soft Comput. 40, 355–363 (2007)
12. Mikki, S.M., et al.: Quantum particle swarm optimization for electromagnetics. IEEE Trans. Antennas Propag. 54(10), 2764–2775 (2006)
13. Wang, J., et al.: Quantum-behaved particle swarm optimization with generalized local search operator for global optimization. Lect. Notes Comput. Sci. 4682, 851–860 (2007)
14. Ling-ling, H., et al.: Wavelet fractionally spaced blind equalization algorithm based on QPSO. Comput. Eng. 37(24), 195–197 (2011)

15. Patra, J.C., et.al.: Nonlinear channel equalization with QAM signal using Chebyshev artificial neural network. In: Proceedings of International Joint Conference on Neural Networks, Montreal, Canada, pp. 3214–3219 (2005)
16. Zhao, H., et al.: A novel joint-processing adaptive nonlinear equalizer using a modular recurrent neural network for chaotic communication systems. Neural Netw. **24**, 12–18 (2011)
17. Zhao, H., et al.: An adaptive decision feedback equalizer based on the combination of the FIR and FLNN. Digit. Signal Process. **21**, 679–689 (2011)
18. Zhao, H., et al.: Pipelined functional link artificial recurrent neural network with the decision feedback structure for nonlinear channel equalization. Inf. Sci. **181**, 3677–3692 (2011)
19. Zhao, H., et al.: Adaptively combined FIR and functional link neural network equalizer for nonlinear communication channel. IEEE Trans. Neural Netw. **20**(4), 665–674 (2009)
20. Zhao, H., et al.: Complex-valued pipelined decision feedback recurrent neural network for nonlinear channel equalization. IET Commun. **6**(9), 1082–1096 (2012)
21. Zhao, H., et al.: Adaptive reduced feedback FLNN nonlinear filter for active control of nonlinear noise processes. Sig. Process. **90**(3), 834–847 (2010)
22. Zhao, H., et al.: Nonlinear adaptive equalizer using a pipelined decision feedback recurrent neural network in communication systems. IEEE Trans. on Commun. **58**(8), 2193–2198 (2010)
23. Yogi, S., et.al.: A PSO based functional link artificial neural network training algorithm for equalization of digital communication channels. In: International Conference on Industrial and Information Systems, pp. 107–112 (2010)
24. Lee, C.-H., et al.: Nonlinear systems design by a novel fuzzy neural system via hybridization of electromagnetism-like mechanism and particle swarm optimization algorithms. Inf. Sci. **186**(1), 59–72 (2012)
25. Lin, C.-J., et al.: Image backlight compensation using neuro-fuzzy networks with immune particle swarm optimization. Expert Syst. Appl. **36**, 3–5220 (2009)
26. Lin, C.-J., et al.: Nonlinear system control using self-evolving neural fuzzy inference networks with reinforcement evolutionary learning. New Appl. Soft. Comput. **11**(8), 5463–5476 (2011)
27. Hong, W.-C.: Rainfall forecasting by technological machine learning models. Appl. Math. Comput. **200**(1), 41–57 (2008)
28. Potter, C., et al.: RNN based MIMO channel prediction. Signal Process. **90**(2), 440–450 (2010)
29. Das, G., et al.: Artificial neural network trained by particle swarm optimization for non-linear channel equalization. Expert Syst. Appl. **41**, 3491–3496 (2014)
30. Panigrahi, S.P., et al.: Hybrid ANN reducing training time requirements and decision delay for equalization in presence of co-channel interference. Appl. Soft Comput. **8**, 1536–1538 (2008)
31. Kennedy, J., et.al.: Particle swarm optimization. In: Proceedings of IEEE International Conference on Neural Networks, vol. 4, pp. 1942–1948. Perth, Australia, November-December 1995
32. Shi, et.al.: Modified particle swarm optimizer. In: Proceedings of the IEEE Conference on Evolutionary Computation (ICEC'98), pp. 69–73. Anchorage, Alaska, USA, May 1998

Complexity Analysis of Multiuser Detection Schemes Based on Sphere Decoder for MIMO Wireless Communication System

Sakar Gupta and Sunita Gupta

Abstract The computational and hardware complexity of sphere decoding algorithm is suboptimal for multiple-input multiple-output (MIMO) systems. In this research paper, we have modified the sphere decoding algorithm and proposed the new sphere decoder that reduces computational complexity and number of flops count with compared to sphere decoding algorithm. The progress is realized using a new characterization for sphere radius by varying the values adaptively. Simulation results show the new algorithm has a low bit error ratio (BER) degradation compared to the generalized sphere decoding algorithm and the total number of nodes visited during the radius search is optimal for the proposed sphere algorithm with compare to sphere decoder and ML decoder.

Keywords MIMO · ZF (zero forcing) · MMSE (minimum mean square error estimation) · ML (maximum likelihood) · BER (bit error rate) · SNR (signal to noise ratio) · SD (sphere decoder)

1 Introduction

The increasing demand for high data rates in wireless communications due to emerging new technologies makes wireless communications an exciting and challenging field. In order to support the required large throughputs, wireless communication systems must employ signaling formats and receiver algorithms that provide a significant increase in spectrum efficiency and capacity over current systems. Recently, a lot of research on multiple-input multiple-output (MIMO)

S. Gupta (✉)
ECE Department, Poornima College of Engineering, Sitapura, Jaipur, India
e-mail: sakargupta@gmail.com

S. Gupta
CSE Department, Swami Keshvanand Institute of Technology Management & Gramothan, Jaipur, India
e-mail: drsunitagupta2016@gmail.com

© Springer Nature Singapore Pte Ltd. 2018   587
M. Pant et al. (eds.), *Soft Computing: Theories and Applications*,
Advances in Intelligent Systems and Computing 583,
https://doi.org/10.1007/978-981-10-5687-1_53

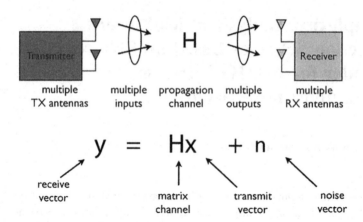

Fig. 1 Multiple-input multiple-output system (MIMO)

systems has been performed. MIMO wireless technology is one possible solution that can meet the demands of future generation wireless communications In contrast to conventional wireless systems, in which there are single transmitting and single receiving antenna, multiple-input multiple-output systems use array of several antennas at both ends of the wireless link, all operating at the same frequency and at the same time. Multiple-input multiple-output schemes can be split into two, depending on the aim to be achieved. The first is spatial diversity, which is aimed at increasing the performance of the wireless communication system by transmitting different representations of the same data stream on different transmitter antennas. The second focus of this research paper is Spatial multiplexing MIMO, which achieves spectral efficiency by transmitting independent data streams on different transmitter branches simultaneously and at the same carrier frequency (Fig. 1).

Decoding techniques for space time codes include zero forcing (ZF), Minimum mean square error (MMSE), and maximum likelihood (ML) decoder. ML detector considered as the best one among these as its performance is higher. The major constraint with ML detector is that it includes higher computational complexity and high order constellation. To lower the complexity, sphere decoding technique is can be easily used which has near ML performance with low complexity.

Multiple-inputs multiple-outputs is planned to be used in mobile radio telephone standards such as 3G and 4G standards. In general, multiple-input multiple-output gives a significant increase in spectral efficiency, improves the reception, and allows for a better reach and rate of transmission. All upcoming 4G systems will also employ multiple-input multiple-output technology.

In this research paper, we will focus on different steps which include the system model in Sect. 2, in Sect. 3 the sphere decoder is explained, and in Sect. 4 the complexity of different decoder is explained and the simulation results in Sect. 5.

The simulation part includes firstly the Signal to Noise Ratio (E_b/N_0) and bit error rate (BER) comparisons for Conventional MIMO Detection Schemes, minimum mean square error estimation (MMSE), ZF, ML detection and sphere decoding techniques for flat Rayleigh fading channel. Secondly, there is an entire number of visiting nodes for the sphere decoder, ML decoder and the proposed sphere decoding techniques with the different antenna configurations. The comparisons are done initially for the 16-QAM modulation technique. Also, the generalized sphere decoder is used which is based on the Schnarr–Euchner (SE) method.

2 System Model

MIMO systems increase system capacity and spectrum efficiency by exploiting the spatial dimension through the use of multiple antennas at both the transmitter and receiver sides of the wireless link. Ideally, the data rates of multiple-input multiple-output systems grow linearly with the number of transmit antennas [1]. However, this increase in system capacity and spectral efficiency come at the cost of increased computational complexity in the receiver. MIMO receivers which are broadly classified into linear and nonlinear detection techniques [2]. There are several Multiuser detection techniques by which we can determine the type of the receiver which can be used appropriately for the detection purpose. The multiuser receiver is basically divided into two categories viz. Optimal MMSE technique and the suboptimal technique. Here our main concern is the suboptimal technique which further categorizes into two parts which are known as Linear and Nonlinear Receivers. Linear receivers consist of the decorrelators and the MMSE (minimum mean square estimation) receivers. Linear signal detection techniques treats all the transmitted signals as interference except for the desired stream from the target transmit antenna. Therefore, interference signals from other transmitting antennas are minimized or nullified in the course of detecting the desired signal from the target transmitting antenna. Accordingly, the received signal vector Y is multiplied with the filter matrix H obtained from channel matrix H and then followed by a parallel decision on all the layers. Mathematically, using the narrowband system Eq. (1), the linear detection process can be expressed as follows

$$X_{\text{est}} = GY \tag{1}$$

ZF Linear Detector: This is the basic algorithm; It works as a standard equalizer where the inverse of the channel frequency response is applied to the received signal. This hypothetically sounds capable but in a realistic situation, it is very susceptible to noise as the inverse of the received noise is also applied to the signal, since the channel response includes noise. Therefore, the ZF algorithm is very superior for noiseless channels as it would effectively eliminate all inter-symbol interference (ISI), but it is unrealistic for a noisy channel as it would amplify the noise experienced at the receiver Obviously, in order to utilize this algorithm, the channel knowledge is required at the receiver which adds to system complexity. Therefore, with respect to MIMO systems, the estimate, of the received signal, $\bar{y}$ can be written as:

$$\overline{y_{ZF}} = (H^H H)^{-1} H^H H = H^+ y \tag{2}$$

Here H^+ refers to the Moore–Penrose inverse which is essentially a pseudo-inverse of the matrix, H. It is evident that the data seen by both receivers is made up of signals from both transmit antennas.

The MMSE detection method estimates the transmitted vector by minimizing the mean square error between the real transmitted symbol vector and the output of the linear detector as shown in Eq. (3)

$$\varepsilon^2 = E\{(x - x_{est})^H (x - x_{est})\} \tag{3}$$

The estimate that minimizes the above expression is obtained from the following equation

$$X_{est} = G_{MMSE} Y \tag{4}$$

where, the MMSE filter matrix G_{MMSE} is given by,

$$G_{MMSE} = (\sigma_n^2 I_{N_t} + H^H H)^{-1} H^H \tag{5}$$

The estimation errors of different layers correspond to the main diagonal elements of the error covariance matrix [3].

$$\emptyset_{MMSE} = E\{(X_{est} - X)(X_{est} - X)^H\} = \sigma_n^2 (\sigma_n^2 I_{N_t} + H^H H)^{-1} \tag{6}$$

3 Sphere Decoder

The conventional MIMO detection techniques under the category of linear and nonlinear detection methods were discussed. Even though it was said that better performances could be obtained by using nonlinear methods, still the performance of these methods is suboptimal when compared with the other detection techniques. Mathematically, the optimal MIMO detection scheme is maximum likelihood detection (MLD) [4]. In order words to decrease the computational complexity of *Maximum Likelihood*, sphere decoding algorithm (SDA) has been proposed. SDA are divided into two types. One is breadth first sphere decoding (BF SD) and another is depth first sphere decoding (DF-SD).

The MLD method estimates the transmitted vector X as to the maximum likelihood principle where the received vector is compared with the entire possible transmitted vector (which is modified by the channel H). Mathematically, the idea is to find a vector X_j for which the conditional probability $P(X_j/Y)$ is maximized (with $1 < j \leq \Omega$), where Ω denotes the ensemble of the possible transmitted vectors.

Considering, a MIMO system with (N_t, N_r) antenna configuration and employing M-point constellation,

$$\Omega = M^{N_t} \tag{7}$$

The maximum likelihood solution is given below:

$$x_{\text{ml}} = \arg \min \|y - Hx_j\|^2 \quad \text{here } x \in \{x_1 \cdots x_\Omega\} \tag{8}$$

The theory of the sphere decoding [5] is to limit the calculation of feasible code words by assuming only those which are within a sphere centered at the received signal vector. In a simple way to find out the nearest lattice point to the expected signal within a sphere radius is sphere decoding in which each codeword is represented by a lattice point in a lattice field.

To consider the SDA, we are using QR decomposition on the channel coefficients. It is expressed as $H = QR$ here Q is a unitary orthogonal matrix and R is refer as an upper triangular matrix whose diagonal possesses real valued positive entries [6].

$$y = Hx + n \tag{9}$$

$$\bar{x}_{\text{ml}} = \arg \min \|y - R\bar{x}\|^2 \tag{10}$$

The manner at which an SDA works is basically solving (10) where it can be clearly seen that in order to achieve correct results, the utilized SDA algorithm needs to completely identify all the constellation points that exist within the hyper-sphere with a radius, d_r centered around a received vector point, x. Using (9), a relationship can be established between d_r and x as seen in (10). Therefore, a lattice point $H\bar{x}$ would exist in a sphere of radius, d_r if and only if the condition of (11) is met [5].

$$d_r^2 \geq \|y - H\bar{x}\|^2 \tag{11}$$

4 Complexity Analysis of MIMO Detection Techniques

MIMO techniques will only become part of future generation wireless systems. A complexity analysis and comparison will be carried out for the most promising MIMO algorithms [7]. This allows to estimate the potential cost of such systems and to identify possible bottle necks for the hardware implementation. Even though there is no real consent in the digital communication system on how accurately to interpret the idea of complexity, it is normally defined as the total number of additions and multiplications [8]. Which compute the estimate of the transmitted

Table 1 Summary of the complexity of preamble processing for MIMO detection techniques

Detection techniques	Complexity of preamble processing		
	Number of R_{ADDS}	Number of M_{MULS}	Number of E_{ADDS}
ZF	$4N_t^3 + N_t^2(8N_r - 2) - 2N_tN_r$	$4N_t^3 + 8N_t^2N_r$	$44N_t^3 + 88N_t^2N_r - 2N_t^2 - 2N_tN_r$
MMSE	$4N_t^3 + N_t^2(8N_r - 2) - 2N_tN_r + N_t$	$4N_t^3 + 8N_t^2N_r$	$44N_t^3 + 88N_t^2N_r - 2N_t^2 - 2N_tN_r + N_t$
ML	$2M^2N_r \frac{M^{N_t}-1}{M-1} + 2N_rN_tM$	$4N_rN_tM$	$2M^2N_r \frac{M^{N_t}-1}{M-1} + 42N_rN_tM$
DF-SD	$4N_t^3 + 12N_t^2N_r - 3N_t^2 - 4N_tN_r$	$4N_t^3 + 12N_t^2N_r + N_t$	$44N_t^3 + 132N_t^2N_r - 3N_t^2 - 4N_tN_r + 10N_t$

Table 2 Summary of the complexity of payload processing for MIMO detection techniques

Detection techniques	Complexity of preamble processing		Number of E_{ADDS}
	Number of R_{ADDS}	Number of M_{MULS}	
ZF	$2N_tN_r + 2N_t(N_r-1)N_t\log_2(M)$	$4N_tN_r$	$44N_tN_r + 2N_t + N_t\log_2(M)$
MMSE	$2N_tN_r + 2N_t(N_r-1)N_t\log_2(M)$	$4N_tN_r$	$44N_tN_r - 2N_t + N_t\log_2(M)$
ML	$4N_rM^{N_t} - 1$	$2N_rM^{N_t}$	$24N_rM^{N_t} - 1$
DF-SD	$4N_tN_r - 2N_t + 6M + \sum_{i=1}^{N_t-1}[4(N_t-i+1)+8M]$ P_{i+1}	$4N_tN_r + 6M + \sum_{i=1}^{N_t-1}[4(N_t-i+1)+6M]$ P_{i+1}	$44N_tN_r - 2N_t + 66M + \sum_{i=1}^{N_t-1}[44(N_t-i+1)+68M]$ P_{i+1}

vector x or the execution time of the algorithm when implemented on some definite platform. The complexity of the detectors will only be computed in terms of additions, multiplications, which is another measure proportional to the running time [9] (Tables 1 and 2).

5 Simulations

Figure 2 shows the bit error rate (BER) characteristics of the conventional MIMO detection techniques in context with SNR for a 4 × 4 model that executes in a flat Rayleigh fading environment. In this QAM, a modulation technique is used and there is absence of channel coding. The BER comparison for Sphere Decoder is shown in the figure with ML, ZF, and MMSE in respect to their Signal to Noise Ratio.

Visited nodes for different antenna configuration: Figure 3 also includes the total number of nodes that have been visited. It is used as a point of reference to examine how much complexity is added to the system by utilizing the modified radius in the proposed SD (PSD). This reference has been applied to Figs. 4 and 5 where different values of m, N_t, N_r have been deployed for different system configurations to determine the robustness of the Proposed SD.

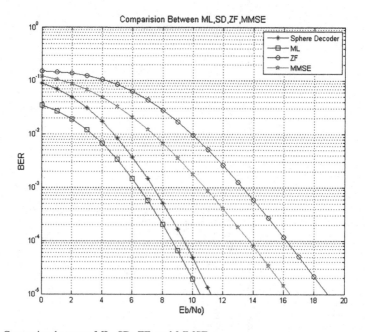

Fig. 2 Comparing between ML, SD, ZF, and MMSE

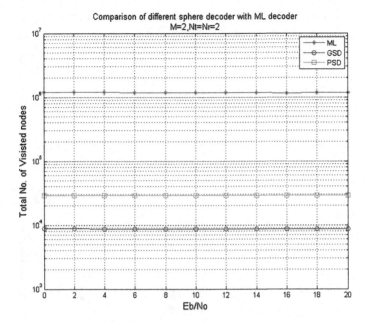

Fig. 3 Comparing the total number of visiting nodes with $m = 2$, $N_t = N_r = 2$

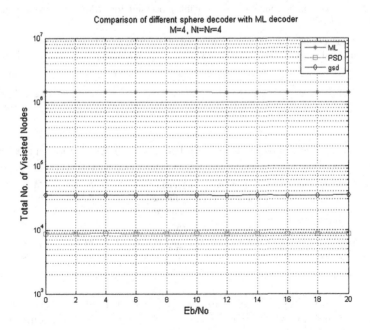

Fig. 4 Comparing the total number of visiting nodes with $m = 4$, $N_t = N_r = 4$

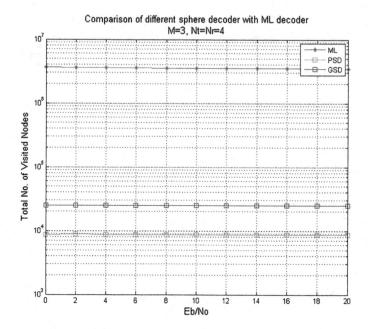

Fig. 5 Comparing the total number of visiting nodes with $m = 3$, $N_t = N_r = 4$

Table 3 Summary of the complexity of preamble processing for MIMO detection techniques for $N_t = N_r = 4$, $M = 2$

	Mean of real multiplication	Mean of real addition	Mean of Estimated addition	No. of flops
ML	320	128	1600	1088
PSD	1040	1028	11,320	4148
SD	1040	1028	11,320	4148

Table 4 Summary of the complexity of payload processing for MIMO detection techniques for $N_t = N_r = 4$, $M = 2$

	Mean of real multiplication	Mean of real addition	Mean of estimated addition	No. of flops	Max. no. of node visited
ML	255	128	1535	893	$10^{6.5}$
SD	196	123	1285	743	$10^{4.2}$
PSD	148	104	1188	548	$10^{3.9}$

It is clear from Figs. 3, 4, and 5 that the PSD adds a slight complexity Decreases in comparison with the SD. This is visualized by the increased amount of nodes visited for the different scenarios evaluated. Tables 3 and 4 have been drafted below to include the total number of visited nodes and the total number of flops counted for different detection schemes.

6 Complexity Analysis Between Different Decoders

PSD achieves a tremendous decrease in the average complexity as it describe in table No. 4 for 4 * 4 antennas, there we consider the mean of real multiplication, there is a significant reduction in proposed sphere decoder, also when the mean of real addition is used into account, the proposed sphere decoder perform better performance. Also, the number of flops counted decreases in the case of PSD as compared to the sphere decoder. In the proposed sphere decoder, the maximum numbers of visited nodes are also reduced in context with the sphere decoder.

7 Conclusion

In this research paper, we explained a novel constant radius modified sphere decoder algorithm for decoding MIMO transmission, based on the mean of real multiplications, counting flops, analyzing a number of visited nodes. Also, the complexity analysis of ML Decoder, Sphere Decoder, and the Proposed Sphere Decoder has been done. It shows a near optimal solution with ML decoder. In future wireless communication technologies, it will be more possible to incorporate a very huge amount of transmitting and receiving antennas. In LTE systems and WIMAX, standards have been previously discussed for future user terminals and base stations with huge number of antenna arrays. This type of system, in general, with high rate QAM modulation scheme will make today's MIMO ML decoding algorithms a very problematic issue for the future hardware implementation. The proposed sphere decoder algorithm allows the employment of the sphere decoder to a small section of the matrices, allowing the accommodation of huge antenna arrays featuring a large amount of spatial streams.

References

1. Wolniansky, P.W., Foschini, G.J., Golden, G.D., Valenzuela, R.A.: V-BLAST: an architecture for realizing very high data rates over the rich-scattering wireless channel. In: Proceedings of International Symposium on Signals, Systems, and Electronics (ISSSE '98), pp. 295–300 (1998)
2. Jang, H., Lee, H., Nooshabadi, S.: Reduced-complexity orthotope sphere decoding for multiple-input multiple-output antenna system. 978-1-4244-7773-9/10 IEEE (2010)

3. Wubben, D., Bohnke, R., Kuhn, V., Kammeyer, K.D.: MMSE extension of VBLAST based on sorted QR decomposition. IEEE 58th Vehicular Technology Conference, vol. 1, pp. 508–512 (2003)
4. Hassibi, B., Vikalo, H.: On the sphere-decoding algorithm I. Expected complexity. IEEE Trans. Signal Process. **53**, 2806–2818 (2005)
5. Hosseini, F.E., Moghaddam, S.S.: Controlling initial and final radii to achieve a low-complexity sphere decoding technique in MIMO channels. Int. J. Antennas Propag. (Hindawi Publishing Corporation), Article ID 192964, 9p. doi:10.1155/2012/192964(2012)
6. Schnarr, C.P., Euchner, M.: Lattice basis reduction: improved practical algorithms and solving subset sum problems. Math. Program. **66**, 181–191 (1994)
7. Shariat-Yazdi, R., Kwasniewski, T.: Low complexity sphere decoding algorithms. IEEE ISWCS (2008)
8. Golub, G.H., Van Loan, C.F.: Matrix computations. Johns Hopkins University Press, Baltimore (1996)
9. Younis, A., Di Renzo, M., Mesleh, R., Haas, H.: Sphere decoding for spatial modulation. IEEE International Conference on Communications (ICC), Japan (2011)

Question Classification Using a Knowledge-Based Semantic Kernel

Mudasir Mohd and Rana Hashmy

Abstract Question classification is a significant component of any question answering system and plays a vital role in the overall accuracy of the QA system, and the key to the accuracy of a question classifier depends on the set of features extracted. Generally, for all text classification problems, the data is represented in a vector space model with bag-of-words (BOW) approach. But despite the simplicity of BOW approach, it suffers some serious drawbacks like it cannot handle synonymy or polysemy and does not take into account the semantic relatedness between the words. In this paper, we propose knowledge-based semantic kernel that uses WordNet as its knowledge-base and a semantic relatedness measure SR. We experimented with five machine learning algorithms viz. Nearest Neighbors (NN), Naive Bayes (NB), Decision Tree (DT), and Support Vector Machines (SVM) to compare the results. For SVM we experimented with linear kernel and the proposed semantic kernel represented by SVMSR.

Keywords Ambiguity problem · Semantic relatedness (SR) · Semantic compactness (SCM) · Semantic path elaboration (SP) · WordNet

1 Introduction

The search engines or the current information retrieval systems available at present return whole document in response to a user query when the user may not be all the time interested in the whole document. This problem got solved with the advances in the research of Question answering technology that identifies and extracts exact answers to the user queries posed in natural languages, from a large collection of text documents. Since the user is interested in exact answers rather than whole document that contains it, thus, question answering systems are more focused for

M. Mohd (✉) · R. Hashmy
Department of Computer Science, University of Kashmir, Hazratbal, Srinagar 190006,
Jammu and Kashmir, India
e-mail: mudie.mohammad@gmail.com

© Springer Nature Singapore Pte Ltd. 2018
M. Pant et al. (eds.), *Soft Computing: Theories and Applications*,
Advances in Intelligent Systems and Computing 583,
https://doi.org/10.1007/978-981-10-5687-1_54

example for a query like "Who was the first women prime minister of India" should return "Indira Gandhi" and not documents containing "Women," "Prime Minister," and "India."

Question classification is the defining phase of the question answering system firstly because the total correctness of the QA system depends heavily on the accuracy of the underlying question classifier. It has been proved by the results obtained from the error analysis of the question answering systems that 36.4% of the errors are due to misclassification of questions. And second, because of its ability to eliminate candidate answers that are irrelevant to the question thus reducing the search space.

There are two common approaches to question classification:

- Surface pattern identification-based question classification.
- Semantic categorization-based question classification.

Surface pattern identification-based approach classifies the questions as the set of word-based patterns and answers are fetched based on these patterns. Such type of question classification strategy suffers from the finite capability to extract answers that belong to irrelevant classes. While as semantic categorization-based question classifiers use external knowledge-base like WordNet to classify the questions taking care of synonymy and polysemy.

While text categorization has been researched better than question classification, which is relatively a new field. With there being very small difference that is "what, when, is, of, the," etc. are neglected in text classification while these words are important in question classification.

2 Related Works

The major conference in the area of question answering is TREC. The body of question answering research associated with our work emphasizes on the automatic question classification. Question classification method proposed by Ravichandran and Hovy [1] does not depend on external knowledge-base but classifies questions on the different sets of surface patterns. Li and Roth [2] used a diverse feature set consisting of both syntactic features like parts of speech tags and semantic features like named entities to achieve a performance of 84.2%. [3] proposed support vector-based question classification with a tree kernel used for finding the syntactic information about the question. [4] contributed the concept of informer a short two to three word phrase present in the question that can be used to accurately classify a particular question. They used a meta learning model to classify informers and then combine the features of the predicted informer with more general features into a single large feature vector and used linear support vector machines to classify. This approach achieved an accuracy of 86.2%. [5] derived features from head words of the principal noun phrases in the question (such as WordNet) hypernyms achieving an accuracy of about 89.2%. The machine learning approach used by R C Balabantaray et al. [6] showed good improvement in the baseline accuracy.

3 Question Classification

This component of the question answering system is concerned with mapping of questions into various semantic categories with the intension that this classification system, along with other constraints on the answer, will be employed by different upcoming processes of the system to choose a correct answer among the various candidates. We have used a two-layered question hierarchy which contains six coarse grain classes (ABBREVIATION, ENTITY, DESCRIPTION, HUMAN, LOCATION, and NUMERIC VALUE) and fifty fine-grained categories, Table 1 shows the details. Each coarse-grained category contains a non-overlapping set of fine-grained categories.

4 The Ambiguity Problem

The absence of a vivid boundary between classes has given rise to ambiguity problem in the question classification process. Due to this problem the classification of a specific question has become ambiguous. Consider a situation.

Who do you buy groceries for?

The answer to such a question can be a pet animal or a human family member, thus it is hard to categorize a question of this type into a particular class and there are chances that there will be misclassification involved in the process if we do so. In order to avoid such problem, assigning multiple class labels to a single question is permissible in our classifiers. The method of assigning multiple class labels is

Table 1 Coarse and fine-grained question categories

Coarse	Fine
ABBR	abbreviation, expansion
DESC	Definition, description, manner, reason
ENTY	Animal, body, color, creation, currency, disease/medical Event, food, instrument, language, letter, other, plant Product, religion, sport, substance, symbol, technique Term, vehicle, word
HUM	Description, group, individual, title
LOC	City, country, mountain, other, state
NUM	Code, count, date, distance, money, order, other

satisfactory then only one label is assigned, because in upcoming processing steps all the classes can be applied thereby preserving any loss. But to simplify the experimentation process, we assume that one question resides in only one category. That is to say, an ambiguous question is labeled with its most probable category.

We use machine learning to methods for question classification because it has got advantages over manual classification. The construction of manual methods is a cumbersome task that requires the analysis of a large number of questions. Moreover, mapping questions into fine grain classes require the use of specific words (syntactic and semantic information) and therefore an explicit representation of the mapping can be very large. On the contrary, in our learning (using machine learning) approach we can define only a small number of "types" of features, which can then be expanded in a data-driven way to a potentially large number of features (Cumby and Roth, 2000), relying on the ability of the learning process to handle it. It is hard to imagine writing explicitly a classifier that depends on thousands or more features. Finally, a learned classifier is more flexible because it can be trained on a new taxonomy in a small amount of time.

5 Dataset and Features

We used the UIUC questions dataset that is publicly available. This dataset contains approximately 5500 training questions from TREC 8, 9 QA track, and 500 testing questions from TREC 10 QA track. The dataset has been manually labeled by UIUC according to the coarse and fine grain categories in Table 1. We take two types of feature simple bag-of-words (BOW) and bag-of-words with semantic relatedness (BOW with SR).

Since there are some major short comings of simple bag-of-words approach. Despite its ease of use, it cannot handle word synonymy and polysemy and thus does not take into account semantic relatedness. In this paper, we overcome these short comings of the bag-of-words by adding a WordNet-based semantic relatedness measure called SR for a pair of words into our semantic kernel. The proposed measure uses the TF-IDF weighing scheme, thus we create a semantic kernel that combines both statistical as well as semantic information from the text. Thus, we present a semantic smoothing matrix and kernel for text classification, based on a semantic relatedness measure that takes into account all of the available semantic relations in WordNet by using the semantic relatedness measure SR. From our experimental results, we prove that by embedding a semantic relatedness measure through a semantic kernel results in increased accuracy in question classification.

6 Semantic Relatedness Measure

Semantic relatedness using WordNet considers two component measures path length and path depth. The two measures are combined to represent the semantic relatedness (SR) between two terms. The path length is represented by compactness (SCM) and the path depth is represented by semantic path elaboration (SPE). The semantic relatedness between any pair of terms in the WordNet is given by the following:

Definition 1 *Given a word thesaurus W, a weighting scheme for edges that assigns a weight $e \in \{0,1\}$ for each edge, a pair of senses $S = (s_1, s_2)$, and a path of length l connecting the two senses, then the semantic compactness of S is given by the following:*

$$\text{SCM}(S, W) = \prod_{i=1}^{l} e^i$$

where $e_1, e_2, ..., e_l$ are the path's edges. If $s_1 = s_2$ SCM $(S, W) = 1$. If there is no path between s_1 and s_2 SCM$(S, W) = 0$.

Semantic compactness considers the path length and has values in the set [0, 1]. Higher compactness between senses declares higher semantic relatedness and larger weight are assigned to stronger edges. The basic idea behind edge weighting is that some edges provide stronger semantic connections than others. All the paths are weighted and the path with maximum weight is selected.

Definition 2 *Given a word thesaurus W, and a pair of senses $S = (s_1, s2)$ where s_1, $s_2 \in W$ and $s_1 \neq s_2$ and a path of length l between the two senses, then the semantic path elaboration is given by the following:*

$$\text{SPE}(S, W) = \prod_{i=1}^{l} \frac{2di.di + 1}{(di + di + 1)d_{\max}}$$

where d_i is the depth of sense s_i in W and d_{max} is the maximum depth of W. If $s_1 = s_2$, and $d_1 = d_2 = d$, $SPE(S, W) = \frac{d}{d\max}$. If there is no path between s_1 and s_2 then $SPE(S, W) = 0$.

Semantic path elaboration is the second parameter that affects the semantic relatedness between the terms. The standard method of measuring the depth of a sense node in WordNet is by checking for hypernym/hyponym hierarchical relation for the noun and adjective parts of speech and hypernym/ troponym for the verb parts of speech.

Definition 3 *Given a word thesaurus W, a pair of terms $T = (t_1, t_2)$, and all pairs of senses $S = (s_1, s_2)$, where s_1, s_2 senses of t_1, t2, respectively. The semantic relatedness of T is defined as follows:*

$$\text{SR } (T, S, W) = \max\{\text{SCM}(S, W) \cdot \text{SPE}/(S, W)\}.$$

SR between two terms t_i, t_j where $t_i \equiv t_j \equiv t$ and $t \notin W$ is defined as 1. If $t_i \in W$ but $t_j \notin W$, or $ti \notin W$ but $t \in W$, SR is defined as 0.

7   Semantic Relatedness-Based Semantic Kernel

Given a question q, we know it is represented in simple bag-of-words representation as follows:

$$\emptyset(q) = [\mathrm{tf} - \mathrm{idf}(t_1, q), \mathrm{tf} - \mathrm{idf}(t_2, q), \ldots, \mathrm{tf} - \mathrm{idf}(t_N, q)]^T \in R^N$$

where tf-idf (ti, q) is the TF-IDF weight of term ti in a particular question, and N is the total number of terms (e.g., words) in the dictionary (the superscript T denotes the transpose operator). In the above expression, the function $\emptyset(q)$ represents the question q as a TF-IDF vector. The tf-idf representation implements a down weighting of irrelevant terms as well as highlighting potentially discriminative ones, but nonetheless is still not capable of recognizing when two terms are semantically related. It is therefore not able to establish a connection between two questions that share no terms, even when they address the same topic through the use of synonyms. The only way that this can be achieved is through the introduction of semantic similarity between terms. So in order to do that we have to enrich the bag-of-words representation with some semantic information or in other words, we have to transform the tf-idf matrix into a semantic relatedness matrix (let us call this matrix as M) using the semantic relatedness measure SR defined above. The ijth element of M is given by SR (T, S, W) which quantifies the semantic relatedness between terms T (ti, tj). Thus M is an NxN matrix with 1's in the principle diagonal. Thus, this smoothing matrix can be used to transform the questions vectors in such a way that semantically related questions are brought closer together in the transformed (feature) space. Mathematical representation of this semantically enriched bag-of-words is as follows:

$$^\wedge(q) = [(q)^{TM}]^T$$

This new feature space can be as such used for many classification purposes but since we want to use this feature space inside a kernel function. The general representation of a kernel function is as follows:

$$K(X, Z) = \emptyset(X)^T(Z)$$

where X and Z are vectors. The kernel function simply computes the inner product of these vectors. In our case, the questions are represented as vectors so our kernel function will be as follows:

$$K\left(q_i, q_j\right) = ^\wedge (q_i)^T(q_i) = (q_i)^{TMMT}\left(q_j\right)$$

Now in order to prove that this is a valid kernel the Gram matrix (G) formed from the above kernel function should hold the Mercer's conditions. Mercer's conditions are satisfied only when the Gram matrix is positive semi-definite. The proof that the matrix G formed by the kernel function is indeed a positive semi-definite with the outer matrix product MMT has been proved already.

8 Experimental Results

We performed our experiments on the UIUC dataset for question classification. The data is preprocessed using tokenization and TF-IDF matrix construction, it is noteworthy to say that we did not use the stop word removal technique as our questions are small documents comprising of only a few terms and we assume all the terms are important for correct classification. We employed four machine learning algorithms namely Nearest Neighbors (NN), Naïve Bayes (NB), Decision Tree (DT), and Support Vector Machines (SVM) with the bag-of-words approach and compared them with support vector machine with semantic smoothing of semantic relatedness approach. We performed our experiments using WEKA (for NN, NB, DT, SVM) and SVM Light package (for SVM with semantic smoothing). The results of our experiments are given in Tables 2 and 3. We used the default parameters for the different machine learning algorithms where the subscripts with SVM algorithm viz. Linear and SR mean linear kernel and semantic relatedness kernel, respectively

The table shows that by using the semantic kernel for the support vector machines the accuracy of the classification gets increased and in fact is better than SVM with a linear kernel. We have tested the SVM on the poly kernel and RBF 1 kernel as well as we found that there is no significant difference from linear kernel so we decided to include the only linear kernel.

In summary, the experimental results showed that by using a semantic kernel in SVM out-performs the other method including SVM with a linear kernel.

Table 2 Question classification accuracy using different machine learning algorithms under the coarse grain category definition	Classifier	Accuracy (%)
	NN	75.6
	NB	77.4
	DT	84.2
	SVM_{linear}	85.8
	SVM_{SR}	91.9

Table 3 Question classification accuracy using different machine learning algorithms under the fine grain category definition	Classifier	Accuracy (%)
	NN	68.4
	NB	58.4
	DT	77.0
	SVM_{linear}	80.2
	SVM_{SR}	86.41

9   Conclusions

In this paper, we presented a semantic kernel approach to question classification by enriching the BOW with semantic information. We evaluated the impact on question classification using SVM with the semantic kernel on the commonly used question classification dataset. We found that by using SVM with semantic kernel there is a significant improvement in the performance

References

1. Hovy, E., Gerber, L., Hermjakob, U., Lin, C., Ravichandran, D.: Towards semantics-based answer pinpointing. In: Proceedings of the DARPA Human language Technology Conference (HLT). San Diego, CA (1999)
2. Li, X., Roth, D.: Learning question classifiers. In: Proceedings of the 19th International Conference Computational Linguistics (COLING'02) (2002)
3. Gabrilovich, E., Markovitch, S.: Computing semantic relatedness using Wikipedia-based explicit semantic analysis. In: Proceedings of the 20th IJCAI, pp. 1606–1611. Hyderabad, India (2007)
4. Vijay, K., Manning, C.D.: An effective two-stage model for exploiting non-local dependencies in named entity recognition. In: Proceedings of the 21st International Conference on Computational Linguistics and the 44th annual meeting of the Association for Computational Linguistics. Association for Computational Linguistics (2006)
5. Huang, Z., Thint, M., Qin, Z.: Question classification using head words and their hypernyms. In: Proceedings of the Conference on Empirical Methods in Natural Language Processing (EMNLP '08). Association for Computational Linguistics, Stroudsburg, PA, USA, pp. 927–936 (2008)
6. Balabantaray, R.C., Mudasir M., Nibha S.: Multi-class emotion classification: A new approach. Int. J. Appl. Inf. Syst. (IJAIS) – ISSN: 2249-0868 Foundation of Computer Science FCS, New York, USA Volume 4–No.1, September 2012

Integrating Six-Sigma and Theory of Constraint for Manufacturing Process: A Case Study

Chiranjib Bhowmik, Sachin Gangwar and Amitava Ray

Abstract The aim of this paper is to combine theory of constraint and six-sigma methodology for evaluation of production system performance. This paper proposes an integrated theory of constraint and six-sigma methodology that increase the product quality of a manufacturing organization. In this work, initially, production constraint is identified, and thereafter, the process capability analysis of production process satisfies the subordination process of an integrated TOC and six-sigma executive philosophy. The re-engineering of the system elevates the constraint of the production unit that enhances the product quality by minimizing the defective production rates. This research is pertinent to any of the production house in which manufactured goods quality diminishes the throughput of the organization.

Keywords Six-sigma · Theory of constraints · Statistical process control
Quality management

1 Introduction

Six-sigma is mostly a commerce process enhancement methodology in which sigma speaks for an arithmetical proportion of inconstancy in the process. Although six-sigma have created a benchmark in prevailing and inherent customers that any product obtain by this methodology will follow the path of good quality. The dominant feature of six-sigma is that it does not allow the best utilization of its

C. Bhowmik (✉)
Department of Mechanical Engineering, NIT Silchar, Silchar, Assam, India
e-mail: chiranjibbhowmik18@gmail.com

S. Gangwar
Department of Mechanical Engineering, JKLU, Jaipur, India
e-mail: sachingangwar@jklu.edu.in

A. Ray
JGEC, Jalpaiguri, West Bengal, India
e-mail: amitavaray.siliguri@gmail.com

© Springer Nature Singapore Pte Ltd. 2018
M. Pant et al. (eds.), *Soft Computing: Theories and Applications*,
Advances in Intelligent Systems and Computing 583,
https://doi.org/10.1007/978-981-10-5687-1_55

607

current resources, which can be a painful issue for a small company. Six-sigma usage an immovable analytical mechanism to fight in opposition to flippancy inside any organization. The leading objective of this measure is to diagnose the censorious consumer demand, legalize the renovation, and promote it. According to Gitlow and Levine [1] the most prevailing scenario pursued by the unit is the DMAIC (Define, Measure, Analyse, Improve, Control), this recipe follows a protocol that, the venture unit knows that, what and when the arithmetical tools should be used, as well as alternative gadget that are worn to perceive the preferences of several scheme confer to the disparity and consequence of the organization. Lean manufacturing and six-sigma strategy also have been used for continuous improvement of any organization [2, 3] and an analytical hierarchy process (AHP) model was developed for optimal industrial result too [4]. For reliable decision-making for sustainable development, six-sigma also plays a vital role [5, 6]. According to Plotkin [7] the strategy hauls the crucial manufacturing, engineering, and transactional processes of integrated structure through the five transformational phases [8].

TOC was developed by Eliyahu M. Goldratt, he explained a system through a chain that, the chain is never strong as its weakest link. The first and foremost objective of theory of constraints is to convert production arrange method to an executive doctrine that highlights on continuous improvement. According to Rahman [9] the theory of constraint (TOC) suggests that, the weakest link gets found by the combination of strategic, tactic and operation level people, respectively, and controlled to a point that everything works around it. TOC evolve a set of policies and methodologies to diagnose and amend such constraints. At the beginning of the modern era TOC process was tested in management appoint stage and later on several areas such as operations, finance, and measures [10]. In order to implement the TOC in any industry a strategy is needed which consists of five focusing steps suggested by Goldratt et al. [11], Ray et al. [12, 13], Wang et al. [14] and Bhowmik [10].

In this paper, an integrated strategy is shown for any organization, that have the urgency to develop not only the products and services, but also their ability and productivity issues in terms of quality as well described by Jin et al. [15]. An integrated model will demonstrate the path, to rendezvous environments where monetary resources are limited in six-sigma and theory of constraint project that will clarify the issues with a spacious scale of any product that returns very little dividends to the business described by Gupta [16, 17]. This paper highlights the potential combination of two distinct strategies, Theory of constraint (TOC) and six-sigma for improving mechanized system performance. Using these integrated strategies any organization can improve their throughput in terms of quality. The efficiency of the integrated strategies is examined and suggestion of applying this advance approach is presented.

2 Materials and Method

The raw material received for the operation is of specification C40/65 sq. and cut weight also maintained as per specification ranging from 2400 to 2500 kg. The raw material was found acceptable and material cutting was performed in shearing machine. In shearing machine the billets of the raw material are made and are taken to next step for treatment. In the next step heat treatment of the billet is performed for further operations. During heat treatment, billet is heated in oil fired batch type furnace (temperature ranging from 1150 to 1200 °C) and monitoring the temperature by using digital optical pyrometer. The furnace operates maintaining the temp within the specification by adjusting oil and air valve. No discrepancy was observed in billet heating process. After heating is done, processed billet is taken for the next operation—Hot Forging. In this operation upsetting of heated billets reduces the billet thickness and stretches the length to suit the die mold. After the forging when the billet is approximately fit to be put in mold the next operation of block formation is performed so that while placing in the blocker impression the billet is correctly filled up for finisher impression. After placing the blocker impression, finisher impression is then given the stroke to form the impression. Once the impressions are made on the billet, raw product is ready which is then taken to the last finishing operation of trimming, piercing and coining. During trimming excess flash is removed. One end flashes shearing other end fine. Then, unfinished product is taken for piercing. During this operation an eccentric bore was formed and here no defect was observed. Piercing dimensions were checked and found fine. After this final coining operation was performed and in this operation, bend was removed but some flatness was noticed on I-section and simultaneously thickness undersize was observed up to 0.5 mm. The various stages of process implementation discussed above and depicted in Fig. 1.

Although six-sigma and theory of constraint have distinct executive ideology, various industries use one methodology in their business or use both methodologies to figure out their demands. On the one hand, we have six-sigma that can solve complex problems that require deep solutions, and on the other hand we have, TOC management philosophy that can make public bottlenecks in the system and expand them. Nave [18] suggested that, most prevailing assimilation recycled by these two

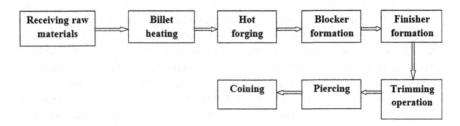

Fig. 1 Simplified process chart for connecting rod manufacturing

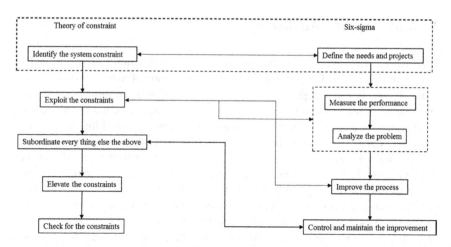

Fig. 2 TOC & DMAIC an integrated improvement methodology

regulations, which consist in identifying the constraint that block the improvement for any organization of global performance. Phase 1 of the integrated framework is equal, such as meeting customer needs or increasing the throughput depicted in Fig. 2. Phase 2 measures the current performance and analyze the root cause needed for improvement. These two phases of six-sigma are involved in the character of TOC by exploiting the selected process. Once the root causes are identified in phase 3 of the integrated approach technique for the process improvement. The purpose is to consume the current ability of the process without incurring supplementary resources expenditures. Phase 4 acts as the safeguard for preceding steps are accurately followed by the rest of the system. Phase 5 has been taken from TOC for the purpose to enhance the quality and fulfill the consumer needs of the organization. Precondition for sixth phase ensures the observant of dynamic humor of the production arrangement and steadily invigilate the occurrence of recent constraints [8, 19, 20].

3 A Case Study

An Asian-based company produces connecting rod SQL for other company. Cut weight to be maintained as per engineering specification given as 2.400–2.500 kg and cut weight maintained and found acceptable. The company had analyzed external failure cost for factory and faced most of customer complaint and rejection in SQL connecting rod component so in this paper, taken the component for complete process study and analyze the problems by using an integrated TOC and DMAIC methodology. The business case for this study has been proposed as a result of the enhancement of quality and the management people was deal with either shutting down the plant or eradicate the non-value added process to raise the capacity without incurring new principal investment. After joining various TOC seminars the managers decide to integrate the TOC and six-sigma to boost their upgrading stab. The project team felt that TOC philosophy could help them with a focus on global system improvement. The unit evoked by reviewing the system map to resolve the potential hindrance in the progress, an in-depth observation of the process SQL connecting rod manufacturing system has been considered to identify the causes lie in the system. After the broad study, the project team identified that the connecting rod has I-section flat instead of radius, flash shearing on I-section and thickness undersized on I-section up to 0.5 mm. The highlighted integrated methodology was adopted to make the improvement.

3.1 Phase 1: Identify the System Constraint and Determine the Process to Be Improved

According to TOC and six-sigma executive philosophy, the first stage of ongoing enhancement associates single out the constraint(s) that block the desired ambition of generating capital and pleasurable patron commitment. Historically, the executive team enforced distinct renovation strategy based on opportunities described by the unit. The unit not at all criticized an enhancement opportunity in terms of quality from arrangement facet which focuses on overall advancement. Subsequently, the project unit determined to single out the adjacent enhancement that would margin regional recovery to comprehensive prosperity.

The unit contrived a thorough interpretation of the plant development from the point of throughput. Brainstorming sessions has been organized between experts of strategic level people, tactical level people, and operational level people. These three level people decided that in man, material, machine, policy, etc., the system constraint(s) lies in between them and give an implication that the connecting rod I-section flat instead of radius according to cause and effect analysis shown in Fig. 3; flash shearing on I-section and thickness undersized on I-section up to 0.5 mm. According to the executive people 1500 nos. are rejected due to I-section

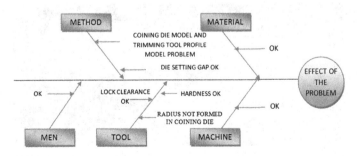

Fig. 3 Identification of root cause

flat therefore, improving trimming quality and coining process are the hindrance and conscript as the leading objective for improvement in this paper. In a typical cause and effect diagram show that, the main effect is most of the products are getting rejected because of flatted I-section and is situated at the "fish head". The elements of the chattels are then stick out onward the "cartilage" such as material, method, machine, tool, and men, respectively and the sub causes are coining die model and trimming tool profile model problem, die setting gap, hardness, radius not formed in coining die, lock clearance, etc. It is a team brainstorming tool used to diagnose probable root causes of the problem. After confirming trimming quality and coining operations as the bottleneck, the team initially decided to change the process setup, however, cardinal compulsion revetment the plant, it is not viable to enhance. The executive people pronounced to feat the constraint or escalate the usage of prevailing technique comparatively than to generate new cardinal venture in advanced development as suggested by TOC perception.

3.2 Phase 2: Measure Current Performance and Identify the Root Cause

This stage measures the performance of elected process, to analyze and single out the essence causes of that particular problem to reduce the rejection rate by using process capability analysis. In this research, the reaction magnitude access, which scrutinize the productivity of the mechanism estimating the defect rate in terms of defect per million, the management team convert the defect rates, to distinguish the method performance using integrated methodology. In order to meet the customer need the management team agreed to disciple the flawed standard into "machine capacity index (MCI)." MCI was figure out based on the ratio of the number of merchandise formed, that meet the engineering specification and crest capacity [8]. The company believes that MCI will communicate well to meet the throughput in terms of customer requirement shown in Fig. 4; 75% of MCI value was measure over the period of sixth month period this indicates that 75% of crest capacity was

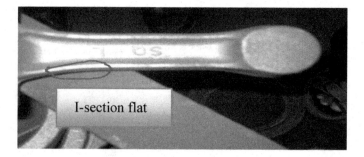

Fig. 4 Rejection due to I-section flat

utilized to meet the quality. On the other words 25% of machining process is either wasted due to breakdown, wait for material and setup. The management team reviewed the process and found that, due to improper trimming and coining process the connecting rods did not attain the specification.

3.3 Phase 3: Exploit the Constraint by Improving the Process

This stage counsels the management team to take specific actions against, formerly diagnosed essence sources in order to perk up the performance of the tabbed process operations. The management squad sure that, all small enhancements made must increase the process capability. As a measure of the wherewithal or capability analysis, the executive team wind up that, they remodeled the die, loaded in 3 Ton hammer press and forged 2000 nos. components has been produced after rectification of coining die and trimming tool. Radius checked on I-section and found that the entire 2000 nos. are fine and no flat thickness undersize on I-section shown in Fig. 5.

3.3.1 Calculations

Specification limit [21];
Upper specification limit (USL) = 15
Lower specification limit (LSL) = 14
Sub-sample size [22], $n = 2$
Number of sub-samples, $N = 25$
Averages $R^- = \sum R/N = 0.14$
Mean of averages $R^= = \sum X/N = 14.48$

Fig. 5 Rectification of coining die and connecting rod

R-Chart calculation;
Upper control limit (UCL) = $D4 * R^- = 0.457$
Lower control limit (LCL) = $D3 * R^- = 0.000$
X-Bar Chart calculation;
Upper control limit (UCL) = $X^= + A2 * R = 14.75$
Lower control limit (LCL) = $X^= - A2 * R = 14.42$
Standard deviation, $\delta = R^-/d2 = 0.124$
Process capability, $Cp = (USL - LSL)/(6*\delta) = 1.34$
Process capability index, $Cpi = (X^= - LSL)/(3*\delta) = 1.29$
Process capability index, $Cpu = (USL - X^=)/(3 * \delta) = 1.39$; $Cpk = MIN (Cpi, Cpu) = 1.29$
Result: $Cp = 1.34$ and $Cpk = 1.29$ as we know that $Cp > 1$, the process is capable of meeting the specification limit found from Bhowmik and Ray [23].

3.4 Phase 4: Subordinate Everything Rather Than the Above 3 Phases to Sustain the Improvement

Already the recent development is in position the target shifted to monitoring the practice, assuring that, the renovation are uninterrupted discussed by Mackey et al. [24], Ftonen and Pass [25], Plotkin [7]. This phase is so designed to visualize and examine the advanced operation circumstances via statistical process control (SPC) tool. In this study, SPC tool helps to monitor the process and I-section flat has been eliminated as shown in Fig. 6, to standardize the trimming and coining process and no customer rejection after implementation of an integrated improvement methodology.

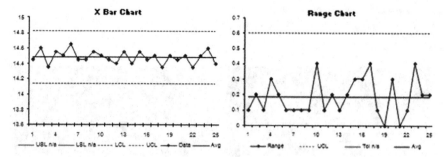

Fig. 6 X-bar and R-chart of the process

3.5   Phase 5: Elevate the Constraints

Since the first four aspects were capable to enhance the development momentously so there is no need to elevate the constraints. If remodeling of the coining die and trimming tool increases the throughput of the system, the decision would be acceptable and satisfactory.

3.6   Phase 6: Check for Next Constraints

The ongoing renovation venture was originated and completed on March 18, 2009. For this purpose, last stride clinch that the enhancements and alternation retrain engaged on future to persuade that it will perfectly bolster [10] in any system or organization. Once the trimming and coining processes were enhanced and turn into tolerable, it would not be the hindrance of the manufacturing unit. The most crucial mechanism is the adoption of ISO standards, EURO standards to an identical scheme for a smaller corporation that can achieve the enhancements by this integrated methodology. The management team continued to analyze the production performance and customer needs to revel probable contemporary constraints that would be a purpose of advanced improvement strategy in terms of quality [10, 20].

4   Conclusion and Future Scope

Various methods have been proposed to deal with the manufacturing problems. An efficient assessment philosophy is essential for the desired model. This paper put forward by describing the fundamentals of six-sigma methodology and the theory of constraint(s) philosophy [10] is worn for retrieve the maximum throughput. Growth of an industry depends on positive and proficient decision-making process. Momentous renovation is attained with the integrated approach when correlated with the results by charts. Statistical process control can apply to each phase of manufacturing and business unit. Once the SPC technique is applied during production process to overcome the defect rate then capability analysis advice to regulate the tolerance restraints and engineering specifications for any production process and it can be adopted not only in production period but also to the machine and machine tool [26]. In this study, X-chart and R-bar chart has been constructed for monitoring the production process. A real-life production problem has been selected and the results of the problem indicates that, the production unit ameliorated staggeringly from its insistence on comprehensive renovation shepherd by an integrated TOC and six-sigma philosophy, project team officials were competent to privileged continuous enhancement [8] had a superior impact on the operation level performance, i.e., to enhance the quality. This work can be further investigated by extending the domain of risk for sustainable environment. The reliability estimation of short-term and long-term cash flow and investment calculation is yet another area of research.

References

1. Gitlow, H.S., Levine, D.M.: Six Sigma for green belts and champions, ISBN-10 X, 13117262. Editorial Pearson Education (2005)
2. Indrawati, S., Ridwansyah, M.: Manufacturing continuous improvement using lean Six Sigma: an iron ores industry case application. Procedia Manuf. **4**, 528–534 (2015)
3. Marzagão, D.S.L., Carvalho, M.M.: Critical success factors for Six Sigma projects. Int. J. Project Manage. **34**(8), 1505–1518 (2016)
4. Alhuraish, I., Robledo, C., Kobi, A.: Assessment of lean manufacturing and Six Sigma operation with decision making based on the analytic hierarchy process. IFAC-PapersOnLine **49**(12), 59–64 (2016)
5. Cherrafi, A., Elfezazi, S., Chiarini, A., Mokhlis, A., Benhida, K.: The integration of lean manufacturing, Six Sigma and sustainability: a literature review and future research directions for developing a specific model. J. Clean. Prod. **139**, 828–846 (2016)
6. Mitchell, E.M., Kovach, J.V.: Improving supply chain information sharing using design for Six Sigma. Investigaciones. Eur. Res. Manag. Bus. Econ. **22**, 147–154 (2016)
7. Plotkin, H.: Six Sigma: what it is and how to use it. Harv. Manag. Update **4**(8), 6–7 (1999)
8. Badiru, A.B., Ehie, I.C., Sawhney, R.: Integrating Six Sigma and lean manufacturing for process improvement. In: Handbook of Industrial and Systems Engineering, 2nd edn, pp. 323–336. CRC Press (2013)
9. Rahman, S.U.: Theory of constraints: a review of the philosophy and its applications. Int. J. Oper. Prod. Manag. **18**(4), 336–355 (1998)
10. Bhowmik, C.: Optimization of process parameter using theory of constraints. J. Mat. Sci. Mech. Eng. **1**(2), 41–44 (2014)
11. Goldratt, E.M., Cox, J., Whitford, D.: The Goal: A Process of Ongoing Improvement, 3rd edn. North River Press, Great Barrington (2004)
12. Ray, A., Sarkar, B., Sanyal, S.K.: An improved theory of constraints. Int. J. Accou. Info. Manag. **16**(2), 155–165 (2008)
13. Ray, A., Sarkar, B., Sanyal, S.: The TOC-based algorithm for solving multiple constraint resources. IEEE Trans. Eng. Manag. **57**(2), 301–309 (2010)
14. Wang, J.Q., Zhang, Z.T., Chen, J., Guo, Y.Z., Wang, S., Sun, S.D., Qu, T., Huang, G.Q.: The TOC-based algorithm for solving multiple constraint resources: a re-examination. IEEE Trans. Eng. Manag. **61**(1), 138–146 (2014)
15. Jin, K., Hyder, A.R., Elkassabgi, Y., Zhou, H. Herrera, A.: Integrating the theory of constraints and Six Sigma in manufacturing process improvement. In: Proceedings of World Academy of Science, Engineering and Technology 37, (2009)
16. Gupta, M.: Activity-based throughput management in a manufacturing company. Int. J. Prod. Res. **39**(6), 1163–1182 (2001)
17. Gupta, M.: Constraints management–recent advances and practices. Int. J. Prod. Res. **41**(4), 647–659 (2003)
18. Nave, D.: How to compare Six Sigma, lean and the theory of constraints. Qual. Prog. **35**(3), 73 (2002)
19. Ehie, I., Sheu, C.: Integrating Six Sigma and theory of constraints for continuous improvement: a case study. J. Manuf. Tech. Manag. **16**(5), 542–553 (2005)
20. Nahavandi, N., Parsaei, Z., Montazeri, M.: Integrated framework for using TRIZ and TOC together: a case study. Int. J. Bus. Innov. Res. **5**(4), 309–324 (2011)

21. Khlebnikova, E.: Statistical tools for process qualification. J. Valid. Tech. **18**(2), 60 (2012)
22. Morimune, K., Hoshino, Y.: Testing homogeneity of a large data set by bootstrapping. Math. Comput. Simul. **78**(2), 292–302 (2008)
23. Bhowmik, C., Ray, A.: The application of theory of constraints in industry: a case study. Int. J. Ext. Res. **2**, 21–29 (2015)
24. Mackey, J., Noreen, E., Smith, D.: The theory of constraints and its implications for management accounting (1955)
25. Ftonen, B., Pass, S.: Focused management: a business-oriented approach to total quality management (1994)
26. Motorcu, A.R., Güllü, A.: Statistical process control in machining, a case study for machine tool capability and process capability. Mat. Design. **27**(5), 364–372 (2006)

SumItUp: A Hybrid Single-Document Text Summarizer

Iram Khurshid Bhat, Mudasir Mohd and Rana Hashmy

Abstract Summarization task helps us to represent significant portion of the original text in concise manner, while preserving its information content and overall meaning. Summarization approach can either be abstractive or be extractive. Our system is concerned with the hybrid of both the approaches. Our approach uses semantic and statistical features to generate the extractive summary. We have used emotion described by text as semantic feature. Emotions play an important part in describing the emotional affinity of the user and sentences that have implicit emotional content in them are thus important to the writer and thus should be part of the summary. The generated extractive summary is then fed to the Novel language generator which is a combination of WordNet, Lesk algorithm and part-of-speech tagger to transform extractive summary into abstractive summary, resulting in a hybrid summarizer. We evaluated our summarizer using DUC 2007 data set and achieved significant results compared to the MS Word.

Keywords Abstract summary · Extract summary · Hybrid summarization
Machine learning techniques · Emotion analysis · WordNet

1 Introduction

Due to the information overload, gigantic amount of available information is stored in textual databases and this is growing at an exponential pace. Thus the information overload huge amount of textual information is generated leading to usage of information retrieval processes and thus information retrieval has found many

I.K. Bhat (✉)
Department of Information Technology, Central University of Kashmir,
Ganderbal, India
e-mail: bhatiramkhurshid@gmail.com

M. Mohd · R. Hashmy
Department of Computer Science, University of Kashmir, Srinagar 190006,
Jammu and Kashmir, India

applications in the text mining approach. Text mining is used to derive various analogies from the unstructured text. Among the various techniques of information retrieval we have text summarization, which is used to process unstructured text into meaningful information. It condenses formless input text into shorter version while preserving its information content and overall meaning.

Summary is the shorter version of text(s) that is generated from one or more texts. Summary contains significant portions of text that contain overall information and meaning of the original text, thus summary is the shorter and concise version of the original text presented in the meaningful and readable manner. Text summarization can be classified into three broad categories: extractive summarization, abstractive summarization; and hybrid summarization. An extractive summarization [1] method consists of selecting salient sentences, paragraphs, etc. from the original document and concatenating them into shorter form. How important a sentence is can be described by the presence of statistical and linguistic features present in it. An abstractive summarization [1] attempts to develop an understanding of the main concepts in a document and then express those concepts in clear natural language. Abstractive summarization makes use of linguistic features to find and examine text and then use these methods find the new concepts and expressions to best describe it by generating a new shorter text that conveys the most important information from the original text document.

This paper is concerned with hybrid of both approaches (Extract as well as Abstract). In our hybrid approach initially an extractive summary is generated using a generic method of sentence scoring. Sentences are ranked on the basis of statistical features such as sentence length, sentence position, frequency (TF-IDF), noun phrase and verb phrase, proper noun, aggregate cosine similarity, and cue-phrases. We have also included a semantic feature which is emotion. Emotions play an important role in describing the emotional affinity of writer and thus sentences which have implicit emotional words in them are important to a writer and thus are important to be included in summary. Sentences with highest ranks and least similarity are extracted. This extractive summary is then fed to our novel language generator, which uses WordNet, Lesk Algorithm and P.O.S tagger, to transform extractive summary into the abstractive one.

This paper is structured as follows: Sect. 2 presents the relevant literature work. Section 3 explains the features used for extraction of salient sentences and methodology. Section 4 presents the summarization algorithm. Section 5 describes analysis and results. In the conclusions, an account of the contribution made is presented together with lines for further work.

2 Related Work

Sentence extraction came into existence in 1950s, when the initial work on automatic text summarization was started by *Luhn*. For identification of salient sentences he alluded the use of frequency of occurrence of the words; words that occur frequently in the text are descriptive or topic words and the sentences that contain theses words are the salient sentences, therefore should be enclosed in the summary [2].

Luhn's work was elaborated by *Edmundson* by bringing forth that several features may indicate salient sentences. Features used by him to assign ranks to the sentences in a scientific article were: (1) word frequency {count of the word's occurrence in the text}, (2) count of the title words or heading section words in the sentence, (3) sentence position with respect to the entire text and the section, (4) the number of cue-phrases such as "in conclusion, in summary" [2].

Paice's work became foundation for the language generation techniques in summarization. He emphasized on the problem in sentence extraction of unintentionally choosing those sentences for the summary that refer to the sentences not present in the summary [2].

Ample amount of the work done in text summarization area has concentrated on extractive summarization approach. According to this approach salient sentences are selected from the original text to form summary [3]. Linear combination of statistical features is used to find descriptive words and key phrases or sentences [4]. Multiple machine learning techniques, which made use of training corpus, are used for various features to get implemented in order to rank sentences [5].

The abstractive summarization approach does not deal with the method of salient sentence extraction for summary generation, instead authors aims to generate summary by writing significant information in the text in their own words. Usually Language generation and compression techniques are required [6].

Very first summarization work began with the summarization of single documents. Single-document summarization approach generates summaries of the input text pertaining to only one single document. As time elapsed, the information content on the web started growing at exponential pace, and the need for multi-document summarization was experienced. Multi-document summarization approach tries to generate summaries from more than one input documents on the same event or topic [3].

On the basis of the content, summaries are classified as indicative and informative summaries. Summaries that give general idea or information about original text are called Indicative Summaries. These summaries are used to determine the aboutness of the input text. Summaries that give information about the topics discussed in the original text instead of the original text itself are called informative summaries [7]. Mohd et al, used several statistical features to summarize the text and produced effiecent summaries of the text [8].

In the domain of natural language processing (NLP) automatic text summarization is a familiar task. Researchers achieved successful results in sentence extraction based text summarization. But researchers still dream of accurate abstractive summarizers [9]. As the abstract summaries do not copy the sentences from input text, abstract approach needs almost full interpretation of the input text in order to generate new sentences. These new sentences can be less redundant [9].

To find a good summary lot of work done, but to decide the quality of the summary still a challenging task. Traditionally evaluation of summarization involves human judgments of different quality metrics, for example, coherence,

conciseness, grammaticality, readability, and content. Later ROUGE was introduced for automatic evaluation of summaries and its evaluations. ROUGE stands
for Recall-Oriented Understudy for Gisting Evaluation. It includes several automatic evaluation methods that measure the similarity between summaries [10].

3   Methodology

Combination of the best features of both the approaches will lead us to the scalable,
reliable and more efficient system. Our summarization procedure can be divided
into five STAGES:

- Preprocessing.
- Generating Ranked Sentences.
- Normalizing Values and Finding Total Score.
- Using Cosine Similarity to Remove Redundancy.
- Making Abstract Summary (Fig. 1).

3.1   Preprocessing

The real-world data needs to be preprocessed into the appropriate form before the
sentence scoring is performed. Data preprocessing improves the quality of the data,
thereby helping to improve the quality of the summary. Therefore, for the quality

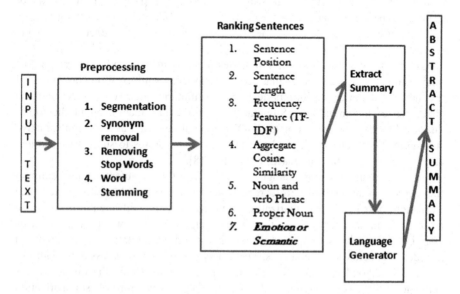

Fig. 1 Diagrammatic representation our system

summary, it is essential to preprocess the input text. In our approach preprocessing is taken into account in following four simple steps.

1. *Segmentation*: The first step of preprocessing is the segmentation, in which the text to be summarized is fragmentized into sentences and then into words in order to rank them.

2. *Synonym Removal*: For sentence ranking the term frequency factor has proven to be critical. Term frequency is the count of occurrence of the given term in the text. Authors usually use synonyms (different lexemes with the same meaning) in their texts to mean the same thing example *big, large, ample, huge, and jumbo*. So it becomes necessary that all the synonyms appearing in the text are replaced with the same word, in order to achieve correct term frequencies for every word.

3. *Removing Stop Words*: Highly frequent words which do not play any role in identifying the important sentences of the text are called stop words. Stop words include *determiners, prepositions and auxiliary verbs, or common domain words*. Such words are not indicative of topicality [2], therefore these words need to be removed. Our implementation makes use of stop list (list of highly frequent words) to eliminate stop words.

4. *Word Stemming*: The process of removing the affixes from words like plural "*s*" from nouns, the "*ing*" from verbs, etc. and collapsing it to a single root form is called stemming. The outcome of the stemming process is a word called the *stem* of the word on which stemming was applied. For example, without stemming, the words like *inflect, inflectable, inflected, inflection and inflective* will be considered as distinct terms having separate term frequencies. The proposed system employs the *Porter stemming algorithm* for stemming purpose.

3.2 Generating Ranked Sentences

Sentence scoring is performed on the preprocessed text. In our approach generation of ranked sentences is based on the combination of statistical and semantic features.

- **Statistical Features**.

1. *Sentence Length*: *Edumpson* suggested that importance of the sentence is directly proportional to its length [11]. Therefore in our implementation number of words in every sentence is considered to be its length.

2. *Sentence Position*: *Edumpson* proposed multiple features that indicated sentence importance and one of them was position of the sentence in the document [2]. Later it was also proposed that leading sentences of an article are important [12]. Our implementation uses Eq. (1) to determine the position of the sentence.

$$\text{Sentence Position} = 1 - \frac{S_i - 1}{N} \quad (1 < S_i < N) \tag{1}$$

where N is total number of sentences in the document and S_i is the sentence number of the ith sentence. According to this rule the leading sentence gets highest rank.

3. *Frequency (TF-IDF)*: The earliest work done by *Luhn* alluded the use of frequency of occurrence of the words to identify the salient sentences in a given document. As word frequency plays a crucial role in determining the importance of the sentences, our model uses *TF-IDF* (TF stands for term frequency, IDF for inverse document frequency) feature. Defined below in Eq. (2):

$$\text{TF}_i * \text{IDF}_i = f(w) * \log\left(\frac{bg}{bg(w)}\right) \tag{2}$$

Here,

TF_i = term frequency of the ith word in the document.
IDF_i = inverse document frequency of ith word in the document.
$f(w)$ is the frequency count of the ith word in given text
bg = total number of background documents taken.
$bg(w)$ = is number of background documents that contain the ith word.

4. *Noun Phrase and Verb Phrase*: The sentence containing a noun or a verb phrase is taken into account as an important sentence and therefore should be enclosed in the summary [13]. Our implementation used *Stanford-POS tagger* to tag the input text. Tagging process is used assign the various parts-of-speech like *determiners (DT)*, *verbs (VBZ)*, *adverbs (ADVB)*, *nouns (NN)*, *adjectives (JJ)*, *coordinating conjunction (CC)*, etc. to each word in the input text. Tagger also assigns the respective labels to the syntactically correlated expressions as *adjective phrase (AP)*, *verb phrase (VB)*, *Noun phrase (NP)*, etc. After tagging each sentence is allotted a numeric variable "*nvp*" whose value is the count of noun and verb phrases contained by it.

5. *Proper Noun*: Proper noun is another important feature, sentences containing proper nouns are observed to be more important and therefore have greater chances of being included in the summary [14]. Therefore to achieve this feature the above mentioned POS tagger is used to determine the number of proper nouns in every sentence.

6. *Aggregate Cosine Similarity*: Aggregate cosine similarity for a given sentence is the addition of its cosine similarities with other all sentences in input text. It is given by Eq. (3).

$$\text{sim}(S_i S_j) = \sum_{k=1}^{n} W_{ik} * W_{jk} \tag{3}$$

$$\text{Aggregate Cosine Similarity}(s_i) = \sum_{j=1, j \neq i}^{n} \text{sim}(S_i S_j) \qquad (4)$$

where

W_{ik} = TF-IDF of the kth word in the ith sentence.

The cosine similarity between two sentences can be represented as $S_i = [W_{i1}, W_{i2}, \ldots, W_{im}]$ and $S_j = [W_{j1}, W_{j2}, \ldots, W_{jm}]$; according to this rule cosine similarity between W_{i1} (first word of sentence "i") and W_{j1} (first word of sentence "j") is calculated and similarly for the rest words of the sentence. The sum of cosine similarity measure for words present in the sentence represents the cosine similarity of that particular sentence. Our implementation uses the below mentioned rule which represents the Standard Cosine Similarity.

$$\text{sim}(S_i S_j) = \frac{\left(\sum_{k=1}^{n} W_{ik} * W_{jk} \right)}{\sum_{k=1}^{n} W_{ik}^2 \sum_{k=1}^{n} W_{jk}^2} \quad i, j = 1 \text{ to } n \qquad (5)$$

7. *Cue-phrases*: This feature was used for the first time in 1968 (*Edmundson*). The sentences that begin with phrases like "the paper describes", "in conclusion" "in summary", "our investigation", and emphasizes such as "the best", "the most important", "in" "particular", "according to the study", "significantly", "important", "hardly", "impossible" as well as domain-specific bonus phrases terms can help in identifying the important contents of an input text [15–17]. In our implementation a cue-list is used to identify the cue-phrases in a given sentence. Therefore, each sentence is allotted a numeric variable "CP" whose value is the total count of cue-phrases contained by it.

- **Semantic Feature or Emotion Feature**.

Emotions play important role in human intelligence, rational decision making, social interaction, perception, memory, learning, creativity, and thus helps in identifying the salient sentences. Our implementation used seven emotion class labels (*positive, negative, fear, joy, surprise, hate and disgust*) as seed words. Then we retrieved their synsets from WordNet which resulted in a lexical database of emotion words, which was then used to identify the emotions in the text. After identifying the emotions, each sentence is assigned a value '*emo*', where '*emo*' is equal to the number of emotion words in a given sentence.

3.3 Normalizing Values and Finding Total Score

Normalization of values means scaling values so as to fall within a small specified range such as "−1.0 to 1.0" or "0.0 to 1.0" or to standardize values measured on different scales to one notionally common scale. In our implementation, for every feature to make its value normalized following methods are used (*sentence position feature does not need to be normalized*):

- *Normalizing Sentence Length Values*: Value for the sentence length feature of each sentence is normalized by computing.

$$\text{sLen}'_i = \frac{\text{sLen}_i}{\text{sLen}_{\max}} \tag{6}$$

where

sLen_i = sentence length of the *i*th sentence.
$\text{sLen}_{\max}$ = sentence length value of the sentence having maximum sentence length value.
And sLen_i' = is the normalized sentence length value of the *i*th sentence.

- *Normalizing Frequency (TF-IDF) Values*: Value for the frequency feature of each sentence is normalized by computing

$$(\text{tf} * \text{idf})'_i = \frac{(\text{tf} * \text{idf})_i}{(\text{tf} * \text{idf})_{\max}} \tag{7}$$

where

$(\text{tf} * \text{idf})_i$ = term frequency-inverse document frequency value of the *i*th sentence.
$(\text{tf} * \text{idf})_{\max}$ = term frequency-inverse document frequency value of the sentence having maximum term frequency-inverse document frequency value.
And $(\text{tf} * \text{idf})_i'$ = normalized term frequency-inverse document frequency value of the *i*th sentence.

- *Normalizing Noun phrase and verb Phrase Values*: Normalized values for the noun and verb phrase feature of each sentence is obtained by using the below mentioned formula,

$$\text{nvp}'_i = \frac{\text{nvp}_i}{\text{nvp}_{\max}} \tag{8}$$

where

nvp_i = noun-verb phrase value of the *i*th sentence.
$\text{nvp}_{\max}$ = noun-verb phrase value of the sentence having maximum noun and verb phrase value.
And nvp_i' = normalized noun-verb phrase value of the *i*th sentence.

- *Normalizing Proper noun Values*: Value for the proper noun feature of each sentence is normalized by computing

$$PN'_i = \frac{PN_i}{PN_{max}} \tag{9}$$

where

PN_i = proper noun value of the ith sentence.
PN_{max}=proper noun value of the sentence having maximum proper noun value.
And PN_i' = normalized proper noun value of the ith sentence.

- *Normalizing Aggregate Cosine Similarity Values*: Value for the aggregate cosine similarity feature of each sentence is normalized by computing

$$ACS(S_i)' = \frac{ACS(S_i)}{ACS(S_{max})} \tag{10}$$

where

$ACS(S_i)$ = aggregate cosine similarity of the ith sentence.
$ACS(S_{max})$ = aggregate cosine similarity value of the sentence having maximum aggregate cosine similarity value.
And $ACS(S_i)'$ = is the normalized aggregate cosine similarity value of the ith sentence.

- *Normalizing Cue-phrases Values*: Normalized values for the cue-phrase feature of each sentence is obtained by using the below mentioned formula,

$$CP'_i = \frac{CP_i}{tCP} \tag{11}$$

where

CP_i = Count of the cue-phrases in the ith sentence.
tCP = Total number of cue-phrases in the text document.
And CP_i' = Normalized cue-phrase score of the ith sentence.

- *Normalizing Emotion Values*: Value for the emotion feature of each sentence is normalized by computing

$$emo'_i = \frac{emo_i}{emoT} \tag{12}$$

where,

emo$_i$ = Number of emotion words in the ith sentence.
emoT = Total number of emotion words in the text document.
And emo$_i'$ = Normalized emotion score of the ith sentence.

After normalization, we calculate the **total score** for each sentence by adding the values obtained for every feature. This total score for a given sentence represents its **rank.** On the basis of this rank sentences are opted for the extract summary, higher the rank greater are the chances for selection.

The total score for the given sentence is obtained by computing:

$$\text{TotalScore}(s_i) = \text{position}_i + \text{sLen}_i' + (\text{tf} * \text{idf})_i' + \text{nvp}_i' + \text{PN}_i' + \text{ASC}(S_i)' + \text{CP}_i' + \text{emo}_i' \tag{13}$$

where

position$_i$ = position of the ith sentence.
sLen$_i'$ = normalized sentence length value of the ith sentence.
(tf * idf)$_i'$ = normalized term frequency-inverse document frequency value of the ith sentence.
nvp$_i'$ = normalized noun and verb phrase value of the ith sentence.
PN$_i'$ = normalized proper noun value of the ith sentence.
ACS(S_i)$'$ = normalized aggregate cosine similarity value of the ith sentence.
CP$_i'$ = normalized cue-phrase score of the ith sentence.
emo$_i'$ = normalized emotion score of the ith sentence.

3.4 Redundancy Removing

Input text can include more than one sentences containing same information. There are chances that these sentences get good enough ranks to be included in the summary, because of which summary content becomes redundant. Therefore for the quality summary redundancy needs to be removed. Our implementation uses cosine similarity measure to remove redundant sentences in the summary.

To remove redundancy, the sentence with the highest rank is added in summary. The next sentence will be added in the summary only if the cosine similarity value (new sentence, summary) is smaller than predefined threshold θ. By executing this step we achieve summary having minimal redundancy but the sequence of the sentences is lost. Therefore, in order to uphold the sequence, sentences need to be rearranged according to the initial index.

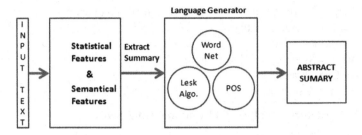

Fig. 2 The language generator

3.5 Making Abstract Summary

We have designed a novel language generator which is a combination of WordNet, Lesk algorithm and Parts-Of-Speech tagger. Using this language generator we are trying to generate abstract summary from the extracted salient sentences. When extract summary is fed to the language generator, words are selected so that they can be replaced by appropriate synonyms to make it abstract. Using WordNet we get the Synsets for the particular word. Then Lesk algorithm is used for word-sense-disambiguation to get synonym of the word to be replaced. The resultant word retrieved using Lesk is checked for its part of speech. If its pos tag matches to pos of the word to be replaced then the original word is substituted by the retrieved word, if not, procedure is repeated until we get the proper substitution. Figure 2 represents the block diagram of the novel language generator.

4 Algorithm

Stage 1: Preprocessing

Input: Text Documents
Output: Preprocessed Text

Step 1: Segmentation.
Step 2: Synonym removal.
Step 3: Removing Stop Words.
Step 4: Word Stemming.

Stage 2: Generating Ranked Sentences.

Input: Preprocessed Text
Output: Ranked Sentences

Step 1: Score the sentence given with 8 different measures.

F_1: $sLen_i$ = Sentence Length
F_2: sPos = Sentence Position

F_3: $(tf * idf)_I$ = Term Frequency-Inverse Document Frequency
F_4: nvp_i = Noun and Verb Phrase
F_5: PN_i = Proper Noun
F_6: $ACS(S_i)$ = Aggregate cosine similarity
F_7: CP_i = Cue-Phrase
F_8: emo_i = Emotion Score

Stage 3: Normalizing Values And Finding Total Score.

Input: Values That are Not Normalized.
Output: Normalized Values And Their Sum In Sorted Order.

Step 1: Apply normalization rules.

F_1': $sLen_i'$ = *Normalized* Sentence Length
F_2': $sPos'$ = *Normalized* Sentence Position
F_3': $(tf * idf)_i'$ = *Normalized* Term Frequency-Inverse Document Frequency
F_4': nvp_i' = *Normalized* Noun and Verb Phrase
F_5': PN_i' = *Normalized* Proper Noun
F_6': $ACS(S_i)'$ = *Normalized* Aggregate cosine similarity
F_7': CP_i' = *Normalized* Cue-Phrase
F_8': emo_i' = *Normalized* Emotion Score

Step 2: Add all the features for every sentence. This sum represents the rank of the given sentence.

$$\text{TotalScore} = \sum_{n=1}^{8} F_n' \tag{14}$$

Step 3: Sorting on the basis of *Total Score*, sentence having the highest total score is the most important sentence.

Stage 4: Using Cosine Similarity To Remove Redundancy.

Input: Sorted sentences.
Output: Salient sentences.

Step 1: Summary = NULL and Threshold value is 0.15.
Step 2: Summary = sentence having highest rank
Step 3: For i = 1 to (total sentences) if [Similarity (Summary, i^{th} sentence) < θ]
Then Summary = Summary + i^{th} sentence
Step 4: In order to uphold the sequence, rearrange the sentences according to their initial index.

Stage 5: Making Abstract Summary.

Input: Extracted Salient sentences.
Output: Abstract Summary.

Step 1: extracted sentences are fed to the novel language generator to transform them into Abstract summary.

Table 1 Values of different measures and their total score for sentences

Sentence no.	Sentence position	Sentence length	Frequency (TF-IDF)	Aggregate cosine similarity	Noun and verb phrase + proper noun	Cue-phrases	Emotion	Total score
0.0	0.19999999	0.36111111	0.37625497	0.19828724	0.38095238	0.0	0.0	1.516605705
1.0	0.86666666	0.16666666	0.10087043	1.0	0.14285714	0.0	0.0	2.277060907
2.0	0.46666567	0.5:777778	0.42192206	0.22089630	0.47619048	0.0	0.0	2.113453273
3.0	0.33333337	0.38888889	0.48018890	0.19192888	0.28571429	0.0	0.0	1.580054292
4.0	0.73333333	0.41666666	0.28143911	0.30877509	0.33333333	0.0	0.4	2.473547533
5.0	0.66666667	0.38888889	0.50937505	0.49024595	0.57142857	0.0	0.0	2.626605126
6.0	1.06666667	0.41666667	0.77616642	0.17505527	0.57142857	0.0	0.0	3.005983593
7.0	1.0	0.88888888	0.88025405	0.21592447	1.0	0.0	0.0	3.985067414
8.0	0.53333333	0.30555556	0.46761594	0.25518146	0.28571429	0.0	0.4	1.47400576
9.0	0.93333333	0.44444444	0.52642453	0.16966249	0.33333333	0.0	0.0	2.407198134
10.0	0.26666667	0.36111111	0.33497640	0.26515444	0.23809524	0.0	0.0	1.466003862
11.0	0.13333333	0.47222222	0.31967674	0.39737532	0.38095238	0.0	0.0	1.703559994
12.0	0.4	0.25	0.22419370	0.42306457	0.23809524	0.0	0.0	1.53535351:
13.0	0.6	1.0	1.0	0.25112163	0.95238095	0.0	0.0	3.803502584
14.0	0.8	0.27777777	0.52624173	0.18974744	0.33333333	0.0	0.2	2.327100283

5 Analysis and Results

We evaluate our single-document hybrid summarization system using (DUC, 2007) dataset. Form (DUC, 2007) dataset we took 15 news articles as input. For each input document our system generated summary which was about 35% of the original document. We compared our system generated summaries with the MS Word generated summaries. The precision of the summaries was calculated with respect to human summary for both the systems. Group of people were given the 15 input documents and were asked to generate their respective summaries.

Table 1 shows different features used in our system for sentence ranking, their values and the total score obtained. Figure 3 is the screenshot of our system. The details of the comparison of our system with MS Word summarization are given in Table 2.

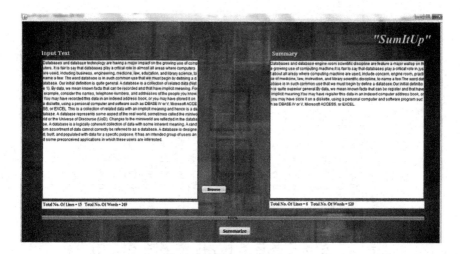

Fig. 3 Screenshot of our system

Table 2 Comparison of summaries generated by our system with the summaries generated by MS Word

Input document	Sentences selected by our system	Sentences selected by MS word	Sentences selected by human judgement	Precision of MS word	Precision of our system
InputDoc1	1,3,4,9,12	1,3,5,10,12	1,2,3,9,14	0.4	0.6
InputDoc2	2,4,5,6, 1O	1,4,5,8,12	1,2,4,5,15	0.6	0.6
InputDoc3	1,6,9,12,19	2,5,9,13,19	1,6,9,13,20	0.4	0.4
InputDoc4	1,8,9,13,19	1,9,13,14,16	1,8,11,13,19	0.6	0.4
InputDoc5	1,6,8,10,14	3,6,8,14,20	2,6,10,14,20	0.6	0.6

(continued)

Table 2 (continued)

Input document	Sentences selected by our system	Sentences selected by MS word	Sentences selected by human judgement	Precision of MS word	Precision of our system
InputDoc6	1,2,7,10,13	2,5,8,12,14	1,2,7,13,18	0.2	0.8
InputDoc7	1,2,4,11,16	4,10,11,15,19	1,2,4,11,19	0.6	0.8
InputDoc8	1,4,11,12,20	1,3,8,12,20	1,2,3,4,20	0.6	0.6
InputDoc9	1,2,16,18,19,20	1,8,13,14,17	1,4,8,14,19	0.4	0.6
InputDoc10	1,3,5,8,11	5,8,9,10,12	1,2,3,8,14	0.2	0.6
InputDoc11	3,4,10,12,19	3,7.10,12,18	3,4,7,15,19	0.6	0.4
InputDoc12	1,3,6,8,13	4,10,12,14,16	1,3,6,16,19	0.2	0.6
InputDoc13	1,2,6,10,18	1,5,9,12,19	2,6,9,12,18	0.4	0.6
InputDoc14	1,2,3,7,17	1,2,6,15,18	1,2,3,14,18	0.4	0.4
InputDoc15	1,3,5,13,15	3,5,11,16,18	1,3,5,12,15	0.4	0.8

6 Conclusion and Future Work

In this paper we explained a hybrid approach to a single-document summarization. Our approach was the hybrid of extraction and abstraction. We first generated extract summary using statistical and novel semantic (emotion) features. We used emotion feature as emotion plays a critical role in identifying the salient sentences to be included in the extract summary. Then using novel language generator extract summary was transformed into abstract summary. We evaluated our system by comparing its summaries and MS Word generated summaries with the human generated summaries and achieved significant results. In most of the cases the relevance rate of our system was more than MS Word.

Further we will evaluate our system using ROUGE.

References

1. Balabantaray, R.C., Sahoo, D.K., Sahoo, B., Swain, M.: Text summarization using term weights. Int. J. Comput. Appl. **38**(1), 10–14 (2012)
2. Nenkova, A., McKeown, K.: Automatic summarization. Found. Trends® Inf. Retr. **5**(3), 235–422 (2011)
3. Kupiec, J., et al.: A trainable document summarizer. In: Proceedings of the 18th Annual International ACM SIGIR Conference on Research and Development in Information Retrieval, pp. 68–73 (1995)
4. Luhn, H.P.: The automatic creation of literature abstracts. IBM J. Res. Dev. **2**, 159–165 (1958)
5. Larsen, B.: A trainable summarizer with knowledge acquired from robust NLP techniques. Adv. Autom. Text Summ., 71 (1999

6. Das, D., Martins, A.F.: A survey on automatic text summarization. In: Literature Survey for the Language and Statistics II course at CMU, vol. 4, pp. 192–195 (2007)
7. Saggion, H., Poibeau, T.: Automatic text summarization: Past, present and future. In: Multi-source, Multilingual Information Extraction and Summarization, pp. 3–21. Springer (2013)
8. Mohd, M., Shah, M.B., Bhat, S.A., Kawa, U.B., Khanday, H.A., Wani, A.H., Wani, M.A., Hashmy, R.: Sumdoc: A Unified Approach for Automatic Text Summarization pp. 333–343. In: Proceedings of Fifth International Conference on Soft Computing for Problem Solving. Springer, Singapore. (2016)
9. Genest, P.E., Lapalme, G.: Framework for abstractive summarization using text- to-text generation. In: Proceedings of the Workshop on Monolingual Text-To-Text Generation, pp. 64–73 (2011)
10. Lin, C.Y.: Rouge: a package for automatic evaluation of summaries. In: Proceedings of Workshop Text Summarization Branches Out (WAS 2004), no. 1, pp. 25–26 (2004)
11. Edmundson, H., Wyllys, R.: Automatic abstracting and indexing—survey and recommendations. Commun. ACM 4(5), 226–234 (1961)
12. Baxendale, P.B.: Machine-made index for technical literature: an experiment. IBM J. Res. Dev. 2(4), 354–361 (1958). doi:10.1147/rd.24.0354
13. Kulkarni, A.R.: An automatic text summarization using feature terms for relevance measure, Dec 2002
14. Ferreira, R., De Souza Cabral, L., Lins, R.D., Pereira E Silva, G., Freitas, F., Cavalcanti, G.D. C., Lima, R., Simske, S.J., Favaro, L.: Assessing sentence scoring techniques for extractive text summarization. Expert Syst. Appl. 40(14), 5755–5764 (2013)
15. Gupta, P., Pendluri, V.S., Vats. I.: Summarizing text by ranking text units according to shallow linguistic features, pp. 1620–1625. In: 13th International Conference on Advanced Communication Technology (2011)
16. Prasad, R.S., Uplavikar, N.M., Wakhare, S.S., Jain, V.Y, Tejas, A.: Feature based text summarization. Int. Adv. Comput. Inf. Res. 1 (2012)
17. Kulkarni, U.V., Prasad, R.S.: Implementation and evaluation of evolutionary connectionist approaches to automated text summarization. J. Comput. Sci., pp. 1366–1376 (2010, Science Publications)

A Novel Locally and Globally Tuned Biogeography-based Optimization Algorithm

Parimal Kumar Giri, Sagar S. De and Satchidananda Dehuri

Abstract *Biogeography-Based Optimization (BBO)* is a nature-inspired meta-heuristic algorithm, which uses the idea of the migration strategy of animals or other species for solving complex optimization problems. In BBO, adaptation of the intensification and diversification for solving complex optimization problem is a challenging task. Migration and mutation operators are two imperative features that largely affect the performance and computational efficiency in BBO, which maintains both exploration and exploitation of existing approaches. In this paper, an innovative migration operator has been introduced in BBO, which inherit the features from a nearest neighbor of the local best individual to be migrated to the globally best individual of the pool and we name it as "*Locally and Globally Tuned BBO (LGBBO)*". We have carried out an extensive numerical evaluation on ten benchmark functions to measure the efficiency of the proposed method. The experimental study confirms that LGBBO is better than canonical and blended BBO in terms of accuracy and convergence time to locate the global optimal solution.

Keywords Island · Habitats · Immigration · Emigration · Exploitation Exploration

P.K. Giri (✉)
National Institute of Science & Technology, Berhampur 761008, India
e-mail: parimal.6789@gmail.com

S.S. De
S.N, Bose National Centre for Basic Sciences, Kolkata 700098, India
e-mail: sagar.s.de@gmail.com

S. Dehuri
Department of I&CT, Fakir Mohan University, Balasore 756019, India
e-mail: satchi.lapa@gmail.com

© Springer Nature Singapore Pte Ltd. 2018
M. Pant et al. (eds.), *Soft Computing: Theories and Applications*,
Advances in Intelligent Systems and Computing 583,
https://doi.org/10.1007/978-981-10-5687-1_57

1 Introduction

The knowledge of biogeography can be traced back to the work of the nineteenth century by naturalists such as Darwin and Wallace [1, 2]. In the early 1960s, MacArthur and Wilson begin working together on mathematical models of bio-geography, the work culminating with the classic 1967 work "The Theory of Island Biogeography" [3]. Their concentration was primarily observant on the distribution of biological species surrounded by neighboring islands along with the geo-temporal revolution. They were attracted to mathematical models of bio-geography to describe speciation (the evolution of new species), the migration of species (animals, fish, birds, or insects) between islands, and the extinction of species.

BBO algorithm introduced by Simon in 2008 [4] was motivated by the theory of island biogeography. The novel idea of original BBO algorithm is based on the principle of migration strategy of biological genesis for solving complex opti-mization problem by maintaining a population of candidate solutions. In BBO, the components' involvement in arrangements is equal to the species' movement in biogeography. Migration model imitate species migration among islands, which provides a recombination way for candidate solutions to interact with each other so that the properties of the population can be improved by keeping the best solutions from previous generation. In BBO, a global optimum solution is one with low Habitat Suitability Index (HSI) that can share their features with poor habitat. This can be achieved only by migrating Suitability Index Variables (SIVs) from emi-grating habitats to immigrating habitats. The original BBO is based on linear migration model [4], and the way to perk up algorithms' performance, several other popular novel migration models are introduced. Motivated by the migration mechanisms of ecosystems and its mathematical model, various extensions to BBO are proposed for achieving information sharing by species migration.

In this paper, we propose a novel technique for migration for enhancing the performance of BBO. The migration operator combines the features from a locally best nearest neighborhood of the individual to be migrated with globally best individual of the pool. Thereby, the LGBBO mimic the species distribution under local best and global best optimum solution, and thus achieves a much better balance between exploration (global search) and exploitation (local search). In Sect. 2, the overviews of BBO and its improvements have been summarized. In Sect. 3 the LGBBO technique has been discussed. The numerical benchmarks are working to test the proposed migration operators and the results are compared with previous work in Sect. 4. In Sect. 5, the conclusions and future research directions are discussed.

2 Review of Biogeography-based Optimization

In BBO, algorithm initializes with population of candidate solutions that are called habitats. Habitats with a high HSI can support many species, whereas low HSI habitats support only a few species. Low HSI habitats can improve their HSI by accepting new features from more attractive habitats in the adaptation process. BBO migration is a probabilistic operator that adjusts each habitat Hi by taking SIVs from a higher HSI habitat. In [4], Simon proposed a migration model which is expressed in Eq. (1). Each habitat has its own probabilistic operators based on HSI as emigration rate (μ) and immigration rate (λ) to define the migration rate for next generation. The migration rates are directly related to the number of species in a habitat. Thus the migration process increases the diversity of the habitat and contributes the likelihood of which information to be shared between the species. The emigration and immigration rates can be calculated in Eq. (2) as follows when there are k species in the habitat:

$$H_i(\text{SIV}) \leftarrow H_j(\text{SIV}) \tag{1}$$

$$\mu_k = \frac{Ek}{S_{\max}} \text{ and } \lambda_k = I(1 - \mu_k), \tag{2}$$

where E is the maximum emigration rate, I is the maximum immigration rate, and $S_{\max}$ is the largest achievable number of species that the habitat can support. We have shown the emigration and immigration rates in Fig. 1 as straight-line model of species large quantity in a single habitat gives us a general description of the process of emigration and immigration [5]. If there are no species on the island, then the emigration rate is zero. The equilibrium number of species is S_0, at which point the emigration and immigration rates are equal.

Fig. 1 Illustration of two candidate solutions: S_1 is a relatively poor solution, while S_2 is a relatively good solution

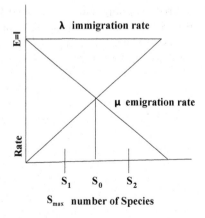

S_1 in Fig. 1 represents a low HSI solution, while S_2 represents a high HSI solution. Thus, the immigration rate λ_1 for S_1 will be higher than the immigration rate λ_2 for S_2. Similarly, the emigration rate μ_1 for S_1 will be lower than the emigration rate μ_2 for S_2. After migration procedure, the mutation operator is used to enhance the diversity of the population to get better solutions. In BBO, mutation is also a probabilistic operator which is used for modifying one or more randomly selected SIV of a solution based on its priori probability of existence P_i. It changes a habitat's SIV randomly based on mutation rate m_i in Eq. (3) that is calculated using solution probability:

$$m_i = M_{\max}\left(1 - \frac{P_i}{P_{\max}}\right), \tag{3}$$

where m_i and $M_{\max}$ are the user-defined parameters of mutation rate and the maximum mutation rate, and $P_{\max}$ is the maximum probability of species count.

2.1   Canonical BBO

Simon [4] uses linear migration model in BBO, which means that λ and μ are the linear functions of solution fitness and are normalized to the range [0, 1]. The pseudocode of canonical BBO is given below.

2.2   Variants of BBO

It is evident from the literature study on BBO that the improvements are based on a modification of migration process only. Guo et al. [6] investigated the migration model by stochastic approaches and exposed the relationship between migration rates and algorithm's performance. Ma et al. [7] and Ma and Simon in [8] proposed a variation of migration model of BBO by using Markov theory. After central BBO, Ergezer et al. [9] proposed oppositional BBO (OBBO) by employing opposition-based learning (OBL). The migration operation of OBBO can be expressed in Eq. (4) as

---

Algorithm 1: Pseudocode of BBO

Initialize: $E = 1$, $I = 1$, $M_{max} = 1$, *Population size, Maximum iteration*;
while *Termination criteria is not satisfied* **do**
    Compute *HSI* for each habitat and sort them in according order;
    for *each habitat (solution)* **do**
        Calculate λ_i, μ_i, m_i and p_i;
        for *each SIV* **do**
            Generate a random number *rand* $\in [0, 1]$;
            if *rand* $< \lambda_i$ **then**
                Select $H_i(SIV)$ through *roulette wheel* with probability $\propto \lambda_i$;
                if $H_i(SIV)$ *selected* **then**
                    Select a habitat $H_j(SIV)$ randomly $\propto \mu_j$;
                    if $H_j(SIV)$ *selected* **then**
                        $H_i(SIV) \leftarrow H_j(SIV)$
                  end
                end
            end
            if *rand* $< m_i$ **then**
                Replace any *SIVs* in $H_i(SIV)$ for mutation;
            end
        end
    end
end

$$OH_i(\text{SIV}) \leftarrow \text{Min} + \text{Max} - H_i(\text{SIV}). \tag{4}$$

In [4], the original BBO employs a linear migration model, after that, in addition to the linear migration model, Ma et al. [7] explored the performance of six migration models which are inspired by the science of biogeography generalizing the equilibrium species count of biogeography theory and showed that the sinusoidal migration model outperforms other models. In [10], Ma et al. proposed a blended migration operator (BMO) shown in Eq. (5), which is inspired from [7]:

$$H_i(\text{SIV}) \leftarrow \alpha H_i(\text{SIV}) + (1 - \alpha)H_i(\text{SIV}), \tag{5}$$

where $\alpha \in [0, 1]$ is a random or deterministic value. In [10], Ma and Simon extend the work of [8] to propose a uniform blended migration operator (UBMO) and nonuniform blended migration operator (NUBMO). They investigate the setting of α in an experimental work to test the results and conclude that a proper value of α, say $\alpha = 0.5$, performs better than a large or a small value of α, say $\alpha = 0.0$ and 0.8, respectively. Feng et al. in [11] proposed the Heuristic Migration Operator (HMO) as shown in Eq. (6), where the parameter $\beta \in [0, 1]$ and $F(.)$ is a fitness function:

$$H_i(\text{SIV}) \leftarrow H_i(\text{SIV}) + \beta(H_j(\text{SIV}) - H_i(\text{SIV})), F(H_j) \geq F(H_i). \tag{6}$$

Gong et al. [12] have proposed popular hybrid alpha-heuristic approach called DE/BBO algorithm in order to balance the exploration of DE and the exploitation of BBO effectively. Siarry and Ahmed-Nacer in [13] proposed a hybridize BBO with different kinds of evolutionary algorithms such as Ant Colony Optimization (ACO) and Artificial Immune Algorithm (AIA) in two different ways. Simon proposed that the canonical approach of BBO has some weakness on its exploration [4]. In total immigration-based BBO, λ_k, is used to decide whether a whole solution should immigrate. If a solution is selected to be immigrated, all the composing features will involve in immigration. Xiong et al. [14, 15] proposed a polyphyletic migration operator; here the current candidate will learn from another solution to extract best features. The migration operation can be expressed in Eq. (7) as follows. When immigration satisfied then,

$$H_i(\text{SIV}) = \begin{cases} H_j(\text{SIV}) + k(H_j(\text{SIV}) - H_i(\text{SIV})), & \text{if migration satisfied} \\ H_j(\text{SIV}), & \text{Otherwise} \end{cases} \quad (7)$$

Simon et al. [16] have proposed an idea of the multi-parent migration operator came from the multi-operators in GA and DE. Orthogonal migration operator was introduced in [17] by employing an orthogonal crossover rule. In standard BBO, if a solution is not selected to be immigrating, migration operator does not run. Thus, Li et al. [18] proposed a new variant of BBO known as Perturb Biogeography-Based Optimization (PBBO) that is used to select a neighborhood solution to update the current one. Based on the previous researches, we are inspired to propose novel and effective migration operator to find the global best feasible optimal solution by using the nearest best solution. Zheng et al. [19] proposed a new variation of BBO, named eco-geography-based optimization (EBO), which regards the population of islands as an ecological system with a local topology.

3 Proposed Work

Recall that BBO is natured inspired algorithm and motivated by the geographical distribution of organisms which involves the study of the migration of biological species between habitats. From subsection 2.2 it is evident that researchers contributed various migration models of BBO with significant results in performance. However, they have their own merits and demerits. To avoid some of the pitfalls of the existing BBO, the proposed LGBBO adapted a novel migration operation which is strongly inspired by the learning mechanisms of school children. The idea is centered on the learning mechanism of a weaker class student for adapting knowledge from stronger students. In nature, it is very often noticed that a weaker student is directly influenced by a student who is better in local context rather than global context. In other words, a weak individual try to adapt best features from the best individual from their nearest neighbor instead of adapting features from best individual of the pool (i.e., global best). However, moving with this strategy a habitat may trap in local optima; therefore, in this approach the combined effort of local best habitat from a predefined size of neighborhood and a global best is explored to uncover the global optimal solution. The model is presented in Eq. (8):

$$H_i(\text{SIV}) \leftarrow \alpha NN(H_i(\text{SIV})) + (1 - \alpha)H_j(\text{SIV}), \tag{8}$$

where the parameter α is named as immaturity index, to represent the island immaturity of the geographical system (population), which is inversely proportional to the invasion resistance of the system. The nearest neighbor of habitat $NN(H_i(SIV))$ can be defined in Eq. (9) as

$$NN(H_i(\text{SIV})) \leftarrow H_{(i \leq r?1:i-r)}(\text{SIV}), \tag{9}$$

where r is the radius of neighborhood. Since the HSIs are sorted in manner, the best nearest neighbor habitat can be found at $(i - r)$. When $i \leq r$ then the best habitat has been chosen as the locally best.

Algorithm 2: Proposed Migration Algorithm (LGBBO)

Initialize: $E = 1$, $I = 1$, $m_{max} = 1$, *Population size* (N_p), *Maximum iteration*;
Create a random set of habitats (populations) $H_1, H_2, \ldots, H_{N_p}$;
while *Termination criteria is not satisfied* **do**
 Compute HSI (fitness);
 Compute λ, μ, p_{mut} and mutation rate m_i for each habitat α HSI ;
 Generate a random number $rand \in (0, 1)$;
 for *each habitat from best to worst according to their HSI values* **do**
  Select a habitat $H_i(SIV)$ probabilistically $\propto \lambda_i$;
  if $rand < \lambda_i$ *and* $H_i(SIV)$ *selected* **then**
   Select an $H_k(SIV)$ as locally best to $H_i(SIV)$ using $NN(H_i(SIV))$;
   Select an habitat $H_j(SIV)$ probabilistically $\propto \mu_j$;
   if $rand < \mu_j$ *and* $H_j(SIV)$ *selected* **then**
    Generate a constant $\alpha \in [0, 1]$;
    for *each SIVs (solution features)* **do**
     $H_i(SIV) \leftarrow \alpha NN(H_i(SIV)) + (1 - \alpha)H_j(SIV)$;
    end
   end
  end
 end
 Select an $SIV(H_i)$ based on mutation probability proportional to p_i;
 if $rand < m_i$ **then**
  Replace $SIV(H_i)$ with randomly generated SIV;
 end
end

---

4   Experimental Works

The focus of this section is to evaluate the efficiency of the nearly developed model. Hence to accomplish the objectives, this section is divided into three Sects. 4.1, 4.2, and 4.3.

4.1   Test Functions and Environments

The task of any good global optimization algorithm is to find globally optimal or at least sub-optimal solutions. The objective functions could be characterized as continuous, differentiable, unimodal, multimodal, separable, and regular. Table 1 presents the details of the well-established 10 benchmark functions and their features that are used to test the performance of the proposed migration model as LGBBO and the results are compared with other developed models like canonical BBO and BBBO. The more details about benchmark functions can be found in [4]. The simulation has been done in an Octa Core i7 x64 CPU with 8GB 1600FSB RAM. We use R programming on LINUX platform for the analysis.

4.2   Parameter Setup

In order to compare the performances of BBO and BBBO with proposed LGBBO, a series of experiments on benchmark functions are carried out to test the efficiency. For initializing the LGBBO, the maximum species count, the maximum migration

Table 1 Benchmark functions and their features

Function	Range/domain	Optimum solution	Features	
f_{01}: Sphere	$[-100, 100]^{30}$	0	U, S, R, C, D	C: Continuous
f_{02}: Schaffer2	$[-100, 100]^{30}$	0	U, S, IR, DC, ND	DC: Discontinuous
f_{03}: Powell's	$[-4, 5]^{30}$	0	U, NS, R, C, D	U: Unimodal
f_{04}: Levy's	$[-100, 100]^{30}$	0	U, NS, R, C, D	M: Multimodal
f_{05}: Schwefel's	$[-100, 100]^{30}$	0	M, S, IR, C, D	D: Differentiable
f_{06}: De Jung	$[-65.5, 65.5]^{30}$	0	M, NS, R, C, D	ND: Nondifferentiable
f_{07}: Rosenbrock	$[-30, 30]^{30}$	0	U, NS, R, C, D	R: Regular
f_{08}: Rastrigin's	$[-5.12, 5.12]^{30}$	0	M, S, R, C, D	IR: Irregular
f_{09}: Ackley's	$[-32, 32]^{30}$	0	M, NS, R, C, D	S: Separable
f_{10}: Griewank's	$[-600, 600]^{30}$	0	M, NS, R, C, D	NS: Non-separable

rates, the maximum mutation rate, and an elitism parameter are defined. 100 habitats and 250 maximum iterations with initial mutation probability of 0.1 have been considered. An α value of 0.15, 0.25, 0.35, and 0.5 has been tested for the BBBO and LGBBO. For the LGBBO a fraction of 0.1, 0.15, and 0.2 habitats (i.e., $100 * 0.1 = 10$ habitats) has been chosen as the neighbors.

4.3 Results and Analysis

Table 2 presents the simulation result obtained by BBO, BBBO ($\alpha = 0.25$), and LGBBO ($nndist = 0.1$, $\alpha = 0.25$) for the 10 benchmark functions over 50 independent runs. The table shows the comparative result of best (min), mean, and standard deviation (Std.) values over the iterations.

Figure 2 provides a graphical view of the comparison of BBO, BBBO, and LGBBO. In Fig. 2, the cost over generation has been plotted for the benchmark functions over 50 independent runs. The plot has been built using generation-wise boxplot for three types of BBO. The comparative results of these functions indicate that LGBBO performs significantly better than other BBOs. The experimental results illustrate that LGBBO has the superior searching ability to other methods both on convergence speed and accuracy.

Table 2 Simulation statistics obtained by BBO, BBBO ($\alpha = 0.25$), and LGBBO ($nndist = 0.1$, $\alpha = 0.25$) on 10 bench mark functions over 50 independent run

Matric	Function	BBO	BBBO	LGBBO	Function	BBO	BBBO	LGBBO
Mean	f_{01}	1.09E-06	3.02E-10	3.30E-13	f_{06}	6.72E-07	3.21E-11	2.14E-13
Std.		8.52E-07	3.32E-10	5.14E-13		5.21E-07	3.43E-11	5.88E-13
Min		2.39E-07	1.72E-11	1.20E-15		6.68E-08	4.10E-14	1.19E-17
Mean	f_{02}	1.25E-06	8.54E-11	9.46E-15	f_{07}	1.43E-06	1.72E-11	2.35E-13
Std.		1.54E-06	1.28E-10	1.10E-14		1.44E-06	3.42E-11	6.16E-13
Min		9.87E-08	1.89E-13	5.65E-16		2.04E-07	2.53E-15	3.09E-16
Mean	f_{03}	9.19E-07	6.34E-11	3.44E-13	f_{08}	8.16E-07	1.44E-11	3.47E-13
Std.		9.44E-07	1.29E-10	6.63E-13		7.65E-07	2.19E-11	9.90E-13
Min		2.68E-08	1.44E-12	3.97E-17		7.71E-08	1.15E-12	2.11E-16
Mean	f_{04}	9.13E-07	6.91E-12	2.75E-12	f_{09}	7.05E-07	1.05E-11	5.14E-13
Std.		7.97E-07	8.32E-12	8.32E-12		6.90E-07	1.49E-11	7.17E-13
Min		1.85E-07	1.87E-13	2.73E-16		9.33E-08	1.89E-14	1.65E-15
Mean	f_{05}	4.93E-07	1.86E-11	5.51E-13	f_{10}	1.51E-06	4.34E-11	2.03E-13
Std.		3.64E-07	1.13E-11	1.28E-12		1.63E-06	6.70E-11	3.18E-13
Min		8.57E-08	2.03E-12	8.57E-17		1.13E-07	1.52E-13	4.98E-16

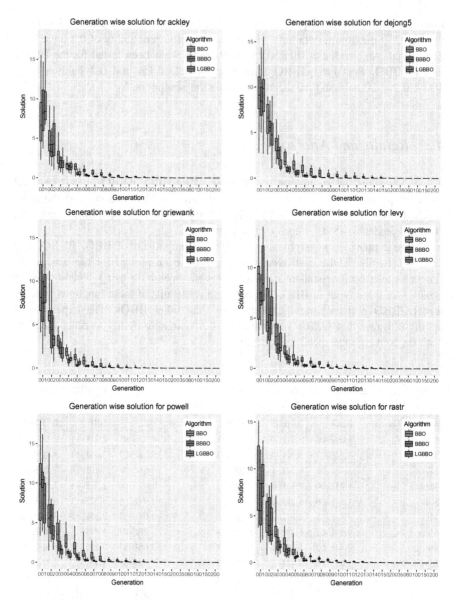

Fig. 2 Graphical presentations of the generation-wise cost for the benchmark functions Ackley, Dejong, Griewank, Levy, Powell, and Raster. The simulation has been carried out for 30-D over 50 independent runs. The box plot indicates generation-wise mean and standard deviation for BBO, BBBO, and LGBBO

5 Conclusions and Future Research Directions

In this paper, to eradicate the deficiencies of the canonical BBO algorithm, we proposed a new BBO by using a new migration method called LGBBO. By using 10 benchmark test functions including unimodal and multimodal functions, we provide a comparative study of LGBBO with canonical BBO and BBBO. An experimental result shows that LGBBO algorithm has strong local best and global best searching ability. It has improved the convergence speed and convergence precision; therefore, it is very effective to solve complex functions optimization problems. Our future research direction includes (i) performance evaluation of LGBBO in domains like financial engineering and big data analysis, (ii) convergence analysis, and (iii) many more specific problems of computational finance.

References

1. Darwin, C.: The Origin of Species. Gramercy, New York (2005)
2. Wallace, A.: The Geographical Distribution of Animals (two volumes). Adamant Media Corporation, Boston (2005)
3. MacArthur, R.H., Wilson, E.O.: The theory of Island Biogeography. Princeton University Press, Princeton (1967)
4. Simon, D.: Biogeography-based optimization. IEEE Trans. Evol. Comput. **12**(6), 702–713 (2008)
5. Ma, H.: An analysis of the equilibrium of migration models for biogeography-based optimization. Inform. Sci. **180**(18), 3444–3464 (2010)
6. Guo, W., Wang, L., Wu, Q.: An analysis of the migration rates for biogeography-based optimization. Inform. Sci. **254**, 111–140 (2014)
7. Ma, H., Simon, D., Fei, M., Xie, Z.: Variations of biogeography-based optimization and Markov analysis. Inform. Sci. **220**, 492–506 (2013)
8. Ma, H., Simon, D.: Analysis of migration models of biogeography based optimization using Markov theory. Eng. Appl. Artif. Intell. **24**(6), 1052–1060 (2011)
9. Ergezer, M., Simon, D., Du, D.: Oppositional biogeography based optimization. In: 2009 IEEE International conference on systems, man and cybernetics (SMC 2009), 1,9, 1009–1014, (2009)
10. Ma, H., Simon, D.: Blended biogeography-based optimization for constrained optimization. Eng. Appl. Artif. Intell. **24**(3), 517–525 (2011)
11. Feng, Q., Liu, S., Tang, G., Yong, L., Zhang, J.: Biogeography based optimization with orthogonal crossover. Math. Problem. Eng. **353969**, 1–20 (2013)
12. Gong, W., Cai, Z., Ling, X.: DE/BBO: a hybrid differential evolution with biogeography-based optimization for global numerical optimization. Soft. Comput. **15**(4), 645–665 (2011)
13. Siarry, P., Boussaid, I., Chatterjee, A., Ahmed-Nacer, M.: Hybridizing biogeography-based optimization with differential evolution for optimal power allocation in wireless sensor networks. IEEE Trans. Veh. Tech. **60**(5), 2347–2353 (2011)
14. Xiong, G., Li, Y., Chen, J., Shi, D., Duan, X.: Polyphyletic migration operator and orthogonal learning aided biogeography based optimization for dynamic economic dispatch with valve point effects. Engergy Convers. Manag. **80**, 457–468 (2014)

15. Xiong, G., Shi, D., Duan, X.: Enhancing the performance of biogeography-based optimization using polyphyletic migration operator and orthogonal learning. Comput. Oper. Res. **41**(5), 125–139 (2014)
16. Simon, D., Rarick, R., Ergezer, M., Du, D.: Analytical and numerical comparisons of biogeography-based optimization and genetic algorithms. Inform. Sci. **181**(7), 1224–1248 (2011)
17. Ma, H., Fei, M., Ding, Z., Jin, J.: Biogeography-based optimization with ensemble of migration models for global numerical optimization. IEEE World Congress on Computational Intelligence, pp. 2981–2988. Brisbane, Australia (2012)
18. Li, X., Wang, J., Zhou, J., Yin, M.: A perturbs biogeography based optimization with mutation for global numerical optimization. Appl. Math. Comput. **218**(2), 598–609 (2011)
19. Zheng, Y., Feng Ling, H., Yun Xue, J.: Eco-geography based optimization, Enhancing biogeography-based optimization with eco-geographic barriers and differentiations. Comput. Oper. Res. **50**(4), 115–127 (2014)

Improved Meta-Heuristic Technique for Test Case Prioritization

**Deepak Panwar, Pradeep Tomar, Harshvardhan Harsh
and Mohammad Husnain Siddique**

Abstract Time is a very critical factor for decision of cost of any software. The cost and validity of software is based on the quality and quantity of the existing test cases. A large number of software testing approaches are available having both advantages and limitations. Original test cases are supposed to be reused and the new test cases have to be supplemented in regression testing of the updated software. For effective and efficient test case prioritization, the techniques like test case prioritization and test case selection were introduced for scheduling test cases and implementing test cases for fulfillment of some particular criteria. Test case prioritization supports the most useful test cases to execute first by making software testing cost-effective and efficiently covering most extreme number of faults in least time. But test case prioritization requires huge time and effort. This paper has proposed an improved meta-heuristic technique (Ant Colony Optimization) algorithm to find the best optimal path by prioritization of test cases.

Keywords ACO · Antennae · Test case prioritization · Software quality assurance

1 Introduction

Software quality ensures functionality and features of a software product by reflecting its ability to satisfy stated or implied needs based on functional requirements and specifications. Till now, software has been considered a product and in quality sense, all the attributes are expected to be improved. But the definition of quality is not complete at this point. Quality control focuses on finding

D. Panwar (✉) · H. Harsh · M.H. Siddique
Amity School of Engineering and Technology, Amity University, Noida, Rajasthan, India
e-mail: deepakpanwar0@gmail.com

P. Tomar
School of I.C.T., Gautam Buddha University, Greater Noida, India
e-mail: parry.tomar@gmail.com

© Springer Nature Singapore Pte Ltd. 2018
M. Pant et al. (eds.), *Soft Computing: Theories and Applications*,
Advances in Intelligent Systems and Computing 583,
https://doi.org/10.1007/978-981-10-5687-1_58

647

and removing defects, and the main purpose of quality assurance is to verify that applicable procedures and standards are being followed [1].

ISO 9126 model has been adopted for standardizing software quality assurance activities. ISO 9126 is an international quality software standard from ISO. It provides a set of six quality characteristics with their definitions. The evaluation processes associated with the qualities of software are provided by ISO 9126 which are used during evaluation of the quality of the software products and specification of their requirements throughout their life cycle. The Six quality characteristics of ISO 9126 are Functionality, Usability, Maintainability, Reliability, Efficiency and Portability. Artificial Intelligence (AI) is related with the computer-based exploration methods and the fields defined in Computational Intelligence (CI). Understanding of the principles which make intelligent behavior possible, in artificial systems or nature is the central goal of the Computational Intelligence. Testing is necessary because it is an imperative part of the software advancement life cycle. Software testing measures the quality of software in terms of test run, the execution time of the processes related to the software, number of defects found and the system covered by the tests. Testing is referred as execution of a project with the plan of discovering deficiencies [2]. Regression testing is performed in case of bug-fixing or whenever there is need to incorporate any new requirement in the software. IEEE software glossary defines regression testing [3] that "Regression testing is the particular retesting of a framework or segment to confirm that alterations have not created unintended impacts and that the framework or segment still conforms to its particular prerequisites." Software testing is one of the most promising areas of research in Artificial Intelligence (AI) [4–6]. Ant Colony Optimization is one such meta-heuristic technique of AI [7, 8]. ACO algorithm is one of the swarm intelligence algorithms. The ant colony is always able to find the shortest path for food from its home. These ants use pheromone to mark some favorable path by depositing it on the ground that should be followed by other members of the colony [9]. They use food trial pheromone and antenna to be in touch with each other. Ants use antenna for information sharing by colliding with each other and touching each other with their antenna. After the proposal of first ACO algorithm, it had become one of the important means to solve optimization problems related to artificial intelligence and it attracted attention of a large number of scholars and researchers. ACO is utilized as another approach to take care of time limitation prioritization issue.

During last few years Ant Colony Optimization approach has been used by many authors to automatic test sequence generation for state-based software testing [10]. The paper introduces an enhanced Ant Colony Optimization (I-ACO) calculation for reordering the test cases in time-constrained environment by using regression test selection technique which selects an appropriate number of test cases from a test suite that might expose a fault in the modified programs.

2 Related Work

The essential point of different proposed methods for software testing is to discover greatest deficiencies in least time. Amid the most recent couple of years Ant Colonies (AC's) have been utilized as a broadly useful heuristic to take care of combinatorial issue like travelling salesman problem, sequential ordering, vehicle routing, data mining, telecommunication networks and other NP-hard problems [8, 11–20]. Several ACO applications are available now for a wide range of different optimization problems. NP-hard problems hold the majority of these applications, that is related to the problems for which the best algorithm which guarantees to identify an optimal solution have exponential time worst case complexity. Apart from TSP, other NP-hard problems were also considered. The broad categories of these problems are *routing problem, assignment problem, scheduling problem, and subset problem*. In addition, on other problem emerging fields like bioinformatics and machine learning, ACO has been successfully applied.

In this paper, researchers have proposed an improved Ant Colony Optimization (I-ACO) Algorithm for prioritization of test cases in time-constrained environment.

3 ACO Concepts for Solving Combinatorial Optimization Problem

Ant Colony Optimization (ACO) Algorithm was proposed by M. Dorogo in 1991. ACO is based on the foraging achievement technique of the ant colonies. The real ants are able to find out the shortest path between the food and their nest by the pheromone information exploitation. The standard ACO algorithm was introduced by the example of Travelling Salesman Problem.

In this paper, to show optimization, the modified ACO algorithm has been proposed using Traveling Salesman Problem (TSP). TSP is the NP-hard problem to find the minimum tour cost by visiting each city once. Let $A = \{(i, j): i, j \in V\}$ be the each edge, $d(i, j) = d(j, i)$ be a cost measure connected with edge $(i, j) \in A$ and $V = \{1,..., n\}$ be a set of cities. The working function of TSP is:

$$minD = \sum_{i=1}^{n-1} d(i, i+1) + d(n, 1) \tag{1}$$

where $d(i, j)(i, j = 1, 2,..., n)$ is the tour length form city i to city j. Node i is the current location of the ant and node j is the position where the ant can reach. The probability p_{ij} of the ant's move from node i to node j is given by:

$$p_{ij}(k) = \begin{cases} \dfrac{\tau_{ij}^{\alpha} \eta_{ij}^{\beta}}{\sum_{j \in C} \tau_{ik}^{\alpha} \eta_{ik}^{\beta}} & \text{If } j \in C \\ 0 & \text{otherwise} \end{cases} \tag{2}$$

where C is expressing the set of nodes that the kth ant can select the next node. Selection probability of following node is based on both the pheromone quantity as well as heuristic information. The heuristic information and pheromone information values are controlled by α and β. The heuristic information η_{ij} is:

$$\eta_{ij} = \frac{1}{d_{ij}} \tag{3}$$

where d_{ij} is the distance between the nodes. The pheromone τ_{ij}, associated with the edge joining the nodes i and j, is updated as:

$$\tau_{ij} \leftarrow (1 - \rho) \cdot \tau_{ij} + \sum_{k=1}^{n} \Delta \tau_{ij}^{k} \tag{4}$$

where n is the number of ants and ρ is the evaporation rate and τ_{ij}^{k} is the quantity of pheromone laid by the ant k on the edge (i, j) and is given by:

$$\Delta \tau_{ij}^{k} = \begin{cases} Q/L_k & \text{if the edge } (i,j) \text{ is used by ant } k \\ 0 & \text{otherwise} \end{cases} \tag{5}$$

where Q is constant and L_k is the tour length constructed by ant k.

4 Proposed Algorithm

In this section, the paper proposes an enhanced Ant Colony Optimization (I-ACO) algorithm in which the modification of heuristic information increases the probability of finding optimal path towards next nodes and solves test case prioritization problem. The proposed algorithm is based on Simple ACO (S-ACO) in which ants move in two modes: Forward mode and Backward mode. As ants move in forward mode from source (nest) to destination (food) and in the other way around in reverse mode the collision between ants occur. During collision, ants share information by using their antenna [21, 22]. This shared information leads to increase in heuristic information, on which the probability of choosing the next node is based. The heuristic information is modified by additional information gained by collision between ants moving from source to destination.

Let the modified heuristic information is given by η'_{ij}, which is obtained by adding $rand()$ with β.

$$\eta'_{ij} = \eta_{ij}^{\beta + rand()} \tag{6}$$

Here, *rand()* signifies the additional information gained by collision of ants antennas with other ant's antennas and β is a constant.

For node j, if $\eta_{ij} \leq \eta'_{ij}$ then the heuristic information is increased and also the probability of finding best path by the ant is increased which will lead to obtaining the best cost of the path with minimum iterations. As the number of iterations will be decreased, the consumption of time to get best cost also decreases. Therefore, for $\eta_{ij} \leq \eta'_{ij}$.

$$\frac{\tau_{ij}^{\alpha} \eta_{ij}^{\beta}}{\sum_{j \in C} \tau_{ik}^{\alpha} \eta_{ik}^{\beta}} \leq \frac{\tau_{ij}^{\alpha} \eta_{ij}^{\beta + rand()}}{\sum_{j \in C} \tau_{ik}^{\alpha} \eta_{ik}^{\beta + rand()}} \tag{7}$$

Thus, we have

$$p_{ij}(\alpha, \beta) \leq p_{ij}(\alpha, \beta + rand()) \tag{8}$$

The above equation signifies that the probability of finding the best path increases due to modification in heuristic information by summation of additional information (i.e., *rand()*) with constant value of β.

This paper compares the performance of standard ACO algorithm and I-ACO algorithm in terms of the best cost of path and number of iterations. Traveling Salesman Problem (TSP) is an NP-hard problem which was used to demonstrate standard ACO [9]. The I-ACO algorithm was trained and validated for TSP using Matlab 7.9.0. The I-ACO was trained and validated to obtain best cost of path traversed by ants in minimum number of iterations. Figures 1 and 2 show the output of standard ACO applied on TSP. In standard ACO the parameters α and β are set to 1, i.e., $\alpha = 1$ and $\beta = 1$.

Fig. 1 Plot graph for standard ACO at $\beta = 1$

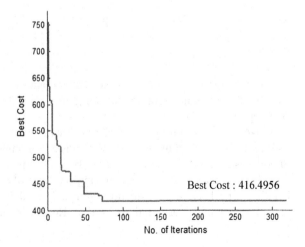

Fig. 2 Path traversed by ant for standard ACO at $\beta = 1$

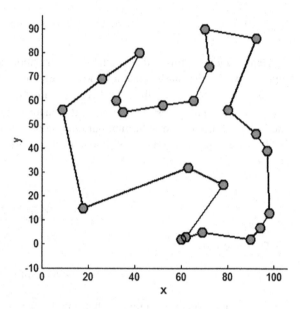

Fig. 3 Plot graph for I-ACO at $\beta = 1 + rand()$

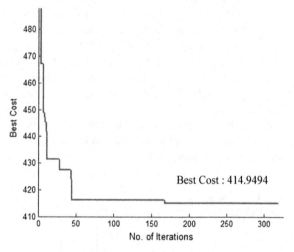

The I-ACO was trained and validated to obtain best cost of path traversed by ants in minimum number of iterations. Figures 1 and 2 show the output of standard ACO applied on TSP.

In I-ACO algorithm the parameters are set, i.e., $\alpha = 1$, $\beta = 1 + rand()$, evaporation rate of 5%, total number of iteration is 320, number of ants is 40 and number of nodes is 22. At $\beta = 1 + rand()$, gives best result in minimum iteration. Figures 3 and 4 show the output of I-ACO having best cost of path in minimum number of iteration. The heuristic information with additional information ($\beta = 1 + rand()$) has reduced number of iterations in TSP. Reduction in no. of iteration decreases execution time.

Fig. 4 Path traversed by ant
for I-ACO at $\beta = 1 + rand()$

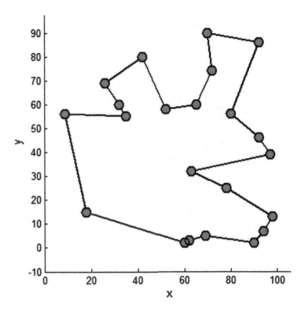

4.1 Modeling Test Case Prioritization Using Improved Ant Colony Optimization

1. The ant knows initial pheromone and initial heuristic at node S.
2. The next node is chosen according to the initial pheromone and heuristic information. Heuristic information is increased by additional information gained by collision between ants moving in forward mode (i.e., from node S to D) and backward mode (i.e., from node D to S).
3. After each collision the heuristic information is increased.
4. A new ant will start moving towards destination node D after one ant has completed the iteration and this ant will give the information by collision of its antennae with the new ant's antenna.

4.2 Example

This paper has used triangle classification problem to generate its DD graph for test case prioritization. The program demonstrated in this paper is to specify whether the given triangle is an equilateral, isosceles or scalene triangle. The code is given as follows:

```
include<stdio.h>
include<conio.h>

void main ()
    {
    double a.b.c;
    int valid=0;
    clrscr();
    printf("enter first side of triangle");
    scanf("%f",&a);
    printf("enter second side of triangle");
    scanf("%f",&b);
    printf("enter third side of triangle");
    scanf("%f",&c);
    if (a>0 && a<50 && b>0 && b<50 && c>0 &&c<50)
        {
                if ((a+b)>c && (b+c)>a && (c+a)>b)
                {
                        valid=1;
                }
                else
                {
                        valid= -1;
                }
        }
    if ( valid==-1)
        {
                printf("\n Invalid triangle ");

        }
    else if ( valid == 1)
        {
                if ((a == b) & (b!=c) || (b==c) & (c!=a) ||
                (c==a) &(a!=b))
                {
                        Printf("Isosceles triangle");

                }
                else if ((a==b) && (b==c))
                {
                        printf("right angled triangle");

                }
                else  if((a!=b) && (b!=c))
                {
                        printf("acute angled triangle");

                }
        }
    else
        {
                printf("\n input values are out of range ");
        }
    getch();
}
```

Fig. 5 DD path graph for the
given triangle problem

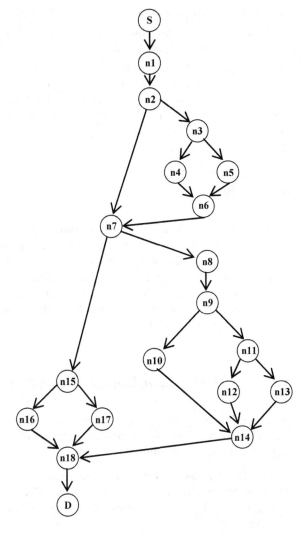

The decision-to-decision path graph for the triangle problem program is given in Fig. 5. The number of independent paths in DD path graph as per given in Fig. 5 is 7 because its cyclometic complexity is 7. The different independent paths and their execution times are given below in Table 1. In I-ACO, it is assumed that the execution time for test case depends on number of edges to be traversed by each ant. In test case T1, there are 7 edges where each edge is having execution time equal to 1. So, execution time for T1 is 7 and similarly, test cases T2, T3, T4, T5, T6 and T7 have execution time 7, 10, 10, 12, 9 and 12 respectively.

Table 1 Independent paths and execution times for DD path graph

Test cases	Independent paths	Execution time
T1	S, n1, n2, n7, n15, n17, n18, D	7
T2	S, n1, n2, n7, n15, n16, n18, D	7
T3	S, n1, n2, n7, n8, n9, n11, n12, n14, n18, D	10
T4	S, n1, n2, n7, n8, n9, n11, n13, n14, n18, D	10
T5	S, n1, n2, n3, n5, n6, n7, n8, n9, n10, n14, n18, D	12
T6	S, n1, n2, n7, n8, n9, n10, n14, n18, D	9
T7	S, n1, n2, n3, n4, n6, n7, n8, n9, n10, n14, n18, D	12

5 Results

In this section, the paper presents the results of the proposed algorithm for the validation of its effectiveness. The I-ACO algorithm traverses each node and then prioritizes the paths according to the weight of the path. There are seven test cases in the regression test suite shown in Table 1. These test cases are prioritized in such a way that all the ten faults are covered in minimum test case execution time. Table 3 indicates the number of faults detected and the execution time of the test cases to find out those faults.

5.1 I-ACO Algorithm Example Validation

After drawing DD path graph and independent path, the researchers have traversed the move of an Ant upon and then prioritization of paths for testing. The I-ACO algorithm is applied as follows.

1. Initially, it is assumed that initial pheromone (τ_0) value is set to any random value. In the given example, it is set to 0.095 and initial heuristic information (η_0) value is equal to the cyclometric complexity of the DD path graph. Here, it is set to 7.
2. After each edge is covered, 5% pheromone is evaporated from that edge.
3. After traversal of node (i.e., moving towards n1), if the next node is not the destination node (i.e., D) then the above steps are again followed from ant's current node.
4. At node n2, the probability of choosing next node (i.e., n3 or n7) is equal and also neither n3 nor n7 are destination node, so, above steps are again repeated.
5. The new pheromone τ_{new} is calculated by including the measure of pheromone left after evaporation and measure of pheromone stored after traversal of way. The modification in heuristic information is calculated as follows:

$$\Delta\eta_{ij} \leftarrow \eta_o/L_i^K, \text{ then}$$

$$\eta_{new} \leftarrow \eta_o + \Delta\eta_{ij}, \text{ then}$$

6. The required heuristic (η'_{ij}) value is calculated as follows:

$$\eta'_{ij} \leftarrow \eta_{new}^{\beta + rand()}$$

where, $\beta = 1$ and $rand()$ $[0–1] = 0.2$.

7. Above steps are repeated until all the nodes in the DD path graph are traversed.
8. After traversing all nodes of the DD path graph, calculate the weight of all the independent paths of the test cases. For example, weight of the independent path in test case T1 (S, n1, n2, n7, n15, n17, n18, D)

   (a) weight(edge) = new pheromone value(τ_{new}) * updated heuristic value (τ_{new})
   (b) Weight(Test Case) = sum of weight of all edges of test

   Case/Execution time of test case
   Similarly, find weight of all other edges of the test cases and sum all the weights, i.e., path weight of test case W(T1) = {w(s − n1) + w(n1 − n2) + w(n2 − n7) + w(n7 − n15) + w(n15 − n17) + w(n17 − n18) + w(n18 − D)}/
   7 = 129.07/7 = 18.43

   The test case having maximum path weight will have highest priority and all the test cases will be executed according to their prioritization. Table 2 describes all the nine moves of an Ant to traverse all the nodes of DD path graph for the prioritization of test cases.

   Table 4 shows the weight of the test case paths and their prioritization according to their weights.

5.2 Comparison

This paper has compared the mentioned example with four different ordering of test cases: *Original Ordering, Random Ordering, Reverse Ordering and Optimal Ordering*. The ordering of test cases with respect to their prioritization approaches is shown in Table 5. The prioritization of test cases in *I-ACO Order* covers all the ten faults in minimum execution time which is 23 whereas in Optimal Ordering, execution time to cover faults is 29.

Table 2 Moves of an ant for traversal of all nodes of DD path graph

S. No.	Path traverse	L_i^k	τ_0	$\tau_e \leftarrow (1-\rho)*\tau_0$	$\Delta\tau_{ij} = 1/L_i^k$	$\tau_{new} \leftarrow \tau_e + \Delta\tau_{ij}$	η_0	$\Delta\eta_{ij} \leftarrow \eta_0/L_i^k$	$\eta_{new} \leftarrow \eta_0 + \Delta\eta_{ij}$	Modified heuristic (rand) $\eta'_{ij} \leftarrow \eta_{new}^{\beta + rand()}$ () = 0.2
			Initial pheromone	Evaporation (5%)			Initial heuristic			
1	S–n1	12	0.095	0.09	0.083	0.173	7	0.583	7.583	11.37
	n1–n2	12	0.095	0.09	0.083	0.173	7	0.583	7.583	11.37
	n2–n3	12	0.095	0.09	0.083	0.173	7	0.583	7.583	11.37
	n3–n4	12	0.095	0.09	0.083	0.173	7	0.583	7.583	11.37
	n4–n6	12	0.095	0.09	0.083	0.173	7	0.583	7.583	11.37
	n6–n7	12	0.095	0.09	0.083	0.173	7	0.583	7.583	11.37
	n7–n8	12	0.095	0.09	0.083	0.173	7	0.583	7.583	11.37
	n8–n9	12	0.095	0.09	0.083	0.173	7	0.583	7.583	11.37
	n9–n10	12	0.095	0.09	0.083	0.173	7	0.583	7.583	11.37
	n10–n14	12	0.095	0.09	0.083	0.173	7	0.583	7.583	11.37
	n14–n18	12	0.095	0.09	0.083	0.173	7	0.583	7.583	11.37
	n18–D	12	0.095	0.09	0.083	0.173	7	0.583	7.583	11.37
2	S–n1	7	0.173	0.164	0.142	0.306	7.583	1.08	8.66	13.33
	n1–n2	7	0.173	0.164	0.142	0.306	7.583	1.08	8.66	13.33
	n2–n7	7	0.095	0.09	0.142	0.232	7	1	8	12.1257
	n7–n15	7	0.095	0.09	0.142	0.232	7	1	8	12.1257
	n15–n16	7	0.095	0.09	0.142	0.232	7	1	8	12.1257
	n16–n18	7	0.095	0.09	0.142	0.232	7	1	8	12.1257
	n18–D	7	0.095	0.09	0.142	0.232	7	1	8	12.1257
3	S–n1	7	0.306	0.2907	0.142	0.432	8.66	1.2371	9.8971	15.65
	n1–n2	7	0.306	0.2907	0.142	0.432	8.66	1.2371	9.8971	15.65
	n2–n7	7	0.232	0.2204	0.142	0.362	8	1.1428	9.1428	14.2329
	n7–n15	7	0.232	0.2204	0.142	0.362	8	1.1428	9.1428	14.2329
	n15–n17	7	0.095	0.09	0.142	0.232	7	1	8	12.1257

(continued)

Table 2 (continued)

S. No.	Path traverse	L_i^k	τ_0 Initial pheromone	$\tau_e \leftarrow (1-\rho)*\tau_0$ Evaporation (5%)	$\Delta\tau_{ij} = 1/L_i^k$	$\tau_{new} \leftarrow \tau_e + \Delta\tau_{ij}$	η_0 Initial heuristic	$\Delta\eta_{ij} \leftarrow \eta_o/L_i^k$	$\eta_{new} \leftarrow \eta_o + \Delta\eta_{ij}$	$\eta'_{ij} \leftarrow \eta_{new}^{\beta+rand()}$ Modified heuristic (rand) () = 0.2)
	n17–n18	7	0.095	0.09	0.142	0.232	7	1	8	12.1257
	n18–D	7	0.232	0.2204	0.142	0.362	8	1.1428	9.1428	14.2329
4	S–n1	13	0.432	0.41	0.077	0.487	9.897	0.76	10.65	17.09
	n1–n2	13	0.432	0.41	0.077	0.487	9.897	0.76	10.65	17.09
	n2–n3	13	0.173	0.164	0.077	0.241	7.583	0.58	8.16	12.4173
	n3–n5	13	0.095	0.09	0.077	0.167	7	0.5384	7.5384	11.29
	n5–n6	13	0.095	0.09	0.077	0.167	7	0.5384	7.5384	11.29
	n6–n7	13	0.173	0.164	0.077	0.241	7.583	0.58	8.16	12.4173
	n7–n8	13	0.173	0.164	0.077	0.241	7.583	0.58	8.16	12.4173
	n8–n9	13	0.173	0.164	0.077	0.241	7.583	0.58	8.16	12.4173
	n9–n11	13	0.095	0.09	0.077	0.167	7	0.5384	7.5384	11.29
	n11–n12	13	0.095	0.09	0.077	0.167	7	0.5384	7.5384	11.29
	n12–n14	13	0.095	0.09	0.077	0.167	7	0.5384	7.5384	11.29
	n14–n18	13	0.173	0.164	0.077	0.241	7.583	0.58	8.16	12.4173
	n18–D	13	0.362	0.344	0.077	0.421	9.142	0.7023	9.8451	15.5547
5	S–n1	7	0.487	0.4626	0.142	0.604	10.65	1.5214	12.1714	20.0635
	1–n2	7	0.487	0.4626	0.142	0.604	10.65	1.5214	12.1714	20.0635
	n2–n7	7	0.362	0.34	0.142	0.482	9.142	1.306	10.4488	16.7062
	n7–n15	7	0.362	0.34	0.142	0.482	9.142	1.306	10.4488	16.7062
	n15–n16	7	0.232	0.22	0.142	0.362	8	1.1428	9.1428	14.2329
	n16–n18	7	0.232	0.22	0.142	0.362	8	1.1428	9.1428	14.2329
	n18–D	7	0.421	0.3999	0.142	0.541	9.845	1.4064	11.2515	18.2579

(continued)

Table 2 (continued)

S. No.	Path traverse	L_i^k	τ_0 Initial pheromone	$\tau_e \leftarrow (1-\rho) * \tau_0$ Evaporation (5%)	$\Delta\tau_{ij} = 1/L_i^k$	$\tau_{new} \leftarrow \tau_e + \Delta\tau_{ij}$	η_0 Initial heuristic	$\Delta\eta_{ij} \leftarrow \eta_0/L_i^k$	$\eta_{new} \leftarrow \eta_0 + \Delta\eta_{ij}$	Modified heuristic (rand) $\eta'_{ij} \leftarrow \eta_{new}^{\beta+rand()}$ () = 0.2
6	S–n1	12	0.604	0.5743	0.083	0.657	12.17	1.014	13.1854	22.08
	n1–n2	12	0.604	0.5743	0.083	0.657	12.17	1.014	13.1854	22.08
	n2–n3	12	0.241	0.2289	0.083	0.311	8.16	0.68	8.84	13.6691
	n3–n4	12	0.173	0.164	0.083	0.247	7.583	0.631	8.215	12.51
	n4–n6	12	0.173	0.164	0.083	0.247	7.583	0.631	8.215	12.51
	n6–n7	12	0.241	0.2289	0.083	0.311	8.16	0.68	8.84	13.6691
	n7–n8	12	0.241	0.2289	0.083	0.311	8.16	0.68	8.84	13.6691
	n8–n9	12	0.241	0.2289	0.083	0.311	8.16	0.68	8.84	13.6691
	n9–n10	12	0.173	0.164	0.083	0.247	7.583	0.631	8.215	12.51
	n10–n14	12	0.173	0.164	0.083	0.247	7.583	0.631	8.215	12.51
	n14–n18	12	0.241	0.2289	0.083	0.311	8.16	0.68	8.84	13.6691
	n18–D	12	0.541	0.5148	0.083	0.597	11.251	0.937	12.1891	20.09
7	S–n1	7	0.657	0.6244	0.142	0.766	13.185	1.88	15.0654	25.91
	n1–n2	7	0.657	0.6244	0.142	0.766	13.185	1.88	15.0654	25.91
	n2–n7	7	0.482	0.4579	0.142	0.599	10.448	1.49	11.9388	19.6
	n7–n15	7	0.482	0.4579	0.142	0.599	10.448	1.49	11.9388	19.6
	n15–n17	7	0.232	0.22	0.142	0.362	8	1.142	9.142	14.23
	n17–n18	7	0.232	0.22	0.142	0.362	8	1.142	9.142	14.23
	n18–D	7	0.597	0.5679	0.142	0.709	12.189	1.74	13.9291	23.5888

(continued)

Table 2 (continued)

S. No.	Path traverse	L_i^k	τ_0 Initial pheromone	$\tau_e \leftarrow (1-\rho)*\tau_0$ Evaporation (5%)	$\Delta\tau_{ij} = 1/L_i^k$	$\tau_{new} \leftarrow \tau_e + \Delta\tau_{ij}$	η_0 Initial heuristic	$\Delta\eta_{ij} \leftarrow \eta_0/L_i^k$	$\eta_{new} \leftarrow \eta_0 + \Delta\eta_{ij}$	$\eta'_{ij} \leftarrow \eta_{new}^{\beta+rand()}$ Modified heuristic (rand() = 0.2)
8	S–n1	7	0.766	0.728	0.142	0.87	15.0654	2.15	17.215	30.415
	n1–n2	7	0.766	0.728	0.142	0.87	15.0654	2.15	17.215	30.415
	n2–n7	7	0.599	0.5699	0.142	0.711	11.9388	1.705	13.64	23
	n7–n15	7	0.599	0.5699	0.142	0.711	11.9388	1.705	13.64	23
	n15–n16	7	0.362	0.3439	0.142	0.485	9.1428	1.306	10.4488	16.704
	n16–n18	7	0.362	0.3439	0.142	0.485	9.1428	1.306	10.4488	16.704
	n18–D	7	0.709	0.6744	0.142	0.816	13.929	1.9898	15.9189	27.688
9	S–n1	13	0.87	0.8265	0.077	0.903	17.215	1.317	18.586	33.34
	n1–n2	13	0.87	0.8265	0.077	0.903	17.215	1.317	18.586	33.34
	n2–n3	13	0.311	0.2963	0.077	0.373	13.669	1.05	14.7191	25.203
	n3–n5	13	0.167	0.158	0.077	0.235	7.538	0.5798	8.1182	12.34
	n5–n6	13	0.167	0.158	0.077	0.235	7.538	0.5798	8.1182	12.34
	n6–n7	13	0.311	0.296	0.077	0.373	13.669	1.05	14.7191	25.203
	n7–n8	13	0.311	0.296	0.077	0.373	13.669	1.05	14.7191	25.203
	n8–n9	13	0.311	0.296	0.077	0.373	13.669	1.05	14.7191	25.203
	n9–n11	13	0.167	0.158	0.077	0.235	11.29	0.8684	12.1584	20.03
	n11–n13	13	0.173	0.164	0.077	0.241	7	0.538	7.538	11.29
	n13–n14	13	0.173	0.164	0.077	0.241	7	0.538	7.538	11.29
	n14–n18	13	0.311	0.296	0.077	0.373	13.669	1.05	14.7191	25.203
	n18–D	13	0.816	0.7755	0.077	0.852	15.918	1.2245	17.1434	30.2633

Table 3 Faults detected by test cases and time for their execution

Test case/faults	T1	T2	T3	T4	T5	T6	T7
F1			⊕			⊕	
F2		⊕					
F3				⊕	⊕	⊕	⊕
F4		⊕		⊕			
F5	⊕		⊕				
F6	⊕				⊕		⊕
F7		⊕	⊕				
F8			⊕			⊕	⊕
F9		⊕		⊕			
F10					⊕	⊕	
Number of faults	2	4	4	3	3	4	3
Execution time	7	7	10	10	12	9	12

Table 4 Weight and priority of test cases

Test case	Weight	Priority
T1	18.43	2
T2	19.28	1
T3	13.906	5
T4	14.074	4
T5	12.08	7
T6	15.19	3
T7	12.11	6

Table 5 Prioritization approaches by ordering of test cases

Original ordering	Random ordering	Reverse ordering	Optimal ordering	I-ACO ordering
T1	T5	T7	T2	T2
T2	T2	T6	T3	T1
T3	T4	T5	T5	T6
T4	T3	T4	T4	T4
T5	T1	T3	T6	T3
T6	T6	T2	T7	T7
T7	T7	T1	T1	T5

6 Conclusion

This work suggested an improved Ant Colony Optimization technique for the test cases prioritization using the additional information to achieve the goal or even find an optimal solution. And here the goal is to satisfy the stakeholders while they are the customers or the producers. So the main focus is on software quality assurance. Proposed algorithm is based on the information sharing capability of ants though there antennas means the base of this improved algorithm is heuristic function. With the help of additional information in the form of increasing heuristic function value it is easy to find the maximum number of faults by executing minimum numbers of test cases in comparisons to previous defined algorithms for test cases prioritization. So, this algorithm recommends a pivotal use of nature inspired algorithm in the field of software quality assurance.

References

1. Jolate, P.: CMM in Practice. Person Education Pte. Ltd., Delhi (2004)
2. Mayers, G.J.: The Art of Software Testing. Wiley, USA (1997)
3. I.S.G. of Software Engg. Terminology, IEEE Standards Collection, IEEE Std 610.12 1990 (1990)
4. Pedrycz, W., Peters, J.F.: Comput. Intell. Softw. Eng. World Scientific, Singapore (1998)
5. McMinn, P.: Search-based software test data generation: a survey. Softw. Test. Verif. Reliab. **14**, 212–223 (2004)
6. Harman, M.: The current state and future of search based software engineering. In: International Conference on Software Engineering. Future of Software Engineering, pp. 324–357. IEEE Computer Society Press, Washington, DC, USA (2007)
7. Dorigo, M., Stutzle, T.: Ant Colony Optimization. MIT press, USA (2005)
8. Ayari, K., Bouktif, S., Antoniol, G.: Automatic mutation test input data generation via ant colony. In: Genetic and Evolutionary Computation Conference, pp. 1074–1081, London, UK (2007)
9. Dorigo, M., Birattari, M., Stutzle, T.: Ant Colony Optimization. IEEE Comput. Intell. Mag. **1** (4), 28–39 (2006)
10. Li, H.Z., Peng, L.A.M.: An ant colony optimization approach to test sequence generation for state-based software testing. Proceedings of the Fifth International Conference on Quality Software(QSIC'05). pp. 255-264 (2005)
11. Caro, G.D., Dorigo, M.: AntNet: distributed stigmergetic control and communication networks. J. Artif. Intell. Res. **9**, 317–365 (1998)
12. Dorigo, M., Maniezzo, V., Colorni, A.: Ant System: optimization by a colony of cooperating agents. IEEE Trans. Syst. Man Cybern. Part B Cybern. **26**(1), 29–41 (1996)
13. Gomez, O., Baren, B.: Omicron ACO: a new ant colony optimization algorithm. CLEI Electron. J. **8**(1), 1–8 (2005)
14. Huaizhong, L., Lam, C.P.: Software test data generation using ant colony optimization. Proc. World Acad. Sci. Eng. Technol. **1**, 1–4 (2005)
15. Li, L., Ju, S., Zhang, Y.: Improved ant colony optimization for the traveling salesman problem. In: Proceedings of 1st International Conference on Intelligent Computation Technology and Automation, pp. 76–80 (2008)

16. Parpinelli, R.S., Lopes, H.S., Freitas, A.A.: Data mining with an ant colony optimization algorithm. IEEE Trans. Evol. Comput. **6**(4), 321–332 (2002)
17. Gambardella, L.M., Dorigo, M.: Ant colony system hybridized with a new local search for the sequential ordering problem. INFORMS J. Comput. **12**(3), 237–255 (2000)
18. Zhao, P., Zhang, X.: New ant colony optimization for the knapsack problem. In: Proceedings of the 7th International Conference on Computer-Aided Industrial Design and Conceptual Design, pp. 1–3 (2006)
19. Zhang, W.J.: Selforganizology: a science that deals with self-organization. Netw. Biol. **3**(1), 1–14 (2013)
20. Zhang, W.J.: Self-Organization: Theories And Methods. Nova Science Publishers, New York, USA (2013)
21. Lopez-Riquelme, G., Malo, E.A., Lopaz, C., Fanyal-Moles, M.L.: Antennal alfactory sensitivity in response to task-related odours of the ant Atta maxicana. Physiol. Entomol. **31** (4), 353–360 (2006)
22. Styrsky, J.D., Ebunks, M.D.: Ecological consequences of interactions between ants and honeydew producing insects". Proc. Biol. Sci. **274**(1067), 151–164 (2007)

Modified Gbest Artificial Bee Colony Algorithm

Pawan Bhambu, Sangeeta Sharma and Sandeep Kumar

Abstract Artificial Bee Colony (ABC) algorithm is considered an efficient nature inspired algorithm to solve continues unconstraint optimization problems. It was developed by taking inspiration from food foraging behavior of honey bee. To get better speed of convergence in ABC, Zhu and Kwong anticipated an enhanced version of ABC, namely Gbest-guided ABC (GABC). But, both the algorithms, i.e., GABC and ABC could not perform well for solving constraints optimization problems. In this paper, a variant of GABC is proposed that may be able to solve constraint optimization problems as well as unconstraints optimization problems. In the proposed variant, namely modified GABC (MGABC), a strategy is proposed which adjusts the step size of the solutions, iteratively during the global optima search process. The competence and toughness of the newly anticipated MGABC are measured by testing it over 8 real-world complex optimization problems. The simulated results are compared with ABC, best-so-far ABC, GABC and modified ABC.

Keywords Constraint optimization problems · Nature inspired algorithm
Swarm intelligence · Optimization techniques · Soft computing

1 Introduction

The Artificial Bee Colony (ABC) algorithm was developed in 2005 by Karaboga [1]. The bee colony includes two types of swarm of bees, namely employed bees and unemployed bees. The employed bees gather nectar from the sources of food. If food source associated to a bee get exhausted then that bee is converted to the scout

P. Bhambu · S. Sharma
Department of Computer Science & Engineering, Arya College of Engineering & IT,
Jaipur, India

S. Kumar (✉)
Faculty of Engineering & Technology, Jagannath University, Jaipur, India
e-mail: sandpoonia@gmail.com

© Springer Nature Singapore Pte Ltd. 2018
M. Pant et al. (eds.), *Soft Computing: Theories and Applications*,
Advances in Intelligent Systems and Computing 583,
https://doi.org/10.1007/978-981-10-5687-1_59

bee. The scout bee searches the food sources in different directions. The employed bee and onlooker bee phases are accountable for the intensification of the solutions in the feasible zone while scout bee phase is responsible for diversification of the solutions. For efficient searching, a proper balance is required between intensification and diversification properties in any swarm intelligence based algorithm. In ABC, the position update process in both the phases is highly depends on random components therefore, it always tends to explore the search space at the cost of chance to skip true solution.

Researchers are continuously improving the performance of ABC algorithm while trying to establish a proper balance between the intensification and diversification properties. In order to improve the exploitation, Gao et al. [2] proposed a new search strategy in which the bee searches just in proximity of the best solutions from preceding round by improving position update equation of ABC. Bansal et al. assimilated local search stratagem in ABC to improve the intensification potential of it [3]. Similar efforts are done by Sharma et al. in [4–6]. Banharnsakun et al. [7] suggested three key amendments in the best-so-far selection in ABC algorithm (BSFABC): The best-so-far method (collect information in relation to the finest solutions established up to now), an adaptable search radius (change radius of search in each iteration), and an objective value-based comparison in ABC (most of the algorithm compare fitness of function). Karaboga and Akay [8] used a different method in ABC strategy to get rid of constrained optimization problems. The newly anticipated strategy was named as Modified ABC (MABC) algorithm. In this sequence, in 2010, Zhu and Kwong [9] anticipated a new variant of ABC algorithm that is Gbest-guided ABC (GABC). The GABC algorithm improves the exploitation capabilities of basic ABC technique using information about best feasible solution in whole swarm that is termed as global best (Gbest) solution and modifies the position update equation. In case of GABC all the entities inspired from the global best solution that leads to problem of stagnation. The proposed GABC performed well for the unconstraints optimization problems but fails to establish its competitiveness to solve the constraints optimization problems. Therefore, in this paper, a new variant of GABC algorithm, namely Modified GABC (MGABC) is presented for the constraints optimization problems. The proposed strategy is tested over 8 problems. Through intensive statistical analysis, it is claimed that the MGABC is a competitive variant of ABC algorithm while solving complex real-world problems.

Section 2 of this paper explains basic ABC algorithm in detail. Section 3 explains the proposed MGABC algorithm. Section 4 discuss result for MGABC and analyze performance of the newly anticipated with the help of experiments. Section 5 concludes the paper.

2 Artificial Bee Colony (ABC) Algorithm

The key steps of ABC algorithm are summarized as follow:
Initialization of the swarm

$$X_{ij} = X_{minj} + rand[0, 1](X_{maxj} - X_{minj}) \tag{1}$$

Here the ith food source in the swarm represented by x_i, the lower and upper bounds of x_i in jth dimension are $x_{min\ j}$ and $x_{max\ j}$, respectively and $rand[0, 1]$ is an evenly scattered arbitrary number in the range [0, 1].
Employed bee phase

$$X_{newij} = X_{ij} + \phi_{ij}(X_{ij} - X_{kj}) \tag{2}$$

This phase updates position of ith candidate solution. Here $k \in \{1, 2, ..., SN\}$ and $j \in \{1, 2, ...,D\}$ are arbitrarily selected indicators and $k \neq i.\varphi_{ij}$ is an arbitrary number in the range $[-1, 1]$.
Onlooker bee phase: Based on fitness of food sources onlooker bees decides that which food sources are most feasible and select a solution with probability $prob_i$. Here $prob_i$ may be decided with the help of fitness (there may be some other):

$$prob_i(G) = 0.9 \times \frac{fitness_i}{max\ fit} + 0.1 \tag{3}$$

Here $fitness_i$ represents the fitness value of the ith solution and max fit represents the highest fitness among all the solutions. Onlooker bees also select a solution and update it according to their probability of selection.
Scout bee phase: In this phase a food source considered as discarded and replaced by arbitrarily generated new solution if its location is not getting updated for a certain number of cycles.

3 Modified Gbest ABC Algorithm

It is proved in literature that the ABC algorithm is performing well for continues unconstraints optimization problems. The performance of ABC algorithm is further enhanced by Zhu and Kwong [9] in their proposed variant of ABC, namely GABC algorithm. But due to inefficiency of ABC to solve the constraints optimization problems, Karaboga and Akay [8] developed a variant of ABC for constraints optimization problems, namely modified ABC (MABC). In similar way, in this paper, a variant of GABC algorithm, namely Modified GABC (MGABC) is proposed to deal with the constraints as well as unconstraint problems. In the proposed MGABC algorithm, position update process of GABC algorithm is modified in employed as well as onlooker bee phase. In GABC, individuals update the positions using following equation:

$$xnew_{ij} = x_{ij} + \phi_{ij}(x_{ij} - x_{kj}) + \psi_{ij}(G_j - x_{ij}) \tag{4}$$

Here, ϕ_{ij} is a universal random number $\in [-1, 1]$, Ψ_{ij} is an evenly distributed arbitrary number in $[0, C]$ a positive constant C, G_j is the jth dimension of the current best solution, x_k is a neighboring solution, and x_i is the solution going to modify its position. It may be easily observed from position update equation of GABC (refer Eq. 4) that the solutions will be attracted towards the current best solution during the position update process. This may speed up the convergence speed with fear of premature convergence and stagnation. We know that a proper balance is required between the diversification and intensification properties of any search algorithm for better search process and diversification of the solutions to cover the search space is achieved by large step size whereas intensification of the solutions in the identified search area is achieved using small step size. To balance the step size of the solutions, the proposed MGABC modify the position update equations of employed as well as onlooker bee phase as follows:

(1) In employed bee phase of MGABC, following position update process is applied:

for $j = 1$ to D do

if$(prob_i \leq \psi_{ij})$ then

$$xnew_{ij} = x_{ij} + \phi_{ij}(x_{ij} - x_{kj}) + \frac{(1 - t)}{T} \times \psi_{ij}(y_j - x_{ij})$$

else

$$xnew_{ij} = x_{ij}$$

Here, t is a current iteration counter whereas T is the total iterations. In this solution search process, the part $B(\psi_{ij}(y_j - x_{ij}))$ influences to the solutions towards the global optima, hence improves the convergence speed while part $A(\phi_{ij}(x_{ij} - x_{kj}))$ includes the stochastic nature in the search process. As, in this process, in one iteration, depending on the probability, which is a function of fitness, multiple dimensions of the solution are changed as well as iteratively, the weightage to part B is reduced, hence, it will get better the diversification potential of the MGABC algorithm.

(2) In Onlooker bee phase of MGABC, following position update process is applied:

Select a random dimension $j = \cup(1, D)$

$$xnew_{ij} = x_{ij} + \frac{(1 - t)}{T} \times \phi_{ij}(x_{ij} - x_{kj}) + \psi_{ij}(y_j - x_{ij})$$

Here, symbols have their usual meaning. In this solution search process, weight to the part B is iteratively reduced, i.e., now search process is more biased by the part A. Therefore, this will get better the intensification potential of the MGABC algorithm.

4 Experimental Results

The performance of the MGABC algorithm is evaluated in this section.

4.1 Considered Test Problems

To authenticate the efficiency of the projected MGABC, 8 complex optimization problems (f_1-f_8) among which 05 problems $(f_1, f_2, f_4, f_6, f_7)$ are unconstraints problems and (f_3, f_5, f_8) are constraints problems, are selected to check the performance of the anticipated MGABC. The considered problems are as follows:

Problem 1 (Neumaier 3 Problem (NF3)) This is a minimization problem of 10 dimensions as shown below:

$$f_1(x) = \sum_{i=1}^{D} (x_i - 1)^2 - \sum_{i=2}^{D} x_i x_i - 1$$

The search boundary for the variables is $[-D^2, D^2]$. The best known solution is $f(\overline{0}) = -(D \times (D+4)(D-1))/6.0$. The acceptable error for a successful run is fixed to be 1.0E−01 [10].

Problem 2 (Colville Function) This is a minimization problem of four dimensions as shown below:

$$f_2(x) = 100(x_2 - x_1^2)^2 + (1 - x_1)^2 + 90(x_4 - x_3^2)^2 + (1 - x_3)^2$$
$$+ 10.1[(x_2 - 1)^2 + (x_4 - 1)^2] + 19.8(x_2 - 1)(x_4 - 1)$$

The search boundary for the variables is $[-10, 10]$. The most feasible solution is $f(\overline{1}) = 0$. The acceptable error for a successful run is fixed to be 1.0E−05 [10].

Problem 3 (Compression Spring) The compression spring problem [11] concerns with minimization of the compression spring's weight, subject to some constraints. There are three main design variables for a compression spring: the diameter of wire x_1, the mean diameter of coil x_2, and the number of active coils x_3. A simple explanation of compression spring problem explained here. This problem mathematically formulated as follow:

$x_1 \in \{1, ..70\}$ with granularity $1, x_2 \in [0.6, 3], x_3 \in [0.207, 0.5]$ with granularity 0.001

and four constra int $g_1 := \dfrac{8c_f F_{max} x_2}{\pi x_3^3} - S \leq 0, g_2 := l_f - l_{max} \leq 0,$

$g_3 := \sigma_p - \sigma_{pm} \leq 0, g_4 := \sigma_w - \dfrac{F_{max} - F_p}{K} \leq 0$

where

$c_f = 1 + 0.75 \dfrac{x_3}{x_2 - x_3} + 0.615 \dfrac{x_3}{x_2}, F_{max} = 1000, S = 189000, l_f = \dfrac{F_{max}}{K} + 1.05(x_1 + 2)x_3,$

$l_{max} = 14, \sigma_p = \dfrac{F_p}{K}, \sigma_{pm} = 6, F_p = 300, K = 11.5 \times 10^6 \dfrac{x_3^4}{8x_1 x_2^3}, \sigma_w = 1.25$

The most feasible solution is $f(7, 1.386599591, 0.292) = 2.6254$. Here tolerable error is fixed to be 1.0E−04.

Problem 4 (Moved axis parallel hyper-ellipsoid This is a minimization problem of 30 dimensions as shown below:

$$f_4(x) = \sum_{i=1}^{D} 5ix_i^2$$

The search boundary for the variables is [−5.12, 5.12]. The best known solution is $f(x) = 0$; $x(i) = 5i$, $i = 1, ... D$. The acceptable error for a successful run is fixed to be 1.0E−15 [10].

Problem 5 (Pressure Vessel design without Granularity) The problem of pressure vessel design formulated as follows:

$$f_5(x) = 0.6224x_1 x_3 x_4 + 1.7781x_2 x_3^2 + 3.1611x_1^2 x_4 + 19.84x_1^2 x_3$$

subject to

$$g_1(x) = 0.0193x_3 - x_1 \leq 0, g_2(x) = 0.00954x_3 - x_2 \leq 0,$$

$$g_3(x) = 750 \times 1728 - \pi x_3^2 \left(x_4 + \frac{4}{3}x_3\right) \leq 0$$

where x_1, x_2, x_3 and x_4 are thickness of shell, thickness of spherical head, radius of cylindrical shell and shell length. The search boundaries for the variables are $1.125 \leq x_1 \leq 12.5$, $0.625 \leq x_2 \leq 12.5$, $1.0 \times 10^{-8} \leq x_3 \leq 240$ and $1.0 \times 10^{-8} \leq x_4 \leq 240$. The most feasible solution is $f(1.125, 0.625, 58.29016, 43.69266) = 7197.729$ [12]. Acceptable error for a successful run is fixed to be 1.0E−05.

Problem 6 (Lennard–Jones) The Lennard–Jones (LJ) problem is minimization problem. It minimizes a type of potential energy of a set of N atoms. In case of LJ problem the dimension of the search space is 3 N, as ith atom's position X_i has three coordinates. Practically the ones of the X_i form coordinates of a point X. Precisely, it can be written as $X = (X_1, X_2, ...,X_N)$, and we have then

$$f_6(x) = \sum_{i=1}^{N-1} \sum_{j=i+1}^{N} \left(\frac{1}{\left\| X_i - X_j \right\|^{2\alpha}} - \frac{1}{\left\| X_i - X_j \right\|^{\alpha}} \right)$$

Here $N = 5$, $\alpha = 6$, and $[-2, 2]$ is considered as search space [13].

Problem 7 (Frequency-Modulated sound wave) It is a 6D optimization problem where the vector to be optimized is $\overline{X} = \{a_1, w_1, a_2, w_2, a_3, w_3\}$. The problem is to produce a sound same as objective. Its minimum value is $f(X_{\text{sol}}) = 0$. The objective sound waves are given as:

$$y(t) = a_1 \sin(w_1 t\theta + a_2 \sin(w_2 t\theta + a_3 \sin(w_3 t\theta)))$$

$$y_0(t) = (1.0) \sin((5.0)t\theta - (1.5) \sin((4.8)t\theta + (2.0) \sin((4.9)t\theta))),$$

respectively, where $\theta = 2\pi/100$ and the parameters are defined in the range $[-6.4, 6.35]$. A run is considered as successful if error is less than $1.0E-05$. The fitness function is defined as follows:

$$f_7(x) = \sum_{i=0}^{100} (y(t) - y_0(t))^2$$

Problem 8 (Welded beam design optimization problem) The welded beam design problem required to minimize cost, subject to some constraints [14]. The main goal is to reduce the cost of fabricating the welded beam subject to constraints on bending stress σ, shear stress τ, buckling load P_c, end deflection δ, and side constraint. x_1, x_2, x_3 and x_4 are four important design variables.

$$f_8(x) = 1.1047x_1^2 x_2 + 0.04822x_3 x_4(14.0 + x_2)$$

Subject to:

$$g_1(x) = \tau(x) - \tau_{\max} \leq 0, g_2(x) = \sigma(x) - \sigma_{\max} \leq 0, g_3(x) = x_1 - x_4 \leq 0,$$
$$g_4(x) = \delta(x) - \delta_{\max} \leq 0, g_5(x) = P - P_c(x) \leq 0$$

$$0.125 \leq X_1 \leq 5, 0.1 \leq X_2, X_3 \leq 10 \text{ and } 0.1 \leq X_4 \leq 5$$

where

$$\tau(x) = \sqrt{\tau'^2 - \tau'\tau''\frac{x_2}{R} + \tau''^2}, \tau' = \frac{P}{\sqrt{2}x_1x_2}, \tau'' = \frac{MR}{J}, M = P\left(L + \frac{x_2}{2}\right), \sigma(x) = \frac{6PL}{x_4x_3^2},$$

$$\delta(x) = \frac{6PL^3}{Ex_4x_3^2}, P_c(x) = \frac{4.013Ex_3x_4^3}{6L^2}\left(1 - \frac{x_3}{2L}\sqrt{\frac{E}{4G}}\right), R = \sqrt{\frac{x_2^2}{4} + \left(\frac{x_1 + x_2}{2}\right)^2},$$

$$J = \frac{2}{\sqrt{2}x_1x_2\left[\frac{x_2^2}{4} + \left(\frac{x_1 + x_2}{2}\right)^2\right]}$$

P = 6000 lb, L = 14 in., δ_{max} = 0.25 in., σ_{max} = 30,000 psi, τ_{max} = 13600 psi, $E = 30 \times 106$ psi, $G = 12 \times 106$ psi.

The most suitable solution is f(0.205730, 3.470489, 9.036624, 0.205729) = 1.724852. The maximum permissible error is 1.0E−01.

4.2 Experimental Setting

The considered test problems (TP) are solved using a penalty function strategy. The penalty function approach redefine the original problem to by adding a penalty term leads defining unconstrained optimization problem in case of constraints breach as described here:

$$f(x) = f(x) + \beta \tag{5}$$

where $f(x)$ is the basic function value and β is the penalty term (here 10^3). The experimental results of MGABC analyzed by comparing results with MABC [15], BSFABC [7], GBestABC [9] and basic ABC. Comparison based on success rate (SR), mean error (ME), standard deviation (SD) and average number of function evaluations (AFE). For the experiments, following parameter setting is adopted.

- Simulations = 100, Number of solutions SN = 25 [6],
- Limit = $D \times SN$ [9], C = 1.5 [18], α = 30.
- The terminating criteria: Acceptable error or achieve upper limit of function evaluations (predefined as 200000),

The parameter settings for other algorithms taken from their original paper.

4.3 Analysis of Experimental Results

Numerical results for problems f_1–f_8 are shown in Tables 1, 2, 3 and 4 by means of the recommended experimental settings as discussed in previous subsection. After in depth analysis of results, it can be observed that MGABC algorithm is superior

Table 1 Comparison based on AFE, TP: test problem

TP	ABC	BSFABC	GABC	MABC	MGABC
f_1	198449.49	200022.9	127452	199685.01	39512.26
f_2	198798.6	163406.61	149232.24	148585.62	15162.67
f_3	194506.44	199868.29	178018.13	181138.43	107684.61
f_4	62508.5	71352	59902	39639.5	28612
f_5	200024.75	200031.95	200026.39	200025.02	200023.36
f_6	70769.21	153149.56	200031.9	97169.96	91955.94
f_7	200032.98	200031.63	199853.34	190675.86	112249.61
f_8	200018.02	53813.47	31698.69	126909.03	4329.3

then other competitive algorithms as reduce ME (thus it is more accurate), improve SR (thus more reliability) and reduce AFE (results in more efficient algorithm). In order to prove that newly proposed algorithm is better than other algorithm a few additional analytical test like the Mann–Whitney U rank sum test, boxplots analysis and acceleration rate (AR) [16] have been done.

4.4 Statistical Analysis

The numerical results of MGABC are compared with basic ABC, BSFABC, GABC and MABC based on AFE, ME, SD and SR to prove that performance of MGABC is better than these considered algorithms. It can be easily observed from the results shown in Table 1 that MGABC costs less in terms of AFE on 6 test functions ($f_1, f_2,$ f_3, f_4, f_7 and f_8) among all the considered algorithms. For f_6, AFE of MGABC is less that the all the considered algorithm except ABC. The f_5 problem could not be solved by any of the considered algorithms; therefore the algorithms are compared on basis of mean error. It is clear from Table 3 that the error reported by MGABC is significantly low than the other considered algorithms for this problem. The boxplots [17] for AFE have been depicted in Fig. 1. Boxplot shown in Fig. 1 can proficiently characterize the experimental distribution of outcomes. Figure 1 demonstrates that MGABC algorithm is very efficient.

It can be easily observed from box plots that MGABC is very efficient in comparison to ABC, BSFABC, GABC and MABC, i.e., MGABC's has different results from the other considered algorithms. In order to prove that new algorithm is really good a non-parametric statistical test named Mann–Whitney U rank sum [18], a non-parametric test performed for AFE. This test compares non-Gaussian data. Here level of significance is taken 0.05 (α). Test performed between MGABC–ABC, MGABC–BSFABC, MGABC–GABC, and MGABC–MABC. The outcomes of the Mann–Whitney U rank sum test for the AFE of 100 runs are shown in Table 5. In Table 5 the important variation observed first, i.e., is there significant difference between two data sets. If the null hypothesis is accepted then sign '=' appears and when the null hypothesis is rejected, then perform comparison

Table 2 Comparison based on SR, TP: test problem

TP	ABC	BSFABC	MABC	GABC	MGABC
f_1	3	0	96	2	100
f_2	2	38	50	46	100
f_3	4	1	22	20	74
f_4	100	100	100	100	100
f_5	0	0	0	0	0
f_6	100	89	0	83	95
f_7	0	0	1	14	59
f_8	0	99	100	65	100

Table 3 Comparison based on ME, TP: test problem

TP	ABC	BSFABC	MABC	GABC	MGABC
f_1	7.95E−01	3.53E+00	1.00E−01	1.51E+00	9.87E−02
f_2	1.63E−01	2.24E−02	1.29E−02	1.63E−02	8.18E−03
f_3	1.38E−02	2.91E−02	4.70E−03	8.69E−03	1.93E−03
f_4	9.24E−16	6.62E−16	9.19E−16	9.27E−16	9.14E−16
f_5	1.85E+01	2.30E+01	1.55E+01	5.40E+00	2.68E−01
f_6	8.52E−04	9.89E−04	4.51E−01	1.09E−03	9.05E−04
f_7	5.63E+00	9.95E+00	2.69E+00	3.51E+00	4.46E+00
f_8	2.45E−01	9.56E−02	9.47E−02	9.98E−02	9.46E−02

Table 4 Comparison based on SD, TP: test problem

TP	ABC	BSFABC	MABC	GABC	MGABC
f_1	6.71E−01	3.87E+00	1.83E−02	1.72E+00	2.24E−03
f_2	9.42E−02	2.17E−02	9.94E−03	1.25E−02	1.84E−03
f_3	1.11E−02	6.06E−03	5.89E−03	9.56E−03	3.02E−03
f_4	7.65E−17	2.44E−16	7.48E−17	7.45E−17	7.75E−17
f_5	1.11E+01	2.01E+01	9.74E+00	3.36E+00	1.81E−01
f_6	1.44E−04	7.00E−04	1.46E−01	6.98E−04	2.25E−04
f_7	5.27E+00	5.11E+00	3.09E+00	5.06E+00	5.43E+00
f_8	7.99E−02	5.13E−03	4.90E−03	1.06E−02	4.79E−03

between the AFEs. If MGABC takes less AFE then put '+' sign and put '−' sign for more AFE than the other considered algorithms. In Table 5, '+' represent that MGABC is considerably superior and '−' demonstrate that MGABC is notably of inferior quality. Out of 32 there are 27 '+' signs in Table 5. It indicates that the outcomes of MGABC are efficient than ABC, MABC, GABC, and BSFABC for measured test problems.

Fig. 1 AFE representation through boxplots

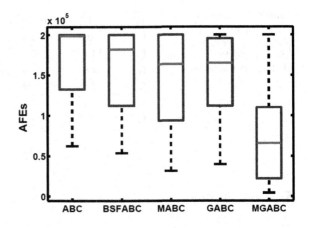

Table 5 The Mann–Whitney U rank sum test for AFE, TP: test problem

TP	Mann–Whitney U rank sum test with MGABC			
	ABC	BSFABC	GABC	MABC
f_1	+	+	+	+
f_2	+	+	+	+
f_3	+	+	+	+
f_4	+	+	+	+
f_5	=	=	=	=
f_6	−	+	+	+
f_7	+	+	+	+
f_8	+	+	+	+

Next, the convergence speed of MGABC measured. If AFE for some algorithm is small it means that it has high convergence speed. The convergence speed calculated by measuring the AFEs. The acceleration rate (AR) is used to compare speed of convergence, which is defined in this way; it depends on AFE of two algorithms ALGO and MGABC:

$$AR = \frac{AFE_{ALGO}}{AFE_{MGABC}} \tag{6}$$

where ALGO $\in$ {ABC, BSFABC, GABC, and MABC} and AR > 1 indicate that MGABC is quicker other. The AR is measured for AFEs of the considered algorithms using (6). Table 6 demonstrates a comparison between MGABC and ABC, MGABC and BSFABC, MGABC and GABC, and MGABC and MABC in terms of AR. Table 6 demonstrates that the speed of convergence for MGABC is good than measured algorithms for most of the functions.

Table 6 Acceleration rate (AR) of MGABC as compared to the ABC, BSFABC, GABC, and MABC, TP: test problems

TP	ABC	BSFABC	GABC	MABC
f_1	5.02	5.06	3.23	5.05
f_2	13.11	10.78	9.84	9.80
f_3	1.81	1.86	1.65	1.68
f_4	2.18	2.49	2.09	1.39
f_5	1.00	1.00	1.00	1.00
f_6	0.77	1.67	2.18	1.06
f_7	1.78	1.78	1.78	1.70
f_8	46.20	12.43	7.32	29.31

5 Conclusions

This paper presents an improved variant of Gbest-Guided ABC (GABC) algorithm, namely Modified GABC (MGABC). The proposed variant is developed in view to solve the constraints as well as unconstraints optimization problems. In the proposed variant, position update process of GABC algorithm in employed as well as in onlooker bee phase is modified. In employed bee, to introducing fluctuations, multiple dimensions are allowed to update in a single iterations. The proposed is tested over 8 constraints as well as unconstraints optimization problems and compared with ABC, MABC, BSFABC and GABC algorithms. Through statistical analysis, it is proved that MGABC is a competitive variant of GABC algorithm.

References

1. Karaboga, D.: An idea based on honey bee swarm for numerical optimization. Technical Report TR06, Erciyes University Press, Erciyes (2005)
2. Gao, W.F., Liu, S.Y.: A modified artificial bee colony algorithm. Comput. Oper. Res. **39**(3), 687–697 (2012)
3. Bansal, J.C., Sharma, H., Arya, K.V., Nagar, A.: Memetic search in artificial bee colony algorithm. Soft. Comput. **17**(10), 1911–1928 (2013)
4. Sharma, H., Bansal, J.C., Arya, K.V.: Opposition based lvy flight artificial bee colony. Memet. Comput. **5**(3), 213–227 (2013)
5. Sharma, H., Bansal, J.C., Arya, K.V.: Power law-based local search in artificial bee colony. Int. J. Artif. Intell. Soft Comput. **4**(2–3), 164–194 (2014)
6. Sharma, H., Bansal, J.C., Arya, K.V., Yang, X.S.: Lvy flight artificial bee colony algorithm. Int. J. Syst. Sci. **47**(11), 2652–2670 (2016)
7. Banharnsakun, A., Achalakul, T., Sirinaovakul, B.: The best-so-far selection in artificial bee colony algorithm. Appl. Soft Comput. **11**(2), 2888–2901 (2011)
8. Karaboga, D., Akay, B.: A modified artificial bee colony (ABC) algorithm for constrained optimization problems. Appl. Soft Comput. **11**(3), 3021–3031 (2011)
9. Zhu, G., Kwong, S.: Gbest-guided artificial bee colony algorithm for numerical function optimization. Appl. Math. Comput. **217**(7), 3166–3173 (2010)

10. Ali, M.M., Khompatraporn, C., Zabinsky, Z.B.: A numerical evaluation of several stochastic algorithms on selected continuous global optimization test problems. J. Glob. Optim. **31**(4), 635–672 (2005)

11. Onwubolu, G.C., Babu, B.V.: New Optimization Techniques In Engineering, vol. 141. Springer (2013)

12. Wang, X., Gao, X.Z., Ovaska, S.J.: A simulated annealing-based immune optimization method. In: Proceedings of the International and Interdisciplinary Conference on Adaptive Knowledge Representation and Reasoning, pp. 41–47. Porvoo, Finland (2008)

13. Clerc, M.:. List based PSO for real problems. http://clerc.maurice.free.fr/pso/ListBasedPSO/ListBasedPSO28PSOsite29.pdf. Accessed 16 July 2012

14. Ragsdell, K.M., Phillips, D.T.: Optimal design of a class of welded structures using geometric programming. J. Eng. Ind. **98**(3), 1021–1025 (1976)

15. Akay, B., Karaboga, D.: A modified artificial bee colony algorithm for real-parameter optimization. Inf. Sci. **192**, 120–142 (2012)

16. Rahnamayan, S., Tizhoosh, H.R., Salama, M.M.: Opposition-based differential evolution. IEEE Trans. Evol. Comput. **12**(1), 64–79 (2008)

17. Williamson, D.F., Parker, R.A., Kendrick, J.S.: The box plot: a simple visual method to interpret data. Ann. Int. Med. **110**(11), 916–921 (1989)

18. Mann, H.B., Whitney, D.R.: On a test of whether one of two random variables is stochastically larger than the other. The Ann. Math. Stat. **18**(1), 50–60 (1947)

Mathematical Study of Queue System with Impatient Customers Under Fuzzy Environment

Reeta Bhardwaj, T.P. Singh and Vijay Kumar

Abstract Queuing models with impatient customers have wide variant of applications in modeling such as business and industries. Impatient customers affect the progress and sustainability of any business. Therefore, it is necessary to study the behavior of such customers. In the present study, authors analyze a queue system having two queues in series with reneging customers. In real-world scenario it has been observed that the arrival pattern, service rates and reneging rates of customers are uncertain, i.e., in linguistic form may be slow, moderate, fast, etc. The amount of uncertainty in a system can be reduced by using fuzzy logic because it offers better capabilities to handle linguistic uncertainties by modeling vagueness and unreliability of information. Hence, fuzzy concept has been introduced which results the queue characteristics as a fuzzy expression. It has also been assumed that the service rates as well as the reneging rates depend upon their respective numbers. A numerical illustration is given at the end to clarify the model.

Keywords Fuzzy logic · Reneging rate · Queuing · Impatient customer Stochastic behavior

1 Introduction

The particular interest in this paper is reneging behavior of customers in fuzzy situation. In queuing literature, a customer is said to have reneged if it joins a system but departs without completely receiving service. In our daily life, reneging

R. Bhardwaj (✉) · V. Kumar
Amity University Haryana, Gurgaon 122413, Haryana, India
e-mail: bhardwajreeta84@gmail.com

V. Kumar
e-mail: vkb1605@gmail.com

T.P. Singh
Yamuna Institute of Engineering & Technology, Yamunanagar, India
e-mail: tpsingh78@yahoo.com

© Springer Nature Singapore Pte Ltd. 2018
M. Pant et al. (eds.), *Soft Computing: Theories and Applications*,
Advances in Intelligent Systems and Computing 583,
https://doi.org/10.1007/978-981-10-5687-1_60

is a commonly observed phenomenon. Kelly [1] finds insignificant place for impatient customers with random service in a queue system. But in modern society and in many real-world situations the impatient customers play a significant role in a service system. Practical situations occurs such as involving impatient telephone switch board customers, emergency patients attended by critical care professionals and the inventory system with storage of valuable goods and so on. To any system analyst, the implications are unfavorable as reneging implies not only economic loss in business but also psychological dissatisfaction among customers. Unsatisfied customers are often likely to spread the rumors or words about organization which brings ripple effect in longer time.

Haight [2] first of all considered the effect of reneging phenomenon on simple queue model. Ancker and Gafarian [3] carried out an early study on Markovian queue system and obtained the probability of balking, reneging customers and loss rate of the system. The reneging concept was incorporated in serial queue network by Singh [4] and obtained a transient solution of the model with stochastic parameters. Mishra and Mishra [5] explored the cost analysis of machine interference queue model with balking and reneging effect. Singh and Tyagi [6] discussed the behavioral aspect of a stochastic serial queue model and made an attempt to find the effect of reneging rate on variation of queue length and the probabilities variations with respect to reneging. Bhardwaj et al. [7] made an analytical study of service surrender queue system with fuzzy arithmetic. Kumar [8] discussed performance analysis of a markovian queuing system with reneging and retention of reneged customers. Recently Tyagi et al. [9] and Singh et al. [10] discussed a serial queue model with balking and reneging connected with non-serial queue channel.

In the present study authors assumed that all parameters of model such as arrival rate, service rate and reneging rates are fuzzy in nature. The assumptions of fuzzy parameters are closer to real-world scenario. The present study is an extension work [9] in the sense that fuzzy logic has been introduced in modeling which results in driving queue characteristics as a fuzzy expression. The triangular membership function has been used to derive and analyze the characteristics of the system. The problem is solved on the basis of α cut. The numerical illustration demonstrates how the parameters of fuzzy queue model influence the optimal service rate of the system. The model has been dealt in literature in stochastic environment but in this case the environment is uncertain, having information deficiency, fragmentary and vague. The importance of study shows that it is practically more significant, reliable in a real-world scenario. The amount of uncertainty in the system has been reduced by using fuzzy logic because it offers better capabilities to handle linguistic uncertainties by modeling vagueness and unreliability of information.

1.1 Fuzzy Logic

A fuzzy set is simply an extension of classical set in which characteristic function is permitted to have any values between 0 and 1. The concept of fuzzy logic was first of all given by Zadeh [11]. The membership function for a fuzzy set A on the universe of discourse X is defined as $\mu_A : X \rightarrow [0, 1]$ where each element of X is mapped to a value between 0 and 1. This value is called degree of membership, quantifies the grade of membership of the element in X to fuzzy set A.

1.2 Symbol Used

$\tilde{\lambda}$ Fuzzy arrival rate

$\tilde{\mu_1}$ Fuzzy service rate for server S_1 and fuzzy arrival rate for server S_2

$\tilde{\mu_2}$ Fuzzy service rate for server S_2

$\tilde{r_1}$ Fuzzy reneging rate at S_1

$\tilde{\rho}$ Fuzzy busy time of the server

$\tilde{L}$ Expected number of customers in fuzzy system

1.3 Triangular Membership Function

A triangular membership function is specified by three parameters $<a, b, c>$ as follows:

$$\mu_{\tilde{A}}(x) = \begin{cases} x - a/b - a & a \le x \le b \\ c - x/c - b & b \le x \le c \\ 0 & \text{otherwise} \end{cases}$$

The parameters with $a < b < c$ determine the x-coordinates of the three corners of underlying triangular membership function.

α-cut: Let us define a fuzzy set $\tilde{A}$ on R and $0 \le \alpha \le 1$. Then **α-cuts** of the fuzzy set $\tilde{A}$ is the crisp set A contains all the elements of universe of discourse X whose membership grade in A are greater than or equal to specified value α. i.e., $A_\alpha = \{x/\mu_A(x) \ge \alpha, x \in X\} = \left\{L_{\tilde{A}}(\alpha), U_{\tilde{A}}(\alpha)\right\}$ where $L_{\tilde{A}}(\alpha)$ and $U_{\tilde{A}}(\alpha)$ represent the lower bound and upper bound of the α-cut of $\tilde{A}$, respectively.

2 Model Description

The mathematical model consists of servers S_1 and S_2 in series that serve customers in phase 1 and phase 2, respectively. The model follows the assumption that customers in phase 2 cannot be attended for the service at S_2 until he or she completes

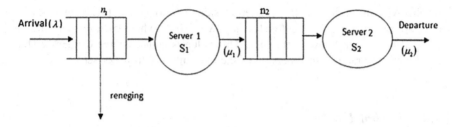

Fig. 1 Queuing model

the services provided in phase 1 at channel S_1. It is also considered that customers follow a Poisson stream with mean rate λ and form a queue in front of S_1, if it is busy. The output of S_1 becomes the input of S_2. It is also assumed that the service time at each service channel distributed exponentially with parameters μ_{n_1} and μ_{n_2} where μ_{n_1} and μ_{n_2} are defined by $\mu_{n_1} = \mu_1 n_1$ and $\mu_{n_2} = \mu_2 n_2$, i.e., μ_{n_1} and μ_{n_2} are directly proportional to their respective queue numbers n_1 and n_2 where μ_1 and μ_2 are known as proportionality constants (Fig. 1).

The customer leaves the system at S_1 after realizing that waiting time is intolerable for him i.e. reneges from queue at S_1. Considering the system as a whole, reneging will be a function of the system state, i.e., if the system is in state n_1, the reneging rate for the system is r_{n_1} where $r_{n_1} = r_1 n_1$ where r_1 is known as proportionality constant.

2.1 Solution Technique

  i First, we form the differential difference equations of the said model on the basis of probability argument. To solve equations we apply generating function technique and partial differential difference equation due to Lagrange.

  ii. The transient solution of the problem has been found on the basis of work done by [9].

 iii. We convert the transient solution into steady-state form and mean queue size of the system has been calculated through well known statistical formulae.

 iv. Then we fuzzified the model by taking the parameters in triangular form. Using α cut and fuzzy arithmetic, the mean queue length of the system is calculated. The inputs in Fuzzy are received by crisp input using fuzzification. Through α-cut, fuzzification methods are to be deployed. Initially, the rule base was formed in data base. These collectively are called knowledge base. Finally defuzzification process is carried out in order to produce crisp output. On the basis of these rules, suitable decisions are made in the decision making unit.

2.2 Application

The practical situation of the above models can be seen where service is done in sequential parts, e.g., assembly line of an industry in two stages or in a bank with two counters S_1 and S_2 where S_1, S_2 deal with customers in phase 1 and phase 2. In phase 1 customer handover cheque and get a token for payment for his phase 2 service. Customers with long awaited line, leave the queue at the first stage.

3 Formation of Queue Equations

Define $P_{n_1,n_2}(t)$ the joint probability at time t where $n_1 \geq 0$ and $n_2 \geq 0$ are units waiting in front of server S_1 and S_2 respectively. We shall start the analysis by attempting to find the state probabilities. Applying Markov process theory and connecting the various state probabilities at time $t + \delta t$ letting $\delta t \to 0$, the following set of time dependent differential difference equations for the model are obtained in the simplified form as

$$P'_{n_1,n_2}(t) = -(\lambda + \mu_1 n_1 + r_1 n_1 + \mu_2 n_2)P_{n_1,n_2}(t) + \lambda P_{n_1-1,n_2}(t)$$
$$+ \mu_1(n_1+1)P_{n_1+1,n_2-1}(t) + r_1(n_1+1)P_{n_1+1,n_2}(t) + \mu_2(n_2+1)P_{n_1,n_2+1}(t) \quad (1)$$

For $n_1 = 0$, $n_2 > 0$

$$P'_{0,n_2}(t) = -(\lambda + \mu_2 n_2)P_{0,n_2}(t) + \mu_1 P_{1,n_2-1}(t) + \mu_2(n_2+1)P_{0,n_2+1}(t) \quad (2)$$

For $n_2 = 0$, $n_1 > 0$

$$P'_{n_1,0}(t) = -(\lambda + \mu_1 n_1 + r_1 n_1)P_{n_1,0}(t) + \lambda P_{n_1-1,0}(t) + r_1(n_1+1)P_{n_1+1,0}(t) + \mu_2 P_{n_1,1}(t) \quad (3)$$

For $n_2 = 0$, $n_1 = 0$

$$P'_{0,0}(t) = -\lambda P_{0,0}(t) + \mu_2 P_{0,1}(t) \quad (4)$$

Using initial condition

$$P_{n_1,n_2}(t) = \begin{cases} 1 & (n_1, n_2 > 0) \\ 0 & \text{otherwise} \end{cases} \quad (5)$$

4   Solution of the Model

To solve equations from (1) to (4), we apply generating function technique as

$$G(x, y, t) = \sum_{n_1=0}^{\infty} \sum_{n_2=0}^{\infty} P_{n_1,n_2}(t) x^{n_1} y^{n_2} \tag{6}$$

$$G_{n_2}(x, t) = \sum_{n_1=0}^{\infty} P_{n_1,n_2}(t) x^{n_1} \tag{7}$$

By using Eqs. (6) and (7) becomes

$$G(x, y, t) = \sum_{n_2=0}^{\infty} G_{n_2}(x, t) y^{n_2} \tag{8}$$

Using generating function technique, and on the basis of solution methodology adopted by [4, 9], the ultimate solution of Eqs. (1)–(4) in transient state is stated as:

$$
P_{n_1,n_2}(t) = \frac{1}{n_1!} \left\{ \frac{\lambda}{\mu_2(\mu_1 + r_1)} \mu_1 \left(1 - e^{-\mu_2 t} + \frac{\mu_2 e^{-\mu_2 t}}{\mu_2 - \mu_1 - r_1} \right) \right\}^{n_1} \cdot \frac{1}{n_2!}
$$
$$
\left\{ \frac{\lambda}{\mu_2(\mu_1 + r_1)} \mu_2 \left(1 - e^{-(\mu_1 + r_1)t} + \frac{\mu_1 e^{-(\mu_1 + r_1)t}}{\mu_2 - \mu_1 - r_1} \right) \right\}^{n_2}
$$
$$
\exp\left[\frac{-\lambda}{\mu_2(\mu_1 + r_1)} \left[\mu_1 \left(1 - e^{-\mu_2 t} + \frac{\mu_2 e^{-\mu_2 t}}{\mu_2 - \mu_1 - r_1} \right) \right] + \mu_2 \left(1 - e^{-(\mu_1 + r_1)t} + \frac{\mu_1 e^{-(\mu_1 + r_1)t}}{\mu_2 - \mu_1 - r_1} \right) \right], \quad n_1, n_2 \geq 0
$$
$$\tag{9}$$

This is the joint probability at any time t, which is known as the required transition probability of the system.

4.1   Mean Queue Size

Let the mean queue size of the system be denoted by L. Then L as statistical formula is defined as

$$L = \sum_{n_1=0}^{\infty} \sum_{n_2=0}^{\infty} (n_1 + n_2) P_{n_1,n_2}(t)$$

Now putting $P_{n_1,n_2}(t)$ from Eq. (9), we have

$$
L = \frac{\lambda}{\mu_2(\mu_1 + r_1)} \left\{ \mu_1 \left(1 - e^{-\mu_2 t} + \frac{\mu_2 e^{-\mu_2 t}}{\mu_2 - \mu_1 - r_1} \right) \right\} + \frac{\lambda}{\mu_2(\mu_1 + r_1)} \left\{ \mu_2 \left(1 - e^{-(\mu_1 + r_1)t} + \frac{\mu_1 e^{-(\mu_1 + r_1)t}}{\mu_2 - \mu_1 - r_1} \right) \right\}
$$
$$\tag{10}$$

4.2 Steady-State Solution

The steady-state solution of the model can be found by letting $t \to \infty$. Hence, when $t \to \infty$, Eq. (9) gives us the corresponding steady-state probability, as

$$P_{n_1, n_2} = \frac{1}{n_1!} \left\{ \frac{\lambda \mu_1}{\mu_2(\mu_1 + r_1)} \right\}^{n_1} \frac{1}{n_2!} \left\{ \frac{\lambda \mu_2}{\mu_2(\mu_1 + r_1)} \right\}^{n_2} \exp \left[\frac{-\lambda(\mu_1 + \mu_2)}{\mu_2(\mu_1 + r_1)} \right] \quad (11)$$

Also the mean queue size L of the system in the steady-state case is as

$$L = \frac{\lambda(\mu_1 + \mu_2)}{\mu_2(\mu_1 + r_1)} \quad (12)$$

4.3 Special Cases

Case 1: When customers provided a constant service rate by servers with parameter $\mu_1 = \mu_2 = \mu$

$$L = \frac{2\lambda}{\mu + r_1}$$

Case 2: Mean system size with no reneging can be similarly derived by reconstructing the steady-state equation. In that case we have

$$L = \frac{\lambda(\mu_1 + \mu_2)}{\mu_1 \mu_2}$$

Case 3: As the customer's has random patient time, this implies that the reneging rate of the system would depend upon the state of system. The average reneging rate is given by

$$\text{Avg.r.r}_1 = \sum_{n_1=1}^{\infty} \sum_{n_2=1}^{\infty} (n_1 + n_2) r_1 P_{n_1 n_2}$$

$$= r_1 \sum_{n_1=1}^{\infty} \sum_{n_2=1}^{\infty} (n_1 + n_2) P_{n_1 n_2}$$

$$= r_1 \frac{\lambda(\mu_1 + \mu_2)}{\mu_2(\mu_1 + r_1)}$$

In system management, the customers who renege represent business loss. Authors' main interest is to find out the proportion of customers lost. The formula for this is given by

$$= \frac{\text{Avereage reneging rate}}{\lambda}$$

$$= \frac{r_1(\mu_1 + \mu_2)}{\mu_2(\mu_1 + r_1)}$$

4.4 Fuzzified Queue Model

In real scenario, it has been found that parametric distribution may only be characterized subjectively. Arrival, service pattern and reneging rate of customers are more suitably described in linguistic variable such as slow, moderate and fast. It is better to fuzzify the model. Fuzzy queue models are potentially more realistic than commonly used stochastic models.

$$\tilde{L} = \frac{\tilde{\lambda}(\tilde{\mu}_1 + \tilde{\mu}_2)}{\tilde{\mu}_2(\tilde{\mu}_1 + \tilde{r}_1)}$$

Using α-cut

$$\tilde{\lambda} = (3, 5, 7), \tilde{\mu}_1 = (9, 10, 11), \tilde{\mu}_2 = (14, 16, 18), \tilde{r}_1 = (4, 5, 6)$$

$$\lambda(\alpha) = (2\alpha + 3, 7 - 2\alpha), \mu_1(\alpha) = (\alpha + 9, 11 - \alpha),$$
$$\mu_2(\alpha) = (2\alpha + 14, 18 - 2\alpha), r_1(\alpha) = (\alpha + 4, 6 - \alpha)$$

Now using fuzzy arithmetic on above fuzzy numbers the mean queue length is given by (Fig. 2)

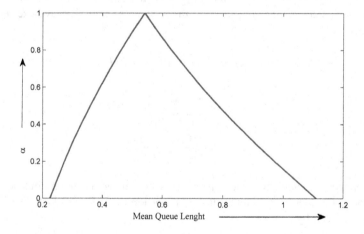

Fig. 2 Mean queue length versus α

Table 1 Mean queue length versus α

α	$L_\alpha = \left(L_\alpha^{\text{lower}}, L_\alpha^{\text{upper}}\right)$
0.0	(0.225, 1.11)
0.1	(0.249, 1.041)
0.2	(0.274, 0.971)
0.3	(0.301, 0.905)
0.4	(0.330, 0.843)
0.5	(0.360, 0.785)
0.6	(0.392, 0.730)
0.7	(0.426, 0.679)
0.8	(0.462, 0.630)
0.9	(0.500, 0.584)
1.0	(0.541, 0.541)

$$L_\alpha = \left[\frac{(2\alpha+3)(3\alpha+23)}{(18-2\alpha)(17-2\alpha)}, \frac{(7-2\alpha)(29-3\alpha)}{(2\alpha+14)(2\alpha+13)} \right] = \left[L_\alpha{}^{\text{lower}}, L_\alpha{}^{\text{upper}} \right]$$

5 Analysis of Table and Graph

With the help of MATLAB, we perform α cuts of arrival rate, service rate and reneging rate and fuzzy queue length at eleven distinct α levels $0.0, 0.1, \ldots, 1.0$. Crisp intervals for fuzzy mean queue length at different possibilistic α levels are presented in Table 1. α cut represent the possibility that the performance measure fuzzy expected length will lies in the associated range. Specially, $\alpha = 0$ the performance measure could appear and for $\alpha = 1$, the performance measure is likely to be. The most likely value of mean queue length L is 0.541 and its value is impossible to fall outside the range of 0.225 and 1.11. The above information will be very useful for designing a queuing system.

Similarly by using appropriate formulae, other performance measures such as fuzzy expected waiting time in queue and fuzzy waiting time in system can be calculated.

6 Conclusion

The present study emphasis on how fuzzy set theory is applicable to a queue model with impatient customers. The present model can be used in operations and service mechanism for evaluating the system performance. As the input parameters of the models are fuzzy in nature, output will also be in fuzzy. Moreover, the problem can also be transformed into crisp problem. We find that on increasing the value of α, the lower bound increases and upper bound decreases and therefore uncertainty in parameters decreases and the system become closer to crisp value, i.e., we reach closer to certainty. The α cut limits the spread of membership function on the universe of discourse. Hence, the level of uncertainty is controlled and it discards low possibilities in the domain. The other queue characteristics such as fuzzy queue length and waiting time can also be converted into linguistic variable such as short, medium and large.

References

1. Kelly, F.P.: Reversibility and Stochastic Network. Wiley, New York (1979)
2. Haight, F.A.: Queuing with reneging. Biometric **52**, 186–197 (1959)
3. Ancker Jr., C.J., Gafarian, A.V.: Some queuing problems with balking and reneging-II. Oper. Res. **11**(6), 928–937 (1963)
4. Singh, T.P., Kumar, A.: On two queues in series with reneging. J. Indian Soc. Stat. Oper. Res. **6**, 31–39 (1985)
5. Mishra, S.S., Mishra, V.: The cost analysis of machine interference model with balking reneging and spares. Opsearch **41**, 35–46 (2004)
6. Singh, T.P., Tyagi, A.: Cost analysis of queue a system with impatient customer. Arya Bhatta J. Math. Inf. **6**(2), 309–312 (2014)
7. Bhardwaj, R., Singh, T.P, Kumar, V.: Analysis of service surrender queue model in fuzzy system. Int. J. Math. Oper. Res. (2017, in press)
8. Kumar, R.: Performance analysis of a markovian queuing system with reneging and retention of reneged customers. Advertising and Branding: Concepts, Meth., Tools, and App. **30**, 686–694 (2017)
9. Tyagi, A., Singh, T.P., Saroa, M.S.: Stochastic analysis of queue network with impatient customer. Int. J. Math. Sci. Eng. Appl. **8**(1), 337–347 (2014)
10. Singh, S., Singh, M., Taneja, G.: The study of serial channels connected to non-serial channel with feedback and balking in serial channels and reneging both types of channels. Arya Bhatta J. Math. Inf. **7**(2), 407–418 (2015)
11. Zadeh, L.A.: Fuzzy sets. Inf. Control **8**(3), 338–353 (1965)

Multi-item Fixed Charged Solid Shipment Problem with Type-2 Fuzzy Variables

Dhiman Dutta and Mausumi Sen

Abstract A multi-item fixed charged solid shipment model with criterion, e.g., shipment penalty, amounts, demands and carriages as type-2 triangular fuzzy variables with condition on few components and carriages is proposed here. With the critical value (CV) based contractions of corresponding type-2 fuzzy parameters, a nearest interval approximation model applying generalized credibility measure for the constraints is introduced for this particular model. An example is provided to explain the model and is then worked out by applying a gradient-based optimization—generalized reduced gradient technique (applying LINGO 16). The sensitivity analysis of the model is also given to illustrate the model.

Keywords Multi-item solid shipment problem · Fixed charge shipment problem Type-2 fuzzy sets · Nearest interval approximation

1 Introduction

The solid shipment (transportation) model (SSM) is an exclusive form of linear programming model where we deal with condition of sources, stations and carriages. The classical shipment model is an exclusive form of solid shipment model if only one type of carriage is taken under consideration. During the shipment movement due to complex situation, a few important criterions in the SSM are always treated as unclear variables to fit the realistic positions. There are cases to form a shipment plan for the later months; the amount of quantity at every origin, the requirement at every station and the carriage quantity are much necessary to be determined by experienced knowledge or probability statistics as a result of no definite data. It is much better to explore this issue by applying fuzzy or stochastic

D. Dutta (✉) · M. Sen
Department of Mathematics, National Institute of Technology Silchar, Silchar, India
e-mail: dhimanduttabigm@gmail.com

M. Sen
e-mail: senmausumi@gmail.com

© Springer Nature Singapore Pte Ltd. 2018
M. Pant et al. (eds.), *Soft Computing: Theories and Applications*,
Advances in Intelligent Systems and Computing 583,
https://doi.org/10.1007/978-981-10-5687-1_61

optimization models. It is difficult to predict the exact shipment cost for a sure time period. Shipment model is associated with additional costs along with shipping cost. These locked costs may be due to toll charges, road taxes etc. In this case it is called fixed charge shipment model. Fuzzy set theory is the one of the popular approaches to deal with uncertainty. Type-2 fuzzy sets were proposed by [12] as a development of type-1 fuzzy sets [11]. Type-2 fuzzy sets have membership functions as type-1 fuzzy sets. The advantage of type-2 fuzzy sets is that they are helpful in a few cases where it is uncertain to find the definite membership functions for fuzzy sets. Multi-item SSM is a model of shipping multiple components from multiple sources to multiple destinations over a few carriages. While transporting a few components from source, a situation may arise when not all brands of components can be shipped over all brands of carriages because of quality of components (e.g., liquid, breakable). Multi-item solid fixed charge shipment model (SFCSM) with condition on carriages is a model of shipping goods to a few destinations over a particular carriage with additional fixed charge for that particular route. The main motivation of this paper is to study solid shipment model with type-2 fuzzy parameters. The solid shipment model in fuzzy environment has been discussed by many researchers [2, 3, 10]. In [3] the author has studied fixed charged shipment problem and in [4] the author has described the several items solid shipment model with conditions on carriages. We have proposed the model multi-item solid fixed charged shipment model with conditions on carriages and taken type-2 triangular fuzzy variables to solve the model.

2 Preliminaries

Definition 1 ([1]) A pair (η', q', Pos) is termed as a possibility space, where η' is non-empty set of points, q' is power set of η' and $\text{Pos} : \eta' \mapsto [0, 1]$, called possibility measure defined as

1. $\text{Pos}(\phi) = 0$ and $\text{Pos}(\eta') = 1$;
2. For any $\{A_k | k \in K\} \subset \eta', \text{Pos}(\cup A_k) = \sup_k \text{Pos}(A_k)$.

Definition 2 ([8]) The possibility measure of a fuzzy event $\{\tau' \in C'\} C' \subset \mathbb{R}$ is explained as $\text{Pos}\{\tilde{\tau}' \in C'\} = \sup_{y' \in C'} \mu_{\tilde{\tau}'}(y')$, where $\mu_{\tilde{\tau}'}(y')$ is defined as a possibility distribution of τ'.

Definition 3 ([5]) The necessity (Nec) and credibility measure (Cr) of a normalized fuzzy variable $\left(\sup_{y' \in \mathbb{R}} \mu_{\tilde{\tau}'}(y') = 1\right)$ is defined as follows

1. $$Nec\left\{\tilde{\tau}' \in B'\right\} = 1 - Pos\left\{\tilde{\tau}' \in B'^c\right\} = 1 - \sup_{y' \in B'^c}\mu_{\tilde{\tau}'}(y').$$

2. $$\mathrm{Cr}\{\tilde{\tau}' \in B'\} = (\mathrm{Pos}\{\tilde{\tau}' \in B'\} + \mathrm{Nec}\{\tilde{\tau}' \in B'\})/2.$$

Definition 4 ([7]) For a possibility space $(\eta', q', \mathrm{Pos})$ a regular fuzzy variable (RFV) $\tilde{\tau}'$ is denoted by $\eta' \mapsto [0, 1]$, in the notion that for every $s' \in [0, 1]$, one has $\{\delta' \in \eta' | \mu_{\tilde{\tau}'}(\delta') \le s'\} \in q'$.

Definition 5 ([7]) If $(\eta', q', \mathrm{Pos})$ is a fuzzy possibility space then a type-2 fuzzy variable (T2 FV) $\tilde{\tau}'$ is expressed as $\eta' \mapsto \mathbb{R}$, such that for any $s' \in \mathbb{R}$, the set $\{\delta' \in \eta' | \mu_{\tilde{\tau}'}(\delta') \le s'\} \in q'$.

Definition 6 ([12]) A type-2 fuzzy set $\tilde{D}$ explained on the universe of discourse Z is described by a membership function $\tilde{\mu}_{\tilde{D}} : \mapsto F([0, 1])$ and is expressed by the following set notation: $\tilde{D} = \left\{(z, \tilde{\mu}_{\tilde{D}}(z)) : z \in Z\right\}$.

***Example 1* ([6])** A type-2 triangular fuzzy variable $\tilde{\tau}'$ is expressed by $\left(r'_1, r'_2, r'_3; \theta'_l, \theta'_r\right)$, where r'_1, r'_2, r'_3 are real numbers and θ'_l, θ'_r are two criterions defining the degree of ambiguity that $\tilde{\tau}'$ takes a value x and the secondary possibility distribution function $\tilde{\mu}_{\tilde{\tau}'}(x')$ of $\tilde{\tau}'$ is denoted as

$$\tilde{\mu}_{\tilde{\tau}'}(x') = \begin{cases} (\frac{x'-r'_1}{r'_2-r'_1} - \theta'_l \frac{x'-r'_1}{r'_2-r'_1}, \frac{x'-r'_1}{r'_2-r'_1}, \frac{x'-r'_1}{r'_2-r'_1} + \theta'_r \frac{x'-r'_1}{r'_2-r'_1}), & \text{if } x' \in \left[r'_1, \frac{r'_1+r'_2}{2}\right]; \\ (\frac{x'-r'_1}{r'_2-r'_1} - \theta'_l \frac{r'_2-x'}{r'_2-r'_1}, \frac{x'-r'_1}{r'_2-r'_1}, \frac{x'-r'_1}{r'_2-r'_1} + \theta'_r \frac{r'_2-x'}{r'_2-r'_1}), & \text{if } x' \in \left(\frac{r'_1+r'_2}{2}, r'_2\right]; \\ (\frac{r'_3-x'}{r'_3-r'_2} - \theta'_l \frac{x'-r'_2}{r'_3-r'_2}, \frac{r'_3-x'}{r'_3-r'_2}, \frac{r'_3-x'}{r'_3-r'_2} + \theta'_r \frac{x'-r'_2}{r'_3-r'_2}), & \text{if } x' \in \left(r'_2, \frac{r'_2+r'_3}{2}\right]; \\ (\frac{r'_3-x'}{r'_3-r'_2} - \theta'_l \frac{r'_3-x'}{r'_3-r'_2}, \frac{r'_3-x'}{r'_3-r'_2}, \frac{r'_3-x'}{r'_3-r'_2} + \theta'_r \frac{r'_3-x'}{r'_3-r'_2}), & \text{if } x' \in \left(\frac{r'_2+r'_3}{2}, r'_3\right]. \end{cases}$$

2.1 Critical Values for RFVs

The different forms of critical values (CV) [9] of a RFV $\tilde{\tau}'$ is defined below.

(i)   The optimistic CV of $\tilde{\tau}'$, denoted by $\mathrm{CV}^*[\tau']$, is defined as

$$\mathrm{CV}^*[\tilde{\tau}'] = \sup_{\alpha' \in [0,1]}[\alpha' \wedge \mathrm{Pos}\{\tilde{\tau}' \ge \alpha'\}].$$

(ii) The pessimistic CV of $\tilde{\tau}'$, denoted by $\mathrm{CV}_*[\tilde{\tau}']$, is defined as

$$\mathrm{CV}_*[\tilde{\tau}'] = \sup_{\alpha' \in [0,1]}[\alpha' \wedge \mathrm{Nec}\{\tilde{\tau}' \ge \alpha'\}].$$

(iii) The CV of $\tilde{\tau}'$, $\mathrm{CV}^*[\tilde{\tau}']$, is defined as

$$\mathrm{CV}\left[\tilde{\tau}'\right] = \sup_{\alpha' \in [0,1]} \left[\alpha' \wedge \mathrm{Cr}\left\{\tilde{\tau}' \geq \alpha'\right\}\right].$$

3 CV-Based Reduction Approach for T2 FVs

CV-based reduction approach introduced by Qin et al. [9] which reduces a T2 FV to a type-1 fuzzy variable. Let $\tilde{\tau}'$ be a T2 FV with secondary membership function $\tilde{\mu}_{\tilde{\tau}'}(y)$. The approach is to propose the CVs as a defining value for RFV, $\tilde{\mu}_{\tilde{\tau}'}(y)$, i.e., $\mathrm{CV}^*[\tilde{\mu}_{\tilde{\tau}'}(y)], \mathrm{CV}_*[\tilde{\mu}_{\tilde{\tau}'}(y)]$ or $\mathrm{CV}[\tilde{\mu}_{\tilde{\tau}'}(y)]$. Then these are accordingly called optimistic CV reduction, pessimistic CV reduction and CV reduction approach.

Theorem 1 [9] Suppose that $\tilde{\tau}' = (s_1', s_2', s_3'; \eta_l', \eta_r')$ be a type-2 triangular fuzzy variable. Then we have:

(i) The reduction of $\tilde{\tau}$ to $\tilde{\tau}_1'$ applying the optimistic CV reduction approach has the consecutive possibility distribution

$$\tilde{\mu}_{\tilde{\tau}_1'}(x') = \begin{cases} \dfrac{(1+\eta_r')(x'-s_1')}{s_2'-s_1'+\eta_r'(x'-s_1')}, & \text{if } x' \in \left[s_1', \dfrac{s_1'+s_2'}{2}\right]; \\[2mm] \dfrac{(1-\eta_r')x'+\eta_r's_2'-s_1}{s_2'-s_1'+\eta_r'(s_2'-s_1')}, & \text{if } x' \in \left(\dfrac{s_1'+s_2'}{2}, s_2'\right]; \\[2mm] \dfrac{(-1+\eta_r')x'-\eta_r's_2'+s_3'}{s_3'-s_2'+\eta_r'(x'-s_2')}, & \text{if } x' \in \left(s_2', \dfrac{s_2'+s_3'}{2}\right]; \\[2mm] \dfrac{(1+\eta_r')(s_3'-x')}{s_3'-s_2'+\eta_r'(s_3'-x')}, & \text{if } x' \in \left(\dfrac{s_2'+s_3'}{2}, s_3'\right]. \end{cases}$$

(ii) The reduction of $\tilde{\tau}'$ to $\tilde{\tau}_2'$ applying the pessimistic CV reduction approach has the consecutive possibility distribution

$$\mu_{\tilde{\tau}_2'}(x') = \begin{cases} \dfrac{x'-s_1}{s_2'-s_1'+\eta_l'(x'-s_1')}, & \text{if } x' \in \left[s_1', \dfrac{s_1'+s_2'}{2}\right]; \\[2mm] \dfrac{x'-s_1}{s_2'-s_1'+\eta_l'(s_2'-x')}, & \text{if } x' \in \left(\dfrac{s_1'+s_2'}{2}, s_2'\right]; \\[2mm] \dfrac{s_3'-x'}{s_3'-s_2'+\eta_l'(x'-s_2')}, & \text{if } x' \in \left(s_2', \dfrac{s_2'+s_3'}{2}\right]; \\[2mm] \dfrac{s_3'-x'}{s_3'-s_2'+\eta_l'(s_3'-x')}, & \text{if } x' \in \left(\dfrac{s_2'+s_3'}{2}, s_3'\right]. \end{cases}$$

(iii) The reduction of $\tilde{\tau}'$ to $\tilde{\tau}_3'$ applying the CV reduction approach has the consecutive possibility distribution

$$
\mu_{\tilde{\tau}_3'}(x') = \begin{cases} \dfrac{(1+\eta_r')(x'-s_1')}{s_2'-s_1'+2\eta_r'(x'-s_1')}, & \text{if } x' \in \left[s_1', \frac{s_1'+s_2'}{2}\right]; \\[2mm] \dfrac{(1-\eta_l')x'+\eta_l's_2'-s_1'}{s_2'-s_1'+2\eta_l'(s_2'-x')}, & \text{if } x' \in \left(\frac{s_1'+s_2'}{2}, s_2'\right]; \\[2mm] \dfrac{(-1+\eta_l')x'-\eta_l's_2'+s_3'}{s_3'-s_2'+2\eta_l'(x'-s_2')}, & \text{if } x' \in \left(s_2', \frac{s_2'+s_3'}{2}\right]; \\[2mm] \dfrac{(1+\eta_r')(s_3'-x')}{s_3'-s_2'+2\eta_r'(s_3'-x')}, & \text{if } x' \in \left(\frac{s_2'+s_3'}{2}, s_3'\right]. \end{cases}
$$

4 Nearest Interval Approximation of Continuous Type-2 Fuzzy Variables

Kundu et al. [4] proposed the interval approximation of type-2 fuzzy variables by applying the α cut of the optimistic, pessimistic and credibilistic approximation of type-2 triangular fuzzy variables given by theorem 1. Lastly, using interval approximation method to these α cuts estimated crisp intervals are obtained which are given below:

(i) **Applying α cut of the optimistic CV-based reduction (optimistic interval approximation):** The nearest interval approximation of $\tilde{\tau}$ is calculated as $[C_L, C_R]$ where,

$$
C_L = C_{L_1} + C_{L_2} \tag{1}
$$

$$
c_{L_1} = \frac{s_1 + \eta_r s_1}{\eta_r} \ln\left(\frac{1+\eta_r}{1+0.5\eta_r}\right) - \frac{s_2 - s_1 - \eta_r s_1}{\eta_r^2}\left[0.5\eta_r - (1+\eta_r)\ln\left(\frac{1+\eta_r}{1+0.5\eta_r}\right)\right],
$$

$$
c_{L_2} = \frac{-s_1 + \eta_r s_2}{\eta_r}\ln(1 - 0.5\eta_r) + \frac{s_2 - s_1 - \eta_r s_2}{\eta_r^2}\left[0.5\eta_r + (1 - \eta_r)\ln(1 - 0.5\eta_r)\right].
$$

$$
C_R = C_{R_1} + C_{R_2} \tag{2}
$$

$$
c_{R_1} = \frac{s_3 + \eta_r s_3}{\eta_r}\ln\left(\frac{1+\eta_r}{1+0.5\eta_r}\right) + \frac{s_3 - s_2 + \eta_r s_3}{\eta_r^2}\left[0.5\eta_r - (1+\eta_r)\ln\left(\frac{1+\eta_r}{1+0.5\eta_r}\right)\right],
$$

$$
c_{R_2} = \frac{-s_3 + \eta_r s_2}{\eta_r}\ln(1 - 0.5\eta_r) - \frac{s_3 - s_2 - \eta_r s_2}{\eta_r^2}\left[0.5\eta_r + (1 - \eta_r)\ln(1 - 0.5\eta_r)\right].
$$

(ii) **Applying α cut of the pessimistic CV-based reduction (pessimistic interval approximation):** The nearest interval approximation of $\tilde{\tau}$ is calculated as $[C_L, C_R]$, where

$$C_L = C_{L_1} + C_{L_2} \tag{3}$$

$$c_{L_1} = -\frac{s_1}{\eta_r}\ln(1+0.5\eta_r) - \frac{s_2 - s_1 - \eta_l s_1}{\eta_l^2}[0.5\eta_l + \ln(1+0.5\eta_r)],$$

$$c_{L_2} = \frac{s_1}{\eta_l}\ln\left(\frac{1+\eta_l}{1+0.5\eta_l}\right) + \frac{s_2 - s_1 + \eta_l s_2}{\eta_l^2}\left[0.5\eta_l - \ln\left(\frac{1+\eta_l}{1+0.5\eta_l}\right)\right].$$

$$C_R = C_{R_1} + C_{R_2} \tag{4}$$

$$c_{R_1} = -\frac{s_3}{\eta_l}\ln(1-0.5\eta_r) + \frac{s_3 - s_2 + \eta_l s_3}{\eta_l^2}[0.5\eta_l + \ln(1-0.5\eta_r)],$$

$$c_{R_2} = \frac{s_3}{\eta_l}\ln\left(\frac{1+\eta_l}{1+0.5\eta_l}\right) + \frac{s_3 - s_2 - \eta_l s_2}{\eta_l^2}\left[0.5\eta_l - \ln\left(\frac{1+\eta_l}{1+0.5\eta_l}\right)\right].$$

(iii) **Applying α cut of the CV reduction (credibilistic interval approximation)**:
The nearest interval approximation of $\tilde{\tau}$ is calculated as $[C_L, C_R]$ where,

$$C_L = C_{L_1} + C_{L_2} \tag{5}$$

$$c_{L_1} = \frac{s_1 + \eta_r s_1}{2\eta_r}\ln(1+\eta_r) - \frac{s_2 - s_1 - 2\eta_r s_1}{4\eta_r^2}[\eta_r - (1+\eta_r)\ln(1+\eta_r)],$$

$$c_{L_2} = \frac{s_1 - \eta_l s_2}{2\eta_l}\ln(1+\eta_l) + \frac{s_2 - s_1 - 2\eta_l s_2}{4\eta_l^2}[\eta_l - (1-\eta_l)\ln(1+\eta_l)].$$

$$C_R = C_{R_1} + C_{R_2} \tag{6}$$

$$c_{R_1} = \frac{s_3 + \eta_r s_3}{2\eta_r}\ln(1+\eta_r) + \frac{s_3 - s_2 + 2\eta_r s_3}{4\eta_r^2}[\eta_r - (1+\eta_r)\ln(1+\eta_r)],$$

$$c_{R_2} = \frac{s_3 - \eta_l s_2}{2\eta_l}\ln(1+\eta_l) - \frac{s_3 - s_2 - 2\eta_l s_2}{4\eta_l^2}[\eta_l - (1-\eta_l)\ln(1+\eta_l)].$$

5 Model: Multi-item Solid Fixed Charged Shipment Problem with Conditions on Carriages

Suppose that $k(k = 1, 2, \ldots, K)$ different modes of carriages are necessary to transport l components from m sources $O_i(i = 1, 2, \ldots, m)$ to n stations $D_j(j = 1, 2, \ldots, n)$. In addition to that there are a few conditions on a few particular components and carriages so that a few components cannot be shipped over a few carriages. Let us assume J_k as the set of components which can be shipped over carriages. We use representation $p'(= 1, 2, \ldots, l)$ to stand for the components.

The solid fixed charge shipment model (SFCSM) is linked with two types of penalties, unit shipment penalty for shipping unit goods from origin i to station j and a fixed penalty for the direction (i,j). We develop a multi-item solid fixed charged shipment model (MISFCSM) with m sources, n stations, k carriages, shipment penalty and locked penalty criterion as T2 FVs as follows:

$$\text{Min Z} = \sum_{p'=1}^{l} \sum_{i=1}^{m} \sum_{j=1}^{n} \sum_{k=1}^{K} d_{ijk}^{p'} \left(c_{ijk}^{p'} x_{ijk}^{p'} \right) + e_{ijk}^{p'} y_{ijk}^{p'}$$

$$\text{subject to } \sum_{j=1}^{n} \sum_{k=1}^{K} d_{ijk}^{p'} x_{ijk}^{p'} \le a_i^{p'}, i = 1, 2, \ldots, m; p' = 1, 2, \ldots, l,$$

$$\sum_{i=1}^{m} \sum_{k=1}^{K} d_{ijk}^{p'} x_{ijk}^{p'} \le b_j^{p'}, j = 1, 2, \ldots, n; p' = 1, 2, \ldots, l, \qquad (7)$$

$$\sum_{p'=1}^{l} \sum_{i=1}^{m} \sum_{j=1}^{n} d_{ijk}^{p'} x_{ijk}^{p'} \le f_k, k = 1, 2, \ldots, K,$$

$$\sum_{i=1}^{m} x_{ijk}^{p} \ge 0, \forall i,j,k,p', y_{ijk}^{p'} = \begin{cases} 1, & \text{if } x_{ijk}^{p} > 0; \\ 0, & \text{otherwise.} \end{cases}$$

where d_{ijk}^{p} is defined as $d_{ijk}^{p'} = \begin{cases} 1, & \text{if } p' \in I_K \forall i,j,k,p'; \\ 0, & \text{otherwise.} \end{cases}$

Here, $x_{ijk}^{p'}$ is the decision variable representing the supply of p'-th thing shipped from source i to station j, $e_{ijk}^{p'}$ is the type-2 fuzzy fixed cost linked with direction (i,j) for the objective Z. The unit shipment cost $c_{ijk}^{p'}$ for the objective Z, total supply of p'-th thing $a_i^{p'}$ at i-th source, total claim of p'-th thing $b_j^{p'}$ at j-th station and total quantity f_k of k-th carriage are all T2 FVs.

6 Solution Procedure (Applying Nearest Interval Approximation)

We consider $c_{ijk}^{p'}, e_{ijk}^{p'} a_i^{p'}, b_j^{p'}$ and f_k are all jointly free type-2 triangular fuzzy variables defined by $c_{ijk}^{p'} = \left(c_{ijk}^{p'1}, c_{ijk}^{p'2}, c_{ijk}^{p'3}; \theta_{l,ijk}^{p'}, \theta_{r,ijk}^{p'} \right)$, $e_{ijk}^{p'} = \left(e_{ijk}^{p'1}, e_{ijk}^{p'2}, e_{ijk}^{p'3}; \theta_{l,ijk}^{p'}, \theta_{r,ijk}^{p'} \right)$, $a_i^{p'} = \left(a_i^{p'1}, a_i^{p'2}, a_i^{p'3}; \theta_{l,i}^{p'}, \theta_{r,i}^{p'} \right)$, $b_j^{p'} = \left(b_j^{p'1}, b_{ijk}^{p'2}, b_j^{p'3}; \theta_{l,j}^{p'}, \theta_{r,j}^{p'} \right)$ and $f_k = \left(f_k^1, f_k^2, f_k^3; \theta_{l,k}, \theta_{r,k} \right)$. We find nearest interval approximations (credibilistic interval approximation) of $c_{ijk}^{p'}, e_{ijk}^{p'} a_i^{p'}, b_j^{p'}$ and f_k from (5) and (6) and suppose these are $\left[c_{ijkL}^{p'}, c_{ijkR}^{p'} \right], \left[e_{ijkL}^{p'}, e_{ijkR}^{p'} \right]$, $\left[a_{iL}^{p'}, a_{iR}^{p'} \right]$, $\left[b_{jL}^{p'}, b_{jR}^{p'} \right]$ and $[f_{kL}, f_{kR}]$ correspondingly. The nearest interval approximation of the model is given below:

$$Z = \sum_{p'=1}^{l}\sum_{i=1}^{m}\sum_{j=1}^{n}\sum_{k=1}^{K} d_{ijk}^{p'}\left(\left[c_{ijkL}^{p'}, c_{ijkR}^{p'}\right]x_{ijk}^{p'}\right) + \left[e_{ijkL}^{p'}, e_{ijkR}^{p'}\right]y_{ijk}^{p'},$$

$$\text{subject to } \sum_{j=1}^{n}\sum_{k=1}^{K} d_{ijk}^{p'}x_{ijk}^{p'} \le \left[a_{iL}^{p'}, a_{iR}^{p'}\right], i = 1, 2, \ldots, m; p' = 1, 2, \ldots, l,$$

$$\sum_{i=1}^{m}\sum_{k=1}^{K} d_{ijk}^{p'}x_{ijk}^{p'} \le \left[b_{jL}^{p'}, b_{jR}^{p'}\right], j = 1, 2, \ldots, n; p' = 1, 2, \ldots, l, \qquad (8)$$

$$\sum_{p'=1}^{l}\sum_{i=1}^{m}\sum_{j=1}^{n} d_{ijk}^{p'}x_{ijk}^{p'} \le [f_{kL}, f_{kR}], k = 1, 2, \ldots, K,$$

$$\sum_{i=1}^{m} x_{ijk}^{p} \ge 0, \forall i, j, k, p', \quad y_{ijk}^{p'} = \begin{cases} 1, & \text{if } x_{ijk}^{p} > 0; \\ 0, & \text{otherwise.} \end{cases}$$

where d_{ijk}^{p} is defined as $d_{ijk}^{p'} = \begin{cases} 1, & \text{if } p' \in I_K \forall i, j, k, p'; \\ 0, & \text{otherwise.} \end{cases}$

6.1 Deterministic Form

Kun et al. [4] introduced deterministic forms of the dubious constraints applying the notion of possibility degree of interval number [14] defining definite degree by which one interval is greater or lesser than another and denoted the left hand side expressions of the origin, station and carriage quantity constraints of the model by $S_i^{p'}, D_j^{p'}$ and E_k correspondingly. The possibility degree of achievement of these constraints are represented as

$$P_{S_i^{p'} \le [a_{iL}^{p'}, a_{iR}^{p'}]} = \begin{cases} 1, & S_i^{p'} \le a_{iL}^{p'} \\ \frac{a_{iR}^{p'} - S_i^{p'}}{a_{iR}^{p'} - a_{iL}^{p'}}, & a_{iL}^{p'} < S_i^{p'} \le a_{iR}^{p'}; \\ 0, & S_i^{p'} > a_{iR}^{p'}. \end{cases}$$

$$P_{D_j^{p'} \ge [b_{jL}^{p'}, b_{jR}^{p'}]} = \begin{cases} 0, & D_j^{p'} < b_{jL}^{p'} \\ \frac{D_j^{p'} - b_{jL}^{p'}}{b_{jR}^{p'} - b_{jL}^{p'}}, & b_{jL}^{p'} < D_j^{p'} \le b_{jR}^{p'}; \\ 1, & D_j^{p'} > b_{jL}^{p'}. \end{cases}$$

$$P_{E_k \le [f_{kL}, f_{kR}]} = \begin{cases} 1, & E_k \le f_{kL} \\ \frac{f_{kR} - E_k}{f_{kR} - f_{kL}}, & f_{kL} < E_k \le f_{kR}; \\ 0, & E_k > d_{iR}^{p'}. \end{cases}$$

The constraints are allowed to be satisfied with a few predetermined possibility degree levels $\alpha_i^{p'}, \beta_j^{p'}$ and $\gamma_k (0 < \alpha_i^{p'}, \beta_j^{p'}, \gamma_k \le 1)$, respectively, i.e.,

$P_{S_i^{p'} \le [d'_{iL}, d'_{iR}]} \ge \alpha_i^{p'}, P_{D_j^{p'} \ge [b'^{p'}_{jL}, b'^{p'}_{jR}]} \ge \beta_j^{p'}, P_{E_k \le [f_{kL}, f_{kR}]} \ge \gamma_k$, and then the identical deterministic inequalities of the various constraints are given below:

$$S_i^{p'} \le d_{iR}^{p'} - \alpha_i^{p'} \left[d_{iR}^{p'} - d_{iL}^{p'} \right], i = 1, 2, \ldots, m; p' = 1, 2, \ldots, l, \tag{9}$$

$$D_j^{p'} \ge b_{jL}^{p'} + \beta_j^{p'} \left[b_{jR}^{p'} - b_{jL}^{p'} \right], j = 1, 2, \ldots, n; p' = 1, 2, \ldots, l, \tag{10}$$

$$E_k \le f_{kR} - \gamma_k [f_{kR} - f_{kL}], k = 1, 2, \ldots, K. \tag{11}$$

We get the lowest objective function value (say $\underline{Z}$) and the highest objective function ($\bar{Z}$) for the interval penalties $\left[c_{ijkL}^{p'}, c_{ijkR}^{p'} \right], \left[e_{ijkL}^{p'}, e_{ijkR}^{p'} \right]$, by working out the two models:

$$\bar{Z} = \max_{c_{ijkL}^{p'} \le c_{ijk}^{p'} \le c_{ijkR}^{p'}, e_{ijkL}^{p'} \le e_{ijk}^{p'} \le e_{ijkR}^{p'}} \sum_{p'=1}^{l} \sum_{i=1}^{m} \sum_{j=1}^{n} \sum_{k=1}^{K} d_{ijk}^{p'} \left(\left[c_{ijkL}^{p'}, c_{ijkR}^{p'} \right] x_{ijk}^{p'} \right) + \left[e_{ijkL}^{p'}, e_{ijkR}^{p'} \right] y_{ijk}^{p'}$$

$$\tag{12}$$

$$\underline{Z} = \min_{c_{ijkL}^{p'} \le c_{ijk}^{p'} \le c_{ijkR}^{p'}, e_{ijkL}^{p'} \le e_{ijk}^{p'} \le e_{ijkR}^{p'}} \sum_{p'=1}^{l} \sum_{i=1}^{m} \sum_{j=1}^{n} \sum_{k=1}^{K} d_{ijk}^{p'} \left(\left[c_{ijkL}^{p'}, c_{ijkR}^{p'} \right] x_{ijk}^{p'} \right) + \left[e_{ijkL}^{p'}, e_{ijkR}^{p'} \right] y_{ijk}^{p'}$$

$$\tag{13}$$

subject to the above constraints (9)–(11) for both cases. Now we evaluate the optimal solutions by dealing with the above problems and (12)–(13) mutually as two objective problems and using fuzzy linear programming [13] given below: Suppose that

$$Z_1 = \sum_{p'=1}^{l} \sum_{i=1}^{m} \sum_{j=1}^{n} \sum_{k=1}^{K} d_{ijk}^{p'} \left(c_{ijk}^{p''} x_{ijk}^{p'} \right) + e_{ijk}^{p''} y_{ijk}^{p'} \text{ and } Z_2 = \sum_{p'=1}^{l} \sum_{i=1}^{m} \sum_{j=1}^{n} \sum_{k=1}^{K} d_{ijk}^{p'} \left(c_{ijk}^{p'''} x_{ijk}^{p'} \right) +$$

$e_{ijk}^{p'''} y_{ijk}^{p'}$ so that $Z_1 \left(x_{ijk}^{p''} \right) = \underline{Z}$ and $Z_2 \left(x_{ijk}^{p''} \right) = \bar{Z}$.

We search for the lower and upper bound for the two objectives as $L_1 = Z_1 \left(x_{ijk}^{p''} \right), U_1 = Z_1 \left(x_{ijk}^{p'''} \right)$, and $L_2 = Z_1 \left(x_{ijk}^{p'''} \right), U_2 = Z_1 \left(x_{ijk}^{p''} \right)$ respectively.

We have then constructed the next two membership function for the objective functions correspondingly as

$$\mu_1(Z_1) = \begin{cases} 1, \text{if } Z_1 \le L_1; \\ \frac{U_1 - Z_1}{U_1 - L_1}, \text{if } L_1 < Z_1 < U_1 \text{ ; and} \\ 0, \text{if } Z_1 \ge U_1. \end{cases}$$

$$\mu_2(Z_2) = \begin{cases} 1, \text{if } Z_2 \le L_2; \\ \frac{U_2 - Z_2}{U_2 - L_2}, \text{if } L_2 < Z_2 < U_2 \\ 0, \text{if } Z_2 \ge U_2. \end{cases}$$

We finally work out the model

$$\text{Max } \delta$$

$$\text{subject to } \delta \le \mu_t(\bar{Z}_t) = \frac{U_t - \bar{Z}_t}{U_t - L_t}, t = 1, 2. \tag{14}$$

and the constraints of (9)–(11); $0 \le \delta \le 1$.

We find the optimal solution, say x_{ijk}^p by working out the two objectives Z_1, Z_2 with some degree $\delta = \delta^*$(say) and results of the the objectives gives the range of the objective value, $[\underline{Z}^*, \bar{Z}^*]$.

7 Numerical Model

The proposed problem is illustrated numerically in this section. The proposed approach ability is solved numerically by taking one example of the model. Consider the model with sources (i = 1, 2, 3), stations (j = 1, 2, 3), carriage (k = 1, 2, 3, 4) and components (p' = 1, 2, 3). Suppose that $J_1 = \{1, 2\}, J_2 = \{1, 2, 3\}, J_3 = \{3\}, J_4 = \{1, 2, 3\}$. The credibilistic interval approximations of the triangular type-2 fuzzy parameters are calculated using (5) and (6). The shipment costs and fixed costs for this model are given in Tables 1–6. The supplies, demands and carriage capacities are the next data:

$a_1^1 = [22.4651, 25.0232], a_1^2 = [26.9808, 29.0192], a_1^3 = [24.982, 27.527],$

$a_2^1 = [26.9861, 30.0278], a_2^2 = [21.9722, 25.5208], a_2^3 = [33.4874, 35.5042],$

$a_3^1 = [27.4935, 28.5065], a_3^2 = [22.9768, 25.0232], a_3^3 = [23.9768, 27.0465],$

$b_1^1 = [10.4935, 13.0043], b_1^2 = [15.4974, 16.5026], b_1^3 = [16.4921, 19.0053],$

$b_2^1 = [11.9898, 14.0102], b_2^2 = [12.5096, 13.9808], b_2^3 = [12.4804, 15.013],$

$b_3^1 = [10.9891, 13.5164], b_3^2 = [9.9882, 11.5059], b_3^3 = [12.9887, 16.0057],$

$e_1 = [34.9898, 36.5051], e_2 = [47.473, 49.509], e_3 = [28.987, 31.5196],$

$e_4 = [41.4847, 44.0102].$

Table 1 c^1_{ijk}

i/j	1	2	3	k
1	[3.4975, 4.5025]	[4.5028, 5.4972]	[3.4974, 5.5079]	
2	[3.4926, 5.0148]	[5.4977, 6.5023]	[5.4898, 6.5102]	1
3	[3.0159, 4.9841]	[4.5, 5.5]	[5.0049, 6.9951]	
1	[5.9898, 7.5051]	[6.4974, 7.5026]	[7.4977, 8.5023]	
2	[8.4866, 10.5045]	[7.5, 8.5]	[8.0049, 9.4975]	2
3	[1.5, 2.5]	[1.5079, 2.4921]	[2.5048, 3.4952]	
1	[9.4969, 11.0061]	[9.4931, 10.5069]	[11.5059, 12.4941]	
2	[9.5074, 10.9852]	[11.5, 12.5]	[11.5102, 12.9795]	4
3	[11.5, 12.5]	[11.5, 12.5]	[12.4935, 14.5022]	

Table 2 c^2_{ijk}

i/j	1	2	3	k
1	[5.5, 7]	[4.4975, 7.0099]	[5.9898, 8.0102]	
2	[4, 6]	[7.5, 8.5]	[7.5042, 8.9916]	1
3	[7.0041, 8.9959]	[7.9808, 10.0192]	[9.4884, 11.0232]	
1	[5.4821, 7.0359]	[1.5025, 2.9951]	[5.4974, 6.5026]	
2	[5.9852, 7.5074]	[3.9947, 6.0053]	[4.4936, 5.5064]	2
3	[9.498, 10.502]	[7.4955, 10.0179]	[7.502, 10.4898]	
1	[1.5092, 2.9816]	[4.5, 5.5]	[2.5051, 3.4949]	
2	[3.5028, 4.4972]	[3.0205, 4.9795]	[1.9898, 4.0102]	4
3	[8.0041, 9.498]	[3.9954, 6.0046]	[5.9959, 8.0041]	

Table 3 c^3_{ijk}

i/j	1	2	3	k
1	[7.9808, 10.0192]	[7.5051, 8.4949]	[5.5025, 6.4975]	
2	[8.4974, 10.5079]	[10.4958, 12.0084]	[6.4978, 8.0043]	2
3	[4.5074, 5.9852]	[5.5048, 6.9905]	[4.4978, 6.0043]	
1	[10.5048, 11.4952]	[5.5074, 6.9852]	[6.0046, 7.9954]	
2	[4.5145, 5.4855]	[10.5143, 11.9714]	[9.4974, 10.5026]	3
3	[10.4921, 11.5079]	[11.4941, 12.5059]	[12.4812, 13.5188]	
1	[2.5153, 4.4949]	[4.0049, 5.4975]	[4.9951, 6.5025]	
2	[10.9898, 12.5051]	[10.9796, 13.5051]	[12.5069, 14.4977]	4
3	[6.5672, 8.4776]	[8.0489, 9.4755]	[7.4898, 10.502]	

The corresponding deterministic forms of all the constraints are attained using (9)–(11) by taking $\alpha^{p'}_i = 0.7, \beta^{p'}_i = 0.7, \gamma^{p'}_i = 0.7$. We get minimum and maximum

Table 4 e_{ijk}^1

i/j	1	2	3	k
1	[4.5102, 5.4898]	[3.9943, 5.5028]	[5.9947, 7.5026]	
2	[3.0148, 4.4926]	[2.0232, 3.9768]	[5.0102, 7.9796]	1
3	[3.9898, 5.5051]	[5.009, 6.4955]	[6.5065, 7.4935]	
1	[3.9795, 6.0205]	[4.0296, 6.4926]	[4.9659, 7.5085]	
2	[6.9947, 8.5026]	[7.9947, 9.5026]	[8.991, 10.5045]	2
3	[9.9852, 11.5074]	[11.4975, 12.5025]	[12.5116, 13.4884]	
1	[7.5074, 8.9852]	[8.0057, 9.9943]	[8.9947, 11.0053]	
2	[9.9752, 12.0248]	[10.9905, 13.0095]	[12.0046, 13.9954]	4
3	[4.0109, 5.9891]	[5.0118, 6.9882]	[6.0128, 7.9872]	

Table 5 e_{ijk}^2

i/j	1	2	3	k
1	[4.0232, 5.4884]	[5.0232, 6.4884]	[6.018, 7.491]	
2	[7.0109, 8.4945]	[8.5051, 9.4949]	[9.5051, 10.4949]	1
3	[9.9091, 11.5045]	[10.982, 12.509]	[11.9905, 13.5048]	
1	[6.0148, 7.4926]	[7.018, 8.491]	[8.013, 9.4935]	
2	[9.4955, 10.5045]	[10.4884, 11.5116]	[11.4952, 12.5048]	2
3	[12, 13.5]	[13, 14.5]	[13.0106, 15.4974]	
1	[4.5051, 5.4949]	[6.0053, 7.4974]	[6.9898, 8.5051]	
2	[2.9905, 4.5048]	[3.991, 5.5045]	[4.4874, 6.5042]	4
3	[5.5222, 7.4926]	[7, 8.5]	[7.4874, 9.5042]	

Table 6 e_{ijk}^3

i/j	1	2	3	k
1	[1.5026, 2.4974]	[2.5074, 3.4926]	[3.498, 4.502]	
2	[4.5074, 5.4926]	[5.4977, 6.5023]	[5.4935, 7.5022]	2
3	[6.4921, 8.5026]	[7.9951, 9.5025]	[8.9954, 10.5023]	
1	[4, 5.5]	[4.4921, 6.5026]	[5.4847, 7.5051]	
2	[8.9905, 10.5048]	[9.991, 11.5045]	[10.4921, 12.5026]	3
3	[14.0046, 15.4977]	[13.4885, 16.5023]	[15.9959, 17.502]	
1	[2.9894, 5.5026]	[4.5208, 6.4931]	[5.5349, 7.4884]	
2	[6.5, 8.5]	[8.0232, 9.4884]	[9.0192, 10.4904]	4
3	[9.9954, 11.5023]	[10.4931, 12.5023]	[11.9954, 13.5023]	

possible values of the objective function by solving (12), (13) and solutions are given in Table 7.

Table 7 Best possible result of the model

Shipment cost	Shipment amount
$\bar{Z} = 446.5881$	$x^1_{131} = 12.7532$, $x^1_{311} = 6.2114$, $x^1_{312} = 6.0379$, $x^1_{322} = 13.4041$, $x^2_{211} = 2.1481$, $x^2_{122} = 13.5394$, $x^2_{114} = 14.0529$, $x^2_{234} = 11.0506$, $x^3_{332} = 15.1006$, $x^3_{123} = 8.6067$, $x^3_{213} = 6.759$, $x^3_{114} = 11.4923$, $x^3_{124} = 5.6465$.
$\underline{Z} = 661.3566$	$x^1_{111} = 10.616$, $x^1_{312} = 1.6351$, $x^1_{322} = 13.4041$, $x^1_{332} = 12.7582$, $x^2_{122} = 11.3913$, $x^2_{114} = 16.201$, $x^2_{224} = 2.1481$, $x^2_{234} = 11.0506$, $x^3_{332} = 8.8951$, $x^3_{123} = 7.6161$, $x^3_{213} = 18.2513$, $x^3_{124} = 6.6371$, $x^3_{134} = 6.2055$.
$Z = [451.1117, 661.309]$ $\delta = 0.8394297$	$x^1_{131} = 12.7582$, $x^1_{312} = 12.2511$, $x^1_{322} = 13.4041$, $x^2_{211} = 2.1481$, $x^2_{122} = 13.5394$, $x^2_{114} = 14.0529$, $x^2_{234} = 11.0506$, $x^3_{332} = 8.8892$, $x^3_{123} = 2.3953$, $x^3_{133} = 6.2114$, $x^3_{213} = 12.9704$, $x^3_{114} = 5.2809$, $x^3_{124} = 11.8579$.

Table 8 Variations in shipment cost for distinct credibility levels

$\alpha^{p'}_i$	$\beta^{p'}_j$	γ_k	Shipped amount	Shipment cost
0.8	0.7	0.7	126.8095	[451.4395, 661.4945]
0.9	0.7	0.7	126.8095	[451.7675, 661.6801]
0.7	0.8	0.7	128.7212	[459.7955, 672.1176]
0.7	0.9	0.7	130.6327	[468.4781, 682.9246]
0.7	0.7	0.8	126.8095	[451.899, 661.9889]
0.7	0.7	0.9	126.8095	[452.6864, 662.6688]

Here, $L_1 = 446.5881$, $U_1 = 474.7602$, and $L_2 = 661.3566$, $U_2 = 665.459$ and the compromise optimal solution of (14) using LINGO 16 solver, based upon GRG technique are given in Table 7. The nearest interval range of objective value is $[451.1117, 661.309]$.

8 Sensitivity Analysis

A sensitivity analysis of the model is given to view the effectiveness and reasonably accuracy of the crisp equivalent form and solution methods of the given problem. The changes of the shipment cost for different generalized credibility levels for the origin, station and carriage constraints of the problem are given in Table 8.

From Table 8 it can be seen that for fixed $\alpha^{p'}_i$, γ_k and different $\beta^{p'}_j$, the shipment cost, shipment amount increases with the increased value of the crisp amount of the demands. The shipment amount remains same for different $\alpha^{p'}_i$, γ_k and fixed $\beta^{p'}_j$. The cause of the increase of shipment cost is due to the decrease of capacity of the defuzzified amount of the supply.

9 Conclusion

In this paper, we have projected and worked out a multi-item solid fixed charge shipment model with type-2 triangular fuzzy variables. A nearest interval approximation approach is used to solve the model applying LINGO 16 solver. The approach can be extended in other decision-making models in distinct areas with type-2 fuzzy parameters.

References

1. Chen, S.M., Wang, C.Y.: Fuzzy decision making systems based on interval type-2 fuzzy sets. Inf. Sci. **242**, 1–21 (2013)
2. Kundu, P., Kar, S., Maiti, M.: Multi-objective solid transportation problems with budget constraint in uncertain environment. Int. J. Syst. Sci. **45**(8), 1668–1682 (2014)
3. Kundu, P., Kar, S., Maiti, M.: Fixed charge transportation problem with type-2 fuzzy variables. Inf. Sci. **255**, 170–186 (2014)
4. Kundu, P., Kar, S., Maiti, M.: Multi-item solid transportation problem with type-2 fuzzy parameters. Appl. Soft Comput. **31**, 61–80 (2015)
5. Liu, B., Liu, Y.K.: Expected value of fuzzy variable and fuzzy expected value models. IEEE Trans. Fuzzy Syst. **10**, 445–450 (2002)
6. Liu, B.: Theory and Practice of Uncertain Programming, UTLAB, 3rd edn. (2009). http://orsc.edu.cn/liu/up.pdf
7. Liu, Z.Q., Liu, Y.K.: Type-2 fuzzy variables and their arithmetic. Soft Comput. **14**, 729–747 (2010)
8. Nahmias, S.: Fuzzy variable. Fuzzy Sets Syst. **1**, 97–101 (1978)
9. Qin, R., Liu, Y.K., Liu, Z.Q.: Methods of critical value reduction for type-2 fuzzy variables and their applications. J. Comput. Appl. Math. **235**, 1454–1481 (2011)
10. Yang, L., Liu, L.: Fuzzy fixed charge solid transportation problem and algorithm. Appl. Soft Comput. **7**, 879–889 (2007)
11. Zadeh, L.A.: Fuzzy Sets Inf. Control **8**, 338–353 (1965)
12. Zadeh, L.A.: The Concept of a linguistic variable and its application to approximate reasoning-I. Inf. Sci. **8**, 199–249 (1975)
13. Zimmermann, H.-J.: Fuzzy programming and linear programming with several objective functions. Fuzzy Sets Syst. **1**, 45–55 (1978)
14. Zhang, Q., Fan, Z., Pan, D.: A ranking approach for interval numbers in uncertain multiple attribute decision making problems. Syst. Eng. Theory Pract. **5**, 129–133 (1999)

Capturing Performance Requirements of Real-Time Systems Using UML/MARTE Profile

Disha Khakhar and Ashalatha Nayak

Abstract Performance is a critical parameter for successful development of real-time systems since their correctness is based not only on logical behavior, but also on timeliness of output. Traditional software development methods use 'fix it later' approach where focus is on correctness of the software and performance considerations are made in the later phases of software development. If performance problems are discovered at this point, software is modified to fix performance issues. This technique does not work well for time critical systems. In order to address this problem, Model Driven Software Performance Engineering (MDSPE) approach is used to include performance analysis in the early stage of software development life cycle. Performance parameters are associated with UML model elements using UML profile for MARTE, to capture software requirement in the design phase. Annotated UML model is transformed to various performance model in order to perform analysis. By evaluating the performance model, it is possible to obtain various performance output parameters using simulation and analytical techniques. These parameters will help in evaluating alternative design for the same system. Currently, there are no approaches that investigate the issue of annotation of existing UML diagrams with MARTE profile. The proposed research focuses on capturing performance requirements by annotation of UML models using UML profile for MARTE.

Keywords UML MARTE profile · Real-time system · UML sequence diagram
Software performance engineering · Performance parameters

D. Khakhar (✉) · A. Nayak
Department of Computer Science and Engineering, Manipal Institute of Technology,
Manipal University, Manipal 576104, India
e-mail: disha.p.khakhar@gmail.com

A. Nayak
e-mail: asha.nayak@manipal.edu

© Springer Nature Singapore Pte Ltd. 2018
M. Pant et al. (eds.), *Soft Computing: Theories and Applications*,
Advances in Intelligent Systems and Computing 583,
https://doi.org/10.1007/978-981-10-5687-1_62

1 Introduction

Usually, performance of software is tested later in the development phase. Unsatisfactory outcomes may lead to changes in implementation or in the worst case, and may require changes in the architecture design itself. Therefore, analyzing software performance requirement is important for evaluating various design alternatives. This can be achieved by integrating performance analysis in the early phase of software development life cycle. Unified Modeling Language (UML) is the most common way to model software requirements. OMG has standardized UML profile for Modeling and Analysis of Real-Time and Embedded systems (MARTE) in 2008 [1]. The UML profile for MARTE provides a standard for annotating UML model with performance parameters. UML/MARTE model acts as a basis for development of performance model. Performance estimation is done by evaluating software performance based on its performance model. This process involves building the performance model for performance critical scenarios followed by its evaluation using analytical and simulation techniques.

In the approach by Street et al. [2], performance modeling and analysis techniques were used for the design of object-oriented software system. It highlights the lessons learned while implementing this approach using UML profile for Schedulability, Performance and Time (SPT) along with Colored Petri Nets (CPN). An important lesson learnt was that performance analysis technique should be known prior to modeling so that only required tags are filled in the performance model.

In the approach by Traore et al. [3], Model Driven Software Performance Engineering (MDSPE) approach is presented where performance analysis is integrated with functional analysis. This paper focuses on Performance Annotation step to encapsulate performance characteristics into the software design model using UML profile for SPT.

The approach by Middleton et al. [4] describes their experience of using UML profile for MARTE to model systems with stochastic behavior using PapyrusUML editor. Demanthieu et al. [5], have highlighted how key features of MARTE profile can be used to model behavior of real-time systems. A case study related to real-time and embedded systems was developed using MARTE adopted specification. Depending on the system to be modeled (whether it is distributed), one may deal with physical or logical time. MARTE has an advantage over SPT by introducing stereotypes for time observation with explicit reference to clocks, which are useful in design of these systems.

In the approach by Akshay KC et al. [6], a case study of ATM system is taken up and UML sequence diagrams are annotated with MARTE profile to capture properties of sequential and simultaneous transactions. Data race was detected by permutation algorithm to find valid scenarios of messages retrieved from the combined fragment.

UML sequence diagram can model behavior of a system but they cannot be evaluated due to lack of formal semantics. Therefore, after the annotation of UML models with MARTE profile, UML model should be transformed to performance model for estimation of performance parameters followed by evaluation of alternative design based on performance. Currently, there are no approaches that investigate the issue of annotation of existing UML diagrams with MARTE profile. As a result, there is no systematic procedure followed for the annotation of UML diagrams, and all the existing papers directly work on transformed UML MARTE diagrams for analysis. However, during design stage, UML diagrams are adopted as de facto standard which can later be transformed for annotation purpose to capture performance requirements. Our main motivation is to capture the software performance requirements of a real-time system, in the early phase of software development life cycle, by modeling the performance requirements using UML profile for MARTE. This paper describes the use of MARTE stereotypes by taking up a case study of a time critical system.

The paper is structured as follows. Section 2 introduces the UML profile for MARTE along with description and usage of its stereotypes required to capture performance requirements. A case study of time critical systems is presented in Sect. 3 along with its UML modeling using sequence diagram. Section 4 describes the method involved in the transformation process of UML diagrams to UML/MARTE diagrams using the case study presented in Sect. 3, followed by conclusion of the research in Sect. 5.

2 Background

This section introduces the main concept of capturing performance requirement in UML model using UML profile for MARTE along with description and usage of its stereotypes.

2.1 Introduction to UML Profile for MARTE

The UML profile for MARTE adds capabilities to UML for model-driven development of Real-Time and Embedded Systems (RTES). MARTE defines the foundations for model-based description of real-time and embedded systems characterized by timing constraints, concurrency, etc. These core concepts are then refined for the purpose of modeling and analysis. Modeling part provides the support for specification of real-time and embedded characteristics to system design. Analysis part provides facilities to annotate the model with information required to perform specific analysis.

2.2 Profile Architecture of MARTE

The profile is structured around two main concerns, one to model the features of RTES and the other to annotate model so as to support analysis of system properties. These concepts provide generic description of real-time and embedded characteristics generalized under MARTE foundation. Diagrammatic representation of basic MARTE architecture is shown in Fig. 1.

Each package consists of various sub-packages which addresses specific concerns like timing, resource allocation etc.

MARTE Foundation Model—It is a shared package which addresses common concerns while describing use of concurrent resources and time. It consists of following sub-packages:

- **Non-Functional Properties Modeling**—Application properties are grouped into two categories: functional properties, which are concerned with what the system does at run-time; and Non-Functional Properties (NFPs), which describe how well the system performs its functions. NFPs provide information about various properties such as throughput, overhead, delays, memory usage, etc. This package provides mechanism for specification of NFPs in UML/MARTE model.
- **Time Modeling**—Real-time systems are associated with timing constraints. This package provides framework for representing time and time-related concepts of real-time system in UML/MARTE model.

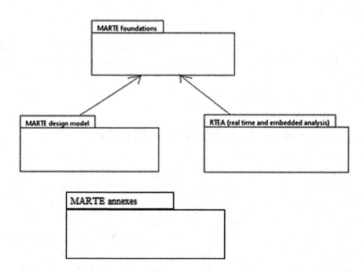

Fig. 1 Basic architecture of MARTE profile

- **General Resource Modeling**—It specifies how to describe resource model at system level. It includes features which are required for dealing with modeling of both software (operating system, etc.) and hardware (memory unit, communication channel, etc.).

MARTE Design Model—It defines the MARTE concepts for model-based design of RTES. It consists of following sub-packages:

- **High-Level Application Modeling**—It provides high-level modeling concepts to deal with real-time and embedded features modeling.
- **Software Resource Modeling (SRM)**—There are two approaches to the design of RTES applications: sequential based design approach and multitask based design approach. Applications designed with multitasking approach have specific execution mechanism on platforms requiring specific execution support. This support provides a set of resources and services for real-time features of an application. It is possible to describe the structure of such support by using modeling artifacts specified by SRM.

Real-Time and Embedded Analysis—It is focused on model-based analysis. It does not define new analysis technologies, but additional information for annotation of models for analysis. It consists of following sub-packages:

- **Generic Quantitative Analysis Modeling**—The generic analysis domain includes specialized domains in which the analysis is based on the software behavior, such as performance, availability, etc. Quantitative analysis (i.e. analysis of non-functional properties (NFPs)) techniques determine the values of 'output NFPs' based on data provided as 'input NFPs'.
- **Performance Analysis Modeling**—It describes the analysis of temporal properties of soft real-time systems, including web-based services, multimedia, networked services, etc. for which performance measures are statistical such as mean throughput or delay.

MARTE Annexes—Annexes contain useful information about various value specification languages provided by MARTE profile. Value specification language deals with specification of parameters, expressions, relationship between different variables in textual form.

2.3 MARTE Stereotypes

The UML profile for MARTE defines a set of stereotypes which allows us to map model elements to characteristics of real-time system. Stereotypes are associated with attributes that gives values for properties which are needed in order to carry out the analysis. Stereotypes used for the proposed research are summarized in Table 1.

Table 1 MARTE stereotypes

Stereotype	Description	Attributes
≪PaStep≫	A step is a unit of a scenario. It is a basic sequential execution step on a host processor	execTime—time for execution (response time minus any initial scheduling delays)
≪PaCommStep≫	A CommStep is an operation which conveys a message from one locale to another (e.g.: response from server to client). The message conveyance may be executed by a combination of host middleware and network services	msgSize—the size of the message to be transmitted by the step
≪RtService≫	It can specify the real-time features described by its attributes	concPolicy—concurrency policy used for the real-time service (reader/writer) isAtomic—when true, implies that the RtService executes as one indivisible unit, non-interleaved with other RtServices
≪Acquire≫	It is used to acquire a protected resource	isBlocking—if true it indicates that any attempt to acquire the resource may result in a blocking situation if it is not available. If false it indicates the unavailability of the protected resource will not block the caller but it will be returned as part of the service results instead
≪Release≫	It is used to free an acquired protected resource	nil
≪GaWorkload Event≫	It is a stream of events that initiate system level behavior	pattern—this attribute defines a pattern of arrival events. It can be periodic, aperiodic, irregular, etc.
≪SWConcurrent Resource≫	This resource defines entities, which may execute concurrent instructions while providing an executing context to a routine	nil
≪PaLogical Resource≫	A PaLogicalResource is a resource that can be acquired and released explicitly by AcqStep or RelStep. It may be a single unit resource, as a mutex or exclusive lock, or have multiple units, as a buffer pool or an access token pool	poolSize—the number of units of the resource
≪SharedDataCom Resource≫	It defines specific resource used to share the same area of memory among concurrent resources. They allow concurrent resources to exchange safely information by reading and writing the same area in memory	nil

3 Case Study and Its UML Modeling

To illustrate the modeling elements introduced above, a case study named "CPU Allotment" is presented in this section. First, the requirements are stated through problem description followed by modeling of the system. System modeling includes:

- Identifying use-case to depict system functionality
- Sequence diagram to understand system behavior for each use-case

3.1 Description of CPU Allotment Case Study

CPU Allotment is a web-based application that allows students to reserve CPU in advance every time they want to practice in lab. They can register to obtain login credentials required for reserving CPU, and each student can reserve only one CPU per slot. This web-based application allows a student to select date and slot corresponding to which available CPUs are displayed. The required CPU can be reserved by selecting and clicking on confirm button.

3.2 Modeling the System Using Use-Case Diagram

The above requirements can be captured by a use-case diagram shown in Fig. 2, which depicts system functionalities by means of use-cases.

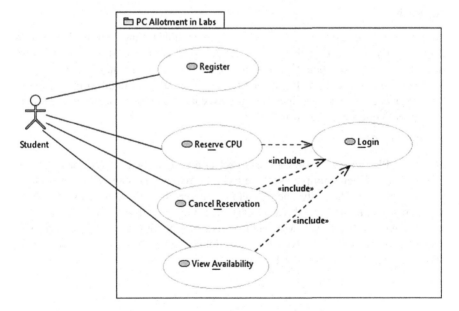

Fig. 2 Use-case diagram for CPU allotment

The use-cases are described in brief below:

- Register—Allows a first time user to obtain login credentials
- Login—Verifies whether a student is allowed to use system functionalities
- Reserve CPU—Allows a student to reserve CPU for selected slot
- Cancel reservation—Allows a user to cancel reservation for a particular slot
- View Availability—System displays available CPUs for selected slot

From the use-cases, Register and Reserve CPU are identified as performance critical. For Register use-case, whenever a student tries to set his/her UserID the system should verify that it is unique in real time. When multiple users are trying to reserve CPU concurrently, the system should allot CPU to only one user. Hence, these two use-cases involve concurrency and performance criteria for the system to respond in real time for correct functioning.

3.3 Modeling the System Using Sequence Diagram

It is assumed that application runs concurrently serving the request of users by creating a new thread for each incoming request. A database which maintains reservation/user information is shared by concurrently executing threads. Multiple read operations are allowed while only one thread can write to the database at a time. A thread cannot write while some other thread is reading from the database. Therefore, before writing, each thread has to acquire the lock and release it after writing. Sequence diagram depicting system behavior is shown Figs. 3 and 4.

A. Register

Register use-case allows a user to set Login credentials for the website. The system should verify, in real time that the userID provided is unique. If a user with same userID already exists, then it should prompt the user to change it. Once uniqueness is verified, entry corresponding to the new user should be added to the database. The sequence of messages involved for executing Register use-case scenario is shown as Sequence Diagram in Fig. 3.

B. Reserve CPU

Reserve CPU use-case allows a user to select a CPU to be reserved for a specific date and slot. When multiple users are trying to reserve the same CPU, the system should make sure that only one of them is successful. This is done by restricting write operation by multiple threads to database. Only one thread is allowed to write at a time while other threads will wait in a queue for gaining access to database. The sequence of messages involved for executing Reserve CPU use-case scenario is shown as sequence diagram in Fig. 4.

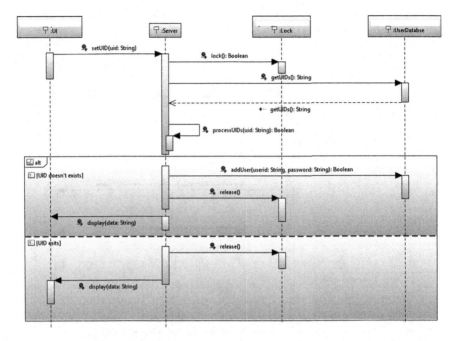

Fig. 3 Sequence diagram for Register

4 Methodology

Using UML diagrams as input, performance requirement of the system is captured by annotating UML model with MARTE profile. The process of adding stereotype labels to UML model for expressing performance (or similar quantitative) concepts is called as annotation. This process involves following steps:

1. Identifying performance critical scenarios of the system.
2. Selecting the stereotypes required to map UML model elements to characteristics of the system.
3. Defining values for stereotype attributes (tagged values) which represents quantitative properties of the system.
4. Associating stereotype labels, along with tagged values to elements of UML model.

The output of this step is annotated UML/MARTE model depicting system behavior and its real-time characteristics. Since UML lacks formal semantics, it is not possible to apply mathematical techniques directly to evaluate performance. Therefore, a transformation to performance model is required for analysis.

This section explains how each step of methodology is implemented for the case study described in Sect. 3. We begin with describing how stereotypes are applied to UML models to obtain annotated UML/MARTE Sequence Diagram.

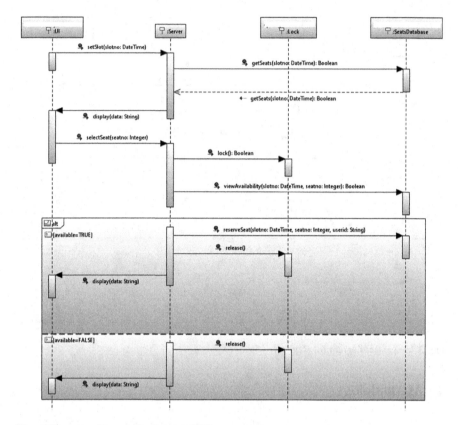

Fig. 4 Sequence diagram for Reserve CPU

Referring to Table 1, the User interface is labeled as «SwConcurrent Resource» since concurrent users interact with the application through it. Lock is annotated with «PaLogicalResource» with pool size as 1 because only one thread can acquire it at a time. Processing step annotated with «PaStep» indicates that this step is performance critical with 'execTime' attribute denoting time required for execution of this step. Function call to services which require response in real time are annotated with «RtService» stereotype. «RtService» is associated with the type of operation (read or write) being performed by using 'concPolicy' attribute. If the required operation is writing, then it is to be performed atomically specified by 'isAtomic' attribute. Response from server to client is annotated with «PaComm Step» because it is sent via a communication channel between them. Database is associated with «sharedDataCommResource» stereotype since it is shared among multiple concurrently executing threads. Annotated sequence diagram with MARTE profile for use-cases Reserve CPU and Register is shown in Figs. 5 and 6.

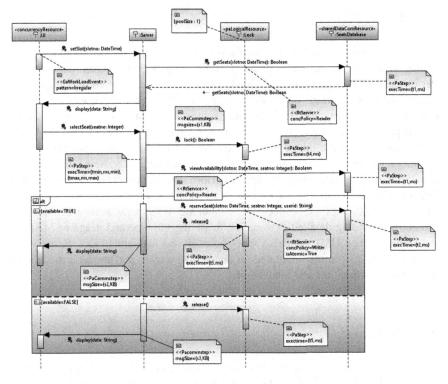

Fig. 5 Annotated sequence diagram for Reserve CPU use-case

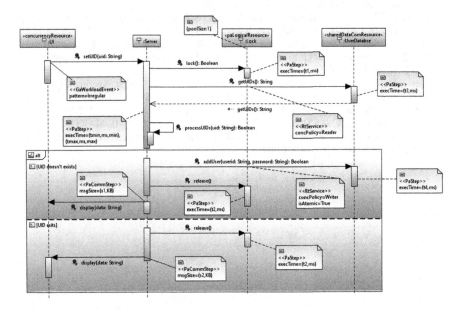

Fig. 6 Annotated sequence diagram for Register use-case

5 Conclusion

In order to demonstrate the transformation of UML diagrams to capture performance requirements, a case study of time critical system has been stated. Performance critical scenarios identified in the given case study are modeled using UML. Use-case diagrams depicting system functionalities and sequence diagrams depicting system behavior are designed. Capturing performance requirement in the design phase has increasingly become critical for real-time applications. MARTE profile allows construction of models that may be used to make quantitative predictions regarding real-time and embedded characteristics of systems, by allowing annotation of UML diagrams, to capture performance requirements. Therefore, it is important to understand the procedure involved in annotation of UML diagrams. This paper addresses the concern of annotating UML diagrams with MARTE profile by providing detailed explanation involving the type and usage of MARTE stereotypes. Since UML models cannot be analyzed mathematically, future work will include transforming annotated sequence diagrams to performance models for performance estimation and evaluation.

References

1. OMG.: The UML profile for Modeling and Analysis of Real Time and Embedded System (MARTE). Available: http://www.omgmarte.org/
2. Street, J.A., Pettit, R.G.: Lessons learned applying performance modeling and analysis techniques. In: Ninth IEEE International Symposium on Object and Component-Oriented Real-Time Distributed Computing (ISORC'06), pp. 7 (2006)
3. Traore, I., Woungang, I., Ahmed, A., Obaidat, M.S.: Software Performance Modeling using the UML: a Case Study, J. Networks, 7(1), 4–20 (2012)
4. Middleton, S.E., Servin, A., Zlatev, Z., Nasser, B., Papay, J., Boniface, M.: Experiences using the UML profile for MARTE to stochastically model post-production interactive applications. In: eChallenges e-2010 Conference, pp. 1–8 (2010)
5. Demathieu, S., Thomas, F., AndrÃ c, C., GÃ crard, S., Terrier, F.: First experiments using the UML profile for MARTE. In: 1th IEEE International Symposium on Object and Component-Oriented Real-Time Distributed Computing (ISORC), pp. 50–57 (2008)
6. Akshay, K.C., Nayak, A., Muniyal, B.: Modeling data races using UML/MARTE profile. In: International Conference on Advances in Computing, Communications and Informatics (ICACCI 2014), pp. 238–244 (2014)

On the Security of Certificateless Aggregate Signature Scheme in Vehicular Ad Hoc Networks

Pankaj Kumar and Vishnu Sharma

Abstract Certificateless aggregate signature scheme is a very effective technique for ad hoc networks such as vehicular ad hoc network. An aggregate scheme aggregates the individual signature, which reduces the computational overhead and useful in the areas, where bandwidth limitation is a major issue. Recently, Malhi and Batra proposed a certificateless aggregate signature scheme for vehicular ad hoc networks and proved the scheme mathematically secure against possible types of security attacks. In this paper, we present the cryptanalysis of the scheme while applying some concrete attack "honest but curious". The additional contribution to this paper is an improvement of the certificateless aggregated signature scheme.

Keywords Cryptography · CLAS · CLS · Digital signature

1 Introduction

The recent enlightenments in the arena of information technology make the security most crucial part of any network infrastructure design. Cryptographic techniques are providing more attractive ways to secure the systems. This is absolutely more and more difficult for the scenarios, where devices are not static. In ad hoc networks, nodes are moving independently and communicating with dedicated short range waves. Bandwidth limitation is one of the fundamental issues in ad hoc networks. On the same line, the Vehicular Ad hoc NETworks (VANETs), where nodes are moving at a relatively high speed, countenance similar issue [1, 2]. Vehicular ad hoc network is a specialized category of mobile ad hoc networks,

P. Kumar (✉) · V. Sharma
School of Computing Science and Engineering, Galgotias University, Uttar Pradesh, India
e-mail: pkumar240183@gmail.com

V. Sharma
e-mail: vishnusharma97@gmail.com

© Springer Nature Singapore Pte Ltd. 2018
M. Pant et al. (eds.), *Soft Computing: Theories and Applications*,
Advances in Intelligent Systems and Computing 583,
https://doi.org/10.1007/978-981-10-5687-1_63

consisting of mainly three components: vehicles, moving at a high speed communicating with the road side unit (RSU) situated along side of the road and an on-board unit installed in the vehicle. Primarily, VANET faces the basic problem of security such as authentication, non-repudiation, privacy, integrity and availability to protect the network. Due to the varying movement of nodes, limited bandwidth is also a foremost challenge here.

Cryptography is an important tool to provide a secure atmosphere to these networks against the above discussed security requirements. Broadly, there are two directions to solve the purpose. These are commonly known as symmetric key cryptography and public key cryptography. The higher communication cost in symmetric key cryptography fails the purpose of using resource constrained sensors in this technology. Public key cryptography uses two types of the keys, public key and private key for every single user. For bonding of these keys, we need a certificate which leads to the problem of certification overhead.

Shamir [3] presents the solution of this certification problem and suggests a novel idea namely, identity-based cryptography. In identity-based cryptography, users are free to select their public key while private key is generated by the third party. This third party is commonly known as private key generator (PKI). Several identity-based schemes have been presented with efficient results but the whole idea is experiencing a rare but possible key escrow problem. Since, private key of the user is generated by the PKI in the ID based scheme, the possibility of PKI being malicious cannot be denied and hence the PKI can forge the messages on behalf of legitimate user.

Certificateless signature scheme (CLS) provides the solution of key escrow problem. Al riyami and Peterson [4] introduced the concept of certificateless signature scheme first time in 2003. Several other certificateless signature schemes and cryptanalysis were presented [5–9]. Later, Boneh [10] presented an idea of aggregate signature scheme which takes input of the all individual signature holders, compress them and produce an aggregate signature as an output. Xiong et al. [11, 12] introduced two efficient certificateless aggregate signature (CLAS) schemes, using constant pairing, for mobile computation and vehicle ad hoc networks. Horng et al. [13] also proposed their CLAS scheme for VANETs. Various CLAS schemes were presented by the researchers in recent years [11–22]. Few schemes were cryptanalized and their improved versions have been presented [23–26]. Shen [25] found Chen et al.'s CLAS scheme [22] insecure by applying concrete attacks with adversary A_1 and adversary A_2, where adversary A_1 has the authority to replace the public key of user but cannot access the master key of KGC and adversary A_2 can access the master key of KGC but cannot change public key of the user. Shim [26] proved that the Zhang and Zhang CLAS scheme [18] is insecure in random oracle model. He et al. [24] also proved that the Xiong et al.'s scheme [11] is insecure.

2 Review of Malhi and Batra [20] CLS Scheme

Set up: Taking input as a security parameter 1^k where $k \in N$, KGC generates two groups G_1 and G_2 of same order q with generator P and a bilinear pairing.

KGC sets a master key $s \in Z_q^*$ and a public key of KGC as $P_{\text{pub}} = sP$. KGC selects two hash functions $H_2 : \{0, 1\}^* \rightarrow Z_q^*$ and $H_3 : \{0, 1\}^* \rightarrow Z_q^*$, with message space $M = \{0, 1\}^*$. Every RSU chooses a secret key $y_i \in Z_q^*$ and computes their corresponding public key as $P_{\text{rsu}_i} = y_i P$. The KGC publishes the system parameter list $\{G_1, G_2, e, P, P_{\text{pub}}, H_2, H_3, P_{\text{rsu}_1}, P_{\text{rsu}_2} \ldots, P_{\text{rsu}_i}\}$.

Vehicle registration:

1. RTA selects a cryptographic function $H_1 : \{0, 1\}^* \rightarrow G_1$
2. Register the vehicle Identity ID_i with RTA as $Q_{ID_i} = H_1(ID_i) \in G_1$, where vehicle identity space is $\{0, 1\}^*$.

PartialKeyGen: KGC takes as parameter list, master key and Q_{ID_i}, then KGC select user's partial private key $pp_i = sQ_{ID_i}$, where $Q_{ID_i} = H_1(ID_i) \in G_1$.

UserKeyGen: The Vehicle with Identity Q_{ID_i} chooses a random number $x_i \in Z_q^*$ and sets it as a secret key of corresponding vehicle with public key of vehicle as $P_i = x_i P$.

PseudonymGen: An autonomous network formed with 4 RSUs in the scheme and produces the pseudonym of the vehicle in two parts $PS1_j$ and $PS2_j$, such that $PS_j = PS1_j + PS2_j$. Select a random number $a_j \in Z_q^*$ for RSU_i and set $PS1_j = a_j Q_{ID_i}$. Second part of pseudonym $PS2_j = a_j T_j$ where $T_j = H_3(PS1_j) \in Z_q^*$

Finally, the complete pseudonym will be $PS_j = PS1_j + PS2_j$.

Sign: After taking a message $m_k \in M$, the partial private key pp_i, the secret key x_i with vehicle identity Q_{ID_i} and the corresponding public key P_i, the user generates the signature as follows:

(i)   Select a random number $r_i \in Z_q^*$, and compute $U_i = r_i P \in G_1$
(ii) Compute $h_{ijk} = H_2(m_k, PS1_j, P_i, U_i) \in Z_q^*$
(iii) Compute $V_{ijk} = pp_i.PS2_j + h_{ijk} r_i P_{\text{pub}} + h_{ijk} x_i P_{\text{rsu}_i}$
(iv) Output (U_i, V_{ijk}) as a signature on m_k

Verify: for verification of signature (U_i, V_{ijk}) of message m_k

(i)   Compute $h_{ijk} = H_2(m_k, PS1_j, P_i, U_i)$, $T_j = H_3(PS1_j)$
(ii) Verify $e(V_{ijk}, P) = e(PS1_j T_j + h_{ijk} U_i, P_{\text{pub}}) \, e(h_{ijk} P_i, P_{\text{rsu}_i})$
(iii) If the above verify equation satisfies then accept otherwise reject the signature.

Aggregate Sign: The aggregator node collects all individual signatures σ_i with corresponding identity ID_i, with pseudonyms pp_i corresponding to the vehicle's public key p_i on message m_i as input and creates an aggregate signature

$\{\sigma_1 = (U_1, V_1), \sigma_2 = (U_2, V_2)\ldots\sigma_n = (U_n, V_n)\}$ on messages $\{m_1, m_2, \ldots, m_n\}$. The aggregate signature can be computed as $V = \sum_{i=1}^{n} V_i$ and an aggregate signature pair is $\sigma = (U_1, U_2, \ldots U_n, V)$.

Aggregate Verify: The verifier can verify the aggregate signature using the following steps:

Compute $h_i = H_2(m_i, PS1_i, P_i, U_i)$, $T_i = H_3(PS1_i)$ for $i \in [1, n]$

Verify the following equation

$$e(V, P) = e\left(\sum_{i=1}^{n} [PS1_i.T_i + h_i.U_i], P_{pub}\right) e\left(\sum_{i=n}^{n} h_i.P_i, P_{rsu}\right)$$

If it satisfies then accept otherwise reject the aggregate signature.

3 Cryptanalysis of Malhi and Batra [20] CLS Scheme

Malhi and Batra [20] proposed an efficient certificateless aggregate signature and proved that it is secure against adaptive chosen message and identity attacks. In this section, we demonstrate that it is not secure against type-2 adversary attack.

Since A_2 is the attacking adversary who knows the master key of KGC and with the help of master key, the adversary A_2 can find partial private key of user by $pp_i = sQ_{ID_i}$ where $Q_{ID_i} = H_1(ID_i) \in G_1$.

A_2 can query the sign oracle and extract a valid signature (U_i, V_{ijk}) on the message m_k

i)   Select a random number $r_i \in Z_q^*$ and compute $U_i = r_i P \in G_1$

ii) Compute $h_{ijk} = H_2(m_k, PS1_j, P_i, U_i) \in Z_q^*$, $T_j = H_3(PS1_j)$

iii) Compute $V_{ijk} = pp_i.PS2_j + h_{ijk}r_i P_{pub} + h_{ijk}x_i P_{rsu_i}$

Now, A_2 can intercept information $T_{ijk} = h_{ijk}^{-1}(V_{ijk} - pp_i.PS2_j)$ with $h_{ijk}.h_{ijk}^{-1} \equiv 1$.

Choose another message m_k', A_2 computes $U' = U$ and $h_{ijk}' = H_2(m_k', PS1_j, P_i, U_i') \in Z_q^*$.

Then, compute $V_{ijk}' = pp_i.PS2_j + h_{ijk}' T_{ijk}$.

Now, A_2 outputs the signature (U_i', V_{ijk}') on the message m_k'.

$$T_{ijk} = h_{ijk}^{-1}(V_{ijk} - pp_i.PS2_j)$$

$$T_{ijk} = h_{ijk}^{-1}\left(pp_i.PS2_j + h_{ijk}r_i P_{pub} + h_{ijk}x_i P_{rsu_i} - pp_i.PS2_j\right)$$
$$= r_i P_{pub} + x_i P_{rsu_i}$$

$$V'_{ijk} = pp_i.PS2_j + h'_{ijk}T_{ijk}$$
$$= pp_i.PS2_j + h'_{ijk}(r_iP_{pub} + x_iP_{rsu_i})$$
$$= pp_i.PS2_j + h'_{ijk}r_iP_{pub} + h'_{ijk}x_iP_{rsu_i}$$

Correctness:

$$e(V'_{ijk}, P) = e(pp_i.PS2_j + h'_{ijk}r_iP_{pub} + h'_{ijk}x_iP_{rsu_i}, P)$$
$$= e(s.Q_{ID_i}.a_j.PS1_j, P)\,e(h'_{ijk}r_iP, P_{pub})\,e(h'_{ijk}x_iP, P_{rsu_i})$$
$$= e(PS1_j.T_j, sP)\,e(h'_{ijk}U_i, P_{pub})\,e(h'_{ijk}P_i, P_{rsu_i})$$
$$= e(PS1_j.T_j, P_{pub})\,e(h'_{ijk}U_i, P_{pub})\,e(h'_{ijk}P_i, P_{rsu_i})$$
$$= e(PS1_j.T_j + h'_{ijk}U_i, P_{pub})\,e(h'_{ijk}P_i, P_{rsu_i})$$

4 Cryptanalysis of Malhi and Batra [20] CLAS Scheme

Adversary A_2 performs the following steps to forge the certificateless aggregate signature.

Since A_2 knows the master key of KGC and he can find partial private key of user by $pp_i = sQ_{ID_i}$ where $Q_{ID_i} = H_1(ID_i) \in G_1$

A_2 queries the sign oracle and extract a valid signature (U_i, V_i) on the message m_i where

(i) Compute $U_i = r_iP \in G_1$, where $r_i \in Z_q^*$
(ii) Compute $h_i = H_2(m_i, PS1_i, P_i, U_i) \in Z_q^*$, $T_i = H_3(PS1_i)$
(iii) Compute $V_i = pp_i.PS2_j + h_ir_iP_{pub} + h_ix_iP_{rsu}$

Now A_2 can find $T_i = h_i^{-1}(V_i - pp_i.PS2_i)$ where h_i satisfying $h_i.h_i^{-1} \equiv 1$. Choose any m'_i, A_2 computes $U' = U$ and $h'_i = H_2(m'_i, PS1_i, P_i, U'_i) \in Z_q^*$. Then compute $V'_i = pp_i.PS2_i + h'_iT_i$.

A_2 outputs the valid signature (U'_i, V'_i) on the message m'_i.

$$T_i = h_i^{-1}(V_i - pp_i.PS2_i)$$

$$T_i = h_i^{-1}(pp_i.PS2_i + h_ir_iP_{pub} + h_ix_iP_{rsu} - pp_i.PS2_i)$$
$$= r_iP_{pub} + x_iP_{rsu}$$

A_2 collect the entire individual signature $V = \sum_{i=1}^{n} V_i'$ and provides an aggregate signature $\sigma' = (U_1', U_2', \ldots, U_n', V)$ with corresponding public key p_i on messages $\{m_1, m_2, \ldots, m_n\}$.

$$V = \sum_{i=1}^{n} pp_i.PS2_j + \sum_{i=1}^{n} h_i'.T_i$$
$$= \sum_{i=1}^{n} pp_i.PS_j + \sum_{i=1}^{n} h_i'(r_i P_{\text{pub}} + x_i P_{\text{rsu}})$$
$$= \sum_{i=1}^{n} pp_i.PS2_j + \sum_{i=1}^{n} h_i' r_i P_{\text{pub}} + \sum_{i=1}^{n} h_i' x_i P_{\text{rsu}}$$

A_2 can verify the signature by the following equation.

$$e(V, P) = e\left(\sum_{i=1}^{n} [PS1_i.T_i + + h_i.U_i'], P_{\text{pub}} \right) e\left(\sum_{i=1}^{n} h_i'.P_i, P_{\text{rsu}} \right)$$

5 Improvement of Malhi and Batra [20] CLAS Scheme

Set up: Given input as a security parameter 1^k where $k \in N$ then KGC generates an additive cyclic group and multiplicative group G_2 of same order q with generator P the bilinear pairing. The KGC generates a master key $s \in Z_q^*$ and set public key of KGC as $P_{\text{pub}} = sP$. The KGC generates two different hash functions $H_2 : \{0,1\}^* \rightarrow Z_q^*$ and $H_3 : \{0,1\}^* \rightarrow Z_q^*$, $H_4 : \{0,1\}^* \rightarrow Z_q^*$ where message space is $M = \{0,1\}^*$. Each RSU sets a secret key of $y_i \in Z_q^*$ and then a corresponding public key is $P_{\text{rsu}_i} = y_i P$. The KGC declares the system parameter list as $\{G_1, G_2, e, P, P_{\text{pub}}, H_2, H_3, H_4, P_{\text{rsu}_1}, P_{\text{rsu}_2}, \ldots, P_{\text{rsu}_i}\}$. The next following algorithms Partialkeygen, Userkeygen, PseudonymGen are same as above CLS scheme.

Sign: Given a message $m_k \in M$, a partial private key pp_i, a secret key x_i with vehicle identity Q_{ID_i}, a corresponding public key P_i as input and generates the signature as follows:

(i)   Select a random $r_i \in Z_q^*$, and compute $U_i = r_i P \in G_1$
(ii) Compute $h_{ijk} = H_2(m_k, PS1_j, P_i, U_i) \in Z_q^*$, $t_{ijk} = H_3(m_k, PS1_j, P_i, U_i) \in Z_q^*$
(iii) Compute $V_{ijk} = pp_i.PS2_j + h_{ijk} r_i P_{rsu_i} + t_{ijk} x_i P_{rsu_i}$
(iv) Output (U_i, V_{ijk}) as a signature on m_k.

Verification: For signature verification (U_i, V_{ijk}) of message m_k, the verifier takes the following action.

(i) Compute $h_{ijk} = H_2(m_k, PS1_j, P_i, U_i), T_j = H_3(PS1_j) \in Z_q^*, t_{ijk} = H_4(m_k, PS1_j, P_i, U_i)$
$\in Z_q^*$

(ii) Verify $e(V_{ijk}, P) = e(PS1_j T_j, P_{pub}) e(h_{ijk}.U_i + t_{ijk}.P_i, P_{rsu_i})$

(iii) If it satisfies then accept otherwise reject the signature

Correctness:

$$
\begin{aligned}
e(V_{ijk}, P) &= e(pp_i.PS2_j + h_{ijk}.r_i.P_{rsu_i} + t_{ijk}.x_i.P_{rsu_i}, P) \\
&= e(s.Q_{ID_i}.a_j T_j, P) e(h_{ijk}.r_i.P, P_{rsu_i}) e(t_{ijk}.x_i.P, P_{rsu_i}) \\
&= e(PS_1.T_j, sP) e(h_{ijk}.U_i, P_{rsu_i}) e(t_{ijk}.P_i, P_{rsu_i}) \\
&= e(PS_1.T_j, P_{pub}) e(h_{ijk}.U_i, P_{rsu_i}) e(t_{ijk}.P_i, P_{rsu_i}) \\
&= e(PS_1.T_j, P_{pub}) e(h_{ijk}.U_i + t_{ijk}.P_i, P_{rsu_i})
\end{aligned}
$$

Aggregate Verify: Verifier can verify the aggregate signature by the following steps:

(1) Compute $h_i = H_2(m_i, PS1_i, P_i, U_i), T_i = H_3(PS1_i)$ for $i \in [1, n]$

(2) Compute $t_i = H_4(m_i, PS1_i, P_i, U_i) \in Z_q^*$

Verify the following equation

$$
e(V, P) = e\left(\sum_{i=1}^{n} PS1_i.T_i, P_{pub}\right) e\left(\sum_{i=1}^{n} (h_i.U_i + t_i P_i, P_{rsu})\right)
$$

6 Conclusion

Recently, Malhi and Batra introduced a certificateless aggregate signature scheme which is applicable in vehicle ad hoc networks. In this paper, we presented a cryptanalysis of their CLAS scheme and found it insecure against type-2 adversary. In type-2 adversary, KGC plays the role of malicious adversary and knows the partial private key of the user. Thereafter, we proposed a modification of their CLAS scheme and our primary focus is on elimination of security flaws in the victim scheme.

References

1. Khan, A.R.: Analysis the channel allocation for removing the traffic problems from the roaming systems. Int. J. Adv. Res. Eng. Appl. Sci. 3(7), 74–86 (2014)
2. Khan, A.R.: Possible solution for traffic in roaming system. Int. J. Adv. Res. Eng. Appl. Sci. 3 (8), 1–15 (2014)

3. Shamir, A.: Identity based cryptosystems and signature schemes, Crypto'84, LNCS 196, pp. 47–53. Springer-Verlag, Santa Barbara, USA (1984)
4. Al-Riyami, S., Paterson, K.: Certificateless public key cryptography, Asiacrypt' 03, LNCS 2894, pp. 452–473. Springer-Verlag (2003)
5. Sharma, G., Bala, S., Verma, A. K.: On the security of certificateless signature schemes. Int. J. Distrib. Sens. Netw. 2013, Article ID 102508, 6 p. (2013) Hindawi. doi:10.1155/2013/102508
6. Bala, S., Sharma, G., Verma, A.K.: Cryptanalysis of certificateless signature scheme. Int. Inform. Inst. J **16**(11), 7827–7830 (2013)
7. Sharma, G., Verma, A.K.: Breaking the RSA based certificateless signature scheme. Int. Inf. Inst. J **16**(11), 7831–7836 (2013)
8. Sharma, G., Bala, S., Verma, A.K.: An Improved RSA based certificateless signature scheme for wireless sensor networks. Int. J. Netw. Secur. (IJNS) **18**(1), 82–89 (2016)
9. Sharma, G., Bala, S., Verma, A.K.: Extending certificateless authentication for wireless sensor networks: a novel insight. Int. J. Comput. Sci. Issues **10**(6), 167–172 (2013)
10. Boneh, D., Gentry, C., Lynn, B., Shacham, H.: Aggregate and verifiably encrypted signatures from bilinear maps, Eurocrypt 2003, LNCS 2656, pp. 416–432. Springer-Verlag, Poland (2003)
11. Xiong, H., Guan, Z., Chen, Z., Li, F.: An efficient certificateless aggregate signature with constant pairing computations. Inf. Sci. **219**, 225–235 (2013)
12. Xiong, H., Qianhong, W., Zhong, C.: Strong security enabled certificateless aggregate signatures applicable to mobile computation. In: Third International Conference on Intelligent Networking and Collaborative Systems, pp. 92–99 (2011)
13. Horng, S., Tzeng, P., Wang, X., Li, T., Khan, M.K.: An efficient certificateless aggregate signature with conditional privacy-preserving for vehicular sensor networks. Inf. Sci. **317**, 48–66 (2015)
14. Huang, X., Mu, Y., Susilo, W., Wong, D.S., Wu, W.: Certificateless signatures: new schemes and security models. Comput. J. **55**(4), (2012)
15. Eslami, Z., Pakniat, N.: Certificateless aggregate signcryption: security model and a concrete construction secure in the random oracle model. J. King Saud Univ. Comput. Inform. Sci. **26**, 276–286 (2014)
16. Zhang, Z., Wong, D.S., Xu, J., Feng, D.: Certificateless public-key signature: security model and efficient construction. In: Applied Cryptography and Network Security, Lecture Notes in Computer Science, vol. **3989**, pp. 293–308. Springer, Germany (2006)
17. Chen, Y.C., Horng, G., Liu, C., Tsai, Y., Chan, C.: Efficient certificateless aggregate signature scheme. J. Electron. Sci. Technol. **10**(3), 209–214 (2012)
18. Zhang, L., Zhang, F.: A new certificateless aggregate signature scheme. Comput. Commun. **32**(6), 1079–1085 (2009)
19. Zhang, L., Qin, B., Wu, Q., Zhang, F.: Efficient many-to-one authentication with certificateless aggregate signatures. Comput. Netw. **54**(14), (2010)
20. Malhi, A.K., Batra, S.: An efficient certificateless aggregate signature scheme for vehicular ad-hoc networks. Discrete Math. Theor. Comput. Sci. DMTCS, **17**(1) (2015)
21. Zhang, L., Zhang, F.: A new certificateless aggregate signature scheme. Comput. Commun. **32**, 1079–1085 (2009)
22. Chen, Y., Tso, R., Mambo, M., Huang, K., Horng, G.: Certificateless aggregate signature with efficient verification. Secur. Commun. Netw. (2015)
23. Zhang, F., Shen, L., Wu, G.: Notes on the security of certificateless aggregate signature schemes. Inf. Sci. **287**, 32–37 (2014)
24. He, D., Miaomiao, T., Jianhua, C.: Insecurity of an efficient certificateless aggregate signature with constant computations. Inf. Sci. **268**, 458–462 (2014)
25. Shen, H., Chen, J., Shen, J., Debiao, H.: Cryptanalysis of a certificateless aggregate signature scheme with efficient verification. Secur. Commun. Netw. (2016)
26. Shim, K.A.: On the Security of a certificateless aggregate signature scheme. IEEE Commun. Lett., **15**(10), (2011)

Fuzzy Transportation Problem with Generalized Triangular-Trapezoidal Fuzzy Number

Rajesh Kumar Saini, Atul Sangal and Om Prakash

Abstract The shortcoming of an existing method for comparing the proposed new method for generalized triangular-trapezoidal fuzzy numbers (TTFN) are pointed out in this paper. Here the proposed ranking method is used for solving unbalanced fuzzy transportation problem (UFTP). A comparison is set between the optimal fuzzy and crisp solutions by using Vogel's approximation method (VAM) and improved zero suffix method (IZSM) after balancing by existing and proposed minima row–column method (MRCM). The effectiveness of proposed method is illustrated by a numerical example and setting a comparison among optimal solutions.

Keywords Generalized triangular-trapezoidal fuzzy numbers · Unbalanced transportation problem · Minima row–column method

1 Introduction

In transportation problem if costs, supply and demand are fuzzy quantities, then it called fuzzy transportation problem (FTP) [1, 2]. Hitchcock [3] originally developed the basic transportation problem. O'heigeartaigh [4] proposed an algorithm for solving TP where the capacities and requirements are fuzzy sets with linear are triangular membership functions. Recently Kaur and Kumar [5] proposed a new

R.K. Saini (✉)
Department of Mathematics, BU, Jhansi, U.P, India
e-mail: rksaini.bu@gmail.com

A. Sangal
Department of Management Studies, Sharda University, Greater Noida, U.P, India
e-mail: atul.sangal@rediffmail.com

O. Prakash
Department of Mathematics, IIT, Patna, Bihar, India
e-mail: om@iitp.ac.in

© Springer Nature Singapore Pte Ltd. 2018
M. Pant et al. (eds.), *Soft Computing: Theories and Applications*,
Advances in Intelligent Systems and Computing 583,
https://doi.org/10.1007/978-981-10-5687-1_64

method for solving FTP using ranking function. For different authors' approaches to solve FTP see [4, 6–15].

Before any action to be taken by a decision maker, an FN must be ranked. A real number may be linearly related as either $\leq$ or $\geq$; however, this type of inequality does not exist in an FN. Since an FN is represented by possibility distribution and can overlap with each other, so it is difficult to determine exactly and clearly weather a fuzzy number is $\leq$ or $\geq$ to another one. So that for ordering of an FN, which maps each FN on the real line, the ranking method is more efficient.

Kaufmann and Gupta [16] in 1985 proposed normalization processes to convert an FN into an NFN, by which solve the real-life problems by obtaining NFN but they pointed out that there is a serious disadvantage in the proposed normalization process see [17–20]. The present problem contains a proposed new ranking method for converting generalized TTFN into crisp numbers.

2 Preliminaries

In this section some basic definitions remarks and arithmetic operations are reviewed.

Definition 2.1 In a crisp set $A \in X$ the characteristic function $\mu_{\tilde{A}}$ assigns values either 0 or 1 to each member in X. If characteristic function $\mu_{\tilde{A}}$ generalized as $\mu_{\tilde{A}} : X \rightarrow [0,1]$, i.e., $0 \leq \mu_{\tilde{A}} \leq 1$ then ordered pair of a number $x \in X$ with $\mu_{\tilde{A}}$ is called FN where $\mu_{\tilde{A}}$ is called the membership function or membership grade of $x \in X$ and the set $\tilde{A} = \{(x, \mu_{\tilde{A}}(x)) : x \in X\}$ is called a fuzzy set.

Definition 2.2 For $a_1, a_2, a_3, a_4, a_5, a_6, a_7 \in R$, where $a_1 \leq \cdots \leq a_7$ a fuzzy set on R is generalized if its membership function as follows:

(i) $\mu_{\tilde{A}} : R \rightarrow [0,1]$ is continuous
(ii) $\mu_{\tilde{A}} = 0$, for all $x \in (-\infty, a_1] \cup [a_7, \infty)$
(iii) $\mu_{\tilde{A}}(x)$ is strictly increasing on $[a_1, a_2] \cup [a_3, a_4]$ and strictly decreasing on $[a_4, a_5] \cup [a_6, a_7]$
(iv) $\mu_{\tilde{A}}(x) = \Omega$ for all $0 \leq \Omega \leq 1$.

Remark 2.1 On the basis of collected data by an expert is a generalized TTFN $\tilde{A} = (a_1, a_2, a_3, a_4, a_5, a_6, a_7; \Omega)$.

Definition 2.3 For $r \in [0, w]$ and $t \in [w, \Omega]$ the generalized TTFN is $\tilde{A}_{\Omega} = (l_1(r), s_1(t), s_2(t), l_2(r))$, shown in Fig. 1.

Remark 2.2 For normality $\Omega = 1$.

Fig. 1 Generalized TTFN

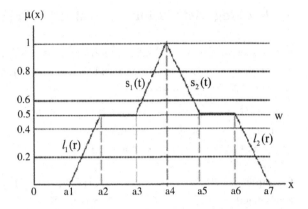

Definition 2.4 A generalized TTFN, denoted by $\tilde{A} = (a_1, a_2, a_3, a_4, a_5, a_6, a_7; \Omega)$ where $a_1, a_2, a_3, a_4, a_5, a_6, a_7$ and w are real numbers and its membership function is defined as

$$
\mu_{\tilde{A}}(x) = \begin{cases}
0, & \text{for } x < a_1 \\
w\{(x - a_1)/(a_2 - a_1)\}, & \text{for } a_1 \leq x \leq a_2 \\
w, & \text{for } a_2 \leq x \leq a_3 \\
w + (\Omega - w)\{(x - a_3)/(a_4 - a_3)\}, & \text{for } a_3 \leq x \leq a_4 \\
1, & \text{for } x = a_4 \\
\Omega + (\Omega - w)\{(a_5 - x)/(a_5 - a_4)\}, & \text{for } a_4 \leq x \leq a_5 \\
w, & \text{for } a_5 \leq x \leq a_6 \\
w\{(a_7 - x)/(a_7 - a_6)\}, & \text{for } a_6 \leq x \leq a_7 \\
0, & \text{for otherwise}
\end{cases}
$$

where $0 < w < \Omega$.

Remark 2.3 If $w = 0$, the generalized TTFN reduces to the TFN (a_3, a_4, a_5) and if $w = 1$, it reduces to TPFN (a_1, a_2, a_6, a_7).

Definition 2.5 A generalized TTFN $\tilde{A}_\Omega$ is the ordered quadruple $\tilde{A}_\Omega = (l_1(r), s_1(t), s_2(t), l_2(r))$, for $r \in [0, w]$ and $t \in [w, \Omega]$, where

$$
l_1(r) = w\left(\frac{r - a_1}{a_2 - a_1}\right), \qquad s_1(t) = w + (\Omega - w)\left(\frac{t - a_3}{a_4 - a_3}\right),
$$
$$
s_2(t) = \Omega + (\Omega - w)\left(\frac{a_5 - t}{a_5 - a_4}\right), \qquad l_2(r) = w\left(\frac{a_7 - r}{a_7 - a_6}\right).
$$

Remark 2.4 Here $\tilde{A}_\Omega$ a continuous function.

Definition 2.6 If $\tilde{A}_\Omega$ be a TTFN, then the α-cut of $\tilde{A}_\Omega$ is

$$
[\tilde{A}_\Omega]_\alpha = \{x/\tilde{A}_\Omega \geq \alpha\} = \begin{cases}
l_1(\alpha), l_2(\alpha) & \text{for } \alpha \in [0, w) \\
s_1(\alpha), s_2(\alpha) & \text{for } \alpha \in [w, \Omega)
\end{cases}
$$

3   Ranking Method for Generalized TTFN

In this section, a new method is proposed for the ranking of generalized TTFN.

Let $\tilde{A} = (a_1, a_2, a_3, a_4, a_5, a_6, a_7; \Omega_1)$ and $\tilde{B} = (b_1, b_2, b_3, b_4, b_5, b_6, b_7; \Omega_2)$ be two generalized TTFN, then

(i) $\text{rank}(\tilde{A}) >_R \text{rank}(\tilde{B})$, whenever $\tilde{A} >_R \tilde{B}$,

(ii) $\text{rank}(\tilde{A}) <_R \text{rank}(\tilde{B})$, whenever $\tilde{A} <_R \tilde{B}$

(iii) $\text{rank}(\tilde{A}) = \text{rank}(\tilde{B})$, whenever $\tilde{A} \approx_R \tilde{B}$

Example 3.1 If $\tilde{A} = (1, 1, 1, 1, 1, 1, 1; \Omega_1)$, $\tilde{B} = (1, 1, 1, 1, 1, 1, 1; \Omega_2)$ be two generalized   TTFN,   then $\tilde{A} > \tilde{B}$ if $\Omega_1 > \Omega_2$, $\tilde{A} < \tilde{B}$ if $\Omega_1 < \Omega_2$ and $\tilde{A} \approx \tilde{B}$ if $\Omega_1 \approx \Omega_2$.

Example 3.2 Let $\tilde{A} = (-1, 1, 2, 5, 6, 7, 8; \Omega_1)$, $\tilde{B} = (-1, 1, 4, 5, 6, 7, 8; \Omega_2)$ be two generalized TTFN, then $\tilde{A} \approx \tilde{B}$ for all values of Ω_1, Ω_2.

Chen and Chen [21] in 2009, pointed out a new method for the ranking of generalized FN. Kaur and Kumar [5] also using the proposed ranking method, which satisfy all the condition of Wang and Kerre [22] method for generalized FN to solve the complex FTP in real life. For details see [23–25]. In this section, a new method is proposed for the ranking of generalized TTFN.

4   Method for Finding the Values $\mathcal{R}(\tilde{A})$ and $\mathcal{R}(\tilde{B})$

Let $\tilde{A} = (a_1, a_2, a_3, a_4, a_5, a_6, a_7; \Omega_1)$ and $\tilde{B} = (b_1, b_2, b_3, b_4, b_5, b_6, b_7; \Omega_2)$ be two generalized TTFN, then use the following steps to find the values of $\mathcal{R}(\tilde{A})$ and $\mathcal{R}(\tilde{B})$. Choose $\Omega = \min\{\Omega_1, \Omega_2\}$

$$\mathcal{R}(\tilde{A}) = \frac{1}{2} \int\limits_0^w \{l_1(x) + l_2(x)\} dx + \frac{1}{2} \int\limits_w^\Omega \{s_1(y) + s_2(y)\} dy$$

where

$$l_1(x) = a_1 + \frac{(a_2 - a_1)}{w} x, \qquad l_2(x) = a_6 + \frac{(a_7 - a_6)}{w} x,$$
$$s_1(x) = a_3 + \frac{(a_4 - a_3)}{(\Omega - w)} (x - w), \qquad s_2(x) = a_5 + \frac{(a_5 - a_4)}{(\Omega - w)} (x - w)$$

Thus

$$\mathcal{R}(\tilde{A}) = \frac{1}{2} \int_0^w \left\{ \left(a_1 + \frac{(a_2 - a_1)}{w} x \right) + \left(a_7 - \frac{(a_7 - a_6)}{w} x \right) \right\} dx$$

$$+ \frac{1}{2} \int_w^\Omega \left\{ \left(a_3 + \frac{(a_4 - a_3)}{(\Omega - w)} (y - w) \right) + \left(a_5 - \frac{(a_5 - a_4)}{(\Omega - w)} (y - \Omega) \right) \right\} dy$$

$$\Rightarrow \mathcal{R}(\tilde{A}) = \frac{1}{2} \left[\frac{w}{2} \{ (a_1 + a_2 + a_6 + a_7) \} + \frac{(\Omega - w)}{2} \{ (a_3 + 2a_4 + a_5) \} \right]$$

For $w = \Omega/2$,

$$\mathcal{R}(\tilde{A}) = \frac{\Omega}{8} [a_1 + a_2 + a_3 + 2a_4 + a_5 + a_6 + a_7]$$

Similarly

$$\mathcal{R}(\tilde{B}) = \frac{\Omega}{8} [b_1 + b_2 + b_3 + 2b_4 + b_5 + b_6 + b_7], \quad \text{where } \Omega \in [0, 1].$$

Example 4.1 Let $\tilde{A} = (-2, 0, 2, 4, 5, 7, 9; 0.7)$, $\tilde{B} = (-3, -1, 0, 1, 3, 5, 7; 0.8)$ and $\tilde{C} = (0, 1, 2, 3, 4, 6, 8; 0.6)$. Now to compute $\mathcal{R}(\tilde{A})$, $\mathcal{R}(\tilde{B})$ and $\mathcal{R}(\tilde{C})$, choose $\Omega = \min\{\Omega_1, \Omega_2\}$, for $\alpha = 0.4$, we have $\mathcal{R}(\tilde{A}) = 1.66$, $\mathcal{R}(\tilde{B}) = 0.63$ and $\mathcal{R}(\tilde{C}) = 1.31$, which implies that $\tilde{A} >_R \tilde{C} >_R \tilde{B}$ also $(\tilde{A} \oplus \tilde{B}) >_R (\tilde{C} \oplus \tilde{B})$ and $(\tilde{A} \ominus \tilde{B}) >_R (\tilde{B} \ominus \tilde{B})$. Thus the proposed method satisfy all the conditions of ranking function as in [3, 11, 31].

5 Some Operations

Let $\tilde{A} = (a_1, a_2, a_3, a_4, a_5, a_6, a_7; \Omega_1)$ and $\tilde{B} = (b_1, b_2, b_3, b_4, b_5, b_6, b_7; \Omega_2)$ be two generalized TTFN defined on universal set R, then

(i) $\tilde{A} \oplus \tilde{B} = (a_1 + b_1, a_2 + b_2, a_3 + b_3, a_4 + b_4, a_5 + b_5, a_6 + b_6, a_7 + b_7; \min(\Omega_1, \Omega_2))$

(ii) $\tilde{A} \ominus \tilde{B} = (a_1 - b_7, a_2 - b_6, a_3 - b_5, a_4 - b_4, a_5 - b_3, a_6 - b_2, a_7 - b_1; \min(\Omega_1, \Omega_2))$

(iii) $\tilde{A} \otimes \tilde{B} = (t_1, t_2, t_3, t_4, t_5, t_6, t_7; \Omega)$
where

$t_1 = \min(a_1 \otimes b_1, a_1 \otimes b_7, a_7 \otimes b_1, a_7 \otimes b_7)$, $t_2 = \min(a_2 \otimes b_2, a_2 \otimes b_6, a_6 \otimes b_2, a_6 \otimes b_6)$

$t_3 = \min(a_3 \otimes b_3, a_3 \otimes b_5, a_5 \otimes b_3, a_5 \otimes b_5)$, $t_4 = a_4 \otimes b_4$

$t_5 = \max(a_3 \otimes b_3, a_3 \otimes b_5, a_5 \otimes b_3, a_5 \otimes b_5)$, $t_6 = \max(a_2 \otimes b_2, a_2 \otimes b_6, a_6 \otimes b_2, a_6 \otimes b_6)$

$t_7 = \max(a_1 \otimes b_1, a_1 \otimes b_7, a_7 \otimes b_1, a_7 \otimes b_7)$, $\Omega = \min(\Omega_1, \Omega_2)$

(iv) $\tilde{A} \div \tilde{B} = \tilde{A} \otimes \frac{1}{\tilde{B}} = (a_1, a_2, a_3, a_4, a_5, a_6, a_7; \Omega_1) \otimes \left(\frac{1}{b_1}, \frac{1}{b_2}, \frac{1}{b_3}, \frac{1}{b_4}, \frac{1}{b_5}, \frac{1}{b_6}, \frac{1}{b_7}; \Omega_2 \right)$

(v) $\lambda \tilde{A} = \begin{cases} (\lambda a_1, \lambda a_2, \lambda a_3, \lambda a_4, \lambda a_5, \lambda a_6, \lambda a_7; \Omega) & \text{if } \lambda > 0, \\ (\lambda a_7, \lambda a_6, \lambda a_5, \lambda a_4, \lambda a_3, \lambda a_2, \lambda a_1; \Omega) & \text{if } \lambda < 0. \end{cases}$

6 MRCM Method

The minimum row–column method (MRCM) is the generalization of method in [26] for TTFN. The following steps are used to balance the UFTP by MRCM:

Step 6.1 Balance the given UFTP as follows:

$$\tilde{a}_{m+1} = \sum_{i=1}^{m} \tilde{a}_i \quad \text{and} \quad \tilde{b}_{n+1} = \sum_{i=1}^{n} \tilde{a}_i \oplus \text{excess supply}.$$

or

$$\tilde{b}_{n+1} = \sum_{i=1}^{n} \tilde{b}_i \quad \text{and} \quad \tilde{a}_{m+1} = \sum_{i=1}^{m} \tilde{b}_i \oplus \text{excess demand}.$$

The unit transportation costs are taken as follows:

$$\tilde{c}_{i(n+1)} = \min_{1 \le j \le n} (\tilde{c}_{ij}), 1 \le i \le m, \quad \tilde{c}_{(m+1)j} = \min_{1 \le i \le m} (\tilde{c}_{ij}), 1 \le j \le n,$$

$$\tilde{c}_{ij} = \tilde{c}_{ji}, 1 \le i \le m, 1 \le j \le n, \quad \text{and} \quad \tilde{c}_{(m+1)(n+1)} = \min_{1 \le i \le m} (0, 0, 0, 0, 0, 0, 0; 0.8).$$

Step 6.2 After balancing the problem convert the fuzzy values into crisp values by using given ranking function.

Step 6.3 Obtain optimal crisp as well as fuzzy solution of the FTP (step 6.1).

Step 6.4 Assuming $\tilde{u}'_{m+1} = (0, 0, 0, 0, 0, 0, 0; 0.8)$ for $\tilde{u}'_i \oplus \tilde{v}'_j = \tilde{c}'_{ij}$.

Step 6.5 According to MRCM, $\tilde{u}'_i = \tilde{u}_i$, and $\tilde{v}'_j = \tilde{v}_j$, for $1 \le i \le m, 1 \le j \le n$, obtain only central rank zero duals, after that in terms of original sources S_i and destinations D_j find the fuzzy optimal solution of the problem. Increase the value of the basic variables $\tilde{x}_{(m+1)p}$ and $\tilde{x}_{q(n+1)}$ in the cell with $\min_{i \in I} \tilde{c}_{ip}$ and $\min_{j \in J} \tilde{c}_{qj}$ [26].

We use VAM [8] and IZSM [25] for finding an optimal solution to the UFTP of generalized TTFN.

7 Mathematical Formulation

Consider an UFTP with generalized TTFN having m sources S_i, for $(0 \leq i \leq m)$ with fuzzy availability a_1, a_2, a_3, …, a_m and n destinations D_n for $(0 \leq j \leq n)$ with fuzzy demand b_1, b_2, b_3, …, b_n. The transportation problem $(P1)$ is:

$$(P1) \quad \text{Minimize} \quad \sum_{i=1}^{m} \sum_{j=1}^{n} \tilde{c}_{ij} \otimes \tilde{x}_{ij}$$

$$\text{subject to} \quad \sum_{j=1}^{n} \tilde{x}_{ij} \succcurlyeq \tilde{a}_i, (1 \leq i \leq m)$$

$$\sum_{i=1}^{m} \tilde{x}_{ij} \succcurlyeq \tilde{b}_j, (1 \leq i \leq m), \; \tilde{x}_{ij} \succcurlyeq 0, \quad \text{for } 0 \leq \Omega \leq 1.$$

where

$$\tilde{x}_{ij} \approx \left(\tilde{x}_{ij}^1, \tilde{x}_{ij}^2, \tilde{x}_{ij}^3, \tilde{x}_{ij}^4, \tilde{x}_{ij}^5, \tilde{x}_{ij}^6, \tilde{x}_{ij}^7; \Omega \right), \quad \tilde{c}_{ij} \approx \left(\tilde{c}_{ij}^1, \tilde{c}_{ij}^2, \tilde{c}_{ij}^3, \tilde{c}_{ij}^4, \tilde{c}_{ij}^5, \tilde{c}_{ij}^6, \tilde{c}_{ij}^7; \Omega \right)$$

$$\tilde{a}_i \approx \left(\tilde{a}_i^1, \tilde{a}_i^2, \tilde{a}_i^3, \tilde{a}_i^4, \tilde{a}_i^5, \tilde{a}_i^6, \tilde{a}_i^7; \Omega \right) \quad \tilde{b}_j \approx \left(\tilde{b}_j^1, \tilde{b}_j^2, \tilde{b}_j^3, \tilde{b}_j^4, \tilde{b}_j^5, \tilde{b}_j^6, \tilde{b}_j^7; \Omega \right).$$

The necessary and sufficient condition for the solution of $(P1)$ is that $\sum_{i=1}^{m} \tilde{a}_i \approx \sum_{j=1}^{n} \tilde{b}_j$.

8 Numerical Example

Consider an UFTP with three sources say S_1, S_2, S_3 and three destinations D_1, D_2, D_3, Let us choose the initial transportation Table 1 where each value of fuzzy cost, fuzzy demand and fuzzy supply is generalized TTFN with $\Omega = 0.8$.

Solution: For balance the UFTP in Table 1, adding a dummy source (S_4) with supply equal to $\sum_{j=1}^{3} \tilde{b}_j - \sum_{i=1}^{3} \tilde{a}_i$ with transportation costs as zero generalized TTFN. The balanced TP in tabular form is shown in Table 2 as follows:

Table 1 Transportation problem

(3, 5, 7, 11, 13, 14, 16; 0.8)	(−3, −1, 1, 2, 3, 7, 9; 0.8)	(−4, −3, 0, 1, 3, 5, 7; 0.8)	(4, 8, 12, 18, 24, 30, 36; 0.8)
(−3, −1, 1, 2, 3, 7, 9; 0.8)	(4, 8, 12, 15, 18, 22, 26; 0.8)	(4, 8, 12, 18, 24, 30, 36; 0.8)	(3, 5, 7, 11, 13, 14, 16; 0.8)
(−2, 0, 2, 4, 5, 7, 10; 0.8)	(3, 5, 7, 11, 13, 14, 16; 0.8)	(−3, −1, 1, 2, 3, 7, 9; 0.8)	(0, 1, 3, 7, 11, 14, 17; 0.8)
(7, 12, 18, 25, 32, 38, 43; 0.8)	(−1, 3, 6, 9, 12, 15, 17; 0.8)	(0, 2, 3, 5, 8, 12, 15; 0.8)	

Table 2 Balanced TP in tabular form

$D\rightarrow$ $S\downarrow$	D_1	D_2	D_3	FS
S_1	(3, 5, 7, 11, 13, 14, 16; 0.8)	(−3, −1, 1, 2, 3, 7, 9; 0.8)	(−4, −3, 0, 1, 3, 5, 7; 0.8)	**(4, 8, 12, 18, 24, 30, 36; 0.8)**
S_2	(−3, −1, 1, 2, 3, 7, 9; 0.8)	(4, 8, 12, 15, 18, 22, 26; 0.8)	(4, 8, 12, 18, 24, 30, 36; 0.8)	**(3, 5, 7, 11, 13, 14, 16, 0.8)**
S_3	(−2, 0, 2, 4, 5, 7, 10; 0.8)	(3, 5, 7, 11, 13, 14, 16; 0.8)	(−3, −1, 1, 2, 3, 7, 9; 0.8)	**(0, 1, 3, 7, 11, 14, 17; 0.8)**
S_4	(0, 0, 0, 0, 0, 0, 0; 0.8)	(0, 0, 0, 0, 0, 0, 0; 0.8)	(0, 0, 0, 0, 0, 0, 0; 0.8)	**(−63, −41, −21, 3, 30, 51, 68; 0.8)**
FD	**(7, 12, 18, 25, 32, 38, 43; 0.8)**	**(−1, 3, 6, 9, 12, 15, 17; 0.8)**	**(0, 2, 3, 5, 8, 12, 15; 0.8)**	

According to the definition of TTFN, the ranking function $\mathcal{R}(\tilde{A})$ of $\tilde{A}$ is calculated as:

$$\mathcal{R}(\tilde{A}) = \frac{\Omega}{8}[a_1 + a_2 + a_3 + 2a_4 + a_5 + a_6 + a_7]$$

$$\mathcal{R}(c_{11}) = \mathcal{R}(3, 5, 7, 11, 13, 14, 16; 0.8) = \frac{0.8}{8}[3 + 5 + 7 + 2x11 + 13 + 14 + 16] = 8$$

$\mathcal{R}(c_{12}) = 2, \ \mathcal{R}(c_{13}) = 1, \ \mathcal{R}(c_{21}) = 2, \ \mathcal{R}(c_{22}) = 12, \ \mathcal{R}(c_{23}) = 15, \ \mathcal{R}(c_{31}) = 3,$

$\mathcal{R}(c_{32}) = 8, \ \mathcal{R}(c_{33}) = 2, \ \mathcal{R}(a_1) = 15, \ \mathcal{R}(a_2) = 8, \ \mathcal{R}(a_3) = 5.5, \ \mathcal{R}(a_4) = 3,$

$\mathcal{R}(b_1) = 20, \ \mathcal{R}(b_2) = 7, \ \mathcal{R}(b_3) = 5, \ \mathcal{R}(c_{41}) = 0 = \mathcal{R}(c_{42}) = \mathcal{R}(c_{43}),$

The fuzzy solution of TP by VAM is given in Table 3.

The fuzzy optimal cost is z = (−641, −331, −37, 117, 349, 711, 1164) and the corresponding crisp cost is 144.9.

Table 3 Fuzzy solution of (P1)

$D\rightarrow$ $S\downarrow$	D_1	D_2	D_3	Fuzzy supply
S_1	(3, 5, 7, 11, 13, 14, 16; 0.8) **(−28, −19, −8, 4, 15, 25, 37; 0.8)**	(−3, −1, 1, 2, 3, 7, 9; 0.8) **(−1, 3, 6, 9, 12, 15, 17; 0.8)**	(−4, −3, 0, 1, 3, 5, 7; 0.8) **(0.2, 3, 5, 8, 12, 15; 0.8)**	(4, 8, 12, 18, 24, 30, 36; 0.8)
S_2	(−3, −1, 1, 2, 3, 7, 9; 0.8) **(3, 5, 7, 11, 13, 14, 16; 0.8)**	(4, 8, 12, 15, 18, 22, 26; 0.8)	(4, 8, 12, 18, 24, 30, 36; 0.8)	(3, 5, 7, 11, 13, 14, 16; 0.8)
S_3	(−2, 0, 2, 4, 5, 7, 10; 0.8) **(0, 1, 3, 7, 11, 14, 17; 0.8)**	(3, 5, 7, 11, 13, 14, 16; 0.8)	(−3, −1, 1, 2, 3, 7, 9; 0.8)	(0, 1, 3, 7, 11, 14, 17; 0.8)
S_4	(0, 0, 0, 0, 0, 0, 0; 0.8) **(−63, −41, −21, 3, 30, 51, 68; 0.8)**	(0, 0, 0, 0, 0, 0, 0; 0.8)	(0, 0, 0, 0, 0, 0, 0; 0.8)	(−63, −41, −21, 3, 30, 51, 68; 0.8)
FD	(7, 12, 18, 25, 32, 38, 43; 0.8)	(−1, 3, 6, 9, 12, 15, 17; 0.8)	(0, 2, 3, 5, 8, 12, 15; 0.8)	

Table 4 Balance the UFTP in Table 1 by applying MRCM

D_1	D_2	D_3	D_4	FS
(3, 5, 7, 11, 13, 14, 16; 0.8)	(−3, −1, 1, 2, 3, 7, 9; 0.8)	(−4, −3, 0, 1, 3, 5, 7; 0.8)	(−4, −3, 0, 1, 3, 5, 7; 0.8)	**(4, 8, 12, 18, 24, 30, 36; 0.8)**
(−3, −1, 1, 2, 3, 7, 9; 0.8)	(4, 8, 12, 15, 18, 22, 26; 0.8)	(4, 8, 12, 18, 24, 30, 36; 0.8)	(−3, −1, 1, 2, 3, 7, 9; 0.8)	**(3, 5, 7, 11, 13, 14, 16; 0.8)**
(−2, 0, 2, 4, 5, 7, 10; 0.8)	(3, 5, 7, 11, 13, 14, 16; 0.8)	(−3, −1, 1, 2, 3, 7, 9; 0.8)	(−3, −1, 1, 2, 3, 7, 9; 0.8)	**(0, 1, 3, 7, 11, 14, 17; 0.8)**
(−3, −1, 1, 2, 3, 7, 9; 0.8)	(−3, −1, 1, 2, 3, 7, 9; 0.8)	(−4, −3, 0, 1, 3, 5, 7; 0.8)	(0, 0, 0, 0, 0, 0, 0; 0.8)	**(7, 14, 22, 36, 48, 58, 69; 0.8)**
(7, 12, 18, 25, 32, 38, 43; 0.8)	**(−1, 3, 6, 9, 12, 15, 17; 0.8)**	**(0, 2, 3, 5, 8, 12, 15; 0.8)**	**(−61, −37, −8, 33, 69, 99, 132; 0.8)**	

Similarly by IZSM the fuzzy optimal solution is

$$x_{11} = (-26, -16, -6, 7, 22, 32, 40; 0.8), \qquad x_{12} = (-1, 3, 6, 9, 12, 15, 17; 0.8),$$
$$x_{13} = (-68, -49, -27, 2, 29, 53, 58; 0.8), \qquad x_{21} = (3, 5, 7, 11, 13, 14, 16; 0.8),$$
$$x_{31} = (0, 1, 3, 7, 11, 14, 17; 0.8), \qquad \text{and} \quad x_{43} = (-63, -41, -21, 3, 30, 51, 68; 0.8)$$

The fuzzy cost z = (−1025, −498, −140, 147, 503, 1014, 1513) and the crisp cost is 166.1 (Tables 4, 5).

The fuzzy optimal cost by VAM is z = (−623, −272, −23, 91, 265, 710, 1191; 0.8) and the corresponding crisp cost is 143.0 (Table 6).

The fuzzy optimal cost by IZSM is z = (−676, −297, −18, 91, 283, 814, 1356; 0.8) and the corresponding crisp cost is 164.4.

Table 5 The fuzzy optimal solutions by VAM after applying MRCM

D_1	D_2	D_3	D_4	FS
(3, 5, 7, 11, 13, 14, 16; 0.8)	(−3, −1, 1, 2, 3, 7, 9; 0.8) **(−1, 3, 6, 9, 12, 15, 17; 0.8)**	(−4, −3, 0, 1, 3, 5, 7; 0.8) **(0, 2, 3, 5, 8, 12, 15; 0.8)**	(−4, −3, 0, 1, 3, 5, 7; 0.8) **(−28, −19, −8, 4, 15, 25, 37; 0.8)**	(4, 8, 12, 18, 24, 30, 36; 0.8)
(−3, −1, 1, 2, 3, 7, 9; 0.8) **(3, 5, 7, 11, 13, 14, 16; 0.8)**	(4, 8, 12, 15, 18, 22, 26; 0.8)	(4, 8, 12, 18, 24, 30, 36; 0.8)	(−3, −1, 1, 2, 3, 7, 9; 0.8)	(3, 5, 7, 11, 13, 14, 16; 0.8)
(−2, 0, 2, 4, 5, 7, 10; 0.8) **(0, 1, 3, 7, 11, 14, 17; 0.8)**	(3, 5, 7, 11, 13, 14, 16; 0.8)	(−3, −1, 1, 2, 3, 7, 9; 0.8)	(−3, −1, 1, 2, 3, 7, 9; 0.8)	(0, 1, 3, 7, 11, 14, 17; 0.8)
(−3, −1, 1, 2, 3, 7, 9; 0.8) **(−26, −16, −6, 7, 22, 32, 40; 0.8)**	(−3, −1, 1, 2, 3, 7, 9; 0.8)	(−4, −3, 0, 1, 3, 5, 7; 0.8)	(0, 0, 0, 0, 0, 0, 0; 0.8) **(−33, −18, 0, 29, 54, 74, 95; 0.8)**	(7, 14, 22, 36, 48, 58, 69; 0.8)
(7, 12, 18, 25, 32, 38, 43; 0.8)	(−1, 3, 6, 9, 12, 15, 17; 0.8)	(0, 2, 3, 5, 8, 12, 15; 0.8)	(−61, −37, −8, 33, 69, 99, 132; 0.8)	

Table 6 The fuzzy optimal solutions by IZSM after applying MRCM

D_1	D_2	D_3	D_4	Fuzzy supply
(3, 5, 7, 11, 13, 14, 16; 0.8)	(−3, −1, 1, 2, 3, 7, 9; 0.8) **(−1, 3, 6, 9, 12, 15, 17; 0.8)**	(−4, −3, 0, 1, 3, 5, 7; 0.8)	(−4, −3, 0, 1, 3, 5, 7; 0.8) **(−13, −7, 0, 9, 18, 27, 37; 0.8)**	(4, 8, 12, 18, 24, 30, 36; 0.8)
(−3, −1, 1, 2, 3, 7, 9; 0.8) **(3, 5, 7, 11, 13, 14, 16; 0.8)**	(4, 8, 12, 15, 18, 22, 26; 0.8)	(4, 8, 12, 18, 24, 30, 36; 0.8)	(−3, −1, 1, 2, 3, 7, 9; 0.8)	(3, 5, 7, 11, 13, 14, 16; 0.8)
(−2, 0, 2, 4, 5, 7, 10; 0.8) **(−15, −11, −5, 2, 8, 12, 17; 0.8)**	(3, 5, 7, 11, 13, 14, 16; 0.8)	(−3, −1, 1, 2, 3, 7, 9; 0.8) **(0, 2, 3, 5, 8, 12, 15; 0.8)**	(−3, −1, 1, 2, 3, 7, 9; 0.8)	(0, 1, 3, 7, 11, 14, 17; 0.8)
(−3, −1, 1, 2, 3, 7, 9; 0.8) **(−26, −14, −3, 12, 30, 44, 55; 0.8)**	(−3, −1, 1, 2, 3, 7, 9; 0.8)	(−4, −3, 0, 1, 3, 5, 7; 0.8)	(0, 0, 0, 0, 0, 0, 0; 0.8) **(−98, −64, −26, 24, 69, 106, 145; 0.8)**	(7, 14, 22, 36, 48, 58, 69; 0.8)
(7, 12, 18, 25, 32, 38, 43; 0.8)	(−1, 3, 6, 9, 12, 15, 17; 0.8)	(0, 2, 3, 5, 8, 12, 15; 0.8)	(−61, −37, −8, 33, 69, 99, 132; 0.8)	

9 Conclusion

With the help of numerical example in Table 1, a comparison is settled between the fuzzy optimal solutions (with corresponding crisp cost) obtained by VAM and IZSM of UFTP by existing and proposed MRCM in Table 7. The proposed MRCM for balancing the UFTP of generalized TTFN is easy to understand, apply and obtaining the fuzzy optimal solution for a real-life problem.

Table 7 Comparison of fuzzy optimal solutions with corresponding crisp solutions

Existing method for (UFTP)	
VAM	IZSM
The optimal fuzzy cost is $z = (−641, −331, −85, 117, 349, 711, 1164)$ Corresponding crisp cost $Z = 144.9$	The optimal fuzzy cost is $z = (−1050, −498, −140, 147, 503, 1014, 1513)$ Corresponding crisp cost $Z = 166.1$
Proposed method for (UFTP)	
VAM	IZSM
The optimal fuzzy cost is $z = (−623, −272, −23, 91, 265, 710, 1191)$ Corresponding crisp cost $Z = 143$	The optimal fuzzy cost is $z = (−676, −297, −18, 91, 283, 814, 1356)$ Corresponding crisp cost $Z = 164.4$

References

1. Bellman, R.E., Zadeh, L.A.: Decision-making in a fuzzy environment. Manage. Sci. **17**, B141–B164 (1970)
2. Zadeh, L.A.: Fuzzy sets. Inf. Control **08**, 338–353 (1965)
3. Hitchcock, F.L.: The distribution of a product from several sources to numerous localities. J. Math. Phys. **20**, 224–230 (1941)
4. Oheigeartaigh, H.: A fuzzy transportation algorithm. Fuzzy Set Syst. 235–243 (1982)
5. Kaur, A., Kumar, A.: A new method for solving fuzzy transportation problems using ranking function. Appl. Math. Model. **35**, 5652–5661 (2011)
6. Chanas, S., Kuchta, D.: A concept of the optimal solution of the transportation problem with fuzzy cost coefficients. Fuzzy Sets Sys. **82**, 299–305 (1996)
7. Chen, S.H.: Operations on fuzzy numbers with function principal. Tamkang J. Manage. Sci. **6**, 13–25 (1985)
8. Jain, R.: Decision-making in the presence of fuzzy variables. IEEE Trans. Syst. Man Cybern. **6**, 698–703 (1976)
9. Kaufmann, A.: Introduction to the theory of fuzzy sets, vol. I. Academic Press, New York (1976)
10. ÓhÉigeartaigh, M.: A fuzzy transportation algorithm. Fuzzy Sets Syst. **8**, 235–243 (1982)
11. Nagoor, A., Gani, K., Razak, A.: Two stage fuzzy transportation problem. J. Phys. Sci. **10**, 63–69 (2006)
12. Pandian, P., Natarajan, G.: A new algorithm for finding a fuzzy optimal solution for fuzzy transportation problem. Appl. Math. Sci., **4**, 79–90 (2010)
13. Pandian, P., Natrajan, G.: An optimal more-for-less solution to fuzzy transportation problems with mixed constraints. Appl. Math. Sci. **4**, 1405–1415 (2010)
14. Saini, R.K., Sangal, A., Prakash, O.: Unbalanced transportation problems in fuzzy environment using centroid ranking technique. Int. J. Comput. Appl. (0975–8887), **110** (11), 27–33 (2015)
15. Liu, S.T., Kao, C.: Solving fuzzy transportation problems based on extension principle. Eur. J. Oper. Res. **153**, 661–674 (2004)
16. Kaufmann, A., Gupta, M.M.: Introduction to fuzzy arithmetic: theory and applications. Van Nostrand Reinhold, New York (1985)
17. Basirzadeh, H., Abbasi, R.: A new approach for ranking fuzzy numbers based on α-cuts, JAMI. J. Appl. Math. Inf. **26**, 767–778 (2008)
18. Gani, A.N., Samuel, A.E., Anuradha, D.: Simplex type algorithm for solving fuzzy transportation problem. Tamsui Oxford J. Math. Sci. **27**, 89–98 (2011)
19. Hsieh C.H., Chen S.H., Similarity of generalized fuzzy numbers with graded mean integration representation. In: Proceedings 8th International Fuzzy System Association World Congress, Vol. 2, pp. 551–555. Taipei, Taiwan (1999)
20. Malini, S.U., Kennedy F.C.: An Approach for solving fuzzy transportation problem using octagonal fuzzy numbers. Appl. Math. Sci. 7(54), 2661–2673 (2013)
21. Chen, S.M., Chen, J.H.: Fuzzy risk analysis based on the ranking generalized fuzzy numbers with different heights and different spreads. Expert Syst. Appl. **36**, 6833–6842 (2009)
22. Wang, X., Kerre, E.E.: Reasonable properties for the ordering of fuzzy quantities (I). Fuzzy Sets Syst. **118**, 375–385 (2001)
23. Abbasbandy, S., Asady, B.: Ranking of fuzzy numbers by sign distance. Inf. Sci. **176**, 2405–2416 (2006)
24. Cheng, C.H.: A new approach for ranking fuzzy numbers by distance method. Fuzzy sets syst. **95**, 307–317 (1998)
25. Liou, T.S., Wang, M.J.: Ranking fuzzy number with integral value. Fuzzy Sets Syst, **50**, 247–255 (1992)

26. Rani, D., Gulati, T.R., Kumar, A.: A method for unbalanced transportation problems in fuzzy environment. Sadhana (IAS) **39**, 573–581 (2014)
27. Chanas, S., Kolodziejczyk, W., Machaj, A.: A fuzzy approach to the transportation problem. Fuzzy Sets Syst. **13**, 211–221 (1984)

Review Study of Navigation Systems for Indian Regional Navigation Satellite System (IRNSS)

Kanta Prasad Sharma and Ramesh C. Poonia

Abstract Indian Regional Navigation System (IRNSS) is an independent regional navigation system being developed by the Indian Space Research Organization (ISRO). IRNSS is known as (NAVigation with Indian Constellation) NAVIK. IRNSS is helpful for navigation in space, land mass and ocean mass in Indian border and exceeding range 1500 km beyond Indian borders. It will be useful for precise timing, mapping, geodetic data capture, terrestrial navigation, visual and voice navigation for drivers. This paper proposes, why IRNSS can be useful for navigation in Indian subcontinent?

Keywords IRNSS · GPS · GAGAN · ISRO · SPS · TDS

1 Introduction

The art of finding the path from one place to another place is called navigation. Navigation has been a major scientific and technological challenge in the twenty-first century. A growing number of GNSS users demand highly correct position with minimal latency delay. China has deployed a GNSS named COMPASS, familiar with BeiDou Navigation Satellite System (BDS) as of February 2011. It has offered navigation services for customers in China and neighbor regions. It is providing services in Asia-Pacific region since 2012. Now, China has exported its system to many Asian countries. China has vision to cover 70–80% of the domestic market by 2020 and create strategic research work on the navigation projects for enhanced investments to spur economic growth [1, 2].

Nevertheless, the use GLONASS and GPS measuring lead to significantly better results with the observation in short time. For the GNSS (GPA + GLONASS) user

K.P. Sharma (✉) · R.C. Poonia
Amity Institute of Information Technology, Amity University, Rajasthan, India
e-mail: tokpsharma@gmail.com

R.C. Poonia
e-mail: rameshcpoonia@gmail.com

© Springer Nature Singapore Pte Ltd. 2018
M. Pant et al. (eds.), *Soft Computing: Theories and Applications*,
Advances in Intelligent Systems and Computing 583,
https://doi.org/10.1007/978-981-10-5687-1_65

simultaneous work in open sky areas, which extend around 60% satellite availability as compared to GPS scenario with high accuracy and fast conversion [3]. The design of GLONASS is likewise to GPS except that each satellite broadcasts its own particular frequency with the same code FDMA (Frequency Division Multiple Access) strategy. GLONASS gives two levels of services military user compared to civilian users. A Satellite Based Augmentation System (SBAS) is familiar as GAGAN (GPA Aided GEO Augmentation System), jointly implemented by ISRO and Airport Authority of India (AAI). GAGAN have two downlinks L1 and L5. It is implemented on three phases such as Technical Demonstration System (TDS), Initial Experimental Phase (IEP) and Final Operational Phase (FOP) [4]. The space segment of GAGAN with the form of a dual frequency signal, GPS compatible payload is planned to be flown on India's GSAT-8 and GSAT-10 satellites. The ground segment working with eight Indian Reference Stations (INRESs), one Indian Master Control Center, one Indian Land Uplink Station and related navigation software and communication links have been installed.

The preliminary system acceptance test has indicated that, the position accuracies available are good or not. The GPS augmented system GAGAN enhances reliability, safety, meteorological information and reduces delays, air traffic control. GAGAN provides better accuracy, reliability and availability through the integration of internal information.

The (Indian Regional Navigation Satellite System) IRNSS is a regional independent system with seven satellites are familiar as NAVIK [5], which would be further extended into eleven or more satellites in future. It will enhance the country image in the international arena of space sciences. It will be very useful for national security besides its civilian use. The network and navigation services will provide Indian mass and exceed up to 1,500 km around the country borders. IRNSS satellites are equipped with a laser retroreflector array to enable highly accurate distance measurements. The IRNSS will provide accurate position navigation and time services on different platforms with all time availability under all weather conditions.

2 Related Work

ISRO developed navigation based satellite system. IRSS is a constellation of seven satellites. It will provide accurate position greater than 20 m over Indian land and Ocean due to availability of all seven satellites [6, 7]. According to S. Sayeenath 2013, from Constellation of seven satellites, three are in GEO orbit at 32.50, 83° and 131.5 °E and four satellites are in GSO orbit at 55° and 111.75 °E. Indian launcher PLSV launches IRNSS satellites. First satellite successfully launched on July 1, 2013 and seventh satellite was successfully launched in orbit on April 28, 2016 [8–13]. The system will provide two types of services like SPS (Standard Position Service) RS (Restricted Service). SPS services are available for common users and RS restricted for security defense uses. SPS services are using BPSK (Byi—Phase Shift Key) Modulation and RS will employ Binary Offset Carrier [BOC (5, 2)] [14].

The transmission process doing using L5 Band (1176.45 MHz) and S Band (2492.028 MHz) provides right hand circular polarized signal by array antenna, Therefore users can select single and dual frequency mode for microwave signals with high frequency rate [7].

The IRNSS 1A satellite was successfully launched on July 1, 2013 in Geo Orbits [15]. It provides high frequency using C, S and L5 band for navigation and ranging signals. IRNSS 1B was successfully launched on 4 April 2014 in Geo orbit. It provides navigation and ranging signal with high bandwidth signal for users. IRNSS 1C, 1D, and 1E was launched into orbit on November 10, 2014, March 28, 2015, and January 20, 2016, respectively. Similarly, IRNSS 1F and 1G satellites are successfully launched on March 10, 2016 and April 28, 2016 with high bandwidth frequency for single and dual services [13]. All satellites are working properly with high CDMA and laser ranging signals. Finally, The IRNSS is available for navigation of accurate position in the Indian region.

3 Indian Regional Navigation Satellite System Architecture Model

The IRNSS constellations of seven satellites are working with L5 band and S band frequency for SPS and RS services to the users [16]. IRNSS is operating in three segments such as *Space, Ground and User* Segment.

(i) **Space Segment**—IRNSS satellite carry a navigation payload in a redundant configuration. It consists of three GEO satellites and four GSO satellites routed around the earth. It uses separate C band transponder for precise CDMA ranging with payload configuration. IRNSS Payload transmits navigation, timing information in L5 band and timing information in S band to Ground Segment (Fig. 1).

Constellation of seven satellite three in GEO orbit at 32.50, 83° and 131.5 °E and four satellites in GSO orbit at 55° and 111.75 °E (two in each plane).

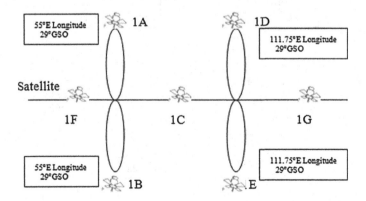

Fig. 1 Space segment

It carries a navigation payload in a redundant configuration using C band for precise CDMA (Code Division Multiple Access) ranging and S band for navigation, timing information satellites.

(ii) **Ground Segment**—Ground segment of IRNSS has main systems for controlling the satellite constellation and will consist of the IRNSS Spacecraft Control Facility, IRNSS Navigation Control features, IRNSS Range and Integrity Monitoring Stations, ranging stations, a timing center, IRNSS TTC and uplink stations, and the IRNSS. IRNSS Ground Segment Elements: IRNSS Satellite Control Facility, IRNSS TTC and Land Uplink Stations, IRNSS Satellite Control Centre, IRNSS Range and Integrity Monitoring Stations, IRNSS Navigation Control Facility, IRNSS Data Communication Network with 17 IRIMS sites. IRIMS stations will be providing signals across the country for orbit determination and Ionospheric modeling. Four ranging stations, disjointed by wide and long baselines, will provide two-way CDMA ranging [11]. The IRNSS timing center is working with highly onboard clocks for timing signals and entire data will manage through communication links at navigation center then it will process and transmit the signals to the satellites (Fig. 2).

IRNSS work on the network of 21 ranging stations geographically allocated region across India. They provide data for the orbit searching of IRNSS satellites and monitoring of the navigation signals. The data from the ranging and monitoring stations is sent to the data processing solution at INC where it is processed to generate the navigation messages. The navigation information is then communicated from INC to the IRNSS satellites through the spacecraft control facility at Hassan/Bhopal, India.

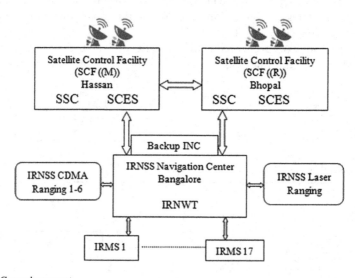

Fig. 2 Ground segment

Fig. 3 Space segment

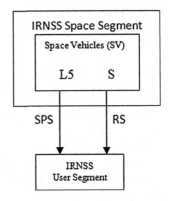

(iii) **User Segment**—It consists of IRNSS based receiver operation with single frequency (L5 or S band) and Dual frequency (L5 and S band) services [17]. Network time is managed by atomic clock at the Indian navigation center (INC) (Fig. 3).

IRNSS [18] users can find out the accurate position with the timing information embedded in the navigation signal, which received from the IRNSS satellites. The timing information is being broadcast in the navigation signal is derived from the atomic clock on board the IRNSS satellite. The IRNWT (IRNSS Network Time) is determining from onboard clock composed of the cesium and hydrogen maser in atomic clocks at the INC (Indian Navigation Centre) ground stations. High bandwidth services are received from satellite to IRNSS users. L5 and S band transmits signals for civilian and restricted operations. RS services are restricted for unauthorized user [19].

4 Comparative Analysis of Satellites System

SNS (Satellite Navigation System) are categorized on three sub-navigation systems such as GSNS (Global Satellite Navigation System), RSNS (Regional Satellite Navigation System) and SBAS (Satellite Based Augmentation System) [20] (Fig. 4).

Different Global Satellite Navigation Systems are working for navigation with world wide services such as GPS (USA), GLONASS (Russia), Galileo (Europe) and Compass (China). Navigation process at Regional label is working independent with RSNS (Regional Satellite Navigation System) of different countries such as China (BeiDou), Japan (QZSS), and India (IRNSS). SBAS (Satellite Based Augmentation System) navigation process is dependent with GPS. These satellites working in USA (WAAS), Europe (EGNOS), Japan (MSAS) and India (GAGAN). Table 1, shown working status of GNSS systems.

Fig. 4 Satellite navigation
system

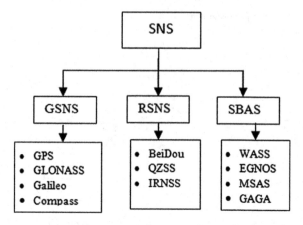

Table 1 Satellite constellation system [21]

System	GPS	GLONASS	Compass	Galileo	IRNSS
Owner	United States	Russian	China	European Union	India
Coding	CDMA	FDMA, CDMA	CDMA	CDMA	CDMA, TDD
Orbital altitude	20,180 km, (12,540 mi)	19,130 km, (11,890 mi)	21,150 km, (13,140 mi)	23,222 km, (14,429 mi)	36,000 km, (22,000 mi)
Period	11 h 58 min	11 h 16 min	12 h 38 min	14 h 5 min	23 h 42 min
Per sidereal delay	2	17/8	17/9	17/10	
Satellites	32	24	35	22	7
Frequency	1.57542 GHz (L1 signal) 1.2276 GHz (L2 signal)	1.602– 1.246 (L3 signal) GHz (SP)	1.561098 GHz (B1), 1.589742 GHz (B1-2), 1.20714 GHz (B2), 1.26852 GHz (B3), E5	1.164– 1.215 GHz (E5a and E5b), 1.260– 1.300 GHz (E6), 1.559– 1.592 GHz (E2-L1-E11)	L5- ba1164.45– 1188.45 MHz S-band 2483.5– 2500 MHz
Status	Operational	Operational	Operational	Operational	Operational

Satellites working in circular orbits at an altitude of 23,150 km, each satellite completes the orbit in 12 h 38 min, means that for a stationary observation the same satellite is visible at the similar point in the sky in every nine sidereal days. Since there are nine satellites in each orbit plane, satellite positions in the sky get repeated each. So that it provide high bandwidth signals with dual frequency services to the China users. Similarly, Galileo satellites working in orbit range 23,222 km. Each satellite complete the orbit in 14 h 5 min, means that for a stationary observation the same satellite is visible at the same points in the sky every 10 sidereal delay.

IRNSS system is working with seven working satellites in orbit of 36,000 km. Each satellite complete the orbit in 23 h 43 min, means that the satellites visible at the same point in the sky. Four satellites are in a circular orbit at 55° and 111.75 °E at GSO orbit and three in GEO orbit. It provides high frequency signal with dual service (SPS, RS) for the users in Indian region. SPS services are available for civilian's uses and RS services can not access by unauthorized users due to security purpose.

5 Conclusion and Future Work

The IRNSS provides services for precise navigation on the ground, mountainous and oceanic mass for enhanced security tremendously. IRNSS is able to provide a vast spectrum of services to the users and resolve issues of GPS navigation problems relate to low signal and noise ratio (SNR), spectrum size, radio frequency disturbances, ionospheric influence, jamming, spoofing, and satellite segment malfunction in India. IRNSS provide working navigation applications on BHUVAN, allow users 2D/3D presentation of Earth with the help of IRNSS satellites. Indian GPS is providing services in commutation, navigation, disaster etc. IRNSS provided significant growth of the country in the field of science and technology and enhance the economical growth in India.

References

1. Cooper, S., et al.: A Kalman filter model for GPS navigation of land vehicles. In: Intelligent Robots and Systems' 94. Advanced Robotic Systems and the Real World, IROS, pp. 157–163 1994
2. Bonnor, N., et al.: A brief history of global navigation satellite systems. J. Navig. 65(01), 1–14 (2012)
3. Koyuncu, H., et al.: A survey of indoor positioning and object locating systems. Int. J. Comput. Sci. Netw. Secur. 10(5), 121–128 (2010)
4. Fernandez-Prades, C., et al.: Satellite radio localization from GPS to GNSS and beyond: Novel technologies and applications for civil mass market. Proc. IEEE 99(11), 1882–1904 (2011)
5. Bhuvan Port: http://bhuvan.nrsc.gov.in/bhuvan_links.php#. Accessed 02 Aug 2016
6. IRNSS: http://www.isro.gov.in/irnss-programme. Accessed 18 July 2016
7. Dow, J.M., et al.: The international GNSS service in a changing landscape of global navigation satellite systems. J. Geodesy 83(3–4), 91–198 (2009)
8. Tornatore, V., et al.: First considerations on the feasibility of GNSS observations by the VLBI technique & measuring the future. In: Proceedings of the Fifth IVS General Meeting, p. 439 (2008)
9. Plett, M., et al.: Free-space optical communication link across 16 kilometers over the Chesapeake Bay to a modulated retroreflector array. Opt. Eng. 47(4), 045001 (2008)

10. Nadeem, T., et al.: Traffic view: traffic data dissemination using car-to-car communication. ACM SIGMOBILE Mob. Comput. Commun. Rev. **8**(3), 6–19 (2004)
11. Rizos, C. etal.: The future of global navigation satellite systems. Technical documents, University of New South Wales (2007)
12. Yao, Z., et al.: ACE-BOC: dual-frequency constant envelope multiplexing for satellite navigation. IEEE Trans. Aerosp. Electron. Syst. **52**(1), 466–485 (2016)
13. IRNSS: http://www.isro.gov.in/Spacecraft/irnss-1a. Accessed 02 Aug 2016
14. Saikiran, B., et al.: IRNSS architecture and applications. KIET Int. J. Comm. Electron **1**(3), 21–27 (2013)
15. Thoelert, S., et al.: IRNSS-1A: signal and clock characterization of the Indian regional navigation system. GPS Solutions **18**(1), 147–152 (2014)
16. Nadarajah, N., et al.: Assessing the IRNSS L5-signal in combination with GPS, Galileo, and QZSS L5/E5a-signals for positioning and navigation. GPS Solutions **20**(2), 289–297 (2016)
17. Rizos, C., et al.: Experimental results of location: a high accuracy indoor positioning system. In: Indoor Positioning and Indoor Navigation (IPIN), International Conference, pp. 1–7 (2010)
18. Saikiran, B., et al.: IRNSS architecture and applications. KIET Int. J. Comm. Electron **1**(3), 21–27 (2016)
19. Montenbruck, O., et al.: IRNSS orbit determination and broadcast ephemeris assessment, ION ITM, pp. 185–193 (2015)
20. Karthick, M., et al.: Design of multiband fractal antenna for IRNSS and GAGAN applications. In: IEEE Innovations in Information, Embedded and Communication Systems (ICIIECS), pp. 1–4 (2015)
21. Carreno-Luengo, H., et al.: Experimental evaluation of GNSS-reflectometry altimetric precision using the P (Y) and C/A signals. IEEE J. Sel. Top. Appl. Earth Obs. Remote Sens. **7** (5), 1493–1500 (2014)

Adaptation of Binary Pigeon-Inspired Algorithm for Solving Multidimensional Knapsack Problem

Asaju La'aro Bolaji, Balogun Sulaiman Babatunde
and Peter Bamidele Shola

Abstract The multidimensional knapsack problem (MKP) is among the complex optimization problems that have number of practical applications in engineering domain. Several solution methods such as local search and population-based search algorithms have been proposed to solve the MKP. In this paper, a new Binary pigeon-inspired optimization (Binary PIO) algorithm is proposed for solving MKP with focus on 0–1 MKP. For the purpose evaluation, the benchmark instances drawn from the OR library are utilized to test the performance of the proposed binary method. Experimentally, it is concluded that the application of Binary PIO offers promising performance when compared with existing techniques from the literature.

Keywords Multidimensional knapsack problem · Pigeon-inspired optimization
Biologically inspired algorithms · Population-based methods

1 Introduction

The role of combinatorial optimization cannot be overemphasized in the field of artificial intelligence and operational research with sole aim of solving several difficult combinatorial optimization problems. The multidimensional knapsack problem (MKP) as a classical example has its numerous practical applications in

A.L. Bolaji (✉)
Department of Computer Science, Faculty of Pure and Applied Sciences,
Federal University Wukari, Wukari, Taraba State, Nigeria
e-mail: lbasaju@fuwukari.edu.ng

B.S. Babatunde · P.B. Shola
Department of Computer Science, Faculty of Communication
and Information Sciences, University of Ilorin, Ilorin, Nigeria
e-mail: aceslimz@gmail.com

P.B. Shola
e-mail: shola.pb@unilorin.edu.ng

© Springer Nature Singapore Pte Ltd. 2018
M. Pant et al. (eds.), *Soft Computing: Theories and Applications*,
Advances in Intelligent Systems and Computing 583,
https://doi.org/10.1007/978-981-10-5687-1_66

different areas such as cryptography [1–3], warehouse location problem, production scheduling problem, assignment problem and reliability problem [4]. The main objective of the MKP as a resource allocation model, is to select a subset of objects that produce the highest profit, while satisfying the set of constraints on knapsack capacities. Due to the NP-hard nature of the MKP, numerous approaches have been developed by the researchers in the domain which are grouped into: exact and approximation methods. The exact method like branch and bound method [5], dynamic programming [6], etc., could generate the exact solutions; however, its time complexity increases exponentially with the size of the problem, whereas the approximation method could produce a near-optimal solution within reasonable computational times when compared to the mathematical method [7, 8].

The emergence of nature-inspired computation as a class of approximation method led to the introduction of numerous algorithmic techniques by the researchers to tackle different formulations of MKP. This is probably due to their robust search ability in handling the problems of high dimensionality more effectively and efficiently. Some examples of nature-inspired algorithms that have been utilized to solve MKPs include artificial algae algorithm [9], ant algorithms [10], artificial bee colony algorithm [11], cuckoo search algorithm [12], fruit fly optimization algorithm (FOA) [13], genetic algorithm [14], harmony search algorithm [15], and particle swarm optimization [16]. Similarly, hybrid and other metaheuristic approaches have also been recently utilized to tackle MKPs can be found in Refs. [17–19]. Among newly introduced nature-inspired algorithms that have shown more promising behavior is the pigeon-inspired optimization (PIO) algorithm, an easy but powerful population-based search technique inspired by the homing pigeons. Thus, this motivated the researchers to investigate its performance when utilized to tackle the complex optimization problems like MKP.

Pigeon-inspired optimization (PIO) is a new class of biologically inspired optimization algorithm proposed by Duan and Qiao [20]. It is a member of swarm intelligence optimizer which belongs to the population-based algorithm inspired by the movement of pigeons. Homing pigeons can easily locate their homes based on three homing components which are magnetic field, sun and landmarks. In the context of this algorithm, map and compass model is formulated according to magnetic field and sun, while presentation of landmark operator model is done based on landmarks [21]. Note that pigeons perceived the magnetic field through the nose from the magnetic particles taken to the brain by trigeminal nerve [22]. It was observed that the pigeons probably employ different navigational tools during different parts of their journey. They may depend more on compass-like tools at the initial stage of the journey, while in the middle and sometimes switch to landmarks in order reassess their routes to makes corrections [23]. The successes of the PIO algorithm have been recorded when utilized to tackle some complex optimization problems like orbital spacecraft formation reconfiguration [24], unmanned aerial vehicles [25–27], and it has proven to have a better or comparable performance with some of the existing algorithms like artificial bee colony, genetic algorithm, particle swarm optimization, and so on [28]. However, none of the studies have adopted its binary formulation to tackle MKP problems. Therefore, this paper presents a binary pigeon-inspired optimization (Binary PIO) algorithm for solving binary optimization problems, especially the 0–1 MKP. The performance of the proposed Binary PIO is evaluated using the standard MKP benchmark datasets published by OR Library. Experimental results and comparisons demonstrate the effectiveness of the proposed algorithm for the MKP.

2 Multidimensional Knapsack problem (MKP) Formulation

The MKP consists of a set of m knapsacks with a set of given capacities $C = \{c_0, c_1, \ldots, c_{m-1}\}$ and a set of n entities $e = \{e_0, e_1, \ldots, e_{n-1}\}$. The binary variables $Y_i(i = 0, 1, \ldots, n - 1)$ correspond to the chosen items to be carried in m knapsacks. The Y_i assumes value of 1 if entity i is in the knapsack and 0 otherwise. Each item e_i has an associated profit $P_i \geq 0$ and weight $W_{ij} \geq 0$ for each knapsack j. The goal is to find the best combination of n entities by maximizing the sum of profits P_i multiplied by the binary variable Y_i, which is mathematically represented as shown in Eq. (1).

$$\max \left(\sum_{i=0}^{n-1} (P_i \times Y_i) \right) \tag{1}$$

Their constraints are the capacity $C_j \geq 0$ of each knapsack. Therefore, the sum of the values of Y_i multiplied by W_{ij} must be less than or equal to C_j as given in Eq. (2)

$$\sum_{i=0}^{m-1} (W_{ij} \times Y_i) \leq C_j \tag{2}$$

Note that this formulation is adopted from [29].

3 Pigeon-Inspired Optimization Algorithm

PIO is a novel nature-inspired, metaheuristic algorithm that has been utilized for solving global optimization problems. It is based on imitating the natural homing pigeon behavior. Recently, studies on pigeon's behaviors have shown that the pigeon can follow their paths using some landmark properties like main roads, railways, and rivers rather than head for their destination directly. The two mathematical models employed to summarize the migration of pigeons are: map and compass operator, and landmark operator.

3.1 Map and Compass Operator

Naturally, the map and compass operator assist pigeons to locate their path and determine the direction. Analogously, in the map and compass operator, the definition of the rules with which the path X_i and the velocity V_i of pigeon i, and the positions and velocities in a D-dimension search space are updated in each iteration. The formulation of the new position X_i and velocity V_i of pigeon i at the tth iteration is given in Eqs. 3 and 4 as follows:

$$V_i(t) = V_i(t-1) \cdot e^{-rt} + \text{rand} \cdot \left(X_g - X_i(t-1)\right) \tag{3}$$

$$X_i(t) = X_i(t-1) + V_i(t) \tag{4}$$

where r represent the map and compass factor, rand is a random number between 0 and 1, and X_g represent the current global best paths that is obtained by comparing all the paths among all the pigeons.

3.2 Landmark Operator

In the landmark operator, the total number of pigeons is reduced to half N_P in every generation, where pigeons in the lower half of the line sorted by fitness values are abandoned. This due to the fact that they are believed to be far from the destination and unfamiliar with the landmarks. Let $X_c(t)$ represent the center of some pigeon's path at the tth iteration, and that every pigeon can fly straight to the destination. The position updating rule for pigeon i at tth iteration can be formulated as shown in Eq. 5:

$$N_P(t) = \frac{N_P(t-1)}{2} \tag{5}$$

$$X_c(t) = \frac{\sum X_i(t) \cdot \text{fitness}(X_i(t))}{N_P \sum \text{fitness}(X_i(t))} \tag{6}$$

$$X_t(t) = X_i(t-1) + \text{rand} \cdot (X_c(t) - X_i(t-1)) \tag{7}$$

where fitness () is the object value (i.e., quality) of the pigeon individual. Note that for the minimization problems, then the fitness cost is given as shown in Eq. 8:

$$f(X_i(t)) = \frac{1}{f(X_i(t)) + \varepsilon} \tag{8}$$

whereas the fitness cost for maximization problem is given in Eq. 9 as:

$$f(X_i(t)) = f(X_i(t)). \tag{9}$$

4　Proposed Binary Pigeon-Inspired Optimization

This section presents the proposed binary version of the classical PIO described previously such that more problem specific knowledge is considered in the constraint handling part. The solution approach utilized in proposed binary version is divided into two stages as given in the next subsections.

4.1　Solution Representation and Objective Function

The initial step of adapting the PIO for solving any given problem is to formulate a suitable representation model, i.e., a way to represent pigeon in the PIO population. Note that for the problem under consideration (i.e., MKP), 0–1 binary representation is an obvious choice since it represents the underlying 0–1 integer variables. Therefore, in this research, a n-bit binary string representation is employed, where n is the number of variables in the MKP, a value of 0 or 1 at the ith path means that $Y_i = 0$ or 1 in the MKP solution, respectively. Figure 1 shows the binary representation of each pigeon (i.e., solution) in the search space for the MKP.

It is worthy of mentioning that in some case the solution generated based on the above procedure may not be feasible. Thus in order to avoid the utilization of such solution in the population, a penalty function method is employed. Therefore, the fitness cost function is given in Eq. 10 as follows:

$$\text{fitness}\,(S) = \sum_{i=0}^{n-1} p_i s\,[i] \times \text{Pen}\,[i] \tag{10}$$

4.2　Binary PIO for the MKP

In binary PIO, the initial step to initialize the parameters and each pigeon is initialized with the random path and velocity. Then, the fitness cost of each pigeon is calculated in order to determine the current best path of the pigeon. During the

Fig. 1 MKP solution representation

i	0	1	2	3	4	-------	n-1
Y_i	0	1	0	1	1	-------	0

search process, each pigeon explores the solution search space by adjusting its flight in accordance its own personal best position or by flying toward the global best path which is determined by comparing all the paths of the pigeons using the Map and Compass operator. This search process is repeated until the map and compass operator generation reaches the maximum. In the landmark operator, the fitness costs of the pigeons are ranked and the center of the pigeon is discovered and pigeons with the lowest fitness costs will move toward those highest fitness costs. The best path and the fitness costs are stored. The search process in this phase is repeated until landmark iteration l-iteration reaches its maximum. Algorithm 1 shows the search process of the binary PIO.

Algorithm 1: Binary PIO algorithm

1: Initialization of Binary PIO parameters: N, D, t_{1max}, t_{2max}.
2: Initialize the path and the velocity of each pigeon. {Initialized the population of pigeons}
3: Calculate the fitness cost of each pigeon and compare the individual best fitness costs
   of all the pigeons to obtain the global best path {Evaluate each pigeon}
4: **while** $t \leq t1max$ **do**
5:    Operate the map and compass operator. Update the velocities and paths of each
      pigeon, and update the global best path.
6: **end while**
7: **while** $t \leq t2max$ **do**
8:    Operate the landmark operator. Sort all pigeons according their fitness costs. Those pigeons with low fitness
      costs will follow those with high fitness cost using Eq. (5). Then determine the center of all pigeons based on
      Eq. (6), and this center is the desirable destination. Adjust all pigeons fly directions in accordance with the Eq.
      (7). Memorize the best pigeon parameters and the best fitness cost.
9: **end while**

---

5 Experimental Results and Discussion

The proposed Binary PIO is implemented in this section is coded in VB.net and run on a PC with Intel core i3 1.9 GHz, 4 GB memory running on Windows 10. The MKP dataset provided by the OR library is utilized for the evaluation of the proposed binary technique. These datasets include the MKNAPCB 1 and MKNAPCB 4 datasets that belong to class of big instances of the MKP library. The conducted experiments are designed to investigate the effectiveness of Binary PIO on the MKP where each instance is run for 10 times. The parameter settings are: pigeon size = 150, map and compass = 0.5, and iteration is fixed to 1500.

Similarly, Table 1 shows the results obtained by Binary PIO in comparison with previous techniques from the literature. These techniques are for Binary Cuckoo Search algorithm BCS [12], Standard binary PSO with penalty function, PSO-P [30], and the quantum inspired Cuckoo search QICSA [31]. Note that the best results are highlighted in bold. It is apparent that the Binary PIO is able to obtain feasible solutions for all datasets. Similarly, it is able to obtain high quality solutions in comparison with previous methods. For example, the Binary PIO achieved first rank in all instances of the MKNAPCB 1 and MKNAPCB 4 datasets except in

Table 1 The best results achieved by the Binary PIO and other methods

Benchmak name	Dataset size	Best known	BCS	PSO-P	QICSA	BPIO
mknapcb1	5.100.01	24381	**23510**	22525	23416	23494
	5.100.02	24274	22938	22244	22880	**232277**
	5.100.03	23551	22518	21822	22525	**22942**
	5.100.04	23534	22677	22057	22167	**22895**
	5.100.05	23991	23232	22167	22854	**23502**
mknapcb4	10.100.01	23064	21841	20895	21796	**22237**
	10.100.02	22801	21708	20663	21348	**22203**
	10.100.03	22131	20945	20058	20961	**21614**
	10.100.04	22772	21395	20908	21377	**22236**
	10.100.05	22751	21453	20488	21251	**22157**

mknapcb1 5.100.01 where the proposed algorithm achieved second best. This proved that Binary PIO is good template solution which could be adopted to tackle other complex problem by the researchers in the domain.

6   Conclusion and Future Work

In this paper, a new binary pigeon-inspired optimization algorithm called Binary PIO is proposed for solving the 0–1 multidimensional knapsack problem (MKP). In the binary PIO, the continuous nature of the classical pigeon-inspired optimization algorithm is modified using the binary representation to cope with the search space nature of the MKP. In the aim to verify and prove the performance of the proposed algorithm, it is evaluated on some MKP benchmarks taken from OR Library. Experimental results show that the proposed BPIO is better than existing algorithms that worked on the problem. Based on this promising result, our future direction is to further study its performance when employed to solve more complicated optimization problems.

References

1. Chor, B., Rivest, R.L.: A knapsack-type public key cryptosystem based on arithmetic in finite fields. IEEE Trans. Inf. Theory **34**(5), 901–909 (1988)
2. McAuley, A.: A new trapdoor knapsack public key cryptosystem. In: Advances in Cryptology. Proceedings of EUROCRYPT 84. A Workshop on the Theory and Application of Cryptographic Techniques-Paris, France, vol 209, 9–11 Apr 1984. Springer (2007) 150
3. Laih, C.S., Lee, J.Y., Harn, L., Su, Y.K.: Linearly shift knapsack public-key cryptosystem. IEEE J. Sel. Areas Commun. **7**(4), 534–539 (1989)

4. Martello, S., Toth, P.: Knapsack Problems: Algorithms and Computer Implementations. Wiley, USA (1990)
5. Shih, W.: A branch and bound method for the multiconstraint zero-one knapsack problem. J. Oper. Res. Soc. **30**, 4 (1979)
6. Toth, P.: Dynamic programming algorithms for the zero-one knapsack problem. Computing **25**(1), 29–45 (1980)
7. Hussain, T.S.: An introduction to evolutionary computation. Tutorial presentation. CITO Researcher Retreat, 12–14 May 1998
8. Jourdan, L., Basseur, M., Talbi, E.G.: Hybridizing exact methods and metaheuristics: a taxonomy. Eur. J. Oper. Res. **199**(3), 620–629 (2009)
9. Zhang, X., Wu, C., Li, J., Wang, X., Yang, Z., Lee, J.M., Jung, K.H.: Binary artificial algae algorithm for multidimensional knapsack problems. Appl. Soft Comput. **43**, 583–595 (2016)
10. Ke, L., Feng, Z., Ren, Z., Wei, X.: An ant colony optimization approach for the multidimensional knapsack problem. J. Heuristics **16**(1), 65–83 (2010)
11. Sundar, S., Singh, A., Rossi, A.: An artificial bee colony algorithm for the 0–1 multidimensional knapsack problem. In: Contemporary Computing, pp. 141–151. Springer, Heidelberg (2010)
12. Gherboudj, A., Layeb, A., Chikhi, S.: Solving 0-1 knapsack problems by a discrete binary version of cuckoo search algorithm. Int. J. Bio-Inspir. Comput. **4**(4), 229–236 (2012)
13. Pan, W.T.: A new fruit fly optimization algorithm: taking the financial distress model as an example. Knowl. Based Syst. **26**, 69–74 (2012)
14. Raidl, G.R., Gottlieb, J.: Empirical analysis of locality, heritability and heuristic bias in evolutionary algorithms: a case study for the multidimensional knapsack problem. Evol. Comput. **13**(4), 441–475 (2005)
15. Kong, X., Gao, L., Ouyang, H., Li, S.: A simplified binary harmony search algorithm for large scale 0–1 knapsack problems. Expert Syst. Appl. **42**(12), 5337–5355 (2015)
16. Bansal, J.C., Deep, K.: A modified binary particle swarm optimization for knapsack problems. Appl. Math. Comput. **218**(22), 11042–11061 (2012)
17. Haddar, B., Khemakhem, M., Hanafi, S., Wilbaut, C.: A hybrid quantum particle swarm optimization for the multidimensional knapsack problem. Eng. Appl. Artif. Intell. **55**, 1–13 (2016)
18. Zhang, B., Pan, Q.K., Zhang, X.L., Duan, P.Y.: An effective hybrid harmony search-based algorithm for solving multidimensional knapsack problems. Appl. Soft Comput. **29**, 288–297 (2015)
19. Tasgetiren, M.F., Pan, Q.K., Kizilay, D., Suer, G.: A differential evolution algorithm with variable neighborhood search for multidimensional knapsack problem. In: Proceedings of 2015 IEEE Congress on Evolutionary Computation (CEC), pp. 2797–2804. IEEE (2015)
20. Duan, H., Qiao, P.: Pigeon-inspired optimization: a new swarm intelligence optimizer for air robot path planning. Int. J. Intell. Comput. Cybern. **7**(1), 24–37 (2014)
21. Zhang, B., Duan, H.: Predator-prey pigeon-inspired optimization for UAV three-dimensional path planning. In: Advances in Swarm Intelligence, pp. 96–105. Springer, Heidelberg (2014)
22. Mora, C.V., Davison, M., Wild, J.M., Walker, M.M.: Magneto-reception and its trigeminal mediation in the homing pigeon. Nature **432**(7016), 508–511 (2004)
23. Guilford, T., Roberts, S., Biro, D., Rezek, I.: Positional entropy during pigeon homing II: navigational interpretation of bayesian latent state models. J. Theor. Biol. **227**(1), 25–38 (2004)
24. Zhang, S., Duan, H.: Gaussian pigeon-inspired optimization approach to orbital spacecraft formation reconfiguration. Chin. J. Aeronaut. **28**(1), 200–205 (2015)
25. Hao, R., Luo, D., Duan, H.: Multiple UAVs mission assignment based on modified pigeon-inspired optimization algorithm. In: IEEE Chinese Guidance, Navigation and Control Conference (CGNCC), pp. 2692–2697 (2014)
26. Li, C., Duan, H.: Target detection approach for UAVs via improved pigeon-inspired optimization and edge potential function. Aerosp. Sci. Technol. **39**, 352–360 (2014)

27. Zhang, S., Duan, H.: Multiple UCAVs target assignment via bloch quantum-behaved pigeon-inspired optimization. In: IEEE 34th Chinese Control Conference (CCC), pp. 6936–6941 (2015)
28. Qiu, H., Duan, H.: Multi-objective pigeon-inspired optimization for brushless direct current motor parameter design. Sci. China Technol. Sci. **58**(11), 1915–1923 (2015)
29. Andre', L., Parpinelli, R.S.: A binary differential evolution with adaptive parameters applied to the multiple knapsack problem. In: Nature-Inspired Computation and Machine Learning, pp. 61–71. Springer, Heidelberg (2014)
30. Kong, M., Tian, P.: Apply the particle swarm optimization to the multidimensional knapsack problem. In: International Conference on Artificial Intelligence and Soft Computing, pp. 1140–1149. Springer, Heidelberg (2006)
31. Layeb, A.: A novel quantum inspired cuckoo search for knapsack problems. Int. J. Bio-Inspired Comput. **3**(5), 297–305 (2011)

Performance Evaluation and Statistical Analysis of AUR-Chord Algorithm with Default Working of Structured P2P Overlay Network

Anil Saroliya, Upendra Mishra and Ajay Rana

Abstract In today's scenario, where data streaming and file sharing (audio, video, jpeg, etc) are the most performed tasks over the internet, the several characteristic functionalities of peer-to-peer network for resource distribution makes its usage much more than the initially used client–server-based network. Peer-to-peer network requires less time compared to client–server approach for distributing data over the network because at an instance of time any node can behave as client and server in accordance to the requirement, i.e., there exists no predefined behavior/state of a node. The distribution of nodes and their scalable linking increases the churn rate in P2P network which as a result affects various parameters like data availability and data security as the higher churn rate creates an overhead on the data transaction activities. In this paper, statistical analysis and simulation results depict that improved algorithm give a highly efficient response which minimize the incomplete lookup as compared to earlier one.

Keywords P2P networks · Chord protocol · Structure overlay network
Churn rate · Performance evaluation

A. Saroliya (✉) · U. Mishra
Amity School of Engineering & Technology,
Amity University Rajasthan, Jaipur, India
e-mail: anilsaroliya1@gmail.com

U. Mishra
e-mail: umishra@jpr.amity.edu

A. Rana
Amity University Uttar Pradesh, Noida, India
e-mail: ajay_rana@amity.edu

© Springer Nature Singapore Pte Ltd. 2018
M. Pant et al. (eds.), *Soft Computing: Theories and Applications*,
Advances in Intelligent Systems and Computing 583,
https://doi.org/10.1007/978-981-10-5687-1_67

1 Introduction

In Distributed Networking Environment usage of DHT (distributed hash table)-oriented protocols like CHORD algorithm, makes data sharing much easier and faster. If by any chance one is able to initiate an inspection to fill the gaps, that will conceptually decrease the distance between the computing devices which are located much far from each other in distributed networks. Nowadays, most of the scientific researchers are thus inclined toward the motive of providing information to the end-user with less delay in time and at lower costs. Thus all such requirements make P2P network very useful as it uses overlay protocols (based on distributed hash tables) for data distribution. DHT based overlay networks comprises of multiple clusters, where each cluster contains many nodes which are logically connected to each other by holding their peer information in terms of key (a unique resource identifier) and domain. Using this peer information in P2P network, the tasks involved in locating the network resource becomes much easier than the methodologies used in client–server networks. In P2P networks maintaining the security during transaction is also an important aspect [1]. Since, DHT protocols mainly focus on locating a resource in the P2P network, thus other additional tasks like downloading the required data or storing data on any P2P network node can be efficiently performed by using proposed infrastructures like cloud-based infrastructure.

This paper is structured in subsequent sections: Sect. 2 explains the detailed working of proposed cloud infrastructure with P2P network while Sect. 3 presents the experimental setup and simulation method in brief. Section 4 explains the Performance analysis between AUR-Chord and Chord Algorithm. In Sect. 5, comparison of proposed and existing (default) model done through statistic analysis. Finally, Sect. 6 provides the conclusion of the paper.

2 Cloud Infrastructure with P2P Network

To reduce the high churn rate available in Chord protocol [2] of P2P networking, in this work cloud computing [3] infrastructure is included so that resource distribution among peers can be possible in a secure way and node compromising can be avoid The presence of cloud infrastructure can introduce a trust factor in existing P2P network which were missing earlier.

Figure 1 portrays the functioning of cloud infrastructure with peer-to-peer network [4]. It can be simply observed that allocation of resources can be performed easily to the actual desired node of the network. The above solution is known as AUR-Chord, its name is based on the combination of the first name or last name starting letter of the author (of this paper). Resource lookup functionality of the cloud-enabled solution (AUR-Chord) is shown in Fig. 2.

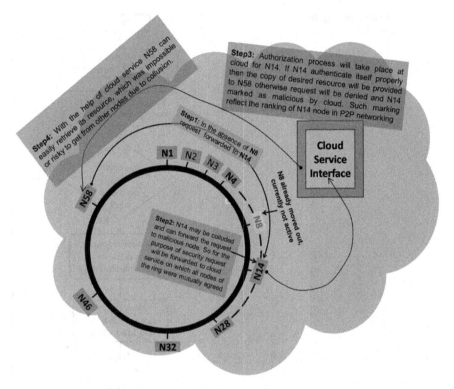

Fig. 1 Resources allocation with cloud Infrastructure in P2P networking (AUR-chord)

3 Experimental Setup and Simulations

A virtual environment is simulated to study the impact of AUR-Chord over basic chord on the basis of number of uncompromised and compromised nodes. In this environment, some randomly connected nodes are placed in unidirectional circular graph. Vertices in circular graph represents computing nodes in the network and assigned with a unique ID, which is known as node identifier (Node-ID). All the recourse ids (which will be shared or exchange by P2P nodes) are associated with particular Node-ID for which it is responsible. The Node-IDs and Resource-IDs indicate the position of the node and the respective resources in the P2P chord network. In above simulated comparisons, some results have been identified.

Following section will statistically analyze the working of AUR-Chord and default chord routing algorithms based on number of successful and unsuccessful lookups and hop counts.

Fig. 2 Lookup processing in
AUR-chord model

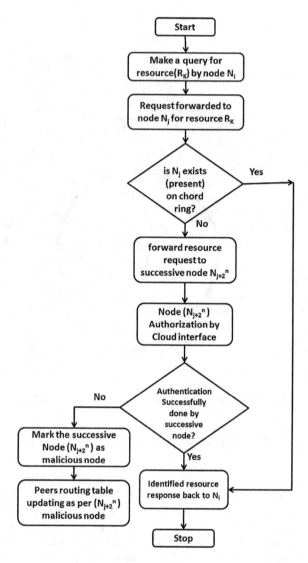

4 Performance Analysis Between AUR-Chord and Chord Algorithm

AUR-Chord reduces the effect of high churn rate [5] on structured P2P network which were designed through Chord algorithm (base model). Table 1 shows the comparative analysis [6] in between AUR-Chord model (AM) and Base (Chord) model (BM).

In Table 1, MCN means malicious colluding nodes, CL means complete lookups, ICL means incomplete lookups and HC means hop count, shows that the mean

Table 1 Comparisons between AUR-chord and basic chord algorithm

Values	No. of good nodes	MCN	CL	ICL	All HC
BM mean	16,200	1670	59.252	40.747	510.89
AM mean	16,200	1670	92.747	7.252	3891.81
BM SD	8573.21	857.32	4.185	4.185	12.273
AM SD	8616.84	861.684	3.08	3.08	1452.98

Here *BM* means basic chord model, *AM* means AUR-chord model and *SD* means standard deviations

of good and malicious nodes are same in basic and AUR both cases but mean complete lookups of AUR Model are (92.747 + 3.0816) and for basic model are (59.252 + 4.185), whereas mean incomplete lookups of AUR Model are (7.252 + 3.0816) and for basic model are (40.747 + 12.273). Mean Hop-Count for basic model is (511 + 4.185) and Mean Hop-Count for AUR model is (3895.33 + 1452.980).

The statistical analysis [7] of data reveals that complete lookups in AUR Model are greater than basic model and incomplete lookups are less in AUR Model. It shows the good performance of AUR Model as compared to earlier basic model. Some security aspects are included in our AUR-Chord model due to Hop-Count increase. Table 1 shows Hop-Count for AUR model are higher than basic model which does not indicate negative results but for security reason we would ignore this aspect for our updated AUR model.

5 Comparison of Models: Data Analysis

Data samples used in this research paper are independent. The measurements are applied on two different models and if the values in one sample reveal no information about those of the other sample, then the samples are independent.

In a normal distribution, the mean is equal to the median. Also, about 68% of the data is within one standard deviation of the mean, 95% is within two standard deviations, and 99% is within three standard deviations. The data set which has been used in this work has these properties, and then these samples come from a population that is normally distributed (Tables 2 and 3).

Table 2 Basic model data analysis

Mean	SD	Median
16,341	8688.4	16,350
1685.8	871.7591	1685
59.22	4.198557	58
40.79	4.207485	42
510.89	12.32301	511.5

Table 3 AUR model data analysis

Mean	SD	Median
16,341	8606.616	16,350
1685.8	861.6844	1685
92.69	3.119489	93
7.27	3.071061	7
3891.81	1446.053	3466.5

6 Comparison of Models: Statistical Analysis

For testing the hypothesis, the mean Incomplete and complete lookups were analyzed statistically [8] to determine whether there was a significant difference in the mean data. The data was evaluated using t-test [8] statistical method and the result obtained is presented in tables. In this statistical analysis, the two general levels are used to mean something is superiorly adequate to be understood, are 0.05 and 0.01. This means that the result has a 95 and 99% chance of being true. "0.05," meaning that the result has a 5% (0.05) chance of not being true, which is the opposite of a 95% chance of being true.

6.1 Comparison of Complete Lookups in Basic and AUR Approaches

Figure 3 represents the comparison of successful lookups (complete lookups) of AUR-Chord and basic chord model. Following test explain more details about above-mentioned comparisons (Table 4).

For this comparison, a two-tailed test is used to test the significance of difference between two means. The t values for df(100 + 100 − 2 = 198) are 1.97 at 0.05 and 2.6 at 0.01 level of significance. Since the obtained t value 3.22 is greater than the

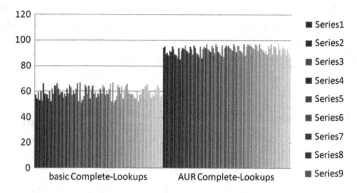

Fig. 3 Comparison of complete lookups in basic and AUR model

Table 4 Two tailed test results after differentiating the complete lookups of AUR and basic model

Approaches	Mean	SD	*t* value
Basic complete lookups	59.22	4.198	3.22
AUR complete lookups	92.69	3.119	Df = 198

table value necessary for the rejection of the null hypothesis at 0.05 and 0.01 level for degree of freedom 198, it is determined that there is a considerable difference in the mean complete lookups of basic and AUR approach; AUR approach is good as compared to basic because the mean of AUR complete lookups is greater than basic complete lookups.

6.2 Comparison of Incomplete Lookups in Basic and AUR approaches

Figure 4 represents the comparison of incomplete lookups of AUR-Chord and basic chord model. Following test explains more details about the above-mentioned comparisons (Table 5).

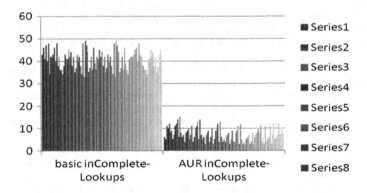

Fig. 4 Comparison of incomplete lookups in basic and AUR model

Table 5 Two tailed test results after differentiating the incomplete lookups of AUR and basic model

Approaches	Mean	SD	*t* value
Basic incomplete lookups	40.747	4.185	8.43
AUR incomplete lookups	7.252	3.0816	Df = 198

Again the t values for df(100 + 100 − 2 = 198) are 1.97 at 0.05 and 2.6 at 0.01 level of significance. Since the obtained t value 8.43 is greater than the table value necessary for the rejection of the null hypothesis at 0.05 and 0.01 level for degree of freedom 198, again it is interpreted that there is a considerable difference in the mean incomplete lookups of basic and AUR approach; AUR approach is good as compared to basic because the mean of AUR incomplete lookups is quite lesser than basic incomplete lookups.

7 Conclusion

The Chord-enabled P2P networks are very flexible network due to the higher churn rate, particular resource identification and its authenticity is always risky for an end-user. In this paper, through simulation, we warily examined the performance of the AUR-Chord algorithm with basic chord algorithm for structured peer-to-peer resource distribution networks. The statistical results shows, that with the help of rooted cloud infrastructure on P2P network success of lookup completion increases as compare to traditional approach.

References

1. Saroliya, A., Mishra, U., Rana, A.: A pragmatic analysis of peer to peer networks and protocols for security and confidentiality. Int. J. Comput. Corp. Res. 2(6) (2012)
2. Stoica, I., Morris, R., Karger, D., Kaashoek, M.F., Balakrishnan, H.: Chord: a scalable peer-to-peer lookup service for internet applications. In Proceedings of ACM SIGCOMM'01, San Diego, California (2001)
3. Sriram, I., Khajeh-Hosseini, A.: Research agenda in cloud technologies. In: Computing Research Repository. Cornell University, USA (2010)
4. Saroliya, A., Rana, A., Mishra, U., Vaibhav, K.: Efficient distribution of resources with cloud service in P2P networks at the time of higher churn rate. In: NSACCS-04 Proceedings of the 2015 Seminar on Advances in Computer Networks, Communications and Security, AUMP Gwalior, India. ISBN:978-93-85000-09-6
5. Hoang, G.N., Chan, H.N., et al.: Performance improvement of chord distributed hash table under high churn rate. In: International Conference on Advanced Technologies for Communications, pp. 191–196 (2009)
6. Kurkwski, S., Camp, T., Colagrosso, M.: MANET simulation studies: the incredible. Mob. Comput. Commun. Rev. 9(4), (50–61) (2005)
7. Garrett, H.E.: Statistics in Psychology and Education. Longman, Green and Co. (1926)
8. Kothari C.R.: Research Methodology. New Age International(P) Ltd. (1985). ISBN:978-81-224-24881

Multiscale LMMSE-Based Statistical Estimation for Image Denoising

Rajesh Kumar Gupta, Milind Anand and Anita Bai

Abstract Denoising of a legitimate image depraved by the additive white Gaussian noise (AWGN) is a famous problem in image processing. Thresholding does the extraction from noisy wavelet coefficients using denoising method by reserving the major coefficients and adjusting the remaining to zero. In this paper, extraction of Gaussian noise from vociferous image is done. The proposed scheme in this paper outmatches some present denoising method by using linear minimum mean square error (LMMSE) based maximum aposteriori (MAP) estimation. Some parameters are altered to alienatet the noise proficiently, such as variance of the classical MMSE estimator of the noisy wavelet coefficients in the neighborhood window. Each and every procedure is analyzed separately and experiments are operated to evaluate their performance in view of peak signal-to-noise ratio (PSNR). Improved results are obtained for highly corrupted inartificial images.

Keywords Image denoising · Additive Gaussian noise · Discrete wavelet transform · Maximum aposteriori (MAP) estimation · Linear minimum mean square error (LMMSE)

R.K. Gupta (✉)
Department of Computer Science & Engineering, National Institute of Technology, Rourkela, India
e-mail: rajeshgupta.cse@gmail.com

M. Anand · A. Bai
Department of Computer Science & Engineering, Visvesvaraya National Institute of Technology, Nagpur, India
e-mail: milindjune1988@gmail.com

A. Bai
e-mail: anita.bai@students.vnit.ac.in

© Springer Nature Singapore Pte Ltd. 2018
M. Pant et al. (eds.), *Soft Computing: Theories and Applications*,
Advances in Intelligent Systems and Computing 583,
https://doi.org/10.1007/978-981-10-5687-1_68

1   Introduction

Digital images are usually depraved by various types of noises such as Gaussian noise [1] during the process of obtainment, devolution, and retrieval from storage media. These images are congregated by different types of sensors. Noise depraves both videos and images. During the image obtainment phase, interference noise likely to be produced because of faulty instrument employed in image processing. The motive of denoising algorithm is to extract such noise. Image denoising [1] is required as a noisy image is not pleasant to see. Therefore, it is prerequisite to extract the noise from the image before postliminary processing. Pattern recognition [2] deals in denoised image to work efficiently and effectively. Irregular and uncorrelated noise pattern are not compressible. So keeping all these factors in knowledge, denoising plays a vital role in enhancing the quality of image. The wavelet transform [3]-based approach have become more popular and artistical in last decades, as it can extract noise easily and effectively from the noisy images [4–11]. Input image signal is decomposed into multiple scales using wavelet transform based approach that depicts various time frequency components of the original image [3, 12, 13].

2   Motivation

Recently, lot of research work have been taken place in the field of image enhancement based on denoising algorithms and it keeps the curiosity alive and finds a scope in future. Now a days, a lot of research is available for wavelet transforms, i.e., continuous wavelet transform (CWT) and discrete wavelet transform (DWT) to amplify the resolution. So lifting wavelet transform is used for improvement of resolution of images, after gets motivated in this particular area of research. In this work, an endeavor is made to come up with a new algorithm based on lifting wavelet transform.

2.1   Performance Measurement

Images are debased by Gaussian noise and then it is denoised using different methods. For the performance measurement of wavelet-based denoising algorithm, the following method is used:

1. PSNR (peak signal-to-noise ratio)

$$MSE = \frac{\sum_{I,J}(A_1(i,j) - A_2(i,j))^2}{I \times J} \qquad (1)$$

$$PSNR = 10\log_{10}\frac{r^2}{MSE} \qquad (2)$$

where $I \times J$ is the size of the image and r is the most extreme variance in the image data type.

3 Multiscale LMMSE-Based Statistical Estimation for Image Denoising

In this paper, Multiscale LMMSE-based image denoising method is introduced, which follows statistical estimations. This proposed method extracts Gaussian noise from digital images using some parameters called locally estimated variance. A rectification is done in the parameter estimation of signal and noise variance commensurate with the LMMSE estimator. LMMSE is an optimal predictor under the chimera of Gaussian noise, for the clean wavelet coefficient. The work introduced in this paper is common in approach of LMMSE [14]. In spite of that, the proposed idea and approach is different than statistical estimation model with parameter Estimation. Without using downsampling, first of all, decomposition of input image took place in wavelet domain. Then each and every wavelet coefficients are combined at one time. Initially, signal and noise variance is estimated for each wavelet coefficient using statistical estimation technique (MAP and ML estimation) from the local neighborhood window and then linear minimum mean squared error (LMMSE) estimation is applied (Fig. 1).

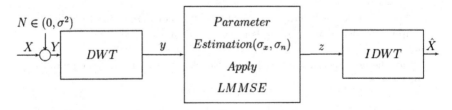

Fig. 1 Illustrated diagram of proposed method

3.1 LMMSE of Wavelet Coefficients

Under the chimera of Gaussian noise [1, 6], locally estimated variance is used by LMMSE method. Gaussian noise is an optimal predictor for clean wavelet coefficient. In this paper, LMMSE is applied instead of soft thresholding [12]. Suppose original image X is corrupted with additive Gaussian noise η from below equation

$$y = x + n \tag{3}$$

After enforcing wavelet transform to the noisy image Y at scale j and $y(i, j)$ is wavelet coefficient of Y. Estimated noise variance is denoted by $\hat{\sigma}_n^2$ and $\hat{\sigma}_x^2$ denotes signal variance. The LMMSE of wavelet coefficient of original image $x(i, j)$ is estimated and signal variance is

$$\hat{x} = \left(\frac{\hat{\sigma}_x^2}{\hat{\sigma}_x^2 + \hat{\sigma}_n^2} \right) y \tag{4}$$

where M is dimension of X $(0 < i, j \leq M)$.

3.2 Denoising Algorithm

We proposed a denoising algorithm to remove Gaussian noise from digital images which provide better denoised output compared to the following noise removal techniques in terms of PSNR. This algorithm optimizes the estimation approach and gives better quality of the image.

Data: Image contaminated with additive Gaussian noise
Result: Estimated noiseless image
Decompose the noisy image using Discrete wavelet transform(DWT) into
   wavelet domain up to J^{th} decomposition level;
for *Each subband (i.e. HH, HL, and LH) in decomposition level j* **do**
   Parameters Estimation signal variance and noise variance σ_x^2 and σ_n^2
      ,respectively;
   Apply LMMSE estimator to obtain modified noiseless wavelet coefficients.;
   up to J^{th} decomposition levels repeat steps (3) and (4);
end
Restore the denoised data using inverse wavelet transform from the modified
   coefficients;

3.3 Parameter Estimation

To dissociate the noiseless wavelet coefficients, parameters are estimated for noisy coefficients. MAP and ML estimation [15] is used for signal and noise variance calculation and then in place of thresholding, LMMSE method is exercised. This plan is concentrated on size of the image for every subband and noise variance. Results are improved in terms of PSNR when a more adaptive threshold is taken and local features of the signal are considered. The parameter σ_n is generally evaluated with the scaled Median Absolute Deviation (MAD) estimator [16]. In DWT, image is decomposed into multiple subband. Wavelet details coefficients ψ^{HH_1} of the finest decomposition level are concerned only to the noise. The MAD estimator makes use of the median of absolute value of ψ^{HH_1} coefficients to estimate noise variance. MAD is defined in Eq. 5.

$$\hat{\sigma}_m = \frac{\text{MAD}(|\psi^{HH_1}|)}{0.6745} \tag{5}$$

where MAD is the Median Absolute Deviation.

For the extraction of more noiseless coefficient of noisy wavelet coefficient for each decomposition levels, weighting factor $g(k)$ is introduced. Equation for estimation of noise variance and signal variance is as follows:

$$\hat{\sigma}_n^2 = g(k)\hat{\sigma}_m^2 \tag{6}$$

$$= 2^{\frac{k}{2k+1}}\hat{\sigma}_m^2 \tag{7}$$

where value of k is taken as 2.

The signal variance of noiseless image is estimated as follows:

$$\hat{\sigma}_x^2 = \begin{cases} \sigma_y^2 - \hat{\sigma}_m^2, & \text{if } \sigma_y^2 > \hat{\sigma}_m^2 \\ 0, & \text{otherwise} \end{cases} \tag{8}$$

$$\sigma_y^2 = \frac{1}{M.N} \sum_{m=1}^{M} \sum_{n=1}^{N} w^2(m, n) \tag{9}$$

Table 1 PSNR (dB) image denoising performance for lena and cameraman images

Images	Noise levels	Denoising algorithms				
		Neighshrink	Bivariate	GIDNWC	IAWDMBNC	MLMMSE
Lena	10	33.21	34.18	33.65	33.83	34.36
	20	28.56	30.50	29.22	29.56	30.74
	30	26.06	28.33	26.74	27.44	28.83
	50	23.10	25.55	24.01	24.33	26.64
	100	22.10	21.20	22.30	22.40	23.81
Cameraman	10	32.23	34.12	33.45	33.67	34.16
	20	26.44	30.23	27.42	27.76	30.40
	30	24.23	27.89	25.04	25.46	27.33
	50	21.44	24.92	22.67	23.55	24.91
	100	19.23	20.70	19.67	20.54	22.05

4 Experiment and Results

Experiments are done in matlab environment. This proposed technique is applied on various images taken from USC-SIPI image database [17]. This technique is applied on miscellaneous images like Lena and Cameraman. Table 1 indicates the peak signal-to-noise ratio (PSNR) results of the five methods on the two benchmark images corrupted by different levels of additive Gaussian noise. Two-standard gray scale images (Lena, Cameraman) of size 256×256 are contaminated with zero mean Gaussian noise ($\sigma = 10; 20; 30; 50;$ and 100) and then denoised using various methods, including the one proposed in this paper. In this paper, Figs. 2 and 3 describe the performance of denoised technique for $\sigma = 10$ Gaussian noise. Fig. 4 shows denoising performance comparison of lena and cameraman images. Experiments show that the proposed scheme outperforms some existing denoising methods. Although the proposed scheme works well for images Lena and Cameraman, it can be seen from Table 1.

(a)   (b)

(c)   (d)

(e)   (f)

Fig. 2 **a** Noisy Lena with ($\sigma = 20$), **b** denoised using neighshrink (28.56 dB), **c** denoised using Bivariate (30.50 dB), **d** denoised using GIDNWC (29.22 dB), **e** denoised using IAWDMBNC (29.56dB), **f** denoised image using MLMMSE (30.74 dB)

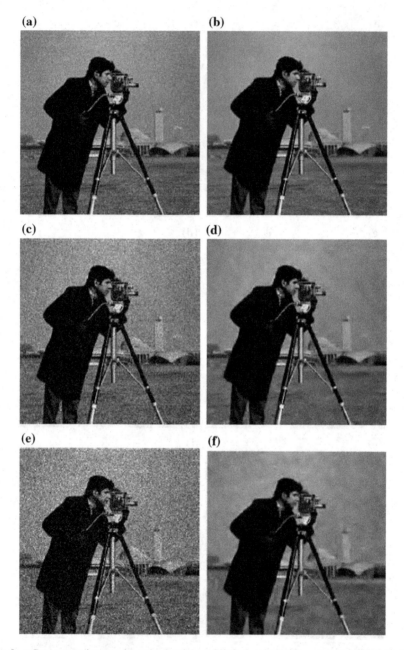

Fig. 3 **a** Cameraman image with noise level ($\sigma = 20$), **b** denoised using neighshrink (26.44 dB), **c** denoised using Bivariate (30.23 dB), **d** denoised using GIDNWC (27.42 dB), **e** denoised using IAWDMBNC (27.76 dB), **f** denoised using MLMMSE (30.40 dB)

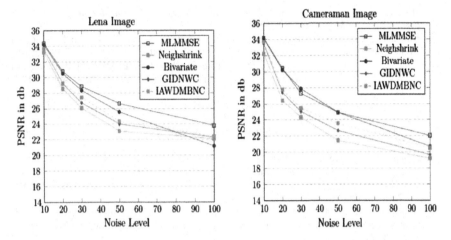

Fig. 4 Denoising performance comparison

5 Conclusion

In this paper, a novel parameter estimation technique for LMMSE estimator is proposed for image denoising. Using statistical estimation model, multiple estimation expression and various experimental results are obtained. This idea impart properly satisfying outcomes in PSNR and results good as compare to existing algorithm as cataloged in the table in experimental results. Although the proposed scheme outperforms other popular denoising schemes for both of the images. This work can be further enhanced to denoise the other type of images, as well, like RGB, Indexed and Binary images. Use of AI techniques will lead to the optimal solution directly with more efficiency and less tedious work.

References

1. Rafael, C.G., Richard, E.W.: Digital image processing. In: Pearson Prenctice Hall (2007). ISBN 0-13-168728-X
2. http://en.wikipedia.org/wiki/Pattern_Recognition
3. Mallat, S.: A wavelet tour of signal processing. Academic press (1999)
4. Zhang, D., Bao, P., Wu, X.: Multiscale LMMSE-based image denoising with optimal wavelet selection. IEEE Trans. Circuits Syst. Video Technol. **15**(4), 469–481 (2005)
5. Chang, S.G., Yu, B., Vetterli, M.: Adaptive wavelet thresholding for image denoising and compression. IEEE Trans. Image Process. **9**(9), 1532–1546 (2000)
6. Om, H., Biswas, M.: A generalized image denoising method using neighbouring wavelet coefficients. SIViP 1–10 (2013)
7. Sendur, L., Selesnick, I.W.: Bivariate shrinkage with local variance estimation. IEEE Signal Process. Lett. **9**(12), 438–441 (2002)

8. Hou, Z.: Adaptive singular value decomposition in wavelet domain for image denoising. Pattern Recogn. **36**(8), 1747–1763 (2003)
9. Ruikar, S., Doye, D. D.: Image denoising using wavelet transform. In: 2nd International Conference on Mechanical and Electrical Technology (ICMET), pp 509–515, IEEE (2010)
10. Liua, J., Caicheng S., Meiguo G.: Image denoising based on BEMD and PDE. In: 3rd International Conference on Computer Research and Development (ICCRD). Vol. 3, IEEE (2011)
11. Wang, Z. et al.: Image quality assessment: from error visibility to structural similarity. IEEE Trans. Image Process. **13**(4), 600–612 (2004)
12. Donoho, D.L.: Denoising by soft-thresholding. IEEE Trans. Inf. Theor. **41**(3), 613–627 (1995)
13. Donoho, D.L.: Unconditional bases are optimal bases for data compression and for statistical estimation. Appl. Comput. Harmonic Anal. **1**(1), 100–115 (1993)
14. Om, H., Biswas, M.: MMSE based map estimation for image denoising. Opt. Laser Technol. **57**, 252–264 (2014)
15. Boubchir, L., Boualem, B.: Wavelet denoising based on the MAP estimation using the BKF prior with application to images and EEG signals. IEEE Trans. Signal Process. **61**(8), 1880–1894 (2013)
16. Donoho, D.L., Johnstone, J.M.: Ideal spatial adaptation by wavelet shrinkage. Biometrika **81**(3), 425–455 (1994)
17. Weber, A.: USC-SIPI Image Database. [Online] Available at: http://sipi.usc.edu/database/database.php
18. Donoho, D.L., Johnstone, I.M.: Adapting to unknown smoothness via wavelet shrinkage. J. Am. Stat. Assoc. **90**(432), 1200–1224 (1995)

Coarse-to-Fine Registration of Remote Sensing Optical Images Using SIFT and SPSA Optimization

Sourabh Paul and Umesh C. Pati

Abstract Sub-pixel accuracy is the vital requirement of remote sensing optical image registration. For this purpose, a coarse-to-fine registration algorithm is proposed to register the remote sensing optical images. The coarse registration operation is performed by the scale invariant feature transform (SIFT) approach with an outlier removal method. The outliers are removed by the random sample consensus (RANSAC) algorithm. The fine registration process is performed by maximizing the mutual information between the input images using the first-order simultaneous perturbation stochastic approximation (SPSA) along with the second-order SPSA. To verify the effectiveness of the proposed method, experiments are performed using three sets of optical image pairs.

Keywords Image registration · Scale invariant feature transform (SIFT) Simultaneous perturbation stochastic approximation (SPSA)

1 Introduction

Remote sensing image registration is the process of geometrical alignment of remotely sensed images captured by same or different sensors. Image registration methods can be classified as intensity-based methods [1] and feature-based methods [2]. In the intensity-based method, the pixels intensities of the image pair are used to measure the similarity between the images. The similarity measurement matrices used for intensity-based methods are maximum likelihood, least square matching, cross-correlation, mutual information, etc. In the intensity-based method, an optimization technique is utilized with a similarity matrix to speed up the registration

S. Paul (✉) · U.C. Pati
Department of Electronics and Communication, National Institute
of Technology Rourkela, Rourkela 769008, India
e-mail: sourabhpaul26@gmail.com

U.C. Pati
e-mail: ucpati@nitrkl.ac.in

© Springer Nature Singapore Pte Ltd. 2018
M. Pant et al. (eds.), *Soft Computing: Theories and Applications*,
Advances in Intelligent Systems and Computing 583,
https://doi.org/10.1007/978-981-10-5687-1_69

operation. In feature-based methods, the robust features such as corner, edges, contours, etc., are extracted and the matching operation is performed.

Scale invariant feature transform (SIFT) [3] is a well-known approach used for the feature-based image registration. SIFT is a very effective approach to extract the distinctive invariant features and it can be used to perform the matching operation between the images. Regardless of its robustness, some problems still arise such as the uneven distribution of the matched features and the existence of outliers in the matched pairs. So, different modified SIFT algorithms have been proposed in the past to improve the performance of SIFT approach. Goncalves et al. [4] proposed an automatic image registration algorithm (IS-SIFT) using image segmentation and SIFT. In [4], the bivariate histogram is utilized to remove the outliers obtained from the SIFT feature matching operation. Sedaghat et al. [5] presented a uniform robust scale invariant feature transform (UR-SIFT) algorithm to uniformly distribute the extracted features. In [5], uniform SIFT features are matched through cross-matching technique to increase the number of matched pairs. Gong et al. [6] developed a coarse-to-fine registration scheme using SIFT and mutual information. In [6], coarse registration was performed by the standard SIFT with reliable outlier method and the fine registration was implemented by maximizing the mutual information using a modified Marquardt–Levenberg optimization technique. Zhang et al. [7] proposed a coarse-to-fine registration algorithm to register the large size very high resolution images using SIFT, Oriented FAST, and Rotated BRIEF (ORB) feature matching. In [8, 9], improved SIFT based matching is performed to register the remote sensing images.

The mutual information is a popular similarity matrix used for the intensity-based image registration methods. Cole-Rodes et al. [10] presented a first-order simultaneous perturbation stochastic approximation (SPSA) optimization with the maximization of mutual information criterion to register the remote sensing images. But, the first-order SPSA converges slowly if the result is very close to the optimum solution. In order to increase the computational speed of the optimization technique, Cole-Rodes et al. [11] utilized a second-order SPSA algorithm. Suri et al. [12] presented a coarse-to-fine registration algorithm to register the IKONOS and TERRA SAR images through image segmentation and maximization of mutual information criterion using the first-order SPSA optimization technique.

Although a variety of feature-based and intensity-based registration algorithms have been proposed to register the remotely sensed optical images, sub-pixel accuracy is still a vital challenge in remote sensing image registration. In this paper, we have proposed a coarse-to-fine registration method to obtain the sub-pixel accuracy between the registered images. Motivated by methods [6, 7], we have performed the coarse registration by using the SIFT feature matching. The matches obtained by SIFT matching contain many outliers. Random sample consensus (RANSAC) [13] is implemented to refine the matches. The fine registration scheme is conducted by maximizing the mutual information with the first-order and second-order SPSA optimization approach. Inspired by the accuracy and the convergence rate of the algorithms [10, 11], the first-order and second-order SPSA are utilized for the fine registration process.

Rest of the paper is organized as follows: Sect. 2 presents the initial coarse registration method followed by Sect. 3, which provides the fine registration algorithm. The simulation results are discussed in Sect. 4. Finally, Sect. 5 offers a conclusion.

2 Coarse Registration

The coarse registration process is performed by standard SIFT matching with RANSAC-based outlier removal process. The SIFT algorithm is implemented for feature extraction and matching of the remote sensing optical images. The algorithm consists of five major steps:

2.1 Scale-Space Extrema Detection

The extrema points are detected in each scale of every octave of the difference of gaussian (DOG) images by comparing every pixel to its eight surrounding neighbors of current scale, nine neighbors in the upper and lower scales. A particular key-point is selected if it is larger or smaller than all of its 26 neighbors.

2.2 Key-Points Localization

The location and scale of every feature point is estimated by 3D quadratic function. For eliminating the poorly localized features principal curvature analysis is performed. The features with the contrast value less than 0.03 (proposed by Lowe [3]) are discarded.

2.3 Orientation Assignment

The local gradients directions are estimated to compute the dominant orientation of a feature point. Therefore, the feature points are invariant to any rotation of the image.

2.4 Key-Points Descriptor Formation

The gradient magnitudes and orientations in a 16 × 16 location grid around the key-point location are computed to form the 128 elements descriptor.

2.5 Feature Matching

It is performed by calculating the minimum Euclidean distance between the descriptors of input images. But, directly using this minimum distance criteria produces many false matches. So, Lowe [3] proposed d_{ratio} factor to improve the correct rate. The ratio of the Euclidean distance of first nearest neighbor to the second nearest neighbor is defined as d_{ratio}. By choosing a high d_{ratio} value, the number of matched pairs can be increased, but the correct rate decreases [6]. A very small value of d_{ratio} can improve the correct rate, but in this case, the number of matches decrease. So, a proper selection of d_{ratio} is an important factor for remote sensing image registration. In our proposed method, the value of d_{ratio} is set to 0.6. But, a number of outliers still exists in the matched pairs set. So, RANSAC [13] algorithm is implemented to remove the remaining outliers.

3 Fine Registration

The fine registration is performed by maximizing the mutual information using the first-order SPSA along with the second-order SPSA optimization techniques. In [11], Cole-Rodes et al. proposed a switching scheme between first-order and second-order SPSA to register the remote sensing images in a multiresolution framework. The second-order SPSA provides a fast convergence, whereas the first-order is more robust to get a solution which is close to the optimum one from a further away point. The algorithm is very effective when the input images have low distortion between them. But, it fails to register the images if a large distortion occurs between the images. Our proposed coarse registration method coarsely aligns the input images. Using our coarse registration process, a very low distortion can be obtained between coarsely registered images. So, we have utilized SPSA in the fine registration process to improve the accuracy of registration.

3.1 Mutual Information

Mutual information is the measurement of relative entropy between two functions. Let A and B are the two input images for registration. Then, mutual information between A and B is defined as

$$\text{MI} = H(A) + H(B) - H(A, B) \tag{1}$$

where $H(A, B)$ is the joint entropy and $H(A)$ and $H(B)$ represent the marginal entropy of A and B, respectively.

3.2 Multiresolution Approach

To speed up the fine registration operation the SPSA optimization is used in a multiresolution framework. The reference and sensed images are decomposed through the Simoncelli steerable filter. The low pass bands of the filter are iteratively registered from the coarsest level to the finest level by maximizing the mutual information using the SPSA optimization. In our proposed method, we have used four levels of decomposition and the image size gets halved in each level starting from finest to coarsest level. The values of transformation parameters, which provides a maximum mutual information between the reference and sensed image, is used as an initial transformation for the next finer level.

3.3 First-Order SPSA

SPSA algorithm was introduced by Spall [14] to find a root of the multivariate gradient equation using gradient approximation. It is popularly used to solve the optimization problem in remote sensing image registration [11]. In the first-order SPSA, the update rule for the transformation parameters is given as

$$\theta_{n+1} = \theta_n + a_n g_n \tag{2}$$

where the gradient vector $g_n = [(g_n)_1 (g_n)_2 \ldots (g_n)_p]^T$ for the p-dimensional parameter space is estimated as

$$(g_n)_i = \frac{L(\theta_n + c_n \Delta_n) - L(\theta_n - c_n \Delta_n)}{2c_n (\Delta_n)_i} \quad \text{for } i = 1, 2, \ldots, p \tag{3}$$

where L is the objective function that has to be optimized.

According to Bernoulli's distribution, every element $(\Delta_n)_i$ of the vector (Δ_n) takes a value of +1 or −1. a_n and c_n are the positive sequences of the form

$$a_n = \frac{a}{(A + n + 1)^\alpha} \quad \text{and} \quad c_n = \frac{c}{(n+1)^\gamma} \quad \text{where } 0 < \gamma < \alpha < 1 \tag{4}$$

where A, a and c are the constants of the optimization. As suggested in [10], these parameters values are $A = 100$, $c = 0.5$, $a = 12$, $\alpha = 0.602$, and $\gamma = 0.101$.

3.4   Second-Order SPSA

The first-order SPSA optimization converges slowly when it is very closer to the optimum solution. In [15], Spall presented a second-order SPSA optimization technique to speed up thwhere the gradient vectore convergence. The update rule for the second-order SPSA optimization algorithm is given as

$$\theta_{n+1} = \theta_n + a_{2n}\overline{H}_n^{-1} g_n, \quad \overline{H}_n = f_n(\hat{H}_n) \tag{5}$$

$$\hat{H}_n = \frac{n}{n+1}\hat{H}_{m-1} + \frac{1}{n+1}H_n \tag{6}$$

where H_n is the per-iteration estimate of the Hessian of L, and a_{2n} is a constant. Equation (6) shows a recursive calculation of the per-iteration Hessian estimate $\hat{H}$. In Eq. (5), f_n is the transformation function for which $\hat{H}$ becomes invertible. Further details can be found in [15]. In our proposed method, the value of a_{2n} is set to 0.5.

3.5   Transformation Model

Let f_r (u', v') and f_s (u, v) are the reference and sensed images, respectively. To register the images, a geometric transformation is estimated so that image f_s (u, v) gets aligned perfectly with f_r (u', v'). So, the transformation gives the spatial relationship between the image pair. Here, affine transformation model is used as a geometric transformation function. The affine model is given as

$$\begin{bmatrix} u' \\ v' \\ 1 \end{bmatrix} = \begin{bmatrix} a_{11} & a_{12} & \alpha_x \\ a_{21} & a_{22} & \alpha_y \\ 0 & 0 & 1 \end{bmatrix} \begin{bmatrix} u \\ v \\ 1 \end{bmatrix} \tag{7}$$

where $(a_{11}, a_{12}, a_{21}a_{22})$ combinedly represent the rotation, scale, and shear differences and (α_x, α_y) are the translation parameters.

4   Simulation and Analysis

Three sets of optical images are selected for the experimental result analysis. The images are taken from USGS Earth Explorer (http://earthexplorer.usgs.gov). We have compared our proposed method with the other two remote sensing optical image registration algorithms named as IS-SIFT [4] and UR-SIFT [5]. All the experiments are performed on an Intel core i7-4770, 3.40 GHz CPU and 4 GB of physical memory computer using MATLAB 2014a.

The first pair of images is obtained from LANDSAT Enhanced Thematic Mapper Plus (ETM+) sensor (resolution 30 m) on July 10, 2001 at the region of Baltimore. A section of 500 × 500 pixels in band 4 (0.76–0.90 μm) is taken as reference image and a segment of the same size in the band 5 (1.55–1.75 μm) with five degrees simulated rotation and translation of −10 pixels in X-direction and −20 pixels in Y-direction, is selected as sensed image. Figure 1a, b show the reference image and the sensed image, respectively. The second pair of images is taken from the OrbView-3 on June 13, 2004 (resolution 1 m) which cover the region of Barcelona. A section of size 800 × 800 pixels taken is considered as the reference image and a segment with the same size with 10° simulated rotation and translation of 20 pixels in X-direction and −35 pixels in Y-direction, is selected as sensed image. Figure 2a, b show the reference image and the sensed image, respectively. The third pair of images is taken from the Earth Observer-1 Advanced Land Imager (ALI) on May 4, 2013 over an area of Chesapeake Bay. A section of size 256 × 256 pixels taken is taken as the reference image and a segment with the same size with 15° simulated rotation and translation of 20 pixels in X-direction and −10 pixels in Y-direction, is selected as sensed image. Figure 3a and b show the reference image and the sensed image, respectively.

Table 1 shows the transformation parameters comparison between IS-SIFT [4], UR-SIFT [5] and the proposed method for different image pairs. Table 2 presents mutual information (MI), root mean square error (RMSE) and associated computational time comparison between different methods. It can be clearly observed that the proposed method provides less RMSE value compared to the other methods and the registration parameters are very close to the ground truth values in the case of our proposed algorithm. The results obtained using the UR-SIFT and IS-SIFT algorithms are acceptable but, still accuracy is less than the proposed method. IS-SIFT method is less accurate than the other two algorithms because the number of matches is less. The d_{ratio} criterion of IS-SIFT eliminates a number of correct correspondences. Although the number of matches obtained in coarse registration scheme of our proposed method is less than the UR-SIFT, still accuracy is

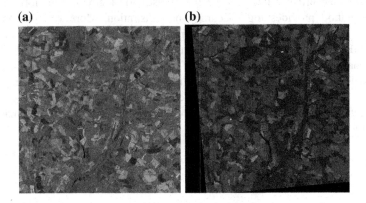

Fig. 1 Images of Baltimore, USA **a** reference image, **b** sensed image

(a)   **(b)**

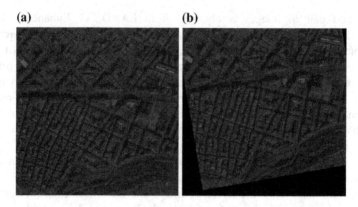

Fig. 2 Images at the region of Barcelona in Spain **a** reference image, **b** sensed image

(a)   **(b)**

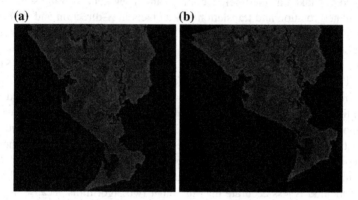

Fig. 3 Images at the region of Chesapeake Bay **a** reference image, **b** sensed image

comparatively better in our method. The transformation obtained from the coarse registration is further refined by the fine registration method. Therefore, the proposed method provides high accuracy in registration. Moreover, the proposed method takes less computational time compared to IS-SIFT and UR-SIFT methods. In IS-SIFT, significant computational time is required for the segmentation the input images. The uniform distribution of features and cross-matching algorithm of UR-SIFT need more computational time than the proposed method. The registered images obtained by using the proposed method are shown in Fig. 4. From the visual representation, it is clear that the edges and the regions of the registered images are perfectly aligned.

Table 1 Registration parameters (a_{11}, a_{12}, a_{21}, a_{22}, α_x, α_y) comparison between different methods

Pair	Method	a_{11}	a_{12}	a_{21}	a_{22}	α_x	α_y
1	Ground truth	0.9962	−0.0872	0.0872	0.9962	10.00	20.00
	IS-SIFT	0.9943	−0.0915	0.0917	0.9945	8.91	20.75
	UR-SIFT	0.9958	−0.0860	0.0862	0.9956	10.84	20.67
	Proposed method	0.9959	−0.0878	0.0879	0.9960	9.43	20.29
2	Ground truth	0.9848	−0.1736	0.1736	0.9848	−20.00	−35.00
	IS-SIFT	0.9844	−0.1740	0.1739	0.9841	−20.17	−34.43
	UR-SIFT	0.9846	−0.1738	0.1734	0.9850	−20.07	−34.91
	Proposed method	0.9847	−0.1737	0.1735	0.9849	−20.03	−35.04
3	Ground truth	0.9659	−0.2588	0.2588	0.9659	20.00	−10.00
	IS-SIFT	0.9604	−0.2508	0.2511	0.9608	20.78	−10.87
	UR-SIFT	0.9626	−0.2524	0.2531	0.9629	20.56	−10.67
	Proposed method	0.9637	−0.2542	0.2548	0.9641	20.42	−10.45

Table 2 RMSE, MI and the associated computational time comparison between different methods

Pair	Method	RMSE	MI	Time (s)
1	Ground truth	–	1.0587	–
	IS-SIFT	0.9230	1.0381	36
	UR-SIFT	0.6701	1.0403	65
	Proposed method	0.3205	1.0432	32
2	Ground truth	–	1.2000	–
	IS-SIFT	0.4324	1.1877	112
	UR-SIFT	0.0315	1.1984	121
	Proposed method	0.0285	1.1987	105
3	Ground truth	–	0.8200	–
	IS-SIFT	0.8190	0.7681	22
	UR-SIFT	0.6133	0.7941	32
	Proposed method	0.4246	0.8072	17

(a) **(b)** **(c)**

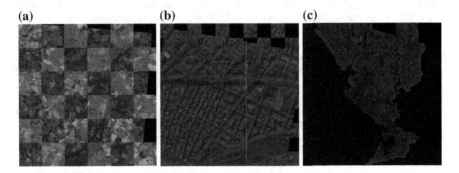

Fig. 4 Checkerboard mosaiced image of **a** pair 1, **b** pair 2, and **c** pair 3

5 Conclusion

In this paper, we have proposed a coarse-to-fine registration method to register remote sensing optical images. The coarse registration is performed by using the standard SIFT approach with RANSAC-based outlier removal technique. The matched features obtained by using SIFT approach are refined through RANSAC algorithm. So, the result of coarse registration is very close to the optimum solution. The transformation parameter values obtained in the coarse registration scheme are used as an initial solution for the SPSA optimizer in the fine registration scheme. The fine registration is performed by the maximization of mutual information using the SPSA optimization in a multiresolution framework. The simulation result shows that the proposed method provides comparatively higher accuracy in registration of remote sensing optical images compared to the other existing algorithms.

References

1. Zitova, B., Flusser, J.: Image registration methods: a survey. Image Vis. Comput. **21**, 977–1000 (2003)
2. Brown, L.G.: A survey of image registration techniques. ACM Comput. Surv. **24**(4), 325–376 (1992)
3. Lowe, D.G.: Distinctive image features from scale-invariant keypoints. Int. J. Comput. Vision **60**(2), 91–110 (2004)
4. Goncalves, H., Corte-Real, L., Goncalves, J.A.: Automatic image registration through image segmentation and SIFT. IEEE Trans. Geosci. Remote Sens. **49**(7), 2589–2600 (2011)
5. Sedaghat, A., Mokhtarzade, M., Ebadi, H.: Uniform robust scale-invariant feature matching for optical remote sensing images. IEEE Trans. Geosci. Remote Sens. **49**(11), 4516–4527 (2011)
6. Gong, M., Zhao, S., Jiao, L., Tian, D., Wang, S.: A novel coarse-to-fine scheme for automatic image registration based on SIFT and mutual information. IEEE Trans. Geosci. Remote Sens. **52**(7), 4328–4338 (2014)
7. Zhang, Y., Zhou, P., Ren, Y., Zou, Z.: GPU-accelerated large-size VHR images registration via coarse-to-fine matching. Comput. Geosci. **66**, 54–65 (2014)
8. Wu, Y., Ma, W., Gong, M., Su, L., Jiao, L.: A novel point matching algorithm based on fast sample consensus for image registration. IEEE Trans. Geosci. Remote Sens. Lett. **12**(1), 43–47 (2015)
9. Sedaghat, A., Ebadi, H.: Remote sensing image matching based on adaptive binning SIFT descriptor. IEEE Trans. Geosci. Remote Sens. **53**(10), 5283–5293 (2015)
10. Cole-Rhodes, A.A., Johnson, K.L., LeMoigne, J., Zavorin, I.: Multiresolution registration of remote sensing imagery by optimization of mutual information using a stochastic gradient. IEEE Trans. Image Process. **12**(12), 1495–1511 (2003)
11. Cole-Rhodes, A.A., Johnson, K.L., LeMoigne, J.: Image registration using a 2nd order stochastic optimization of mutual information. Proc. IGARS. **6**, 4038–4040 (2003)
12. Suri, S., Reinartz, P.: Mutual-information-based registration of TerraSAR-X and Ikonos imagery in urban areas. IEEE Trans. Geosci. Remote Sens. **48**(2), 939–949 (2010)
13. Fischler, M.A., Bolles, R.C.: Random sample consensus: a paradigm for model fitting with applications to image analysis and automated cartography. Commun. ACM **24**(6), 381–395 (1981)

14. Spall, J.C.: Multivariate stochastic approximation using a simultaneous perturbation gradient approximation. IEEE Trans. Automat. Contr. **37**, 332–341 (1992)
15. Spall, J.C.: Accelarated second-order stochastic optimization using only function measurements. Proc. DAC. 1417–1424 (1997)

Review on Human Activity Recognition Using Soft Computing

Rashim Bhardwaj, Kirti Dang, Subhash Chand Gupta
and Sushil Kumar

Abstract In this paper, different techniques for human action recognition has been used such as spiking neural network, genetic algorithm, evolutionary algorithm, state-of-the-art method etc. The comparison of all these techniques has been done to recognise the human activities and after comparison best technique will be selected which works well for recognising the human activities in normal and abnormal situations. Machine learning concept is also included in this paper such as K-NN classifier, neural networks, hidden markov model, spiking neural network, BCM etc. Research studies states that every technique works well in one situation where on the other hand when complex situation comes some of the technique doesn't work. To deal with such type of situation GRN technique is used. The dataset which is considered by authors is UCF50 dataset. Performance of Differential evolution algorithm over genetic algorithm is the future work to recognise the human activities.

Keywords Bag of features · SLFN · OSLEM · K-NN classifier
Image processing · EEMD · SBMLR · FLS-SVM · Accelerometer
GRN

R. Bhardwaj (✉) · K. Dang · S.C. Gupta · S. Kumar
Amity School of Engineering & Technology, Amity University Uttar Pradesh,
Noida, India
e-mail: bhardwajgauri7@gmail.com

K. Dang
e-mail: dangkirti@gmail.com

S.C. Gupta
e-mail: scgupta@amity.edu

S. Kumar
e-mail: kumarsushiliitr@gmail.com

1 Introduction

From the previous years, various researchers have been attracted towards the topic which is recognising human activities in videos. With the advancement of technology, there is possibility that crime can take birth from any root. As it is clear that video camera is used in different areas such as entertainment, healthcare and public areas like railway station and roads. To identify the crime before it occurs, human activity recognition is necessary. Various techniques have been used to recognise the human activities such as genetic algorithm, neural networks, spiking neural networks.

This paper is organised into four sections. Section 1 covers the introductory part. the literature review will be covered in Sect. 2. In Sect. 3, experimental results. conclusion and future work is mentioned in Sect. 4.

2 Literature Review

In order to start the literature review, some of the research questions have been formulated.

Research Questions:

RQ1: Why there is a need to Detect Human activities in videos?
RQ2: Describe Various Techniques used for Detecting?
RQ3: To Identify the Future work?

Meng et al. [1] discussed that it is difficult to detect human activity in videos from various scenes. The main difficulty from the above detection is how to separate the temporal and spatial features. To overcome this problem, a new technique was proposed by the author which is a model of BCM-based spiking neural network which is the advanced version of conventional neural network. GRN is used to regulate their weights. The main motive of this paper is to make a classifier model which can separate the temporal features automatically and we have to extract only the spatial features. Results proved that GRM-BCM network is very effective for recognising human activities online as compared to other algorithms. Implementation of various evolutionary algorithms to improve their performance is the future work.

Xie et al. [2] proposed that the recognition of human activities based on sensors can be improved by genetic programming. The fitness function used in GP is very sensitive it changes with change in environment. GP based classifier does not work well when multiple human activities are present. Therefore, the main objective of this paper is to modify the GP classifier so that it overcomes the issues such as fitness accuracy, use of online time series for normalisation. The dataset which was taken by the author to implement it is the sensor-based data collected from the smartphone. To handle the human activity transition is the future work.

Kunwar et al. [3] discussed the time lapse technique to capture the scenes at slower rate and when it has to reply back or present the scenes. These capture scenes will appear at normal rates. For normal activities like sunrise and sunset, periodic time lapse techniques are used where for uncertain activities a-periodic time lapse technique is used. The main objective of this paper is to use the differential time lapse video one of the classifications of a-periodic video. The selection method in differential time lapse video may be pixel to pixel, edge detection technique, etc. From the above review, authors conclude that differential time lapse is the best way to visualise the slowly moving activities.

Valle and Starostenko [4] discussed that the main objective of this paper is to recognise the humans while walking, running and some other activities. Recognition of human activities at a high precession is still a problem, the approach which is used to analyse the full body activities is neural networks. The disadvantage of this proposed approach is that it is difficult to adjust the parameters on neural network while working with multiple activities. These discussed disadvantages are the future work.

Meng et al. [5] proposed the model spiking neural network, from the studies it is proved that spiking neural network is more powerful than normal neural networks. As it solves the complex real world problems, the combination of GRN-BCM is used in this paper. The comparison of GRN-BCM is compared with various models some of them are hidden markov model, single BCM, etc. After comparing their performance GRN-BCM is applied on two dataset of human behaviour recognition. From this scenario, results proved that use of GRN-BCM model is very difficult to recognise the human behaviour as compared to the state-of-the-art method.

Wu et al. [6] discussed that video camera's is used for monitoring the human activities in different environment, due to this there is demand of system analysis to deal with large amount of videos. In this paper, radial basis function neural network approach is used which is trained by local generalisation error model. All the features from various videos are extracted by the use of action bank and these extracted feature vectors are used as input of RBFNN. Results prove that it works well for human action recognition as compared to SVM.

Vishwakarma et al. [7] discussed that there are various simple and robust method which can be used to detect the abnormal activities so that it can be monitored at early stages. Change in the direction of human body is used for the detection of abnormal activities then these activities are classified by K-NN classifier. This method is highly accepted due to their high human activities recognition rate. If no any activity is detected then it is considered as normal video otherwise abnormal.

Chen et al. [8] presented the algorithm for the recognition of human activities by the use of wearable sensors. The main techniques which are used in this paper are EEMD, SBMLR, FLS-SVM. From the sensor-based data, the features will be extracted and further processed by EEMD which results in reducing the dimensions. Then further FLS-SVM will be used to deal with the reduced features to detect the human activities. The results depicts that it works well on large scale dataset.

Jeroudi et al. [9] discussed that the recognition of human activities is the necessary for all real world problems. In this paper, inertial sensor is used to collect all the data from the sensors because to map the performance of activity with the corresponding data is not possible. For this, OSELM is used to train the SLFN (single layer feed forward network) where results proves 82.05% average accuracy. Their future work is to further investigate the type of activation function.

Nguyen et al. [10] presented the use of ISA (independent subspace analysis) with three-layer neural networks to learn the features of the videos. The features which are learned from the videos using ISA are passed to BOF which is used for recognising human interactions. Results proved that convolution neural network is able to learn the features which include complex activities as well as real world activities.

Wawrzyniak et al. [11] discussed that the clustering approach is used to overcome the problem of Human activity recognition using the motion data. Accelerometer and gyroscope are the main sources for collecting the data. Tracking system is used to monitor the position of the human.

ljjina et al. [12] presented the paper on recognising human activities using genetic algorithm and convolution neural networks. The weights which will be initialized on neural network will become the solutions for genetic algorithm. To train the neural network, gradient descent approach is used during the evaluation of fitness function from Chromosomes. The global and local search abilities are explored to find the solution which is closer to the global optimum. This proves that combining the evidences of classifier using GA help us to improve the performance. The future work can be improvising the performance of the person adjoining.

3 Experimental Results

After reviewing each paper, experiment results which are analysed are listed below. In [12], ljjina et al. (2016) analysed the performance by checking the experimental results. In this experiment, genetic algorithm runs for five generations whose seed value is 0–5000. Crossover probability taken here is 0.8 where mutation probability is 0.01. The probability which is used here is low mutation probability. The main motive of this implementation is to find the optimal solution using small number of generations and population size with the help of steepest descent algorithm. Results of UCF50 dataset are analysed on the basis of fivefold cross validation and the best result of fivecross fold validation in the forms of graph are as follows.

Figures 1 and 2 show the proper selection of GA parameters.

Results analysed in the Figs. 3 and 4 shows that the extreme learning machine gives better results. Classification accuracy so far achieved is 99.98% [2]. To check the performance of GRN-BCM, KTH dataset is used to conduct the experiment. Results can be analysed by analysing the behaviour of different classes. On the KTH dataset, 400 videos are tested. Results prove that the behaviour recognition rate on KTH dataset is 82.5% [3]. "Differential time lapse video generation

Fig. 1 Best values decreases

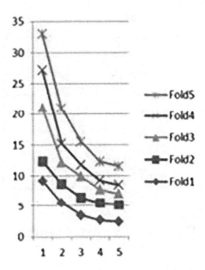

Fig. 2 Mean value decreases

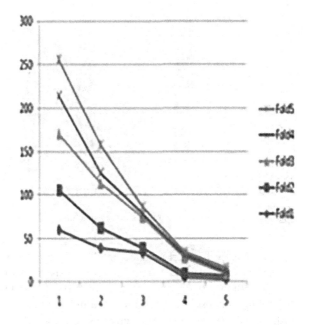

algorithm" is proposed; in this algorithm, edge detection technique is used. Results show that the performance of this algorithm does not work properly for real-time videos. To overcome this limitation, already stored files has been used [4]. In this paper, the recognition rate is very high. To measure it more accurately, a new pattern which is not participated in CNN networks will be taken. By doing this experiment, results show 87% accuracy [5].

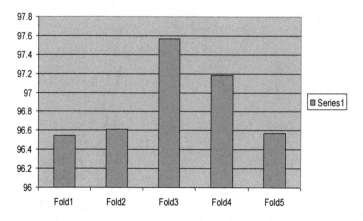

Fig. 3 Average results of neural network classifier

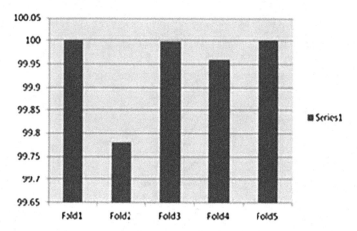

Fig. 4 Average results of extreme learning machine

To compare the performance of GRN-BCM networks, various state-of-the-art algorithms are selected. Comparative analysis shows better results [6]. The proposed method provides an accuracy of 98.61%. On testing further with different dataset the proposed algorithm resulted in higher performance [7]. In this paper, in order to detect the abnormal activities, methodologies such as background subtraction, Silhouette extraction and Feature extraction-nearest neighbour are used. Results show that this method is feasible only if the background is stationary, when the abnormal activity occurs [8]. In this paper, experiments include comparison of extracted features with several other features. Results show that FLS-SVM achieves recognition rate of 93.43%, LS-SVM with recognition rate of 92.83% and SVM with the rate of 90.03%. This concludes that FLS-SVM is more efficient than any other above-stated algorithm [9]. This paper states that when the number of neurons increases in artificial neural network, the accuracy increases hand in hand. Experimental results show that using 210 neurons and 71 features, 82% accuracy is achieved. In this paper, method works well only when human is in a moving state. It gives poor results while the human is at stationary position [10]. Two types of experiments were conducted to verify the performance of three-layer neural network and pooling layer. In the first case, it achieves accuracy of 90% and on the second it goes to 85%. The limitation of this paper is that the ISA proposed method lacks the ability for spatially and temporal activities. This limitation can be taken as the future work [11]. In this paper, results conclude that SVM classifier and LDA classifier combined together produces slightly poor scores [12]. In this paper, using genetic algorithm for human activity recognition achieves an accuracy of 99.98%.

4 Conclusion and Future Scope

During the detailed review of human activity recognition, the main focus is to find the different techniques which were used to recognise the human activities. The need of human activity detection in today's world is very important because of the crime and frauds are increasing day by day. In order to control these crimes, video cameras should be present at every corner of the street. The techniques which are used to recognise the human activities are genetic algorithm, neural networks, simple machine learning methods and evolutionary algorithms. The performance of genetic algorithm and evolutionary algorithm is compared.

Various papers have been reviewed along with their experiments. In this, performance of each and every method is compared that uses genetic algorithm to recognise the human activity. The experiment results of genetic algorithm that uses extreme learning machine gives better results compared to the neural network classifier. During this experimental results analysis, the main issues which was faced by the authors of this paper were over fitting of training data and possibility of solution getting stuck at the local minimum.

The future work from this detailed review can be the implementation of differential evolution to recognise the human activities along with neural networks. Differential evolution is a technique which is used to improve candidate solution by considering search over large space. It has four stages: (1) initialisation, (2) Mutation, (3) Recombination, (4) selection. On the basis of the best fit solution neural network will be trained. If the best solution is not found, then the process continuous with mutation and goes through various stages until the fittest solution is not found. Studies proved that differential evolution gives better performance as compare to other algorithm.

References

1. Meng, Y., Jin, Y., Yin, J., Conforth, M.: Human activity detection using spiking neural networks regulated by a gene regulatory network. In: International Joint Conference on Neural Networks (IJCNN), pp. 1–6 (2010)
2. Xie, F., Qin, A.K., Song, A., Ciesielski,V.: Sensor-based activity recognition with improved GP-based classifier. In: IEEE Congress on Evolutionary Computation, pp. 3043–3050 (2013)
3. Kunwar, K., Patil, A.S., Sagar, A.: Real time application to generate the differential time lapse video with edge detection. In: Nirma University International Conference on Engineering (NUiCONE), pp. 1–6 (2012)
4. Valle, E.A., Starostenko, O.: Recognition of human walking/running actions based on neural network. In: 10th International Conference on Computing Science and Automatic Control (CCE), pp. 239–244 (2013)
5. Meng, Y., Jin, Y., Yin, J.: Modeling activity-dependent plasticity in BCM spiking neural networks with application to human behaviour recognition. IEEE Trans. Neural Netw. **22**, 1952–1966 (2011)
6. Wu, Z.M., Wing, Y., Ng, W.: Human action recognition using action bank and RBFNN trained by L-GEM. In: International Conference on Wavelet Analysis and Pattern Recognition, pp. 30–35 (2014)
7. Vishawkarma, D.K., Kapoor, R., Maheshwari, R., Kapoor, V., Raman, S.: Recognition of abnormal human activity using the changes in orientation of silhouette in key frames. In: 2nd International Conference on Computing for Sustainable Global Development (INDIACom), pp. 336–341 (2015)
8. Chen, Y., Guo, M., Wang, Z.: Improved algorithm for human activity recognition using wearable sensors. In: 8th International Conference on Advanced Computational Intelligence (ICACI), pp. 248–252 (2016)
9. Jeroudi, Y.A., Ali, M.A., Latief, M., Akmeliawati, R.: Online sequential extreme learning machine algorithm based human activity recognition using inertial data. In: 10th Asian control conference, pp. 1–6 (2016)
10. Nguyen, N., Yoshitaka, A.: Human interaction recognition using independent subspace analysis algorithm. In: IEEE International Symposium, pp. 40–46 (2014)
11. Wawrzyniak, S., Niemiro, W.: Clustering approach to the problem of human activity recognition using motion data. In: Federated conference on Computer science and information System, pp. 411–416 (2015)
12. Ijjina, E.P., Chalavadi, K.M.: Human action recognition using genetic algorithms and convolutional neural networks. Sci. Direct, 1–4 (2016)

Solving Sudoku Puzzles Using Evolutionary Techniques—A Systematic Survey

Deepti Bala Mishra, Rajashree Mishra, Kedar Nath Das and Arup Abhinna Acharya

Abstract Sudoku puzzle is a game which takes the form of an N × N matrix. It requires the players to organize the number sequences from 1 to N in the submatrices of the original matrix in such a way that no numbers are reused in each sub matrices and also the numbers are not reused in each column and rows. It is mainly based on the number replacement game and is a combinatorial puzzle. Several evolutionary techniques such as Genetic algorithm, Particle Swarm Optimization, Ant Colony Optimization, and Artificial Bee Colony Optimization are used for solving, rating, and generating Sudoku Puzzles. This research paper presents a survey of solving Sudoku Puzzles using different evolutionary technique-based hybridized algorithms and analyze the results, i.e., success rates found in solving the puzzles of different levels such as Easy, Medium, Challenging, Hard, Evil, and Super Hard.

Keywords Sudoku puzzle · Genetic Algorithm (GA) · Particle Swarm Optimization (PSO) · Ant Colony Optimization (ACO) · Artificial Bee Colony Optimization (ABCO)

D.B. Mishra (✉) · A.A. Acharya
School of Computer Engineering, KIIT University, Bhubaneswar 751024, India
e-mail: dbm2980@gmail.com

A.A. Acharya
e-mail: aacharyafcs@kiit.ac.in

R. Mishra
School of Applied Sciences, KIIT University, Bhubaneswar 751024, India
e-mail: rajashreemishra011@gmail.com

K.N. Das
NIT Silchar, Silchar, Assam, India
e-mail: kedar.iitr@gmail.com

© Springer Nature Singapore Pte Ltd. 2018
M. Pant et al. (eds.), *Soft Computing: Theories and Applications*,
Advances in Intelligent Systems and Computing 583,
https://doi.org/10.1007/978-981-10-5687-1_71

791

1 Introduction

Nowadays, Sudoku Puzzles are one of the most popular and addictive game in all over the world. In 1979, the first Sudoku Puzzle was published in a puzzle magazine of USA. The word Sudoku means to a "Single Number," which is a combination of two Japanese words Su (Number) and Duko (Single) [1]. It is mainly based on the logic that numbers are placed in a N $\times$ N matrix with 1 to N^2 digits in such a way that each of these numbers should present exactly once in all the subgrids, columns, and rows including the fixed cells which are never be changed. A set of unique numbers should contain in all the rows, columns, and submatrices with no repetitions in a particular sequence. So there exists a unique solution to a particular Sudoku puzzle that satisfies all the given constraints. It is a combinatorial puzzle with N $\times$ N subgrids of order N. The puzzle can be partially filled $N^2 \times N^2$ grids. These types of problems are non trivial problems and are also known as NP-complete. A single player can play this game by partially filling a $N^2 \times N^2$ grid with the digits ranging from 1 to N^2. Basically, the puzzle consists of 9 $\times$ 9 subdivided by 3 $\times$ 3 subgrids in which there are some prefilled cells with constant digits that will never changed or moved [2]. It is a discrete optimization problem where among several discrete values there is a unique solution to the problem is found. Puzzles are also called as NP-Hard or NP-complete problems. Figure 1 shows an example of Sudoku puzzle [1] of the order 9 $\times$9 with 26 given numbers and Fig. 2 shows the solution for the puzzle. Given number of Fig. 1 remains constant with their same position. In Fig. 2, the solution contains the numbers from 1 to 9 in such a way that the numbers are present exactly once in each row, column and sub square.

Fig. 1 A Sudoku puzzle with 26 givens

	8			3		4		
				5				1
					4	5	8	
	5	7				2		9
9								4
	3		4			6	5	
	7	9	2					
5					6			
		6		4			2	

Fig. 2 Solution to the Sudoku puzzle of Fig. 1

7	8	5	9	3	1	4	6	2
2	4	3	8	5	6	9	7	1
6	9	1	7	2	4	5	8	3
4	5	7	6	1	2	3	9	8
9	6	8	5	7	3	2	1	4
1	3	2	4	9	8	6	5	7
3	7	9	2	8	5	1	4	6
5	2	4	1	6	7	8	3	9
8	1	6	3	4	9	7	2	5

This paper presents a survey of how different types of evolutionary techniques such as GA, PSO, and ACO (meta-heuristic method) have been efficiently used to solve Sudoku Puzzles of different difficulty levels as Easy, Medium, Hard, and Super Hard. Further, the paper is divided into five sections. Section 1 presents the Introduction to Sudoku Puzzles, Sect. 2 presents the background of different Evolutionary techniques, Sect. 3 contains related previous work in the field of evolutionary techniques-based Sudoku solver, Sect. 4 shows the performance and results of discussed algorithms. Finally, Sect. 5 gives some conclusions of the work done and future work is outlined.

2 An Introduction to Evolutionary Algorithm

Evolutionary algorithm based on biological behavior or evolution of population, which can be used to solve many complex and real-life problems by producing high quality test data automatically [3, 4].

This algorithm based on the principle as survival of the fittest and models some natural phenomena like genetic inheritance and Darwinian strife for survival, constitute an interesting category of modern heuristic search [3].

2.1 Genetic Algorithm (GA)

GA has emerged as a practical, robust optimization technique and search method and it is inspired by the way nature evolves species using natural selection of the fittest individuals. The algorithm was developed by John Holland in United States. The solution to a specific problem can be solved by a population of chromosomes. It is the best way to solve optimization problems by searching for good genes and applying the different genetic operators like selection, crossover, mutation, and Elitism to the specific fitness function according to the problem [4, 5].

2.2 Particle Swarm Optimization Algorithm (PSO)

PSO is a search-based optimization technique that studies the social behavior of bird flocking or fish schooling. The algorithm can converged very speedily to an optimal solution. The best solution can be found by a number of particles constituting a swarm, moving around in a particular real valued N-dimensional search space and adjusting their flying according to own and other's flying experience [6]. Particles are always keeping track for personal best solution, denoted by p-best and the best value of any particle, denoted by g-best. The particles are called as potential solution [7, 8].

2.3 Ant Colony Optimization Algorithm (ACO)

Ant Colony Optimization (ACO) is a distributed meta-heuristic algorithm, inspired by biological behaviors of real-world ants mainly used to solve many optimization problems. The Ant System Algorithm was first proposed by Marco Dorigo et al. in 1991 to solve combinatorial discrete optimization problems [9, 10]. In this algorithm, the optimization problem is represented as a graph and the artificial ants move around the paths of the graph repeatedly to find the best solution. Ants select their paths according to the higher pheromone levels of the graph edges. For each iterations, possible solutions are created and finally evaluate the best quality solution using a heuristic measure [10].

2.4 Artificial Bee Colony Optimization Algorithm (ABCO)

Artificial Bee Colony (ABC) was introduced by Karabora in 2005. This algorithm is used to solve different optimization problems with real parameter and this algorithm uses scouts and foragers as initial populations. In the first step, the food sources are initialized, scouts are searching food sources until the stopping criteria fulfilled and forager finds the best quality food source by exploiting the corresponding scouts [11, 12].

3 Related Works

This section provides a systematic literature review on different evolutionary techniques like GA, PSO, and ACO used to solve NP-Hard problem as Sudoku Puzzles.

Moraglio et al. [13], proposed a genetic algorithm with product geometrical crossover by combining the new geometric crossovers with the preexisting geometric crossovers in a simple way to solve Sudoku Puzzles. They found their proposed method can solve easy Sudoku efficiently, but medium and superior difficulties puzzles have not been solved. The success rate of the algorithm is 14–72%. Their proposed fitness function is to maximize the total sum of the unique numbers present in each column, row, and sub grid. The maximum global fitness is less than the value shown in Eq. (1).

$$\Delta F = 216 \tag{1}$$

Mantere and Koljonen [14–17], have used different evolutionary algorithms (EA) such as genetic algorithm (GA), cultural algorithm (CA), ant colony optimization (ACO), and Hybrid Genetic Algorithm/Ant Colony Optimization

(HGA/ACO) algorithm to solve Sudoku Puzzles. Genetic algorithm-based idea is innovatively valuable and novelty but it gives a slow rate of convergence. They proposed a new elitism and sorting strategy which is an intelligent problem-based sorting strategy and the results show the ability of high improvement in the solving speed of Sudoku Puzzles. They found the results that HGA/ACO was more efficient than other evolutionary methods. Their fitness function shown in Eq. (2).

$$
\begin{aligned}
&\textbf{If Best} \, (\textbf{generation}_i) = \textbf{Best}(\textbf{generation}_{i-1}) \\
&\textbf{then Value}(\textbf{Best}) + \, = \textbf{1}; \\
&\textbf{A} = \{\, \textbf{1}, \textbf{2}, \textbf{3}, \textbf{4}, \textbf{5}, \textbf{6}, \textbf{7}, \textbf{8}, \textbf{9} \}, \textbf{g}_{i3} = |\textbf{A} - \textbf{x}_i| \, \textbf{and} \, \textbf{g}_{j3} = |\textbf{A} - \textbf{x}_j|
\end{aligned} \tag{2}
$$

In [16], they again proposed a Cultural algorithm (CA) by adding a space to the genetic algorithm and found solving efficiency was increased in comparison to genetic algorithm that they have used previously.

Again in [17], the author introduced an improved hybridized ant colony optimization/genetic algorithm to solve Sudoku Puzzles. The proposed algorithm was based on sorting process of populations and elitism process to improve their previous version of evolutionary algorithm and the author found the speed of solving Sudoku Puzzles has improved significantly.

Hereford and Gerlach [6], developed a hybridized of the PSO algorithm to solve a Sudoku puzzle as Integer PSO (IPSO). It performs better on the Sudoku Puzzles than the general PSO by several parameter adjustments. In their work, the solution of the puzzle have represented by a list of integers, where N is the length of the list and that is also the number of unknowns in the puzzle. They used the velocity update equation for their modified PSO shown in Eq. (3).

$$
\begin{aligned}
&\textbf{v}_{n+1} = \textbf{v}_n + \textbf{c1} * \textbf{r1} * (p\textbf{best}_n - \textbf{p}_n) + \textbf{c2} * \textbf{r2} * (l\textbf{best}_n - \textbf{p}_n) \\
&\textbf{p}_{n+1} = \textbf{1}, \textbf{if rand}() < \textbf{S}(\textbf{v}_{n+1}) \\
&\textbf{p}_{n+1} = \textbf{0}, \textbf{otherwise}
\end{aligned} \tag{3}
$$

Pacurib et al. [11], proposed an improved a hybridized form of the Artificial Bee Colony algorithm for solving Sudoku Puzzles. The authors compared their algorithm with GA-based Sudoku solver and they found ABC-based algorithm performs outstanding for all types of Sudoku Puzzles with optimal solution. The authors used the fitness function shown in Eq. (4).

$$
\textbf{fit}_i = 1/1 + f_i \tag{4}
$$

where $\text{fit}_{i,}$ represents the fitness of the feasible solution of ith bee and for every feasible solution the penalty value is $f_{i,}$ where the Penalty value is the total number of missing values in each row and column and the optimal solution is achieved when the fitness value is 1 and penalty value is 0. Their proposed Sudoku solver is very efficient in finding the optimal solution of the puzzle for every time.

Sato and Inoue [4], proposed a genetic algorithm based on effective building blocks in genetic operations to solve Sudoku problems. They proposed a stronger local search function in mutation which gives a higher improved optimum solution. They used the fitness function shown in Eq. (5).

$$f(x) = \sum_{i=1}^{9} g_i(x) + \sum_{j=1}^{9} h_j(x) \tag{5}$$

where $g_i(x)$ and $h_j(x)$ represents the number of entries in a specified row and column and the maximum value of $f(x)$ will be 162. Their proposed crossover method and improved local search functions of mutation are very effective by increasing the solution rate.

Sato et al. [5], developed a GA model based on coarse grained to solve Sudoku Puzzles. The authors used a genetic operation which links with building blocks using GPU (Graphics Processing Units) in applications for parallel processing. This model can solve the processing time problem by countering the dependent individuals. In a few second of processing, they found 100% correct solution for extremely difficult problems. They used the same fitness function as shown in Eq. (5).

Nath et al. [1], introduced Ret-GA called as Retrievable Genetic Algorithm for solving Sudoku puzzle. The proposed algorithm is based on creating a blue print matrix in which the value s as 0 or 1 is assigned to the entries in Sudoku puzzle whether the entry is given or not, respectively. Their algorithm starts creating the blue print matrix of the same size as the Sudoku puzzle. They used the fitness function for each individual as shown in Eq. (6).

$$f(i,j,k,l) = \begin{cases} 0, & \textbf{if}(i,j) = (k,l) \\ 1, & \textbf{otherwise} \end{cases} \tag{6}$$

where (i, j) and (k, l) refer to two entries of N × N Sudoku puzzle. The authors performed Ret-GA on a set of 75 new puzzles with difficulty rating as Easy, Medium and Hard. They found when the difficulty level changed from Easy to Hard the success rate decreases with the increasing of the average number of generations and the standard deviation. In their work they ordered the difficulty level of Sudoku puzzle from Easy to Hard as Easy < Medium < Hard by taking into consideration, the success rate and average number of generations for Ret-GA. They again compared with another nine puzzles taken from [14], to check the effectiveness of their proposed Ret-GA. Finally, they concluded that the algorithm can solve easy puzzles with lesser number of generations with 100% success rate as compared to existing GA, and it is not able to solve difficult and super difficult puzzles with 100% success rate but the success rate is always higher than GA.

Deng and Li [18], have proposed one improved genetic algorithm (IGA) to enhance the convergence speed of the genetic algorithm, by improving the different operators of GA for higher reliability, better stability, and quicker convergence speed of solved puzzles. In their proposed method, the crossover and mutation probability are kept constant at the time of optimization, which affect the stability and practicability of the algorithm and restrict the convergence speed to a particular degree. The authors again proposed a hybrid genetic algorithm (HGA) [19], in which all the operators of Genetic algorithm have been improved depending on the futures of Sudoku Puzzles. After simulation the convergence rate and stability of the algorithm gives very good results for easy and challenging Sudoku but it is not so satisfactory for difficult and super difficult Sudoku Puzzles. The authors used the fitness function as in Eq. (7).

$$w = \sum_{i=1}^{9} (r_i + c_i) \qquad (7)$$

where r_i, c_i represents the repeated integers in ith row and column, respectively. The optimal solution will achieved when w becomes 0.

Waiyapara et al. [20], proposed a node-based coincidence algorithm(NB-COIN) to solve easy medium and hard level Sudoku Puzzles, which combines the features of NHBSA (node histogram-based sampling algorithms) and coincidence algorithms. Their fitness function is same as Eq. (5). They found NB-COIN gives out performance result for solving multimodal combinatorial optimization problems.

Kamei and Nakano [7], proposed to solve simplified Sudoku and complete Sudoku. They found their proposals gives better results than the theoretical value and PSO is able to solve simplified Sudoku of 15 blank grids. They overcome this problem by modifying the exploration process which again reexplore from another initial state if the results are quasi-optimum. They found the modified PSO can solve complete Sudoku with more than 17 blank grids. Their fitness function is shown in Eq. (8) and they taken the small value as the good result.

Redding et al. [21], proposed a method known as WoAC (Wisdom of Artificial Crowds) with GA to constrain the solution space after a certain generation by aggregating the agreement between the members of the population. A time variant mutation operation is implemented in the proposed algorithm and found puzzles of Easy, Medium, Hard, and Evil difficult can solved in 2 min only. They compared the performance of solving hard and evil difficulty puzzles with GA method and GA + WoAC method and found their method shows computational efficient and effective results by reducing the number of generations. Finally, they conclude that by applying WoAC the time to find a solution for easy and medium difficulty Sudoku can be reduced up to 50%.

$$f = \sum_{C=1}^{9} |45 - \mathbf{line}_C| + \sum_{R=1}^{9} |45 - \mathbf{line}_R| + \sum_{G=1}^{9} |45 - \mathbf{subgrid}_G| + \left|405 - \sum_{\alpha=1}^{81} \mathbf{grid}_\alpha\right|$$
$$(8)$$

Wang et al. [3], proposed the filtered operation for mutation to solve Sudoku Puzzles, where the two cells of subblocks are swapped and filtered from the given cells by taking the candidate arrays and check whether at least one of the two values is correct entry or not. They have taken eight different levels of puzzles in their computational experiments and found 100% success by taking appropriate GA parameters. They used the fitness function shown in Eq. (9).

$$\mathbf{fit} = \sum_{i=1}^{9} (r_i + c_i) \tag{9}$$

where c_i and r_i represents the number of repeated integers on the ith column and ith row, respectively. The optimal solution for is achieved when fit = 0.

Singh and Deep [2], presented a new algorithm as MA_PSO_M based on the modified rules of PSO coupled with a carefully designed mutation operator within the framework of cell-like P-systems. They defined a search space for solving the Sudoku problem. Their proposed algorithm is very efficient and reliable for easy and medium difficulty levels and it also performs very well for the hard and evil difficultly levels of puzzles by using an additional deterministic phase in which there may exists few empty cells with only single probable candidates. They found the results are better than other similar attempt of PSO-based membrane algorithm and it successfully improved the performance.

After an extensive study about the different techniques of Sudoku Puzzle solving, we came to learn that the evolutionary techniques are vastly used for solving different kinds of Sudoku Puzzle by hybridizing the original evolutionary techniques. Table 1 shows a brief summary of different evolutionary algorithms used for solving Sudoku Puzzles and the success rates for different levels of Puzzles in comparison to other technique, found in the previous works to till date.

Table 1 A brief summary of different evolutionary algorithms used to solve Sudoku Puzzles and their success rates in comparison to other techniques

Authors	Proposed algorithm	Compared with	No. of puzzles taken	Success rate (Easy, Challenging, Medium, Hard and Evil)
Moraglio et al. [13], 2006	Geometric crossover GA	Hill climber and mutation	5	–
Mantere et al. [14–17], 2006–2013	GA, CA, ACO and HGA/ACO	GA, NB-COIN	10	Easy-100%, Challenging-30% and Hard-4–6%
Hereford et al. [6], 2008	IPSO	PSO and $\mu + \lambda$ ES	50	Easy-100%, Medium-100% and Hard-0–30%
Pacurib et al. [11], 2009	ABCO	GA	3	Easy, Medium and Hard-100%

(continued)

Table 1 (continued)

Authors	Proposed algorithm	Compared with	No. of puzzles taken	Success rate (Easy, Challenging, Medium, Hard and Evil)
Sato et al. [4], 2010	GA	GA	6	Easy-100%, Medium-100%, Hard-96–100%
Sato et al. [5], 2011	GPU acceleration	GA	66	Easy, Medium and Hard-100%
Nath et al. [1], 2012	RET-GA	GA	75	Easy-100%, Medium-67–98%, Hard-12–40%
Deng et al. [18], 2013	HGA	IGA and GA	14	Easy 80–100%, Challenging (Medium)-60–80% Difficult (Hard)-17%
Waiyapara et al. [20], 2013	NB-COIN	GA	6	Easy, Medium and Hard-100%
Kamei et al. [7], 2014	Improved PSO	PSO	50	Easy-100%, Medium-100%, Hard-40%
Redding et al. [21], 2015	WoAC + GA	GA	–	–
Wang et al. [3], 2015	GA + Filtered Mutation	GA with random mutation	8	Easy and Medium, Hard-100% and Evil-100%
Singh et al. [8], 2016	MA-PSO-M	PSO	60	Easy-40–100%, Medium-0–100%, Hard-50–100% and Evil-0–100%

4 Results and Discussions

Figure 3 shows the performance and success rate of discussed algorithms for Easy Level Sudoku Puzzles. It was found that most of the proposed algorithms based on evolutionary techniques can successively solve Sudoku Puzzles of Easy Levels. Figures 4 and 5 represent the success rate of previously proposed algorithm for Medium and Hard Level Sudoku Puzzles, respectively. For medium level puzzles, all proposed algorithms gives 100% success rates except two or three the algorithms. Although we are taking the upper range of the success rate, it is found that most of the algorithms can solve both easy and medium level puzzles successively, but in case of hard level puzzles the proposed algorithms can solve for a success rate from 6 to 100%.

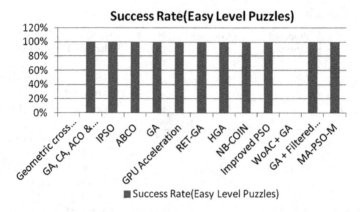

Fig. 3 Success rate for easy level Sudoku

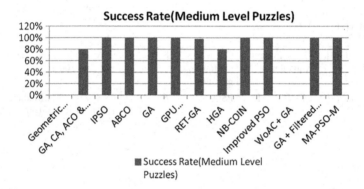

Fig. 4 Success rate for medium level Sudoku

Fig. 5 Success rate for hard level Sudoku

5 Conclusion and Future Work

This review paper gives us a brief idea how different types of evolutionary techniques such as GA, PSO, and ACO have been efficiently used to solve different types of Sudoku Puzzles, leveled by Easy, Medium, Challenging, Hard, Evil, and Super Hard. The results and performance of solving these Puzzles using evolutionary techniques are very effective and efficient and the overall analysis shows that many of the discussed algorithms could successfully solve the easy and medium level puzzles, but in case of the Hard level puzzles, the success rates gradually decreases except some proposed algorithm like ABCO, proposed by Pacurib et al. [11], GPU acceleration with Genetic operation, developed by Sato et al. [5], GA with filtered mutation, proposed by Wang et al. [3] and NB-COIN algorithm, proposed by Waiyapara et al. [20] which gives 100% success rate.

In future, we want to extend the result analysis of discussed algorithms and it is planned to design an efficient evolutionary-based algorithm to solve the different levels Sudoku Puzzles.

References

1. Das, K.N., Bhatia, S., Puri, S., Deep, K.: A retrievable GA for solving Sudoku puzzles. Technical report (2012)
2. Sabuncu, I.: Work-in-progress: solving Sudoku puzzles using hybrid ant colony optimization algorithm. In: 2015 1st International Conference on Industrial Networks and Intelligent Systems (INISCom), pp. 181–184. IEEE (2015)
3. Wang, Z., Yasuda, T., Ohkura, K.: An evolutionary approach to Sudoku puzzles with filtered mutations. In: 2015 IEEE Congress on Evolutionary Computation (CEC), pp. 1732–1737. IEEE (2015)
4. Sato, Y., Inoue, H.: Solving Sudoku with genetic operations that preserve building blocks. In: Proceedings of the 2010 IEEE Conference on Computational Intelligence and Games, pp. 23–29. IEEE (2010)
5. Sato, Y., Hasegawa, N., Sato, M.: GPU acceleration for Sudoku solution with genetic operations. In: 2011 IEEE Congress of Evolutionary Computation (CEC), pp. 296–303. IEEE (2011)
6. Hereford, J.M., Gerlach, H.: Integer-valued Particle Swarm Optimization applied to Sudoku puzzles. In: Swarm Intelligence Symposium (SIS 2008), pp. 1–7. IEEE (2008)
7. Kamei, K., Nakano, M. An approach to search the solution of a puzzle game by Particle Swarm Optimization. In: 2014 Joint 7th International Conference on and Advanced Intelligent Systems (ISIS), 15th International Symposium on Soft Computing and Intelligent Systems (SCIS), pp. 75–80. IEEE (2014)
8. Singh, G., Deep, K.: A new membrane algorithm using the rules of Particle Swarm Optimization incorporated within the framework of cell-like P-systems to solve Sudoku. Appl. Soft Comput. 45, 27–39 (2016)
9. Schiff, K.: An ant algorithm for the Sudoku problem. J. Autom. Mob. Robot. Intell. Syst. 9 (2015)
10. Asif, M., Baig, R.: Solving NP-complete problem using ACO algorithm. In: International Conference on Emerging Technologies (ICET 2009), pp. 13–16. IEEE (October 2009)

11. Pacurib, J.A., Seno, G.M.M., Yusiong, J.P.T.: Solving Sudoku puzzles using improved artificial bee colony algorithm. In: 2009 Fourth International Conference on Innovative Computing, Information and Control (ICICIC), pp. 885–888. IEEE (2009)
12. Kaur, A., Goyal, S.: A survey on the applications of bee colony optimization techniques. Int. J. Comput. Sci. Eng. **3**(8), 3037 (2011)
13. Moraglio, A., Togelius, J., Lucas, S.: Product geometric crossover for the Sudoku puzzle. In: 2006 IEEE International Conference on Evolutionary Computation, pp. 470–476. IEEE (2006)
14. Mantere, T., Koljonen, J.: Solving and rating Sudoku puzzles with genetic algorithms. In: New Developments in Artificial Intelligence and the Semantic Web, Proceedings of the 12th Finnish Artificial Intelligence Conference STeP, pp. 86–92 (2006)
15. Mantere, T., Koljonen, J.: Solving, rating and generating Sudoku puzzles with GA. In: 2007 IEEE Congress on Evolutionary Computation, pp. 1382–1389. IEEE (2007)
16. Mantere, T., Koljonen, J.: Solving and analyzing Sudokus with cultural algorithms. In: 2008 IEEE Congress on Evolutionary Computation (IEEE World Congress on Computational Intelligence), pp. 4053–4060. IEEE (2008)
17. Mantere, T.: Improved ant colony genetic algorithm hybrid for Sudoku solving. In: 2013 Third World Congress on Information and Communication Technologies (WICT), pp. 274–279. IEEE (2013)
18. Li, Y., Deng, X.: Solving Sudoku puzzles based on improved genetic algorithm. Jisuanji Yingyong yu Ruanjian **28**(3), 68–70 (2011)
19. Deng, X.Q., Da Li, Y.: A novel hybrid genetic algorithm for solving Sudoku puzzles. Optim. Lett. **7**(2), 241–257 (2013)
20. Waiyapara, K., Wattanapornprom, W., Chongstitvatana, P.: Solving Sudoku puzzles with node based Coincidence algorithm. In: 2013 10th International Joint Conference on Computer Science and Software Engineering (JCSSE), pp. 11–16. IEEE (2013)
21. Redding, J., Schreiver, J., Shrum, C., Lauf, A., Yampolskiy, R.: Solving NP-hard number matrix games with Wisdom of Artificial Crowds. In: Computer Games: AI, Animation, Mobile, Multimedia, Educational and Serious Games (CGAMES), pp. 38–43. IEEE (2015)

Energy-Efficient Straight Robotic Assembly Line Using Metaheuristic Algorithms

Janardhanan Mukund Nilakantan, S.G. Ponnambalam
and Peter Nielsen

Abstract This paper focuses on the implementation of metaheuristic algorithms to solve straight robotic assembly line balancing problem with an objective of maximizing line efficiency by minimizing the energy consumption of the assembly line. Reduction in the energy consumption is of high importance these days due to the need of creating environmental friendly industries and also due to the increase in the cost of energy. Due to the availability of different types of robots in the market, there is a necessity of selecting efficient set of robots to perform the tasks in the assembly line and optimizing the efficiency of its usage in the line effectively. Two well-known metaheuristic algorithms: particle swarm optimization (PSO) and differential evolution (DE) are implemented to solve due to the NP-hard nature of the problem. Proposed algorithms are tested on the benchmark problems available in the literature and the detailed comparative results are presented in this paper. It can be seen that proposed DE algorithm could obtain better results when compared with PSO from the experimental study.

Keywords Energy efficiency · Metaheuristics · Robotic assembly line · Balancing

J.M. Nilakantan (✉) · P. Nielsen
Department of Mechanical and Manufacturing Engineering,
Aalborg University, Aalborg, Denmark
e-mail: mnj@m-tech.aau.dk

P. Nielsen
e-mail: peter@m-tech.aau.dk

S.G. Ponnambalam
School of Engineering, Monash University Malaysia,
Bandar Sunway 47500, Malaysia
e-mail: sgponnambalam@monash.edu

© Springer Nature Singapore Pte Ltd. 2018
M. Pant et al. (eds.), *Soft Computing: Theories and Applications*,
Advances in Intelligent Systems and Computing 583,
https://doi.org/10.1007/978-981-10-5687-1_72

803

1   Introduction

Robots are extensively used in the assembly line in the recent years due to several advantages over the manual assembly lines such as high manufacturing flexibility, high productivity, and ability for production of good quality products [1]. These robotic assembly line systems are highly cost intensive and consumption of energy is one of the major expenses encountered by the industries which use them extensively. Due to the increase in the energy prices and concern over the effects of carbon dioxide emission during electricity production throughout the world, there is an increased requirement among industries and corporations to minimize the energy consumption [2]. It is reported by that cost of energy consumption in a car manufacturing process is around 9–12% of the total manufacturing cost. If a 20% reduction in the energy consumption could be achieved, the total manufacturing cost can be reduced by 2–2.5% [3]. Due to significant economic importance, industries always try for improving the efficiency and cost effectiveness of the assembly line operations [4]. Assembly line operations are performed as the last step of production process, where the final assembly of the product is carried out by performing set of specific tasks at each workstation and the product moves between different workstations through a conveyor system in a predetermined manner. Assembly lines have been operated manually as well as automated assembly lines using robots are used extensively in the industries. When using assembly lines, an important decision and planning problem is encountered. This is termed as assembly line balancing (ALB) problems. This problem mainly aims at assigning tasks to workstations in such a way the workstations are not overloaded and tasks assignment are in a balanced manner [5].

As discussed earlier, usage of robots in assembly line are increasing and effectively balancing an assembly line with robots is of utmost importance due to the cost intensive character of such assembly line systems. Different robots with different specifications and efficiencies are available in the market. Hence, proper allocation of robots is critical for the performance of the assembly line. Robotic assembly line balancing (RALB) problem aims at allocating tasks to the workstations and selecting the best-fit robot to perform these tasks. By doing so, efficiency and productivity of the robotic assembly line is improved. RALB problems are more complex than the traditional ALB problems [1]. Researchers over the years have reported different variants of RALB problems. Two types of robotic assembly lines are used in the industry (straight and U-shaped). The major objective considered in such robotic assembly line balancing problems are minimizing number of workstations (RALB-I), minimizing cycle time (RALB-II), minimizing energy consumption and maximizing efficiency [6]. Most of the research was focused on RALB-I and RALB-II problems and from the literature study, it can be seen that very minimal attention has been given on minimizing energy consumption of manufacturing systems. It can be seen that no work has been reported in maximizing line efficiency by minimizing energy consumption of a straight robotic assembly line. By employing an energy-efficient manufacturing system, the energy consumption of the assembly can be reduced significantly [7]. Due to the NP-hard nature of the problem, researchers are utilizing extensively metaheuristic algorithms to obtain near optimal results in a short computational time [8]. This study proposes to utilize two well-known metaheuristic algorithms such as particle swarm optimization (PSO) and differential evolution (DE) to maximize line efficiency by minimizing the energy consumption of a straight robotic assembly line. This study can be a reference guide for industries for choosing an efficient method for reducing energy consumption and maximize the efficiency of a straight robotic assembly line system. The remaining of the paper is organized as follows: Sect. 2 describes the RALB problem along with assumptions considered in the study. In Sect. 3, the detailed implementation of the metaheuristics to solve the problem is presented. Section 4 presents the comparative results of the experimental study on the metaheuristics and Sect. 5 concludes the outcome of this study.

2 Problem Description and Assumptions

This study considers a straight robotic assembly line where different assembly tasks are performed at each workstation using best-fit robot to produce/assembly a given product. The assembly line consists of set of robots and workstations which are connected using an efficient material handling system (e.g., conveyor belt). The algorithm proposed mainly aims at minimizing the energy consumption of a straight robotic assembly line at each work station and in turn tries to maximize the line efficiency of the considered assembly line. A pictorial representation of a typical a straight robotic assembly line is depicted in Fig. 1. In this type of assembly line, workstations are arranged in a straight line and to perform the tasks in the workstations, different types of robots are allocated. This paper aims to find the energy-efficient robots to these workstations in such a way that the total energy consumption is minimized.

The following assumptions are considered for the RALB problem addressed in this paper.

Fig. 1 Straight robotic assembly line

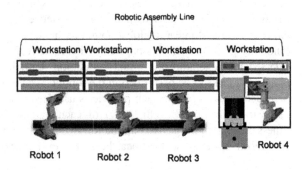

1. Power consumption of the robots are taken from the literature. These power consumption values of the robots are used to calculate the energy consumption.
2. Tasks cannot be subdivided to more than one workstation. But any tasks can be assigned to any workstation and performed by any robot.
3. Number of workstations is equal to number of robots.
4. Time for material handling, loading and unloading time are included in the tasks time.
5. All robots are available without any limitations.
6. A single unique product is assembled.
7. Planning horizon and maintenance operation is not considered in this study.

3   Metaheuristic Algorithms Design to Solve RALB Problem

Assembly line balancing problems are classified in the category of NP-hard problems [9]. Over the years, different researchers have extensively utilized metaheuristic algorithms to solve ALB problems. The problem addressed in this paper also falls under this category. For obtaining fast and feasible solution for this type of problems, two well-known metaheuristics (particle swarm optimization and differential evolution) are applied to solve this RALB problem with an objective of maximizing line efficiency by minimizing total energy consumption of the assembly line. These metaheuristics are selected due to their simplicity in adapting to this problem, faster convergence and lesser parameters to fine tune [10, 11].

3.1   Particle Swarm Optimization (PSO)

PSO algorithm is developed based on the social behavior of bird flocking or fish schooling [12]. PSO has been utilized extensively by researchers for solving engineering optimization problems due to its simplicity in implementation and ability to converge quickly to a reasonably good solution [10]. PSO procedure starts with a set of particles (swarm) randomly generated in the search space. Every particle in the population flies in the solution space by adjusting the velocity and position. Particle uses its own flying experience along with the partners flying experience to reach near to global optimum. In general, initial population for PSO is randomly generated. In this paper, few heuristic rules reported in the literature are selected for generating the initial population. Particle structure is a string which consists of tasks to be performed in the robotic assembly line arranged based on the precedence relationship. Each particle is assigned with a random velocity. Fitness value (maximizing line efficiency) of each particle is evaluated. Procedure for fitness evaluation is discussed in the following section. Each particle remembers the best result achieved so far (local best) and exchanges the information with other particles to find out the best particle (global best) in the swarm. For every iteration, velocity and position of each particle is updated. Using the velocity vector, each particle updates it position toward the global optimum solution. A flowchart of the functioning of the PSO algorithm is presented in Fig. 2.

Using Eq. (1), the velocity is updated.

$$v_i^{t+1} = v_i^t + \underbrace{c_1 * [U_1 * ({}^{lo}P_i^t - P_i^t)]}_{cognitivepart} + \underbrace{c_2 * [U_2 * (G^t - P_i^t)]}_{socialpart} \tag{1}$$

where U_1 and U_2 are random numbers between 0 and 1, c_1 and c_2 are known as acceleration coefficients, v_i^t is the initial velocity, ${}^{lo}P_i^t$ is the local best, G_t is the global best solution at generation 't' and P_i^t is the current particle position. Transposition rule is used for updating the velocity due to the characteristic of the problem. Using the updated velocity, position of the particle is updated. This updating mechanism helps to find the position of the new particle based on Eq. (2).

$$P_i^{t+1} = P_i^t + v_i^{t+1} \tag{2}$$

In this paper, number of velocity pairs are generated randomly and fixed based on the number of tasks in the problem considered. For instance, for problems with 20 tasks, the number of velocity pairs is fixed as 4. In the velocity equation,

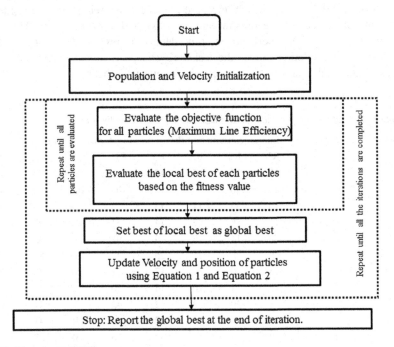

Fig. 2 Flowchart of PSO

coefficient values c_1, U_1 and c_2, U_2 acts like a probability percentage which will decide the number of pairs to be selected for performing the transpositions to form the updated velocity. Detailed implementation of the velocity and position update procedure is available in [13]. The velocity and position update is performed for a set of generations and the procedure completes when it reaches a predetermined number of generations.

3.1.1 Fitness Evaluation

A heuristic procedure based on the consecutive procedure proposed by Levitin et al. [1] is adopted to find fitness value of this problem. This heuristic aims at allocating tasks and robots to the workstations with an objective of minimizing the energy consumption of the straight assembly line. In case of straight robotic assembly line, tasks are allocated based on the sequence of tasks generated based on the heuristic rules discussed in the previous section. These tasks are arranged in such way that they meet the precedence relationship. Robots are checked for allotment to the workstation to perform the tasks allocated in it. Initial energy consumption E_0 is calculated initially and allocation of the tasks to the workstation is based on E_0 value. Maximum allocation of tasks based on the E_0 is attempted. Robot which performs the allotted tasks with minimum energy consumption is allotted. If all tasks cannot be allotted with E_0, E_0 is incremented by 1 and this procedure is repeated until all tasks and robots are allotted. A feasible task sequence (1-3-2-4-5-6-7-9-8-10-11) which meets the precedence relations is considered. Problem considered for the illustration is 11 task and 4 workstations. The robot energy consumption and performance time details are presented in Table 1 along with precedence relationship of the tasks. Key steps involved in evaluating the fitness value of the particle (line efficiency) is presented here.

Table 1 Energy consumption and performance time for 11 tasks by 4 robots

Tasks	Precedence	Energy consumption (kJ)				Performance time units			
		R1	R2	R3	R4	R1	R2	R3	R4
1	–	20	15	15	17	81	37	51	49
2	1	27	40	27	15	109	101	90	42
3	1	16	32	11	18	65	80	38	52
4	1	13	16	27	14	51	41	91	40
5	1	23	14	10	9	92	36	33	25
6	2	19	26	25	25	77	65	83	71
7	3, 4, 5	13	20	12	17	51	51	40	49
8	6	13	17	10	15	50	42	34	44
9	7	11	30	12	12	43	76	41	33
10	8	11	18	12	27	45	46	41	77
11	9, 10	19	15	25	30	76	38	83	87

(i) Initially E_0 is calculated as follows.

$$E_0 = \left[\sum_{j=1}^{N_a} \min_{1 \le i \le N_r} e_{hi}/N_w \right] \tag{3}$$

Here N_a is the total number of tasks, N_w is the total number of workstations, N_r is the total number of robots and e_{hi} is the energy consumption by robot h to perform task i. For example, consider E_0 value is calculated as 35.

(ii) The first station is opened and the procedure allocates the tasks based on the sequence in the order of its occurrence and checks if any one of the robots can perform the allocated tasks within E_0.

(iii) If the present workstation is fully allotted with tasks next workstation is open and the procedure attempts to allocate the remaining tasks. Until all tasks are assigned, Steps 2 and 3 are to be repeated.

(iv) If the procedure cannot assign all the tasks within the given E_0, E_0 is incremented by one and until all tasks gets allocated, Steps 2 and 3 are repeated.

(v) Now, all the stations are allocated with certain set of tasks, the best available robot to perform these tasks based on the energy consumption is selected and allocated.

(vi) Workstation time of all the stations are calculated for the allotted tasks and robots.

(vii) Using the cycle time (maximum workstation time) and workstation times, using Eq. (4), line efficiency is calculated.

$$LE = \frac{\sum_{k=1}^{N_w} S_k}{N_w * C} * 100 \tag{4}$$

Figure 3 displays the tasks and robot allocation for given sequence of tasks. Energy consumed and workstation times calculated are presented. Based on the objective of minimizing the energy consumption, allocation of tasks and workstation times are calculated for a straight robotic assembly line. The workstation times are used to calculate the line efficiency. Line efficiency is the direct indication of the efficiency of a given assembly line. Straight line robotic assembly line's efficiency is calculated as follows: LE = (89 + 107 + 171 + 171)/(4 * 171) * 100 = 78.65%.

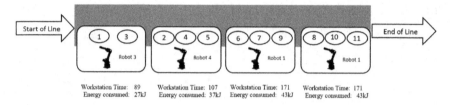

Fig. 3 Workstation times and energy consumption in straight RALB

3.2 Differential Evolution (DE)

Differential evolution algorithm is another well-known metaheuristic algorithm proposed by Storn and Price [14] for solving engineering optimization problems. DE has been widely used due to its simplicity in implementation and less parameters to be fine-tuned. DE has been used recently for solving job shop scheduling and design optimization problems [11]. DE is comparable to genetic algorithm. However, the differences are on the mechanism of mutation and crossover operators. DE also starts with a set of population called target vectors. Each target vector undergoes mutation, crossover, and selection operations and these operations are used to produce a new population with better quality individuals. Each target vector undergoes mutation operation and produces a set of donor vector. Crossover operation is performed with the target vectors and donor vectors to create a trial vector. Based on the fitness value (refer Sect. 3.1.1) of every trial vector and their corresponding target vector, vectors for next generation are selected. Procedure completes when a predetermined number of generations is met. In DE, mutation operator is the primary operator. Three individuals in the population are selected randomly for the mutation process. Donor vector is created using the target vectors which are selected. Perturbation is based on the Eq. (5). Here F is the mutation scaling factor and $x_{r1,g}$, $x_{r2,g}$ and $x_{r3,g}$ are the randomly selected target vectors.

$$y_{ig} = x_{r1,g} + F(x_{r2,G} - x_{r3,G}), \text{ where } i = 1, \ldots, 5 \tag{5}$$

Next operation to be performed is the crossover. This paper uses OX (order crossover) operator for the crossover. Detailed implementation of the mutation and crossover operation can be found in [6]. Next operation to be performed is the selection. This operation selects the population for the next generation using the target vector and trial vectors. The vector which has better fitness value is copied for the next generation. In this paper, fitness value is based on the objective of better efficiency of the straight robotic assembly line.

Table 2 Parameters for PSO and DE for RALB problem

Parameters for PSO	Parameters for DE
Population size: 25	Population size: 25
Number of iterations: 30	Number of iterations: 30
Acceleration coefficients: $c_1 = 1$, $c_2 = 2$	Mutation factor: 0.5 Crossover rate: 0.9

3.3 Parameters for PSO and DE

Performance of PSO and DE mainly depends on the selection of parameters. In this paper, parameters are selected based on the ability to obtain good solution quality in an acceptable time span. They are selected based on pilot test conducted on three datasets and the best combination of parameters is chosen for testing the remaining problems. Table 2 summarizes the parameters utilized for solving the considered RALB problem.

4 Experimental Results

To test the performance of the proposed PSO and DE algorithm, a set of benchmark instances available from the literature are considered. A total of 32 test problems ranging from 25 to 297 tasks are available in the literature [15]. The proposed algorithms are coded in C++ and are tested on Intel core i5 processor (2.3 GHz). Table 3 reports the results obtained for both the algorithms. Solutions reported are the average of 10 runs. Straight robotic assembly line is evaluated using PSO and DE with the objective of maximizing the line efficiency when the energy consumption is minimized. Energy data is used for the allocation of tasks and robots, using the allocation details the workstation time and cycle time of the assembly line are calculated. The workstation time and cycle time is used for calculating the line efficiency of the assembly line.

The results reported are the best solutions obtained for both the algorithms for the objective of maximizing the line efficiency as shown in Table 3. Column I presents the problem evaluated, Column II, III, and IV presents the line efficiency (LE-%), energy consumption (EC-kilojoules) and cycle time (CT) of the best solution evaluated using PSO algorithm. Column V, VI, and VII presents the results obtained using DE algorithm. When comparing the results between PSO and DE, it is observed that DE produces results with an average improvement in the line efficiency of 7.8% for small size datasets (up to 70 task problems) and the average improvement in the line efficiency for large size datasets is found to be 8.1%. The results reported for line efficiency is calculated by minimizing the energy consumption of the workstation. It is observed that the energy consumption is lower for 21 datasets and for the other datasets PSO reported lower energy consumption.

Table 3 Experimental results for 32 RALB problems

Dataset	Proposed PSO			Proposed DE			Dataset	Proposed PSO			Proposed DE		
	LE	EC	CT	LE	EC	CT		LE	EC	CT	LE	EC	CT
25-3	98.1	489	521	**98.4**	502	536	89-8	75.4	4956	562	**77.0**	4,922	562
25-4	85.3	353	342	**90.3**	343	329	89-12	80.0	5509	438	**81.3**	5,499	438
25-6	88.0	357	207	**88.0**	357	207	89-16	83.2	4906	288	**92.6**	4,887	265
25-9	74.5	235	143	**76.4**	242	147	89-21	72.5	4150	236	**74.3**	4,182	232
35-4	**90.2**	1037	518	80.9	1045	516	111-9	84.11	7149	529	**92.2**	7,131	674
35-5	76.7	900	444	**96.4**	890	357	111-13	82.1	7030	396	**89.0**	7,137	391
35-7	77.5	1000	343	**85.8**	989	314	111-17	88.3	6857	280	**92.1**	6,877	266
35-12	**68.7**	688	153	73.9	687	130	111-22	72.3	6630	255	**87.7**	6,667	215
53-5	**74.1**	2680	605	75.7	2680	590	148-10	94.6	9798	688	**95.8**	9,798	678
53-7	69.9	1970	439	**90.6**	1970	345	148-14	88.7	10524	461	**90.7**	10,645	466
53-10	80.7	2186	308	**83.6**	2148	263	148-21	75.3	10084	384	**83.5**	9,927	335
53-14	73.6	2016	202	**79.8**	2050	183	148-29	57.1	8334	317	**68.3**	8,508	263
70-7	94.2	4093	620	**94.4**	4191	633	297-19	78.3	24518	809	**88.7**	24,351	688
70-10	88.5	3046	290	**90.3**	3106	302	297-29	**87.4**	24554	458	83.8	24,404	528
70-14	67	3815	338	**83.2**	3766	256	297-38	84.2	22485	409	**86.1**	22,482	348
70-19	63.1	3243	232	**71.2**	3225	204	297-50	64.8	21089	348	**80.4**	21,050	336

Table 4 Average computational time of PSO and DE for straight RALB

Problems	Average computational time	
	PSO	DE
25	9	8
35	23	20
53	41	36
70	73	67
89	95	92
111	288	281
148	550	540
297	1700	1739

25 datasets reported same or better cycle time for DE when compared with PSO. Even though cycle time is low for 7 datasets for PSO, line efficiency is still better for DE. This could be due to the variation of workstation times which affects the line efficiency. Table 4 presents the average computational time for both the algorithms. DE is able to produce results at a faster rate than PSO for almost all datasets. However, computational time for 297 tasks problem is high for DE. This could be due to the large size of the problem and repeated selection procedure in DE.

5 Conclusion

RALB problem with an objective of maximizing line efficiency by minimizing energy consumption is presented. Particle Swarm Optimization and differential evolution algorithm is proposed to solve the proposed problem and benchmark problems available are used for evaluating the performance. Reducing energy consumption is very important due to serious environmental impacts and increasing energy cost. This paper will be an important addition to the literature since most of the works reported focused on objectives such as cycle time and cost minimization. From the results obtained it can be observed that DE algorithm reports better line efficiency than PSO algorithm. Detailed results obtained using these two algorithms are also reported. In terms of computational time also, proposed DE could obtain solutions in less computational time. In future, maintenance operation could be considered and new models could be studied for other types of robotic assembly line. Other alternative metaheuristic algorithms can also be experimented.

References

1. Levitin, G., Rubinovitz, J., Shnits, B.: A genetic algorithm for robotic assembly line balancing. Eur. J. Oper. Res. **168**(3), 811–825 (2006)
2. Mouzon, G., Yildirim, M.B.: A framework to minimise total energy consumption and total tardiness on a single machine. Int. J. Sustain. Eng. **1**(2), 105–116 (2008)
3. Fysikopoulos, A., et al.: An empirical study of the energy consumption in automotive assembly. Procedia CIRP **3**, 477–482 (2012)
4. Rubinovitz, J., Bukchin, J., Lenz, E.: RALB–A heuristic algorithm for design and balancing of robotic assembly lines. CIRP Ann. Manuf. Technol. **42**(1), 497–500 (1993)
5. Kumar, D.M., Assembly line balancing: a review of developments and trends in approach to industrial application. Glob. J. Res. Eng. **13**(2), 1–23 (2013)
6. Nilakantan, J.M., et al.: Differential evolution algorithm for solving RALB problem using cost-and time-based models. Int. J. Adv. Manuf. Technol. **89**(1–4), 311–332 (2017)
7. Chryssolouris, G.: Manufacturing Systems: Theory and Practice. Springer, New York (2013)
8. Gao, J., et al.: An efficient approach for type II robotic assembly line balancing problems. Comput. Ind. Eng. **56**(3), 1065–1080 (2009)
9. Scholl, A., Becker, C.: State-of-the-art exact and heuristic solution procedures for simple assembly line balancing. Eur. J. Oper. Res. **168**(3), 666–693 (2006)
10. Rameshkumar, K., Suresh, R., Mohanasundaram, K.: Discrete particle swarm optimization (DPSO) algorithm for permutation flowshop scheduling to minimize makespan. In: International Conference on Natural Computation. Springer (2005)
11. Wang, G.-G., et al.: A novel improved accelerated particle swarm optimization algorithm for global numerical optimization. Eng. Comput. **31**(7), 1198–1220 (2014)
12. Kennedy, J., Eberhart, R.: Particle swarm optimization. In: Proceedings of the IEEE International Conference on Neural Networks (1995)
13. Nilakantan, J.M., et al.: Bio-inspired search algorithms to solve robotic assembly line balancing problems. Neural Comput. Appl. **26**(6), 1379–1393 (2015)
14. Storn, R., Price, K.: Differential evolution—a simple and efficient heuristic for global optimization over continuous spaces. J. Glob. Optim. **11**(4), 341–359 (1997)
15. Nilakantan, J.M., Huang, G.Q., Ponnambalam, S.: An investigation on minimizing cycle time and total energy consumption in robotic assembly line systems. J. Clean. Prod. **90**, 311–325 (2015)

Soft Computing of Credit Risk of Bond Portfolios

Ravindran Ramasamy, Bavani Chandra Kumar
and Siti Balqis Mohd Saldi

Abstract Non-traded financial assets like bonds and loans have credit risk quantification methods and models which are still under debate for their effectiveness. Quantifying credit risk in portfolio context is all the more challenging as it incorporates covariance and weights. This article quantifies credit risk prevalent in non-traded financial assets by applying Transition Probabilities and Mahalanobis distance in the portfolio context. Bonds are rated by rating agencies and when the credit rating drops the bonds lose value as there is higher risk for the investors. The Investors and lenders have to assess the collective loss expected by them if rating migrates in portfolio context. This portfolio credit risk is affected by two parameters covariance and proportion of funds invested (weight) in a particular bond. We chose a real bond portfolio invested by a Malaysian mutual fund company and demonstrate the complex computations through soft computing by a MATLAB algorithm compiled by us. We computed the credit risk by classical weighted average method and also by Mahalanobis Distance method. The results show the classical method underestimate the credit risk leading to suboptimal hedging.

Keywords Credit risk · Mahalanobis distance · Portfolio · Transition probabilities Hedging

R. Ramasamy (✉) · B.C. Kumar · S.B.M. Saldi
Graduate School of Business, University Tun Abdul Razak, Kuala Lumpur, Malaysia
e-mail: ravindran@unirazak.edu.my

B.C. Kumar
e-mail: bavani@unirazak.edu.my

S.B.M. Saldi
e-mail: siti.balqis@unirazak.edu.my

© Springer Nature Singapore Pte Ltd. 2018
M. Pant et al. (eds.), *Soft Computing: Theories and Applications*,
Advances in Intelligent Systems and Computing 583,
https://doi.org/10.1007/978-981-10-5687-1_73

1 Introduction

Banks are worried over their nonperforming assets and loans. Though the banks are very cautious while lending, still due to borrower-specific problems, they face the problem of default [3, 19]. The banks not only lose the money but also they lose the chance of further lending. The country's economy is another victim where scarce capital is not well utilised. To reduce the capital adequacy reserve [2] insisted under Basel III the banks and financial institutions set up special purpose vehicles and convert the loans and bonds into structured products like asset-backed securities [10], collateralised debt obligations [7, 12]. When the borrowers default the structured products are affected. The institutional investors who invest funds in the structured products and asset-backed securities are another group of victim of these defaults [10]. The regulators perplexed over this problem of default and impose more and more rules and regulations to prevent this anomaly, which makes the bond market more rigid and inflexible [17]. Countless rules will not help to solve this problem of default as it is bound to occur while lending [8]. The credit risk management through hedging is the solution, but it is not free but with significant costs to the banks in terms of manpower, technology, monitoring and host of other related activities.

1.1 Valuation of Loans and Bonds

Financial institutions like banks lend loans and invest funds in bonds [28]. A significant portion of the surplus funds of the banks are invested in bonds, structured products, asset-backed securities and derivatives [22]. Normally, these financial assets are not traded in the market and this makes difficult to find their arm's length value which determines drop in value. The accounting rules after the introduction Sarbanes Oxley Act changed the reporting of investments especially in the financial assets [13]. They are to be reported at current market values, if a ready market is available for them. Otherwise, they are to be valued with the relevant inputs available from the market. The profits and losses result from this valuation are to be reported in the current year income statement and they cannot be deferred under any pretext. This reporting rule significantly increases the volatility of the income of investor banks which affects the earning per share and ultimately results in volatility in share prices prevailing in the market which is undesirable in finance theory [23]. To avoid the volatility hedging is recommended [9]. To mitigate this volatility, hedging is recommended which is known as risk management.

1.2 Nontrading Nature of Financial Assets

Financial assets are broadly classified under four heads such as shares, bonds, units and derivatives. Shares have established markets for their trading in the form of over the counter (OTC), regular stock market and also the modern online trading in computer portal. As these provide a very efficient trading mechanism where trading is easy and the prices are determined by the market forces. But for the other three financial assets, though market exists for them, they are rarely traded as there is no excitement as in shares [18]. The long-term loans are another financial asset for which there is no market at all unless they are converted into structured products [5]. Finding fair values for these assets are really challenging not only because of absence of market inputs but also due to non-availability of correct models and methods. There is no standardised model or method recommended by the regulators. Significant variation exists among the practitioners in valuing these assets which ultimately defeats the objective of uniform reporting of financial results all over the globe [6].

1.3 Significance of This Study

Many attempts have been made to quantify the credit risk with several models, but all have their own merits and demerits [1]. The individual valuation of financial assets are more or less standardised, but when come to portfolio there is no standard model. It is all the more challenging as the existing models have to incorporate two additional parameters covariance and the proportion of funds invested in every security in portfolio context [25]. The mean–variance–covariance model applied in share portfolio, if applied in bond market it is incompatible as these assets have no time series market prices to capture variance and covariance [15]. Hence, this study applies the transition probabilities of rating migration and the yield rates of financial assets depending on the present status of their credit rating [25]. This accurate expected loss will lead to correct amount of hedging which will bring down the cost of hedging and also protect the portfolio loss. This will reduce volatility in earnings per share and also in share prices. The bankers, investors, regulators and practitioners like investments analysts can use this model. This article contributes academically also to the existing literature, to research students and also lecturers to enhance their knowledge.

1.4 Organisation of This Paper

Section 1 introduces the topic, Sect. 2 reviews existing literature in this area. Section 3 gives the methodology adopted and describes data, Sect. 4 discusses the results obtained and their implications, and finally Sect. 5 concludes this article.

2 Literature Review

Many researchers have attempted to estimate the risks present in lending. The interest rate risk, market risk, credit risk and liquidity risk, etc., are covered by the researchers. The interest rate risk and market risks are fairly straightforward. The challenging risk is the credit risk which is difficult to quantify because of non-availability data [14]. The loans are not tradable and it is difficult to estimate the mean and variance of the value movements of these loans [26]. The bonds also join this complexity as the bond market is not so active in many countries [4]. Literature suggests three methods prominently applied in credit risk estimation [21]. The internal method is more or less similar to the model discussed in this article. The credit risk metrics proposed by the JPMorgan is another method which is popular among practitioners [24]. This model uses the transition probability to calculate the expected values and standard deviation. The third model is the Merton model which uses the Black Scholes framework to estimate the default risk and credit risk [27]. All these models end in estimating the credit risk of individual loans and bonds. Only a few studies go to portfolio models [16]. This is due to lack of data, models, and software to accomplish the computations. Most of the computations are based on matrices which are difficult to handle in excel and in other software. The MATLAB software is promising, but the algorithm and programme writing, testing and debugging poses challenge. The portfolio credit risk computation requires powerful soft computing which we have applied in this article.

3 Methodology for Estimating Credit Risk

All financial assets are valued by discounting the future cash flows expected from the asset. It is universal to assume the cash flows are uninterrupted, which is rarely true. Many loans and bonds' repayments are interrupted as the borrowers' cash flows are not regular due to ups and downs in business [23]. In addition, their poor budgeting practices and poor financial management leads to defaults. Moreover, the values of these assets fall when there is a rating downgrade.

3.1 Rating and Credit Risk

Credit ratings are mandatory and should be obtained from rating agencies before going to the market for funds. Even for long-term loans the lenders now insists the borrowers to get rated from rating agencies [11]. The company law imposes mandatory requirement for the issuers of bonds to get rated first before approaching the bond market for funds. The initial rating is assigned by rating agencies on the basis of inputs received from the issuers of bonds. Normally, the ratings such as AAA, AA and A are high grades, BBB and BB are risky but provides higher yields as the risk is high [20]. Rates B and C are on the verge of insolvency and D is already defaulted.

3.2 The Coupon Rate and Cash Flows

Coupon rate is the rate the bond issuer promises to pay at end of every year. In finance, the coupon rate has a limited role of finding cash flows only. In all other decisions, the yield rate is applied. The cash flows are computed by multiplying par value of the bond by its coupon rate. The final year cash flow will be a huge amount as the maturity value of the bond is included in it. For every rating the same cash flows are needed for discounting. If the cash flows are discounted eight fair values will be available each for a rating.

3.3 Dynamic Yield Rates

The yield rate depends normally on demand and supply of funds, prevailing economic conditions and also the base lending rate fixed by the central bank [8]. These yield rates are dynamic not fixed as coupon rate. They are ever changing like any other economic variable. In addition, the yield rate of a bond depending on the financial strengths and weaknesses of the issuer of bond. For instance, if the borrower's credit rating goes down the risk for the lenders increase and they will demand higher yield rates to compensate the higher risk.

3.4 Fair Values at Different Rating

When cash flows are discounted, the present values are generated. Yield rates are used for discounting. They are converted into discounting factors. Since different yield rates are applied in different years the power rule is applied for finding discounting factors. The first cash flow is not discounted hence the factor is taken as one. The second year yield rates are squared, the third year rate is raised to the power of three and so on. Since the portfolio life is assumed to be 5 years, the discounting stopped at year five. The cash flows are divided by the discounting factors to get the present values of cash flows of a bond and finally added to get fair values at eight different rating.

The first year is the current year and need not be discounted, and hence only the future four years' yield rates are computed. These yield rates are forecasted based on certain parameters and it is applicable for all bonds. To get specific bond yield rates, 3% is subtracted and the specific bond's yield rate is added. The adjusted yield rates are used for discounting.

3.5 Transition Probabilities of Ratings

Credit ratings are assigned by rating agencies initially and they do not stay at the same level during the life of the bonds. The initial rating will be revised by the rating agencies based on the new inputs every year. Therefore, the initial ratings are not static and they migrate year by year depending on the situation prevailing in the economy and the financial strengths and weaknesses of the individual company. These rating movements are documented and their percentages of movements are published in the form of transition probabilities. When the credit rating moves upward, the bond gains value and when it drifts downward the expected yield rate increases causing the bond value to fall. This fall in value is credit rate migration risk which is to be precisely quantified and hedged. These probabilities are to be understood row-wise but not column-wise.

3.6 Expected Value Calculation of Single Bond

The bonds fair values computed above are multiplied by the TPs of the rating presently prevailing and added to get fair value otherwise known as expected value.

3.7 Expected Standard Deviation

The expected (mean) value of each bond is subtracted from expected value to get deviations (residuals). These deviations are squared and again multiplied by transition probabilities and added to get variance. The square root of variance is the standard deviation which is the credit risk of each bond.

3.8   Covariance of Bonds

Each bond has eight values at different rating and at different yield rates. These values are converted into covariance by cov function of MATLAB.

3.9   Weight of Bonds

The proportion of funds invested in a bond is computed by dividing the individual current market values by the total current market values of all bonds present in the portfolio.

3.10   Mahalanobis Distance (MD)

The covariance matrix is multiplied by two types of weights. One weight is vertical vector and the other weight is a horizontal vector. Similarly, the deviations which are to be multiplied in left and right while computing MD are scaled by the vertical and horizontal weights, respectively. Then all the three (left deviations, covariance matrix and right deviations) are multiplied and added row-wise and natural logarithm is computed to get a column vector of MD. Each Bond has one MD, this could be interpreted as the relative risk of each bond in the portfolio.

3.11   Expected Loss of Portfolio

The MD column vector is the relative risk of individual Bond. The mean MD and the standard deviation are computed as usual. This MD's mean and standard deviation are applied to compute the portfolio value at 95 and 99% confidence levels and subtracted from mean value to get portfolio credit risk.

3.12   Classical Risk Versus Mahalanobis Risk

We compare the portfolio expected loss estimated by classical method and by MD method. There is an appreciable difference. We conclude that Mahalanobis distance computation is more accurate and more meaningful for hedging and computing capital adequacy required by BASEL III.

3.13 Data

To apply Mahalanobis distance, we selected a real bond portfolio from KAF Investment Bank Berhad consisting 28 different companies bonds from fundssu-permart.com. The transition probabilities, yield rates for different rating and at different years were collected form Bond Pricing Agency of Malaysia. A MATLAB algorithm is developed to estimate the credit risk of the above portfolio. The algorithm incorporates both covariance and proportion of funds invested in a particular bond. The MATLAB algorithm is given in the appendix.

4 Results and Discussion

The bonds included in the portfolio are high graded as the safety of funds is important in any portfolio. The portfolio included three AAA-rated bonds, 24-AA-rated bonds and a single A-rated bond. The cash flows were estimated with coupon rate and added with the maturity value in the final year. With a loop command and with the repmat command (repeat matrix) eight cash flows of similar values for each bond in the matrix form was generated. The yield rates, and the discounting factors, the present vales of cash flows and finally the fair values at each rating for each bond were also computed within the same loop.

Table 1 shows the fair values at eight levels for each bond at each credit rate. If the yield rates are stable, only one fair value will appear for each bond. As the credit rating migrate, the risk also increases or decreases causing the risk to the investors increase or decrease, which leads to demanding risk premiums. The yield rates depending on the risk premiums go up or down causing the bond values drop or increase. Though Co 1 and Co 2 are in the AAA rating initially, their fair values are less than par value. This is due to the risk present in those specific bonds. Co 4 to Co 27 are AA-rated initially, whose value show a mixed pattern. Few bonds show a fair value of more than the par value and few others show lesser value. Co 28 single A-rated initially, though it is riskier than other bonds, still the fair values at each rating level is high due to lesser yield rate expected. In general when the rating is downgraded, the values drop and vice versa. Co 8, Co 21 and Co 4 show higher values even when they are placed under D rating. This is due to the lesser risk premium demanded by the investors. Their cash flows are discounted at a lower yield rate hence they produce higher values.

The initial ratings migrate and may be placed under another rating at end of the year. Over the years, their migrations are recorded which is as follows in terms of percentages (Table 2).

These probabilities are to be understood row-wise but not column-wise. 99.33% of times the AAA-rated bond will stay next year at the same rating and there is a chance of 0.67% of chance that it will migrate to AA rating. Similarly a AA-rated bond, 1.23% may be upgraded to AAA next year and may be downgraded to A by

Table 1 Fair values of each company's bond at different ratings

	AAA	AA	A	BBB	BB	B	C	D
Co 1	97.95	96.89	95.71	93.57	90.15	86.89	74.69	45.09
Co 2	98.13	97.07	95.90	93.76	90.34	87.09	74.90	45.32
Co 3	100.03	98.97	97.80	95.65	92.24	88.98	76.75	47.05
Co 4	102.65	101.77	101.21	99.82	98.93	97.81	90.18	75.05
Co 5	97.27	96.23	95.08	92.97	89.61	86.41	74.41	45.24
Co 6	106.01	104.93	103.73	101.51	98.02	94.68	82.07	51.46
Co 7	105.21	104.15	102.97	100.80	97.39	94.12	81.74	51.66
Co 8	103.91	103.76	103.57	103.21	102.79	102.34	99.23	91.31
Co 9	106.94	105.83	104.59	102.33	98.75	95.34	82.46	51.27
Co 10	97.06	95.99	94.80	92.65	89.20	85.90	73.64	43.91
Co 11	103.56	102.68	102.12	100.75	99.86	98.74	91.15	76.09
Co 12	103.87	103.00	102.44	101.06	100.18	99.06	91.46	76.39
Co 13	99.87	98.84	97.69	95.59	92.25	89.06	77.05	47.82
Co 14	99.18	99.09	98.38	96.29	94.31	92.24	82.31	58.25
Co 15	99.35	98.29	97.11	94.97	91.54	88.28	76.04	46.34
Co 16	99.34	98.28	97.10	94.95	91.53	88.26	76.03	46.32
Co 17	97.11	96.03	94.84	92.67	89.18	85.86	73.50	43.57
Co 18	99.47	99.38	98.66	96.53	94.50	92.39	82.28	57.81
Co 19	96.93	95.90	94.76	92.68	89.36	86.19	74.30	45.38
Co 20	96.78	95.72	94.55	92.43	89.01	85.76	73.63	44.18
Co 21	103.83	103.68	103.49	103.14	102.72	102.27	99.17	91.29
Co 22	99.55	99.45	98.75	96.65	94.66	92.59	82.65	58.54
Co 23	97.72	96.66	95.49	93.35	89.94	86.68	74.51	44.96
Co 24	97.31	96.28	95.14	93.05	89.72	86.55	74.63	45.64
Co 25	99.07	98.06	96.93	94.86	91.58	88.45	76.63	47.81
Co 26	104.53	103.44	102.23	100.00	96.47	93.10	80.42	49.70
Co 27	98.25	97.19	96.02	93.89	90.48	87.23	75.07	45.54
Co 28	105.55	104.46	103.25	101.03	97.52	94.16	81.51	50.82

Table 2 Five year cumulative transition probabilities of credit rating migration

	AAA	AA	A	BBB	BB	B	C	D
AAA	0.9933	0.0067	0	0	0	0	0	0
AA	0.0123	0.9562	0.0226	0.0061	0.0014	0.0007	0.0007	0
A	0.0011	0.0464	0.8862	0.0306	0.0103	0.0011	0.0011	0.0232
BBB	0	0.0152	0.0909	0.6515	0.0455	0.0303	0	0.1666
BB	0	0	0	0.0625	0.625	0.1875	0	0.125
B	0	0	0	0	0.0217	0.8043	0.0436	0.1304
C	0	0	0	0	0	0	0.8205	0.1795
D	0	0	0	0	0	0	0	1

Source Bond rating agency of Malaysia

Table 3 Expected values, expected loss and Mahalanobis distance of each company's bond

	Expected values	Standard deviation	95% CL loss	99% CL loss	Mahalanobis distance	Relative distance (basis points)
Co 1	97.94	0.09	0.14	0.20	34.46	371
Co 2	98.12	0.09	0.14	0.20	34.04	366
Co 3	100.02	0.09	0.14	0.20	35.71	384
Co 4	101.74	0.39	0.65	0.92	33.08	356
Co 5	96.16	0.75	1.24	1.75	33.00	355
Co 6	104.86	0.79	1.30	1.84	32.62	351
Co 7	104.09	0.77	1.27	1.80	33.93	365
Co 8	103.75	0.14	0.23	0.33	32.03	345
Co 9	105.76	0.81	1.33	1.88	34.64	373
Co 10	95.92	0.77	1.27	1.80	34.63	373
Co 11	102.65	0.39	0.65	0.91	32.06	345
Co 12	102.97	0.39	0.65	0.91	32.08	345
Co 13	98.77	0.75	1.24	1.75	34.46	371
Co 14	99.03	0.56	0.93	1.31	33.04	355
Co 15	98.22	0.77	1.27	1.79	31.24	336
Co 16	98.21	0.77	1.27	1.79	32.03	345
Co 17	95.96	0.78	1.28	1.81	34.46	371
Co 18	99.32	0.57	0.95	1.34	29.84	321
Co 19	95.84	0.74	1.23	1.74	31.97	344
Co 20	95.66	0.76	1.26	1.78	34.39	370
Co 21	103.67	0.14	0.23	0.33	35.26	379
Co 22	99.40	0.56	0.93	1.32	32.04	345
Co 23	96.59	0.76	1.26	1.78	33.05	356
Co 24	96.22	0.75	1.23	1.74	33.01	355
Co 25	98.00	0.74	1.22	1.72	32.04	345
Co 26	103.37	0.79	1.31	1.85	32.77	353
Co 27	97.13	0.76	1.26	1.78	33.10	356
Co 28	101.93	7.95	13.12	18.53	34.62	372

2.26% and may migrate to BBB by 0.61% and so on. This rating migration results in higher risk, to the investors, consequently they demand higher risk premiums.

Table 3 shows the Mahalanobis distance in absolute form and in relative form in basis points. One in ten thousand is a basis point. In the classical method the standard deviation is treated as the risk and combined in weighted average to get portfolio risk. The standard deviation for the first three companies which are AAA-rated as 0.09. But the relative distance show different results (371, 366 and 384). It shows that there is no connection between the standard deviations and real risk present in the form of relative Mahalanobis distance. Even absolute MD is not the same. The difference is due to the incorporation of covariance among the bonds and also weights applied in a different way while computing MD. The maximum

Table 4 Credit risk of portfolio under Mahalanobis distance and classical method

	95% Confidence level loss	99% Confidence level loss
MD	2.22	3.15
Classical	1.58	2.23
Difference	0.64	0.92

MD is shown by Co 3 which is AAA-rated and the minimum distance is shown by Co 18 which is AA-rated with a distance of 321. In terms of standard deviation Co 3 and Co 18 shows 0.09 and 0.57, respectively. The results show the weakness of the classical method of computing portfolio credit risk under rating migration (Table 4).

When covariance and weights are applied to estimate the portfolio risk in the MD model the risk increases from 1.58 to 2.22% a difference of 0.64% or 64 points over the classical method. Similarly at 99% confidence level classical method computes the risk as 2.23% while MD shows as 3.15% a difference of 0.92% or 92 points. Though MD slightly over estimates, it is more accurate as it incorporates covariance and weights in different forms which are totally absent in classical method.

5 Conclusion

The credit risk estimation in portfolio context poses challenges as covariance and weights are to be incorporated in the model. The classical method ignores these two parameters and applies weighted average method to compute portfolio risk. We applied both the models to compare the portfolio risk. The classical method underestimates the portfolio risk when compared to MD method which incorporates the additional parameters of covariance and weights. In the absence of time series data, we compute the covariances among the bonds from fair values which are the expected values for bonds if they are placed under different ratings. Basel III applies 99% confidence level in estimating credit risk. If classical method is used 92 points less amount is provided as capital adequacy reserve which is undesirable. An accurate expected loss will not only decrease costs of hedging but also reduce capital requirements under BASEL III. In the hedging decision also a lower amount of loss is hedged which is not at the optimum level, leading to uncovered credit risk exposure. We suggest MD method to optimise hedging and provide correct capital reserve to meet the investor confidence and also for the healthy banking system in the country.

Appendix

MATLAB Programme for Mahalanobis Distance and risk calculation

```
close all; clear all; clc
loadkafdata
valtab=[];

%% Cash Flows and Discounting at appropriate yield rates
fori=1:28
    yr1=[yr(:,1) (yr(:,2:end)-0.03)+rates(i,3)]+1;      % Remove 3% and add yield rates of individual bonds
    yr2=[yr1(:,1:2) yr1(:,3).^2 yr1(:,4).^3 yr1(:,5).^4];  % Discounting factors

if (rates(i,4) < 4)                                    % If the life is less than 4 years

cf=repmat(rates(i,2),1,(rates(i,4)-1));                % Coupon cash flows
    cf1= [cfcf(:,end)+100];                            % Final year cash flow with par value
    cf2=repmat(cf1,8,1);                               % Cash flow column for discounting
val=sum((cf2./yr2(:,1:rates(i,4))),2);                 % Fair Value
valtab=[valtabval];                                    % Store the value in the table

elseif (rates(i,4) > 3 && rates(i,4) < 5)              % If the life is between 3 and 5 years

cf=repmat(rates(i,2),1,(rates(i,4)-1));                % Coupon cash flows
    cf1= [cfcf(:,end)+100];                            % Final year cash flow with par value
    cf2=repmat(cf1,8,1);                               % Cash flow column for discounting
val=sum((cf2./yr2(:,1:4)),2);                          % Fair Value
valtab=[valtabval]:                                    % Store the value in the table




x1=repmat(rates(:,5),1,28);                            % Create weights of 28,28 matrix
x2=x1';                                                % Create transpose of the weight matrix
x3=x1*incovar;                                         % Covariance inverse is scaled by row weights
x4=x3*x2;                                              % Covariance inverse is scaled by both row, column weights
x5=mean(valtab);                                       % Mean value of bond values
x6=repmat((ones(1,28).*x5),8,1);                       % Mean value matrix for subtraction
x7=valtab-x6;                                          % Deviation of bond values
x8=x7*x1;                                              % Left deviation * weights
x9=x7*x2;                                              % Right deviation * weights
md=sum((x8*x4).*x9);                                   % Mahalanobis distance (MD)
md1=log(0.5*md);                                       % Natural logarithm of MD

pmean=mean(md1);                                       % Mean of portfolio values
psd=std(md1);                                          % Standard deviation of portfolio values

loss95=pmean-norminv(0.05,pmean,psd)                   % Mahalanobis Risk 95% confidence level
loss99=pmean-norminv(0.01,pmean,psd)                   % Mahalanobis Risk 99% confidence level

%% Simple weighted average risk computation
trorisk = sd1 * rates(:,5);                            % Classical weighted Average
```

References

1. Aalen, O., Borgan, O., Gjessing, H.K.: Survival and Event History Analysis: A Process Point of View. Springer, New York (2008)
2. Aiyar, Shekhar, Calomiris, C.W., Wieladek, T.: Bank capital regulation: theory, empirics, and policy. IMF Econ. Rev. **63**(4), p955–p983 (2015)
3. Altman, E., Suggitt, H.: Default rates in the syndicated bank loan market: A mortality analysis. J. Bank. Financ. **24**(2), 229–253 (2000)
4. Saunders, A.: Financial Institutions Management, 3rd edn. McGraw-Hill, New York (2000)
5. Asarnow, E., Edwards, D.: Measuring loss on defaulted bank loans: a 24 year study. J. Commer. Lend. **77**(7), 11–23 (1995)
6. Bakshi, G., Madan, D., Zhang, F.: Understanding the role of recovery in default risk models: empirical comparisons and implied recovery rates, FEDS 2001-37 (2001)
7. Basel Committee on Banking Supervision http://www.bis.org/bcbs/basel3.htm?m=3%7C14%7C572 (2013)
8. Bhimani, A., Gulamhussen, M.A., Lopes, da Rocha, S.: The role of financial, macroeconomic, and non-financial information in bank loan default timing prediction. Eur. Acc. Rev. **22**(4), 739–763 (2013)
9. Bielecki, T.R., Rutkowski, M.: Credit risk modeling, valuation and hedging. Springer, Berlin (2004)
10. Bluhm, C.: Structured credit portfolio analysis in credit portfolio management. Chapman and Hall (2007)
11. Brennan, W., McGirt, D., Roche, J., Verde, M.: Bank Loan Ratings, in Bank Loans: Secondary Market and Portfolio Management. Frank J. Fabozzi Associates, New Hope, PA, pp. 57–69 (1998)
12. Calem, P., Lacour-Little, M.: Risk-based capital requirements for mortgage loans. J. Bank. Financ. **28**(3), 647–672 (2004)
13. Carter, K.E.: The joint effect of the sarbanes-oxley act and earnings management on credit ratings. J. Acc. Financ. **15**(4), p77–p94 (2015)
14. CRISIL: Credit risk estimation techniques. https://www.crisil.com/pdf/global-offshoring/Credit_Risk_Estimation_Techniques.pdf
15. Das, S.R., Duffie, D., Kapadia, N., Saita, L.: Common failings: how corporate defaults are correlated. J. Financ. **62**, 93–117 (2007)
16. Duffie, D., Saita, L., Wang, K.: Multi-period corporate default prediction with stochastic covariates. J. Financ. Econ. **83**, 635–665 (2007)
17. Everett, C.: Group membership, relationship banking and loan default risk: the case of online social lending. Bank. Financ. Rev. **7**(2), 15–54 (2015)
18. Friedman, C., Sandow, S.: Ultimate recoveries. Risk **16**(8), 69–73 (2003)
19. Glennon, D., Nigro, P.: Measuring the default risk of small business loans: a survival analysis approach. J. Money Credit Bank. **37**(5), p923–p947 (2005)
20. Helwege, J.: How long do junk bonds spend in default? J. Financ. **54**(1), 341–357 (1999)
21. Lando, D.: Credit Risk Modeling: Theory and Applications. Princeton University Press, Princeton (2004)
22. Mohan, T.P., Croke, J.J., Lockner, R.E., Manbeck, P.C.: Basel III and regulatory capital and liquidity requirements for securitizations. J. Struct. Financ., **17**(4), 31–50 (2012)
23. Ngene, G.M., Kabir Hassan, M., Hippier, III, W.J., Julio, I.: Determinants of mortgage default rates: pre-crisis and crisis period dynamics and stability. J. Hous. Res. **25**(1), 39–64 (2016)
24. Perez Montes, C.: Estimation of regulatory credit risk models. J. Financ. Serv. Res. **48**(2), p161–p191 (2015)
25. Press, W.H., Teukolsky, S.A., Vetterling, W.T., Flannery, B.P.: Numerical Recipes: The Art of Scientific Computing. Cambridge University Press, Cambridge (2007)

26. Seiler, M.J.: Determinants of the strategic mortgage default cumulative distribution function. J. Real Estate Lit. **24**(1), p185–p199 (2016)
27. Switzer, L.N., Wang, J.: Default risk estimation, bank credit risk, and corporate governance. Financ. Mark. Instit. Instrum. **22**(2), 91–112 (2013)
28. Wagner, H.S.: The pricing of bonds in bankruptcy and financial restructuring. J. Fixed Income (June), 40–47 (1996)

A Decision Model to Predict Clinical Stage of Bladder Cancer

Archana Purwar, Sandeep Kumar Singh and Pawan Kesarwani

Abstract An extensive variety of computational techniques and tools have been developed for information investigation in medical domain. In this article, we have exploited those accessible technological headways to predict stage of bladder cancer. Present system of examination involves an invasive procedure called "cystoscopy" to find high risk patients. This unique research work helps in determining contributing factors (demographic as well as pathological) leading to the progression of bladder cancer. The proposed predictive model if used by restorative specialists and professionals will help in eliminating unnecessary cystoscopy in patients with low stage and grade diseases. As additional contribution, this article also validates the performance of our earlier designed hybrid model dealing with missing value imputation (HPM-MI) for the diagnosis of stage of bladder cancer. This model is fit for precise prediction in nearness of vast missing values. The evaluation of this model is examined by means of the classification accuracy, precision, recall and F1 measure. The dataset is collected from Max Super Specialty Hospital (a unit of Balaji Medical and diagnostic research centre), Patparganj, Delhi, retrospectively from records of cancerous patients. The results are found to be 82.39% precise and accurate when compared with actual information.

Keywords Missing value imputation · Prediction model · Data mining
Bladder cancer

A. Purwar (✉) · S.K. Singh
Department of Computer Science and Information Technology, JIIT Noida, Noida, India
e-mail: archana.purwar@gmail.com

S.K. Singh
e-mail: sandeepk.singh@jiit.ac.in

P. Kesarwani
Department of Urology & Renal Transplant, Max Super Speciality Hospital,
Patparganj, New Delhi, India
e-mail: pawanaiims@hotmail.com

© Springer Nature Singapore Pte Ltd. 2018
M. Pant et al. (eds.), *Soft Computing: Theories and Applications*,
Advances in Intelligent Systems and Computing 583,
https://doi.org/10.1007/978-981-10-5687-1_74

1   Introduction

Cancer is one of the main sources of grown-up deaths around the world. In India, the international Agency for Research on Cancer had reported that around 635,000 individuals passed away from cancer in 2008. Death cases of around 8% due to bladder cancer have been reported in other countries whereas around 6% death have occurred in India. Bladder cancer is the fifth largest regular growth in United States (US). Every year, 63,219 fresh cases and 13,180 deaths are reported [1] in US. According to the Indian malignancy registry information in men, it is the ninth most basic disease representing 3.9% of all disease cases. Bladder Cancer is a standout amongst the most well-known urological malignancies that stems from bladder lining. This bladder lining has a mucous layer of surface cells that has the capacity to grow and shrink. It also has a smooth muscle and fibrous layer. Recently, the rate of bladder malignancy is rising continuously. This might be because of the inert impacts of smoking, tobacco misuse and modern cancer-causing agents. Bladder cancer is more prevalent in males as compared to females [2].

Now days, the quick increment in sizes of databases has lead to the need to develop automated tool and techniques for the extraction of knowledge from vast amount of data. The term data mining or knowledge discovery in databases (KDD) has been embraced to find the hidden information or patterns in the data automatically [3]. It encompasses predictive as well as descriptive approaches for effective automated revelation of concealed, unique, valuable, and implicit patterns in vast databases. Formally, data mining is a vital extraction of hidden and most valuable data about data [4]. It consists of various predictive or descriptive models to describe data. These models may be used for forecast and classification of unseen data. Different fields, for example, marketing, customer relationship management, weather forecasting, fraud detection, classification of objects, analyzing social network data, and spam filtering of mails make use of data mining [5]. In addition to these fields, medical domain also experiences the use of data mining techniques to forecast adequacy of surgical systems and medical tests. It also helps in finding association among clinical data [6].

In this research, data of urinary bladder cancer patients undergoing treatment at Max Super Specialty Hospital, Patparganj, Delhi is collected to do retrospective study of 85 patients. The extracted data from the hospital is used to build a decision model to predict the stage of bladder cancer in patients accurately. Currently medical practitioners make use of repeated invasive procedure to find high risk patients. The work reported in this paper take into account demographic as well as pathological contributing factors to judge the progression of bladder cancer. The proposed predictive model (HPM-MI) if used by restorative specialists and professionals will help in eliminating unnecessary cystoscopy in patients with low stage and grade diseases. Moreover, this article also helps practitioner to diagnose of stage of bladder cancer. The residual paper is structured in 4 sections. Section 2 describes about background study of bladder cancer and data mining techniques.

Proposed work is explained in Sect. 3 along with empirical work. Section 4 shows results using different metrics. Finally, Sect. 5 propounds conclusion and future work.

2 Background Study

This section gives background study of data mining methods used for predicting stage of bladder cancer patients. Moreover, it also discusses the previous work done in the predictive classification of bladder cancer.

2.1 Significance of Bladder Cancer

Bladder Cancer is arising at fifth position out of harmful infections in the United States of America with a yearly event of nearly 63,219 fresh cases and 13,180 deaths [6]. Mostly, western nations are also having similar statistics, with occurrence rates of 18–30 fresh cases for every one lakh men. Bladder cancer is one of the main five malignancies in men while females are three times more averse to gain this malady as compared to males. The early superficial disease is detected in 75% of patients if urine sample consists of blood which may be gross or microscopic [7]. The general reappearance rate of sixty five percent and development rate of twenty percent requires long lasting medical surveillance by cystoscopy (i.e. endoscopy of bladder). In spite of the fact that the definite reason for bladder cancer is misty, different variables, for example, introduction to certain sweet-smelling chemicals, remarkably aniline colors and benzidine mixes have been reported as probable causes of this ailment [7, 8]. In addition to this, smoking is also in charge of more than half of the bladder growths found in males and 33% of those obtained in females [2]. More than ninety percent of bladder growths are of the transitional cell type, with squamous cell carcinoma and adenocarinoma forming the rest. Squamous cell carcinoma is present in the patients who suffer from chronic diseases such as bilhariasis [7]. Notable elements involved for exact prediction of bladder growth are the profundity of entrance and level of cell anaplasia (grade) among the transitional cell carcinomas. Carcinoma-in-situ, a mucosal sharp-review injury does not have a surgical procedure because of its diffuse surface spreading conduct. If it is not treated, it will advance to a muscle-intrusive ailment in up to eighty percent cases of bladder disease in 5 years [9].

The intracavitary approach of therapy for shallow bladder cancer is done to avoid the repetition of tumor after effective surgical resection and to destroy remaining ailment, for example, carcinoma-in situ. Different intravesical chemotherapeutic agents, for example, mitomycin, thiotepa, and doxorubicin were the pillar of treatment for a considerable length of time. They accomplished short abatements with sturdy advantage for just 7–14% of patients [10, 11]. In addition to this, other offbeat types of treatment, for example, immunotherapy was acquainted due with unsuitable chemotherapy results [7]. Shah et al. [12] proposed a patient data mining model for bladder tumor patients who were given bacillus Calmette-Guérin (BCG) plus interferon- immunotherapy treatment. In their model, data set was gathered from University of Iowa and from health centers of patients who were given BCG-plus interferon-alpha immunotherapy. The data was utilized to forecast the stage of bladder disease whether it is present or not after the given treatment. Lukka [13] and Li et al. [14] managed the issue of small data set using similarity classifier and mega trend diffusing technique on gene expression data of bladder cancer respectively.

The present system to assess the stage of bladder cancer is dependent on regular check cystoscopy which is an invasive procedure. The development of prediction factors will help us in eliminating unnecessary cystoscopy in patients with low stage and grade diseases. Hence, this paper has collected data of 85 patients who have either undergone either of the two most successful therapy for bladder cancer as transurethral resection of bladder tumor (TURBT) or radical cystectomy which is a further treatment given to patients suffering from T3 and T4 stage of bladder cancer. Data set collected consist of 78 missing entries in 85 patients. Each patient consists of 14 features. The present study has considered 13 features along with 'class attribute' as stage of Bladder Cancer. Prediction of stage of bladder cancer is done using HPM-MI with missing value imputation that is proved to be best in presence of missing values [15].

2.2   Prediction Models

The performance of the HPM-MI proposed in our previous work [15] was evaluated on the real world data set collected from hospital and results are compared with AdaBoost.M1, J48, and Multilayer perceptron (MLP).

2.2.1   HPM-MI

Real world data set under study comprises of various missing values [16] and HPM-MI proposed in [15] turned out to be extremely helpful in prediction of outcome of medical disease especially if it comprises of missing values. This model includes three stages. First stage analyzes and selects imputation approach utilizing most popular approach as K-means clustering. Second stage of the model extracts the correctly classified instances by making use of clustering. Final stage employs MLP for predictive classification. Back propagation algorithm is utilized to train this neural network. MVI is a pivotal step consolidated in this model aimed to locate the best imputation procedure concerning missing information under study. Imputation technique which has generated more condensed grouping of instances is opted to fill in incomplete values present in the data. Consequently, complete data with no missing value is utilized to extract the instances. Correctly classified instances for the data set under study were extracted from best imputed data set. This consequential data set was classified using MLP classifier for classification. One can refer to reported work in [15] for detailed understanding of HPM-MI.

2.2.2 AdaBoost.M1

The AdaBoost (Adaptive Boosting).M1 [17] is accounted for as a viable boosting algorithm to enhance classification accuracies of weak learning algorithms. It measures every instance mirroring its significance and spots the most weights on those instances which are repeatedly wrongly classified by the former classifiers. This strengthens the subsequent learning classifiers to figure out how to focus on instances that are difficult to be accurately classified. At the point when the AdaBoost is adjusted to handle the class imbalance ratio, it has multi fold benefits. One is that AdaBoost can take any of learning algorithm as base classifier. Secondly weight calculation in AdaBoost is equivalent to re-sampling of data that comprises up-sampling as well as down sampling. As a re-sampling technique, it revises the data space naturally taking out the additional learning cost to explore the optimal class distribution. Additionally, re-sampling through weighting tests has little data loss. Moreover, AdaBoost algorithm is stated to be immune to overfitting [18]. Therefore, AdaBoost.M1 is applied for prediction of stage of bladder cancer for real world dataset collected under study [18].

2.2.3 C4.5

Decision tree is standout amongst the most prevalent classification approaches in data mining. Decision tress produces rules that are easy to deduce and recognize. Therefore, classifiers based on decision tree help extraordinarily in splitting in various classes. There are numerous decision tree classifiers such as CART, ID3. C4.5 is an entrenched and generally utilized classifier. C4.5 utilizes information gain ratio as a splitting measure to pick up most discriminatory feature at every progression of decision tree while training model. At every iteration, the information gain ratio is computed and the feature that provides the highest information gain ratio is chosen as best rip. C4.5 works recursively assembling sub-trees as long as (1) it gets set of instances that contain samples of one and only one class (then the leaf node is named by this class), or (2) no feature is left available for split (then the leaf node is named by the majority class), or (3) it is below a predetermined threshold such as tree depth, goodness of split, or the number of instances achieved in the subset [19].

2.2.4   MLP

MLP is identified as feed forward neural network that is allowed to have multiple units of neurons. Researchers have ascertained that MLP consisting of one hidden unit can precisely estimate consistent continuous functions [20]. It comprises of three units. First unit of the network has as many neurons as number of features present in the data set under study. Each and every feature is supplied to the input unit as a neuron. Second unit takes aggregated input from input unit and applies sigmoid function and passes the output to the output unit. This output unit shows the network's classes for given data as 'T1', 'T2', 'T3', and 'T4'. This neural network employs backpropagation algorithm to calculate the optimum weights for training the network.

3   Implementation

3.1   Experimental Data Set

Dataset taken under study was gathered from Max Super Specialty Hospital (a unit of Balaji Medical and diagnostic research centre), Patparganj, Delhi. Data gathering was dependent on familiar as well as unfamiliar features of patients who had bladder cancer. Patients had undergone either transurethral resection of bladder tumor (TURBT) or radical cystectomy. TURBT is the underlying medication of bladder cancer and a staging procedure that helps to locate the cancer. The sample is recovered from surgical treatment and remit to a pathology lab which provides his/her finding with regards to the profundity of attack of the tumor in the bladder wall (T stage). Moreover, grade is also reported as either high or low. Radical cystectomy is an expulsion of the whole bladder, adjacent lymph hubs (Lymphadenectomy), portion of the urethra, and close-by organs that may include growth cells. This information was taken from pathological reports and patients files. It is described by 13 features along with a class attribute shown in Table 1. Various features considered under study are demographic as well as pathological factors like age, sex, smoker and alcohol are the demographic factors. Carcinoma in situ (CIS), Grade (how well developed the cells look under the microscope), Lamina Propria, Muscalaris, lymphovascular invasion (LVI), perineural, no of tumors and aspect of tumor are other factors taken under study. Table 1 shows the name of feature, type of feature and values carried by attribute. Class attribute is "stage" that has four values as T1, T2, T3 and T4. There are 59 and 18 instances of T1 and T2 type respectively. T3 and T4 have 5 and 3 instances. Data set consisting of 85 patients having bladder cancer is used to evaluate the prediction of stage of bladder cancer.

Table 1 Description of data set having 14 attributes

S. No.	Feature name	Type of feature	Values
1	Age	Numerical	Continuous
2	Sex	Categorical	{Male, female}
3	CIS	Categorical	{Seen, not seen}
4	Grade	Categorical	{Low, high}
5	Lamina propria	Categorical	{Present, absent}
6	Muscalaris	Categorical	{Present, absent}
7	LVI	Categorical	{Seen, not Seen}
8	Perineural	Categorical	{Seen, not Seen}
9	Smoker	Categorical	{Yes, no}
10	Alcohol	Categorical	{Yes, no}
11	No. of tumors	Categorical	{Single, multiple}
12	Tumor aspect	Categorical	{Solid, papillary}
13	Mitomycin	Categorical	{Yes, no}
14	Class (stage)	Nominal	{T1, T2, T3, T4}

3.2 Experimental Framework

Two popular data mining tools namely keel [21] and weka [22] were utilized to accomplish mining techniques. Missing value imputation techniques shown in Table 2 are implemented using KEEL. Second column of the table shows the missing value imputation method and third column shows the incorrectly classified instances as a result of Simple K-means clustering. Therefore, WKNN gives the least number of misclassified instances, therefore was opted for filling incomplete values in present the data samples. For the classification, we have divided the data set in training and testing in the ratio of 80:20. Classifiers such as MLP, AdaBoost. M1 and J48 are applied using Weka software.

Table 2 Results of Imputation techniques after Clustering

Imputation method	Incorrectly classified instances (%)
Concept most common	48.24
Fuzzy K-means	54.11
K-means	54.11
K nearest neighbour	14.12
Most common	54.12
Weighted nearest neighbor	**14.12**
Case deletion	36.58

3.3 Evaluation Measures

The following subsection depicts evaluation measures that were utilized to assess the performance of HPM-MI and different classifiers as AdaBoost.M1, C4.5 and MLP.

(a) Confusion matrix

A matrix is ascertained to decipher the outcomes of the classifier known as confusion matrix. The major diagonal elements of a confusion matrix depict the correctly classified instances of each class by respective classifier. Results obtained for HPM-MI on the data set under study are presented Table 3.

(b) Accuracy, Precision and Recall

Accuracy is a prominent measure which alludes to the capacity of the predictive system to accurately anticipate the target feature of different instance [5]. Besides, recall is computed by dividing the number of true positives with the total number of instances that actually belong to the positive class and the precision for a class is termed as a ratio of the number of instances correctly labeled as belonging to the positive class and the sum of true positives and false positives which are instances incorrectly labeled as belonging to the class. Discussed metrics are computed as follows:

$$Accuracy = (TP + TN)/(TP + TN + FP + FN) \tag{1}$$

$$Precision = TP/(TP + FP) \tag{2}$$

$$Recall = TP/(TP + FN) \tag{3}$$

where,
    TP and FP are number of true positives predictions and number of false positives predictions for considered class. TN and FN are number of true negative predictions and false negative predictions for considered class.

(c) F1-Measure

This metric is computed using precision and recall as follows:

$$F1 = 2 * Precision * Recall/(Precision + Recall) \tag{4}$$

where, precision and recall can be computed from Eqs. (2) and (3).

Table 3 Confusion matrix for test data set for HPM-MI

a	b	c	d	Classified as
8	0	0	0	a = T1
0	6	0	0	b = T2
0	1	0	0	c = T3
1	1	0	0	d = T4

Table 4 Values of precision, recall and F1 measure of each class

Class	Precision	Recall	F1
T1	0.889	1	0.941
T2	0.75	1	0.857
T3	0	0	0
T4	0	0	0

Table 5 Classification accuracies of HPM-MI and other classifiers

Classifier	Accuracy
HPM-MI	82.35
AdaBoost.M1	82.35
MLP	76.47
J48	76.47

4 Results and Analysis

The values of precision, recall and F-measure for each class achieved from HPM-MI on the data set under study are shown in Table 4. The instances (patients) belonging to T1 and T2 are correctly classified where as no instance is correctly classified as T3 and T4. The accuracy obtained from the model is 82.35% which is consistent with AdaBoost.M1 and better than MLP and J48 described in Table 5.

5 Conclusions and Future Work

Bladder Cancer is a fatal disease spreading all over the world. This paper has revealed pathological as well as demographic factors of patients suffering from bladder cancer. These factors along with one class attribute are used to diagnose the stage of the bladder cancer. Further, HPM-MI was employed to assess the stage of bladder cancer based on 13 factors. It is demonstrated that accuracy obtained by HPM-Model for the prediction of bladder cancer patient is 82.39%. The results presented in Table 4 shows that precision and recall values of T3 and T4 classes are zero. It means that no instance of this class is correctly classified by the HPM-MI. Future work will analyze possible reasons for not correctly classifying T3 and T4 classes and would develop the strategy for accurate prediction of the patients belonging to T3 and T4 stage. Moreover, the further study in this path will help us in finding the high risk patients so that a strict surveillance protocol could be followed.

Acknowledgements I would like to thank Dr. P.B. Singh for providing the guidance and Mr. Rajesh Saxena for helping me to provide the resources for data collection from Max Super Specialty Hospital (a unit of Balaji Medical and diagnostic research centre), Patparganj, Delhi. This work is submitted and approved by Dr. Indranil Mukopadhyay (Medical Superintendent, Max Super Specialty Hospital, Patparganj, Delhi).

References

1. Jemal, A., et al.: Cancer statistics, 2005. CA Cancer J. Clin. **55**, 10–30 (2005)
2. Wynder, E.L., Goldsmith, R.: The epidemiology of bladder cancer. A second look. Cancer **40**, 1246–1268 (1977)
3. Cunningham, S.J., Holmes, G.: Developing innovative applications in agriculture using data mining (1999)
4. Piatetsky, G.: Knowledge discovery in database: An overview. Knowl. Discov. Database (1991)
5. Han, J., Kamber, M.: Data Mining: Concepts and Techniques (2007)
6. Chapple M.: About.com guide. http://databases.about.Com/od/datamining/g/classification. html
7. Alexandroff, A.B., et al.: BCG immunotherapy of bladder cancer: 20 years on. Lancet **353**, 1689–1694 (1999)
8. Mayo Clinic Staff: Bladder cancer. http://www.mayoclinic.com/invoke.cfm?id=DS00177. Accessed 21 Sept 2004
9. Utz, D., Farrow, G.: Management of carcinoma in situ of the bladder: the case for surgical management. Urol. Clin North Am. **7**, 533–541 (1980)
10. Pawinski, A., et al.: A combined analysis of European Organization for Research and Treatment of Cancer, and Medical Research Councilrandomized clinical trials for the prophylactic treatment of stage TaT1 bladder cancer. J. Urol. **156**, 1934–1941(1996)
11. Lamm, D.L., et al.: Apparent failure of current intravesical chemotherapy prophylaxis to influence the long-term course of superficial transitional cell carcinoma of the bladder. J. Urol. **153**, 1444–1450 (1995)
12. Shah, S.C., Kusiak, A., O'Donnell, M.A.: Patient-recognition data-mining model for BCG-plus interferon immunotherapy bladder cancer treatment. Comput. Biol. Med. **36**, 634–655 (2006)
13. Luukka, P.: Similarity classifier in diagnosis of bladder cancer. Comput. Methods Programs Biomed. **89**, 43–49 (2008)
14. Li, D.-C., et al.: A new method to help diagnose cancers for small sample size. Expert Syst. Appl. **33**, 420–424 (2007)
15. Purwar, A., Singh, S.K.: Hybrid prediction model with missing value imputation for medical data. Expert Syst. Appl. **42**(13), 5621–5631 (2015)
16. Purwar, A., Singh, S.K.: DBSCANI: noise-resistant method for missing value imputation. J. Intell. Syst. **25**(3), 431–440 (2016)
17. Freund, Y., Schapire, R.E.: A Decision-theoretic generalization of on-line learning and an application to boosting. J. Comput. Syst. Sci. **55**, 119–139 (1977)
18. Friedman, J., Hastie, T., Tibshirani, R.: Additive logistic regression: a statistical view of boosting (with discussion and a rejoinder by the authors). Annal. Stat. **28**, 337–407 (2000)
19. Salzberg, S.L.: C4. 5: Programs for machine learning. In: Quinlan, J.R. (ed.) Machine Learning, vol. 16, pp. 235–240. Morgan Kaufmann Publishers inc. (1994)
20. Cybenko, G.: Approximation by superpositions of a sigmoidal function. Math. Control Signals Syst. **2**(4), 303–314 (1989)
21. Alcalá-Fdez, J., et al.: KEEL: a software tool to assess evolutionary algorithms for data mining problems. Soft. Comput. **13**(3), 307–318 (2009)
22. Witten, I.H., et al.: Weka: Practical machine learning tools and techniques with Java implementations (1999)

Dynamic 9 × 9 Substitution-Boxes Using Chaos-Based Heuristic Search

Musheer Ahmad, Farooq Seeru, Ahmed Masihuddin Siddiqi
and Sarfaraz Masood

Abstract Large-sized substitution-boxes tend to provide high security and resistant to some attacks as compared to small sized S-Boxes. However, finding large and efficient $n \times n$ S-boxes ($n > 8$) is an open issue. Here, we propose to put forward a chaos-based heuristic search strategy to generate dynamic 9 × 9 S-boxes. The anticipated strategy has the ability to search optimized S-boxes as the generations are applied. As an instance, the S-box constructed by the proposed strategy is assembled and analyzed against the criterions such as bijectivity, nonlinearity, algebraic degree, differential probability, robustness to differential cryptanalysis, transparency order, etc. The simulation outcomes verify that chaos-based heuristic search strategy is streamlined and has proficiency of synthesizing cryptographically potent 9 × 9 dynamic substitution-boxes capable of exhibiting consistent performance lineaments.

Keywords Large substitution-box · Heuristic search · Nonlinearity
Block ciphers · Chaotic map

M. Ahmad (✉) · S. Masood
Department of Computer Engineering, Faculty of Engineering and Technology,
Jamia Millia Islamia, New Delhi 110025, Delhi, India
e-mail: musheer.cse@gmail.com

F. Seeru
Department of Computer Science & Engineering, Al-Falah School of Engineering
and Technology, Al-Falah University, Faridabad 121004, Haryana, India

A.M. Siddiqi
Department of Computer Science, Shaqra University, Shaqra, Kingdom of Saudi Arabia

© Springer Nature Singapore Pte Ltd. 2018   839
M. Pant et al. (eds.), *Soft Computing: Theories and Applications*,
Advances in Intelligent Systems and Computing 583,
https://doi.org/10.1007/978-981-10-5687-1_75

1 Introduction

The block cryptosystems employ the layers of permutation and substitution for strong design. The two much studied architectures viz. Fiestal network and SP network facilitates the designer to construct strong cryptosystems [1]. Substitution-boxes are prime components for these two networks at substitution layers and meant to carry out required nonlinear transformation and confusion along with the robustness to different cryptographic assaults. These nonlinear components are often deployed as a lookup table. An $n \times n$ S-box takes n input bits and transforms it nonlinearly into n output bits. It is a mapping, from $GF(2^n) \rightarrow GF(2^n)$, that substitutes secretly the n-bit input data to n-bit output data [2]. They can also be viewed as multi-input and multi-output Boolean function, i.e., a 9×9 S-box inherently compose of nine component Boolean functions, where each Boolean function takes 9-bit input and yields 1-bit as output, all nine functions collectively gives 9-bit output bitstream. Consequently, the metrics meant to evaluate Boolean functions are accounted to quantify the security of S-boxes. Substitution-boxes have a central role to play in deciding the security strength of block ciphers. The cryptographic lineaments of S-boxes are of immense significance for the security of cipher systems. A strong S-box with all features concerning a cryptographic system forms the very base of secure block cipher. Factually, the cryptographic lineaments of S-boxes represent the main strength of the corresponding block-based encryption system [3]. Hence, the development of formidable S-boxes is of utmost significance for cryptologists in designing strong cryptosystems.

There have been a number of proposals that are dedicated to the construction of 8×8 S-boxes which is due to the success of AES block cipher and it is S-box [4]. All of them are balanced and whose design primarily based on concepts such as affine transformations, gray coding, power mapping, optimization, chaotic systems, etc. [4–12]. The design of large $n \times n$ S-boxes ($n > 8$) has not been considered significantly. There are few proposals that include MISTY S-box, KASUMI S-box and S-box investigated by Millan et al., they provide the designs and configuration of large S-boxes [13–15]. It has been well investigated that large-sized substitution-boxes tend to offer high security to the cryptosystem and better resistivity to prevailing cryptographic assaults. But, the designing efficient and large S-boxes are very complex problem and an open issue. One of the very basic reasons of this complexity is the size of immensely large search space. Consider the case of a 9×9 S-box, the search space to fetch the optimized and efficient configuration of an S-box where from is (512), which is enormously high and large. Therefore, with an aim to fetch an efficient configuration of an S-box which size is higher than

8 × 8, a chaos-based heuristic search strategy is framed which can construct effective 9 × 9 S-boxes.

The rest of the paper is as follows: Sect. 2 describes the proposed heuristic-based strategy. Simulation results, different metrics, their analyses, and comparative study have been performed in Sect. 3. Section 4 is prepared to conclude the work.

2 Proposed Strategy

A chaotic map is employed as random source in the proposed strategy. Further, the chaotic map facilitates to make the whole system dynamic and key-dependent. It is bounded dynamical system whose permissible values lie in the region of (0, 1) and has governing equation defined as follows [16]:

$$w(i+1) = \begin{cases} \frac{w(i)}{q} & 0 < w(i) \leq q \\ \frac{1-w(i)}{1-q} & q < w(i) < 1 \end{cases}$$

where $w(i)$ defines the state i of chaotic map whose domain and range set is same, the system parameter is q. This map is chaotic for all $q \in (0, 1)$. It is an iterative formula, where next state value $w(i + 1)$ of map is determined by applying iteration over its previous value $w(i)$, keeping q as fixed. Then selected values chosen for w (0) and q initially for execution makes key for dynamic generation of substitution-boxes.

There have been very limited proposals for the designs and investigations of large S-boxes of size 9 × 9. The well-known block ciphers named MISTY and its variant KASUMI have designed and employed static 9 × 9 S-box that are based on the structure of AES S-Box, i.e., using inverse power mapping followed by affine transformation over the galois field $GF(2^9)$. The readers are referred to official document for the detailed description of KASUMI block cipher available at Ref. [14] Like 8 × 8 AES S-box, the static 9 × 9 S-Boxes of MISTY and KASUMI also have excellent nonlinearity and differential probabilities, but exhibits poor lineaments when assessed against some other performance criterions. In this work, the concepts of chaotic system and heuristic search are explored to propose a novel strategy for generating dynamic substitution-boxes of size 9 × 9 that offers better performance when compared with the S-Box of KASUMI. The method of constructing optimized large-sized S-boxes has the following operational steps.

1.   Select the initial values of $w(0)$, q, n_0, gen.
2.   Take local best L_s and global best G_s as empty tables
3.   Apply chaotic map for n_0 times and leave all the values except $w(n_0)$
4.   Take B_i (for $i = 1$ to 3) as empty arrays
5.   Apply chaotic map to get next value random value w
6.   Find an element g such that $0 \leq g \leq 511$ from current w, If it doesn't belong to array B_1 then add g to B_1, else-if check for B_2, else check for B_3
7.   Repeat steps 5 and 6 until all B_i (for $i = 1$ to 3) are completely filled (i.e. $|B_i| = 512$)
8.   Convert all arrays B_i to 16×32 matrices
9.   Proceed as follows:

    if $(nonlinearity(B_1) \geq nonlinearity(B_2))$

    if $(nonlinearity(B_1) \geq nonlinearity(B_3))$

        set $L_s = B_1$

    else

        set $L_s = B_3$

  else

    if $(nonlinearity(B_2) \geq nonlinearity(B_3))$

        set $L_s = B_2$

    else

        set $L_s = B_3$

10.   Update $G_s = L_s$ if $nonlinearity(L_s) \geq nonlinearity(G_s)$
11.   Apply steps 4 to 10 repeatedly for *gen* number of generations

3   Simulation and Assessment

In this section, the performance analysis of two 9×9 S-Boxes, one generated by proposed strategy and other used in KASUMI, are assessed and analyzed to determine the dominance of one over the other. The experimental setup to perform all simulations and analysis is asfollows: $w(0) = 0.7654321$, $q = 0.49876$, $n_0 = 250$ and $gen = 35{,}000$. The final configuration of optimized 9×9 S-box generated by proposed heuristic-based strategy is shown in Table 1. The two S-boxes under consideration are tested, analyzed, and compared against the bijectivity, nonlinearity, algebraic degree, differential probability, robustness, algebraic immunity, transparency order, etc., in the following subsections.

Table 1 Proposed optimized 9 × 9 substitution-box G_s (shown in transposed form)

	0	1	2	3	4	5	6	7	8	9	10	11	12	13	14	15
0	434	98	155	405	3	358	160	217	365	208	390	60	323	26	414	89
1	115	382	250	442	366	483	13	372	360	402	441	262	294	179	95	73
2	492	59	264	187	300	51	510	353	299	398	167	477	406	491	88	205
3	482	348	370	132	120	142	136	288	437	438	303	193	312	99	409	389
4	221	439	315	101	261	325	347	456	404	354	74	436	227	440	417	371
5	68	75	292	8	237	476	199	336	326	495	233	24	275	494	392	367
6	11	83	445	485	38	443	272	1	460	306	79	270	152	244	116	276
7	150	448	140	301	111	112	81	165	239	225	203	100	253	349	64	373
8	289	407	32	338	31	400	43	413	383	76	41	429	423	62	252	307
9	458	420	131	178	118	265	224	489	357	3 88	211	103	341	461	122	54
10	186	506	446	480	308	251	293	135	369	422	487	6	401	37	108	508
11	408	55	378	207	198	226	328	368	271	63	350	82	282	189	267	219
12	385	470	194	469	52	220	269	46	110	247	34	381	23	129	172	340
13	195	196	39	0	430	20	86	90	92	241	258	355	223	222	161	507
14	474	177	197	36	248	493	232	56	209	447	421	463	497	156	94	313
15	317	362	428	22	77	157	185	163	504	148	304	255	184	17	416	236
16	359	393	21	274	19	419	331	213	333	127	488	139	93	47	231	511
17	339	164	450	435	295	285	452	97	147	144	283	201	202	379	25	246
18	42	234	117	230	53	256	394	65	133	130	126	254	345	229	191	216
19	27	412	376	119	374	464	425	18	322	486	4	356	114	321	109	396
20	169	240	125	80	324	58	484	5	2	279	128	143	377	287	286	123
21	45	134	346	260	102	361	206	427	509	145	166	316	124	277	311	159
22	467	176	153	105	397	33	137	310	352	71	320	424	273	455	263	35

(continued)

Table 1 (continued)

	0	1	2	3	4	5	6	7	8	9	10	11	12	13	14	15
23	451	215	146	235	91	158	302	297	479	490	363	290	259	471	309	498
24	67	330	415	410	426	243	433	50	49	242	70	249	168	305	496	375
25	332	104	481	151	87	459	141	468	462	182	210	384	334	296	281	449
26	380	44	170	342	192	113	291	14	228	48	96	16	453	257	505	314
27	138	121	212	466	284	238	188	175	391	457	475	266	473	7	40	329
28	28	30	204	387	344	431	15	85	502	61	454	84	478	499	66	278
29	418	335	181	69	149	268	343	298	327	465	386	403	503	395	472	190
30	319	78	180	351	318	12	200	9	154	218	280	364	174	399	171	57
31	173	500	337	501	107	72	10	245	29	106	183	214	411	432	162	444

3.1 Bijectivity

Definition #1: A 9 × 9 S-Box is said to satisfy the property of bijectivity, if all its nine component Boolean functions are balanced. The Boolean function is balanced if it has same number of 0's and 1's in all its output vectors.

The cryptographic Boolean functions which are not balanced considered as weak and insecure [17]. It has been are verified that each of the component functions of the proposed S-Box are balanced. Moreover, if a 9 × 9 S-Box has unique elements in the range of [0, 511], as the case for S-Box in Table 1, then S-Box is considered as bijectivity.

3.2 Nonlinearity

It is mandatory for cryptographically strong S-boxes to have a high nonlinearity score.

Definition #2: The nonlinearity of a Boolean function f is determined by finding the minimum distance of f to the set of all affine functions [17]. Thus, the component functions of an S-box should have upright nonlinearities scores. The nonlinearity NL_f of any Boolean function f is accounted as

$$NL_f = \frac{1}{2}(2^n - WH_{max}(f))$$

A Boolean function is arrogated as weak if it tends to have low nonlinearity score, where $WH_{max}(f)$ is the Walsh-Hadamard transform of Boolean function f [18]. It has been examined that the maximum nonlinearity of a balanced Boolean function for odd n (=3, 5, 7, 9) equals $2^{n-1} - 2^{(n-1)/2}$, and it is tightly higher than $2^{n-1} - 2^{(n-1)/2}$ as proved by Patterson in [19, 20]. The nonlinearity scores of nine component Boolean functions of proposed optimized 9 × 9 S-box are: 226, 226, 222, 224, 218, 222, 222, 228, 224, which are also listed in Table 2. It is evident from the Table 2 that each of the component functions almost achieved the satisfactory value (min = 218, max = 228, mean = 223.55) which are not too far from the target value of 240.

Table 2 Nonlinearities of Boolean functions f_i of proposed S-Box

Nonlinearities										
Boolean Function	f_1	f_2	f_3	f_4	f_5	f_6	f_7	f_8	f_9	Mean
Score	226	226	222	224	218	222	222	228	224	223.55

3.3 Algebraic Degree

A Boolean function f is usually represented in the form of algebraic normal form (ANF).

Definition #3: The highest degree of this algebraic polynomial (the number of variables in the largest monomial) is called the degree of Boolean function. The degree of S-Box S is defined as [23]:

$$\deg(S) = \max(\deg(f_1), \deg(f_2), \ldots, \deg(f_n))$$

The degree of an S-box should be as high as possible, it has an upper bound of $n-1$ for $n \times n$ S-box (=8 in our case). The algebraic degrees for the two S-boxes under consideration are provided in Table 3.

3.4 Invulnerability Against Differential Cryptanalysis

The differential probability and robustness are the measures evaluated to quantify the resistivity of an S-Box against the differential cryptanalysis (DC). The prevailing attack procedure of DC was suggested by Biham and Shamir, it is concerned with exploiting of imbalance on the input/output distribution to attack S-boxes [21]. Resistance toward differential cryptanalysis can be accomplished if the Exclusive-OR value of each output has equal uniformity with the Exclusive-OR value of each input [21]. If an S-box is closed in input/output probability distribution, then it is resistant against DC. It is desired that the largest value of differential probability (DP) in XOR table (difference distribution table) should be as low as possible.

Table 3 Comparative analysis of KASUMI and proposed 9×9 substitution-boxes

Parameters	KASUMI [14]	Proposed	Dominance
Kasumi S-Box is dominant			
C.1.a. Nonlinearity	240	223.55	*Significant*
C.1.b. Differential probability	0.0039	0.0234	*Slight*
C.1.c. Robustness to DC	0.996	0.977	*Slight*
Proposed S-Box is dominant			
C.2.a. Algebraic degree	2	8	*Significant*
C.2.b. Algebraic immunity	2	4	*Significant*
C.2.c. Transparency order	8.982	8.855	*Slight*
C.2.d. Nature	Static	Dynamic	*Significant*

Definition #4: The DP for a Boolean function $f(x)$ is quantified as:

$$DP_f = \max_{\Delta x \neq 0, \Delta y} \left(\frac{\#\{x \in X | f(x) \oplus f(x \oplus \Delta x) = \Delta y\}}{2^9} \right)$$

where X is the set of all possible input values and the number of its elements is 2^9. It one of the practical attacks that have been successfully utilized to break the DES ciphers. The maximum score of the differential probability in an S-box should be as low enough to resist DC. In our case, the largest DP for S-box in Table 1 is found to be 0.0234 which is quite low.

Definition #5: Let $S = (f_1, f_2, ..., f_n)$ be an $n \times n$ S-box and f_i are its component functions. An S-box S is said to be R-robust against DC [22], if:

$$R = \left(1 - \frac{N}{2^n} \right) \left(1 - \frac{L}{2^n} \right)$$

where L is the largest value in difference distribution table of S and N is the number of nonzero entries in the first column of DDT. The upper bound of robustness of an S-Box is $1 - 2^{-n+1}$ (= 0.996 in our case). Any value which is close to this is considered acceptable and satisfactory like the case of our S-box; the robustness comes out as 0.977. The scores of DP and robustness for two S-boxes are provided in Table 3.

3.5 Algebraic Immunity

Algebraic immunity metric denotes the resistance of an S-box, used in block ciphers, against the algebraic attacks and inversely the effectiveness of the XSL attack on a particular S-box. The procedure of computing the algebraic immunity is explained in [24].

Definition #6: The algebraic immunity AI of an $n \times n$ S-box S is defined in terms of the algebraic immunity AI of the n component functions as follows [25]:

$$AI(S) = \min_{c \in F_2^n} \{ AI(c_1 f_1 \oplus c_2 f_2 \oplus .... \oplus c_n f_n) \}$$

A high score of algebraic immunity is desired to complicate the algebraic attacks on S-boxes. Meaning, an S-Box having high algebraic immunity is considered as better. The algebraic immunity of proposed S-box is 4 which are quite better than the value 2 of KASUMI S-box as evident from Table 3.

3.6 Transparency Order

The vast majority of digital systems are not carefully designed; one can get sensitive data from side channels, for example, power consumption or the timing of operations or the software implementation. The differential power analysis (DPA) is said to be a standout analysis among the most capable strategy against block ciphers to execute side-channel attacks (SCA). Rijindael S-box included in AES cipher is generally focused by cryptanalysts as oracles giving the output corresponding to given information [26]. To evaluate the resistance of S-boxes toward DPA-type of SCA assaults, transparency order (TO) measure is suggested by cryptographers. If an S-box show lower TO score, then the S-box tends to exhibit more resistant against SCA attack, i.e., the count of power traces to identify the correct key will be higher.

Definition #7: According to Prou [26], the transparency order of an S-box S can be defined as:

$$TO(S) = \max_{\beta \in F_2^m} \left(|m - 2H(\beta)| - \frac{1}{2^{2n} - 2^n} \sum_{a \in F_2^{n^*}} \left| \sum_{i=1}^{m} (-1)^{\beta_i} A_{F_i}(a) \right| \right)$$

The terms involved in the formulation can be comprehended in Ref. [26]. An S-box having lower TO score is tend to have better resistivity toward SCA attack than another S-box having higher TO score. The transparency orders for the two S-boxes under analysis are provided in Table 3.

3.7 Comparative Analysis

The scores of mean nonlinearity, differential probability, robustness, algebraic degree, algebraic immunity, and transparency order for the two S-boxes are collected and compared in the Table 3. As far as bijectivity is concern, both the S-boxes satisfy the property of bijectivity. The metrics, opted for comparison are, categorized into two classes, C.1: Metrics where KASUMI S-box has dominance over the proposed S-box and C.2: Metrics where the proposed S-box is dominating over the standard KASUMI 9×9 S-box. The better score is identified by an underline mark. The comparative analysis unveils that the proposed S-box in Table 1 outperforms over the well-known standard KASUMI 9×9 S-box for the security metrics such as algebraic degree, algebraic immunity, transparency order, and nature of their generation. For algebraic degree and algebraic immunity, the proposed S-box is significantly better than the KASUMI S-box. It is slightly better than KASUMI for the case of transparency order. Moreover, the nature of S-box generation cannot be ignored, the S-box of KASUMI is static and fixed, whereas, the proposed strategy of optimized S-box design is dynamic in nature. A different dynamic S-box can be easily generated with a slight variation in any of the key components $w(0)$ or q or n_0 etc., the dynamic nature of generation is one of the major merits of the proposed design similar to the proposals [27–30].

KASUMI S-box has better performance for the metrics such as nonlinearity, differential probability, and robustness to DC. However, the scores of these three metrics for proposed S-box also has satisfactory and acceptable values. The scores for proposed S-box are not that much low unlike KASUMI S-box. For example, algebraic degree and immunity of KASUMI S-box is two and two which is quite low for any cryptographic S-box. Hence, it can be claimed that the proposed S-box design strategy has many lineaments and perform effective in generating optimized 9 × 9 substitution-boxes for strong block ciphers.

4 Conclusion

Here, we have suggested a chaos-based heuristic strategy for the generation of efficient large substitution-boxes of size 9 × 9. The strategy is dynamic in nature and fully key-dependent, i.e., a number of strong S-boxes can be synthesized by slightly varying any of the key values. A chaotic map is employed for random generation of S-box populations in each generation. The strength of proposed S-box is quantified against a number of security metrics and parameters pertinent to S-boxes. The S-box is tested against bijectivity, nonlinearity, DP, robustness to DC, algebraic degree, algebraic immunity, and transparency order. The proposed S-box is also compared with the standard S-Box of KASUMI block cipher to justify the efficacy and effectiveness of proposed method. The simulation and comparison analyses justify the appreciable and honorable security performance of anticipated method.

References

1. Knudsen, L.R., Robshaw, M.: The block cipher companion. Springer Science & Business Media (2011)
2. Stinson, D.R.: Cryptography: theory and practice. CRC press, (2005)
3. Hussain, I., Shah, T., Gondal, M.A., Khan, W.A.: Construction of cryptographically strong 8 × 8 S-boxes. World Appl Sci J. **13**(11), 2389–2395 (2011)
4. Daemen, J., Rijmen, V.: The design of Rijndael: AES – the Advanced Encryption Standard, Springer-Verlag (2002)
5. Lambić, D.: A novel method of S-box design based on chaotic map and composition method. Chaos, Solitons Fractals **58**, 16–21 (2014)

6. Tran, M.T., Bui, D.K., Duong, A.D.: Gray S-box for advanced encryption standard, International Conference on Computational Intelligence and Security, 253–258 (2008)
7. Ahmad, M., Bhatia, D., Hassan, Y.: A Novel Ant Colony Optimization Based Scheme for Substitution Box Design. Procedia Comput. Sci. **57**, 572–580 (2015)
8. Ahmad, M., Chugh, H., Goel, A., Singla, P.: A Chaos Based Method for Efficient Cryptographic S-box Design. In: Thampi, S.M., Atrey, P.K., Fan, C.-I., Pérez, G.M. (eds.) SSCC 2013. CCIS 377, 130–137 (2013)
9. Isa, H., Jamil, N., Z'aba, M.R.: Construction of Cryptographically Strong S-Boxes Inspired by Bee Waggle Dance. New Gener. Comput. **34**(3), 221–238 (2016)
10. Laskari, E.C., Meletiou, G.C., Vrahatis, M.N.: Utilizing Evolutionary Computation Methods for the Design of S-Boxes. In: International Conference on Computational Intelligence and Security, 1299–1302 (2006)
11. Nedjah, N., Mourelle, L.D.M.: Designing substitution boxes for secure ciphers. Int. J. Innovative Comput. Appl. **1**(1), 86–91 (2007)
12. Burnett, L.: Heuristic optimization of Boolean functions and substitution boxes for cryptography. PhD dissertation, Queensland University of Technology (2005)
13. Matsui, M.: New block encryption algorithm MISTY International Workshop on Fast Software Encryption. Lect. Notes Comput. Sci. **1267**, 54–68 (1997)
14. 3rd Generation Partnership Project, Technical Specification Group Services and System Aspects, 3G Security: Specification of the 3GPP Confidentiality and Integrity Algorithms; Document 2: KASUMI Specification, V.3.1.1, (2001)
15. Millan, W., Clark, A., Dawson, E.: Heuristic design of cryptographically strong balanced Boolean functions International Conference on the Theory and Applications of Cryptographic Techniques. Lect. Notes Comput. Sci. **1403**, 489–499 (1998)
16. Li, S., Li, Q., Li, W., Mou, X., Cai, Y.: Statistical properties of digital piecewise linear chaotic maps and their roles in cryptography and pseudo-random coding. In: IMA International Conference on Cryptography and Coding, 205–221 (2001)
17. Cusick, T.W., Stanica, P.: Cryptographic Boolean Functions and Applications. Elsevier, Amsterdam (2009)
18. Hussain, I., Shah, T.: Literature survey on nonlinear components and chaotic nonlinear components of block ciphers. Nonlinear Dyn. **74**(4), 869–904 (2013)
19. Helleseth, T., Klve, T., Mykkelveit, J.: On the covering radius of binary codes. IEEE Trans. Inf. Theory **24**(5), 627–628 (1978)
20. Patterson, N.J., Wiedemann, D.H.: The covering radius of the $[2^{15}, 16]$ Reed-Muller code is at least 16276. IEEE Trans. Inf. Theory **29**(3), 354–356 (1983)
21. Biham, E., Shamir, A.: Differential cryptanalysis of DES-like cryptosystems. J. Cryptol. **4**(1), 3–72 (1991)
22. Seberry, J., Zhang, X.M., Zheng, Y.: Pitfalls in designing substitution boxes Annual International Cryptology Conference. Lect. Notes Comput. Sci. **839**, 383–396 (1994)
23. Isa, H., Jamil, N., Zaba, M. R.: S-box construction from non-permutation power functions. In: Proceedings of the 6th International Conference on Security of Information and Networks, 46–53 (2013)
24. Didier, F., Tillich, J.P.: Computing the algebraic immunity efficiently International Workshop on Fast Software Encryption. Lect. Notes Comput. Sci. **4047**, 359–374 (2006)
25. Wood, C.A.: Large Substitution Boxes with Efficient Combinational Implementations, M.S. Thesis, Rochester Institute of Technology (2013)
26. Prou, E.: DPA Attacks and S-Boxes. In: Proceedings of FSE-2005, Lect. Notes Comput. Sci. **3557**, 424–441 (2005)

27. Ahmad, M., Khan, P.M., Ansari, M.Z.: A simple and efficient keydependent S-box design using Fisher-Yates shuffle technique. In: International Conference on Security in Networks and Distributed Systems, CCIS **420**, 540–550 (2014)
28. Ahmad, M., Haleem, H., Khan, P.M.: A new chaotic substitution box design for block ciphers. In: International Conference on Signal Processing and Integrated Networks, 255–258 (2014)
29. Ahmad, M., Ahmad, F., Nasim, Z., Bano, Z., Zafar, S.: Designing chaos based strong substitution box. In: International Conference on Contemporary Computing, 97–100 (2015)
30. Ahmad, M., Rizvi, D.R., Ahmad, Z.: PWLCM-Based Random Search for Strong Substitution-Box Design. In: Second International Conference on Computer and Communication Technologies, AISC **379**, 471–478 (2016)

Opposition-Based Learning Embedded Shuffled Frog-Leaping Algorithm

Tarun Kumar Sharma and Millie Pant

Abstract Shuffled frog-leaping algorithm (SFLA), a memetic algorithm modeled on the foraging behavior of natural species called frogs. SFLA embeds the features of both particle swarm optimization (PSO) and shuffled complex evolution (SCE) algorithm. It is well documented in literature that SFLA is an efficient algorithm to solve non-traditional optimization problems. However like other memetic algorithms SFLA also limits in convergence rate or shows premature convergence when applied to multifaceted continuous optimization problems. In this study, an opposition-based variant of SFLA named as O–SFLA is proposed. In general, the structure of SFLA, the frog, is divided into memeplexes based on their fitness values where they forage for food. In this study, the opposition-based learning concept is embedded into the memeplexes before the frog initiates foraging. The proposed variant is validated on six optimization benchmark problems taken from literature. Further non-parametric analysis is performed to evaluate the efficacy of the proposal.

Keywords Shuffled frog-leaping algorithm · SFLA
Opposition-based learning · OBL · Global optimization

1 Introduction

The problem of optimization arises in every sphere of human life. As the level of problem increases in terms of complexity, dimension, limited knowledge about the problem domain, feasible region and a like, it becomes difficult to solve such multifaceted problems using traditional optimization methods (e.g., gradient search

T.K. Sharma (✉)
Amity University, Jaipur, Rajasthan, India
e-mail: taruniitr1@gmail.com

M. Pant
IIT Roorkee, Saharanpur Campus, Roorkee, India
e-mail: millidma@gmail.com

© Springer Nature Singapore Pte Ltd. 2018
M. Pant et al. (eds.), *Soft Computing: Theories and Applications*,
Advances in Intelligent Systems and Computing 583,
https://doi.org/10.1007/978-981-10-5687-1_76

methods, etc.). SFLA, introduced in 2003 by Eusuff and Lansey [1, 2], is a recent member of memetic algorithms family. SFLA, like other memetic algorithms, enthused by natural foraging behavior of species. In SFLA, these natural species are frogs. Since introduction, SFLA and its variants have been successfully applied to solve many real-world optimization problems of versatile engineering and management domain. The same can be witnessed from the literature. SFLA embeds the features of both particle swarm optimization (PSO) and shuffled complex evolution (SCE) algorithm. PSO helps in performing local search while SCE helps in global search. In SFLA, like other memetic algorithms, colony of frogs is initialized randomly and in second step the colony is divided into sub colonies or memeplexes.

SFLA performs both exploration by dividing the randomly initialized population of frogs into number of memeplexes and exploitation by performing evolution process in each memeplexes. Although SFLA emerges as successful optimizer, however, it also suffers in global convergence velocity. In this study a modification in the memeplexes is introduced by embedding the concept of opposition-based learning. In general, structure of SFLA, the frogs, is divided into memeplexes based on their fitness values where they forage for food. In this study, the opposition-based learning concept is embedded into each memeplexes before the frog initiates foraging. After each iteration, when information exchange process among memeplexes is performed, OBL is again introduced to improvise global search. The proposal is named as O–SFLA. This modification not only enhances local search mechanism of SFLA but also improves the diversity. The performance of the O–SFLA is evaluated on six well-known benchmark optimization functions referred from literature.

The structure of the study is as follows: Sect. 2 describes SFLA in brief followed by the proposed scheme in Sect. 3. Test bed, Experimental settings and computational results are presented and analyzed non-parametrically in Sect. 4. Section 5 presents the conclusions drawn and future work plan.

2 SFLA Overview

SFLA, introduced by Eusuff and Lansey in 2003, has been successfully implemented in solving many world optimization problems. SFLA performed at-par when compared with GA, PSO and Ant Colony Optimization (ACO) algorithms [1, 2]. Researchers have proposed many improved variants of SFLA and applied to many applications of science, engineering, and management [2–16].

SFLA is a memetic algorithm that is inspired from foraging behaviors of natural species called Frogs. SFLA has the abilities to perform local search process as in PSO algorithm and at the later stage information sharing between memeplexes. The next paragraph details the working principle of SFLA.

Like in other memetic algorithms, a population of frogs (solutions) is randomly initialized in feasible search space. Then frogs are distributed into small colonies called memeplexes. This is done by calculating fitness values using objective function. Now each memeplex contains frogs from different cultures and each frog is supposed to perform local search process in order to modify worst's frog position (in terms of fitness value). This process, within each memeplexes, is performed till some fixed criterion or generations. This process also helps in optimization. Later the information or idea hold by the frogs in each memeplex is then exchanged among other memeplexes through shuffling process. Both local and information exchange process are performed till the fixed number of iterations. In general there are four steps in basic SFLA and are as follows:

A. Initialization of frog population

The population of frogs is randomly initialized between fixed upper and lower bounds, i.e., (ub_j) and (lb_j), using Eq. (1): Set of frogs is presented by F.

$$x_{ij} = lb_i + rand(0, 1) \times (ub_i - lb_i) \tag{1}$$

where $rand$ $(0,1)$ is random number between zero and one which is uniformly distributed; $i = 1, 2, ..., F; j = 1, 2, ..., D$ (dimension).

B. Sorting and division of frogs in memeplexes

The fitness of each frog is calculated and then sorted in descending order. Then division of sorted population is done into m memeplexes such that $F = m$ (memeplexes) $\times n$ (number of frogs in each memeplex). The division of frogs is done such that the frog with higher fitness value will be the part of first memeplex, accordingly the next frog into second memeplex, and following fashion frog $m + 1$ will again move to the first memeplex and so on.

C. Local searching process

Frog with global best (X_{global}), best (X_{best}), and worst (X_{worst}) frogs based on fitness values in each memeplex, are noted. Now, an evolution process is implemented to update the position of worst frogs only using the following Eqs. (2) and (3):

Worst frog's updated position

$$X_{new} = current\,position\,(X_{worst}) + Mov_t \tag{2}$$

$$-Mov_{max} \le Mov_t \le Mov_{max}$$

where

$$Mov_t = rand(0, 1) \times (X_{best} - X_{worst}) \tag{3}$$

where t, Mov_t represents the generations $(1, 2, ..., N_{gen})$ and movement of a frog, respectively. Mov_{max} is the maximum acceptable frog movement in the feasible region. When using above equations, the worst frog position shows significant improvement then the position of worst frog is updated otherwise, the same process using Eqs. (2) and (3) are repeated with respect to the global best frog (i.e., X_{global}

replaces X_{best}). If process shows no improvements, then a new solution is randomly generated using Eq. (1) to update the worst frog. This is an iterative process till the maximum number of generations (N_{gen}).

D. Information exchange or shuffling process

Information exchange is performed among memeplexes or the frogs are sorted and shuffled again to perform the evolution process. This process again continues till the fixed number of iterations.

The searching process and basic SFLA algorithm is depicted in Figs. 1 and 2, respectively.

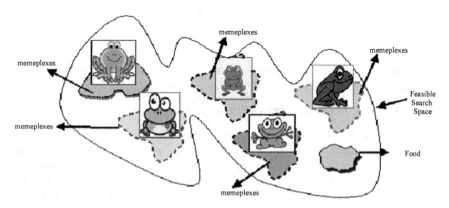

Fig. 1 Frogs performing local search initially then exchanging information with other memeplexes

Algorithm – Basic SFLA

Step 1: Initial population of frog size F is generated and evaluated (fitness value)
Step 2: Distribute the population of frog in m memeplexes where each memeplex contains n fog members such that $F = m \times n$.
Step 3: Start Evolutionary process
Step 4: Set $i = 1$
Step 5: while $i \leq i_{max}$
Step 6: Identify the global best (x_g) position of frog in the population.
Step 7: Identify the worst (x_{worst}) and best (x_{best}) frog in each memeplex.
Step 8: Apply eq. (2) & (3) to generate new frog position and evaluate.
Step 9: **If** $f(x_{new}) < f(x_{worst})$
Step 10: Set $x_{worst} = x_{new}$ and go to Step 13.
Step 11: $x_{best} = x_{global}$. Repeat Step 8 & Step 9.
Step 12: Random position is generated and replaced with x_{worst}.
Step 13: $i = i + 1$
Step 14: End While
Step 15: Population of frog is shuffled.
Step 16: If fixed termination criterion (N_{gen}) is achieved then exit else go to Step 2.

Fig. 2 SFL algorithm

3 Proposed O–SFLA

SFLA has three phases, initialization of frog positions, local search process in each memeplex, and information exchange, i.e., shuffling process. SFLA performs both exploration as well as exploitation. However, because of poor exploration capabilities SFLA sometimes get trapped in local optima that results in poor convergence. In this study, OBL is introduced in the structure of basic SFLA to eliminate its limitations.

OBL
This concept was introduced by Tizhoosh [17, 18]. The opposite positions of frogs can be calculated using Eq. (4):

$$x_{ijOpp} = \alpha_j - \beta_j - x_{ij} \tag{4}$$

where the upper and lower range of search process are represented by α and β, Frogs by $i = 1, 2, ..., F$; and problem's dimension by $j = 1, 2, ..., D$.

Embedded OBL in O–SFLA
In this study, OBL has been employed in local search process of SFLA to improve positions of frogs within memeplex. In each memeplex, opposite positions are generated using OBL (shown in Fig. 3) then elite n positions are selected based on fitness values $(f(n_{pop}) \cup f(n_{OBL\text{-}pop}))$. Further, in order to enhance diversity and convergence one extra step to improve the performance of memeplex is included in the structure of basic SFLA (Step14) in Fig. 4.

The proposed algorithm is discussed in Fig. 4.

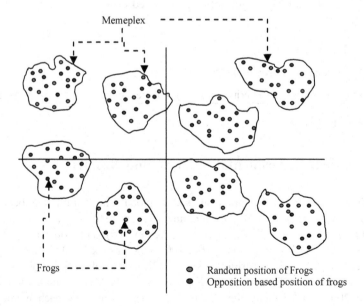

Fig. 3 Generation of opposite position of frogs in each memeplex

Proposed Algorithm: O – SFLA

Step 1: Initial population of frog size N is generated and evaluated (fitness value)

Step 2: Distribute the population of frog in m memeplexes where each memeplex contains n_{pop} fog members such that $F = m$ x n.

Step 3: Start Evolutionary process

Step 4: Set $i = 1$

Step 5: while $i \leq i_{max}$

Step 6: Identify the global best (x_g) position of frog in the population.

Step 7: Apply OBL to generate opposition positions of frogs ($n_{OBL-pop}$) in each memeplex.

Step 8: Select the elite n frogs positions from a set of $n_{pop} \cup n_{OBL-pop}$ based on fitness values.

Step 9: Identify the worst (x_{worst}) and best (x_{best}) frog in each memeplex.

Step 10: Apply eq. (2) & (3) to generate new frog position and evaluate.

Step 11: **If $f(x_{new}) < f(x_{worst})$**

Step 12: Set $x_{worst} = x_{new}$ and go to Step 16 .

Step 13: $x_{best} = x_{global}$. Repeat Step 11 & Step 12.

Step 14: Apply OBL to generate opposite point of x_{worst} i.e. $x_{OBL-worst}$. If $f(x_{OBL-worst}) < f(x_{worst})$ go to Step 16.

Step 15: Random position is generated and replaced with x_{worst}.

Step 16: $i = i + 1$

Step 17: End While

Step 18: Population of frog is shuffled.

Step 19: If fixed termination criterion (N_{gen}) is achieved then exit else go to Step 2.

Fig. 4 O–SFL algorithm

4 Test Bed, Experimental Settings, and Results

To validate efficacy of the proposal, six unimodal and multimodal benchmark functions are referred from the literature. These functions are Sphere (F1), Rosenbrock (F2), Step (F3), Rastrigin (F4), Ackley (F5), and Griekwank (F6). The search space taken for Sphere is [−100, 100], Rosenbrock [−30, 30], Step [−100, 100], Ackley [−32, 32], Rastrigin [−5.12, 5.12], and Griekwank [−600, 600] with the dimension of 30. The optimum value in each case is 0. The population of frog is fixed to 200 with 20 memeplexes. There will be 10 frogs in each memeplexes performing local search. 10 generations are performed in each memeplexes to explore the local region. The maximum movement step is fixed to 0.5 times of the search area. 25 runs are performed to evaluate mean function value and standard deviation. The simulations are performed on Dev C++ with following machine configurations: Intel(R) CPU T1350@1.86 GHz with 1 GB of RAM.

The results of O–SFLA are compared with that of PSO, ABC, SFLA and a recent variant of SFLA, i.e., EOLSFLA [19]. In order to perform unbiased comparison of O–SFLA with the performance of the other evolutionary algorithms, the population size is kept same with their standard parameter values. The comparative results (mean fitness values (Mean) and standard deviation (Std. Dev.)) are given in Tables 1 and 2, respectively, where F_{un} is Function.

It can be analyzed from the simulated results presented in the Table 2 that the O–SFLA outperforms or at-par with the results of most of the algorithms for all functions, except F_4 where EOLSFLA outperforms all the algorithms. Results analysis, in order to perform in-depth analysis of results a statistical analysis [20, 21] is done for testing the algorithms' efficiency. Bonferroni–Dunn [22] test is also done to calculate the significant difference. Critical difference for Bonferroni–Dunn's graph is calculated as (Eq. (5)):

$$CD = Q_\alpha \sqrt{\frac{k(k+1)}{6N}} \tag{5}$$

Table 1 Comparative simulated results in terms of mean of PSO, ABC, SFLA, EOLSFLA and the proposal O–SFLA

F_{un}	PSO	SFLA	ABC	EOLSFLA	O–SFLA
F_1	9.6E−07	6.5E+00	1.2E−12	0.0E+00	0.0E+00
F_2	2.6E+04	3.0E+01	3.0E+01	2.4E+01	1.9E+01
F_3	0.0E+00	1.6E+01	0.0E+00	0.0E+00	0.0E+00
F_4	7.9E+00	5.1E+00	2.4E+07	0.0E+00	4.7E+01
F_5	6.8E−01	7.8E−01	4.0E−06	5.9E−16	2.2E−19
F_6	1.8E−02	1.5E−01	1.0E−09	1.2E−14	1.1E−16

Table 2 Comparative simulated results in terms of standard deviation of PSO, ABC, SFLA, EOLSFLA and the proposal O–SFLA

F_{un}	PSO	SFLA	ABC	EOLSFLA	O–SFLA
F_1	±7.2E−06	±8.3E+01	±5.1E−12	±0.0E+00	±0.0E+00
F_2	±2.6E+05	±0.0E+00	±1.3E+00	±7.5E+01	±3.6E+01
F_3	±0.0E+00	±1.9E+02	±0.0E+00	±0.0E+00	±0.0E+00
F_4	±1.9E+02	±1.3E+12	±2.9E+06	±0.0E+00	±7.0E+01
F_5	±9.2E+00	±3.8E+00	±9.9E−06	±0.0E+00	±3.0E−17
F_6	±1.0E−01	±5.7E−01	±1.0E−08	±1.6E−13	±5.2E−15

Table 3 Friedman's rank

Algorithm	Rank
PSO	3.01
ABC	2.80
SFLA	2.67
EOLSFLA	2.1
O–SFLA	1.98
Critical difference (CD) $\alpha = 0.05$	2.280
Critical difference (CD) $\alpha = 0.10$	2.045

Fig. 5 Bonferroni–Dunn's graph

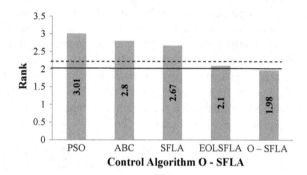

Q_α is critical value for a multiple non-parametric comparison with control [22], k and N represent the total number of algorithms and number of problems considered for comparison respectively. To show two levels of significance at $\alpha = 0.05$ and 0.10, horizontal lines are drawn. Ranks calculated using Friedman's test is presented in Table 3. and a graph is presented in Fig. 5.

Bonferroni–Dunn's test states subsequent significant differences with:

- **O–SFLA as a control algorithm:**

  O–SFLA is better than PSO, ABC, SFLA at $\alpha = 0.05$.
  O–SFLA is better than PSO, ABC, SFLA, and EOLSFLA at $\alpha = 0.10$.

5 Conclusions of the Study and the Future Scope

In this study, a novel variant of SFLA that embeds OBL is proposed and named as O–SFLA. The concept is embedded into the memeplexes before the frog initiates foraging and if the position of the worst frog does not improves then again OBL comes into picture to generate opposite point of worst frog position. The proposal is investigated on six benchmark functions and its performance is tested by comparing it with other algorithms. The efficiency of the proposal is validated using non-parametric test analysis. The results prove the efficiency of the proposal. In future, the proposal will be implemented on a real-world problem of paper and pulp industry.

References

1. Eusuff, M.M., Lansey, K.E.: Optimization of water distribution network design using the shuffled frog leaping algorithm. J. Water Resour. Plan. Manage. **129**, 210–225 (2003)
2. Eusuff, M., Lansey, K., Pasha, F.: Shuffled frog-leaping algorithm: a memetic meta-heuristic for discrete optimization. Eng. Optim. **38**, 129–154 (2006)
3. Ahandani, M.A., Alavi-Rad, H.: Opposition-based learning in the shuffled differential evolution algorithm. Soft. Comput. **16**, 1303–1337 (2012)
4. Li, J., Pan, Q., Xie, S.: An effective shuffled frog-leaping algorithm for multi-objective flexible job shop scheduling problems. Appl. Math. Comput. **218**, 9353–9371 (2012)
5. Tang-Huai, F., Li, L., Jia, Z.: Improved shuffled frog leaping algorithm and its application in node localization of wireless sensor network. Intell. Autom. Soft Comput. **18**, 807–818 (2012)
6. Ahandani, M.A., Alavi-Rad, H.: Opposition-based learning in shuffled frog leaping: An application for parameter identification. Inf. Sci. **291**, 19–42 (2015)
7. Sharma, T.K., Pant, M.: Shuffled artificial bee colony algorithm. Soft Comput. (2016). doi:10.1007/s00500-016-2166-2
8. Sharma, T.K., Pant, M.: Identification of noise in multi noise plant using enhanced version of shuffled frog leaping algorithm, Int. J. Syst. Assur. Eng. Manag. Springer. (2016). doi:10.1007/s13198-016-0466-7
9. Sharma, S., Sharma, T.K., Pant, M., Rajpurohit, J., Naruka, B.: Centroid mutation embedded shuffled leap frog algorithm, Elsevier Procedia Computer Science. **46**, 127–134 (2015)
10. Rajpurohit, J., Sharma, T.K., Nagar, A.: Shuffled frog leaping algorithm with adaptive exploration. In: Proceedings of Fifth International Conference on Soft Computing for Problem Solving Volume 436 of the series Advances in Intelligent Systems and Computing, pp. 595–603 (2015)
11. Liu, C., Niu, P., Li, G., Ma, Y., Zhang, W., Chen, K.: Enhanced shuffled frog-leaping algorithm for solving numerical function optimization problems. J. Intell. Manuf. 1–21 (2015)
12. Aungkulanon, P., Luangpaiboon, P.: Vertical transportation systems embedded on shuffled frog leaping algorithm for manufacturing optimisation problems in industries. Springer Plus. (2016). doi: 10.1186/s40064-016-2449-1
13. Liu, H., Yi, F., Yang, H.: Adaptive grouping cloud model shuffled frog leaping algorithm for solving continuous optimization problems. Comput. Intell. Neurosci. **2016** Article ID 5675349 (2016)
14. Dalavi, A.M., Pawar, P.J., Singh, T.P.: Tool path planning of hole-making operations in ejector plate of injection mould using modified shuffled frog leaping algorithm. J. Comput. Des. Eng. **3**(3), 266–273 (2016)
15. Lei, D., Guo, X.: A shuffled frog-leaping algorithm for hybrid flow shop scheduling with two agents. Expert Syst. Appl. **42**(23), 9333–9339 (2015)
16. Jadidoleslam, M., Ebrahimi, A.: Reliability constrained generation expansion planning by a modified shuffled frog leaping algorithm. Int. J. Electr. Power Energy Syst. **64**, 743–751 (2015)
17. Arshi, S.S., Zolfaghari, A., Mirvakili, S.M.: A multi-objective shuffled frog leaping algorithm for in-core fuel management optimization. Comput. Phys. Commun. **185**(10), 2622–2628 (2014)
18. Tizhoosh, H.R.: Opposition-based learning: a new scheme for machine intelligence. In: Proc. Int. Conf. Comput. Intell. Modeling, Control and Autom., pp. 695–701. Vienna, Austria (2005)
19. Zhao, J., Lv, L.: Shuffled frog-leaping algorithm using elite opposition-based learning. Int. J. of Sens. Netw. **16**(4), 244–251 (2014)
20. Demšar, J.: Statistical comparisons of classifiers over multiple data sets. J. Mach. Learn. Res. **7**, 1–30 (2006)
21. García, S., Herrera, F.: An extension on statistical comparisons of classifiers over multiple data sets for all pairwise comparisons. J. Mach. Learn. Res. **9**, 2677–2694 (2008)
22. Dunn, O.J.: Multiple comparisons among means. J. Am. Stat. Assoc. **56**(293), 52–64 (1961)

Author Index

Printed in the United States
By Bookmasters